Special relativity and quantum theory : a collection of papers on the Poincaré group

Special Relativity and Quantum Theory

Fundamental Theories of Physics

An International Book Series on The Fundamental Theories of Physics: Their Clarification, Development and Application

Special Relativity and Quantum Theory

A Collection of Papers on the Poincaré Group

dedicated to Professor Eugene Paul Wigner on the 50th Anniversary of His Paper on Unitary Representations of the Inhomogeneous Lorentz Group (completed in 1937 and published in 1939)

edited by

M. E. Noz

Department of Radiology,
New York University, U.S.A.

and

Y. S. Kim

Department of Physics and Astronomy,
University of Maryland, U.S.A.

KLUWER ACADEMIC PUBLISHERS
DORDRECHT / BOSTON / LONDON

Library of Congress Cataloging in Publication Data

Special relativity and quantum theory : a collection of papers on the Poincaré group / edited by M.E. Noz, Y.S. Kim.
p. cm. -- (Fundamental theories of physics)
"Dedicated to Professor Eugene Wigner on the 50th anniversary of his paper on unitary representations of the inhomogeneous Lorentz group (completed in 1937 and published in 1939)."
Includes bibliographies.
ISBN 9027727996
1. Quantum field theory--Congresses. 2. Special relativity (Physics)--Congresses. 3. Poincaré series--Congresses. 4. Lorentz groups--Congresses. 5. Wigner, Eugene Paul, 1902- . I. Noz, Marilyn E. II. Kim, Y. S. III. Wigner, Eugene Paul, 1902- . IV. Series.
QC174.46.S64 1988
530.1'2--dc19 88-13278
ISBN 90-277-2799-6 CIP

Published by Kluwer Academic Publishers,
P.O. Box 17, 3300 AA Dordrecht, The Netherlands.

Kluwer Academic Publishers incorporates
the publishing programmes of
D. Reidel, Martinus Nijhoff, Dr W. Junk and MTP Press.

Sold and distributed in the U.S.A. and Canada
by Kluwer Academic Publishers,
101 Philip Drive, Norwell, MA 02061, U.S.A.

In all other countries, sold and distributed
by Kluwer Academic Publishers Group,
P.O. Box 322, 3300 AH Dordrecht, The Netherlands.

Printed in The Netherlands

Table of Contents

Chapter V: Lorentz-Dirac Deformation in High Energy Physics

Chapter VI: Massless Particles and Gauge Transformations

Chapter VII: Group Contractions

Preface

Special relativity and quantum mechanics are likely to remain the two most important languages in physics for many years to come. The underlying language for both disciplines is group theory. Eugene P. Wigner's 1939 paper on the Unitary Representations of the Inhomogeneous Lorentz Group laid the foundation for unifying the concepts and algorithms of quantum mechanics and special relativity. In view of the strong current interest in the space-time symmetries of elementary particles, it is safe to say that Wigner's 1939 paper was fifty years ahead of its time. This edited volume consists of Wigner's 1939 paper and the major papers on the Lorentz group published since 1939.

This volume is intended for graduate and advanced undergraduate students in physics and mathematics, as well as mature physicists wishing to understand the more fundamental aspects of physics than are available from the fashion-oriented theoretical models which come and go. The original papers contained in this volume are useful as supplementary reading material for students in courses on group theory, relativistic quantum mechanics and quantum field theory, relativistic electrodynamics, general relativity, and elementary particle physics.

This reprint collection is an extension of the textbook by the present editors entitled "**Theory and Applications of the Poincaré Group.**" Since this book is largely based on the articles contained herein, the present volume should be viewed as a continuation of and supplementary reading for the previous work.

We would like to thank Professors J. Bjorken, R. Feynman, R. Hofstadter, J. Kuperzstych, L. Michel, M. Namiki, L.Parker, S. Weinberg, E.P. Wigner, A.S. Wightman, and Drs. P. Hussar, M. Ruiz, F. Rotbart, and B. Yurke for allowing us to reprint their papers. We are grateful to Mrs. M. Dirac and Mrs. S. Yukawa for giving us permission to reprint the articles of Professors P.A.M. Dirac and H. Yukawa respectively.

We wish to thank the Annals of Mathematics for permission to reprint Professor Wigner's historic paper. We thank the American Physical Society, the American Association of Physics Teachers, The Royal Society of London, il Nuovo Cimento and Progress in Theorectical Physics for permission to reprint the articles which appeared in their journals and for which they hold the copyright. The excerpt from *Albert Einstein: Historical and Cultural Perspective: The Centennial Symposium in Jerusalem* is reprinted with permission of Princeton University Press; that from

High Energy Collisions is reprinted with permission of Gordon and Breach Science publisher, Inc. and that from *Aspects of Quantum Theory* with permission of Cambridge University Press.

Introduction

One of the most fruitful and still promising approaches to unifying quantum mechanics and special relativity has been and still is the covariant formulation of quantum field theory. The role of Wigner's work on the Poincaré group in quantum field theory is nicely summarized in the fourth paragraph of an article by V. Bargmann *et al.* in the commemorative issue of the Reviews of Modern Physics in honor of Wigner's 60th birthday [Rev. Mod. Phys. *34*, 587 (1962)], which concludes with the sentences:

> "Those who had carefully read the preface of Wigner's great 1939 paper on relativistic invariance and had understood the physical ideas in his 1931 book on group theory and atomic spectra were not surprised by the turn of events in quantum field theory in the 1950's. A fair part of what happened was merely a matter of whipping quantum field theory into line with the insights achieved by Wigner in 1939".

It is important to realize that quantum field theory has not been and is not at present the only theoretical machine with which physicists attempt to unify quantum mechanics and special relativity. Indeed, Dirac devoted much of his professional life to this important task, but, throughout the 1950's and 1960's, his form of relativistic quantum mechanics was overshadowed by the success of quantum field theory. However, in the 1970's, when it was necessary to deal with quarks confined permanently inside hadrons, the limitations of the present form of quantum field theory become apparent. Currently, there are two different opinions on the difficulty of using field theory in dealing with bound-state problems or systems of confined quarks. One of these regards the present difficulty merely as a complication in calculation. According to this view, we should continue developing mathematical techniques which will someday enable us to formulate a bound-state problem with satisfactory solutions within the framework of the existing form of quantum field theory. The opposing opinion is that quantum field theory is a model that can handle only scattering problems in which all particles can be brought to free-particle asymptotic states. According to this view we have to make a fresh start for relativistic bound-state problems.

These two opposing views are not mutually exclusive. *Bound-state models developed in these two different approaches should have the same space-time symmetry*. It is quite possible that independent bound-state models, if successful in explaining what we see in the real world, will eventually complement field theory. One of the purposes of this book is to present the fundamental papers upon which a relativistic bound-state model that can explain basic hadronic features observed in high-energy laboratories could be build in accordance with the principles laid out by Wigner in 1939.

Wigner observed in 1939 that Dirac's electron has an SU(2)-like internal space-time symmetry. However, quarks and hadrons were unknown at that time. Dirac's form of relativistic bound-state quantum mechanics, which starts from the representations of the Poincaré group, makes it possible to study the O(3)-like little group for massive particles and leads to hadronic wave functions which can describe fairly accurately the distribution of quarks inside hadrons. Thus a substantial portion of hadronic physics can be incorporated into the O(3)-like little group for massive particles.

Another important development in modern physics is the extensive use of gauge transformations in connection with massless particles and their interactions. Wigner's 1939 paper has the original discussion of space-time symmetries of massless particles. However, it was only recently recognized that gauge-dependent electromagnetic four-potentials form the basis for a finite-dimensional non-unitary representation of the little group of the Poincaré group. This enables us to associate gauge degrees of freedom with the degrees of freedom left unexplained in Wigner's work. Hence it is possible to impose a gauge condition on the electromagnetic four-potential to construct a unitary representation of the photon polarization vectors.

Wigner showed that the internal space-time symmetry group of massless particles is locally isomorphic to the Euclidian group in two-dimensional space. However, Wigner did not explore the content of this isomorphism, because the physics of the translation-like transformations of this little group was unknown in 1939. Neutrinos were known only as "Dirac electrons without mass", although photons were known to have spins either parallel or antiparallel to their respective momenta. We now know the physics of the degrees of freedom left unexplained in Wigner's paper. Much more is also known about neutrinos today that in 1939. For instance, it is firmly established that neutrinos and anti-neutrinos are left and right handed respectively. Therefore, it is possible to discuss internal space-time symmetries of massless particles starting from Wigner's E(2)-like little group. Recently, it was observed that the O(3)-like little group becomes the E(2)-like group in the limit of small mass and/or large momentum.

Indeed, group theory has become the standard language in physics. Until the 1960's, the only group known to the average physicist had been the three-dimensional rotation group. Gell-Mann's work on the quark model encouraged physicists to study the unitary groups, which are compact groups. The Weinberg-Salam model enhanced this trend. The emergence of supersymmetry in the 1970's has brought the space-time group closer to physicists. These groups are non-compact, and it is difficult to prove or appreciate mathematical theorems for them.

The Poincaré group is a non-compact group. Fortunately, the representations of this group useful in physics are not complicated from the mathematical point of view.

The application of the Lorentz group is not restricted to the symmetries of elementary particles. The (2 + 1)-dimensional Lorentz group is isomorphic to the two-dimensional symplectic group, which is the symmetry group of homogeneous linear canonical transformations in classical mechanics. It is also useful for studying coherent and squeezed states in optics. It is likely that the Lorentz group will serve useful purposes in many other branches of modern physics.

This reprint volume contains the fundamental paper by Wigner, and the papers on applications of his paper to physical problems. This book starts with Wigner's review paper on relativistic invariance and quantum phenomena. The reprinted papers are grouped into nine chapters. Each chapter starts with a brief introduction.

Chapter I

Perspective View of Quantum Space-Time Symmetries

When Einstein formulated his special theory of relativity in 1905, quantum mechanics was not known. Einstein's original version of special relativity deals with point particles without space-time structures and extension. These days, we know that elementary particles can have intrinsic space-time structure manifested by spins. In addition, many of the particles which had been thought to be point particles now have space-time extensions.

The hydrogen atom was known to be a composite particle in which the electron maintains a distance from the proton. Therefore, the hydrogen atom is not a point particle. The proton had been regarded as a point particle until, in 1955, the experiment of Hofstadter and McAllister proved otherwise. These days, the proton is a bound state of more fundamental particles called the quarks. We still do not know whether the quarks have non-zero size, but assume that they are point particles. We assume also that electrons are point particles. However, it is clear that these particles have intrinsic spins. The situation is the same for massless particles. For intrinsic spins, the Wigner's representation of the Poincaré group is the natural scientific language.

As for nonrelativistic extended particles, such as the hydrogen atom, the present form of quantum mechanics with the probability interpretation is quite adequate. If the proton is a bound state of quarks within the framework of quantum mechanics, the description of a rapidly moving proton requires a Lorentz transformation of localized probability distribution. In addition, this description should find its place in Wigner's representation theory of the Poincaré group.

This Chapter consists of one article by Wigner on relativistic invariance of quantum phenomena, and one article by Dirac. As he said in his 1979 paper, Dirac was concerned with the problem of fitting quantum mechanics in with relativity, right from the beginning of quantum mechanics. Dirac suggests that the ideal mechanics should be both relativistic and deterministic. It would be too ambitious to work with both the relativistic and deterministic problem at the same time. Perhaps the easier way is to deal with one aspect at a time. Then there are two routes to the ideal mechanics, as are illustrated in Figure 1. The current literature indicates that it would be easier to make quantum mechanics relativistic than deterministic. In this book, we propose to study the easier problem first.

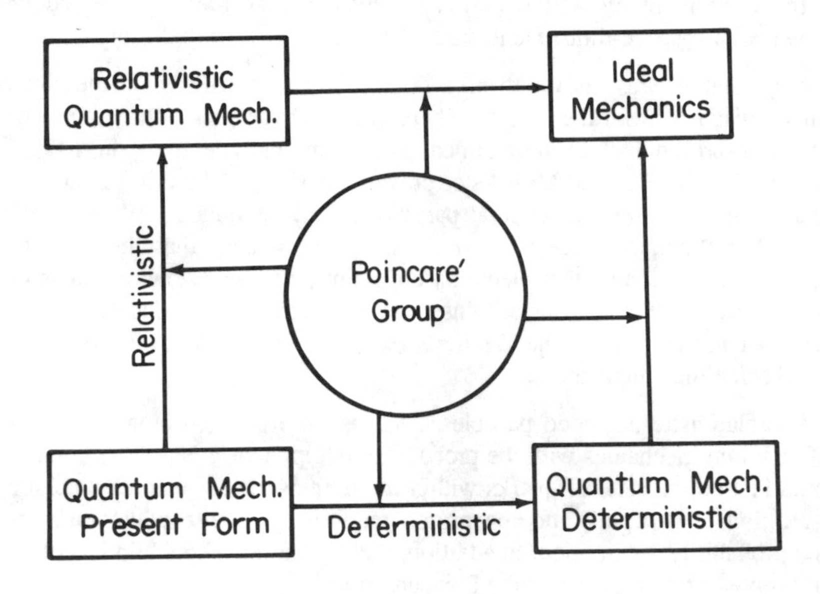

FIG. 1. Two different routes to the ideal mechanics. Covariance and determinism are the two main problems. In approaching these problems, there are two different routes. In either case, the Poincaré group is likely to be the main scientific language.

Reprinted from REVIEWS OF MODERN PHYSICS, Vol. 29, No. 3, 255–268, July, 1957
Printed in U. S. A.

Relativistic Invariance and Quantum Phenomena*

EUGENE P. WIGNER
Palmer Physical Laboratory, Princeton University, Princeton, New Jersey

INTRODUCTION

THE principal theme of this discourse is the great difference between the relation of special relativity and quantum theory on the one hand, and general relativity and quantum theory on the other. Most of the conclusions which will be reported on in connection with the general theory have been arrived at in collaboration with Dr. H. Salecker,[1] who has spent a year in Princeton to investigate this question.

The difference between the two relations is, briefly, that while there are no conceptual problems to separate the theory of special relativity from quantum theory, there is hardly any common ground between the general theory of relativity and quantum mechanics. The statement, that there are no conceptual conflicts between quantum mechanics and the special theory, should not mean that the mathematical formulations of the two theories naturally mesh. This is not the case, and it required the very ingenious work of Tomonaga, Schwinger, Feynman, and Dyson[2] to adjust quantum mechanics to the postulates of the special theory and this was so far successful only on the working level. What is meant is, rather, that the concepts which are used in quantum mechanics, measurements of positions, momenta, and the like, are the same concepts in terms of which the special relativistic postulate is formulated. Hence, it is at least possible to formulate the requirement of special relativistic invariance for quantum theories and to ascertain whether these requirements are met. The fact that the answer is more nearly *no* than *yes*, that quantum mechanics has not yet been fully adjusted to the postulates of the special theory, is perhaps irritating. It does not alter the fact that the question of the consistency of the two theories can at least be formulated, that the question of the special relativistic invariance of quantum mechanics by now has more nearly the aspect of a puzzle than that of a problem.

This is not so with the general theory of relativity. The basic premise of this theory is that coordinates are only auxiliary quantities which can be given arbitrary values for every event. Hence, the measurement of position, that is, of the space coordinates, is certainly not a significant measurement if the postulates of the general theory are adopted: the coordinates can be given any value one wants. The same holds for momenta. Most of us have struggled with the problem of how, under these premises, the general theory of relativity can make meaningful statements and predictions at all. Evidently, the usual statements about future positions of particles, as specified by their coordinates, are not meaningful statements in general relativity. This is a point which cannot be emphasized strongly enough and is the basis of a much deeper dilemma than the more technical question of the Lorentz invariance of the quantum field equations. It pervades all the general theory, and to some degree we mislead both our students and ourselves when we calculate, for instance, the mercury perihelion motion without explaining how our coordinate system is fixed in space, what defines it in such a way that it cannot be rotated, by a few seconds a year, to follow the perihelion's apparent motion. Surely the x axis of our coordinate system could be defined in such a way that it pass through all successive perihelions. There must be some assumption on the nature of the coordinate system which keeps it from following the perihelion. This is not difficult to exhibit in the case of the motion of the perihelion, and it would be useful to exhibit it. Neither is this, in general, an academic point, even

* Address of retiring president of the American Physical Society, January 31, 1957.

[1] This will be reported jointly with H. Salecker in more detail in another journal.

[2] See, e.g., J. M. Jauch and F. Rohrlich, *The Theory of Protons and Electrons* (Addison-Wesley Press, Cambridge, Massachusetts, 1955).

Reprinted from *Rev. Mod. Phys.* **29**, 255 (1957).

though it may be academic in the case of the mercury perihelion. A difference in the tacit assumptions which fix the coordinate system is increasingly recognized to be at the bottom of many conflicting results arrived at in calculations based on the general theory of relativity. Expressing our results in terms of the values of coordinates became a habit with us to such a degree that we adhere to this habit also in general relativity where values of coordinates are not *per se* meaningful. In order to make them meaningful, the mollusk-like coordinate system must be somehow anchored to space-time events and this anchoring is often done with little explicitness. If we want to put general relativity on speaking terms with quantum mechanics, our first task has to be to bring the statements of the general theory of relativity into such form that they conform with the basic principles of the general relativity theory itself. It will be shown below how this may be attempted.

RELATIVISTIC QUANTUM THEORY OF ELEMENTARY SYSTEMS

The relation between special theory and quantum mechanics is most simple for single particles. The equations and properties of these, in the absence of interactions, can be deduced already from relativistic invariance. Two cases have to be distinguished: the particle either can, or cannot, be transformed to rest. If it can, it will behave, in that coordinate system, as any other particle, such as an atom. It will have an intrinsic angular momentum called J in the case of atoms and spin S in the case of elementary particles. This leads to the various possibilities with which we are familiar from spectroscopy, that is spins 0, $\frac{1}{2}$, 1, $\frac{3}{2}$, 2, $\cdots$ each corresponding to a type of particle. If the particle cannot be transformed to rest, its velocity must always be equal to the velocity of light. Every other velocity can be transformed to rest. The rest-mass of these particles is zero because a nonzero rest-mass would entail an infinite energy if moving with light velocity.

Particles with zero rest-mass have only two directions of polarization, no matter how large their spin is. This contrasts with the $2S+1$ directions of polarization for particles with nonzero rest-mass and spin S. Electromagnetic radiation, that is, light, is the most familiar example for this phenomenon. The "spin" of light is 1, but it has only two directions of polarization, instead of $2S+1=3$. The number of polarizations seems to jump discontinuously to two when the rest-mass decreases and reaches the value 0. Bass and Schrödinger[3] followed this out in detail for electromagnetic radiation, that is, for $S=1$. It is good to realize, however, that this decrease in the number of possible polarizations is purely a property of the Lorentz transformation and holds for any value of the spin.

There is nothing fundamentally new that can be said about the number of polarizations of a particle and the principal purpose of the following paragraphs is to illuminate it from a different point of view.[4] Instead of the question: "Why do particles with zero rest-mass have only two directions of polarization?" the slightly different question, "Why do particles with a finite rest-mass have more than two directions of polarization?" is proposed.

The intrinsic angular momentum of a particle with zero rest-mass is parallel to its direction of motion, that is, parallel to its velocity. Thus, if we connect any internal motion with the spin, this is perpendicular to the velocity. In case of light, we speak of transverse polarization. Furthermore, and this is the salient point, the statement that the spin is parallel to the velocity is a relativistically invariant statement: it holds as well if the particle is viewed from a moving coordinate system. If the problem of polarization is regarded from this point of view, it results in the question, "Why can't the angular momentum of a particle with finite rest-mass be parallel to its velocity?" or "Why can't a plane wave represent transverse polarization unless it propagates with light velocity?" The answer is that the angular momentum *can* very well be parallel to the direction of motion and the wave *can* have transverse polarization, but these are not Lorentz invariant statements. In other words, even if velocity and spin are parallel in one coordinate system, they do not appear to be parallel in other coordinate systems. This is most evident if, in this other coordinate system, the particle is at rest: in this coordinate system the

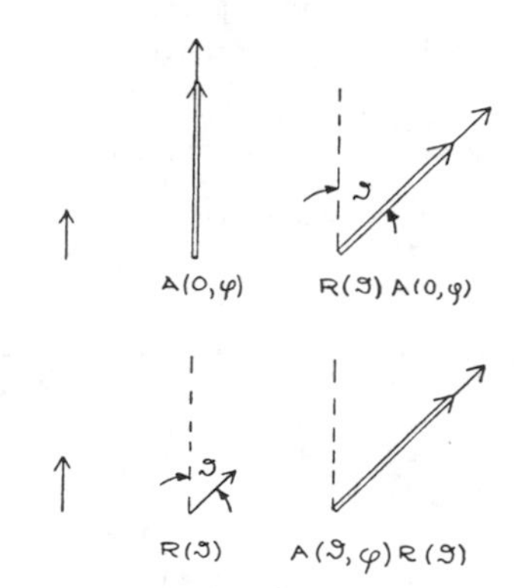

FIG. 1. The short simple arrows illustrate the spin, the double arrows the velocity of the particle. One obtains the same state, no matter whether one first imparts to it a velocity in the direction of the spin, then rotates it $(R(\vartheta)A(0,\varphi))$, or whether one first rotates it, then gives a velocity in the direction of the spin $(A(\vartheta,\varphi)R(\vartheta))$. See Eq. (1.3).

[3] L. Bass and E. Schrödinger, Proc. Roy. Soc. (London) **A232**, 1 (1955).

[4] The essential point of the argument which follows is contained in the present writer's paper, Ann. Math. **40**, 149 (1939) and more explicitly in his address at the Jubilee of Relativity Theory, Bern, 1955 (Birkhauser Verlag, Basel, 1956), A. Mercier and M. Kervaire, editors, p. 210.

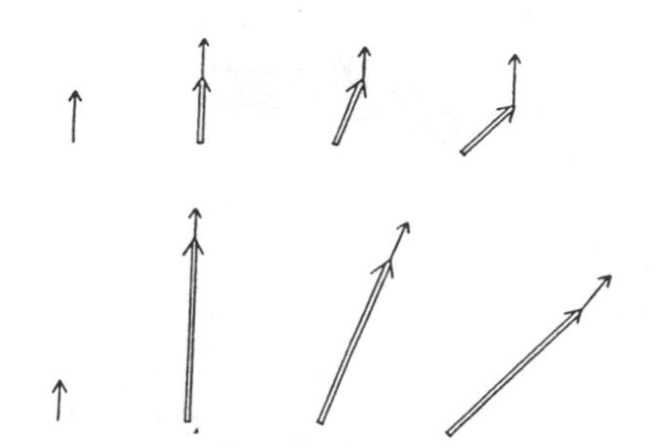

FIG. 2. The particle is first given a small velocity in the direction of its spin, then increasing velocities in a prependicular direction (upper part of the figure). The direction of the spin remains essentially unchanged; it includes an increasingly large angle with the velocity as the velocity in the perpendicular direction increases. If the velocity imparted to the particle is large (lower part of the figure), the direction of the spin seems to follow the direction of the velocity. See Eqs. (1.8) and (1.7).

angular momentum should be parallel to nothing. However, every particle, unless it moves with light velocity, can be viewed from a coordinate system in which it is at rest. In this coordinate system its angular momentum is surely not parallel to its velocity. Hence, the statement that spin and velocity are parallel cannot be universally valid for the particle with finite rest-mass and such a particle must have other states of polarization also.

It may be worthwhile to illustrate this point somewhat more in detail. Let us consider a particle at rest with a given direction of polarization, say the direction of the z axis. Let us consider this particle now from a coordinate system which is moving in the $-z$ direction. The particle will then appear to have a velocity in the z direction and its polarization will be parallel to its velocity (Fig. 1). It will now be shown that this last statement is nearly invariant if the velocity is high. It is evident that the statement is entirely invariant with respect to rotations and with respect to a further increase of the velocity in the z direction. This is illustrated at the bottom of the figure. The coordinate system is first turned to the left and then given a velocity in the direction opposite to the old z axis. The state of the system appears to be exactly the same as if the coordinate system had been first given a velocity in the $-z$ direction and then turned, which is the operation illustrated at the top of the figure. The state of the system appears to be the same not for any physical reason but because the two coordinate systems are identical and they view the same particle (see Appendix I).

Let us now take our particle with a high velocity in the z direction and view it from a coordinate system which moves in the $-y$ direction. The particle now will appear to have a momentum also in the y direction, its velocity will have a direction between the y and z axes (Fig. 2). Its spin, however, will not be in the direction of its motion any more. In the nonrelativistic case, that is, if all velocities are small as compared with the velocity of light, the spin will still be parallel to z and it will, therefore, enclose an angle with the particle's direction of motion. This shows that the statement that the spin is parallel to the direction of motion is not invariant in the nonrelativisitic region. However, if the original velocity of the particle is close to the light velocity, the Lorentz contraction works out in such a way that the angle between spin and velocity is given by

$$\tan \text{ (angle between spin and velocity)} = (1-v^2/c^2)^{\frac{1}{2}} \sin\vartheta, \quad (1)$$

where ϑ is the angle between the velocity v in the moving coordinate system and the velocity in the coordinate system at rest. This last situation is illustrated at the bottom of the figure. If the velocity of the particle is small as compared with the velocity of light, the direction of the spin remains fixed and is the same in the moving coordinate system as in the coordinate system at rest. On the other hand, if the particle's velocity is close to light velocity, the velocity carries the spin with itself and the angle between direction of motion and spin direction becomes very small in the moving coordinate system. Finally, if the particle has light velocity, the statement "spin and velocity are parallel" remains true in every coordinate system. Again, this is not a consequence of any physical property of the spin, but is a consequence of the properties of Lorentz transformations: it is a kind of Lorentz contraction. It is the reason for the different behavior of particles with finite, and particles with zero, rest-mass, as far as the number of states of polarization is concerned. (Details of the calculation are in Appendix I.)

The preceding consideration proves more than was intended: it shows that the statement "spin and velocity are parallel for zero mass particles" is invariant and that, for relativistic reasons, one needs only *one* state of polarization, rather than *two*. This is true as far as proper Lorentz transformations are concerned. The second state of polarization, in which spin and velocity are antiparallel, is a result of the reflection symmetry. Again, this can be illustrated on the example of light: right circularly polarized light appears as right circularly polarized light in all Lorentz frames of reference which can be continuously transformed into each other. Only if one looks at the right circularly polarized light in a mirror does it appear as left circularly polarized light. The postulate of reflection symmetry allows us to infer the existence of left circularly polarized light from the existence of right circularly polarized light—if there were no such reflection symmetry in the real world, the existence of *two* modes of polarization of light, with virtually identical properties, would appear to be a miracle. The situation is entirely different for particles with nonzero mass. For these, the $2S+1$ directions of polarization follow from the invariance of the theory with respect to proper Lorentz transformations. In particular, if the particle is at rest, the spin will have different orientations with respect to coordinate systems which have different orientations in space. Thus, the existence of all the states of polarization follow from the existence of one, if only the theory is invariant with respect to proper Lorentz transformations. For particles with zero rest-mass, there are only two states of polarization, and even the existence of the second one can be inferred only on the basis of reflection symmetry.

REFLECTION SYMMETRY

The problem and existence of reflection symmetry have been furthered in a brilliant way by recent theoretical and experimental research. There is nothing essential that can be added at present to the remarks and conjectures of Lee, Yang, and Oehme, and all that follows has been said, or at least implied, by Salam, Lee, Yang, and Oehme.[5] The sharpness of the break with past concepts is perhaps best illustrated by the cobalt experiment of Wu, Ambler, Hayward, Hoppes, and Hudson.

The ring current—this may be a permanent current in a superconductor—creates a magnetic field. The Co source is in the plane of the current and emits β particles (Fig. 3). The whole experimental arrangement, as shown in Fig. 3, has a symmetry plane and, if the principle of sufficient cause is valid, the symmetry plane should remain valid throughout the further fate of the system. In other words, since the right and left sides of the plane had originally identical properties, there is no sufficient reason for any difference in their properties at a later time. Nevertheless, the intensity of the β radiation is larger on one side of the plane than the other side. The situation is paradoxical no matter what the mechanism of the effect is—in fact, it is most paradoxical if one disregards its mechanism and theory entirely. If the experimental circumstances can be idealized as indicated, even the principle of sufficient cause seems to be violated.

It is natural to look for an interpretation of the experiment which avoids this very far-reaching conclusion and, indeed, there is such an interpretation.[5a] It is good to reiterate, however, that no matter what interpretation is adopted, we have to admit that the symmetry of the real world is smaller than we had thought. However, the symmetry may still include reflections.

[5] Lee, Yang, and Oehme, Phys. Rev. **106**, 340 (1957).

[5a] The interpretation referred to has been proposed independently by numerous authors, including A. Salam, Nuovo cimento **5**, 229 (1957); L. Landau, Nuclear Phys. **3**, 127 (1957); H. D. Smyth and L. Biedenharn (personal communication). Dr. S. Deser has pointed out that the "perturbing possibility" was raised already by Wick, Wightman, and Wigner [Phys. Rev. **88**, 101 (1952)] but was held "remote at that time." Naturally, the apparent unanimity of opinion does not prove its correctness.

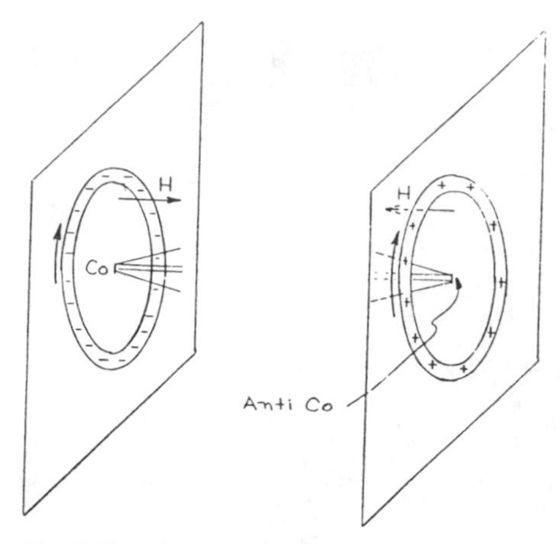

FIG. 3. The right side is the mirror image of the left side, according to the interpretation of the parity experiments[5a] which maintains the reflection as a symmetry element of all physical laws. It must be assumed that the reflection transforms matter into antimatter: the electronic ring current becomes a positronic ring current, the radioactive cobalt is replaced by radioactive anticobalt.

If it is true that a symmetry plane always remains a symmetry plane, the initial state of the Co experiment could not have contained a symmetry plane. This would not be the case if the magnetic vector were polar—in which case the electric vector would be axial. The charge density, the divergence of the electric vector, would then become a pseudoscalar rather than a simple scalar as in current theory. The mirror image of a negative charge would be positive, the mirror image of an electron a positron, and conversely. The mirror image of matter would be antimatter. The Co experiment, viewed through a mirror, would not present a picture contrary to established fact: it would present an experiment carried out with antimatter. The right side of Fig. 3 shows the mirror image of the left side. Thus, the principle of sufficient cause, and the validity of symmetry planes, need not be abandoned if one is willing to admit that the mirror image of matter is antimatter.

The possibility just envisaged would be technically described as the elimination of the operations of reflection and charge conjugation, as presently defined, as true symmetry operations. Their product would still be assumed to be a symmetry operation and proposed to be named, simply, reflection. A few further technical remarks are contained in Appendix II. The proposition just made has two aspects: a very appealing one, and a very alarming one.

Let us look first at the appealing aspect. Dirac has said that the number of elementary particles shows an alarming tendency of increasing. One is tempted to add to this that the number of invariance properties also showed a similar tendency. It is not equally alarming because, while the increase in the number of elementary particles complicates our picture of nature, that of the symmetry properties on the whole simplifies it. Nevertheless the clear correspondence between the invariance properties of the laws of nature, and the symmetry properties of space-time, was most clearly breached by the operation of charge conjugation. This postulated that the laws of nature remain the same if all positive charges are replaced by negative charges and vice versa, or more generally, if all particles are replaced by antiparticles. Reasonable as this postulate appears to us, it corresponds to no symmetry of the space-time continuum. If the preceding interpretation of the Co experiments should be sustained, the correspondence between the natural symmetry elements of space-time, and the invariance properties of the laws of nature, would be restored. It is true that the role of the planes of reflection would not be that to which we are accustomed—the mirror image of an electron would become a positron—but the mirror image of a sequence of events would still be a possible sequence of events. This possible sequence of events would be more difficult to realize in the actual physical world than what we had thought, but it would still be possible.

The restoration of the correspondence between the natural symmetry properties of space-time on one hand, and the laws of nature on the other hand, is the appealing feature of the proposition. It has, actually, two alarming features. The first of these is that a symmetry operation is, physically, so complicated. If it should turn out that the operation of time inversion, as we now conceive it, is not a valid symmetry operation (e.g., if one of the experiments proposed by Treiman and Wyld gave a positive result) we could still maintain the validity of this symmetry operation by reinterpreting it. We could postulate, for instance, that time inversion transforms matter into *meta*-matter which will be discovered later when higher energy accelerators will become available. Thus, maintaining the validity of symmetry planes forces us to a more artificial view of the concept of symmetry and of the invariance of the laws of physics.

The other alarming feature of our new knowledge is that we have been misled for such a long time to believe in more symmetry elements than actually exist. There was ample reason for this and there was ample experimental evidence to believe that the mirror image of a possible event is again a possible event with electrons being the mirror images of electrons and not of positrons. Let us recall in this connection first how the concept of parity, resulting from the beautiful though almost forgotten experiments of Laporte,[6] appeared to be a perfectly valid concept in spectroscopy and in nuclear physics. This concept could be explained very naturally as a result of the reflection symmetry of space-time, the mirror image of electrons being electrons and not positrons. We are now forced to believe that this symmetry is only approximate and the concept of parity, as used in spectroscopy and nuclear physics, is also only approximate. Even more fundamentally, there is a vast body of experimental information in the chemistry of optically active substances which are mirror images of each other and which have optical activities of opposite direction but exactly equal strength. There is the fact that molecules which have symmetry planes are optically inactive; there is the fact of symmetry planes in crystals.[7] All these facts relate properties of right-handed matter to left-handed *matter*, not of right-handed matter to left-handed *antimatter*. The new experiments leave no doubt that the symmetry plane in this sense is not valid for all phenomena, in particular not valid for β decay, that if the concept of symmetry plane is at all valid for all phenomena, it can be valid only in the sense of converting matter into antimatter.

Furthermore, the old-fashioned type of symmetry plane is not the only symmetry concept that is only approximately valid. Charge conjugation was mentioned before, and we are remainded also of isotopic spin, of the exchange character, that is multiplet system, for electrons and also of nuclei which latter holds so accurately that, in practice, parahydrogen molecules can be converted into orthohydrogen molecules only by first destroying them.[8] This approximate validity of laws of symmetry is, therefore, a very general phenomenon—it may be *the* general phenomenon. We are reminded of Mach's axiom that the laws of nature depend on the physical content of the universe, and the physical content of the universe certainly shows no symmetry. This suggests—and this may also be the spirit of the ideas of Yang and Lee—that all symmetry properties are only approximate. The weakest interaction, the gravitational force, is the basis of the distinction between inertial and accelerated coordinate systems, the second weakest known interaction, that leading to β decay, leads to the distinction between matter and antimatter. Let me conclude this subject by expressing the conviction that the discoveries of Wu, Ambler, Hayward, Hoppes, and Hudson,[9] and of Garwin, Lederman, and Weinreich[10] will not remain isolated discoveries. More likely, they herald a revision of our concept of invariance and possibly

[6] O. Laporte, Z. Physik **23**, 135 (1924). For the interpretation of Laporte's rule in terms of the quantum-mechanical operation of inversion, see the writer's *Gruppentheorie und ihre Anwendungen auf die Quantenmechanik der Atmospektren* (Friedrich Vieweg und Sohn, Braunschweig, 1931), Chap. XVIII.

[7] For the role of the space and time inversion operators in classical theory, see H. Zocher and C. Török, Proc. Natl. Acad. Sci. U.S. **39**, 681 (1953) and literature quoted there.

[8] See A. Farkas, *Orthohydrogen, Parahydrogen and Heavy Hydrogen* (Cambridge University Press, New York, 1935).

[9] Wu, Ambler, Hayward, Hoppes, and Hudson, Phys. Rev. **105**, 1413(L) (1957).

[10] Garwin, Lederman, and Weinreich, Phys. Rev. **105**, 1415(L) (1957); also, J. L. Friedman and V. L. Telegdi, *ibid.* **105**, 1681(L) (1957).

of other concepts which are even more taken for granted.

QUANTUM LIMITATIONS OF THE CONCEPTS OF GENERAL RELATIVITY

The last remarks naturally bring us to a discussion of the general theory of relativity. The main premise of this theory is that coordinates are only labels to specify space-time points. Their values have no particular significance unless the coordinate system is somehow anchored to events in space-time.

Let us look at the question of how the equations of the general theory of relativity could be verified. The purpose of these equations, as of all equations of physics, is to calculate, from the knowledge of the present, the state of affairs that will prevail in the future. The quantities describing the present state are called initial conditions; the ways these quantities change are called the equations of motion. In relativity theory, the state is described by the metric which consists of a network of points in space-time, that is a network of events, and the distances between these events. If we wish to translate these general statements into something concrete, we must decide what events are, and how we measure distances between *events*. The metric in the general theory of relativity is a metric in space-time, its elements are distances between space-time points, not between points in ordinary space.

The events of the general theory of relativity are coincidences, that is, collisions between particles. The founder of the theory, when he created this concept, had evidently macroscopic bodies in mind. Coincidences, that is, collisions between such bodies, are immediately observable. This is not the case for elementary particles; a collision between these is something much more evanescent. In fact, the point of a collision between two elementary particles can be closely localized in space-time only in case of high-energy collisions. (See Appendix III.) This shows that the establishment of a close network of points in space-time requires a reasonable energy density, a dense forest of world lines wherever the network is to be established. However, it is not necessary to discuss this in detail because the measurement of the distances between the points of the network gives more stringent requirements than the establishment of the network.

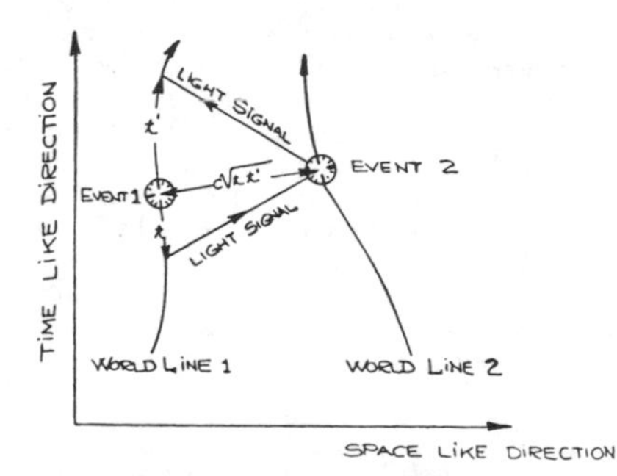

FIG. 4. Measurement of space-like distances by means of a clock. It is assumed that the metric tensor is essentially constant within the space-time region contained in the figure. The space-like distance between events 1 and 2 is measured by means of the light signals which pass through event 2 and a geodesic which goes through event 1. Explanation in Appendix IV.

It is often said that the distances between events must be measured by yardsticks and rods. We found that measurements with a yardstick are rather difficult to describe and that their use would involve a great deal of unnecessary complications. The yardstick gives the distance between events correctly only if its marks coincide with the two events simultaneously from the point of view of the rest-system of the yardstick. Furthermore, it is hard to image yardsticks as anything but macroscopic objects. It is desirable, therefore, to reduce all measurements in space-time to measurements by clocks. Naturally, one can measure by clocks directly only the distances of points which are in time-like relation to each other. The distances of events which are in space-like relation, and which would be measured more naturally by yardsticks, will have to be measured, therefore, indirectly.

It appears, thus, that the simplest framework in space-time, and the one which is most nearly microscopic, is a set of clocks, which are only slowly moving with respect to each other, that is, with world lines which are approximately parallel. These clocks tick off periods and these ticks form the network of events which we wanted to establish. This, at the same time, establishes the distance of those adjacent points which are on the same world line.

Figure 4 shows two world lines and also shows an event, that is, a tick of the clock, on each. The figure shows an artifice which enables one to measure the distance of space-like events: a light signal is sent out from the first clock which strikes the second clock at event 2. This clock, in turn, sends out a light signal which strikes the first clock at time t' after the event 1. If the first light signal had to be sent out at time t before the first event, the calculation given in Appendix IV shows that the space-like distance of events 1 and 2 is the geometric average of the two measured time-like distances t and t'. This is then a way to measure distances between space-like events by clocks instead of yardsticks.

It is interesting to consider the quantum limitations on the accuracy of the conversion of time-like measurements into space-like measurements, which is illustrated in Fig. 4. Naturally, the times t and t' will be well defined only if the light signal is a short pulse. This implies that it is composed of many frequencies and, hence, that its energy spectrum has a corresponding width. As a result, it will give an indeterminate recoil to the second clock, thus further increasing the uncertainty of its momentum. All this is closely related to Heisenberg's uncertainty principle. A more detailed calculation[1] shows that the added uncertainty is of the same order of magnitude as the uncertainty inherent in the nature of the best clock that we could think of, so that the conversion of time-like measurements into space-like measurements is essentially free.

We finally come to the discussion of one of the principal problems—the limitations on the accuracy of the clock. It led us to the conclusion that the inherent limitations on the accuracy of a clock of given weight and size, which should run for a period of a certain length, are quite severe. In fact, the result in summary is that a clock is an essentially nonmicroscopic object. In particular, what we vaguely call an atomic clock, a single atom which ticks off its periods, is surely an idealization which is in conflict with fundamental concepts of measurability. This part of our conclusions can be considered to be well established. On the other hand, the actual formula which will be given for the limitation of the accuracy of time measurement, a sort of uncertainty principle, should be considered as the best present estimate.

Let us state the requirements as follows. The watch shall run T seconds, shall measure time with an accuracy of $T/n=t$, its linear extension shall not exceed l, its mass shall be below m. Since the pointer of the watch must be able to assume n different positions, the system will have to run, in the course of the time T, over at least n orthogonal states. Its state must, therefore, be the superposition of at least n stationary states. It is clear, furthermore, that unless its total energy is at least $\hbar/t$, it cannot measure a time interval which is smaller than t. This is equivalent to the usual uncertainty principle. These two requirements follow directly from the basic principles of quantum theory; they are also the requirements which could well have been anticipated. A clock which conforms with these postulates is, for instance, an oscillator, with a period which is equal to the running time of the clock, if it is with equal probabilty in any of the first n quantum states. Its energy is about n times the energy of the first excited state. This corresponds to the uncertainty principle with the accuracy t as time uncertainty. Broadly speaking, the clock is a very soft oscillator, the oscillating particle moving very slowly and with a rather large amplitude. The pointer of the clock is the position of the oscillating particle.

The clock of the preceding paragraph is still very light. Let us consider, however, the requirement that the linear dimensions of the clock be limited. Since there is little point in dealing with the question in great generality, it may as well be assumed here that the linear dimension shall correspond to the accuracy in time. The requirement $l=ct$ increases the mass of the clock by n^2 which may be a very large factor indeed:

$$m > n^3\hbar t/l^2 = n^3\hbar/c^2 t. \tag{2}$$

For example, a clock, with a running time of a day and an accuracy of 10^{-8} second, must weigh almost a gram—for reasons stemming solely from uncertainty principles and similar considerations.

So far, we have paid attention only to the physical dimension of the clock and the requirement that it be able to distinguish between events which are only a distance t apart on the time scale. In order to make it usable as part of the framework which was described before, it is necessary to *read* the clock and to start it. As part of the framework to map out the metric of space-time, it must either register the readings at which it receives impulses, or transmit these readings to a part of space outside the region to be mapped out. This point was already noted by Schrödinger.[11] However, we found it reassuring that, in the most interesting case in which $l=ct$, that is, if space and time inaccuracies are about equal, the reading requirement introduces only an insignificant numerical factor but does not change the form of the expression for the minimum mass of the clock.

The arrangement to map the metric might consist, therefore, of a lattice of clocks, all more or less at rest with respect to each other. All these clocks can emit light signals and receive them. They can also transmit their reading at the time of the receipt of the light signal to the outside. The clocks may resemble oscillators, well in the nonrelativistic region. In fact, the velocity of the oscillating particle is about n times smaller than the velocity of light where n is the ratio of the *error* in the time measurement, to the *duration* of the whole interval to be measured. This last quantity is the spacing of the events on the time axis, it is also the distance of the clocks from each other, divided by the light velocity. The world lines of the clocks from the dense forest which was mentioned before. Its branches suffuse the region of space-time in which the metric is to be mapped out.

We are not absolutely convinced that our clocks are the best possible. Our principal concern is that we have considered only one space-like dimension. One consequence of this was that the oscillator had to be a one-dimensional oscillator. It is possible that the size limitation does not increase the necessary mass of the clock to the same extent if use is made of all three spatial dimensions.

The curvature tensor can be obtained from the metric in the conventional way, if the metric is measured with sufficient accuracy. It may be of interest, nevertheless, the describe a more direct method for measuring the curvature of space. It involves an arrangement, illustrated in Fig. 5, which is similar to that used for obtaining the metric. There is a clock, and a mirror, at such a distance from each other that the curvature of space can be assumed to be constant in the interven-

[11] E. Schrödinger, Ber. Preuss. Akad. Wiss. phys.-math. Kl. 1931, 238.

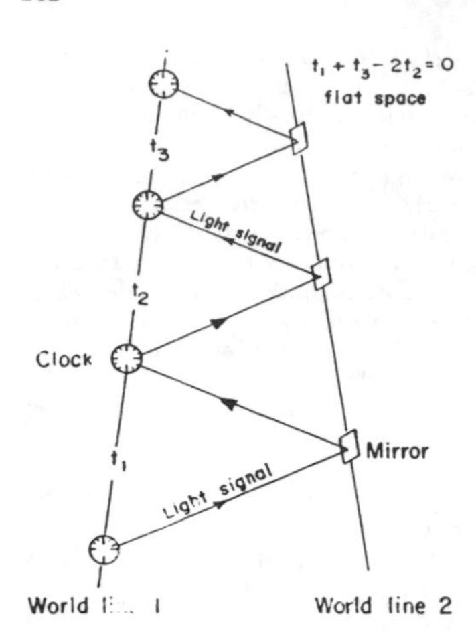

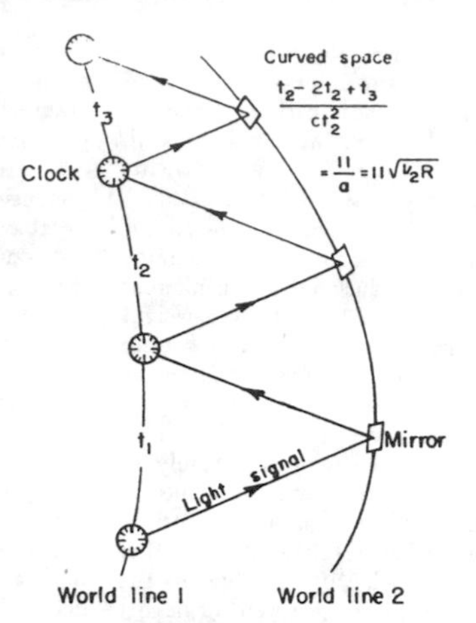

FIG. 5. Direct measurement of the curvature by means of a clock and mirror. Only one space-like dimension is considered and the curvature assumed to be constant within the space-time region contained in the figure. The explanation is given in Appendix V.

ing region. The two clocks need not be at rest with respect to each other, in fact, such a requirement would involve additional measurements to verify it. If the space is flat, the world lines of the clocks can be drawn straight. In order to measure the curvature, a light signal is emitted by the clock, and this is reflected by the mirror. The time of return is read on the clock—it is t_1—and the light signal returned to the mirror. The time which the light signal takes on its second trip to return to the clock is denoted by t_2. The process is repeated a third time, the duration of the last roundtrip denoted by t_3. As shown in Appendix V, the radius of curvature, a, and the relevant component R_{0101} of the Riemann tensor are given by

$$\frac{t_1 - 2t_2 + t_3}{t_2^2} = \frac{11}{a} = 11(\tfrac{1}{2}R_{0101})^{\frac{1}{2}}. \qquad (2)$$

If classical theory would be valid also in the microscopic domain, there would be no limit on the accuracy of the measurement indicated in Fig. 5. If $\hbar$ is infinitely small, the time intervals t_1, t_2, t_3 can all be measured with arbitrary accuracy with an infinitely light clock. Similarly, the light signals between clock and mirror, however short, need carry only an infinitesimal amount of momentum and thus deflect clock and mirror arbitrarily little from their geodesic paths. The quantum phenomena considered before force us, however, to use a clock with a minimum mass if the measurement of the time intervals is to have a given accuracy. In the present case, this accuracy must be relatively high unless the time intervals t_1, t_2, t_3 are of the same order of magnitude as the curvature of space. Similarly, the deflection of clock and mirror from their geodesic paths must be very small if the result of the measurement is to be meaningful. This gives an effective limit for the accuracy with which the curvature can be measured. The result is, as could be anticipated, that the curvature at a *point* in space-time cannot be measured at all; only the average curvature over a finite region of space-time can be obtained. The error of the measurement[1] is inversely proportional to the two-thirds power of the area available in space-time, that is, the area around which a vector is carried, always parallel to itself, in the customary definition of the curvature. The error is also proportional to the cube root of the Compton wavelength of the clock. Our principal hesitation in considering this result as definitive is again its being based on the consideration of only one space-like dimension. The possibilities of measuring devices, as well as the problems, may be substantially different in three-dimensional space.

Whether or not this is the case, the essentially nonmicroscopic nature of the general relativistic concepts seems to us inescapable. If we look at this first from a practical point of view, the situation is rather reassuring. We can note first, that the measurement of electric and magnetic fields, as discussed by Bohr and Rosenfeld,[12] also requires macroscopic, in fact *very* macroscopic, equipment and that this does not render the electromagnetic field concepts useless for the purposes of quantum electrodynamics. It is true that the measurement of space-time curvature requires a finite region of space and there is a minimum for the mass, and even the mass uncertainty, of the measuring equipment. However, numerically, the situation is by no means alarming. Even in interstellar space, it should be possible to measure the curvature

[12] N. Bohr and L. Rosenfeld, Kgl. Danske Videnskab. Selskab Mat.-fys. Medd. 12, No. 8 (1933). See also further literature quoted in L. Rosenfeld's article in *Niels Bohr and the Development of Physics* (Pergamon Press, London, 1955).

in a volume of a light second or so. Furthermore, the mass of the clocks which one will wish to employ for such a measurement is of the order of several micrograms so that the finite mass of elementary particles does not cause any difficulty. The clocks will contain many particles and there is no need, and there is not even an incentive, to employ clocks which are lighter than the elementary particles. This is hardly surprising since the mass which can be derived from the gravitational constant, light velocity, and Planck's constant, is about 20 micrograms.

It is well to repeat, however, that the situation is less satisfactory from a more fundamental point of view. It remains true that we consider, in ordinary quantum theory, position operators as observables without specifying what the coordinates mean. The concepts of quantum field theories are even more weird from the point of view of the basic observation that only coincidences are meaningful. This again is hardly surprising because even a 20-microgram clock is too large for the measurement of atomic times or distances. If we analyze the way in which we "get away" with the use of an absolute space concept, we simply find that we do not. In our experiments we surround the microscopic objects with a very macroscopic framework and observe *coincidences* between the particles emanating from the microscopic system, and parts of the framework. This gives the collision matrix, which is observable, and observable in terms of macroscopic coincidences. However, the so-called observables of the microscopic system are not only not observed, they do not even appear to be meaningful. There is, therefore, a boundary in our experiments between the region in which we use the quantum concepts without worrying about their meaning in face of the fundamental observation of the general theory of relativity, and the surrounding region in which we use concepts which are meaningful also in the face of the basic observation of the general theory of relativity but which cannot be described by means of quantum theory. This appears most unsatisfactory from a strictly logical standpoint.

APPENDIX I

It will be necessary, in this appendix, to compare various states of the same physical system. These states will be generated by looking at the same state—the standard state—from various coordinate systems. Hence every Lorentz frame of reference will define a state of the system—the state as which the standard state appears from the point of view of this coordinate system. In order to define the standard state, we choose an arbitrary but fixed Lorentz frame of reference and stipulate that, in this frame of reference, the particle in the standard state be at rest and its spin (if any) have the direction of the z axis. Thus, if we wish to have a particle moving with a velocity v in the z direction and with a spin also directed along this axis, we look at the particle in the standard state from a coordinate system moving with the velocity v in the $-z$ direction. If we wish to have a particle at rest but with its spin in the yz plane, including an angle α with the z axis, we look at the standard state from a coordinate system the y and z axes of which include an angle α with the y and z axes of the coordinate system in which the standard state was defined. In order to obtain a state in which both velocity and spin have the aforementioned direction (i.e., a direction in the yz plane, including the angles α and $\frac{1}{2}\pi-\alpha$ with the y and z axes), we look at the standard state from the point of view of a coordinate system in which the spin of the standard state is described as this direction and which is moving in the opposite direction.

Two states of the system will be identical only if the Lorentz frames of reference which define them are identical. Under this definition, the relations which will be obtained will be valid independently of the properties of the particle, such as spin or mass (as long as the mass in nonzero so that the standard state exists). Two states will be approximately the same if the two Lorentz frames of reference which define them can be obtained from each other by a very small Lorentz transformation, that is, one which is near the identity. Naturally, all states of a particle which can be compared in this way are related to each other inasmuch as they represent the same standard state viewed from various coordinate systems. However, we shall have to compare only these states.

Let us denote by $A(0,\varphi)$ the matrix of the transformation in which the transformed coordinate system moves with the velocity $-v$ in the z direction where $v=c\tanh\varphi$

$$A(0,\varphi)=\left\|\begin{matrix}1 & 0 & 0\\ 0 & \cosh\varphi & \sinh\varphi\\ 0 & \sinh\varphi & \cosh\varphi\end{matrix}\right\|. \qquad (1.1)$$

Since the x axis will play no role in the following consideration, it is suppressed in (1.1) and the three rows and the three columns of this matrix refer to the y', z', ct' and to the y, z, ct axes, respectively. The matrix (1.1) characterizes the state in which the particle moves with a velocity v in the direction of the z axis and its spin is parallel to this axis.

Let us further denote the matrix of the rotation by an angle φ in the yz plane by

$$R(\vartheta)=\left\|\begin{matrix}\cos\vartheta & \sin\vartheta & 0\\ -\sin\vartheta & \cos\vartheta & 0\\ 0 & 0 & 1\end{matrix}\right\|. \qquad (1.2)$$

We refer to the direction in the yz plane which lies between the y and z axes and includes an angle ϑ with the z axis as the direction ϑ. The coordinate system which moves with the velocity $-v$ in the ϑ direction is obtained by the transformation

$$A(\vartheta,\varphi)=R(\vartheta)A(0,\varphi)R(-\vartheta). \qquad (1.3)$$

In order to obtain a particle which moves in the direction ϑ and is polarized in this direction, we first rotate the coordinate system counterclockwise by ϑ (to have the particle polarized in the proper direction) and impart it then a velocity $-v$ in the ϑ direction. Hence, it is the transformation

$$T(\vartheta,\varphi)=A(\vartheta,\varphi)R(\vartheta)$$

$$=\left\|\begin{matrix}\cos\vartheta & \sin\vartheta\cosh\varphi & \sin\vartheta\sinh\varphi \\ -\sin\vartheta & \cos\vartheta\cosh\varphi & \cos\vartheta\sinh\varphi \\ 0 & \sinh\varphi & \cosh\varphi\end{matrix}\right\| \qquad (1.4)$$

which characterizes the aforementioned state of the particle. It follows from (1.3) that

$$T(\vartheta,\varphi)=R(\vartheta)A(0,\varphi)=R(\vartheta)T(0,\varphi) \qquad (1.5)$$

so that the same state can be obtained also by viewing the state characterized by (1.1) from a coordinate system that is rotated by ϑ. It follows that the statement "velocity and spin are parallel" is invariant under rotations. This had to be expected.

If the state generated by $A(0,\varphi)=T(0,\varphi)$ is viewed from a coordinate system which is moving with the velocity u in the direction of the z axis, the particle will still appear to move in the z direction and its spin will remain parallel to its direction of motion, unless $u>v$ in which case the two directions will become antiparallel, or unless $u=v$ in which case the statement becomes meaningless, the particle appearing to be at rest. Similarly, the other states in which spin and velocity are parallel, i.e., the states generated by the transformations $T(\vartheta,\varphi)$, remain such states if viewed from a coordinate system moving in the direction of the particle's velocity, as long as the coordinate system is not moving faster than the particle. This also had to be expected. However, if the state generated by $T(0,\varphi)$ is viewed from a coordinate system moving with velocity $v'=c\tanh\varphi'$ in the $-y$ direction, spin and velocity will *not* appear parallel any more, *provided the velocity v of the particle is not close to light velocity*. This last proviso is the essential one; it means that the high velocity states of a particle for which spin and velocity are parallel (i.e., the states generated by (1.4) with a large φ) are states of this same nature if viewed from a coordinate system which is not moving too fast in the direction of motion of the particle itself. In the limiting case of the particle moving with light velocity, the aforementioned states become invariant under *all* Lorentz transformations.

Let us first convince ourselves that if the state (1.1) is viewed from a coordinate system moving in the $-y$ direction, its spin and velocity no longer appear parallel. The state in question is generated from the normal state by the transformation

$$A(\tfrac{1}{2}\pi,\varphi')A(0,\varphi)$$

$$=\left\|\begin{matrix}\cosh\varphi' & \sinh\varphi\sinh\varphi' & \cosh\varphi\sinh\varphi' \\ 0 & \cosh\varphi & \sinh\varphi \\ \sinh\varphi' & \sinh\varphi\cosh\varphi' & \cosh\varphi\cosh\varphi'\end{matrix}\right\|. \qquad (1.6)$$

This transformation does not have the form (1.4). In order to bring it into that form, it has to be multiplied on the right by $R(\epsilon)$, i.e., one has to rotate the spin ahead of time. The angle ϵ is given by the equation

$$\tan\epsilon=\frac{\tanh\varphi'}{\sinh\varphi}=\frac{v'}{v}(1-v^2/c^2)^{\frac{1}{2}} \qquad (1.7)$$

and is called the angle between spin and velocity. For $v\ll c$, it becomes equal to the angle which the ordinary resultant of two perpendicular velocities, v and v', includes with the first of these. However, ϵ becomes very small if v is close to c; in this case it is hardly necessary to rotate the spin away from the z axis before giving it a velocity in the z direction. These statements express the identity

$$A(\tfrac{1}{2}\pi,\varphi')A(0,\varphi)R(\epsilon)=T(\vartheta,\varphi'') \qquad (1.8)$$

which can be verified by direct calculation. The right side represents a particle with parallel spin and velocity, the magnitude and direction of the latter being given by the well-known equations

$$v''=c\tanh\varphi''=(v^2+v'^2-v^2v'^2/c^2)^{\frac{1}{2}} \qquad (1.8a)$$

and

$$\tan\vartheta=\frac{\sinh\varphi'}{\tanh\varphi}=\frac{v'}{v(1-v'^2/c^2)^{\frac{1}{2}}}. \qquad (1.8b)$$

Equation (1) given in the text follows from (1.7) and (1.8b) for $v\sim c$.

The fact that the states $T(\vartheta,\varphi)\psi_0$ (where ψ_0 is the standard state and $\varphi\gg1$) are approximately invariant under all Lorentz transformations is expressed mathematically by the equations,

$$R(\vartheta)\cdot T(0,\varphi)\psi_0=T(\vartheta,\varphi)\psi_0, \qquad (1.5a)$$

$$A(0,\varphi')\cdot T(0,\varphi)\psi_0=T(0,\varphi'+\varphi)\psi_0, \qquad (1.9a)$$

and

$$A(\tfrac{1}{2}\pi,\varphi')\cdot T(0,\varphi)\psi_0\rightarrow T(\vartheta,\varphi'')\psi_0, \qquad (1.9b)$$

which give the wave function of the state $T(0,\varphi)\psi_0$, as viewed from other Lorentz frames of reference. Naturally, similar equations apply to all $T(\alpha,\varphi)\psi_0$. In particular, (1.5a) shows that the states in question are invariant under rotations of the coordinate system, (1.9a) that they are invariant with respect to Lorentz transformations with a velocity not too high *in* the direction of motion (so that $\varphi'+\varphi\gg0$, i.e., φ' not too large a negative number). Finally, in order to prove (1.9b), we calculate the transition probability between the states $A(\frac{1}{2}\pi,\varphi')\cdot T(0,\varphi)\psi_0$ and $T(\vartheta,\varphi'')\psi_0$ where ϑ and φ'' are given by (1.8a) and (1.8b). For this, (1.8) gives

$$\begin{aligned}(A(\tfrac{1}{2}\pi,\varphi')\cdot T(0,\varphi)\psi_0,\ T(\vartheta,\varphi'')\psi_0)&=(T(\vartheta,\varphi'')R(\epsilon)^{-1}\psi_0,\ T(\vartheta,\varphi'')\psi_0)\\&=(R(\epsilon)^{-1}\psi_0,\psi_0)\rightarrow(\psi_0,\psi_0).\end{aligned}$$

The second line follows because $T(\vartheta,\varphi'')$ represents a coordinate transformation and is, therefore, unitary. The last member follows because $\epsilon\rightarrow 0$ as $\varphi\rightarrow\infty$ as can be seen from (1.7) and $R(0)=1$.

The preceding consideration is not fundamentally new. It is an elaboration of the facts (a) that the subgroup of the Lorentz group which leaves a null-vector invariant is different from the subgroup which leaves a time like vector invariant[4] and (b) that the representations of the latter subgroup decompose into one dimensional representations if this subgroup is "contracted" into the subgroup which leaves a null-vector invariant.[13]

APPENDIX II

Before the hypothesis of Lee and Yang[14] was put forward, it was commonly assumed that there are, in addition to the symmetry operations of the proper Poincaré group, three further independent symmetry operations. The proper Poincaré group consists of all Lorentz transformations which can be continuously obtained from unity and all translations in space-like and time-like directions, as well as the products of all these transformations. It is a continuous group; the Lorentz transformations contained in it do not change the direction of the time axis and their determinant is 1. The three independent further operations which were considered to be rigorously valid, were

Space inversion I, that is, the transformation $x, y, z\rightarrow -x, -y, -z$, without changing particles into antiparticles.

Time inversion T, more appropriately described by Lüders[15] as *Umkehr der Bewegungsrichtung*, which replaces every velocity by the opposite velocity so that the position of the particles at $+t$ becomes the same as it was, without time inversion, at $-t$. The time inversion T (also called time inversion of the first kind by Lüders[16]) does not convert particles into antiparticles either.

Charge conjugation C, that is, the replacement of positive charges by negative charges and more generally of particles by antiparticles, without changing either the position or the velocity of these particles.[17] The quantum-mechanical expressions for the symmetry operations I and C are unitary, that for T is antiunitary.

The three operations I, T, C, together with their products TC (Lüders' time inversion of the second kind), IC, IT, ITC and the unit operation form a group and the products of the elements of this group with those of the proper Poincaré group were considered to be the symmetry operations of all laws of physics. The suggestion given in the text amounts to eliminating the operations I and C separately while continuing to postulate their product IC as symmetry operation. The discrete symmetry group then reduces to the unit operation plus

$$IC,\ T,\ \text{and}\ ICT, \tag{2.1}$$

and the total symmetry group of the laws of physics becomes the proper Poincaré group plus its products with the elements (2.1). This group is isomorphic (essentially identical) with the unrestricted Poincaré group, i.e., the product of *all* Lorentz transformations with all the displacements in space and time. The quantum mechanical expressions for the operations of the proper Lorentz group and its product with IC are unitary, those for T and ICT (as well as for their products with the elements of the proper Poincaré group) antiunitary. Lüders[16] has pointed out that, under certain very natural conditions, ICT belongs to the symmetry group of every *local* field theory.

APPENDIX III

Let us consider, first, the collision of two particles of equal mass m in the coordinate system in which the average of the sum of their momenta is zero. Let us assume that, at a given time, the wave function of both particles is confined to a distance l in the direction of their average velocity with respect to each other. If we consider only this space-like direction, and the time axis, the area in space-time in which the two wave functions will substantially overlap is [see Fig. 6(a)]

$$a=l^2/2v_{\min}, \tag{3.1}$$

where $v_{\min}$ is the lowest velocity which occurs with substantial probability in the wave packets of the colliding particles. Denoting the average momentum by $\bar{p}$ (this has the same value for both particles) the half-width of the momentum distribution by δ, then $v_{\min}=(\bar{p}-\delta)(m^2+(\bar{p}-\delta)^2/c^2)^{-\frac{1}{2}}$. Since l cannot be below $\hbar/\delta$, the area (3.1) is at least

$$\frac{\hbar^2}{2\delta^2}\frac{(m^2+(\bar{p}-\delta)^2/c^2)^{\frac{1}{2}}}{\bar{p}-\delta}. \tag{3.1a}$$

(Note that the area becomes infinite if $\delta>\bar{p}$.) The

[13] E. Inonu and E. P. Wigner, Proc. Natl. Acad. Sci. U.S. 39, 510 (1953).

[14] T. D. Lee and C. N. Yang, Phys. Rev. 104, 254 (1956). See also E. M. Purcell and N. F. Ramsey, Phys. Rev. 78, 807 (1950).

[15] G. Lüders, Z. Physik 133, 325 (1952).

[16] G. Lüders, Kgl. Danske Videnskab. Selskab Mat.-fys. Medd. 28, No. 5 (1954).

[17] All three symmetry operations were first discussed in detail by J. Schwinger, Phys. Rev. 74, 1439 (1948). See also H. A. Kramers, Proc. Acad. Sci. Amsterdam 40, 814 (1937) and W. Pauli's article in *Niels Bohr and the Development of Physics* (Pergamon Press, London, 1955). The significance of the first two symmetry operations (and their connection with the concepts of parity and the Kramers degeneracy respectively), were first pointed out by the present writer, Z. Physik 43, 624 (1927) and Nachr. Akad. Wiss. Göttingen, Math.-physik. 1932, 546. See also T. D. Newton and E. P. Wigner, Revs. Modern Phys. 21, 400 (1949); S. Watanabe, Revs. Modern Phys. 27, 26 (1945). The concept of charge conjugation is based on the observation of W. Furry, Phys. Rev. 51, 125 (1937).

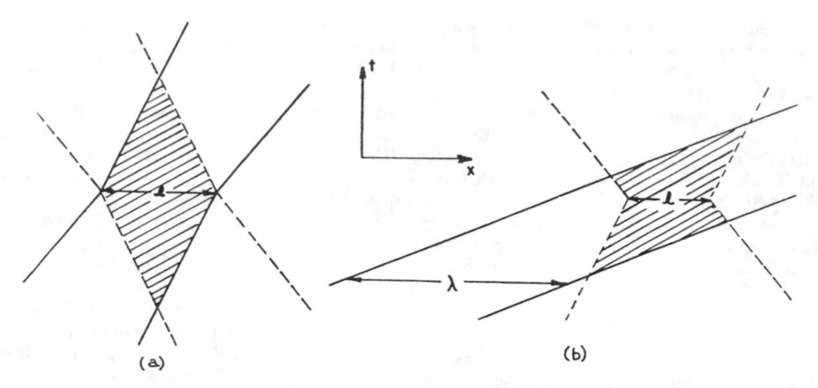

FIG. 6. (a) Localization of a collision of two particles of equal mass. The full lines indicate the effective boundaries of the wave packet of the particle traveling to the right, the broken lines the effective boundaries of the wave packet of the particle traveling to the left. The collision can take place in the shaded area of space-time. (b) Localization of a collision between a particle with finite mass and a particle with zero rest-mass. The full lines, at a distance λ apart in the x direction, indicate the boundary of the particle with zero rest-mass, the broken lines apply to the wave packet of the particle with nonzero rest-mass. The collision can take place in the shaded area.

minimum of (3.1a) is, apart from a numerical factor

$$a_{\min}\approx\frac{\hbar^2}{\bar{p}^3}(m^2+\bar{p}^2/c^2)^{\frac12}\approx\frac{\hbar^2 c}{E^{\frac32}(E+mc^2)^{\frac12}},\tag{3.2}$$

where E is the kinetic energy (total energy minus rest-energy) of the particles.

The kinetic energy E permits the contraction of the wave functions of the colliding particles also in directions perpendicular to the average relative velocity, to an area $\hbar^2c^2/E(E+2mc^2)$. Hence, again apart from a numerical factor, the volume to which the collision can be confined in four dimensional space-time becomes

$$V_{\min}=\frac{\hbar^4c^3}{E^{\frac52}(E+mc^2)^{\frac32}}.\tag{3.3}$$

E is the average kinetic energy of the particles in the coordinate system in which their center of mass is, on the average, at rest. Equation (3.3) is valid apart from a numerical constant of unit order of magnitude but this constant depends on E/mc^2.

Let us consider now the opposite limiting case, the collision of a particle with finite rest-mass m with a particle with zero rest-mass. The collision is viewed again in the coordinate system in which the average linear momentum is zero. In this case, one will wish to confine the wave function of the particle with finite rest-mass to a narrower region l than that of the particle with zero rest-mass. If the latter is confined to a region of thickness λ, [see Fig. 6(b)], its momentum and energy uncertainties will be at least $\hbar/\lambda$ and $\hbar c/\lambda$ and these expressions will also give, apart from a numerical factor, the average values of these quantities. Hence $\bar{p}\approx\hbar/\lambda$. The kinetic energy of the particle with finite restmass will be of the order of magnitude

$$\tfrac12(m^2c^4+(\bar{p}+\hbar/l)^2c^2)^{\frac12}+\tfrac12(m^2c^4+(\bar{p}-\hbar/l)^2c^2)^{\frac12}-mc^2,\tag{3.4}$$

since $\hbar/l$ is the momentum uncertainty. Since $l\leq\lambda$, one can neglect $\bar{p}$ in (3.4) if one is interested only in the order of magnitude. This gives for the total kinetic energy,

$$E\approx\hbar c/\lambda+(m^2c^4+\hbar^2c^2/l^2)^{\frac12}-mc^2,\tag{3.5}$$

while the area in Fig. 6(b) is of the order of magnitude

$$a=(\lambda/c)(l+\Delta v\lambda/c),\tag{3.6}$$

where Δv is the uncertainty in the velocity of the second particle

$$\Delta v=\frac{\bar{p}+\hbar/l}{(m^2+(\bar{p}+\hbar/l)^2/c^2)^{\frac12}}-\frac{\bar{p}-\hbar/l}{(m^2+(\bar{p}-\hbar/l)^2/c^2)^{\frac12}}.\tag{3.6a}$$

This can again be replaced by $(\hbar/l)(m^2+\hbar^2/l^2c^2)^{-\frac12}$.

For given E, the minimum value of a is assumed if the *kinetic energies* of the two particles are of the same order of magnitude. The two terms of (3.6) then become about equal and $l/\lambda\approx(E/(m+E))^{\frac12}$. The minimum value of a, as far as order of magnitude is concerned, is again given by (3.2). Similarly, (3.3) also remains valid if one of the two particles has zero rest-mass.

The two-dimensional case becomes simplest if both particles have zero rest-mass. In this case the wave packets do not spread at all and (3.2) can be immediately seen to be valid. In the four-dimensional case, (3.3) again holds. However, its proof by means of explicitly constructed wave packets (rather than reference to the uncertainty relations) is by no means simple. It requires wave packets which are confined in every direction, do not spread too fast and progress essentially only into one half space (one particle going toward the right, the other toward the left). The construction of such wave packets will not be given in detail. They are necessary to prove (3.2) and (3.3) more rigorously also in the case of finite masses; the preceding proofs, based on the uncertainty relations show only that a and v cannot be *smaller* than the right sides of the corresponding equations. It is clear, in fact, that the limits given by (3.2) and (3.3) would be very difficult to realize, except in the two-dimensional case and for the collision of two particles with zero restmass. In all other cases, the relatively low values of $a_{\min}$ and $V_{\min}$ are predicated on the assumption that the wave packets of the colliding particles are so constituted that they assume a minimum size at the time of the collision. At any rate, (3.2) and (3.3) show that only collisions with a relatively high collision energy, and high energy uncertainty, can be closely localized in space-time.

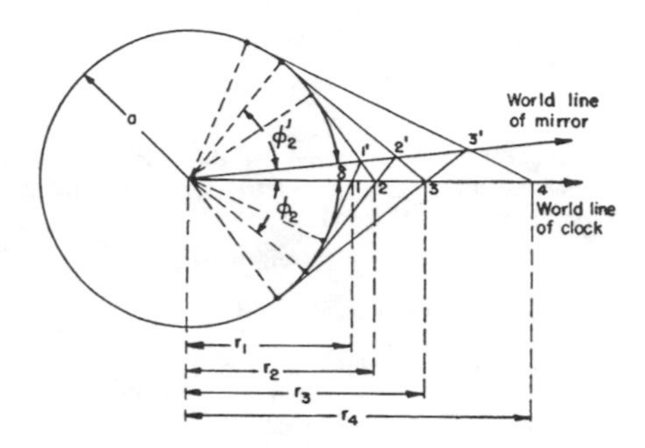

FIG. 7. Analysis of the experiment of Fig. 5. The figure represents a view of the hyperboloid of deSitter space, viewed along its axis. Every point of the plane which is outside the circle corresponds to two points of the deSitter world with one spatial dimension, those with oppositely equal times. The first light signal is emitted at 1, reaches the mirror at 1′, and returns to the clock at 2. The paths of the second and third light signals are 22′3 and 33′4.

APPENDIX IV

Let us denote the components of the vector from event 1 to event 2 by x_i, the components of the unit vector along the world line of the first clock at event 1 by e_i. The components of the first light signal are x_i+te_i, that of the second light signal $x_i-t'e_i$. Hence

$$g^{ik}(x_i+te_i)(x_k+te_k)=0 \tag{4.1}$$

$$g^{ik}(x_i-t'e_i)(x_k-t'e_k)=0. \tag{4.2}$$

Elimination of the linear terms in t and t' by multiplication of (4.1) with t' and (4.2) with t and addition gives

$$2g^{ik}x_ix_k+2tt'g^{ik}e_ie_k=0. \tag{4.3}$$

Since e is a unit vector $g^{ik}e_ie_k=1$ and (4.3) shows that the space-like distance between points 1 and 2 is $(tt')^{\frac{1}{2}}$.

APPENDIX V

Since the measurement of the curvature, described in the text, presupposes *constant curvature* over the space-time domain in which the measurement takes place, we use a space with constant curvature, or, rather, part of a space with constant curvature, to carry out the calculation. We consider only one spatial dimension, i.e., a two-dimensional deSitter space. This will be embedded, in the usual way, in a three-dimensional space[18] with coordinates x, y, τ. The points of the deSitter space then form the hyperboloid

$$x^2+y^2-\tau^2=a^2, \tag{5.1}$$

where a is the "radius of the universe." As coordinates of a point we use x and y, or rather the corresponding polar angles r, ϕ. The metric form in terms of these is

$$ds^2=\frac{a^2}{r^2-a^2}dr^2-r^2d\phi^2. \tag{5.2}$$

Two points of deSitter space correspond to every pair r, ϕ (except $r=a$): those with positive and negative $\tau=(r^2-a^2)^{\frac{1}{2}}$. This will not lead to any confusion as all events take place at positive τ. The null lines (paths of light signals) are the tangents to the $r=a$ circle.

The experiment described in the text can be analyzed by means of Fig. 7. For the sake of simplicity, the clock and mirror are assumed to be "at rest," i.e., their world lines have constant polar angles which will be assumed as 0 and δ, respectively. The first light signal travels from 1 to 1′ and back to 2, the second from 2 to 2′ and back to 3, the third from 3 to 3′ and back to 4. The polar angle of the radius vector which is perpendicular to the first part 22′ of the world line of the second light signal is denoted by ϕ_2. The construction of Fig. 7 shows that angle ϕ_2' which the world line of the mirror includes with the radius vector perpendicular to the second part 2′3 of the second light signal's world line is

$$\phi_2'=\phi_2+\delta. \tag{5.3}$$

The angles ϕ_1, ϕ_1', ϕ_3, ϕ_3' have similar meanings; they are not indicated in the figure in order to avoid overcrowding. For reasons similar to those leading to (5.3), we have

$$\phi_3=\phi_2'+\delta=\phi_2+2\delta \tag{5.3a}$$

$$\phi_1=\phi_2-2\delta \tag{5.3b}$$

$$\phi_4=\phi_3+2\delta=\phi_2+4\delta. \tag{5.3c}$$

[18] See, e.g., H. P. Robertson, Revs. Modern Phys. 5, 62 (1933).

The radial coordinates of the points 1, 2, 3, 4 are denoted by r_1, r_2, r_3, r_4

$$r_i = a/\cos\phi_i. \tag{5.4}$$

The proper time t, registered by the clock, can be obtained by integrating the metric form (5.2) along the world line $\phi=0$ of the clock

$$t = a \ln[r+(r^2-a^2)^{\frac{1}{2}}]. \tag{5.5}$$

Hence, the traveling time t_2 of the second light signal becomes

$$t_2 = a \ln\frac{r_3+(r_3^2-a^2)^{\frac{1}{2}}}{r_2+(r_2^2-a^2)^{\frac{1}{2}}} = a \ln\frac{\cos\phi_2(1+\sin\phi_3)}{\cos\phi_3(1+\sin\phi_3)}. \tag{5.6}$$

Similar expressions apply for the traveling times of the first and third light signals; all ϕ can be expressed by means of (5.3a), (5.3b), (5.3c) in terms of ϕ_2 and δ. This allows the calculation of the expression (3). For small δ, one obtains

$$\frac{t_1-2t_2+t_3}{t_2^2} \approx \frac{11}{a}, \tag{5.7}$$

and Riemann's invariant $R=2/a^2$ is proportional to the square of (5.7). In particular, it vanishes if the expression (3) is zero.

P. A. M. Dirac

THE EARLY YEARS OF RELATIVITY

I AM VERY HAPPY to have this opportunity of talking at the Einstein Symposium because I am a very great admirer of Einstein.

In this Symposium we have heard many historical talks. Historians collect all the documents they can find referring to their subject, assess those documents, collate them, and then give us a detailed account of what happened. I am not going to give you a talk of that nature, because I am not a historian. What I would like to talk to you about is the arrival of relativity as it appeared to someone who lived through it.

I was in England at the time, and for the most part I shall be talking about the coming of relativity to England, although I think it happened very much the same in other Western countries. What I want to emphasize is the tremendous impact that relativity had on the general public. I think that the historians have not emphasized this sufficiently in their talks.

First, I must describe the background. The time I am speaking of is the end of the First World War. That war had been long and terrible. It had been a war in which not very much had happened, from the military point of view. We had battle lines that remained almost static year after year, but we also had tremendous casualties, ever increasing casualties, and nearly everyone had lost close relatives or friends. Then the end of this war came, rather suddenly and unexpectedly, in November 1918. There was immediately an intense feeling of relaxation. It was something dreadful that was now finished. People wanted to get away from thinking about the awful war that had passed. They wanted something new. And that is when relativity burst upon us.

Reprinted from *Albert Einstein: Historical and Cultural Perspectives: The Centennial Symposium in Jerusalem*, (1979).

P. A. M. DIRAC

I can't describe it by other words than by saying that it just *burst* upon us. It was a new idea, a new kind of philosophy, and it aroused interest and excitement in everyone. The newspapers, as well as the magazines, both popular and technical, were continually carrying articles about it. These articles were mainly written from the "philosophical" point of view. Everything had to be considered relatively to something else. Absolutism was just a bad idea that one had to get away from. There was no public radio in those days, so we did not have that kind of propaganda; but all the written material that we had was devoted to bringing out this new idea of relativity. In most of what people wrote there was no real physics involved. The people writing the articles did not understand the physics.

At that time I was sixteen years old and a student of engineering at Bristol University. You might think it was rather unusual for someone only sixteen years old to be a student, but it was not unusual when you consider the times. All the young men had been taken away from the universities to serve in the army. There were some professors left, those who were too old to serve in the army and those who were not physically fit; but they had empty classrooms. So the younger boys were pushed on, as far as they were able to absorb the knowledge, to fill up these empty classrooms. That is how I came to be an engineering student at the time.

I was caught up in the excitement of relativity along with my fellow students. We were studying engineering, and all our work was based on Newton. We had absolute faith in Newton, and now we learned that Newton was wrong in some mysterious way. This was a very puzzling situation. Our professors were not able to help us, because no one really had the precise information needed to explain things properly, except for one man, Arthur Eddington.

Eddington was an astronomer. He had kept in touch with his friend Willem de Sitter, an astronomer in Holland. Holland was neutral at that time, so they could write to each other. De Sitter had kept in contact with Einstein, and in that roundabout way Eddington was in contact with Einstein and had heard all about the development of relativity theory.

At the end of 1918, the end of the war, relativity was not really a new idea. It had dated from 1905, a good many years before the war started. But no one had heard about relativity or about Einstein previously, except for a few specialists at the universities. No one in the engineering faculty at Bristol had heard anything about these things, and it was all completely new to us. Then we had Eddington to explain things.

Eddington was very good at popular exposition. He had a great talent for it, and he applied his talent to explaining the foundations of relativity theory to the general public. He told us about the need to consider that we cannot communicate instantaneously with people at a distance. We could communicate only with the help of light signals and make a note of the time the light signals were sent out and the time we received an answer, and we had to work on that. Eddington also told us about the Michelson-Morley experiment. It was quite an old experiment, but one whose importance had not been previously appreciated. Albert A. Michelson and Edward W. Morley had attempted to determine the velocity of the earth through the ether, but, surprisingly, their experiment failed to give any definite answer. The only way to account for this failure was to suppose a very peculiar behavior of measuring rods and clocks. Moving measuring rods had to be subject to a kind of contraction, called the Lorentz-Fitzgerald contraction. Clocks had to have their rates slowed up when they were moving.

All this was very hard to explain to the general public, but still Eddington and other people wrote innumerable articles about it, doing the best they could. The engineering students were not very much better off. We were told that in some way the absolute scheme of things that we had been using in our engineering studies had to be modified, but there was no very definite way to make the modification. We were not given any definite equations.

Now, Eddington was an astronomer and was very interested in the general theory of relativity, and especially in testing its astronomical consequences. The theory predicts a motion of the perihelion of Mercury. It was easy to work this out and to check that the observations agreed with the Einstein theory. This was the first big triumph of the Einstein theory.

Then there was the question of observing whether light would be deflected when it passed close by the sun, another requirement of the Einstein theory. This is something that can be tested observationally only at a time of total eclipse. Eddington had heard about these things already in 1916, and he immediately set to work to find out if there was a favorable total eclipse coming up soon. He found that there would be a very favorable one occurring in May 1919—very favorable because the time of totality was long and also because the sun was then in a very rich field of stars, so that there would be many stars to observe. Eddington immediately set to work to plan expeditions to make the needed observations. Of course, he knew very well that there would be no hope of making such observations while the war was on, but he was hoping that the war would be over in time. The war was over in November 1918, which was in time, and the eclipse expeditions were sent out. They brought back their photographic plates and measured them up, and they got results in agreement with the Einstein theory. Then the public really went wild. The new theory was proved, and everyone was so excited about it.

I shall mention just one example to illustrate the sort of enthusiasm that prevailed. In a detective story called *The Bishop Murder Case*, an important clue was provided by a piece of paper on which some of the Einstein equations were written down. The theory of relativity was woven into the plot of the story, with the result that the book had a big sale. All the young people were reading it. That shows you the tremendous excitement that pervaded all fields of thought. It has never happened before or since in the history of science that a scientific idea has been so much caught up by the public and has produced so much enthusiasm and excitement.

Yet with all that was written about relativity, we still had very little accurate information. Professor Broad at Bristol gave a series of lectures about relativity, which I attended, but they dealt mostly with philosophical aspects. He did, however, give some information about the geometry of special relativity, in particular about Minkowsky's space, and I began to get some definite information about the theory from him.

We really had no chance to understand relativity properly until 1923, when Eddington published his book, *The Mathematical Theory of Relativity*, which contained all the information needed for a proper understanding of the basis of the theory. This mathematical information was interspersed with a lot of philosophy. Eddington had his own philosophical views, which, I believe, were somewhat different from Einstein's, but developed from them.

But there it was, and it was possible for people who had a knowledge of the calculus, people such as engineering students, to check the work and study it in detail. The going was pretty tough. It was a harder kind of mathematics than we had been used to in our engineering training, but still it was possible to master the theory. That was how I got to know about relativity in an accurate way.

EARLY YEARS OF RELATIVITY

Eddington's book did not give any information about the struggles that Einstein had gone through in order to set up his theory. It just gave us the completed result. I have been very interested in the lectures given by historians at the various Einstein symposia, which enable me to understand better both Einstein's struggles and also his appreciation of the need for beauty in the mathematical foundation. Einstein seemed to feel that beauty in the mathematical foundation was more important, in a very fundamental way, than getting agreement with observation.

This was brought out very clearly in the early work about the theory of the electron. Hendrik Lorentz had set up a theory for the motion of electrons that was in agreement with Einstein's principles, and experiments were made by Walter Kaufmann to see whether this theory was in agreement with observation. The resulting observations did not support the theory of Lorentz and Einstein. Instead, they supported an older theory of the electron, given by Max Abraham. Lorentz was completely knocked out by this result. He bewailed that all his work had gone for nothing. Einstein seems not to have reacted very much to it. I do not know just what he said—that's a question for the historians to decide. I imagine that he said, "Well, I have this beautiful theory, and I'm not going to give it up, whatever the experimenters find; let us just wait and see."

Well, Einstein proved to be right. Three years later the experiments were done again by someone else, and the new experiment supported the Lorentz-Einstein view of the electron. And some years after that, a fault was found in the apparatus of Kaufmann. So it seems that one is very well justified in attaching more importance to the beauty of a theory and not allowing oneself to be too much disturbed by experimenters, who might very well be using faulty apparatus.

Let us return to the general theory of relativity. The observations of the eclipse expeditions supported the theory. Right from the beginning there was agreement. We then had a satisfactory basis for the development of relativity. At that time I was a research student and very much enjoyed the new field of work that was opened up by relativity. One could take some previous piece of work that had been expressed in nonrelativistic language and turn it into the relativistic formalism, get a better understanding of it, and perhaps find enough material to publish a paper.

It was about that time, I think in 1924, that A. H. Compton visited Cambridge and spoke at the Cavendish Laboratory about his experiments on the Compton effect. These experiments involved both the light quantum hypothesis and some of the fundamental relations of special relativity. I can remember very well this colloquium. I found Compton's results very convincing, and I think most of the members of the audience were also convinced by Compton, although there were a few people who held out against the light quantum hypothesis.

The situation was completely changed in 1925, when Werner Heisenberg introduced his new quantum mechanics. This was a theory in which it was soon found out that the basic idea was to have dynamic variables that do not satisfy the commutative law of multiplication. That is to say, $a \times b$ is not equal to $b \times a$. Quite a revolutionary idea, but an idea that rapidly proved to have great success.

Now, in spite of that success, Einstein was always rather hostile to quantum mechanics. How can one understand this? I think it is very easy to understand, because Einstein had been proceeding on different lines, lines of pure geometry. He had been developing geometrical theories and had achieved enormous success. It is only natural that he should think that further problems of physics should be solved by further developments of geometrical ideas. Now, to have $a \times b$ not equal to $b \times a$ is something that does not fit in very well with geometrical ideas; hence his hostility to it.

I first met Einstein at the 1927 Solvay Conference. This was the beginning of the big discussion between Niels Bohr and Einstein, which centered on the interpretation of quantum mechanics. Bohr, backed up by a good many other physicists, insisted that one can use only a statistical interpretation, getting probabilities from the theory and then comparing these probabilities with observation. Einstein insisted that nature does not work in this way, that there should be some underlying determinism. Who was right?

At the present time, one must say that, according to Heisenberg's quantum mechanics, we must accept the Bohr interpretation. Any student who is working for an exam must adopt this interpretation if he is to be successful in his exams. Once he has passed his exams, he may think more freely about it, and then he may be inclined to feel the force of Einstein's argument.

In this discussion at the Solvay Conference between Einstein and Bohr, I did not take much part. I listened to their arguments, but I did not join in them, essentially because I was not very much interested. I was more interested in getting the correct equations. It seemed to me that the foundation of the work of a mathematical physicist is to get the correct equations, that the interpretation of those equations was only of secondary importance.

Right from the beginning of quantum mechanics, I was very much concerned with the problem of fitting it in with relativity. This turned out to be very difficult, except in the case of a single particle, where it was possible to make some progress. One could find equations for describing a single particle in accordance with quantum mechanics, in agreement with the principle of special relativity. It turned out that this theory provided an explanation of the spin of the electron.

Also, one could develop the theory a little further and get to the idea of antimatter. The idea of antimatter really follows directly from Einstein's special theory of relativity when it is combined with the quantum mechanics of Heisenberg. There is no escape from it. With just a single, uniform line of argument one goes right up from special relativity to antimatter.

This was all very satisfactory so long as one considered only a single particle. There remained, of course, the problem of two or more particles interacting with each other. Then one soon found that there were serious difficulties. Applying the standard rules, all one could say was that the theory did not work. The theory allowed one to set up definite equations. When one tried to interpret those equations, one found that certain quantities were infinite according to the theory, when according to common sense they should be finite. That was a very serious difficulty in the theory, a difficulty that still has not been completely resolved.

Physicists have been very clever in finding ways of turning a blind eye to terms they prefer not to see in an equation. They may go on to get useful results, but this procedure is of course very far from the way in which Einstein thought that nature should work.

It seems clear that the present quantum mechanics is not in its final form. Some further changes will be needed, just about as drastic as the changes made in passing from Bohr's orbit theory to quantum mechanics. Some day a new quantum mechanics, a relativistic one, will be discovered, in which we will not have these infinities occurring at all. It might very well be that the new quantum mechanics will have determinism in the way that Einstein wanted. This determinism will be introduced only at the expense of abandoning some other preconceptions that physicists now hold. So, under these conditions I think it is very likely, or at any rate quite possible, that in the long run Einstein will turn out to be correct, even though for the time being physicists have to accept the Bohr probability interpretation, especially if they have examinations in front of them.

There are two other subjects that I would like to talk about. The first concerns an incompleteness in the Einstein theory of gravitation. This Einstein theory gives us field equations, but one cannot solve them and obtain precise results without having some boundary conditions, that is, some conditions that one can use to refer to what space is like at a very great distance. To establish the boundary conditions, we need to have a theory of cosmology, a theory that tells us what the universe as a whole looks like when we smooth out the irregularities caused by stars and galaxies, and so on. Einstein himself realized from the beginning the need for such boundary conditions, and he proposed a model of the universe, a model in which space is limited, although unbounded. That was a new idea for people to get used to, one on which Eddington brought to bear his remarkable powers of explaining to try to get it across to the general public. The idea is not that difficult to comprehend; you can think of the surface of the earth, a region of space in two dimensions that is finite but still unbounded.

Einstein set up a cylindrical model of the universe, but it was soon found that it had to be abandoned because of its static character. Observations showed us that things that are very far away are receding from us with velocities that increase as the distance increases. Einstein's model did not account for that phenomenon. Therefore, it had to be abandoned.

Soon afterward, another model was proposed by de Sitter. De Sitter's model did give correctly the distant matter receding from us, but it gave zero for the average density of matter, which of course is in disagreement with observation. Therefore, this model also had to be abandoned.

Many other models were worked out by Alexander Friedman, Georges Lemaitre, and others, all consistent with Einstein's basic field equations. Among all these other models, I want to call your attention to one that was brought out jointly by Einstein and de Sitter. This model was satisfactory in that it gives distant matter receding from us and also gives a non-zero average density of matter about in agreement with observation. In all its basic features it appears to be satisfactory.

This Einstein–de Sitter joint model I would like to propose as the one that should be generally accepted, because it is the simplest of all the models that are not in some elementary way in disagreement with the facts. We ought to keep to the simplest model until something turns up that causes us to depart from it.

The Einstein–de Sitter model proposes that the universe started with a big bang and will go on expanding forever. Many models claim that the universe will expand to a certain point and then contract again. This is an unneeded complication, and there is no observational evidence supporting it at the present. I feel that such models should not be taken as seriously as the simple Einstein–de Sitter model. The latter has the character that three-dimensional space is flat. It also has the character that the pressure in the smoothed-out universe is very small.

I should mention that all these different models will have their influence on the solutions of the Einstein equations we need in order to account for phenomena occurring in the solar system. But the differences are exceedingly small. They would not affect the wonderful agreement that we have between observation and Einstein's theory in the description of the solar system.

There is one other topic that I would like to talk about: the problem of the unification of the gravitational field and the electromagnetic field. Einstein was very much concerned with this unification. Those were the only two fields that he had to consider. Nowadays, physicists have other fields, but these other fields all involve short-range forces, forces that are significant only for particles extremely close together, lying inside an atomic nucleus. The gravitational force and the electromagnetic force are long-range forces. These forces fall off inversely proportional to the square of the distance. In some ways, they are more important, I think, than the other forces. One feels that there should be a close connection between them.

Einstein had achieved his very great success in accounting for the gravitational force in terms of geometry, and he thought that some generalization of the geometry would bring in the electromagnetic force also. A method of doing this was very soon discovered by Hermann Weyl. Weyl made a rather simple generalization in the geometry of Riemann, which Einstein was using. He supposed that the distance between two neighboring points does not have an absolute value. There is no natural absolute unit to which one could refer it. But one can transport it from one location in space to another, and the equations that govern this transport are such that, if you go around a closed loop and get back to your starting point, the final distance does not agree with the starting value. That led to a generalization of the geometry, which was soon found to provide just what was needed in order to bring in the electromagnetic field.

This seemed to be a very wonderful solution of the problem, but then there was a difficulty. Atomic events do provide a natural scale for measuring distances. You could refer all your distances to this atomic scale, and then there would be no point at all in having the uncertainty in distances introduced by Weyl's geometry. So this very beautiful theory of Weyl was reluctantly abandoned.

Einstein worked for the rest of his life on trying to solve this problem of unifying the gravitational and the electromagnetic fields. He tried one scheme after another. All were unsatisfactory. Other people have joined in this work without achieving any greater success than Einstein did. I have been wondering whether Einstein was limiting his ideas too much in these attempts to unify the gravitational and the electromagnetic fields. It seems to me that it is quite possible that one will have to bring in cosmological effects to arrive at a satisfactory solution of this problem. The way cosmological effects would show up would be this.

Let us accept the Einstein theory as it stands for all problems involving just classical theory, and only when we go over to atomic problems, let us require that some modification is needed. This modification can be expressed by saying that Planck's constant is not really a constant in the cosmological sense but must be considered as varying with the epoch, that is, with the time since the origin of the universe, the "Big Bang." If we have Planck's constant h varying, then we must also have the charge on the electron e varying, because e^2/hc is a dimensionless constant that plays an important role in physics and is observed to have the value 1/137, and it seems to be really constant. Thus, if h is varying, e must vary according to its square root. On this basis one could set up a new theory.

Before introducing such a drastic revision in basic ideas, it is desirable to have some confirmation of it by observation. If you have these atomic constants varying when referred to the Einstein picture, it will mean that atomic clocks will not keep the same time as the time of the Einstein theory. The time of the Einstein theory is the time that governs the motion of the planets around the sun, what astronomers call ephemeris time. So one could look to see whether there is any difference between atomic clocks and clocks based on ephemeris time.

Astronomers have been studying this question. In particular, T. C. Van Flandern, working at the Naval Research Observatory in Washington, has spent many years studying lunar motion, referred to both ephemeris time and atomic time. Lunar motion has been observed with atomic clocks since 1955. However, Van Flandern's results up to the present are still not conclusive. One must wait a little longer to see whether a difference really exists between the two ways of measuring time.

There is another possibility of checking these new ideas about variation of constants that has been followed up by I. I. Shapiro. His method consists in sending radar signals to one of the planets and observing the reflected radar signals, and then timing the to and fro journey with an atomic clock. In effect, he is observing distances in the solar system with atomic clocks, and there should be some discrepancy showing up if the atomic clocks do not keep the same time as ephemeris time.

I was talking to Shapiro just before I came to this Symposium and asked him about the latest information concerning his work. There is a very good chance of observing radar waves reflected from Mars because of the Viking Lander that landed on Mars in 1976, which can be used to send back reflected radar waves. Shapiro said that the time base that he has for the observations with the Viking expedition (just about two years) is not long enough for him to give a definite answer to this question. There were also observations of Mars made previously with a Mariner expedition, which give him a considerably longer time base, eight years instead of two. But he told me that another year or two of work would be needed to evaluate his results before he could answer the question of variation of the constants.

That is the situation at the present time. I am sorry that I cannot offer anything more definite than that. There are hopes that Shapiro will come out with a definite result, and if it is a positive result, we shall have a new basis for looking at the question of the unification of the gravitational and the electromagnetic fields. It might be possible to revive Weyl's geometry, which was abandoned only because it clashed with the distances provided by atomic events. So it seems that if this approach turns out to be correct, the whole theory of unifying these distances with gravitational theory would have to be gone into again. There may very well be greater chances of success.

Chapter II

Representations of the Poincaré Group

This Chapter consists of the fundamental paper of Wigner on the Poincaré group and some of the resulting papers. In his original paper, Wigner constructed subgroups of the Lorentz group whose transformations leave the four-momentum of a given particle invariant. These subgroups are called the little groups. In their 1948 paper, Bargmann and Wigner formulated Wigner's little groups in terms of the infinitesimal generators. The little groups for massive, massless, and imaginary-mass particles are locally isomorphic to the three-dimensional rotation group, the two-dimensional Euclidean group, and the (2 + 1)-dimensional Lorentz group.

In his 1945 paper, Dirac suggested the four-dimensional harmonic oscillator to construct representations of the Lorentz group. In their 1979 papers, Kim, Noz, and Oh constructed a representation of the O(3)-like little group for massive hadrons in the quark model.

Quantum field theory has its place in the development of relativistic quantum theory, and its strength and weakness are well known. As is well known, the Poincaré group is the basic language for quantum field theory, and there are many papers which will indicate this point. In one of his 1964 papers, Weinberg gives a lucid treatment of the role of the representations of the Poincaré group in Feynman diagrams.

ON UNITARY REPRESENTATIONS OF THE INHOMOGENEOUS LORENTZ GROUP

By E. Wigner

(Received December 22, 1937)

Parts of the present paper were presented at the Pittsburgh Symposium on Group Theory and Quantum Mechanics. Cf. Bull. Amer. Math. Soc., 41, p.306, 1935.

Reprinted from *Ann. Math.* **40**, 149 (1939).

1. Origin and Characterization of the Problem

It is perhaps the most fundamental principle of Quantum Mechanics that the system of states forms a *linear manifold*,[1] in which a *unitary scalar* product is defined.[2] The states are generally represented by wave functions[3] in such a way that ϕ and constant multiples of ϕ represent the same physical state. It is possible, therefore, to normalize the wave function, i.e., to multiply it by a constant factor such that its scalar product with itself becomes 1. Then, only a constant factor of modulus 1, the so-called phase, will be left undetermined in the wave function. The linear character of the wave function is called the superposition principle. The square of the modulus of the unitary scalar product (ψ,ϕ) of two normalized wave functions ψ and ϕ is called the transition probability from the state ψ into ϕ, or conversely. This is supposed to give the probability that an experiment performed on a system in the state ϕ, to see whether or not the state is ψ, gives the result that it is ψ. If there are two or more different experiments to decide this (e.g., essentially the same experiment, performed at different times) they are all supposed to give the same result,

[1]The possibility of a future non linear character of the quantum mechanics must be admitted, of course. An indication in this direction is given by the theory of the positron, as developed by P.A.M. Dirac (Proc. Camb. Phil Soc. *30*, 150, 1934, cf. also W. Heisenberg, Zeits. f. Phys. *90*, 209, 1934; *92*, 623, 1934; W. Heisenberg and H. Euler, ibid. *98*, 714, 1936 and R. Serber, Phys. Rev. *48*, 49, 1935; *49*, 545, 1936) which does not use wave functions and is a non linear theory.

[2]Cf. P.A.M. Dirac, *The Principles of Quantum Mechanics*, Oxford 1935, Chapters I and II; J. v. Neumann, *Mathematische Grundlagen der Quantenmechanik*, Berlin 1932, pages 19-24.

[3]The wave functions represent throughout this paper states in the sense of the "Heisenberg picture," i.e. a single wave function represents the state for all past and future. On the other hand, the operator which refers to a measurement at a certain time t contains this t as a parameter. (Cf. e.g. Dirac, *l.c.* ref. 2, pages 115-123). One obtains the wave function $\phi_s(t)$ of the Schroedinger picture from the wave function ϕ_H of the Heisenberg picture by $\phi_s(t) = \exp(-iHt/\hbar)\phi_H$. The operator of the Heisenberg picture is $Q(t) = \exp(iHt/\hbar)Q\exp(-iHt/\hbar)$, where Q is the operator in the Schroedinger picture which does not depend on time. Cf also E. Schroedinger, Sitz. d. Koen. Preuss. Akad. p. 418, 1930.

The wave functions are complex quantities and the undetermined factors in them are complex also. Recently attempts have been made toward a theory with real wave functions. Cf. E. Majorana, Nuovo Cim. *14*, 171, 1937 and P. A. M. Dirac, in print.

i.e., the transition probability has an invariant physical sense.

The wave functions form a description of the physical state, not an invariant however, since the same state will be described in different coordinate systems by different wave functions. In order to put this into evidence, we shall affix an index to our wave functions, denoting the Lorentz frame of reference for which the wave function is given. Thus ϕ_l and $\phi_{l<'}$ represent the same state, but they are different functions. The first is the wave function of the state in the coordinate system l, the second in the coordinate system l'. If $\phi_l = \psi_{l'}$ the state ϕ behaves in the coordinate system l exactly as ψ behaves in the coordinate system l'. If ϕ_l is given, all $\phi_{l'}$ are determined up to a constant factor. Because of the invariance of the transition probability we have

$$|(\phi_l, \psi_l)|^2 = |(\phi_{l'}, \psi_{l'})|^2 \tag{1}$$

and it can be shown[4] that the aforementioned constants in the $\phi_{l'}$ can be chosen in such a way that the $\phi_{l'}$ are obtained from the ϕ_l by a linear unitary operation, depending, of course, on l and l'.

$$\phi_{l'} = D(l', l)\phi_l. \tag{2}$$

The unitary operators D are determined by the physical content of the theory up to a constant factor again, which can depend on l and l'. Apart from this constant however, the operations $D(l', l)$ and $D(l'_1, l_1)$ must be identical if l' arises from l by the same Lorentz transformation, by which l'_1 arises from l_1. If this were not true, there would be a real difference between the frames of reference l and l_1. Thus the unitary operator $D(l', l) = D(L)$ is in every Lorentz invariant quantum mechanical theory (apart from the constant factor which has no physical significance) completely determined by the Lorentz transformation L which carries l into $l' = Ll$. One can write, instead of (2)

$$\phi_{Ll} = D(L)\phi_l. \tag{2a}$$

By going over from a first system of reference l to a second $l' = L_1 l$ and then to a third $l'' = L_2 L_1 l$ or directly to the third $l'' = (L_2 L_1)l$, one must obtain–apart from the above mentioned constant–the same set of wave functions. Hence from

---

[4]E. Wigner, *Gruppentheorie und ihre Anwendungen auf die Quantenmechanik det Atoms-pektren.* Braunschweig 1931, pages 251-254.

$$\phi_{l''} = D(l'', l')D(l', l)\phi_l$$

$$\phi_{l''} = D(l'', l)\phi_l$$

it follows

$$D(l'', l')D(l', l) = \omega D(l'', l) \tag{3}$$

or

$$D(L_2)D(L_1) = \omega D(L_2 L_1), \tag{3a}$$

where ω is a number of modulus 1 and can depend on L_2 and L_1. Thus the D(L) form, up to a factor, a representation of the inhomogeneous Lorentz group by linear, unitary operators.

We see thus[5] that there corresponds to every invariant quantum mechanical system of equations such a representation of the inhomogeneous Lorentz group. This representation, on the other hand, though not sufficient to replace the quantum mechanical equations entirely, can replace them to a large extent. If we knew, e.g., the operator K corresponding to the measurement of a physical quantity at the time t = 0, we could follow up the change of this quantity throughout time. In order to obtain its value for the time $t = t_1$, we could transform the original wave function ϕ_l by $D(l', l)$ to a coordinate system l' the time scale of which begins a time t_1 later. The measurement of the quantity in question in this coordinate system for the time 0 is given–as in the original one–by the operator K. This measurement is identical, however, with the measurement of the quantity at time t_1 in the original system. One can say that the representation can replace the equation of motion, it cannot replace, however, connections holding between operators at one instant of time.

It may be mentioned, finally, that these developments apply not only in quantum mechanics, but also to all linear theories, e.g., the Maxwell equations in empty space. The only difference is that there is no arbitrary factor in the description and the ω can be omitted in (3a) and one is led to real representations instead of representations up to a factor. On the other hand, the unitary character of the representation is not a consequence of the basic assumptions.

---

[5]E. Wigner, *l.c.* Chapter XX.

The increase in generality, obtained by the present calculus, as compared with the usual tensor theory, consists in that no assumptions regarding the field nature of the underlying equations are necessary. Thus more general equations, as far as they exist (e.g., in which the coordinate is quantized, etc.) are also included in the present treatment. It must be realized, however, that some assumptions concerning the continuity of space have been made by assuming Lorentz frames of reference in the classical sense. We should like to mention, on the other hand, that the previous remarks concerning the time-parameter in the observables, have only an explanatory character and we do not make assumptions of the kind that measurements can be performed instantaneously.

We shall endeavor, in the ensuing sections, to determine all the continuous unitary representations up to a factor of the inhomogeneous Lorentz group, i.e., all continuous systems of linear, unitary operators satisfying (3a).[6]

2. Comparison With Previous Treatments and Some Immediate Simplifications

A. Previous treatments

The representations of the Lorentz group have been investigated repeatedly. The first investigation is due to Majorana,[7] who in fact found all representations of the class to be dealt with in the present work excepting two sets of representations. Dirac[8] and Proca[8] gave more elegant derivations of Majorana's results and brought them into a form which can be handled more easily. Klein's work[9] does not endeavor to

[6]The exact definition of the continuous character of a representation up to a factor will be given in Section 5A. The definition of the inhomogeneous Lorentz group is contained in Section 4A.

[7]E. Majorana, Nuovo Cim. *9*, 335, 1932.

[8]P. A. M. Dirac, Proc. Roy. Soc. A. *155*, 447, 1936; Al. Proca, J. de Phys. Rad. *7*, 347, 1936.

[9]Klein, Arkiv f. Matem. Astr. och Fysik, *25A*, No. 15, 1936. I am indebted to Mr. Darling for an interesting conversation on this paper.

derive irreducible representations and seems to be in a less close connection with the present work.

The difference between the present paper and that of Majorana and Dirac lies–apart from the finding of new representations–mainly in its greater mathematical rigor. Majorana and Dirac freely use the notion of infinitesimal operators and a set of functions to all members of which every infinitesimal operator can be applied. This procedure cannot be mathematically justified at present, and no such assumption will be used in the present paper. Also the conditions of reducibility and irreducibility could be, in general, somewhat more complicated than assumed by Majorana and Dirac. Finally, the previous treatments assume from the outset that the space and time coordinates will be continuous variables of the wave function in the usual way. This will not be done, of course, in the present work.

B. Some immediate simplifications

Two representations are physically equivalent if there is a one to one correspondence between the states of both which is 1. invariant under Lorentz transformations and 2. of such a character that the transition probabilities between corresponding states are the same.

It follows from the second condition[5] that there either exists a unitary operator S by which the wave functions $\Phi^{(2)}$ of the second representation can be obtained from the corresponding wave functions $\Phi^{(1)}$ of the first representation

$$\Phi^{(2)} = S\Phi^{(1)} \tag{4}$$

or that this is true for the conjugate imaginary of $\Phi^{(2)}$. Although, in the latter case, the two representations are still equivalent physically, we shall, in keeping with the mathematical convention, not call them equivalent.

The first condition now means that if the states $\Phi^{(1)}$, $\Phi^{(2)} = S\Phi^{(1)}$ correspond to each other in one coordinate system, the states $D^{(1)}(L)\Phi^{(1)}$ and $D^{(2)}(L)\Phi^{(2)}$ correspond to each other also. We have then

$$D^{(2)}(L)\Phi^{(2)} = SD^{(1)}(L)\Phi^{(1)} = SD^{(1)}(L)S^{-1}\Phi^{(2)}. \tag{4a}$$

As this shall hold for every $\Phi^{(2)}$, the existence of a unitary S which transforms $D^{(1)}$ into $D^{(2)}$ is the condition for the equivalence of these two representations. Equivalent representations are not considered to be really different and it will be sufficient to find one sample from every infinite class of equivalent representations.

If there is a closed linear manifold of states which is invariant under all Lorentz transformations, i.e., which contains $D(L)\psi$ if it contains ψ, the linear manifold perpendicular to this one will be invariant also. In fact, if ϕ belongs to the second manifold, $D(L)\phi$ will be, on account of the unitary character of $D(L)$, perpendicular to $D(L)\psi'$ if ψ' belongs to the first manifold. However, $D(L^{-1})\psi$ belongs to the first manifold if ψ does and thus $D(L)\phi$ will be orthogonal to $D(L)D(L^{-1})\psi = \omega\psi$ i.e. to all members of the first manifold and belong itself to the second manifold also. The original representation then "decomposes" into two representations, corresponding to the two linear manifolds. It is clear that, conversely, one can form a representation, by simply "adding" several other representations together, i.e. by considering as states linear combinations of the states of several representations and assume that the states which originate from different representations are perpendicular to each other.

Representations which are equivalent to sums of already known representations are not really new and, in order to master all representations, it will be sufficient to determine those, out of which all others can be obtained by "adding" a finite or infinite number of them together.

Two simple theorems shall be mentioned here which will be proved later (Sections 7A and 8C respectively). The first one refers to unitary representations of any closed group, the second to irreducible unitary representations of any (closed or open) group.

The representations of a closed group by unitary *operators* can be transformed into the sum of unitary representations with matrices of finite dimensions.

Given two non equivalent irreducible unitary representations of an arbitrary group. If the scalar product between the wave functions is invariant under the operations of the group, the wave functions belonging[23] to the first representation are orthogonal to all wave functions belonging to the second representation.

C. Classification of unitary representations according to von Neumann and Murray[10]

Given the operators D(L) of a unitary representation, or a representation up to a factor, one can consider the algebra of these operators, i.e. all linear combinations

$$a_1D(L_1) + a_2D(L_2) + a_3D(L_3) + \cdots$$

of the D(L) and all limits of such linear combinations which are bounded operators. According to the properties of this representation algebra, three classes of unitary representations can be distinguished.

The first class of *irreducible* representations has a representation algebra which contains all bounded operators, i.e. if ψ and ϕ are two arbitrary states, there is an operator A of the representation algebra for which $A\psi = \phi$ and $A\psi' = 0$ if ψ' is orthogonal to ψ. It is clear that the center of the algebra contains only the unit operator and multiply thereof. In fact, if C is in the center one can decompose $C\psi = \alpha\psi + \psi'$ so that ψ' shall be orthogonal to ψ. However, ψ' must vanish since otherwise C would not commute with the operator which leaves ψ invariant and transforms every function orthogonal to it into 0. For similar reasons, α must be the same for all ψ. For irreducible representations there is no closed linear manifold of states, (excepting the manifold of all states) which is invariant under all Lorentz transformations. In fact, according to the above definition, a ϕ' arbitrarily close to any ϕ can be represented by a finite linear combination

$$a_1D(L_1)\psi + a_2D(L_2)\psi + \cdots + a_nD(L_n)\psi.$$

Hence, a closed linear invariant manifold contains every state if it contains one. This is, in fact, the more customary definition for irreducible representations and the one which will be used subsequently. It is well known that all finite dimensional representations are sums of irreducible representations. This is not true,[10] in general, in an infinite number of dimensions.

The second class of representations will be called factorial. For these, the center of the representation algebra still contains only multiples of the unit operator. Clearly, the irreducible representations are all factorial, but not conversely. For finite dimensions, the factorial representations may

[10]F. J. Murray and J. v. Neumann, Ann. of Math. *37*, 116, 1936; J. v. Neumann, to be published soon.

contain one irreducible representation several times. This is also possible in an infinite number of dimensions, but in addition to this, there are the "continuous" representations of Murray and von Neumann.[10] These are not irreducible as there are invariant linear manifolds of states. On the other hand, it is impossible to carry the decomposition so far as to obtain as parts only irreducible representations. In all the examples known so far, the representations into which these continuous representations can be decomposed, are equivalent to the original representation.

The third class contains all possible unitary representations. In a finite number of dimensions, these can be decomposed first into factorial representations, and these, in turn, in irreducible ones. Von Neumann[10] has shown that the first step still is possible in infinite dimensions. We can assume, therefore, from the outset that we are dealing with factorial representations.

In the theory of representations of finite dimensions, it is sufficient to determine only the irreducible ones, all others are equivalent to sums of these. Here, it will be necessary to determine all factorial representations. Having done that, we shall know from the above theorem of von Neumann, that all representations are equivalent to finite or infinite sums of factorial representations.

It will be one of the results of the detailed investigation that the inhomogeneous Lorentz group has no "continuous" representations, all representations can be decomposed into irreducible ones. Thus the work of Majorana and Dirac appears to be justified from this point of view a posteriori.

D. Classification of unitary representations from the point of view of infinitesimal operators

The existence of an infinitesimal operator of a continuous one parametric (cyclic, abelian) unitary group has been shown by Stone.[11] He proved that the operators of such a group can be written as exp(iHt) where H is a (bounded or unbounded) hermitian operator and t is the group parameter.

---

[11] M. H. Stone, Proc. Nat. Acad. *16*, 173, 1930, Ann of Math. *33*, 643, 1932, also J. v. Neumann, ibid, *33*, 567, 1932.

However, the Lorentz group has many one parametric subgroups, and the corresponding infinitesimal operators H_1, H_2, $\cdots$ are all unbounded. For every H an everywhere dense set of functions ϕ can be found such that $H_i\phi$ can be defined. It is not clear, however, that an everywhere dense set can be found to all members of which every H can be applied. In fact, it is not clear that one such ϕ can be found.

Indeed, it may be interesting to remark that for an irreducible representation the existence of one function ϕ to which all infinitesimal operators can be applied, entails the existence of an everywhere dense set of such functions. This again has the consequence that one can operate with infinitesimal operators to a large extent in the usual way.

Proof: Let Q(t) be a one parametric subgroup such that Q(t)Q(t′) = Q(t+t′). If the infinitesimal operator of all subgroups can be applied to ϕ, the

$$\lim_{t=0} t^{-1}(Q(t) - 1)\phi \tag{5}$$

exists. It follows, then, that the infinitesimal operators can be applied to $R\phi$ also where R is an arbitrary operator of the representation: Since $R^{-1}Q(t)\,R$ is also a one parametric subgroup

$$\lim_{t=0} t^{-1}(R^{-1}Q(t)R - 1)\phi = \lim_{t=0} R^{-1}\cdot t^{-1}(Q(t) - 1)R\phi$$

also exists an hence also (R is unitary)

$$\lim_{t=0} t^{-1}(Q(t) - 1)R\phi.$$

Every infinitesimal operator can be applied to $R\phi$ if they all can be applied to ϕ, and the same holds for sums of the kind

$$a_1R_1\phi + a_2R_2\phi + \cdots + a_nR_n\phi. \tag{6}$$

These form, however, an everywhere dense set of functions if the representation is irreducible.

If the representation is not irreducible, one can consider the set N_0 of such wave functions to which every infinitesimal operator can be applied. This set is clearly linear and, according to the previous paragraph, invariant under the operations of the group (i.e. contains every $R\phi$ if it contains ϕ). The same holds for the closed set N generated by N_0 and also of the set P of functions which are perpendicular to all functions of N. In fact, if ϕ_l, is perpendicular to all ϕ_n of N, it is perpendicular also to all $R^{-1}\phi_n$ and, for the unitary character of R, the $R\phi_p$ is perpendicular to all ϕ_n, i.e. is also contained in the set P.

We can decompose thus, by a unitary transformation, every unitary representation into a "normal" and a "pathological" part. For the former, there is an everywhere dense set of functions, to which all infinitesimal operators can be applied. There is no single wave functions to which all infinitesimal operators of a "pathological" representation could be applied.

According to Murray and von Neumann, if the original representation was factorial, all representations into which it can be decomposed will be factorial also. Thus every representation is equivalent to a sum of factorial representations, part of which is "normal," the other part "pathological."

It will turn out again that the inhomogeneous Lorentz group has no pathological representations. Thus this assumption of Majorana and Dirac also will be justified a posteriori. Every unitary representation of the inhomogeneous Lorentz group can be decomposed into normal irreducible representations. It should be stated, however, that the representations in which the unit operator corresponds to every translation have not been determined to date (cf. also section 3, end). Hence, the above statements are not proved for these representations, which are, however, more truly representations of the homogeneous Lorentz group, than of the inhomogeneous group.

While all these points may be of interest to the mathematician only, the new representation of the Lorentz group which will be described in section 7 may interest the physicist also. It describes a particle with a continuous spin.

Acknowledgment. The subject of this paper was suggested to me as early as 1928 by P. M. Dirac who realized even at that date the connection of representations with quantum mechanical equations. I am greatly indebted to him also for many fruitful conversations about this subject, especially during the years 1934/35, the outgrowth of which the present paper is.

I am indebted also to J. v. Neumann for his help and friendly advice.

3. Summary of Ensuing Sections

Section 4 will be devoted to the definition of the inhomogeneous Lorentz group and the theory of characteristic values and characteristic vectors of a homogeneous (ordinary) Lorentz transformation. The discussion will follow very closely the corresponding, well-known theory of the group of motions in ordinary space and the theory of characteristic values of orthogonal transformations.[12] It will contain only a straightforward generalization of the methods usually applied in those discussions.

In section 5, it will be proved that one can determine the physically meaningless constants in the D(L) in such a way that instead of (3a) the more special equation

$$D(L_1)D(L_2) = \pm D(L_1L_2) \tag{7}$$

will be valid. This means that instead of a representation up to a factor, we can consider representations up to the sign. For the case that either L_1 or L_2 is a pure translation, Dirac[13] has given a proof of (7) using infinitesimal operators. A consideration very similar to his can be carried out, however, also using only finite transformations.

For representations with a finite number of dimensions (corresponding to an only finite number of linearly independent states), (7) could be proved also if both L_1 and L_2 are homogeneous Lorentz transformations, by a straightforward application of the method of Weyl and Schreier.[14] However, the Lorentz group has no finite dimensional representation (apart from the trivial one in which the unit operation corresponds to every L). Thus the method of Weyl and Schreier cannot be applied. Its first step is to normalize the indeterminate constants in every matrix D(L) in such a way that the determinant of D(L) becomes 1. No determinant can be defined for general unitary operators.

---

[12]Cf. e.g. E. Wigner, *l.c.* Chapter III. O. Veblen and J.W. Young, *Projective Geometry*, Boston 1917. Vol 2, especially Chapter VII.

[13]P. A. M. Dirac, mimeographed notes of lectures delivered at Princeton University, 1934/35, page 5a.

[14]H. Weyl, Mathem. Zeits. *23*, 271; *24*, 328, 377, 789, 1925; O. Schreier, Abhandl. Mathem. Seminar Hamburg, *4*, 15, 1926; *5*, 233, 1927.

The method to be employed here will be to decompose every L into a product of two involutions L = MN with $M^2 = N^2 = 1$. Then D(M) and D(N) will be normalized so that their squares become unity and D(L) = D(M)D(N) set. It will be possible, then, to prove (7) without going back to the topology of the group.

Sections 6, 7, and 8 will contain the determination of the representations. The pure translations form an invariant subgroup of the whole inhomogeneous Lorentz group and Frobenius' method[15] will be applied in Section 6 to build up the representations of the whole group out of representations of the subgroup, by means of a "little group." In Section 6, it will be shown on the basis of an as yet unpublished work[24] of J.v. Neumann that there is a characteristic (invariant) set of "momentum vectors" for every irreducible representation. The irreducible representations of the Lorentz group will be divided into four classes. The momentum vectors of the

1*st class* are time-like,
2*nd class* are null-vectors, but not all their components will be zero,
3*rd class* vanish (i.e., all their components will be zero),
4*th class* are space-like.

Only the first two cases will be considered in Section 7, although the last case may be the most interesting from the mathematical point of view. I hope to return to it in another paper. I did not succeed so far in giving a complete discussion of the 3rd class. (All these restrictions appear in the previous treatments also.)

In Section 7, we shall find again all known representations of the inhomogeneous Lorentz group (i.e., all known Lorentz invariant equations) and two new sets.

Sections 5, 6, 7 will deal with the "restricted Lorentz group" only, i.e. Lorentz transformations with determinant 1 which do not reverse the direction of the time axis. In section 8, the representations of the extended Lorentz group will be considered, the transformations of which are not subject to these conditions.

---

[15]G. Frobenius, Sitz. d. Koen. Preuss. Akad. p. 501, 1898, I. Schur, ibid, p. 164, 1906; F. Seitz, Ann. of Math. *37*, 17, 1936.

4. Description of the Inhomogeneous Lorentz Group

A.

An inhomogeneous Lorentz transformation L = (a, Λ) is the product of a translation by a real vector a

$$x'_i = x_i + a_i \quad (i=1,2,3,4) \tag{8}$$

and a homogeneous Lorentz transformation Λ with real coefficients

$$x'_i = \sum_{k=1}^{4} \Lambda_{ik} x_k. \tag{9}$$

The translation shall be performed after the homogeneous transformation. The coefficients of the homogeneous transformation satisfy three conditions: (1) They are real and Λ leaves the indefinite quadratic form $-x_1^2-x_2^2-x_3^2+x_4^2$ invariant:

$$\Lambda F \Lambda' = F \tag{10}$$

where the prime denotes the interchange of rows and columns and F is the diagonal matrix with the diagonal elements -1, -1, -1, +1. –(2) The determinant $|\Lambda_{ik}| = 1$ and –(3) $\Lambda_{44} > 0$.

We shall denote the Lorentz-hermitian product of two vectors x and y by

$$\{x, y\} = -\overset{*}{x}_1 y_1 - \overset{*}{x}_2 y_2 - \overset{*}{x}_3 y_3 - \overset{*}{x}_4 y_4. \tag{11}$$

(The star denotes the conjugate imaginary.) If {x, x} < 0 the vector x is called space-like, if {x, x} = 0, it is a null vector, if {x, x} > 0, it is called time-like. A real time-like vector lies in the positive light cone if $x_4 > 0$; it lies in the negative light cone if $x_4 < 0$. Two vectors x and y are called orthogonal if {x, y} = 0.

On account of its linear character a homogeneous Lorentz transformation is completely defined if Λv is given for four linearly independent vectors $v^{(1)}, v^{(2)}, v^{(3)}, v^{(4)}$.

From (11) and (10) it follows that {v, w} = {Λv, Λw} for every pair of vectors v, w. This will be satisfied for every pair if it is satisfied for all pairs $v^{(i)}, v^{(k)}$ of four linearly independent vectors. The reality condition is satisfied if $(\Lambda v^{(i)})^* = \Lambda(v^{(i)*})$ holds for four such vectors.

The scalar product of two vectors x and y is positive if both lie in the positive light cone or both in the negative light cone. It is negative if one lies in the positive, the other in the negative light cone. Since both x and y are time-like $|x_4|^2 > |x_1|^2 + |x_2|^2 + |x_3|^2$; $|y_4|^2 > |y_1|^2 + |y_2|^2 + |y_3|^2$. Hence, by Schwarz's inequality $|x_4^* y_4| > |x_1^* y_1 + x_2^* y_2 + x_3^* y_3|$ and the sign of the scalar product of two real time-like vectors is determined by the product of their time components.

A time-like vector is transformed by a Lorentz transformation into a time-like vector. Furthermore, on account of the condition $\Lambda_{44} > 0$, the vector $v^{(0)}$ with the components 0, 0, 0, 1 remains in the positive light cone, since the fourth component of $\Lambda v^{(0)}$ is Λ_{44}. If $v^{(1)}$ is another vector[16] in the positive light cone $\{v^{(1)}, v^{(0)}\} > 0$ and hence also $\{\Lambda v^{(1)}, \Lambda v^{(0)}\} > 0$ and $\Lambda v^{(1)}$ is in the positive light cone also. The third condition for a Lorentz transformation can be formulated also as the requirement that every vector in (or on) the positive light cone shall remain in (or, respectively, on) the positive light cone.

This formulation of the third condition shows that the third condition holds for the product of two homogeneous Lorentz transformations if it holds for both factors. The same is evident for the first two conditions.

From $\Lambda F \Lambda' = F$ one obtains by multiplying with Λ^{-1} from the left and $\Lambda'^{-1} = (\Lambda^{-1})'$ from the right $F = \Lambda^{-1} F (\Lambda^{-1})'$ so that the reciprocal of a homogeneous Lorentz transformation is again such a transformation. The homogeneous Lorentz transformations form a group, therefore.

One easily calculates that the product of two inhomogeneous Lorentz transformations (b, M) and (c, N) is again an inhomogeneous Lorentz transformation (a, Λ)

$$(b, M)(c, N) = (a, \Lambda) \tag{12}$$

where

$$\Lambda_{ik} = \sum_j M_{ij} N_{jk}; \quad a_i = b_i + \sum_j M_{ij} c_j \tag{12a}$$

[16]Wherever a confusion between vectors and vector components appears to be possible, upper indices will be used for distinguishing different vectors and lower indices for denoting the components of a vector.

or, somewhat shorter

$$\Lambda = MN; \quad a = b + Mc. \tag{12b}$$

B. Theory of characteristic values and characteristic vectors of a homogeneous Lorentz transformation

Linear homogeneous transformations are most simply described by their characteristic values and vectors. Before doing this for the homogeneous Lorentz group, however, we shall need two rules about orthogonal vectors.

[1] *If* $\{v, w\} = 0$ *and* $\{v, v\} > 0$, *then* $\{w, w\} < 0$; *if* $\{v, w\} = 0$, $\{v, v\} = 0$, *then* w *is either space-like, or parallel to* v (*either* $\{w, w\} < 0$, *or* $w = cv$).

Proof:

$$\overset{*}{v}_4 w_4 = \overset{*}{v}_1 w_1 + \overset{*}{v}_2 w_2 + \overset{*}{v}_3 w_3. \tag{13}$$

By Schwarz's inequality, then

$$|v_4|^2 |w_4|^2 \leq (|v_1|^2 + |v_2|^2 + |v_3|^2)(|w_1|^2 + |w_2|^2 + |w_3|^2). \tag{14}$$

For $|v_4|^2 > |v_1|^2 + |v_2|^2 + |v_3|^2$ it follows that $|w_4|^2 < |w_1|^2 + |w_2|^2 + |w_3|^2$. If $|v_4|^2 = |v_1|^2 + |v_2|^2 + |v_3|^2$ the second inequality still follows if the inequality sign holds in (14). The equality sign can hold only, however, if the first three components of the vectors v and w are proportional. Then, on account of (13) and both being null vectors, the fourth components are in the same ratio also.

[2] *If four vectors* $v^{(1)}$, $v^{(2)}$. $v^{(3)}$, $v^{(4)}$ *are mutually orthogonal and linearly independent, one of them is time-like, three are space-like.*

Proof: It follows from the previous paragraph that only one of four mutually orthogonal, linearly independent vectors can be time-like or a null vector. It remains to be shown therefore only that one of them is time-like. Since they are linearly independent, it is possible to express by them any time-like vector

$$v^{(t)} = \sum_{k=1}^{4} \alpha_k v^{(k)}.$$

The scalar product of the left side of this equation with itself is positive and therefore

$$\{\sum_k \alpha_k \mathrm{v}^{(k)},\ \sum_k \alpha_k \mathrm{v}^{(k)}\} \ > \ 0$$

or

$$\sum_k |\alpha_k|^2 \{\mathrm{v}^{(k)},\ \ \mathrm{v}\,Up<(k)>\} \ > \ 0 \qquad (15)$$

and one $\{\mathrm{v}^{(k)}, \mathrm{v}^{(k)}\}$ must be positive. Four mutually orthogonal vectors are not necessarily linearly independent, because a null vector is perpendicular to itself. The linear independence follows, however, if none of the four is a null vector.

We go over now to the characteristic values λ of Λ. These make the determinant $|\Lambda - \lambda 1|$ of the matrix $\Lambda - \lambda 1$ vanish.

[3] *If λ is a characteristic value, λ^*, λ^{-1} and λ^{*-1} are characteristic values also.*

Proof: For λ^* this follows from the fact that Λ is real. Furthermore, from $|\Lambda - \lambda 1| = 0$ also $|\Lambda' - \lambda 1| = 0$ follows, and this multiplied by the determinants of ΛF and F^{-1} gives

$$|\Lambda F| \cdot |\Lambda' - \lambda 1| \cdot |F|^{-1} \ = \ |\Lambda F \Lambda' F^{-1} - \lambda\Lambda| \ = \ |1 - \lambda\Lambda| \ = \ 0,$$

so that λ^{-1} is a characteristic value also.

[4] *The characteristic vectors v_1 and v_2 belonging to two characteristic values λ_1 and λ_2 are orthogonal if $\lambda^*{}_1\lambda_2 \neq 1$.*

Proof:

$$\{\mathrm{v}_1, \mathrm{v}_2\} = \{\Lambda \mathrm{v}_1, \Lambda \mathrm{v}_2\} = \{\lambda_1 \mathrm{v}_1, \lambda_2 \mathrm{v}_2\} = \lambda_1^* \lambda_2 \{\mathrm{v}_1, \mathrm{v}_2\}.$$

Thus if $\{\mathrm{v}_1, \mathrm{v}_2\} \neq 0$, $\lambda_1^*\lambda_2 = 1$.

[5] *If the modulus of a characteristic value λ is $|\lambda| \neq 1$, the corresponding characteristic vector* v *is a null vector and λ itself real and positive.*

From $\{\mathrm{v}, \mathrm{v}\} = \{\Lambda \mathrm{v}, \Lambda \mathrm{v}\} = |\lambda|^2 \{\mathrm{v}, \mathrm{v}\}$ the $\{\mathrm{v}, \mathrm{v}\} = 0$ follows immediately for $|\lambda| \neq 1$. If λ were complex, λ^* would be a characteristic value also. The characteristic vectors of λ and λ^* would be two different null vectors and, because of [4], orthogonal to each other. This is impossible on account of [1]. Thus λ is real and v a real null vector. Then, on account of the third condition for a homogeneous Lorentz transformation, λ must be positive.

[6] *The characteristic value* λ *of a characteristic vector* v *of length null is real and positive.*

If λ were not real, λ^* would be a characteristic value also. The corresponding characteristic vector v^* would be different from v, a null vector also, and perpendicular to v on account of [4]. This is impossible because of [1].

[7] *The characteristic vector* v *of a complex characteristic value* λ *(the modulus of which is 1 on account of* [5]*) is space-like*: $\{v, v\} < 0$.

Proof: λ^* is a characteristic value also, the corresponding characteristic vector is v^*. Since $(\lambda^*)^*\lambda = \lambda^2 \neq 1$, $\{v^*, v\} = 0$. Since they are different, at least one is space-like. On account of $\{v, v\} = \{v^*, v^*\}$ both are space-like. If all four characteristic values were complex and the corresponding characteristic vectors linearly independent (which is true except if Λ has elementary divisors) we should have four space-like, mutually orthogonal vectors. This is impossible, on account of [2]. Hence

[8] *There is not more than one pair of conjugate complex characteristic values, if* Λ *has no elementary divisors. Similarly, under the same condition, there is not more than one pair* λ, λ^{-1} *of characteristic values whose modulus is different from* 1. Otherwise their characteristic vectors would be orthogonal, which they cannot be, being null vectors.

For homogeneous Lorentz transformations which do not have elementary divisors, the following possibilities remain:

(a) There is a pair of complex characteristic values, their modulus is 1, on account of [5]

$$\lambda_1 = \lambda_2^* = \lambda_2^{-1}; \quad |\lambda_1| = |\lambda_2| = 1, \tag{16}$$

and also a pair of characteristic values λ_3, λ_4, the modulus of which is not 1. These must be real and positive:

$$\lambda_4 = \lambda_3^{-1}; \quad \lambda_3 = \lambda_3^* > 0. \tag{16a}$$

The characteristic vectors of the conjugate complex characteristic values are conjugate complex, perpendicular to each other and space-like so that they can be normalized to -1

$$v_1 = \overset{*}{v}_2; \quad \{v_1, v_2\} = \{v_1, \overset{*}{v}_1\} = 0$$
$$\{v_1, v_1\} = \{v_2, v_2\} = -1 \tag{17}$$

those of the real characteristic values are real null vectors, their scalar product can be normalized to 1

$$v_3 = \overset{*}{v}_3 \quad v_4 = \overset{*}{v}_4 \quad \{v_3, v_4\} = 1 \tag{17a}$$
$$\{v_3, v_3\} = \{v_4, v_4\} = 0.$$

Finally, the former pair of characteristic vectors is perpendicular to the latter kind

$$\{v_1, v_3\} = \{v_1, v_4\} = \{v_2, v_3\} = \{v_2, v_4\} = 0. \tag{17b}$$

It will turn out that all the other cases in which Λ has no elementary divisor are special cases of (a).

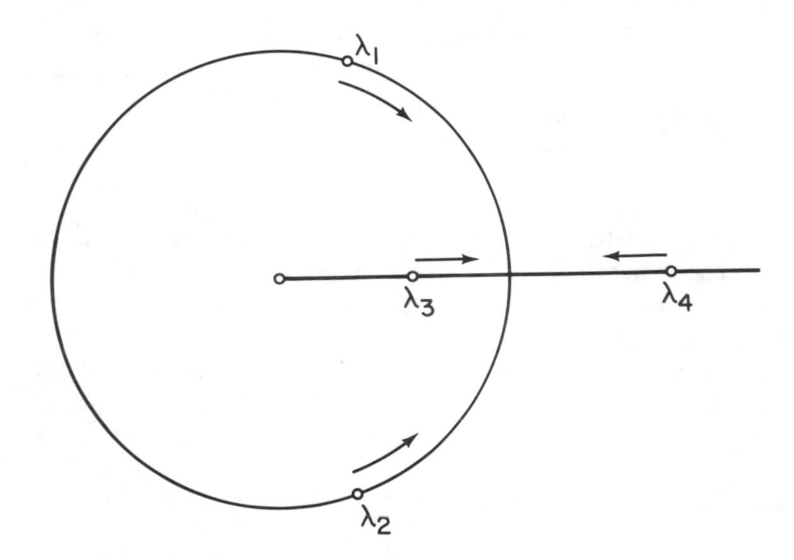

Figure 1: Position of the characteristic values for the general case a) in the complex plane. In case b), λ_3 and λ_4 coincide and are equal 1; in case c), λ_1 and λ_2 coincide and are either +1 or -1. In case d) both pairs $\lambda_3 = \lambda_4 = 1$ and $\lambda_1 = 2 = \neq 1$ coincide.

(b) There is a pair of complex characteristic values λ_1, $\lambda_2 = \lambda_1^{-1} = \lambda_1^*$, $\lambda_1 \neq \lambda_1^*$, $|\lambda_1| = |\lambda_2| = 1$. No pair with $|\lambda_3| \neq 1$, however. Then on account of [8], still $\lambda_3 = \lambda_3^*$ which gives with $|\lambda_3| = 1$, $\lambda_3 = \pm 1$. Since the product $\lambda_1\lambda_2\lambda_3\lambda_4 = 1$, on account of the second condition for homogeneous Lorentz transformations, also $\lambda_4 = \lambda_3 = \pm 1$. The double characteristic value ± 1 has two linearly independent characteristic vectors v_3 and v_4 which can be assumed to be perpendicular to each other, $\{v_3, v_4\} = 0$. According to [2], one of the four characteristic vectors must be time-like and since those of λ_1 and λ_2 are space-like, the time-like one must belong to ± 1. This must be positive, therefore $\lambda_3 = \lambda_4 = 1$. Out of the time-like and space-like vectors $\{v_3, v_3\} = -1$ and $\{v_4, v_4\} = 1$, one can build two null vectors $v_4 + v_3$ and $v_4 - v_3$. Doing this, case (b) becomes the special case of (a) in which the real positive characteristic values become equal $\lambda_3 = \lambda_4^{-1} = 1$.

(c) All characteristic values are real; there is however one pair $\lambda_3 = \lambda_3^*$, $\lambda_4 = \lambda_3^{-1}$, the modulus of which is not unity. Then $\{v_3, v_3\} = \{v_4, v_4\} = 0$ and $\lambda_3 > 0$ and one can conclude for λ_1 and λ_2, as before for λ_3 and λ_4 that $\lambda_1 = \lambda_2 = \pm 1$. This again is a special case of (a); here the two characteristic values of modulus 1 become equal.

(d) All characteristic values are real and of modulus 1. If all of them are +1, we have the unit matrix which clearly can be considered as a special case of (a). The other case is $\lambda_1 = \lambda_2 = -1$, $\lambda_3 = \lambda_4 = +1$. The characteristic vectors of λ_1 and λ_2 must be space-like, on account of the third condition for a homogeneous Lorentz transformation; they can be assumed to be orthogonal and normalized to -1. This is then a special case of (b) and hence of (a) also. The cases (a), (b), (c), (d), are illustrated in Fig. 1.

The cases remain to be considered in which Λ has an elementary divisor. We set therefore

$$\Lambda_e v_e = \lambda_e v_e; \quad \Lambda_e w_e = \lambda_e w_e + v_e. \tag{18}$$

It follows from [5] that either $|\lambda_e| = 1$, or $\{v_e, v_e\} = 0$. We have $\{v_e, w_e\} = \{\Lambda_e v_e, \Lambda_e w_e\} = |\lambda_e|^2\{v_e, w_e\} + \{v_e, v_e\}$. From this equation

$$\{v_e, v_e\} = 0 \tag{19}$$

follows for $|\lambda_e| = 1$, so that (19) holds in any case. It follows then from [6] that λ_e is real, positive and v_e, w_e can be assumed to be real also. The last equation now becomes $\{v_e, w_e\} = \lambda^2{}_e\{v_e, w_e\}$ so that either $\lambda_e = 1$ or $\{v_e, w_e\} = 0$. Finally, we have

$$\begin{aligned}\{w_e, w_e\} &= \{\Lambda_e w_e, \Lambda_e w_e\} \\ &= \lambda^2{}_e\{w_e, w_e\} + 2\lambda_e\{w_e, v_e\} + \{v_e, v_e\}.\end{aligned}$$

This equation now shows that

$$\{w_e, \ v_e\} = 0 \tag{19a}$$

even if $\lambda_e = 1$. From (19), (19a) it follows that w_e is space-like and can be normalized to

$$\{w_e, \ w_e\} = -1. \tag{19b}$$

Inserting (19a) into the preceding equation we finally obtain

$$\lambda_e = 1. \tag{19c}$$

[9] *If* Λ_e *has an elementary divisor, all its characteristic roots are* 1.

From (19c) we see that the root of the elementary divisor is 1 and this is at least a double root. If Λ had a pair of characteristic values $\lambda_1 \neq 1$, $\lambda_2 = \lambda_1^{-1}$, the corresponding characteristic vectors v_1 and v_2 would be orthogonal to v_e and therefore space-like. On account of [5], then $|\lambda_1| = |\lambda_2| = 1$ and $\{v_1, v_2\} = 0$. Furthermore, from $\{w_e, v_1\} = \{\Lambda_e w_e, \Lambda_e v_1\} = \lambda_1\{w_e, v_1\} + \lambda_1\{v_e, v_1\}$ and from $\{v_e, v_1\} = 0$ also $\{w_e, v_1\} = 0$ follows. Thus all the four vectors v_1, v_2, v_e, w_e would be mutually orthogonal. This is excluded by [2] and (19).

Two cases are conceivable now. Either the fourfold characteristic root has only one characteristic vector, or there is in addition to v_e (at least) another characteristic vector v_1. In the former case four linearly independent vectors v_e, w_e, z_e, x_e could be found such that

$$\Lambda_e v_e = v_e \qquad \Lambda_e w_e = w_e + v_e$$

$$\Lambda_e z_e = z_e + w_e \qquad \Lambda_e x_e = x_e + z_e$$

However $\{v_e, x_e\} = \{\Lambda_e v_e, \Lambda_e x_e\} = \{v_e, x_e\} + \{v_e, z_e\}$ from which $\{v_e, z_e\} = 0$ follows. On the other hand

$$\begin{aligned}\{w_e, z_e\} &= \{\Lambda_e w_e, \Lambda_e z_e\} \\ &= \{w_e, z_e\} + \{w_e, w_e\} + \{v_e, z_e\} + \{v_e, w_e\}.\end{aligned}$$

This gives with (19a) and (19b) $\{v_e, z_e\} = 1$ so that this case must be excluded.

(e) There is thus a vector v_1 so that in addition to (18)

$$\Lambda_e v_1 = v_1 \tag{18a}$$

holds. From $\{w_e, v_1\} = \{\Lambda_e w_e, \Lambda_e v_1\} = \{w_e, v_1\} + \{v_e, v_1\}$ follows

$$\{v_e, v_1\} = 0. \tag{19d}$$

The equations (18), (18a) will remain unchanged if we add to w_e and v_1 a multiple of v_e. We can achieve in this way that the fourth components of both w_e and v_1 vanish. Furthermore, v_1 can be normalized to -1 and added to w_e also with an arbitrary coefficient, to make it orthogonal to v_1. Hence, we can assume that

$$v_{14} = w_{e4} = 0; \quad \{v_1, v_1\} = -1; \quad \{w_e, v_1\} = 0. \tag{19e}$$

We can finally define the null vector z_e to be orthogonal to w_e and v_1 and have a scalar product 1 with v_e

$$\{z_e, z_e\} = \{z_e, w_e\} = \{z_e, v_1\} = 0; \quad \{z_e, v_e\} = 1. \tag{19f}$$

Then the null vectors v_e and z_e represent the momenta of two light beams in opposite directions. If we set $\Lambda_e z_e = av_e + bw_e + cz_e + dv_1$ the conditions $\{z_e, v\} = \{\Lambda_e z_e, \Lambda_e v\}$ give, if we set for v the vectors v_e, w_e, z_e, v_1, the conditions $c = 1$; $b = c$; $2ac - b^2 - d^2 = 0$; $d = 0$. Hence

$$\begin{aligned} \Lambda_e v_e &= v_e & \Lambda_e w_e &= w_e + v_e \\ \Lambda_e v_1 &= v_1 & \Lambda_e z_e &= z_e + w_e + 1/2 v_e. \end{aligned} \tag{20}$$

A Lorentz transformation with an elementary divisor can be best characterized by the null vector v_e which is invariant under it and the space part of which forms with the two other vectors w_e and v_1 three mutually orthogonal vectors in ordinary space. The two vectors w_e and v_1 are normalized, v_1 is invariant under Λ_e while the vector v_e is added to w_e upon application of Λ_e. The result of the application of Λ_e to a vector which is linearly independent of v_e, w_e and v_1 is, as we saw, already determined by the expressions for $\Lambda_e v_e$, $\Lambda_e w_e$ and $\Lambda_e v_1$.

The $\Lambda_e(\gamma)$ which have the invariant null vector v_e and also w_e (and hence also v_1) in common and differ only by adding to w_e different multiples γv_e of v_e, form a cyclic group with $\gamma = 0$, the unit transformation as unity:

$$\Lambda_e(\gamma)\Lambda_e(\gamma') = \Lambda_e(\gamma + \gamma').$$

The Lorentz transformation $M(\alpha)$ which leaves v_1 and w_e invariant but replaces v_e by αv_e (and z_e by $\alpha^{-1} z_e$) has the property of transforming $\Lambda_e(\gamma)$

into

$$M(\alpha)\Lambda_e(\gamma)M(\alpha)^{-1} = \Lambda_e(\alpha\gamma). \qquad (+)$$

An example of $\Lambda_e(\gamma)$ and $M(\alpha)$ is

$$\Lambda_e(\gamma) = \begin{bmatrix} 1 & 0 & 0 & 0 \\ 0 & 1 & \gamma & \gamma \\ 0 & -\gamma & 1-\frac{1}{2}\gamma^2 & -\frac{1}{2}\gamma^2 \\ 0 & \gamma & \frac{1}{2}\gamma^2 & 1+\frac{1}{2}\gamma^2 \end{bmatrix} ;$$

$$M(\alpha) = \begin{bmatrix} 1 & 0 & 0 & 0 \\ 0 & 1 & 0 & 0 \\ 0 & 0 & \frac{1}{2}(\alpha+\alpha^{-1}) & \frac{1}{2}(\alpha-\alpha^{-1}) \\ 0 & 0 & \frac{1}{2}(\alpha-\alpha^{-1}) & \frac{1}{2}(\alpha+\alpha^{-1}) \end{bmatrix} .$$

These Lorentz transformations play an important role in the representations with space like momentum vectors.

A behavior like (+) is impossible for finite unitary matrices because the characteristic values of $M(\alpha)^{-1}\Lambda_e(\gamma)M(\alpha)$ and $\Lambda_e(\gamma)$ are the same–those of $\Lambda_e(\gamma\alpha) = \Lambda_e(\gamma)^{\alpha}$ the αth powers of those of $\Lambda_e(\gamma)$. This shows very simply that the Lorentz group has no true unitary representation in a finite number of dimensions.

C. Decomposition of a homogeneous Lorentz transformation into rotations and an acceleration in a given direction

The homogeneous Lorentz transformation is, from the point of view of the physicist, a transformation to a uniformly moving coordinate system, the origin of which coincided at $t = 0$ with the origin of the first coordinate system. One can, therefore, first perform a rotation which brings the direction of motion of the second system into a given direction–say the direction of the third axis–and impart it a velocity in this direction, which will bring it to rest. After this, the two coordinate systems can differ only in a rotation. This means that every homogeneous Lorentz transformation can be decomposed in the following way[17]

[17]Cf. e.g. L. Silberstein, *The Theory of Relativity*, London 1924, p. 142.

$$\Lambda = RZS \tag{21}$$

where R and S are pure rotations, (i.e. $R_{i4} = R_{4i} = S_{i4} = S_{4i} = 0$ for $i \neq 4$ and $R_{44} = S_{44} = 1$, also $R' = R^{-1}$, $S' = S^{-1}$) and Z is an acceleration in the direction of the third axis, i.e.

$$Z = \begin{pmatrix} 1 & 0 & 0 & 0 \\ 0 & 1 & 0 & 0 \\ 0 & 0 & a & b \\ 0 & 0 & b & a \end{pmatrix}$$

with $a^2 - b^2 = 1$, $a > b > 0$. The decomposition (21) is clearly not unique. It will be shown, however, the Z is uniquely determined, i.e. the same in every decomposition of the form (21).

In order to prove this mathematically, we chose R so that in $R^{-1}\Lambda = I$ the first two components in the fourth column $I_{14} = I_{24} = 0$ become zero: R^{-1} shall bring the vector with the components Λ_{14}, Λ_{24}, Λ_{34} into the third axis. Then we take $I_{34} = (\Lambda_{14}^2 + \Lambda_{24}^2 + \Lambda_{34}^2)^{\frac{1}{2}}$ and $I_{44} = \Lambda_{44}$ for b and a to form Z; they satisfy the equation $I_{44}^2 - I_{34}^2 = 1$. Hence, the first three components of the fourth column of $J = Z^{-1}I = Z^{-1}R^{-1}\Lambda$ will become zero and $J_{44} = 1$, because of $J_{44}^2 - J_{14}^2 - J_{24}^2 - J_{34}^2 = 1$. Furthermore, the first three components of the fourth row of J will vanish also, on account of $J_{44}^2 - J_{41}^2 - J_{42}^2 - J_{43}^2 = 1$, i.e. $J = S = Z^{-1}R^{-1}\Lambda$ is a pure rotation. This proves the possibility of the decomposition (21).

The trace of $\Lambda\Lambda' = RZ^2R^{-1}$ is equal to the trace of Z^2, i.e. equal to $2a^2 + 2b^2 + 2 = 4a^2 = 4b^2 + 4$ which shows that the a and b of Z are uniquely determined. In particular $a = 1$, $b = 0$ and Z the unit matrix if $\Lambda\Lambda' = 1$, i.e. Λ a pure rotation.

It is easy to show now that the group space of the homogeneous Lorentz transformations is only doubly connected. If a continuous series $\Lambda(t)$ of homogeneous Lorentz transformations is given, which is unity both for $t = 0$ and $t = 1$, we can decompose it according to (21)

$$\Lambda(t) = R(t)Z(t)S(t). \tag{21a}$$

It is also clear from the foregoing, that R(t) can be assumed to be continuous in t, except for values of t, for which $\Lambda_{14} = \Lambda_{24} = \Lambda_{34} = 0$, i.e. for which Λ is a pure rotation. Similarly, Z(t) will be continuous in t and this will hold even where $\Lambda(t)$ is a pure rotation. Finally, $S = Z^{-1}R^{-1}\Lambda$ will be continuous also, except where $\Lambda(t)$ is a pure rotation.

Let us consider now the series of Lorentz transformations

$$\Lambda_s(t) = R(t)Z(t)^s S(t) \tag{21b}$$

where the b of $Z(t)^s$ is s times the b of $Z(t)$. By decreasing s from 1 to 0 we continuously deform the set $\Lambda_1(t) = \Lambda(t)$ of Lorentz transformations into a set of rotations $\Lambda_0(t) = R(t)S(t)$. Both the beginning $\Lambda_0(0) = 1$ and the end $\Lambda_s(1) = 1$ of the set remain the unit matrix and the sets $\Lambda_s(t)$ remain continuous in t for all values of s. This last fact is evident for such t for which $\Lambda(t)$ is not a rotation: for such t all factors of (21b) are continuous. But it is true also for t_0 for which $\Lambda(t_0)$ is a rotation, and for which, hence $Z(t_0) = 1$ and $\Lambda_s(t_0) = \Lambda_1(t_0) = \Lambda(t_0)$. As $Z(t)$is everywhere continuous, there will be a neighborhood of t_0 in which $Z(t)$ and hence also $Z(t)^s$ is arbitrarily close to the unit matrix. In this neighborhood $\Lambda_s(t) = \Lambda(t)$. $S(t)^{-1}Z(t)^{-1}Z(t)^s S(t)$ is arbitrarily close to $\Lambda(t)$; and, if the neighborhood is small enough, this is arbitrarily close to $\Lambda(t_0) = \Lambda_s(t_0)$.

Thus (21b) replaces the continuous set $\Lambda(t)$ of Lorentz transformations by a continuous set of rotations. Since these form an only doubly connected manifold, the manifold of Lorentz transformations can not be more than doubly connected. The existence of a two valued representation[18] shows that it is actually doubly and not simply connected.

We can form a new group[14] from the Lorentz group, the elements of which are the elements of the Lorentz group, together with a way $\Lambda(t)$, connecting $\Lambda(1) = \Lambda$ with the unity $\Lambda(0) = E$. However, two ways which can be continuously deformed into each other are not considered different. The product of the element "Λ with the way $\Lambda(t)$" with the element "I with the way I(t)" is the element ΛI with the way which goes from E along $\Lambda(t)$ to Λ and hence along $\Lambda I(t)$ to ΛI. Clearly, the Lorentz group is isomorphic with this group and two elements (corresponding to the two essentially different ways to Λ) of this group correspond to one element of the Lorentz group. It is well known,[18] that this group is holomorphic with

[18]Cf. H. Weyl, *Gruppentheorie und Quantenmechanik,* 1st. ed. Leipzig 1928, pages 110-114, 2nd ed. Leipzig 1931, pages 130-133. It may be interesting to remark that essentially the same isomorphism has been recognized already by L. Silberstein, *l.c.* pages 148-157.

the group of unimodular complex two dimensional transformations.

Every continuous representation of the Lorentz group "up to the sign" is a singlevalued, continuous representation of this group. The transformation which corresponds to "Λ with the way $\Lambda(t)$" is that $d(\Lambda)$ which is obtained by going over from $d(E) = d(\Lambda(0)) = 1$ continuously along $d(\Lambda(t))$ to $d(\Lambda(1)) = d(\Lambda)$.

D. The homogeneous Lorentz group is simple

It will be shown, first, that an invariant subgroup of the homogeneous Lorentz group contains a rotation (i.e. a transformation which leaves x_4 invariant).– We can write an arbitrary element of the invariant subgroup in the form RZS of (21). From its presence in the invariant subgroup follows that of $S{\cdot}RZS{\cdot}S^{-1} = SRZ = TZ$. If X_π is the rotation by π about the first axis, $X_\pi Z X_\pi = Z^{-1}$ and $X_\pi TZX_\pi^{-1} = X_\pi TX_\pi X_\pi ZX_\pi = X_\pi TX_\pi Z^{-1}$ is contained in the invariant subgroup also and thus the transform of this with Z, i.e. $Z^{-1}X_\pi TX_\pi$ also. The product of this with TZ is $TX_\pi TX_\pi$ which leaves x_4 invariant. If $TX_\pi TX_\pi = 1$ we can take $TY_\pi TY_\pi$. If this is the unity also, $TX_\pi TX_\pi = TY_\pi TY_\pi$ and T commutes with $X_\pi Y_\pi$, i.e. is a rotation about the third axis. In this case the space like (complex) characteristic vectors of TZ lie in the plane of the first two coordinate axes. Transforming TZ by an acceleration in the direction of the first coordinate axis we obtain a new element of the invariant subgroup for which the space like characteristic vector will have a not vanishing fourth component. Taking this for RZS we can transform it with S again to obtain a new SRZ = TZ. However, since S leaves x_4 invariant, the fourth component of the space like characteristic vectors of this TZ will not vanish and we can obtain from it by the procedure just described a rotation which must be contained in the invariant subgroup.

It remains to be shown that an invariant subgroup which contains a rotation, contains the whole homogeneous Lorentz group. Since the three-dimensional rotation group is simple, all rotations must be contained in the invariant subgroup. Thus the rotation by π around the first axis X_π and also its transform with Z and also

$$ZX_\pi Z^{-1}{\cdot}X_\pi = Z{\cdot}X_\pi Z^{-1}X_\pi = Z^2$$

is contained in the invariant subgroup. However, the general acceleration in the direction of the third axis can be written in this form. As all rotations are contained in the invariant subgroup also, (21) shows that this holds for all elements of the homogeneous Lorentz group.

It follows from this that the homogeneous Lorentz group has apart from the representation with unit matrices only true representations. It follows then from the remark at the end of part B, that these have all infinite dimensions. This holds even for the two-valued representations to which we shall be led in Section 5 equ. (52D), as the group elements to which the positive or negative unit matrix corresponds must form an invariant subgroup also, and because the argument at the end of part B holds for two-valued representations also. One easily sees furthermore from the equations (52B), (52C) that it holds for the inhomogeneous Lorentz group equally well.

5. Reduction of Representations Up to a Factor to Two-Valued Representations

The reduction will be effected by giving each unitary transformation, which is defined by the physical content of the theory and the consideration of reference only up to a factor of modulus unity, a "phase," which will leave only the sign of the representation operators undetermined. The unitary operator corresponding to the translation a will be denoted by T(a), that to the homogeneous Lorentz transformation Λ by d(Λ). To the general inhomogeneous Lorentz transformation then D(a, Λ) = T(a)d(Λ) will correspond. Instead of the relations (12), we shall use the following ones.

$$T(a)T(b) = \omega(a, b)T(a + b) \tag{22B}$$

$$d(\Lambda)T(a) = \omega(\Lambda, a)T(\Lambda a)d(\Lambda) \tag{22C}$$

$$d(\Lambda)d(I) = \omega(\Lambda, I)d(\Lambda I). \tag{22D}$$

The ω are numbers of modulus 1. They enter because the multiplication rules (12) hold for the representative only up to a factor. Otherwise, the relations (22) are consequences of (12) and can in their return replace (12). We shall replace the T(a), d(Λ) by Ω(a)T(a) and Ω(Λ)d(Λ) respectively, for which equations similar to (22) hold, however with

$$\omega(a, b) = 1; \quad \omega(\Lambda, a) = 1; \quad \omega(\Lambda, I) = \neq 1. \tag{22'}$$

A.

It is necessary, first, to show that the undetermined factors in the representation D(L) can be assumed in such a way that the $\omega(a,b)$, $\omega(\Lambda,a)$, $\omega(\Lambda,I)$ become –apart from regions of lower dimensionality–continuous functions of their arguments. This is a consequence of the continuous character of the representation and shall be discussed first.

(a) From the point of view of the physicist, the natural definition of the continuity of a representation up to a factor is as follows. The neighborhood δ of a Lorentz transformation $L_0 = (b, I)$ shall contain all the transformations $L = (a, \Lambda)$ for which $|a_k - b_k| < \delta$ and $|\Lambda_{ik} - I_{ik}| < \delta$. The representation up to a factor D(L) is continuous if there is to every positive number ε, every normalized wave function ϕ and every Lorentz transformation L_0 such a neighborhood δ of L_0 that for every L of this neighborhood one can find an Ω of modulus > 1 (the Ω depending on L and ϕ) such that $(u_\phi, u_\phi) < \varepsilon$ where

$$u_\phi = (D(L_0) - \Omega D(L)\phi). \tag{23}$$

Let us now take a point L_0 in the group space and find a normalized wave function ϕ for which $|(\phi, DL_0\phi)| > 1/6$. There always exists a ϕ with this property, if $|(\phi, D(L_0)\phi)| < 1/6$ then $\psi = \alpha\phi + \beta D(L_0)\phi$ with suitably chosen α and β will be normalized and $|(\psi, D(L_0)\psi)| > 1/6$. We consider then such a neighborhood $\aleph$ of L_0 for all L of which $|(\phi, D(L)\phi)| > 1/12$. It is well known[19] that the whole group space can be covered with such neighborhoods. We want to show now that the $D(L)\phi$ can be multiplied with such phase factors (depending on L) of modulus unity that it becomes strongly continuous in the region $\aleph$.

We shall chose that phase factor so that $(\phi, D(L)\phi)$ becomes real and positive. Denoting then

$$(D(L_1) - D(L))\phi = U_\phi,$$

the (U_ϕ, U_ϕ) can be made arbitrarily small by letting L approach sufficiently near to L_1, if L_1 is in $\aleph$. Indeed, on account of the continuity, as defined above, there is an $\Omega = e^{ik}$ such that $(u, u) < \varepsilon$ if L is sufficiently

[19]This condition is the "separability" of the group. Cf. e.g. A. Haar, Ann. of Math., *34*, 147, 1933.

near to L_1 where

$$u = (D(L_1) - e^{ik}D(L))\phi. \tag{23'}$$

Taking the absolute value of the scalar product of u with ϕ one obtains

$$|(\phi, D(L_1)\phi) - \cos k(\phi, D(L)\phi) - i \sin k(\phi, D(L)\phi)| = |(\phi, u)| \le \sqrt{\varepsilon},$$

because of Schwartz's inequality. If only $\sqrt{\varepsilon} < 1/12$, the k must be smaller than $\pi/2$ because the absolute value is certainly greater than the real part and both $(\phi, D(L_1)\phi)$ and $(\phi, D(L)\phi)$ are real and greater than 1/12.

As the absolute value is also greater than the imaginary part, we

$$\sin k < 12\sqrt{\varepsilon}$$

On the other hand,

$$U_\phi = u + (e^{ik} - 1)D(L)\phi,$$

and thus

$$(U_\phi, U_\phi)^{\frac{1}{2}} \le (u, u)^{\frac{1}{2}} + |e^{ik} - 1| \le \sqrt{\varepsilon} + 2 \sin k/2$$

$$(U_\phi, U_\phi) \le 625\varepsilon.$$

(b) It shall be shown next that if $D(L)\phi$ is strongly continuous in a region and $D(L)$ is continuous in the sense defined at the beginning of this section, then $D(L)\psi$ with an arbitrary ψ is (strongly) continuous in that region also. We shall see, hence, that the $D(L)$, with any normalization which makes a $D(L)\phi$ strongly continuous, is continuous in the ordinary sense: There is to every L_1, ε and *every* ψ a δ so that $(U_\psi, U_\psi) < \varepsilon$ where

$$U_\psi = (D(L_1) - D(L))\psi$$

if L is in the neighborhood δ of L_1.

It is sufficient to show the continuity of $D(L)\psi$ where ψ is orthogonal to ϕ. Indeed, every ψ' can be decomposed into two terms, $\psi' = \alpha\phi + \beta\psi$ the one of which is parallel, the other perpendicular to ϕ. Since $D(L)\phi$ is continuous, according to supposition, $D(L)\psi' = \alpha D(L)\phi + \beta D(L)\psi$ will be continuous also if $D(L)\psi$ is continuous.

The continuity of the representation up to a factor requires that it is possible to achieve that $(u_\psi, u_\psi) < \varepsilon$ and $(u_{\psi+\phi}, u_{\psi+\phi}) < \varepsilon$ where

$$u_\psi = (D(L_1) - \Omega_\psi D(L))\psi, \tag{23a}$$

$$u_{\psi+\phi} = (D(L_1) - \Omega_{\psi+\phi} D(L))(\psi+\phi), \tag{23b}$$

with suitaly chosen Ω's. According to the foregoing, it also is possible to choose L and L_1 so close that $(U_\phi, U_\phi) < \varepsilon$.

Subtracting (23′) and (23a) from (23b) and applying $D(L)^{-1}$ on both sides gives

$$(\Omega_\psi - \Omega_{\psi+\phi})\psi + (1-\Omega_{\psi+\phi})\phi = D(L)^{-1}(u_{\psi+\phi} - u_\psi - U_\phi)$$

The scalar product of the right side with itself is less than 9ε. Hence both $|\Omega_\psi - \Omega_{\psi+\phi}| < 3\varepsilon^{\frac{1}{2}}$ and $|1 - \Omega_{\psi+\phi}| < 3\varepsilon^{\frac{1}{2}}$ or $|1 - \Omega_\psi| < 6\varepsilon^{\frac{1}{2}}$. Because of $U_\psi = u_\psi - (1 - \Omega_\psi)D(L)\psi$, the $(U_\psi, U_\psi)^{\frac{1}{2}} < (u_\psi, u_\psi)^{\frac{1}{2}} + |1 - \Omega_\psi|$ and thus $(U_\psi, U_\psi) < 49\varepsilon$.

This completes the proof of the theorem stated under (b). It also shows that not only the continuity of $D(L)\psi$ has been achieved in the neighborhood of L_0 by the normalization used in (a) but also that of $D(L)\psi$ with every ψ, i.e., the continuity of D(L).

It is clear also that every finite part of the group space can be covered by a finite number of neighborhoods in which D(L) can be made continuous. It is easy to see that the ω of (22) will be also continuous in these neighborhoods so that it is possible to make them continuous, apart from regions of lower dimensionality than their variables have. In the following only the fact will be used that they can be made continuous in the neighborhood of any a, b, and Λ.

B.

(a) We want to show next that all T(a) commute. From (22B) we have

$$T(a)T(b)T(a)^{-1} = c(a,\ b)T(b) \tag{24}$$

where $c(a,\ b) = \omega(a, b)/\omega(b, a)$ and hence

$$c(a,\ b) = c(b,\ a)^{-1}. \tag{24a}$$

Transforming (24) with T(a′) one obtains

$$T(a')T(a)T(b)T(a)^{-1}T(a')^{-1} = c(a,b)T(a')T(b)T(a')^{-1}$$

or

$$\omega(a',a)T(a'+a)T(b)\omega(a',a)^{-1}T(a'+a)^{-1} = c(a,b)c(a',b)T(b)$$

or

$$c(a,b)c(a',b) = c(a+a',b). \tag{25}$$

It follows[20] from (25) and the partial continuity of c(a, b) that

$$c(a, \ b) \ = \ exp(2\pi i\sum_{k=1}^{4} a_k f_k(b)) \tag{26}$$

and, since this is equal to $c(b, a)^{-1} = \exp(-2\pi i\sum b_k f_k(a))$

$$\sum_{k=1}^{4}(a_k f_k(b) \ + \ b_k f_k(a)) \ = \ n(a, \ b) \tag{27}$$

where n(a, b) is an integer. Setting in (27) for b the vector $e^{(\lambda)}$ the λ component of which is 1, all the others zero and for $f_k(e^{(\lambda)}) = -f_{k\lambda}$

$$f_\lambda(a) \ = \ n(a, \ e^{(\lambda)}) \ + \ \sum_k a_k f_{k\lambda},$$

and putting this back into (27) we obtain

$$\sum_{k,\lambda=1}^{4} f_{k\lambda}(a_\lambda b_k + b_\lambda a_k) + \sum_{k=1}^{4} a_k n(b,e^{(k)}) + b_k n(a,e^{(k)}) \ = \ n(a,b). \tag{28}$$

Assuming for the components of a and b such values which are transcendental both with respect to each other and the $f_{k\lambda}$ (which are fixed numbers), one sees that (28) cannot hold except if the coefficient of every one vanishes

$$f_{k\lambda} \ + \ f_{\lambda k} \ = \ 0; \quad n(b, \ e^{(k)}) \ = \ 0, \tag{29}$$

so that (26) becomes

$$c(a, \ b) \ = \ exp(2\pi i \sum_{k,\lambda=1}^{4} f_{k\lambda}, a_\lambda b_k). \tag{30}$$

It is necessary now to consider the existence of an operator d(Λ) satisfying (22C). Transforming this equation with the similar equation containing b instead of a

[20]G. Hamel, Math. Ann. *60*, 460, 1905, quoted from H. Hahn, *Theorie der reellen Funktionen*. Berlin 1921, pages 581-583.

$$d(\Lambda)T(b)d(\Lambda)^{-1}d(\Lambda)T(a)d(\Lambda)^{-1}d(\Lambda)T(b)^{-1}d(\Lambda)^{-1}$$
$$= \omega(\Lambda,\ b)T(\Lambda b)\omega(\Lambda,\ a)T(\Lambda a)\omega(\Lambda,\ b)^{-1}T(\Lambda b)^{-1}$$
$$= \omega(\Lambda,\ a)c(\Lambda b,\ \Lambda a)T(\Lambda a),$$

while the first line is clearly $d(\Lambda)c(b, a)T(a)d(\Lambda)^{-1} = \omega(\Lambda, a)c(b, a)T(\Lambda a)$ whence

$$c(b,\ a) = c(\Lambda b,\ \Lambda a) \tag{31}$$

holds for every Lorentz transformation Λ. Combined with (30) this gives

$$\sum_{k\lambda}(f_{k\lambda}a_k b_\lambda - \sum_{\nu\mu} f_{\nu\mu}\Lambda_{\nu k}\Lambda_{\mu\lambda}a_k b_\lambda) = n'(a,\ b),$$

where $n'(a, b)$ is again an integer. As this equation holds for every a, b

$$f_{k\lambda} = \sum_{\nu\mu} f_{\nu\mu}\Lambda_{\nu k}\Lambda_{\mu\lambda};\quad f = \Lambda' f\Lambda$$

must hold also, for every Lorentz transformation. However, the only form invariant under all Lorentz transformations are multiples of the F of (10). Actually, because of (29), f must vanish and $c(a, b) = 1$, all the operators corresponding to translations commute

$$T(a)T(b) = T(b)T(a). \tag{32}$$

It is well to remember that it was necessary for obtaining this result to use the existence of $d(\Lambda)$ satisfying (22C).

(b) Equation (32) is clearly independent of the normalization of the T(a). If we could fix the translation operators in four linearly independent directions $e^{(1)}, e^{(2)}, e^{(3)}, e^{(4)}$ so that for each of these directions

$$T(ae^{(k)})T(be^{(k)}) = T((a\ +\ b)e^{(k)}) \tag{33}$$

be valid for every pair of numbers a, b, then the normalization

$$T(a_1e^{(1)} + a_2e^{(2)} + a_3e^{(3)} + a_4e^{(4)})$$
$$= T(a_1e^{(1)})T(a_2e^{(2)})T(a_3e^{(3)})T(a_4e^{(4)}) \tag{33a}$$

and (32) would ensure the general validity of

$$T(a)T(b) = T(a\ +\ b). \tag{34}$$

As the four linearly independent directions $e^{(1)}, \cdots, e^{(4)}$ we shall take four null vectors. If e is a null vector, there is, according to section 3, a homogeneous Lorentz transformation[21] Λ_e such that $\Lambda_e e = 2e$.

[21]The index e denotes here the vector e for which $\Lambda_e e = 2e$; this Λ_e has no elementary divisor.

We normalize T (e) so that

$$d(\Lambda_e)T(e)d(\Lambda_e)^{-1} = T(e)^2. \tag{35}$$

This is clearly independent of the normalization of $d(\Lambda_e)$. We further normalize for all (positive and negative) integers n

$$d(\Lambda_e)^n T(e) d(\Lambda_e)^{-n} = T(2^n e). \tag{35a}$$

It follows from this equation also that

$$T(2^n e)^2 = d(\Lambda_e)^n T(e)^2 d(\Lambda_e)^{-n} = d(\Lambda_e)^n d(\Lambda_e) T(e) d(\Lambda_e)^{-1} d(\Lambda_e)^{-n}$$
$$= T(2^{n+1}e). \tag{36}$$

This allows us to normalize for every positive integer k

$$T(k \cdot 2^{-n} e) = T(2^{-n} e)^k \tag{35a}$$

in such a way that the normalizaton remains the same if we replace k by $2^m k$ and n by n + m. This ensures, together with (36), the validity of

$$T(\nu e)T(\mu e) = T((\nu + \mu)e)$$
$$\tag{36a}$$
$$d(\Lambda_e)T(\nu e)d(\Lambda_e)^{-1} = T(2\nu e)$$

for all dyadic fractions ν and μ.

It must be shown that if $\nu_1, \nu_2, \nu_3, \cdots$ is a sequence of dyadic fractions, converging to 0, $\lim T(\nu_i e) = 1$. From $T(a) \cdot T(0) = \omega(a, 0)T(a)$ it follows that T(0) is a constant. According to the theorem of part (A)(b), the $T(\nu e)$, if multiplied by proper constants Ω_ν will converge to 1, i.e., by choosing an arbitrary ϕ, it is possible to make both $(1-\Omega_\nu T(\nu e))\phi = u$ and $(1-\Omega_\nu T(\nu e)) \cdot d(\Lambda_e)^{-1}\phi = u'$ arbitrarily small, by making ν small. Applying $d(\Lambda_e)$ to the second expression, one obtains, for (36a), that $(1 - \Omega_\nu T(2\nu e))\phi = d(\Lambda_e)u'$ is also small. On the other hand, applying $T(\nu e)$ to the first expression one sees that $(T(\nu e) - \Omega_\nu T(2\nu e))\phi = T(\nu e)u$ approaches zero also. Hence, the difference of these two quantities $(1 - T(\nu e))\phi$ goes to zero, i.e. $T(\nu_i e)\phi$ converges to ϕ if $\nu_1, \nu_2, \nu_3, \cdots$ is a sequence of dyadic fractions approaching 0.

Now $\nu_1, \nu_2, \nu_3, \cdots$ be a sequence of dyadic fractions coverging to an arbitary number a. It will be shown then that $T(\nu_i e)$ converges to a multiple of T(ae) and this multiple of T(ae) will be the normalized T(ae). Again, it follows from the continuity that there are such Ω_i that $\Omega_i T(\nu_i e)\phi$ converges to $T(ae)\phi$. The $\Omega_j^{-1}T(\nu_j e)^{-1}\Omega_i T(\nu_i e)\phi$ will converge to ϕ, therefore, as both i and j tend to infinity. However, according to the previous paragraph, $T((\nu_i - \nu_j)e)\phi$ tends to ϕ and thus $\Omega_j^{-1}\Omega_i$ tends to 1. It follows that Ω_i^{-1} converges to a definite number Ω. Hence $\Omega_i^{-1}\cdot\Omega_i T(\nu_i e)\phi$ converges to $\Omega T(ae)\phi$ which will be denoted, henceforth, by T(ae). For the T(ae), normalized in this way, (33) will hold, since if $\mu_1, \mu_2, \mu_3 \cdots$ are dyadic fractions converging to b, we obtain, with the help (36a)

$$T(ae)T(be)\phi = \lim_{i,j=\infty} T((\nu_i + \mu_i)e)\phi = T((a+b)e)\phi.$$

This argument not only shows that it is possible to normalize the $T(ae^{(k)})$ and hence by (33a) the T(a) so that (34) holds for them but, in addition to this, that these T(a) will be continuous in the ordinary sense.

C.

It is clear that (34) will remain valid if one replaces T(a) by $\exp(2\pi i\{a,c\})T(a)$ where c is an arbitrary vector. This remaining freedom in the normalization of T(a) will be used to eliminate the $\omega(\Lambda, a)$ from (22C).

Transforming (22C) $d(\Lambda)T(a)d(\Lambda)^{-1} = \omega(\Lambda, a)T(\Lambda a)$ with d(M) one obtains on the left side $\omega(M, \Lambda)d(M\Lambda)T(a)\omega(M, \Lambda)^{-1}d(M\Lambda)^{-1} = \omega(M\Lambda, a)T(M\Lambda a)$ while the right side becomes $\omega(\Lambda, a)\omega(M, \Lambda a)T(M\Lambda a)$. Hence

$$\omega(M\Lambda, a) = \omega(M, \Lambda a)\omega(\Lambda, a). \tag{37}$$

On the other hand, the product of two equations (22C) with the same Λ but with a and b respectively, instead of a yields with the help of (34)

$$\omega(\Lambda, a)\omega(\Lambda, b) = \omega(\Lambda(a + b)).$$

Hence

$$\omega(\Lambda, a) = exp(2\pi i\{a, f(\Lambda)\}),$$

where $f(\Lambda)$ is a vector which can depend on Λ. Inserting this back into (37) one obtains

$$\{a, f(M\Lambda)\} = \{\Lambda a, f(M)\} + \{a, f(\Lambda)\} + n,$$

$$\{a, f(M\Lambda) - \Lambda^{-1}f(M) - f(\Lambda)\} = n,$$

where n is an integer which must vanish since it is a linear function of a. Hence

$$f(M\Lambda) = \Lambda^{-1}f(M) + f(\Lambda). \tag{38}$$

If we can show that the most general solution of the equation is

$$f(\Lambda) = (\Lambda^{-1} - 1)v_0, \tag{39}$$

where v_0 is a vector independent of Λ, the $\omega(\Lambda, a)$ will become $\omega(\Lambda, a) = \exp(2\pi i\{(\Lambda - 1)a, v_0\})$. Then $\omega(\Lambda, a)$ in (22C) will disappear if we replace $T(a)$ by $\exp(2\pi i\{a, v_0\})T(a)$.

The proof that (39) is a consequence of (38) is somewhat laborious. One can first consider the following homogeneous Lorentz transformations.

$$X(\alpha_1,\gamma_1) = \begin{bmatrix} C_1 & 0 & 0 & S_1 \\ 0 & c_1 & s_1 & 0 \\ 0 & -s_1 & c_1 & 0 \\ S_1 & 0 & 0 & C_1 \end{bmatrix}; \qquad Y(\alpha_2,\gamma_2) = \begin{bmatrix} c_2 & 0 & -s_2 & 0 \\ 0 & C_2 & 0 & S_2 \\ s_2 & 0 & c_2 & 0 \\ 0 & S_2 & 0 & C_2 \end{bmatrix}$$

$$Z(\alpha_3,\gamma_3) = \begin{bmatrix} c_3 & s_3 & 0 & 0 \\ -s_3 & c_3 & 0 & 0 \\ 0 & 0 & C_3 & S_3 \\ 0 & 0 & S_3 & C_3 \end{bmatrix} \tag{40}$$

where $c_i = \cos\alpha_i$; $s_i = \sin\alpha_i$; $C_i = \mathrm{Ch}\gamma_i$; $S_i = \mathrm{Sh}\gamma_i$. All the $X(\alpha, \gamma)$ commute. Let us choose, therefore, two angles α_1, γ_1 for which $1 - X(\alpha_1, \gamma_1)^{-1}$ has a reciprocal. It follows then from (38)

$$X(\alpha,\gamma)^{-1}f(X(\alpha_1,\gamma_1)) + f(X(\alpha,\gamma)) = X(\alpha_1,\gamma_1)^{-1}f(X(\alpha,\gamma)) + f(X(\alpha_1,\gamma_1))$$

or

$$f(X(\alpha,\gamma)) = [1 - X(\alpha_1,\gamma_1)^{-1}]^{-1}[1 - X(\alpha,\gamma)^{-1}]f(X(\alpha_1,\gamma_1)) \tag{41}$$

$$f(X(\alpha,\gamma)) = (1 - X(\alpha,\gamma)^{-1})v_x,$$

where v_x is independent of α, γ. Similar equations hold for the $f(Y(\alpha,\gamma))$ and $f(Z(\alpha,\gamma))$. Let us denote now $X(\pi, 0) = X$; $Y(\pi, 0) = Y$; $Z(\pi, 0) = Z$. These anticommute in the following sense with the transformations (40):

$$YX(\alpha, \gamma)Y = ZX(\alpha, \gamma)Z = X(\alpha, \gamma)^{-1}. \tag{42}$$

From (38) one easily calculates

$$f(YX(\alpha, \gamma)Y) = (YX(\alpha, \gamma)^{-1} + 1)f(Y) + Yf(X(\alpha, \gamma)),$$

or, because of (41) and (42), after some trivial transformations

$$(1 - X(\alpha, \gamma))(1 - Y)(v_X - v_Y) = 0. \tag{43}$$

As α, γ can be taken arbitrarily, the first factor can be dropped. This leaves $(1 - Y)(v_X - v_Y) = 0$, or that the first and third components of v_X and v_Y are equal. One similarly concludes, however, that $(1 - X)(v_Y - v_X) = 0$ and thus that the first three components of v_X, v_Y and also of v_Z are equal.

For $\gamma_1 = \gamma_2 = \gamma_3 = 0$ the transformations (40) are the generators of all rotations, i.e. all Lorentz transformations R not affecting the fourth coordinate. As the 4-4 matrix element of these transformations is 1, the expression $(1 - R^{-1})v$ is independent of the fourth component of v and $(1 - R^{-1})v_X = (1 - R^{-1})v_Y = (1 - R^{-1})v_Z$. It follows from (38) that if $f(R) = (1 - R^{-1})v_X$ and $f(S) = (1 - S^{-1})v_X$, then $f(SR) = (1 - R^{-1} S^{-1})v_X$. Thus $f(R) = (1 - R^{-1})v_X$ is valid with the same v_X for all rotations.

Now

$$\begin{aligned} f(X(\alpha,\gamma)R) &= R^{-1}(1-X(\alpha,\gamma)^{-1})v_X+(1-R^{-1})v_X \\ &= (1-(X(\alpha,\gamma)R)^{-1})v_X. \end{aligned}$$

One easily concludes from (38) that the f(E) corresponding to the unit operation vanishes and $f(\Lambda^{-1}) = -\Lambda f(\Lambda)$. Hence $f(R^{-1}X(\alpha, \gamma)^{-1}) = (1 - X(\alpha,\gamma R)v_X$; and one concludes further that for all Lorentz transformations $\Lambda = RX(\alpha,\gamma)S$, (39) holds with $v_0 = -v_X$ if R and S are rotations. However, every homogeneous Lorentz transformation can be brought into this form (Section 4C). This completes the proof of (39) and thus of $\omega(\Lambda, \alpha) = 1$.

D.

The quantities $\omega(a, b)$ and $\omega(\Lambda, \alpha)$ for which it has just been shown that they can be assumed to be 1, are independent from the normalization of $d(\Lambda)$. We can affix therefore an arbitrary factor of modulus 1 to all the $d(\Lambda)$, without interfering with the normalizations so far accomplished. In consequence hereof, the ensuing discussion will be simply a discussion of the normalization of the operators for the homogeneous Lorentz group and the result to be obtained will be valid for the group also.

Partly because the representations up to a factor of the three dimensional rotation group may be interesting in themselves, but more particularly

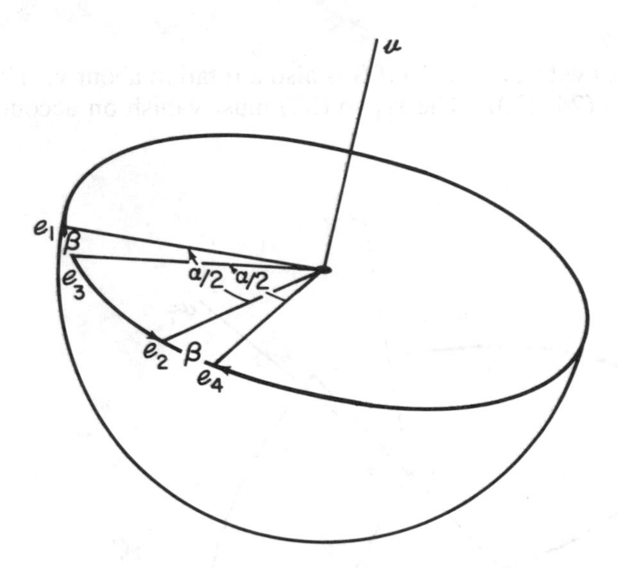

Figure 2:

because the procedure to be followed for the Lorentz group can be especially simply demonstrated for this group, the three dimensional rotation group shall be taken up first.

It is well known that the nomalization cannot be carried so far that $\omega(\Lambda, I) = 1$ in (22D) and there are well known representations for which $\omega(\Lambda, I) = \pm 1$. We shall allow this ambiguity therefore from the outset.

One can observe, first, that the operator corresponding to the unity of the group is a constant. This follows simply from $d(\Lambda)d(E) = \omega(\Lambda, E)d(\Lambda)$. The square of an operator corresponding to an involution is a constant, therefore.

The operator corresponding to the rotation about the axis e by the angle π; normalized so that its square be actually 1, will be denoted by $\tilde{e}$; $\tilde{e}^2 = 1$. The $\tilde{e}$ are–apart from the sign–uniquely defined.

A rotation R about v by the angle α is the product of two rotations by π about e_1 and e_2 where e_1 and e_2 are perpendicular to v and e_2 arises from e_1 by rotation about v and with $\alpha/2$. Choosing for every v an arbitrary e_1 perpendicular to v, we can normalize, therefore

$$d(R) = \pm\tilde{e}_1\tilde{e}_2. \tag{44}$$

Now d(R) commutes with every d(S) if S is also a rotation about v. This is proved in equations (24)-(30). The f_{11} in (30) must vanish on account of (29).

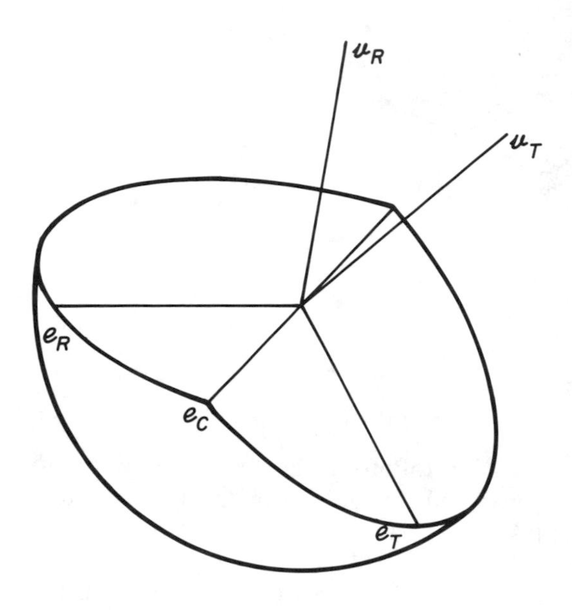

Figure 3:

(Also, both R and S can be arbitrarily accurately represented as powers of a very small rotation about v). Hence, transforming (44) by d(S) one obtains

$$d(R) = \pm d(S)\tilde{e}_1 d(S)^{-1} \cdot d(S)\tilde{e}_2 d(S)^{-1}. \tag{44a}$$

Now $d(S)\tilde{e}_1 d(S)^{-1}$ corresponds to a rotation by π about an axis, perpendicular to v and enclosing an angle β with e_1, where β is the angle of rotation of S. Since the square of $d(S)\tilde{e}_1 d(S)^{-1}$ is also 1, (44a) is simply another way of writing $d(R) = \tilde{e}_3\tilde{e}_4$ as a product of two $\tilde{e}$ and we see that the normalization (44) is independent of the choice of the axis e_1 (Cf. Fig. 2).

For computing d(R)d(T) we can draw the planes perpendicular to the axes of rotation of R and T and use for $d(R) = \tilde{e}_R\tilde{e}_C$ such a development that the axis e_C of the second involution coincide with the intersection line of the above-mentioned planes, while for $d(T) = \tilde{e}_C\tilde{e}_T$ we choose the first involution to be a rotation about this intersection line (Fig. 3). Then, the product

$$d(R)d(T) = \pm\tilde{e}_R\tilde{e}_C\tilde{e}_C\tilde{e}_T = \pm\tilde{e}_R\tilde{e}_T \qquad (45)$$

will automatically have the normalization corresponding to (44). This shows that the operators normalized in (44) give a representation up to the sign.

For the Lorentz group, the proof can be performed along the same line, only the underlying geometrical facts are less obvious. Let Λ be a Lorentz transformation without elementary divisors with the characteristic values $e^{2i\gamma}$, $e^{-2i\gamma}$, $e^{2\chi}$, $e^{-2\chi}$ and the characteristic vectors v_1, $v_2 = \overset{*}{v}_1$, v_3, v_4, as described in section 4B.

We want to make $\Lambda = MN$ with $M^2 = N^2 = 1$. For $\Lambda N = M$, we have $\Lambda N \Lambda N = 1$ and thus $\Lambda N \Lambda = N$. Setting $Nv_i = \sum_k \alpha_{ik} v_k$, we obtain $\Lambda N \Lambda v_i = \sum \lambda_k \alpha_{ik} \lambda_i v_i = \sum \alpha_{ik} v_k$. Because of the linear independence of the v_k this amounts to $\lambda_i \lambda_k \alpha_{ik} = \alpha_{ik}$: all α_{ik} are zero, except those for which $\lambda_i \lambda_k = 1$. As in none of the cases (a), (b), (c), (d) of section 4B is λ_1 or λ_2 reciprocal to one of the last two λ, the vectors v_1 and v_2 will be transformed by N into a linear combination of v_1 and v_2 again, and the same holds for v_3 and v_4. This means the N can be considered as the product of two transformations $N = N_s N_t$, the first in the v_1v_2 plane, the second in the v_3v_4 plane. (Instead of v_1v_2 plane one really should say $v_1 + v_2$, $iv_1 - iv_2$ plane, as v_1 and v_2 are complex themselves. This will be meant always by v_1v_2 plane, etc.). The same holds for M also.

Both N_s and N_t must satisfy the first and third condition for Lorentz transformations (cf. 4A) and both determinants must be either 1, or -1. Furthermore, the square of both of them must be unity.

If both determinants were +1, the N_t had to be unity itself, while N_s could be the unity or a rotation by π in the v_1v_2 plane. Thus v_1, v_2, v_3, v_4 would be characteristic vectors of N itself.

If both determinants are -1 (this will turn out to be the case), N_s is a reflection on a line in the v_1v_2 plane and N_t a reflection in the v_3v_4 plane, interchanging v_3 and v_4. In this case v_1, v_2, v_3, v_4 would not all be characteristic vectors of N.

If v_1, v_2, v_3, v_4 are characteristic vectors of N, they are characteristic vectors of $M = \Lambda N$ also. Then both M and N would be either unity, or a rotation by π in the v_1v_2 plane. If both of them were rotations in the v_1v_2 plane, their product Λ would be the unity which we want to exclude for the present. We can exclude the remaining cases in which the determinants of N_s and N_t are +1 by further stipulating that neither M nor N shall be the unity in the decomposition $\Lambda = MN$.

Hence N is the product of a reflection in the v_1v_2 plane

$$Ns'_\nu = s'_\nu; \quad Ns_\nu = s_\nu \tag{46a}$$

where s_ν and s'_ν are two perpendicular real vectors in the v_1v_2 plane

$$s'_\nu = e^{i\nu}v_1 + e^{-i\nu}v_2; \quad s_\nu = i(e^{i\nu}v_1 - e^{-i\nu}v_2), \tag{46b}$$

and of a reflection in the v_3v_4 plane

$$Nt'_\mu = t'_\mu; \quad Nt_\mu = -t_\mu, \tag{46c}$$

where again t_μ, t'_μ are real vectors in the v_3v_4 plane, perpendicular to each other, t_μ being space-like, t'_μ time-like:

$$t'_\mu = e^{\mu}v_3 + e^{-\mu}v_4; \quad t_\mu = e^{\mu}v_3 - e^{-\mu}v_4. \tag{46d}$$

Thus N becomes a rotation by π in the purely space like $s_\nu t_\mu$ plane. The M can be calculated from $M = \Lambda N$

$$Ms'_\nu = \Lambda Ns'_\nu = \Lambda s'_\nu = e^{i\nu+2i\gamma}v_1 + e^{-i\nu-2i\gamma}v_2$$
$$= \tfrac{1}{2}e^{2i\gamma}(s'_\nu - is_\nu) + \tfrac{1}{2}e^{-2i\gamma}(s'_\nu + is_\nu) = \cos 2\gamma \cdot s'_\nu + \sin 2\gamma \cdot s_\nu$$

$$Ms_\nu = \sin 2\gamma \cdot s'_\nu - \cos 2\gamma \cdot s_\nu \tag{46e}$$

$$Mt'_\mu = \Lambda Nt'_\mu = \Lambda t'_\mu = e^{\mu+2\chi}v_3 + e^{-\mu-2\chi}v_4$$
$$= \tfrac{1}{2}e^{2\chi}(t'_\mu + t_\mu) + \tfrac{1}{2}e^{2\chi}(t'_\mu - t_\mu) = \mathrm{Ch}2\chi \cdot t'_\mu + \mathrm{Sh}2\chi \cdot t_\mu$$

$$Mt_\mu = -\mathrm{Sh}2\chi \cdot t'_\mu - \mathrm{Ch}2\chi \cdot t_\mu.$$

Thus M also becomes a product of two reflections, one in the $v_1v_2 = s'_\nu s_\nu$ the other in the $v_3v_4 = t'_\mu t_\mu$ plane. This completes the decomposition of Λ into two involutions. One of the involutions can be taken to be a rotation by π in an arbitrary space like plane, intersecting both the v_1v_2 and the v_3v_4 planes, as the freedom in choosing ν and μ allows us to fix the lines s_ν, and t_μ arbitrarily in those planes. The involution characterized by (46) will be called $N_{\nu\mu}$ henceforth. The other involution M is then a similar rotation, in a plane, however, which is completely determined once the $s_\nu t_\mu$ plane is fixed. It will be denoted by $M_{\nu\mu}$ (it is, in fact $M_{\nu\mu} = N_{\nu+\gamma\mu+\chi}$). One sees the complete analogy to the three dimensional case if one remembers that γ and χ are the half angles of rotation.

The d(M) and d(N) so normalized that their squares be 1 shall be denoted by $d_1(M_{\nu\mu})$ and $d_1(N_{\nu\mu})$. We must show that the normalization for

$$d(\Lambda) = \pm d_1(M_{\nu\mu})d_1(N_{\nu\mu}) \tag{47}$$

is independent of ν and μ. For this purpose, we transform

$$d(\Lambda) = \pm d_1(M_{00})d_1(N_{00}) \tag{47a}$$

with $d(\Lambda_1)$ where Λ_1 has the same characteristic vectors as Λ but different characteristic values, namely $e^{i\nu}$, $e^{-i\nu}$, e^{μ} and $e^{-\mu}$. Since $\Lambda_1 M_{00}\Lambda_1^{-1} = M_{\nu\mu}$ and $\Lambda_1 N_{00}\Lambda_1^{-1} = N_{\nu\mu}$ we have $d(\Lambda_1)d_1(M_{00})d(\Lambda_1)^{-1} = \omega d_1(M_{\nu\mu})$ where $\omega = \pm 1$, as the squares of both sides are 1. Hence, (47a) becomes if transformed with $d(\Lambda_1)$ just

$$d(\Lambda_1)d(\Lambda)d(\Lambda_1)^{-1} = \pm d_1(M_{\nu\mu})d_1(N_{\nu\mu}). \tag{47b}$$

The normalization (47) would be clearly independent of ν and μ if $d(\Lambda_1)$ commuted with $d(\Lambda)$.

Again, the argument contained in equations (24)-(30) can be applied and shows that

$$d(\Lambda_1)d(\Lambda)d(\Lambda_1)^{-1} = exp(2\pi if(2\gamma\mu - 2\chi\nu))d(\Lambda) \tag{48}$$

holds for every γ, χ, ν, μ. However, the exponential in (48) must be 1 if $\gamma = 0$; $\nu = 2\pi/n$; $\chi = \frac{1}{2}n\mu$ since in this case $\Lambda = \Lambda_1^n$. Thus $\exp(-4\pi^2 if\mu) = 1$ for every μ and $f = 0$ and the left side of (47b) can be replaced by $d(\Lambda)$; the normalization in (47) is independent of ν and μ.

In order to have the analogue of (45), we must show that, having two Lorentz transformations $\Lambda = M_{\nu\mu}N_{\nu\mu}$ and $I = P_{\alpha\beta}Q_{\alpha\beta}$ we can choose ν, μ and α, β so that $N_{\nu\mu} = P_{\alpha\beta}$ i.e. that the plane of rotation $s_\nu t_\mu$ of $N_{\nu\mu}$ coincide with the plane of rotation of $P_{\alpha\beta}$. As the latter plane can be made to an arbitrary space like plane intersecting both the w_1w_2 and the w_3w_4 planes (where w_1, w_2, w_3, w_4 are the characteristic vectors of I), we must show the existence of a space like plane, intersecting all four planes v_1v_2, v_3v_4, w_1w_2, w_3w_4. Both the first and the second pair of planes are orthogonal.

One can show[22] that if Λ and I have no common null vector as characteristic vector, there are always two planes, perpendicular to each other which intersect four such planes. One of these is always space like. It is possible to assume, therefore, that both $N_{\nu\mu}$ and $P_{\alpha\beta}$ are the rotation by π in this plane. Thus

$$\begin{aligned} d(\Lambda)d(I) &= \pm d_1(M_{\nu\mu})d_1(N_{\nu\mu})d_1(P_{\alpha\beta})d_1(Q_{\alpha\beta}) \\ &= \pm d_1(M_{\nu\mu})d_1(Q_{\alpha\beta}), \end{aligned} \tag{49}$$

and $d(\Lambda)d(I)$ has the normalization corresponding to the product of two involutions, neither of which is unity. This is, however, also the normalization adopted for $d(\Lambda I)$. Hence

$$d(\Lambda)d(I) \;=\; \pm d(\Lambda I) \tag{49a}$$

holds if Λ, I and ΛI are Lorentz transformations corresponding to one of the cases (a), (b), (c) or (d) of section 4B and if Λ and I have no common characteristic null vector. In addition to this (49a) holds also, assuming $d(E) = \pm 1$, if any of the transformations Λ, I, ΛI is unity, or if both characteristic null vectors of Λ and I are equal, as in this case the planes v_3v_4 and w_3w_4 and also v_1v_2 and w_1w_2 conincide and there are many space like planes intersecting all.

---

[22]We first suppose the existence of a real plane p intersecting all four planes v_1v_2, v_3v_4, w_1w_2, w_3w_4. If p intersects v_1v_2 the plane q perpendicular to p will intersect the plane v_3v_4 perpendicular to v_1v_2. Indeed, the line which is perpendicular to both p and v_1v_2 (there is such a line as p and v_1v_2 intersect) is contained in both q and v_3v_4. This shows that if there is a plane intersecting all four planes, the plane perpendicular to this will have this property also.

If the plane p–the existence of which we suppose for the time being–contains a time-like vector, q will be space-like (Section 4B, [1]). Both in this case and if p contains only space-like vectors, the theorem in the text is valid. There is a last possibility, that p is tangent to the light cone, i.e. contains only space like vectors and a null vector v. The space-like vectors of p are all orthogonal to v, otherwise p would contain time-like vectors also. In this case the plane q, perpendicular to p will contain v also. The line in which v_1v_2 intersects p is space-like and orthogonal to the vector in which v_3v_4 intersects p. The latter intersection must coincide with v, therefore, as no other vector p is orthogonal to any space-like vector in it. Hence, v is the intersection of p and v_3v_4 and is either v_3 or v_4. One can conclude in the same way that v coincides with either w_3, or w_4 also and we see that if p is tangent to the light cone the two transformations Λ and I have a common null vector as characteristic vector. Thus the theorem in the text is correct if we can show the existence of an arbitrary real plane p intersecting all four planes v_1v_2, v_3v_4, w_1w_2, w_3w_4.

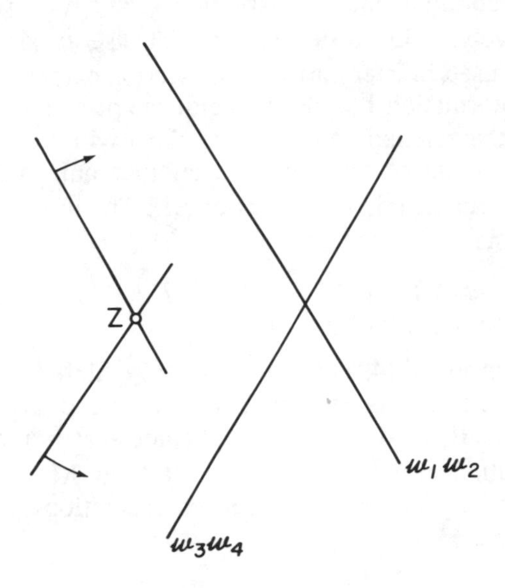

Figure 4: Fig. 4 gives a projection of all lines into the x_1x_2 plane. One sees that there are, in general, two intersecting planes, only in exceptional cases is there only one.

Let us draw a coordinate system in our four dimensional space, the x_1x_2 plane of which is the v_1v_2 plane, the x_3 and x_4 axes having the directions of the vectors $v_3 - v_4$ and $v_3 + v_4$, respectively. The three dimensional manifold M characterized by $x_4 = 1$ intersects all planes in a line, the v_1v_2 plane in the line at infinity of the x_1x_2 plane, the v_3v_4 plane in the x_3 axis. The intersection of M with the w_1w_2 and w_3w_4 planes will be lines in M with directions perpendicular to each other. They will have a common normal through the origin of M, intersecting it at reciprocal distances. This follows from their orthogonality in the four dimensional space.

A plane intersecting v_1v_2 and v_3v_4 will be a line parallel to x_1x_2 through the x_3 axis. If we draw such lines through all points of the line corresponding to w_1w_2, the direction of this line will turn by π if we go from one end of this line to the other. Similarly, the lines going through the line corresponding to w_3w_4 will turn by π in the *opposite* direction. Thus the first set of lines will have at least one line in common with the second set and this line will correspond to a real plane intersecting all four planes v_1v_2, v_3v_4, w_1w_2, w_3w_4. This completes the proof of the theorem referred to in the text.

If Λ and I have one common characteristic null vector, $v_3 = w_3$, the others, v_4 and w_4 respectively, being different, one can use an artifice to prove (49a) which will be used in later parts of this section extensively. One can find a Lorentz transformation J so that none of the pairs I - J; Λ - IJ; ΛIJ - J^{-1} has a common characteristic null vector. This will be true, e.g. if the characteristic null vectors of J are v_4 and another null vector, different from v_3, w_4 and the characteristic vectors of ΛI. Then (49a) will hold for all the above pairs and

$$\begin{aligned} d(\Lambda)d(I) &= \pm d(\Lambda)d(I)d(J)d(J^{-1}) = \pm d(\Lambda)d(IJ)d(J^{-1}) \\ &= \pm d(\Lambda IJ)d(J^{-1}) = \pm d(\Lambda I). \end{aligned}$$

This completes the proof of (49a) for all cases in which Λ, I and ΛI have no elementary divisors. It is evident also that we can replace in the normalization (47) the d_1 by d. One also concludes easily that $d(M)^2$ is in the same representation either +1 for all involutions M, or -1 for every involution. The former ones will give real representations, the latter ones representations up to the sign.

If Λ has an elementary divisor, it can be expressed in the v_e, w_e, z_e, v_1 scheme as the matrix (Cf. equ. (20))

$$\Lambda_e = \begin{bmatrix} 1 & 1 & \frac{1}{2} & 0 \\ 0 & 1 & 1 & 0 \\ 0 & 0 & 1 & 0 \\ 0 & 0 & 0 & 1 \end{bmatrix}$$

and can be written, in the same scheme, as the product of two Lorentz transformations with the square 1

$$\Lambda_e = M_0 N_0 = \begin{bmatrix} 1 & -1 & \frac{1}{2} & 0 \\ 0 & -1 & 1 & 0 \\ 0 & 0 & 1 & 0 \\ 0 & 0 & 0 & -1 \end{bmatrix} \cdot \begin{bmatrix} 1 & 0 & 0 & 0 \\ 0 & -1 & 0 & 0 \\ 0 & 0 & 1 & 0 \\ 0 & 0 & 0 & -1 \end{bmatrix}.$$

We can normalize therefore $d(\Lambda) = \pm\, d(M_0)d(N_0)$. If Λ can be written as the product of two other involutions also $\Lambda_e = M_1N_1$ the corresponding normalization will be identical with the original one. In order to prove this, let us consider a Lorentz transformation J such that neither of the Lorentz transformations J, N_0J, N_1J, $\Lambda_eJ = M_0N_0J = M_1N_1J$ have an elementary divisor. Since the number of free parameters is only 4 in case (e), while 6 for case (a), this is always possible. Then, for (45a)

$$d(M_0)d(N_0)d(J) = \pm d(M_0)d(N_0J) = \pm d(M_0N_0J)$$
$$= \pm d(M_1N_1J) = \pm d(M_1)d(N_1J) = \pm d(M_1)d(N_1)d(J)$$

and thus $d(M_0)d(N_0) = \pm d(M_1)d(N_1)$. This shows also that even if ΛI is in case (e), $\omega(\Lambda, I) = \pm 1$, since (49) leads to the correct normalization.

If $\Lambda = MN$ has an elementary divisor, I not, $d(\Lambda)d(I)$ still will have the normalization corresponding to the product of two involutions. One can find again a J such that neither of the transformations J, J^{-1}, IJ, NIJ, MNIJ, have an elementary divisor. Then

$$d(M)d(N)d(I) = \pm d(M)d(N)d(I)d(J)d(J)^{-1}$$
$$= \pm d(M)d(N)d(IJ)d(J)^{-1} = \pm d(M)d(NIJ)d(J)^{-1}$$
$$= d(\Lambda IJ)d(J^{-1}).$$

The last product has, however, the normalization corresponding to two involutions, as was shown in (49a), since neither ΛIJ, nor J^{-1} is in case (e).

Lastly, we must consider the case when both Λ and I may have an elementary divisor. In this case, we need a J such that neither J, J^{-1}, IJ have one. Then, because of the generalization of (49a) just proved, in which the first factor is in case (e)

$$d(\Lambda)d(I) = \pm d(\Lambda)d(I)d(J)d(J^{-1}) = \pm d(\Lambda)d(IJ)d(J^{-1})$$
$$= \pm d(\Lambda IJ)d(J^{-1})$$

which has the right normalization.

This completes the proof of

$$\omega(\Lambda, I) = \pm 1 \tag{50}$$

for all possible cases, and the normalization of all D(L) of a representation of the inhomogeneous Lorentz group up to a factor, is carried out in such a way that the normalized operators give a representation up to the sign. It is even carried so far that in the first two of equations (22) $\omega = 1$ can be set. We shall consider henceforth systems of operators satisfying (7), or more specifically, (22B) and (22C) with $\omega(a, b) = \omega(\Lambda, a) = 1$ and (22D) with $\omega(\Lambda, I) = \pm 1$.

E.

Lastly, it shall be shown that the renormalization not only did not spoil the partly continuous character of the representation, attained at the first normalization in part (A) of this section, but that the same holds now *everywhere*, in the ordinary sense for T(a) and, apart from the ambiguity of sign, also for $d(\Lambda)$. For T(a) this was proved in part (B)(b) of this section, for $d(\Lambda)$ it means that to every Λ_1, ε and ϕ there is such a δ that *one* of the two quantities

$$((d(\Lambda_1) \mp d(\Lambda))\phi, (d(\Lambda_1) \mp d(\Lambda))\phi) \; < \; \varepsilon \tag{51}$$

if Λ is in the neighborhood δ of Λ_1. The inequality (51) is equivalent to

$$((1 \mp d(\Lambda_0))\phi, (1 \mp d(\Lambda_0))\phi) \; < \; \varepsilon, \tag{51a}$$

where $\Lambda_0 = \Lambda_1^{-1}\Lambda$ now can be assumed to be in the neighborhood of the unity. Thus, the continuity of $d(\Lambda)$ at $\Lambda = E$ entails the continuity everywhere.[23] In fact, it would be sufficient to show that the d(X), d(Y) and d(Z) corresponding to the transformations (40) converge to ± 1, as α, γ approach 0, since one can write every transformation in the neighborhood of the unit element as a product $\Lambda = Z(0, \gamma_3)Y(0, \gamma_2)X(0, \gamma_1)X(\alpha_1, 0)Y(\alpha_2, 0)Z(\alpha_3, 0)$ and the parameters $\alpha_1, \cdots, \phi_3$ will converge to 0 as Λ converges to 1. However, we shall carry out the proof for an arbitrary Λ without an elementary divisor.

For $d(\Lambda)$, equations (46) show that as Λ approaches E (i.e., as ϕ and χ approach zero) both M_{00} and N_{00} approach the same involution, which we shall call K. Let us now consider a wave function $\psi = \phi + d_1(K)\phi$ or, if this vanishes $\psi = \phi - d_1(K)\phi$. We have $d_1(K)\psi = \pm \psi$. If Λ is sufficiently near to unity, $d_1(N_{00})\psi$ will be sufficiently near to $\Omega d_1(K)\psi = \pm \Omega\psi$ and all we have to show is that Ω approaches ± 1. The same thing will hold for $d_1(M_{00})$. Indeed from $d_1(N_{00})\psi - \Omega\psi = u$ it follows by applying $d_1(N_{00})$ on both sides $\psi - \Omega^2\psi = (d_1(N_{00}) + \Omega)u$. As (u, u) goes to zero, Ω must go to ± 1, and consequently, also $d_1(N_{00})\psi$ goes to ψ or to $-\psi$. Applying

[23] J. von Neumann, Sitz. d. Kon. Preuss. Akad. p. 76, 1927.

$d_1(M_{00})$ to this, one sees that $d_1(M_{00})d_1(N_{00})\psi = d(\Lambda)\psi$ goes to $\pm\,\psi$ as Λ goes to unity. The argument given in (A)(b) shows that this holds not only for ψ but for every other function also, i.e. $d\Lambda$ converges to $\pm\,1 = d(E)$ as Λ approaches E. Thus $d(\Lambda)$ is continuous in the neighborhood of E and hence everywhere.

According to the last remark in part 4, the operators $\pm\,d(\Lambda)$ form a single valued representation of the group of complex unimodular two dimensional matrices C. Let us denote the homogeneous Lorentz transformation which corresponds in the isomorphism to C by $\tilde{C}$. Our task of solving the equs. (22) has been reduced to finding all single valued unitary representation[s] of the group with the elements $[a,C] = [a,1]\,[0,C]$, the multiplication rule of which is $[a,C_1]\,[b,C_2] = [a + \tilde{C}_1 b, C_1C_2]$. For the representations of this group $D[a, C] = T(a)d[C]$ we had

$$\begin{aligned} T(a)\,T(b) &= T(a+b) \\ d[C]\,T(a) &= T(\tilde{C}a)\,d[C] \\ d[C_1]\,d[C_2] &= d[C_1C_2] \end{aligned} \qquad (52a)$$

It would be more natural, perhaps, from the mathematical point of view, to use henceforth this new notation for the representations and let the d depend on the C rather than on the $\tilde{C}$ or Λ. However, in order to be reminded on the geometrical significance of the group elements, it appeared to me to be better to keep the old notion. Instead of the equations (22B), (22C), (22D) we have, then

$$T(a)\,T(b) = T(a+b) \qquad (52B)$$

$$d(\Lambda)\,T(a) = T(\Lambda a)\,d(\Lambda) \qquad (52C)$$

$$d(\Lambda)\,d(I) = \pm d(\Lambda I). \qquad (52D)$$

6. Reduction of the Representations of the Inhomogeneous Lorentz Group to Representations of a "Little Group"

This section, unlike the other ones, will often make use of methods, which though commonly accepted in physics, must be further justified from a rigorous mathematical point of view. This has been done, in the meanwhile, by J. von Neumann in an as yet unpublished article and I am much indebted to him for his cooperation in this respect and for his readiness in communicating his results to me. A reference to this paper[24] will be made whenever his work is necessary for making inexact considerations of this section rigorous.

A.

Since the translation operators all commute, it is possible[24] to introduce such a coordinate system in Hilbert space that the wave functions $\phi(p, \zeta)$ contain momentum variables p_1, p_2, p_3, p_4 and a discrete variable ζ so that

$$T(a)\phi(p, \zeta) = e^{i\{p, a\}}\phi(p, \zeta). \tag{53}$$

p will stand for the four variables p_1, p_2, p_3, p_4.

Of course, the fact that the Lorentzian scalar product enters in the exponent, rather than the ordinary, is entirely arbitrary and could be changed by changing the signs of p_1, p_2, p_3.

The unitary scalar product of two wave functions is not yet completely defined by the requirements so far made on the coordinate system. It can be a summation over ζ and an arbitrary Stieltjes integral over the components of p:

$$(\psi, \phi) = \sum_{\zeta} \int \psi(p, \zeta)^* \phi(p, \zeta) df(p, \zeta). \tag{54}$$

The importance of introducing a weight factor, depending on p, for the scalar product lies not so much in the possibility of giving finite but different weights to different regions in p space. Such a weight distribution $g(p, \zeta)$ always could be absorbed into the wave functions, replacing all $\phi(p, \zeta)$ by $\sqrt{g(p,\zeta)}\cdot\phi(p,\zeta)$. The necessity of introducing the $f(p, \zeta)$ lies rather in the possibility of some regions of p having zero weight while, on the other hand, at other places points may have finite weights. On account of the definite metric in Hilbert space, the integral $\int df(p, \zeta)$ over any region r, for any ζ, is either positive, or zero, since it is the scalar product of that function with itself, which is 1 in the region r of p and the value ζ of the discrete variable, zero otherwise.

---

[24]J. von Neumann, Ann. of Math. to appear shortly.

Let us now define the operators

$$P(\Lambda)\phi(p,\ \zeta)\ =\ \phi(\Lambda^{-1}p,\ \zeta). \tag{55}$$

This equation defines the function $P(\Lambda)\phi$, which is, at the point p, ζ, as great as the function ϕ at the point $\Lambda^{-1}p$, ζ. The operator $P(\Lambda)$ is not necessarily unitary, on account of the weight factor in (54). We can easily calculate

$$P(\Lambda)T(a)\phi(p,\ \zeta)\ =\ T(a)\phi(\Lambda^{-1}p,\ \zeta)\ =\ e^{i\{\Lambda^{-1}p,\ a\}}\phi(\Lambda^{-1}p,\ \zeta),$$

$$T(\Lambda a)P(\Lambda)\phi(p,\zeta) = e^{i\{p,\ \Lambda a\}}P(\Lambda)\phi(p,\zeta)\ = e^{i\{p,\Lambda a\}}\phi(\Lambda^{-1}p,\zeta),$$

so that, for $\{\Lambda^{-1}p,\ a\} = \{p,\ \Lambda a\}$, we have

$$P(\Lambda)T(a)\ =\ T(\Lambda a)P(\Lambda). \tag{56}$$

This, together with (52C), shows that $d(\Lambda)P(\Lambda)^{-1} = Q(\Lambda)$ commutes with all T(a) and, therefore, with the multiplication with every function of p, since the exponentials form a complete set of functions of p_1, p_2, p_3, p_4. Thus

$$d(\Lambda)\ =\ Q(\Lambda)P(\Lambda), \tag{57}$$

where $Q(\Lambda)$ is an operator in the space of the ζ alone[24] which can depend, however, on the particular value of p in the underlying space:

$$Q(\Lambda)\phi(p,\zeta)\ =\ \sum_{\eta} Q(P,\Lambda)_{\zeta\eta}\phi(p,\eta). \tag{57a}$$

Here, $Q(p,\ \Lambda)_{\zeta\eta}$ are the components of an ordinary (finite or infinite) matrix, depending on p and Λ. From (57), we obtain

$$\begin{aligned} d(\Lambda)\phi(p,\zeta)\ &=\ \sum_{\eta} Q(p,\ \Lambda)_{\zeta\eta}P(\Lambda)\phi(p,\ \eta) \\ &= \sum_{\eta} Q(p,\ \Lambda)_{\zeta\eta}\phi(\Lambda^{-1}p,\ n). \end{aligned} \tag{57b}$$

As the exponentials form a complete set of functions, we can approximate the operation of multiplication with any function of p_1, p_2, p_3, p_4 by a linear combination

$$f(p)\phi\ =\ \sum_{n} c_n T(a_n)\phi. \tag{58}$$

If we choose f(p) to be such a function that

$$f(p)\ =\ f(\Lambda p) \tag{58a}$$

the operation of multiplication with f(p) will commute with all operations of the group. It commutes evidently with the T(a) and the Q(p, Λ), and on account of (56) and (58), (58a) also with P(Λ). Thus the operation of (58) belongs to the centrum of the algebra of our representation. Since, however, we assume that the representation is factorial (cf. 2), the centrum contains only multiples of the unity and

$$f(p)\phi(p, \zeta) = c\phi(p, \zeta). \tag{58b}$$

This can be true only if ϕ is different from zero only for such momenta p which can be obtained from each other by homogeneous Lorentz transformations, because f(p) needs to be equal to f(p′) only if there is a Λ which brings them into each other.

It will be sufficient, henceforth, to consider only such representations, the wave functions of which vanish except for such momenta which can be obtained from one by homogeneous Lorentz transformations. One can restrict, then, the definition domain of the ϕ to these momenta.

These representations can now naturally be divided into the four classes enumerated in section 3, and two classes contain two subclasses. There will be representations, the wave functions of which are defined for p such that

(1) $\{p, p\} = P < 0$ (3) $p = 0$

(2) $\{p, p\} = P = 0;\ p \neq 0$ (4) $\{p, p\} = < 0.$

The classes 1 and 2 contain two sub-classes each. In the positive subclasses P_+ and 0_+ the time components of all momenta are $p_4 > 0$, in the negative subclasses P_- and 0_- the fourth components of the momenta are negative. Class 3 will be denoted by 0_0. If P is negative, it has no index.

From the condition that d(Λ) shall be a unitary operator, it is possible to infer[24] that one can introduce a coordinate system in Hilbert space in such a way that

$$\int_r df(p, \zeta) = \int_{\Lambda r} df(p, \eta) \tag{59}$$

if $Q(p, \Lambda)_{\zeta\eta} \neq 0$ for the p of the domain r. Otherwise, r is an arbitrary domain in the space of p_1, p_2, p_3, p_4, and Λr is the domain which contains Λp if r contains p. Equation (59) holds for all ζ, η, except for such pairs for which $Q(p, \Lambda)_{\zeta\eta} = 0$. It is possible, hence, to decompose the original representation in such a way that (59) holds within every reduced part.

Neither T(a) nor d(Λ) can have matrix elements between such η and ζ for which (59) does not hold.

In the third class of representations, the variable p can be dropped entirely, and $T(a)\phi(\zeta) = \phi(\zeta)$, i.e., all wave functions are invariant under the operations of the invariant subgroup, formed by the translations. The equation $T(a)\phi(\zeta) = \phi(\zeta)$ is an invariant characterization of the representations of the third class, i.e., a characterization which is not affected by a similarity transformation. Hence, the reduced parts of a representation of class 3 also belong to this class.

Since no wave function of the other classes can remain invariant under all translations, no representation of the third class can be contained in any representation of one of the other classes. In the other classes, the variability domain of p remains three dimensional. It is possible, therefore, to introduce instead of p_1, p_2, p_3, p_4 three independent variables. In the cases 1 and 2 with which we shall be concerned most, p_1, p_2, p_3 can be kept for these three variables. On account of (59), the Stieltjes integral can be replaced by an ordinary integral[24] over these variables, the weight factor being $|p_4|^{-1} = (P + p_1^2 + p_2^2 + p_3^2)^{-\frac{1}{2}}$

$$\{\psi,\phi\} = \sum_{\zeta}\iiint_{-\infty}^{\infty} \psi(p,\zeta)^*\phi(p,\zeta)|p_4|^{-1}dp_1dp_2dp_3. \tag{59a}$$

In fact, with the weight factor $|p_4|^{-1}$ the weight of the domain r i.e., $W_r = \iiint_r |p_4|^{-1}dp_1dp_2dp_3$ is equal to the weight of the domain $W_{\Lambda r}$ as required[25] by (59). Having the scalar product fixed in this way, $P(\Lambda)$ becomes a unitary operator and, hence, $Q(\Lambda)$ will be unitary also.

We want to give next a characterization of the representations with a given P, which is independent of the coordinate system in Hilbert space. It follows from (53), that in a representation with a given P the wave functions $\psi_1, \psi_2, \cdots$ which are different from zero only in a finite domain

[25]The invariance of integrals of the character of (59a) is frequently made use of in relativity theory. One can prove it by calculating the Jacobian of the transformation

$$p'_i = \Lambda_{i1}p_1 + \Lambda_{i2}p_2 + \Lambda_{i3}p_3 + (P + p_1^2 + p_2^2 + p_3^2)^{\frac{1}{2}} \quad (i=1,2,3)$$

which comes out to be $(P + p_1^2 + p_2^2 + p_3^2)^{\frac{1}{2}}(P + p'^2_1 + p'^2_2 + p'^2_3)^{-\frac{1}{2}}$. Equ. (59a) will not be used in later parts of this paper.

of p, form an everywhere dense set, to all elements of which the infinitesimal operators of translation can be applied arbitrarily often

$$\lim_{h=0} h^{-n}(T(he)-1)^n\psi = \lim_{h=0} h^{-n}(e^{ih\{p,e\}}-1)^n\psi = i^n\{p,e\}^n\psi, \quad (60)$$

where e will be a unit vector in the direction of a coordinate axis or oppositely directed to it. Hence for all members ψ of this everywhere dense set

$$\lim_{h=0} \sum_k \pm h^{-2}(T(2he_k) - 2T(he_k) + 1)\psi = (p_1^2 + p_2^2 + p_3^2 - p_4^2)\psi$$
$$= -P\psi, \quad (61)$$

where e_k is a unit vector in (or opposite) the kth coordinate axis and the $\pm$ is + for k = 4, and − for k = 1, 2, 3.

On the other hand, there is no ϕ for which

$$\lim_{h=0} \sum_k \pm h^{-2}(T(2he_k) - 2T(he_k + 1)\phi \quad (61a)$$

if it exists, would be different from $-P\phi$. Suppose the limit in (61a) exists and is $-P\phi + \phi'$. Let us choose then a normalized ψ, from the above set, such that $(\psi,\phi') = \delta$ with $\delta > 0$ and an h so that the expression after the lim sign in (61a) assumes the value $-P\phi + \phi' + u$ with $(u, u) < \delta/3$ and also the expression after the lim sign in (61), with oppositely directed e_k becomes $-P\psi + u'$ with $(u', u') < \delta/3$. Then, on account of the unitary character of T(a) and because of $T(-a) = T(a)^{-1}$

$$(\psi, \sum_k \pm h^{-2}(T(2he_k) - 2T(he_k) + 1)\phi),$$
$$= (\sum_k \pm h^{-2}(T(-2he_k) - 2T(-he_k) + 1)\psi,\phi),$$

or

$$-P(\psi,\phi) + (\psi,\phi') + (\psi,u) = -P(\psi,\phi) + (u',\phi),$$

which is clearly impossible.

Thus if the lim in (61a) exists, it is $-P\phi$ and this constitutes a characterization of the representation which is independent of similarity transformations. Since, according to the foregoing, it is always possible to find wave functions for a representation, to which (61a) can be applied, every reduced part of a representation with a given P must have this same P and no representation with one P can be contained in a representation with an other P. The same argument can be applied evidently to the positive and negative sub-classes of class 1 and 2.

B.

Every automorphism $L \rightarrow L^o$ of the group allows us to construct from one representation D(L) another representation

$$D^o(L) = D(L^o). \tag{62}$$

This principle will allow us to restrict ourselves, for representations with finite, positive or negative P, to one value of P which can be taken respectively, to be $+1_+$ and -1. It will also allow in cases 1 and 2 to construct the representations of the negative sub-classes out of representations of the positive sub-classes.

The first automorphism is $a^o = \alpha a$, $\Lambda^o = \Lambda$. Evidently Equs. (12) are invariant under this transformation. If we set, however,

$$T^o(a)\phi = T(\alpha a)\phi; \quad d^o(\Lambda)\phi = d(\Lambda)\phi,$$

then the occurring p

$$T^o(a)\phi = T(\alpha a)\phi = e^{i\{p,\alpha a\}}\phi = e^{i\{\alpha p,a\}}\phi,$$

will be the p occurring for the unprimed representation, multiplied by α. This allows, with a real positive α, to construct all representations with all possible numerical values of P, from all representation with one numerical value of P. If we take α negative, the representations of the negative sub-classes are obtained from the representations of the positive sub-class.

In case $P = 0_0$ evidently all representations go over into themselves by the transformation (62). In case $P = 0_+$ and $P = 0_-$ it will turn out that for positive α, (62) carries every representation into an equivalent one.

C.

On account of (53) and (56), (57), the equs. (52B) and (52C) are automatically satisfied and the $Q(p, \Lambda)_{\zeta\eta}$ must be determined by (52D). This gives

$$\sum_{\eta\theta} Q(p,\Lambda)_{\zeta\eta} Q(\Lambda^{-1}p,I)_{\eta\theta}\phi(I^{-1}\Lambda^{-1}p,\theta)$$
$$= \pm\sum_{\theta} Q(p,\Lambda I)_{\zeta\theta}\phi(I^{-1}\Lambda^{-1}p,\theta). \tag{63}$$

Since this must hold for every ϕ, one would conclude

$$\sum_{\eta} Q(p,\Lambda)_{\zeta\eta} Q(\Lambda^{-1}p,I)_{\eta\theta} = \pm Q(p,\Lambda I)_{\zeta\theta}. \tag{63a}$$

Actually, this conclusion is not justified, since two wave functions must be considered to be equal even if they are different on a set of measure zero. Thus one cannot conclude, without further consideration, that the two sides of (63a) are equal at every point p. On the other hand,[24] the value of $Q(p,\Lambda)_{\zeta\eta}$ can be changed on a set of measure zero and one can make it continuous in the neighborhood of every point, if the representation is continuous. This allows then, to justify (63a). It follow from (63a) that $Q(p, 1)_{\zeta\eta} = \delta_{\zeta\eta}$.

Let us choose[15] now a basic p_0 arbitrarily. We can consider then the subgroup of all homogeneous Lorentz transformations which leave this p_0 unchanged. For all elements λ, ι of this "little group," we have

$$\sum_{\eta} Q(p_0,\lambda)_{\zeta\eta} Q(p_0,\iota)_{\eta\theta} = \pm Q(p_0,\lambda\iota)_{\zeta\theta}$$

$$q(\lambda)q(\iota) = \pm q(\lambda\iota), \tag{64}$$

where $g(\lambda)$ is the matrix $q(\lambda)_{\zeta\eta} = Q(p_0, \lambda)_{\zeta\eta}$. Because of the unitary character of $Q(\Lambda)$, the $Q(p_0, \Lambda)_{\zeta\eta}$ is unitary matrix and $q(\lambda)$ is unitary also.

If we consider, according to the last paragraph of Section 5, the group formed out of the translations and unimodular two-dimensional matrices, rather than Lorentz transformations, the ± sign in (64) can be replaced by a + sign. In this case, λ and ι are unimodular two-dimensional matrices and the little group is formed by those matrices, the corresponding Lorentz transformations $\tilde{\lambda}$, $\tilde{\iota}$ to which leave p_0 unchanged $\tilde{\lambda}p_0 = \tilde{\iota}p_0 = p_0$.

Adopting this interpretation of (64), one can also see, conversely, that the representation $q(\lambda)$ of the little group, together with the class and P of the representation of the whole group, determines the latter representation, apart from a similarity transformation. In order to prove this, let us define for every p a two-dimensional unimodular matrix $\alpha(p)$ in such a way that the corresponding Lorentz transformation

$$\tilde{\alpha}(p)p_0 = p \tag{65}$$

brings p_0 into p. The $\alpha(p)$ can be quite arbitrary except of being an almost everywhere continuous function of p, especially continuous for $p = p_0$ and $\alpha(p_0) = 1$. Then, we can set

$$d(\alpha(p)^{-1})\phi(p_0,\zeta) = \phi(p,\zeta),$$
$$d(\alpha(p))\phi(p,\zeta) = \phi(p_0,\zeta). \qquad (66)$$

This is equivalent to setting in (58)

$$Q(p,\alpha(p)) = 1 \qquad (66a)$$

and can be achieved by a similarity transformation which replaces $\phi(p, \zeta)$ by $\sum_\eta Q(p_0, \alpha(p)^{-1})_{\zeta\eta}\phi(p, \eta)$. As the matrix $Q(p_0, \alpha(p)^{-1})$ is unitary, this is a unitary transformation. It does not affect, furthermore, (53) since it contains p only as a parameter.

Assuming this transformation to be carried out, (66) will be valid and will define, together with $d(\lambda)$, all the remaining $Q(p, \Lambda)$ uniquely. In fact, calculating $d(\Lambda)\phi(p, \zeta)$, we can decompose Λ into three factors

$$\Lambda = \alpha(p). \quad \alpha(p)^{-1}\Lambda\alpha(\tilde{\Lambda}^{-1}p). \quad \alpha(\tilde{\Lambda}^{-1}p)^{-1} \qquad (67)$$

The second factor $\beta = \alpha(p)^{-1}\Lambda\alpha(\tilde{\Lambda}^{-1}p)$ belongs into the little group: $\tilde{\alpha}(p)^{-1}\tilde{\Lambda}\tilde{\alpha}(\tilde{\Lambda}^{-1}p)p_0 = \tilde{\alpha}(p)^{-1}\tilde{\Lambda}\cdot\tilde{\Lambda}^{-1}p = \tilde{\alpha}(p)^{-1}p = p_0$. We can write, therefore $(\tilde{\Lambda}^{-1}p = p')$

$$\begin{aligned} d(\Lambda)\phi(p,\zeta) &= d(\alpha(p))d(\beta)d(\alpha(p'))^{-1}\phi(p,\zeta) \\ &= d(\beta)d(\alpha(p'))^{-1}\phi(p_0,\zeta) \qquad (67a) \\ &= \sum_\eta q(\beta)_{\zeta\eta}d(\alpha(p')^{-1})\phi(p_0,\eta) = \sum_\eta q(\beta)_{\zeta\eta}\phi(p',\eta). \end{aligned}$$

This shows that all representations of the whole inhomogeneous Lorentz group are equivalent which have the same P and the same representation of the little group. Further than this, the same holds even if the representations of the little group are not the same for the two representations but only equivalent to each other. Let us assume $q_1(\Lambda) = sq_2(\Lambda)s^{-1}$. Then by replacing $\phi(p, \zeta)$ by $\sum_\eta s(\zeta, \eta)\phi(p, \eta)$ we obtain a new form of the representation for which (53) still holds but $q_2(\beta)$ for the little group is replaced by $q_1(\beta)$. Then, by the transformation just described (Eq. (66)), we can bring $d(\Lambda)$ for both into the form (67a). The equivalence of two representations of the little group must be defined as the existence of a *unitary* transformation which transforms them into each other. (Only unitary transformations are used for the whole group, also).

On the other hand, if the representations of the whole group are equivalent, the representations of the little group are equivalent also: the representation of the whole group determines the representation of the little group up to a similarity transformation uniquely.

The representation of the little group was defined as the set of matrices $Q(p_0, \lambda)_{\zeta\eta}$ if the representation is so transformed that (53) and (66a) hold. Having two equivalent representations D and $SDS^{-1} = D^o$ for both of which (53) and (66a) holds, the unitary transformation S bringing the first into the second must leave all displacement operators invariant. Hence, it must have the form (57a), i.e., operate on the ζ only and depend on p only as on a parameter.

$$S\phi(p,\zeta) = \sum_{\eta} S(p)_{\zeta\eta}\phi(p,\eta). \tag{68}$$

Denoting the matrix Q for the two representations by Q and Q^o, the condition $SD(\Lambda) = D^o(\Lambda)S$ gives that

$$\sum_{\eta} S(p)_{\zeta\eta} Q(p,\Lambda)_{\eta\theta} = \sum_{\eta} Q^{o}(p,\Lambda)_{\zeta\eta} S(\Lambda^{-1}p)_{\eta\theta}$$

holds, for every Λ, for almost every p. Setting $\Lambda = \alpha(p_1)$ we can let p approach p_1 in such a way that (68a) remains valid. Since Q is a continuous function of p both $Q(p, \Lambda)$ and $Q^o(p, \Lambda)$ will approach their limiting value 1. It follows that there is no domain in which

$$S(p_1) = S(\alpha(p_1)^{-1}p_1) = S(p_0) \tag{69}$$

would not hold, i.e., that (69) holds for almost every p_1. Since all our equations must hold only for almost every p, the $S(p)_{\zeta\eta}$ can be assumed to be independent of p and (68a) then to hold for every p also. It then follows that the representations of the little group D and D^o are transformed into each other by $S_{\zeta\eta}$.

The definition of the little group involved an arbitrarily chosen momentum vector p_0. It is clear, however, that the little groups corresponding to two different momentum vectors p_0 and p are holomorphic. In fact they can be transformed into each other by $\alpha(p)$: If Λ is an element of the little group leaving p invariant then $\alpha(p)^{-1} \Lambda\alpha(p) = \beta$ is an element of the little group which leaves p_0 invariant. We can see furthermore from (67a) that if Λ is in the little group corresponding to p, i.e. $\Lambda p = p$ then the representation matrix $q(\beta)$ of the little group of p_0, corresponding to β, is identical with the representation matrix of the little group of p, corresponding to $\Lambda = \alpha(p)\beta\alpha(p)^{-1}$. Thus when characterizing a representation of the whole inhomogeneous Lorentz group by P and the representation of the little group, it is not necessary to say which p_0 is left invariant by the little group.

D.

Lastly we shall determine the constitution of the little group in the different cases.

1_+. In case 1_+ we can take for p_0 the vector with the components 0, 0, 0, 1. The little group which leaves this invariant obviously contains all rotations in the space of the first three coordinates. This holds for the little group of all representations of the first class.

0_0. In case 0_0, little group is the whole homogeneous Lorentz group.

-1. In case $P = -1$ the p_0 can be assumed to have the components 1, 0, 0, 0. The little group then contains all transformations which leave the form $-x_2^2 - x_3^2 + x_4^2$ invariant, i.e., is the 2 + 1 dimensional homogeneous Lorentz group. The same holds for all representations with $P < 0$.

0_+. The determination of the little group for $P = 0_+$ is somewhat more complicated. It can be done, however, rather simply, for the group of unimodular two dimensional matrices. The Lorentz transformation corresponding to the matrix $\begin{pmatrix} a & b \\ c & d \end{pmatrix}$ with $ad - bc = 1$ brings the vector with the components x_1, x_2, x_3, x_4, into the vector with the components x_1', x_2', x_3', x_4', where[18]

$$\begin{pmatrix} a & b \\ c & d \end{pmatrix} \begin{pmatrix} x_4+x_3 & x_1+ix_2 \\ x_1-ix_2 & x_4-x_3 \end{pmatrix} \begin{pmatrix} a^* & c^* \\ b^* & d^* \end{pmatrix} = \begin{pmatrix} x'_4+x'_3 & x'_1+ix'_2 \\ x'_1-ix'_2 & x'_4-x'_3 \end{pmatrix}. \tag{70}$$

The condition that a null-vector p_0, say with the components 0, 0, 1, 1 be invariant is easily found to be $|a|^2 = 1$, $c = 0$. Hence the most general element of the little group can be written

$$\begin{pmatrix} e^{-i\beta/2} & (x+iy)e^{i\beta/2} \\ 0 & e^{i\beta/2} \end{pmatrix}, \tag{71}$$

with real x, y, β and $0 \le \beta < 4\pi$. The general element (71) can be written as $t(x, y)\,\delta(\beta)$ where

$$t(x,y) = \begin{pmatrix} 0 & x+iy \\ 0 & 1 \end{pmatrix}; \quad \delta = \begin{pmatrix} e^{-i\beta/2} & 0 \\ 0 & e^{i\beta/2} \end{pmatrix}. \tag{71a}$$

The multiplication rules for these are

$$t(x, y)t(x', y') = t(x + x', \ y + y'), \tag{71b}$$
$$\delta(\beta)t(x, y) = t(x\cos\beta + y\sin\beta, -x\sin\beta + y\cos\beta)\delta(\beta), \tag{71c}$$
$$\delta(\beta)\delta(\beta') = \delta(\beta + \beta'). \tag{71d}$$

One could restrict the variability domain of β in $\delta(\beta)$ from 0 to 2π. As $\delta(2\pi)$ commutes with all elements of the little group, it will be a constant and from $\delta(2\pi)^2 = \delta(4\pi) = 1$ it can be $\delta(2\pi) = \pm 1$. Hence $\delta(\beta + 2\pi) = \pm\delta(\beta)$ and inserting a $\pm$ into equation (71d) one could restrict β to $0 \leq \beta < 2\pi$

These equations are analogous to the equations (52)-(52D) and show that the little group is, in this case, isomorphic with the inhomogeneous rotation group of two dimensions, i.e. the two dimensional Euclidean group.

It may be mentioned that the Lorentz transformations corresponding to t(x,y) have elementary divisors, and constitute all transformations of class e) in 4B, for which $v_e = p_0$. The transformations $\delta(\beta)$ can be considered to be rotations in the ordinary three dimensional space, about the direction of the space part of the vector p_0. It is possible, then, to prove equations (71) also directly.

7. The Representations of the Little Groups

A. Representations of the three dimensional rotation group by unitary transformations.

The representations of the three dimensional rotation group in a space with a finite member of dimensions are well known. There is one irreducible representation with the dimensions 1, 2, 3, 4, $\cdots$ each, the representations with an odd number of dimensions are single valued, those with an even number of dimensions are two-valued. These representations will be denoted by $D^{(j)}$ (R) where the dimension is 2j + 1. Thus for single valued representations j is an integer, for double valued representations a half integer. Every finite dimensional representation can be decomposed into these irreducible representations. Consequently those representations of the Lorentz group with positive P in which the representation of the little group–as defined by (64)–has a finite number of dimensions, can be decomposed into such representations in which the representation of the little group is one of the well known irreducible representations of the rotation group. This result will hold for all representations of the inhomogeneous Lorentz group with positive P, since we shall show that even the infinite dimensional representations of the rotation group can be decomposed into the same, finite, irreducible representations.

In the following, it is more appropriate to consider the subgroup of the two dimensional unimodular group which corresponds to rotations, than the rotation group itself, as we can restrict ourselves to single valued representations in this case (cf. equations (52)). From (70), one easily sees[18] that the condition for $\binom{a\ b}{c\ d}$ to leave the vector with the components 0, 0, 0, 1 invariant is that it shall be unitary. It is, therefore, the two dimensional unimodular unitary group the representations of which we shall consider, instead of the representations of the rotation group.

Let us introduce a discrete coordinate system in the representation space and denote the coefficients of the unitary representation by $q(R)_{k\lambda}$ where R is a two dimensional unitary transformation. The condition for the unitary character of the representation q(R) gives

$$\sum_k q(R)^*_{k\lambda} q(R)_{k\mu} = \delta_{\lambda\mu}; \quad \sum_\lambda q(R)^*_{k\lambda} q(R)_{\nu\lambda} = \delta_{k\nu}, \tag{72}$$

$$\sum_k |q(R)_{k\lambda}|^2 = 1; \quad \sum_\lambda |q(R)_{k\lambda}|^2 = 1. \tag{72a}$$

This show also that $|q(R)_{k\lambda}| \leq 1$ and the $q(R)_{k\lambda}$ are therefore, as functions of R, square integrable:

$$\int |q(R)_{k\lambda}|^2 dR$$

exists if $\int \cdots$ dR is the well known invariant integral in group space. Since this is finite for the rotation group (or the unimodular unitary group), it can be normalized to 1. We then have

$$\sum_k \int |q(R)_{k\lambda}|^2 dR = \sum_\lambda \int |q(R)_{k\lambda}|^2 dR = 1. \tag{73}$$

The $(2j + 1)^{\frac{1}{2}} D^{(j)}(R)_{kl}$ form,[26] a complete set of normalized orthogonal functions for R. We set

[26] H. Weyl and F. Peter, Math. Annal. *97*, 737, 1927.

$$q(R)_{k\lambda} = \sum_{jkl} C^{k\lambda}_{jkl} D^{(j)}(R)_{kl}. \tag{74}$$

We shall calculate now the integral over group space of the product of $D^{(j)}(R)^*_{kl}$ and

$$q(RS)_{k\mu} = \sum_{\lambda} q(R)_{k\lambda} q(S)_{\lambda\mu}. \tag{75}$$

The sum on the right converges uniformly, as for (72a)

$$\sum_{\lambda=N}^{\infty} |q(R)_{k\lambda} q(S)_{\lambda\mu}| \leq \left(\sum_{\lambda=N}^{\infty} |q(R)_{k\lambda}|^2 \sum_{\lambda=N}^{\infty} |q(S)_{\lambda\mu}|^2\right)^{\frac{1}{2}}$$
$$\leq \left(\sum_{\lambda=N}^{\infty} |q(S)_{\lambda\mu}|^2\right)^{\frac{1}{2}}$$

can be made arbitrarily small by choosing an N, independent of R, making the last expression small. Hence, (75) can be integrated term by term and gives

$$\int D^{(j)}(R)^*_{kl} q(RS)_{k\mu} dR = \sum_{\lambda} \int D^{(j)}(R)^*_{kl} q(R)_{k\lambda} q(S)_{\lambda\mu} dR. \tag{76}$$

Substituting $\sum_m D^{(j)}(RS)_{km} D^{(j)}(S^{-1})_{ml}$ for $D^{(j)}(R)_{kl}$ one obtains

$$\sum_m D^{(j)}(S^{-1})^*_{ml} \int D^{(j)}(RS)^*_{km} q(RS)_{k\mu} dR \tag{77}$$
$$= \sum_{\lambda} q(S)_{\lambda\mu} \int D^{(j)}(R)^*_{kl} q(R)_{k\lambda} dR.$$

In the invariant integral on the left of (77), R can be substituted for RS and we obtain, for (74) and the unitary character

$$\sum_m D^{(j)}(S)_{lm} C^{k\mu}_{jkm} = \sum_{\lambda} q(S)_{\lambda\mu} C^{k\lambda}_{jkl}. \tag{78}$$

Multiplying (78) by $D^{(h)}(S)^*_{in}$, the integration on the right side can be carried out term by term again, since the sum over λ converges uniformly

$$\sum_{\lambda=N}^{\infty} |C^{k\lambda}_{jkl} q(S)_{\lambda\mu}| \leq \left(\sum_{\lambda=N}^{\infty} |C^{k\lambda}_{jkl}|^2 \sum_{\lambda=N}^{\infty} |q(S)_{\lambda\mu}|^2\right)^{\frac{1}{2}} \leq \left(\sum_{\lambda=N}^{\infty} |C^{k\lambda}_{jkl}|^2\right)^{\frac{1}{2}}.$$

This can be made arbitrarily small, as even $\sum_{\lambda} \sum_{jkl} (2j+1)^{-1} |C^{k\lambda}_{jkl}|^2$ converges, for (74) and (72a). The integration of (78) yields thus

$$\sum_{\lambda} C_{jkl}^{k\lambda} C_{hin}^{\lambda\mu} = \delta_{jh}\delta_{li}C_{jkn}^{k\mu}. \tag{79}$$

From $q(R)q(E) = q(R)$ follows $q(E) = 1$ and then $q(R^{-1}) = q(R)^{-1} = q(R)^{\dagger}$.

This, with the similar equation for $D^{(j)}(R)$ gives

$$\sum_{jkl} C_{jkl}^{k\lambda} D^{(j)}(R^{-1})_{kl} = q(R^{-1})_{k\lambda} = q(R)_{\lambda k}^{*} \tag{80}$$

$$= \sum_{jkl} C_{jlk}^{\lambda k*} D^{(j)}(R)_{lk}^{*} = \sum_{jkl} C_{jlk}^{\lambda k*} D^{(j)}(R^{-1})_{kl},$$

or

$$C_{jkl}^{k\lambda} = C_{jlk}^{\lambda k*}. \tag{81}$$

On the other hand $q(E)_{k\lambda} = \delta_{k\lambda}$ yields

$$\sum_{jk} C_{jkk}^{k\lambda} = \delta_{k\lambda}. \tag{82}$$

These formulas suffice for the reduction of $q(R)$. Let us choose for every finite irreducible representation $D^{(j)}$ an index k, say $k = 0$. We define then, in the original space of the representation $q(R)$ vectors $v^{(kjl)}$ with the components

$$C_{jkl}^{k1},\ C_{jkl}^{k2},\ C_{jkl}^{k3},\ \cdots .$$

The vectors $v^{(kjl)}$ for different j or l are orthogonal, the scalar product of those with the same j and l is independent of l. This follows from (79) and (81)

$$(v^{(\mu j'l')},\ v^{(kjl)}) = \sum_{\lambda} C_{j'kl'}^{\mu\lambda*} C_{jkl}^{k\lambda} \tag{83}$$

$$= \sum_{\lambda} C_{jkl}^{k\lambda} C_{j'l'k}^{\lambda\mu} = \delta_{jj'}\delta_{ll'}C_{jkk}^{k\mu}.$$

The $v^{(kjl)}$ for all k, j, l, form a complete set of vectors. In order to show this, it is sufficient to form, for every ν, linear combination from them, the ν component of which is 1, all other components 0. This linear combination is

$$\sum_{kjl} C_{jlk}^{\nu k} v^{(kjl)}. \tag{84}$$

In fact, the λ component of (84) is, on account of (79) and (82)

$$\sum_{kjl} C_{jlk}^{\nu k} C_{jkl}^{k\lambda} = \sum_{jl} C_{jll}^{\nu\lambda} = \delta_{\nu\lambda}. \tag{85}$$

However, two v with the same j and l but different first indices k are not orthogonal. We can choose for every j and l, say $l = 0$ and go through the vectors $v^{(1j0)}$, $v^{(2j0)}$, $\cdots$ and, following Schmidt's method, orthogonalize and normalize them. The vectors obtained in this way shall be denoted by

$$w^{(nj0)} = \sum_{\lambda} \alpha_{n\lambda}^{j} v^{(\lambda j0)}. \tag{86}$$

Then, since according to (83) the scalar products $(v^{(kjl)}, v^{(\lambda jl)})$ do not depend on l, the vectors

$$w^{(njl)} = \sum_{\lambda} \alpha_{n\lambda}^{j} v^{(\lambda jl)} \tag{86a}$$

will be mutually orthogonal and normalized also and the vectors $w^{(njl)}$ for all n, j, l will form a complete set of orthonormal vectors. The same holds for the set of the conjugate complex vectors $w^{(njl)*}$. Using these vectors as coordinate axes for the original representation q(R), we shall find that q(R) is completely reduced. The ν component of the vector $q(R)v^{(kjl)*}$ obtained by applying q(R) on $v^{(kjl)*}$ is

$$\sum_{\mu} q(R)_{\nu\mu}(v^{(kjl)*})_{\mu} = \sum_{\mu} q(R)_{\nu\mu} C_{jkl}^{\mu k}. \tag{87}$$

The right side is uniformly convergent. Hence, its product with (2h+1) $\overset{*}{D^{(h)}(R)_{in}}$ can be integrated term by term giving

$$\sum_{\mu} \int (2h+1) D^{(h)}(R)_{in}^{*} q(R)_{\nu\mu} C_{jkl}^{\mu k} dR = \sum_{\mu} C_{hin}^{\nu\mu} C_{jlk}^{\mu k} \tag{88}$$

$$= \delta_{hj} \delta_{ln} C_{jik}^{\nu k}.$$

Thus we have for almost all R

$$\sum_{\mu} q(R)_{\nu\mu}(v^{(kjl)*})_{\mu} = \sum_{i} C_{jik}^{\nu k} D^{(j)}(R)_{il} = \sum_{i} D^{(j)}(R)_{il}(v^{(kji)*})_{\nu}, \tag{88a}$$

or

$$q(R)\, v^{(kjl)*} = \sum_{i} D^{(j)}(R)_{il} v^{(kji)*}. \tag{88b}$$

Since both sides are supposed to be strongly continuous functions of R, (88b) holds for every R. In (86a), for every n, the summation must be carried out only over a finite number of λ. We can write therefore immediately

$$q(R)w^{(njl)*} = \sum_i D^{(j)}(R)_{il} w^{(nji)*}. \tag{89}$$

This proves that the original representation decomposes in the coordinate system of the w into well known finite irreducible representations $D^{(j)}(R)$. Since the w form a complete orthonormal set of vectors, the transition corresponds to a unitary transformation.

This completes the proof of the complete reducibility of all (finite and infinite dimensional) representations of the rotation group or unimodular unitary group. It is clear also that the same consideration applies for all closed groups, i.e., whenever the invariant integral $\int dR$ converges.

The result for the inhomogeneous Lorentz group is: For every positive numerical value of P, the representations of the little group can be, in an irreducible representation, only the $D^{(0)}$, $D^{(\frac{1}{2})}$, $D^{(1)}, \cdots$, both for P_+ and for P_-. All these representations have been found already by Majorana and by Dirac and for positive P there are none in addition to these.

B. Representations of the two dimensional Euclidean group

This group, as pointed out in section 6, has a great similarity with the inhomogeneous Lorentz group. It is possible, again[24], to introduce "momenta", i.e. variables ξ, η and v instead of ζ in such a way that

$$t(x, y)\phi(p_0, \xi, \eta, v) = e^{i(x\xi+y\eta)}\phi(p_0, \xi, \eta, v). \tag{90}$$

Similarly, one can define again operators $R(\beta)$

$$R(\beta)\phi(p_0, \xi, \eta, v) = \phi(p_0, \xi', \eta', v), \tag{91}$$

where

$$\begin{aligned} \xi' &= \xi\cos\beta - \eta\sin\beta, \\ \eta' &= \xi\sin\beta + \eta\cos\beta. \end{aligned} \tag{91a}$$

Then $\delta(\beta)R(\beta)^{-1} = S(\beta)$ will commute, on account of (71c), with t(x, y) and again contain ξ, η as parameter only. The equation corresponding to (57a) is

$$\delta(\beta)\phi(p_0, \xi, \eta, v) = \sum_\omega S(\beta)_{v\omega}\phi(p_0, \xi', \eta', \omega). \tag{92}$$

One can infer from (90) and (92) again that the variability domain of ξ, η can be restricted in such a way that all pairs ξ, η arise from one pair ξ_0, η_0 by a rotation, according (91a). We have, therefore two essentially different cases:

$$\xi^2 + \eta^2 = \Xi \neq 0 \qquad \text{a.)}$$

$$\xi^2 + \eta^2 = \Xi = 0, \qquad \text{i.e.} \quad \xi = \eta = 0. \qquad \text{b.)}$$

The positive definite metric in the ξ, η space excludes the other possibilities of section 6 which were made possible by the Lorentzian metric for the momenta, necessitated by (55).

Case b) can be settled very easily. The "little group" is, in this case, the group of rotations in a plane and we are interested in one and two valued irreducible representations. These are all one dimensional ($e^{is\beta}$)

$$S(\beta) = e^{is\beta} \qquad (93)$$

where s is integer or half integer. These representations were also all found by Majorana and by Dirac. For s = 0 we have simply the equation $\Box\phi = 0$, for $s = \pm \frac{1}{2}$ Dirac's electron equation without mass, for $s = \pm 1$ Maxwell's electromagnetic equations, etc.

In case a) the little group consists only of the unit matrix and the matrix $\begin{pmatrix} -1 & 0 \\ 0 & -1 \end{pmatrix}$ of the two dimensional unimodular group. This group has two irreducible representations, as (1) and (−1) can correspond to the above two dimensional matrix of the little group. This gives two new representations of the whole inhomogeneous Lorentz group, corresponding to every numerical value of Ξ. Both these sets belong to class 0_+ and two similar new sets belong to class 0_-.

The final result is thus as follows: The representations P_{+j} of the first subclass P_+ can be characterized by the two numbers P and j. From these P is positive, otherwise arbitrary, while j is an integer or a half integer, positive, or zero. The same holds for the subclass P_-. There are three kinds of representations of the subclass 0_+. Those of the first kind 0_{+s} can be characterized by a number s, which can be either an integer or a half integer, positive, negative or zero. Those of the second kind $0_+(\Xi)$ are single valued and can be characterized by an arbitrary positive number Ξ, those of the third kind $0'_+(\Xi)$ are double-valued and also can be characterized by a positive Ξ. The same holds for the subclass 0_-. The representations of the other classes (0_0 and P with $P < 0$) have not been determined.

8. Representations of the Extended Lorentz Group

A.

As most wave equations are invariant under a wider group than the one investigated in the previous sections, and as it is very probable that the laws of physics are all invariant under this wider group, it seems appropriate to investigate now how the results of the previous sections will be modified if we go over from the "restricted Lorentz group" defined in section 4A, to the extended Lorentz group. This extended Lorentz group contains in addition to the translations all the homogeneous transformations X satisfying (10)

$$XFX' = F \tag{10'}$$

while the homogeneous transformations of section 4A were restricted by two more conditions. From (10′) it follows that the determinant of X can be +1 or −1 only. If its −1, the determinant of $X_1 = XF$ is +1. If the four-four element of X_1 is negative, that of $X_2 = -X_1$ is positive. It is clear, therefore, that if X is a matrix of the extended Lorentz group, one of the matrices X, XF, −X, −XF is in the restricted Lorentz group. For $F^2 = 1$, conversely, all homogeneous transformations of the extended Lorentz group can be obtained from the homogeneous transformations of the restricted group by multiplication with one of the matrices

$$1, \; F, \; -1, \; -F. \tag{94}$$

The group elements corresponding to these transformations will be denoted by E, F, I, IF. The restricted group contains those elements of the extended group which can be reached continuously from the unity. It follows that the transformation of an element L of the restricted group by F, I, or IF gives again an element of the restricted group. This is, therefore, an invariant subgroup of the extended Lorentz group. In order to find the representations of the extended Lorentz group, we shall use again Frobenius' method.[15]

We shall denote the operators corresponding in a representation to the homogeneous transformations (94) by d(E) = 1, d(F), d(I), d(IF). For deriving the equations (52) it was necessary only to assume the existence of the transformations of the restricted group, it was not necessary to assume that these are the only transformations. These equations will hold, therefore, for elements of the restricted group, in representations of the extended group also. We normalize the indeterminate factors in d(F) and d(I) so that their squares become unity. Then we have d(F)d(I) = ωd(I)d(F) or d(I) = ωd(F)d(I)d(F). Squaring this, one obtains $\omega^2 = \pm 1$. We can set, therefore

$$\begin{aligned} d(IF) &= d(I)d(F) = \pm d(F)d(I) \\ d(F)^2 &= d(I)^2 = 1; \quad d(IF)^2 = \pm 1. \end{aligned} \tag{95}$$

Finally, from

$$d(F)D(L_1)d(F) = \omega(L_1)D(FL_1F) \tag{96}$$

we obtain, multiplying this with the similar equation for L_2

$$\omega(L_1)\omega(L_2) = \omega(L_1L_2)$$

which, gives $\omega(L) = 1$ as the inhomogeneous Lorentz group (or the group used in (52B)-(52D)) has the only one dimensional representation by the unity (1). In this way, we obtain

$$d(F)D(L)d(F) = D(FLF), \tag{96a}$$

$$d(I)D(L)d(I) = D(ILI), \tag{96b}$$

$$d(IF)D(L)d(IF)^{-1} = D(IFLFI). \tag{96c}$$

B.

Given a representation of the extended Lorentz group, one can perform the transformations described in section 6A, by considering the elements of the restricted group only. We shall consider here only such representations of the extended group, for which, after having introduced the momenta, all representations of the restricted group are either in class 1 or 2, i.e. $P \geq 0$ but not 0_0. Following then the procedure of section 6, one can find a set of wave functions for which the operators D(L) of the restricted group have one of the forms, given in section 6 as irreducible representations. We shall proceed, next to find the operator d(F). For the wave functions belonging to an irreducible D(L) of the restricted group, we can introduce a complete set of orthonormal functions $\psi_1(p, \zeta)$, $\psi_2(p, \zeta)$, $\cdots$. We then have

$$D(L)\psi_k(p, \zeta) = \sum_\mu D(L)_{\mu k}\psi_\mu(p, \zeta). \tag{97}$$

The infinite matrices $D(L)_{\mu k}$ defined in (97) are unitary and form a representation which is equivalent to the representation by the operators D(L). The D(L), d(F) are, of course, operators, but the $D(L)_{\mu k}$ are components of a matrix, i.e. numbers. We can now form the wave functions $d(F)\psi_1$, $d(F)_2$, $d(F)\psi_3, \cdots$ and apply D(L) to these. For (96a) and (97) we have

$$\begin{aligned} D(L)d(F)\psi_k &= d(F)D(FLF)\psi_k \\ &= d(F)\sum_\mu D(FLF)_{\mu k}\psi_\mu \\ &= \sum_\mu D(FLF)_{\mu k} d(F)\psi_\mu. \end{aligned} \tag{97a}$$

The matrices $D^o(L)_{\mu k} = D(FLF)_{\mu k}$ give a representation of the restricted group (FLF is an element of the restricted group, we have a new representation by an automorphism, as discussed in section 6B). We shall find out whether $D^o(L)$ is equivalent D(L) or not. The translation operation D^o is

$$T^o(a) = d(F)T(a)d(F) = T(Fa) \tag{98}$$

which, together with (53) shows that D^o has the same P as D(L) itself. In fact, writing

$$U_1\phi(p, \zeta) = \phi(Fp, \zeta) \tag{99}$$

one has $U_1^{-1} = U_1$ and one easily calculates $U_1T^o(a)U_1 = T(a)$. Similarly for $U_1 d^o(\Lambda)U_1$ one has

$$\begin{aligned} U_1 d^o(\Lambda)U_1\phi(p, \zeta) &= U_1 d(F\Lambda F)U_1\phi(p, \zeta) \\ =d(F\Lambda F)U_1\phi(Fp, \zeta) &= \sum_\eta Q(Fp, F\Lambda F)_{\zeta\eta}U_1\phi(F\Lambda^{-1}p, \eta) \\ &= \sum_\eta Q(Fp, F\Lambda F)_{\zeta\eta}\phi(\Lambda^{-1}p, \eta). \end{aligned} \tag{99a}$$

This means that the similarity transformation with U_1 brings $T^o(a)$ into T(a) and $d^o(\Lambda)$ into $Q(Fp, F\Lambda F)P(\Lambda)$. Thus the representation of the "little group" in $U_1 d^o(\Lambda)U_1$ is

$$q^o(\lambda) = Q(Fp_0, F\lambda F).$$

For this latter matrix, one obtains from (67a)

$$q^o(\lambda) = Q(Fp_0, F\lambda F) = q(\alpha(Fp_0)^{-1}F\lambda F\alpha(Fp_0)) = q(\lambda^o) \tag{100}$$

where λ^o is obtained from λ by transforming it with $F\alpha(Fp_0)$.

The representations $D^o(L)$ and $D(L)$ are equivalent if the representation of $q(\lambda)$ is equivalent to the representation which coordinates $q(\lambda^o)$ to λ. The $\alpha(Fp_0)$ is a transformation of the restricted group which brings p_0 into $\alpha(Fp_0)p_0 = Fp_0$. (Cf. (65).) This transformation is, of course, not uniquely determined but if $\alpha(Fp_0)$ is one, the most general can be written as $\alpha(Fp_0)\iota$, where $\iota p_0 = p_0$ is in the little group. For $q(\iota^{-1}\alpha(Fp_0)^{-1}\Lambda\alpha(Fp_0)\iota) = q(\iota^{-1})q(\alpha(Fp_0)^{-1}\Lambda\alpha(Fp_0))q(\iota)$, the freedom in the choice of $\alpha(Fp_0)$ only amounts to a similarity transformation of $q^o(\lambda)$ and naturally does not change the equivalence or non equivalence of $q^o(\lambda)$ with $q(\lambda)$.

For the case P_+, we can choose p_0 in the direction of the fourth axis, with components 0, 0, 0, 1. Then $Fp_0 = p_0$ and $\alpha(Fp_0) = 1$. The little group is the group of rotations in ordinary space and $F\lambda F = \lambda$. Hence $q^o(\lambda) = q(\lambda)$ and $D^o(\Lambda)$ is equivalent to $D(\Lambda)$ in this case. The same holds for the representations of class P_-.

For 0_+ we can assume that p_0 has the components 0, 0, 1, 1. Then the components of Fp_0 are 0, 0, −1, 1. For $\alpha(Fp_0)$ we can take a rotation by π about the second axis and $F\alpha(Fp_0)$ will be a diagonal matrix with diagonal elements 1, −1, 1, 1, i.e., a reflection of the second axis. Thus if λ is the transformation in (70), $\lambda^0 = \alpha(Fp_0)^{-1}F\lambda F\alpha(Fp_0)$ is the transformation for which

$$\lambda^0\begin{pmatrix} x_4 + x_3 & x_1 - ix_2 \\ x_1 + ix_2 & x_4 - ix_3 \end{pmatrix}\lambda^{0\dagger} = \begin{pmatrix} x'_4 + x'_3 & x'_1 - ix'_2 \\ x'_1 + ix'_2 & x'_4 - ix'_3 \end{pmatrix} \tag{101}$$

This is, however, clearly $\lambda^0 = \lambda^*$. Thus the operators of $q^0(\lambda)$ are obtained from the operators $q(\lambda)$ by (cf. (71a))

$$\begin{aligned} t^0(x, y) &= t(x, -y) \\ \delta^0(\beta) &= \delta(-\beta). \end{aligned} \tag{101a}$$

For the representations 0_{+s} with discrete s, the $q^0(\lambda)$ and $q(\lambda)$ are clearly inequivalent as $\delta^0(\beta) = (e^{-is\beta})$ and $\delta(\beta) = (e^{is\beta})$, except for $s = 0$, when they are equivalent. For the representations $0_+(\Xi)$, $0'_+(\Xi)$, the $q^0(\lambda)$ and $q(\lambda)$ are equivalent, both in the single valued and the double valued case, as the substitution $\eta \to -\eta$ transforms them into each other. The same holds for representations of the class 0_-. If $D^0(L)$ and $D(L)$ are equivalent

$$D^{-1}D^0(L)U = D(L), \tag{102}$$

the square of U commutes with all D(L). As a consequence of this, U^2

must be a constant matrix. Otherwise, one could form, in well known manner,[27] an idempotent which is a function of U^2 and thus commutes with D(L) also. Such an idempotent would lead to a reduction of the representation D(L) of the restricted group. As a constant is free in U, we can set

$$U^2 = 1 \tag{102a}$$

C.

Returning now to equation (97a), if $D^0(L) = D(FLF)$ and D(L) are equivalent ($P > 0$ or 0_+, 0_- with continuous Ξ or $s = 0$) there is a unitary matrix $U_{\mu\nu}$, corresponding to U, such that

$$\begin{aligned} \sum_\mu D(FLF)_{k\mu} U_{\mu\nu} &= \sum_\mu U_{k\mu} D(L)_{\mu\nu} \\ \sum_\mu U_{k\mu} U_{\mu\nu} &= \delta_{k\nu}. \end{aligned} \tag{102b}$$

Let us now consider the functions

$$\phi_\nu = \psi_\nu + \sum_\mu U_{\mu\nu} d(F) \psi_\mu. \tag{103}$$

Applying D(L) to these

$$\begin{aligned} D(L)\phi_\nu &= D(L)\psi_\nu + \sum_\mu U_{\mu\nu} D(L) d(F) \phi_\mu \\ &= D(L)\psi_\nu + \sum_\mu U_{\mu\nu} d(F) D(FLF) \psi_\mu \\ &= \sum_\mu D(L)_{\mu\nu}\psi_\mu + \sum_{\mu k} U_{\mu\nu} d(F) D(FLF)_{k\mu} \psi_k \\ &= \sum_\mu D(L)_{\mu\nu}(\psi_\mu + \sum_k U_{k\mu} d(F)\psi_k) = \sum_\mu D(L)_{\mu\nu}\phi_\mu. \end{aligned} \tag{103a}$$

Similarly

---

[27] J. von Neumann, Ann. of Math. *32*, 191, 1931; ref. 2, p. 89.

$$d(F)\phi_\nu = d(F)\psi_\nu + \sum_\mu U_{\mu\nu}\psi_\mu$$
$$= \sum_\mu U_{\mu\nu}(\psi_\mu + \sum_k U_{k\mu}d(F)\psi_k) = \sum_\mu U_{\mu\nu}\phi_\mu. \tag{103b}$$

Thus the wave functions ϕ transform according to the representation in which $D(L)_{\mu\nu}$ corresponds to L and $U_{\mu\nu}$ to d(F). The same holds for the wave functions

$$\phi'_\nu = \psi_\nu - \sum_\mu U_{\mu\nu}d(F)\psi_\mu, \tag{104}$$

except that in this case $(-U_{\mu\nu})$ corresponds to d(F). The ψ_ν and $d(F)\psi_\nu$ can be expressed by the ϕ and ϕ'. If the ψ and $d(F)\psi$ were linearly independent, the ϕ and ϕ' will be linearly independent also. If the $d(F)\psi$ were linear combinations of the ψ, either the ϕ or the ϕ' will vanish.

If we imagine a unitary representation of the group formed by the L and FL in the form in which it is completely reduced out as a representation of the group of restricted transformations L, the above procedure will lead to a reduction of that part of the representation of the group of L and FL, for which D(L) and D(FLF) are equivalent.

If $D(L)_{\mu\nu}$ and $D^o(L)_{\mu\nu}$ are inequivalent, the ψ_k and $d(F)\psi_\nu = \psi'_\nu$ are orthogonal. This is again a generalization of the similar rule for finite unitary representations.[28] One can see this in the following way: Denoting $M_{k\nu} = (\psi_k, \psi'_\nu)$ one has

$$M_{k\nu} = (\psi_k, \psi'_\nu) = (D(L)\psi_k, D(L)\psi'_\nu)$$
$$= \sum_{\mu\lambda} D(L)^*_{\mu k} D^o(L)_{\lambda\nu} M_{\mu\lambda};$$

$$M = D(L)^\dagger M D^o(L).$$

Hence

$$D(L)M = MD^o(L); \qquad M^\dagger D(L) = D^o(L)M^\dagger. \tag{105}$$

From these, one easily infers that $MM^\dagger$ commutes with D(L), and $M^\dagger M$ commutes with $D^o(L)$. Hence both are constant matrices, and if neither of

[28] Cf. e.g. E. Wigner, ref. 4, Chapter XII.

them is zero, M and $M^\dagger$ are, apart from a constant, unitary. Thus D(L) would be equivalent $D^o(L)$ which is contrary to supposition. Hence $MM^\dagger = 0$, $M = 0$ and the ψ are orthogonal to the $d(F)\psi = \psi'$. Together, they give a representation of the group formed by the restricted Lorentz group and F. If they do not form a complete set, the reduction can be continued as before.

One sees, thus, that introducing the operation F ''doubles'' the number of dimensions of the irreducible representations in which the little group was the two dimensional rotation group, while it does not increase the underlying linear manifold in the other cases. This is analogous to what happens, if one adjoins the reflection operation to the rotation groups themselves.[29]

D.

The operations d(I) can be determined in the same manner as the d(F) were found. A complete set of orthonormal functions corresponding to an irreducible representation of the group formed by the L and FL shall be denoted by $\psi_1, \psi_2, \cdots$. For this, we shall assume (97) again, although the D(L) contained therein is now not necessarily irreducible for the restricted group alone but contains, in case of 0_{+s} or 0_{-s} and finite s, both s and –s. We shall set, furthermore

$$d(F)\psi_k = \sum_\mu d(F)_{\mu k}\psi_\mu. \tag{106}$$

We can form then the functions $d(I)\psi_1$, $d(I)\psi_2, \cdots$. The consideration, contained in (97a) shows that these transform according to $D(ILI)_{\mu k}$ for the transformation L of the restricted group:

$$D(L)d(I)\psi_k = \sum_\mu D(ILI)_{\mu k}d(I)\psi_\mu. \tag{106a}$$

Choosing for L a pure translation, a consideration analogous to that performed in (98) shows that the set of momenta in the representation L → D(ILI) has the opposite sign to the set of momenta in the representation D(L). If the latter belongs to a positive subclass, the former belongs to the corresponding negative subclass and conversely. Thus the adjunction of

[29] I. Schur, Sitz. d. kon. Preuss. Akad. pages 189, 297, 1924.

the transformation I always leads to a "doubling" of the number of states, the states of "negative energy" are attached to the system of possible states. One can describe all states $\psi_1, \psi_2, \cdots$, $d(I)\psi_1$, $d(I)\psi_2, \cdots$ by introducing momenta p_1, p_2, p_3, p_4 and restricting the variability domain of p by the condition $\{p, p\} = P$ alone without stipulating a definite sign for p_4.

As we saw before, the $d(I)\psi_1$, $d(I)\psi_2$, are orthogonal to the original set of wave functions $\psi_1, \psi_2, \cdots$. The result of the application of the operations D(L) and d(F) to the $\psi_1, \psi_2, \cdots$ (i.e., the representation of the group formed by the L, FL) was given in part C. The $D(L)d(I)\psi_k$ are given in (106a). On account of the normalization of d(I) we can set

$$d(I)d(I)\psi_k = \psi_k. \tag{106b}$$

For $d(F)d(I)\psi_k$ we have two possibilities, according to the two possibilities in (95). We can either set

$$d(F)d(I)\psi_k = d(I)d(F)\psi_k = \sum_\mu d(F)_{\mu k} d(I)\psi_\mu. \tag{107}$$

or

$$d(F)\cdot d(I)\psi_k = -d(I)d(F)\psi_k = -\sum_\mu d(F)_{\mu k} d(I)\psi_\mu. \tag{107a}$$

Strictly speaking, we thus obtain two different representations. The system of states satisfying (107) could be distinguished from the system of states for which (107a) is valid, however, only if we could really perform the transition to a new coordinate system by the transformation I. As this is, in reality, impossible, the representations distinguished by (107) and (107a) are not different in the same sense as the previously described representations are different.

I am much indebted to the Wisconsin Alumni Research Foundation for their aid enabling me to complete this research.

Madison, Wis.

Group Theoretical Discussion of Relativistic Wave Equations

by V. Bargmann and E.P. Wigner

Princeton University

Read before the Academy, November 18, 1947

Reprinted from *Proc. Nat. Acad. Sci. (U.S.A.)* **34**, 211 (1948).

Introduction.[1]–The wave functions, ψ, describing the possible states of a quantum mechanical system form a linear vector space V which, in general, is infinite dimensional and on which a positive definite inner product (ϕ,ψ) is defined for any two wave functions ϕ and ψ (i.e., they form a Hilbert space). The inner product usually involves an integration over the whole configuration or momentum space and, for particles of higher spin, a summation over the spin indices.

If the wave functions in question refer to a free particle and satisfy relativistic wave equations, there exists a correspondence between the wave functions describing the same state in different Lorentz frames. The transformations considered here form the group of all *inhomogeneous* Lorentz transformations (including translations of the origin in space and time). Let $\psi_{l'}$ and ψ_l be the wave functions of the same state in two Lorentz frames l' and l, respectively. Then $\psi_{l'} = \mathrm{U(L)}\psi_l$, where U(L) is a linear unitary operator which depends on the Lorentz transformation L leading from l to l'. By a proper normalization, U is determined by L up to a factor $\pm$ 1. (For all details the reader is referred to the paper of reference 2, hereafter quoted as (L).) Moreover, the operators U form a single- or double-valued representation of the inhomogeneous Lorentz group, i.e., for a succession of two Lorentz transformations L_1, L_2, we have

$$U(L_2L_1) = \pm\, U(L_2)U(L_1). \tag{1}$$

Since all Lorentz frames are equivalent for the description of our system, it follows that, together with ψ, U(L)ψ is also a possible state viewed from the original Lorentz frame l. Thus, the vector space V contains, with every ψ, all transforms U(L)ψ, where L is any Lorentz transformation.

The operators U may also replace the wave equation of the system. In our discussion, we use the wave functions in the "Heisenberg" representation, so that a given ψ represents the system for all times, and may be chosen as the "Schroedinger" wave function at time 0 in a given Lorentz frame l. To find ψ_{t_0}, the Schroedinger function at time t_0, one must therefore transform to a frame l' for which $t' = t - t_0$, while all other coordinates remain unchanged. Then $\psi_{t_0} = \mathrm{U(L)}\psi$, where L is the transformation leading from l to l'.

A classification of all unitary representations of the Lorentz group, i.e., of all solutions of (1), amounts, therefore, to a classification of all possible relativistic wave equations. Such a classification has been carried out in

(L). Two representations U(l) and $\tilde{U}(L) = VU(L)V^{-1}$, where V is a fixed unitary operator, are equivalent. If the system is described by wave functions ψ, the description by

$$\tilde{\psi}_l = V\psi_l \tag{2}$$

is isomorphic with respect to linear superposition, to forming the inner product of two wave functions, and also to the transition from one Lorentz frame to another. In fact, if $\psi_{l'} = U(L)\psi_l$, then $\tilde{\psi}_{l'} = V\psi_{l'} = \tilde{U}(L)\tilde{\psi}_l$. Thus, one obtains classes of equivalent wave equations. Finally, it is sufficient to determine the irreducible representations since any other may be built up from them.

Two descriptions which are equivalent according to (2) may be quite different in appearance. The best known example is the description of the electromagnetic field by the field strength and the four vector potential, respectively. It cannot be claimed either that equivalence in the sense of (2) implies equivalence in every physical aspect. Thus, two equivalent descriptions may lead to quite different expressions for the charge density or the energy density in configuration space (cf. Fierz,[3]) because (2) only implies global, but not local, equivalence of the wave functions. It should be emphasized, however, that any selection of one among the equivalent systems or the superposition of non-equivalent systems in any particular way involves an explicit or implicit assumption as to possible interactions, the positive character of densities, etc. Our analysis is necessarily restricted to free particles and does not lead to any assertions about possible interactions.

The present discussion is not based on any hypothesis about the structure of the wave equations provided that they be Lorentz invariant. In particular, it is not necessary to assume differential equations in configuration space. But it is a result of the analysis in (L) that every irreducible wave equation is equivalent (in the sense of (2)) to a system of differential equations. For the relation of the present point of view to other treatments of the subjects see reference 11.

In the present note, we shall give, for every representation of (L), a differential equation the solutions of which transform according to that representation. We also will discuss in some detail the infinitesimal operators which generate the irreducible representations determined in (L), and we shall characterize these representations, and hence the covariant differential equations, by certain invariants constructed from the infinitesimal operators. This is of some interest, because the infinitesimal operators are closely related to dynamical variables of the system. L. Garding[4] has recently shown that even in the infinite dimensional case one can rather freely operate with infinitesimal transformations. In particular, it immediately follows from his discussion (although it is not explicitly stated in his note) that the familiar commutation rules remain valid.

1. The Infinitesimal Operators of the Lorentz Group.–The metric tensor is assumed in form $g_{44} = 1$, $g_{11} = g_{22} = g_{33} = -1$, $g_{kl} = 0$ $(k \neq l)$ and $g^{kl} = g_{kl}$. The scalar product of two four vectors a, b will be denoted by $\{a,b\} = a^k b_k$. Both c, the velocity of light, and $\hbar$, Planck's constant divided by 2π, are set equal to 1.

The Infinitesimal Operators p_k and M_{kl}. A translation in the x^k-direction is generated by p_k, a rotation in the $(x^k - x^l)$ plane by $M_{kl} = -M_{lk}$ $(k, l = 1, \cdots, 4)$. These operators are Hermitian, and the unitary operators U which represent the finite Lorentz transformation are obtained by exponentiation; thus $U = \exp(-i\alpha p_k)$ corresponds to a translation by the amount α in the direction x_k. Clearly, p_k are the four momenta of the system, and M_{23}, M_{31}, M_{12} the three components of the total angular momentum. The following commutation rules hold (where $[A, B] = AB - BA$)

$$[M_{kl}, M_{mn}] = i(g_{lm}M_{kn} - g_{km}M_{ln} + g_{kn}M_{lm} - g_{ln}M_{km}), \tag{3a}$$

$$[p_k, p_l] = 0 \qquad [M_{kl}, p_m] = i(g_{lm}p_k - g_{km}p_l). \tag{3b}$$

We now define four operators w_k by

$$(w^1, w^2, w^3, w^4) = (v_{234}, v_{314}, v_{124}, v_{321}), \tag{4a}$$

$$\begin{aligned} v_{klm} &= p_k M_{lm} + p_l M_{mk} + p_m M_{kl} \\ &= M_{lm} p_k + M_{mk} p_l + M_{kl} p_m. \end{aligned} \tag{4b}$$

Note that w_k is a "pseudo-vector," i.e., it is a vector only with respect to Lorentz transformations of determinant 1. By (3),

$$[M_{kl}, w_m] = i(g_{lm}w_k - g_{km}w_l) \quad [p_k, w_l]. \tag{5}$$

It follows from (3) and (5) that the two operators,

$$P = p^k p_k; \qquad W = (1/6)v^{klm}v_{klm} = -w^k w_k, \tag{6}$$

commute with all the infinitesimal operators M_{kl} and p_k. Therefore, they have constant values (i.e., they are multiples of the unit operator) for every irreducible representation of the Lorentz group. (The familiar arguments which establish this for finite dimensional representations can be carried over to the infinite dimensional case. (Cf. V. Bargmann, reference 5, p. 602.))

W may also be written in the form

$$W = 1/2 M_{kl} M^{kl} p_m p^m - M_{km} M^{lm} p^k p_l. \tag{7}$$

(This quantity was first introduced by W. Pauli, cf. Lubánski.[6]) The scalar product $w^k p_k$ vanishes.

2. *Summary of the Results of (L).–(a)* For every irreducible representation the states ψ may be expressed as functions ψ (p, ξ) of the momentum vector p and an auxiliary variable ξ which may assume a finite or an infinite number of values. The momenta p are either all zero, or they vary over the manifold $p^k p_k = \mathrm{P}$, with a constant value P. We confine ourselves to the cases in which $p \neq 0$, and either $\mathrm{P} > 0$ or $\mathrm{P} = 0$, because the remaining cases are unlikely to have direct physical significance.[5].

(b) To every inhomogeneous Lorentz transformation $y^k = \lambda^k{}_l x_l + a_k$ (in vector form: $y = \Lambda x + a$) corresponds a unitary operator U(L) defined by

$$U(L)\psi(p,\xi) = e^{-i\{a,p\}} Q(p, \Lambda)\psi(\Lambda^{-1}p,\xi), \tag{8}$$

where $Q(p,\Lambda)$ is a unitary operator which may depend on p but affects only the variable ξ. The inner product (ϕ,ψ) is obtained by an integration over the manifold $p^k p_k = \mathrm{P}$ and by a summation or integration over the variable ξ.

(c) The subgroup of the homogeneous Lorentz transformations which keep a fixed momentum vector p_0 unchanged is called "little group." (The little groups defined by different vectors p_0 are isomorphic.) The unitary operators $Q(p_0,\Lambda)$ (where $\Lambda p_0 = p_0$) form an irreducible representation of the little group and determine the irreducible representation U(L) of the inhomogeneous Lorentz group.

In all cases the operators M_{kl} have the form $\underset{o}{M}_{kl} + \mathrm{S}_{kl}$, where the

$$\underset{o}{M}_{kl} = i\left(p_k \frac{\partial}{\partial p^l} - p_l \frac{\partial}{\partial p^k}\right) = i(p_k g_{lj} - p_l g_{kj})\frac{\partial}{\partial p_j} \tag{9}$$

act on the variables p and correspond to the orbital angular momenta, while the S_{kl} act on the variables ξ and correspond, to the spin angular momenta. Both M_{kl} and S_{kl} satisfy the commutation rules (3a). Since the $\underset{o}{M}_{kl}$ do not contribute to v_{klm} (cf. (4b)), we have

$$v_{klm} = p_k S_{lm} + p_l S_{mk} + p_m S_{kl}; \qquad [S_{kl}, p_m] = 0. \tag{10}$$

or, introducing the three-dimensional vector operators,

$$\vec{S} = (S_{23}, S_{31}, S_{12}); \quad \vec{S}' = (S_{14}, S_{24}, S_{34}); \quad \vec{p} = (p^1, p^2, p^3); \tag{10a}$$
$$\vec{w} = (w^1, w^2, w^3); \quad w^4 = \vec{p}\cdot\vec{S}; \quad \vec{w} = p^4\vec{S} - (\vec{p}\times\vec{S}').$$

Clearly, M_{kl} may also be replaced by S_{kl} in the expression (7) for W.

For a fixed momentum vector p_o the operators w_k are the infinitesimal generators of the little group. Since $w^k p_k = 0$, only three of them are linearly independent.

3. *Classification of the Irreducible Representations.*– We now turn to a brief summary of the main results, including the characterization of the representations in terms of the operators p and w. A more detailed discussion will follow in the succeeding sections.

The classes found in (L) (§§ 7, 8) are these:

I. P_s. *Particles of finite mass and spin s.–Here* $P = m^2 > 0$. In the rest system of the particle, the momentum vector has only the one non-vanishing component $p^4 = \pm m$, hence, by (10a), $W = m^2 S^2$. The operator $P^{-1}W$ represents the square of the spin angular momentum, and has the value $s(s + 1)$ $(s = 0, 1/2, 1, ...)$ for an irreducible representation. For a given momentum vector there are $2s + 1$ independent states. The representation U(L) is single or double valued according to whether s is integral or half integral. The lowest cases $(s = 0, 1/2, 1)$ correspond to the Klein-Gordon, Dirac and Proca equations, respectively.

II. O_s. *Particles of zero rest mass and discrete spin.*–These representations may be considered limiting cases of the representations P_s for $m \to 0$. Then both P and W are equal to zero, and do not suffice to characterize these representations. For a given momentum vector, there exist 2 independent states if $s \neq 0$ (corresponding to two different states of polarization), and there is only one state if $s = 0$. Right and left circularly polarized states are described by the operator equations $w_k = sp_k$, and $w_k =$

$-sp_k$ respectively, so that the representation O_s is characterized by $P = 0$, $w_k w_l = s^2 p_k p_l$. the lowest cases ($s = 0$, 1/2, 1) correspond to the scalar wave equation, the neutrino equation, and Maxwell's equations, respectively.

III. $O(\Xi)$ and $O'(\Xi)$. *Particles of zero rest mass and continuous spin.*–Here, $P = 0$, $W = \Xi^2$, where Ξ is a real positive number. For a given momentum vector there exist infinitely many different states of polarization, which may be described by a continuous variable. The representation $O(\Xi)$ is single valued, while $O'(\Xi)$ is double valued.

To construct these representations explicitly, we shall select, in each case, one among the equivalent sets of wave equations, define a Lorentz invariant inner product (ϕ,ψ), and prove the operator relations stated above. We shall operate in momentum space; this is particularly simple, because the momenta (but not the coordinates) are defined by the Lorentz group, as infinitesimal translations.

4. The Class P_s.–*(a)* $s = 0$. Here, the variable ξ assumes only one value and may therefore be omitted. Consequently, $Q(p, A) = 1$ (cf. reference 8), and for the little group the trivial one-dimensional representation is obtained. Hence, $S_{kl} = 0$, and $w_k = 0$. The wave equation reduces to $p^k p_k = m^2$; the inner product (ϕ,ψ) is determined by the norm $(\psi,\psi)$of a wave function,

$$(\psi,\psi) = \int |\psi(p)|^2 d\Omega, \quad \text{where} \quad d\Omega = |p^4|^{-1} dp^1 dp^2 dp^3, \tag{11}$$

the integral being extended over both sheets of the hyperboloid $p^k p_k = P = m^2$. The expression (11) is Lorentz invariant, because $d\Omega$ is an invariant volume element in momentum space. For the wave function in configuration space, one finds

$$\psi(x) = (2\pi)^{\frac{-3}{2}} \int e^{-i(p,x)} \psi(p) d\Omega, \tag{12}$$

where x stands for x^1, x^2, x^3, x^4. It is well known that (ψ,ψ) cannot be simply expressed in configuration space, because for the Klein-Gordon equation the density is indefinite, and the integral over the density in configuration space coincides with (11) only if $\psi(p) = 0$ whenever $p^4 < 0$.

(b) $s = \frac{1}{2}N$ with $N = 1, 2, 3, \cdots$. For particles of higher spin we use the equations first derived by Dirac[7] in the form essentially given in reference 8. We use for ξ the N four-valued variables $\zeta_1, \ldots, \zeta_N$ in which the wave

function $\psi(p; \zeta_1, ..., \zeta_N)$ is symmetric. We define for every ζ_ν four-dimensional matrices $\gamma_\nu{}^k$ of the same nature as are used in Dirac's electron theory:

$$\gamma_\nu{}^k\gamma_\nu{}^l + \gamma_\nu{}^l\gamma_\nu{}^k = 2g^{kl}1 \qquad (k,l = 1,2,3,4). \tag{13}$$

The γ with different lower indices ν commute. The γ^1, γ^2, γ^3 are skew Hermitian, γ^4, is Hermitian. The wave equations then are

$$\gamma_\nu{}^k p_k \psi = m\psi \qquad (\nu = 1,2, \cdots ,N). \tag{14}$$

It follows from any of these equations in well known fashion that

$$g^{kl}p_k p_l \psi = p^k p_k \psi = m^2\psi. \tag{14a}$$

The infinitesimal operators of displacement are the p, those of four-dimensional rotation the $\mathrm{M}_{kl} = \underset{o}{M}_{kl} + \mathrm{S}_{kl}$ with $\underset{o}{M}_{kl}$ of (9) and

$$S_{kl} = 1/2i\sum_\nu \gamma_{\nu k}\gamma_{\nu l}(k \neq l), \tag{15}$$

where the

$$\gamma_{\nu k} = g_{kl}\gamma_\nu{}^l \tag{15a}$$

satisfy the same relations (13) as do the $\gamma_\nu{}^k$.

The invariant scalar product is

$$(\psi,\psi) = \int |\sum_\zeta \psi^*\gamma_1{}^4\gamma_2{}^4 \cdots \gamma_N{}^4\psi| d\Omega \tag{16}$$

In fact, (16) is invariant both with respect to the operators $\underset{o}{M}$ and also with respect to the S. The latter condition means that

$$((1 + i\varepsilon S_{kl})\psi, \quad (1 + i\varepsilon S_{kl})\psi) = (\psi,\psi),$$

up to terms with ε^2. This formula can be verified by observing that, if both k and l are space like S_{kl} is a Hermitian matrix and commutes with the product of the γ^4. If either k or l is 4, S_{kl} is skew Hermitian, but anticommutes with product of the γ^4. It follows that (16) is invariant with respect to the proper Lorentz transformations. Its invariance with respect to reflections, etc., can also be shown.

The absolute sign in (16) is necessary to make it positive definite. We now shall give (16) a new form which is based on the set of identities

$$(p_4)^\nu \gamma_\nu{}^4 \cdots \gamma_2{}^4\gamma_1{}^4\psi = m^\nu\psi + A_\nu\psi, \tag{17}$$

where A_ν is a skew Hermitian matrix involving only the first ν of the γ^k (and the p). We can prove (17) best by induction: applying $p_4\gamma_{\nu+1}{}^4$ to (17) gives, by means of (14),

$$(p_4)^{\nu+1}\gamma_{\nu+1}{}^4\gamma_\nu{}^4 \cdots \gamma_2{}^4\gamma_1{}^4\psi = m^\nu p_4\gamma_{\nu+1}{}^4\psi + p_4\gamma_{\nu+1}{}^4 A_\nu\psi \tag{17a}$$
$$= m^{\nu+1}\psi + (-m^\nu p_k\gamma_{\nu+1}{}^k + p_4\gamma_{\nu+1}{}^4 A_\nu)\psi \qquad (k = 1,2,3).$$

The last bracket is $A_{\nu+1}$: it is skew Hermitian and involves only the first $\nu + 1$ of the γ so that (17) is established by induction. Setting $\nu = N$ in (17), multiplying with γ and summing over the ζ yields

$$p_4{}^N\sum_\zeta \psi^*\gamma_1{}^4\gamma_2{}^4 \cdots \gamma_N{}^4\psi = m^N\sum_\zeta |\psi|^2 + \sum_\zeta \psi^* A_N\psi. \tag{17b}$$

Because of the skew Hermitian nature of A_N, the last term is imaginary. Since the two other terms of (17b) are real, they must be equal. As a result, we can write for (16) also

$$(\psi,\psi) = \int |m/p_4|^N\sum_\zeta |\psi|^2 d\Omega. \tag{18}$$

At the same time, (18) permits us to give another form to the scalar product,

$$(\psi,\psi) = \int |p_4|^{-N-1}\sum_\zeta |\psi|^2 dp_1 dp_2 dp_3, \tag{18a}$$

which differs from (18) or (16) by the positive constant m^{-N}. It may be worth noting here that the absolute signs in (16), and in the definition (11) of $d\Omega$ (or in (18a)), can be omitted in case of an odd N. This makes it possible to define a simple positive definite scalar product in coordinate space by means of (12). In particular, for N = 1, (16) (or (18a)) equals the integral of $|\psi^2|$ over ordinary space. In case of even N (integer spin s) no simple positive definite scalar product can be defined in coordinate space.

It is now established that the solutions of (14) form a Lorentz invariant set in which a positive definite scalar product (16) or (18a) can be defined. We shall now determine the representation of 2 to which the solutions belong and will also calculate the invariants P and W.

In order to define a little group, we choose as momentum p_o with the components 0, 0, 0, m. The little group then becomes the group of rotations in ordinary space. If we assume that the γ^4 are diagonal, with diagonal elements 1, 1, -1, -1, equation (14) shows that only those components of ψ can be different from zero which correspond to the first two rows of γ_v. There are 2^N such components, the rest of the 4^N components of ψ must vanish. Even these components will not be independent: as a result of the symmetry of the ψ in the ζ, all components of ψ will be equal in which the same number κ of the N indices ζ correspond to the first row of the γ, the N - κ other indices to the second row. Since κ can assume any of the values between 0 and N, there are N + 1 such components. If $p_4 = -m$, the same considerations will hold, except that the last two rows of γ will play the role which the first two rows play in case of $p_4 = m$.

In order to determine the transformation properties of these $N + 1 = 2s + 1$ independent components under the elements of the little group, we note that the space like $\underset{o}{M}$ give zero if applied to ψ with a purely time like $p = p_0$. We need only to calculate, therefore, the effect of the S_{kl} on ψ. Since, in particular, $1/2i\gamma_1\gamma_2$ commutes with γ^4, but is not identical with it, we can assume that it is diagonal and has the diagonal elements 1/2, -1/2, 1/2, -1/2. If the sum of such $1/2i\gamma_1\gamma_2$ is applied to the component of ψ in which κ of the ζ_v correspond to the first row, N - κ to the second row, this component will be multiplied by $1/2\kappa - 1/2(N - \kappa) = \kappa - s$. Since κ runs from 0 to $N = 2s$, the $M_{12}\psi = S_{12}\psi$ will run from $-s\psi$ to $s\psi$. Hence the representation of the little group in question is $D^{(s)}$, as was postulated.

Because of (10a), W becomes $m^2(S_{23}{}^2 + S_{31}{}^2 + S_{12}{}^2)$ or, since the S_{23}, S_{31}, S_{12} are the infinitesimal operators of $D^{(s)}$, we have $W = m^2 s(s + 1)$ as given[9] in §3. The value of P is m^2 because of (14a).

5. *The Class* O_s *–(a)* $s = 0$. The corresponding discussion in the preceding section may be literally applied to this case, with the exception that $m = 0$ and that the integral (11) is to be extended over the light cone.

(b) The wave equations can be obtained by setting $m = 0$ in (14). The infinitesimal operators continue to be given by (9) and (15). The scalar product must be defined by (18a) because (16) vanishes for all ψ. The invariance of this scalar product follows from the invariance of (18a) for finite mass because, except at $p_1 = p_2 = p_3 = 0$, the wave function is continuous in m.

The essential difference between finite and zero mass is that, in the latter case, not only the infinitesimal operators but also the wave equation are invariant under any one of the operators $<\Gamma>_\nu = i\gamma_\nu^1\gamma_\nu^2\gamma_\nu^3\gamma_\nu^4$. As a result, for m = 0, the linear manifold defined by (14) can be decomposed into invariant manifolds by giving definite values to the $<\Gamma>_\nu$. In particular, we shall be concerned henceforth with the manifold defined by (14) and

$$\Gamma_\nu\psi = \psi \qquad (\nu = 1,2, \cdots, N), \tag{19a}$$

and with the other one for which

$$\Gamma_\nu\psi = -\psi \qquad (\nu = 1,2, \cdots, N) \tag{19b}$$

holds. Both manifolds are invariant under proper Lorentz transformations but go over into each other by reflections: they correspond physically to right and left circular polarization.[10]

Let us now again choose a particular momentum vector p_o in order to define the little group. The covariant components of p_o shall be 0, 0, 1, 1. The wave equations (14) then can be written, after multiplication with γ_ν^3, in the form

$$\gamma_\nu^3\gamma_\nu^4\psi = \psi \qquad (\nu = 1,2, \cdots, N). \tag{20}$$

It is now advantageous to assume that the $\gamma^3\gamma^4$ are diagonal, their diagonal elements being 1, 1, -1, -1. Equation (20) then expresses the fact that ψ for the p_o in question is different from zero only if all ζ have values corresponding to the first two rows of the γ. Since the Γ commute with the $\gamma^4\gamma^3$ but are not identical with them, they may be also assumed to be diagonal, with diagonal elements 1, -1, 1, -1. Hence, in the manifold defined by (20) and (19a) all components of ψ vanish (for $p = p_o$) unless all ζ have values corresponding to the first rows of the γ: the manifold (20), (19a) is one dimensional for given p. The same holds for the manifold defined by (20), (19b) except that in this case $\psi(p_o; \zeta_1, ..., \zeta n)$ differs from zero only if all ζ have values corresponding to the second row of the γ. For given momentum, ψ has only two independent components.

The infinitesimal operators of the little group are M_{12}, M_{13} - M_{14}, M_{23} - M_{24} which leave p_o invariant. The corresponding $\underset{o}{M}$ give again zero if applied to ψ at $p = p_o$. The S corresponding to the second of the above operators (cf. (15), (15a)) is a sum of matrices $1/2i(\gamma_\nu^1\gamma_\nu^3 + \gamma_\nu^1\gamma_\nu^4)$. It vanishes if applied to our ψ as can be seen by applying $\gamma_\nu^1\gamma_\nu^3$ to (20). The same holds for M_{23} - M_{24}. On the other hand, $1/2i\gamma_\nu^1\gamma_\nu^2$ gives $1/2\psi$ if applied to the ψ of (20), (19a), and gives - $1/2\psi$ if applied to the ψ of (20),

(19b). One sees this most easily by applying $1/2i\gamma_v{}^1\gamma_v{}^2$ to (20) and making use of (19). As a result, $M_{12}\psi = \pm 1/2N\psi = \pm s\psi$ for the two manifolds in question: these indeed belong to the representation O_s, of the inhomogeneous Lorentz group.

The value of the invariant P is zero. The above also involves a calculation of the w for the ψ at $p = p_0$: we have $w^3\psi = M_{12}\psi = \pm s\psi$, $w^1\psi = (M_{42} + M_{23})\psi = 0$, $w^2\psi = (M_{31} + M_{14})\psi = 0$, $-w^4\psi = M_{12}\psi = \pm s\psi$. It follows that the value of the second invariant $W = -(w^4)^2 + (w^1)^2 + (w^2)^2 + (w^3)^2$ is also zero for all the manifolds O_s; these cannot be characterized by P and W. However, these manifolds can be characterized by the equation P = 0 with the additional set

$$w_k = sp_k \quad \text{and} \quad w_k = -sp_k, \tag{21}$$

the + applying to (19a), the - to (19b). Both these equations are invariant with respect to proper Lorentz transformations. If reflections are to be included, one can combine them into $w_k w_l = s^2 p_k p_l$.

6. The Class O(Ξ). – Here, the auxiliary is a space like four vector ξ of length l, orthogonal to p. The scalar function $\psi(p,\xi)$ is determined by the equations[11]

$$g^{kl}p_k p_l\psi = 0;\ g^{kl}p_k\xi_l\psi = 0;\ g^{kl}\xi_k\xi_l\psi = -\psi, \tag{22}$$

$$p_k\partial\psi/\partial\xi_k = -i\Xi\psi, \tag{22a}$$

with a real positive constant Ξ. By (22a), for every real number ρ,

$$\psi(p,\xi + \rho p) = e^{-i\rho\Xi}\psi(p,\xi). \tag{23}$$

The infinitesimal operators of displacement are the p_k, those for rotations are the $\underset{o}{M}$ of (9) plus the

$$S_{kl} = i(\xi_k\frac{\partial}{\partial\xi^l} - \xi_l\frac{\partial}{\partial\xi^k}) = i(\xi_k g_{lj} - \xi_l g_{kj})\frac{\partial}{\partial\xi_j} \tag{24}$$

In order to find the invariant scalar product, we introduce, for every vector p on the light cone, two real space like vectors $u^{(1)}(p)$ and $u^{(2)}(p)$ of length one, orthogonal to p and to each other, so that

$$\{u^{(r)}(p),p\} = 0,\quad \{u^{(r)}(p),u^{(s)}(p)\} = \delta_{rs} \quad (r,s = 1,2). \tag{25}$$

Then ξ is a linear combination of p, $u^{(1)}(p)$, $u^{(2)}(p)$,

$$\xi = \alpha p + \beta_1 u^{(1)}(p) + \beta_2 u^{(2)}(p), \tag{26}$$

where α and the β are real. $\{\xi, \xi\} = -1$ implies $\beta_1^2 + \beta_2^2 = 1$, hence $\beta_1 + i\beta_2 = e^{i\tau}$ with a suitable real angle τ. $\psi(p, \xi)$ is therefore a function of p, α, τ,

$$\psi(p,\xi) = \phi(p,\alpha,\tau). \tag{27}$$

The choice of the $u^r(p)$ is, of course, not unique. Let $v^{(r)}(p)$ be another system of vectors which satisfy (25). They may be expressed in the form (26), i.e.,

$$v^{(r)}(p) = \kappa_r p + \sum_s \lambda_{sr} u^{(s)}(p) \qquad (r,s = 1,2).$$

By (25), the matrix λ_{sr} is orthogonal. In terms of the $v^{(r)}$, $\xi = \alpha' p + \sum_r \beta'_r v^{(r)}(p)$, where $\beta'_r = \sum_s \lambda_{sr}\beta_s$. In particular

$$\beta'_1 + i\beta'_2 = e^{i\tau'}; \qquad \tau' = \pm(\tau + \lambda) \tag{28}$$

λ depending on the λ_{sr}. By (23), $|\phi(p, \alpha, \tau)| = |\phi(p, 0, \tau)|$, and we define the norm of ψ by

$$(\psi,\psi) = \int |\phi(p,0,\tau)|^2 d\Omega d\tau. \tag{29}$$

This expression is independent of the choice of the $u^{(r)}$. In fact, let $\phi(p, \alpha, \tau) = \phi'(p, \alpha', \tau')$ where the primed variables refer to another set $v^{(r)}$. Then $|\phi(p, 0, \tau)| = |\phi'(p, \alpha', \tau')| = |\phi(p, 0, \tau')|$, and $|d\tau'/d\tau| = 1$. To prove the Lorentz invariance of (29) we proceed as follows: If a homogeneous Lorentz transformation maps p on $\Lambda^{-1}p$, and ξ on $\Lambda^{-1}\xi$, we may, in particular, choose the $u^{(r)}(p)$ in the new system to be the transforms of the original ones; then the coefficient α, β_1, β_2 in (26), and hence τ remain unchanged, and the integral (29) is invariant.

If we choose as the basic vector again p_o with the components 0, 0, 1, 1, the infinitesimal operators of the little group are again M_{12}, $M_{13} - M_{14}$, $M_{23} - M_{24}$. The $\underset{o}{M}$ parts of these give zero for $p = p_o$, the S parts of the latter two are

$$S_{13} - S_{14} = -i\xi_1\left(\frac{\partial}{\partial\xi_3} + \frac{\partial}{\partial\xi_4}\right) + i(\xi_3 - \xi_4)\frac{\partial}{\partial\xi_1}, \tag{30a}$$

$$S_{23}-S_{24}$$
$$= -i\xi_2(\frac{\partial}{\partial\xi_3} + \frac{\partial}{\partial\xi_4}) + i(\xi_3 - \xi_4)\frac{\partial}{\partial\xi_2}. \tag{30b}$$

Because of (22a), the first term gives, if applied to ψ at $p = p_o$ just $\Xi\xi_1\gamma$ and $\Xi\xi_2\psi$, respectively. The second terms vanish because of the second equation of (22). Hence ψ is not invariant under the "displacements" M_{13} - M_{14} and M_{23} - M_{24} in ξ space, and the sum of the squares of the "momenta" is $(\xi_1{}^2 + \xi_2{}^2)\ \Xi^2 = \Xi^2$ because of the last equation of (22). This is also the value of W, while P = 0.

7. *The Class* $O'(\Xi)$.–Since the discussion of this last case follows the pattern of the preceding section we confine ourselves to stating the main results. We introduce, in addition to the vector ξ, a discrete spin variable ζ which can assume four values. The wave equations become

$$\gamma^k p_k \psi = 0;$$
$$g^{kl} p_k \xi_l \psi = 0; \quad g^{kl} \xi_k \xi_l \psi = -\psi. \tag{31}$$

$$p_k \partial\psi/\partial\xi_k = -i\Xi\psi \tag{31a}$$

The parameters α and τ are introduced as before. The norm is given by

$$(\psi, \psi) = \int p_4{}^{-2} \sum |\phi(p,0,\tau)|^2 dp_1 dp_2 dp_3 d\tau. \tag{32}$$

(Cf. (18a) and (29).) Again $W\psi = \Xi^2\psi$, $P\psi = 0$.

It may be remarked that the scalar product has a simple positive definite form in coordinate space for these equations.[11]

[1] All the essential results of the present paper were obtained by the two authors independently, but they decided to publish them jointly.

[2] Wigner, E.P., *Ann. Math.*, **40**, 149-204 (1939).

[3] Fierz, M., *Helv. Phys. Acta*, **XII**, 3-37 (1939).

[4] Garding, L., *Proc. Nat. Acad. Sci.*, **33**, 331-332 (1947).

[5] Gelfand L., and Neumark, M., *J. Phys. (USSR)*, **X**, 93-94 (1946); Harish-Chandra, *Proc. Roy Soc.* (London), **A, 189**, 372-401 (1947); and Bargmann, V., *Ann. Math.*, *48*, 568-640 (1947), have determined the representations of the homogeneous Lorentz group. These are representations also of the inhomogeneous Lorentz group. In the quantum mechanical interpretation, however, all the states of the corresponding particles are invariant under translations and, in particular, independent of time. It is very unlikely that these

representations have immediate physical significance. In addition, the third paper contains a determination of those representations for which the momentum vectors are space like. These are not considered in the present article as they also are unlikely to have a simple physical interpretation.

[6] Lubánski, J. K. *Physica*, **IX**, 310-324 (1942).

[7] Dirac, P. A., M., *Proc. Roy. Soc.*, **A, 155**, 447-459 (1936).

[8] The literature on relativistic wave equations is very extensive. Beside the papers quoted in reference 11, we only mention the book by the Broglie, L., *Théorie générale des particules à spin* (Paris, 1943), and the following articles which give a systematic account of the subject: Pauli, W., *Rev. Mod. Phys.*, **13**, 203-232 (1941); Bhabha, H.J., *Rev. Mod. Phys.*, **17**, 203-209 (1945); Kramers, H. A., Belinfante, F. J., and Lubánski, J. K., *Physica*, **VIII**, 597-627 (1941). In this paper, the sum of (14) over all ν was postulated; (14a) then has to be added as an independent equation (except for $N = 1$). Reference 11 uses these equations in the form given by Kramers, Belinfante and Lubánski.

[9] One may derive this result in a more elegant way, without specializing the coordinate system. For the sake of brevity, we omit this derivation.

[10] de Wet, J.S., *Phys. Rev.*, **58**, 236-242 (1940), in particular, p. 242.

[11] Wigner, E. P., *Z Physik*, (1947).

Unitary representations of the Lorentz group

By P. A. M. Dirac, F.R.S., *St John's College, Cambridge*

(*Received* 31 *May* 1944)

Certain quantities are introduced which are like tensors in space-time with an infinite enumerable number of components and with an invariant positive definite quadratic form for their squared length. Some of the main properties of these quantities are dealt with, and some applications to quantum mechanics are pointed out.

1. Introduction

Given any group, an important mathematical problem is to get a matrix representation of it, which means to make each element of the group correspond to a matrix in such a way that the matrix corresponding to the product of two elements is the product of the matrices corresponding to the factors. The matrices may be looked upon as linear transformations of the co-ordinates of a vector and then each element of the group corresponds to a linear transformation of a field of vectors. Of special interest are the *unitary* representations, in which the linear transformations leave invariant a positive definite quadratic form in the co-ordinates of a vector.

The Lorentz group is the group of linear transformations of four real variables ξ_0, ξ_1, ξ_2, ξ_3, such that $\xi_0^2-\xi_1^2-\xi_2^2-\xi_3^2$ is invariant. The finite representations of this group, i.e. those whose matrices have a finite number of rows and columns, are all well known, and are dealt with by the usual tensor analysis and its extension spinor analysis. None of them is unitary. The group has also some infinite representations which are unitary. These do not seem to have been studied much, in spite of their possible importance for physical applications.

The present paper gives a new method of attack on these representations, which was suggested by Fock's quantum theory of the harmonic oscillator. It leads to a new kind of tensor quantity in space-time, with an infinite number of components and a positive definite expression for its squared length.

2. Three-dimensional theory

This section will be devoted to some preliminary work applying to the rotation group of three-dimensional Euclidean space. Take an ascending power series

$$a_0+a_1\xi_1+a_2\xi_1^2+a_3\xi_1^3+\ldots \qquad (1)$$

in a real variable ξ_1 with real coefficients a_r. Consider these coefficients to be the co-ordinates of a vector in a certain space of an infinite number of dimensions, and define the squared length of the vector to be

$$\Sigma_0^\infty r!\,a_r^2. \qquad (2)$$

The series (2) must converge for the vector to be a finite one.

Reprinted from *Proc. Roy. Soc. (London)*, **A183**, 284 (1945).

Take two more similar power series $\Sigma_0^\infty b_s \xi_2^s$ and $\Sigma_0^\infty c_t \xi_3^t$ in the real variables ξ_2 and ξ_3 and consider their coefficients to be the co-ordinates of vectors in two more vector spaces, with squared lengths defined by the corresponding formula to (2). Now multiply the three vector spaces together. A general vector in the product space will be a sum of products of vectors of the three original vector spaces, and its co-ordinates A_{rst} can be represented as the coefficients in a power series

$$P = \Sigma_0^\infty A_{rst} \xi_1^r \xi_2^s \xi_3^t \tag{3}$$

in the three variables ξ_1, ξ_2, ξ_3. The squared length of such a vector is

$$\Sigma_0^\infty r!\,s!\,t!\,A_{rst}^2, \tag{4}$$

and two vectors with co-ordinates A_{rst} and B_{rst} have a scalar product

$$\Sigma_0^\infty r!\,s!\,t!\,A_{rst} B_{rst}. \tag{5}$$

If the variables ξ_1, ξ_2, ξ_3 are subjected to a linear transformation, going over into ξ_1', ξ_2', ξ_3', say, the power series (3) will go over into a power series in ξ_1', ξ_2', ξ_3',

$$P = \Sigma A'_{rst} \xi_1'^r \xi_2'^s \xi_3'^t,$$

in which the coefficients A' are linear functions of the previous coefficients A. Thus each linear transformation of the ξ's generates a linear transformation of the coefficients A.

The theorem will now be proved: *A linear transformation of ξ_1, ξ_2, ξ_3 which leaves $\xi_1^2+\xi_2^2+\xi_3^2$ invariant generates a linear transformation of the coefficients A which leaves the squared length* (4) *invariant.* Consider first the infinitesimal transformation

$$\xi_1 = \xi_1' + \epsilon\xi_2', \quad \xi_2 = \xi_2' - \epsilon\xi_1', \quad \xi_3 = \xi_3', \tag{6}$$

which leaves $\xi_1^2+\xi_2^2+\xi_3^2$ invariant, ϵ being a small quantity whose square is negligible. Substituting into (3), one gets

$$P = \Sigma A_{rst}(\xi_1'^r \xi_2'^s + r\epsilon \xi_1'^{r-1} \xi_2'^{s+1} - s\epsilon \xi_1'^{r+1} \xi_2'^{s-1})\, \xi_3'^t.$$

Hence $$A'_{rst} = A_{rst} + (r+1)\,\epsilon A_{r+1,s-1,t} - (s+1)\,\epsilon A_{r-1,s+1,t},$$

in which A_{rst} with a negative suffix is counted as zero. Thus

$$\Sigma r!\,s!\,t!\,A'^2_{rst} = \Sigma r!\,s!\,t!\,[A^2_{rst} + 2(r+1)\,\epsilon A_{rst} A_{r+1,s-1,t} - 2(s+1)\,\epsilon A_{r-1,s+1,t} A_{rst}].$$

The last two terms in the [ ] here cancel, as may be seen by substituting $r-1$ for r in the former and $s-1$ for s in the latter, and hence the squared length (4) is invariant for the transformation (6). Any linear transformation of ξ_1, ξ_2, ξ_3 which leaves $\xi_1^2+\xi_2^2+\xi_3^2$ invariant can be built up from the infinitesimal transformation (6) and similar infinitesimal transformations with ξ_1, ξ_2 and ξ_3 permuted, together with possibly a reflexion $\xi_1 = -\xi_1'$, $\xi_2 = \xi_2'$, $\xi_3 = \xi_3'$, which obviously leaves the squared length (4) invariant, and hence the theorem is proved.

The group of transformations of the ξ's which leave $\xi_1^2+\xi_2^2+\xi_3^2$ invariant is the rotation group in three-dimensional Euclidean space, so the transformations of the coefficients A provide a representation of this rotation group. One may restrict the function P to be homogeneous, of degree u say, and then the representation is a finite one. The coefficients A then form the components of a symmetrical tensor of rank u, the connexion with the usual tensor notation being effected by taking A_{rst} to be $u!/r!s!t!$ times the usual tensor component with the suffix 1 occurring r times, 2 occurring s times and 3 occurring t times, as may be seen from the invariance of expression (3) with (ξ_1, ξ_2, ξ_3) transforming like a vector.

One can make a straightforward generalization of the foregoing theory by introducing other triplets of variables, η_1, η_2, η_3 and $\zeta_1, \zeta_2, \zeta_3$ say, which transform together with ξ_1, ξ_2, ξ_3, and setting up a power series in all the variables. The transformations of the coefficients of these more general power series will provide further representations of the three-dimensional rotation group. If such a more general power series is restricted to be homogeneous, its coefficients will form the components of an unsymmetrical tensor.

3. Four-dimensional theory

Take a descending power series

$$k_0/\xi_0 + k_1/\xi_0^2 + k_2/\xi_0^3 + k_3/\xi_0^4 + \ldots \tag{7}$$

in a real variable ξ_0 with real coefficients k_n. Consider these coefficients to be the co-ordinates of a vector in a certain space of an infinite number of dimensions, and define the squared length of the vector to be

$$\Sigma_0^\infty k_n^2/n!. \tag{8}$$

Multiply this vector space into the vector space of the preceding section. A general vector in the product space will have co-ordinates A_{nrst} which can be represented as the coefficients in a power series

$$Q = \Sigma_0^\infty A_{nrst}\xi_0^{-n-1}\xi_1^r\xi_2^s\xi_3^t \tag{9}$$

in the four variables $\xi_0, \xi_1, \xi_2, \xi_3$. The squared length of such a vector is

$$\Sigma_0^\infty n!^{-1} r!s!t! A_{nrst}^2, \tag{10}$$

and two vectors with co-ordinates A_{nrst} and B_{nrst} have a scalar product

$$\Sigma_0^\infty n!^{-1} r!s!t! A_{nrst} B_{nrst}. \tag{11}$$

The series (7) may be extended backwards to include some terms with non-negative powers of ξ_0, so that coefficients k_n occur with negative n-values, leading to coefficients A_{nrst} with negative n-values. Since $n!$ is infinite for n negative, these new coefficients do not contribute to the squared length of a vector or the scalar product of two vectors. Thus the terms with non-negative powers of ξ_0 should be counted as corresponding to the vector zero, and whether they are present in an expansion or not does not matter.

Now apply a Lorentz transformation to the ξ's,

$$\xi_\mu = \alpha^\nu_\mu \xi'_\nu, \tag{12}$$

the α's satisfying certain conditions so that $\xi_0^2-\xi_1^2-\xi_2^2-\xi_3^2$ is invariant. This makes $\xi_1^r\xi_2^s\xi_3^t$ go over into a finite polynomial in ξ'_0, ξ'_1, ξ'_2, ξ'_3, and ξ_0^{-n-1} go over into

$$(\alpha^0_0\xi'_0+\alpha^1_0\xi'_1+\alpha^2_0\xi'_2+\alpha^3_0\xi'_3)^{-n-1}, \tag{13}$$

which may be expanded in ascending powers of ξ'_1, ξ'_2, ξ'_3 and descending powers of ξ'_0. (The question of the convergence of this expansion is left to the next section, so as not to break the main argument here.) The power series (9) then goes over into a series

$$Q = \Sigma A'_{nrst}\xi_0'^{-n-1}\xi_1'^r\xi_2'^s\xi_3'^t \tag{14}$$

in ascending powers of ξ_1, ξ_2, ξ_3 and descending powers of ξ_0, with coefficients A' which are linear functions of the previous coefficients A. There may be terms in (14) with non-negative powers of ξ_0, but on account of what was said above these can be discarded. If such terms are present in the original series (9), they will not affect any of the coefficients with non-negative n-values in (14).

The Lorentz transformation thus generates a linear transformation of the coefficients A_{nrst} with non-negative n-values. The theorem will now be proved: *The transformation of the coefficients A leaves the squared length* (10) *invariant.*

Consider first the infinitesimal Lorentz transformation

$$\xi_0 = \xi'_0+\epsilon\xi'_1, \quad \xi_1 = \xi'_1+\epsilon\xi'_0, \quad \xi_2 = \xi'_2, \quad \xi_3 = \xi'_3. \tag{15}$$

Substituting into (9), one gets

$$Q = \Sigma A_{nrst}[\xi_0'^{-n-1}\xi_1'^r-(n+1)\epsilon\xi_0'^{-n-2}\xi_1'^{r+1}+r\epsilon\xi_0'^{-n}\xi_1'^{r-1}]\,\xi_2'^s\xi_3'^t.$$

Hence $$A'_{nrst} = A_{nrst}-n\epsilon A_{n-1,r-1,s,t}+(r+1)\,\epsilon A_{n+1,r+1,s,t},$$

in which A_{nrst} with a negative value for r, s or t is counted as zero. Thus

$$\begin{aligned}&\Sigma n!^{-1}r!\,s!\,t!\,A'^2_{nrst}\\ &\quad = \Sigma n!^{-1}r!\,s!\,t!\,[A^2_{nrst}-2n\epsilon A_{nrst}A_{n-1,r-1,s,t}+2(r+1)\,\epsilon A_{n+1,r+1,s,t}A_{nrst}].\end{aligned}$$

The last two terms in the [ ] here cancel, as may be seen by substituting $n+1$ for n in the former and $r-1$ for r in the latter, and hence the squared length (10) is invariant for the transformation (15). Any Lorentz transformation can be built up from infinitesimal transformations like (15) and three-dimensional rotations like those considered in the preceding section, together with possibly a reflexion, which obviously leaves the squared length (10) invariant, and hence the theorem is proved.

The transformations of the coefficients A thus provide a unitary representation of the Lorentz group. The coefficients themselves form the components of a new kind of tensor quantity in space-time. I propose for it the name *expansor*, because of its connexion with binomial expansions. One may restrict the function (9) to be homogeneous and one then gets a simpler kind of expansor, which may be called a *homogeneous expansor*. The analogy with the three-dimensional case suggests that one should look upon a homogeneous expansor as a symmetrical tensor in space-time with the suffix 0 occurring in its components a negative number of times.

The foregoing theory can be generalized, like the three-dimensional theory, by the introduction of other quadruplets of variables, $\eta_0, \eta_1, \eta_2, \eta_3$ and $\zeta_0, \zeta_1, \zeta_2, \zeta_3$ say, which transform together with $\xi_0, \xi_1, \xi_2, \xi_3$. One can then set up a power series in all the variables, ascending in those variables with suffixes 1, 2 and 3, and descending in those variables with suffix 0. The transformations of the coefficients of these more general power series will provide further unitary representations of the Lorentz group, and the coefficients themselves will form the components of more general expansors.

There is another generalization which may readily be made in the theory, namely, to take the values of the index n in (9) to be not integers, but any set of real numbers $n_0, n_0+1, n_0+2, \ldots$ extending to infinity. In the formula (10) for the squared length $n!$ is then to be interpreted as $\Gamma(n+1)$. The terms with negative n-values can no longer be discarded. The expression for the squared length is still positive definite if the minimum value of n is greater than -1, which is the case if the function (9) is homogeneous and its degree is negative. The resulting representation is then still unitary. If, however, the function (9) is homogeneous and its degree is positive, there will be a finite number of negative terms in the expression for the squared length. The resulting kind of representation may be called *nearly unitary*.

4. Some theorems on convergence

If the series (2) is convergent, then (1) is convergent for all values of ξ_1. Similarly, if (4) is convergent, then (3) is convergent for all values of ξ_1, ξ_2, ξ_3. On the other hand, if (8) is convergent, (7) need not be convergent for any value of ξ_0, in which case, of course, it does not define a function of ξ_0. Similarly, if (10) is convergent, (9) need not be convergent for any values for the ξ's. Thus, corresponding to a general expansor A_{nrst}, there need not exist any function Q of the ξ's. However, it will now be proved that *if the series* (9) *is homogeneous and* (10) *is convergent, then* (9) *is absolutely convergent for all values of the ξ's satisfying*

$$\xi_0^2-\xi_1^2-\xi_2^2-\xi_3^2>0. \tag{16}$$

If the series (9) is homogeneous of degree $u-1$, it may be written

$$\Sigma_n \xi_0^{-n-1}\Sigma_S A_{nrst}\xi_1^r\xi_2^s\xi_3^t, \tag{17}$$

where Σ_S means a sum over all values of r, s and t satisfying

$$r+s+t = n+u. \tag{18}$$

With this notation the series (10) may be written

$$\Sigma_n n!^{-1}\Sigma_S r!\,s!\,t!\,A_{nrst}^2. \tag{19}$$

Now apply Cauchy's inequality

$$(x_1y_1+x_2y_2+x_3y_3+\ldots)^2 \leqslant (x_1^2+x_2^2+x_3^2+\ldots)(y_1^2+y_2^2+y_3^2+\ldots) \tag{20}$$

in the following way. Take $(r+s+t)!/r!s!t!$ of the x's in (20) to be each equal to $\xi_1^r\xi_2^s\xi_3^t$ and the corresponding y's to be each equal to $A_{nrst}\,r!s!t!/(r+s+t)!$, and do this for all r, s, t subject to (18) for a fixed value of n. Then (20) becomes

$$(\Sigma_S A_{nrst}\xi_1^r\xi_2^s\xi_3^t)^2 \leqslant (\xi_1^2+\xi_2^2+\xi_3^2)^{n+u}(n+u)!^{-1}\Sigma_S r!s!t!A_{nrst}^2. \tag{21}$$

If the sum with respect to n in (19) converges, there must be some number κ such that

$$n!^{-1}\Sigma_S r!s!t!A_{nrst}^2 < \kappa$$

for all n. Then from (21)

$$|\xi_0^{-n-1}\Sigma_S A_{nrst}\xi_1^r\xi_2^s\xi_3^t| < \kappa^{\frac{1}{2}}|\xi_0|^{u-1}\left\{\frac{n!}{(n+u)!}\right\}^{\frac{1}{2}}\left\{\frac{\xi_1^2+\xi_2^2+\xi_3^2}{\xi_0^2}\right\}^{\frac{1}{2}(n+u)},$$

which shows that (17) is convergent when (16) is satisfied. One may take $\xi_0, \xi_1, \xi_2, \xi_3$ and the A's to be all positive without disturbing the argument, so (17), considered as a quadruple series in n, r, s, t, is absolutely convergent.

Thus for any homogeneous expansor of finite length, there exists a function Q given by (9) defined within the light-cone (16). In transforming this function with a Lorentz transformation (12), it is legitimate to use the expansion of (13) in ascending powers of ξ_1', ξ_2', ξ_3' and descending powers of ξ_0', for the following reason. Suppose ξ_μ is within the light-cone and take for definiteness $\xi_0>0$. Then

$$\alpha_0^0\xi_0'+\alpha_0^1\xi_1'+\alpha_0^2\xi_2'+\alpha_0^3\xi_3'>0. \tag{22}$$

One can change the sign of any of the co-ordinates ξ_1', ξ_2', ξ_3', leaving ξ_0' unchanged and ξ_μ will still lie within the light-cone with $\xi_0>0$, so (22) must still be satisfied. Hence

$$\alpha_0^0\xi_0'-|\alpha_0^1\xi_1'|-|\alpha_0^2\xi_2'|-|\alpha_0^3\xi_3'|>0,$$

which shows that the expansion of (13) in the required manner is absolutely convergent. The use of this expansion in the preceding section is thus justified for the case when (9) is homogeneous, the new coefficients A being determined by the transformed function Q within the light-cone. The justification for (9) not homogeneous then follows since terms of different degree do not interfere. It should be noted that all the foregoing arguments are valid also when the values of n are not integers.

There are some expansors which are invariant under all Lorentz transformations, namely those whose components are the coefficients of $(\xi_0^2-\xi_1^2-\xi_2^2-\xi_3^2)^{-l}$ expanded in ascending powers of ξ_1, ξ_2, ξ_3 and descending powers of ξ_0. For such an expansion to be possible, l must not be a negative integer or zero, but it can be any other real number. Again, there are expansors which transform like ordinary tensors under Lorentz transformations, namely those whose components are the coefficients in the expansion of

$$f(\xi_0,\xi_1,\xi_2,\xi_3)(\xi_0^2-\xi_1^2-\xi_2^2-\xi_3^2)^{-l}, \tag{23}$$

where f is any homogeneous integral polynomial in $\xi_0, \xi_1, \xi_2, \xi_3$, and l is restricted as before. (23) transforms like the polynomial f, and thus like a tensor of order equal to the degree of f. One would expect all these expansors to be of infinite length, as otherwise one could set up a positive definite form for the squared length of a tensor in space-time. A formal proof that they are is as follows.

Suppose the f in (23) is of degree u and express it as

$$f = g_u + \xi_0 g_{u-1} + (\xi_0^2 - \boldsymbol{\xi}^2) g_{u-2} + \xi_0(\xi_0^2 - \boldsymbol{\xi}^2) g_{u-3} + (\xi_0^2 - \boldsymbol{\xi}^2)^2 g_{u-4} + \dots,$$

where $\boldsymbol{\xi}^2 = \xi_1^2 + \xi_2^2 + \xi_3^2$ and each of the g's is a polynomial in ξ_1, ξ_2, ξ_3 only, of degree indicated by the suffix. The successive terms here contribute to (23) the amounts

$$g_u(\xi_0^2 - \boldsymbol{\xi}^2)^{-l} = \Sigma_{m=0}^{\infty} \frac{l(l+1)(l+2)\dots(l+m-1)}{m!} \frac{g_u \boldsymbol{\xi}^{2m}}{\xi_0^{2(m+l)}},$$

$$g_{u-1}\xi_0(\xi_0^2 - \boldsymbol{\xi}^2)^{-l} = \Sigma_{m=0}^{\infty} \frac{l(l+1)(l+2)\dots(l+m-1)}{m!} \frac{g_{u-1} \boldsymbol{\xi}^{2m}}{\xi_0^{2(m+l)-1}},$$

$$g_{u-2}(\xi_0^2 - \boldsymbol{\xi}^2)^{-l+1} = \Sigma_{m=-1}^{\infty} \frac{(l-1)l(l+1)\dots(l+m-1)}{(m+1)!} \frac{g_{u-2} \boldsymbol{\xi}^{2(m+1)}}{\xi_0^{2(m+l)}},$$

$$g_{u-3}\xi_0(\xi_0^2 - \boldsymbol{\xi}^2)^{-l+1} = \Sigma_{m=-1}^{\infty} \frac{(l-1)l(l+1)\dots(l+m-1)}{(m+1)!} \frac{g_{u-3} \boldsymbol{\xi}^{2(m+1)}}{\xi_0^{2(m+l)-1}},$$

and so on. These expansions show that, for large values of m, the terms arising from $g_{u-2}, g_{u-4}, \dots$ are of smaller order than the corresponding terms arising from g_u, and the terms arising from $g_{u-3}, g_{u-5}, \dots$ are of smaller order than the corresponding terms arising from g_{u-1}, so that in testing for the convergence of the series which gives the squared length, only the g_u and g_{u-1} terms need be taken into account. (It will be found that the convergence conditions are not sufficiently critical to be affected by this neglect.) Now express g_u and g_{u-1} as

$$g_u = S_u + \boldsymbol{\xi}^2 S_{u-2} + \boldsymbol{\xi}^4 S_{u-4} + \dots, \quad g_{u-1} = S_{u-1} + \boldsymbol{\xi}^2 S_{u-3} + \boldsymbol{\xi}^4 S_{u-5} + \dots,$$

where the S's are solid harmonic functions of ξ_1, ξ_2, ξ_3, of degrees indicated by the suffixes. Each of them gives a contribution to (23) of the form

$$S_{u-2r}\boldsymbol{\xi}^{2r}(\xi_0^2 - \boldsymbol{\xi}^2)^{-l} = S_{u-2r}\Sigma_{m=0}^{\infty} \frac{(m+l-1)!}{m!(l-1)!} \frac{\boldsymbol{\xi}^{2(m+r)}}{\xi_0^{2(m+l)}} \tag{24}$$

or

$$S_{u-2r-1}\xi_0\boldsymbol{\xi}^{2r}(\xi_0^2 - \boldsymbol{\xi}^2)^{-l} = S_{u-2r-1}\Sigma_{m=0}^{\infty} \frac{(m+l-1)!}{m!(l-1)!} \frac{\boldsymbol{\xi}^{2(m+r)}}{\xi_0^{2(m+l)-1}}. \tag{25}$$

Using the result (41) of the appendix, one finds for the squared lengths of the expansors whose components are the coefficients of (24) and (25), series of the form

$$c\Sigma_m \frac{(m+l-1)!^2}{m!^2(l-1)!^2} \frac{4^{m+r}(m+r)!(m+u-r+\frac{1}{2})!}{(2m+2l-1)!}$$

and

$$c\Sigma_m \frac{(m+l-1)!^2}{m!^2(l-1)!^2} \frac{4^{m+r}(m+r)!(m+u-r-\frac{1}{2})!}{(2m+2l-2)!},$$

respectively. In both these series the ratio of the $(m+1)$th term to the mth is $1+u/m$ for large m, and since u is necessarily positive or zero, the series diverge and the expansors are of infinite length. From the result (40) of the appendix, expansors associated with solid harmonics of different degrees are orthogonal to one another, and hence the total expansor with the coefficients of (23) as components must also be of infinite length.

5. A TRANSFORMATION OF VARIABLES

Up to the present, expansors have always been considered in connexion with certain ξ-functions, the components of the expansor being the coefficients of the ξ-function. In the case of integral n-values, one can make a transformation of variables of the kind which is familiar in quantum mechanics, and get the expansors connected with some other functions, which serves to clarify some of their properties.

Introduce the four operators x_μ, defined by

$$2^{\frac{1}{2}}x_0 = \xi_0 - \partial/\partial\xi_0, \quad 2^{\frac{1}{2}}x_r = \xi_r + \partial/\partial\xi_r \quad (r = 1, 2, 3). \tag{26}$$

Under Lorentz transformations they transform like the components of a 4-vector. The operators $\partial/\partial x_\mu$ may be taken to be

$$2^{\frac{1}{2}}\,\partial/\partial x_0 = \xi_0 + \partial/\partial\xi_0, \quad 2^{\frac{1}{2}}\,\partial/\partial x_r = -\xi_r + \partial/\partial\xi_r, \tag{27}$$

as this gives the correct commutation relations between all the operators. One can now represent ξ-functions with integral n-values by functions of the x's according to the following scheme.

Take as starting point the ξ-function ξ_0^{-1}. It vanishes when operated on by $\partial/\partial\xi_1$, $\partial/\partial\xi_2$, or $\partial/\partial\xi_3$, and also effectively vanishes when multiplied by ξ_0, since the result of the multiplication can be discarded. It must therefore be represented by a function of the x's which vanishes when operated on by $x_r + \partial/\partial x_r$ with $r = 1, 2$ or 3, or by $x_0 + \partial/\partial x_0$, and thus by a multiple of $e^{-\frac{1}{2}(x_0^2+\mathbf{x}^2)}$, where $\mathbf{x}^2 = x_1^2 + x_2^2 + x_3^2$. Take

$$\xi_0^{-1} \equiv \pi^{-1} e^{-\frac{1}{2}(x_0^2+\mathbf{x}^2)}, \tag{28}$$

where the sign $\equiv$ means 'represented by'. Multiply the left-hand side of (28) by the operator $(-\partial/\partial\xi_0)^n\,\xi_1^r\xi_2^s\xi_3^t$ and the right-hand side by the operator equal to it according to (26) and (27). The result is, after dividing through by $n!$,

$$\begin{aligned}\xi_0^{-n-1}\xi_1^r\xi_2^s\xi_3^t &\equiv \pi^{-1}n!^{-1}2^{-\frac{1}{2}(n+r+s+t)}\left(x_0 - \frac{\partial}{\partial x_0}\right)^n \left(x_1 - \frac{\partial}{\partial x_1}\right)^r \left(x_2 - \frac{\partial}{\partial x_2}\right)^s \left(x_3 - \frac{\partial}{\partial x_3}\right)^t e^{-\frac{1}{2}(x_0^2+\mathbf{x}^2)} \\ &= F_{nrst}(x_0, x_1, x_2, x_3),\end{aligned} \tag{29}$$

say. This gives the function F of the x's which represents $\xi_0^{-n-1}\xi_1^r\xi_2^s\xi_3^t$. A general ξ-function is now represented thus,

$$\Sigma A_{nrst}\,\xi_0^{-n-1}\xi_1^r\xi_2^s\xi_3^t \equiv \Sigma A_{nrst}F_{nrst}(x_0, x_1, x_2, x_3). \tag{30}$$

In this way a general expansor A_{nrst} gets connected with a function of the x's.

The chief interest of this connexion is that the law for the scalar product of two expansors becomes very simple when expressed in terms of the functions of the x's. *The scalar product is the integral of the product of the two functions of the x's over all x_0, x_1, x_2 and x_3.* To prove this, first evaluate the integral

$$\int_{-\infty}^{\infty}\left(z-\frac{d}{dz}\right)^m e^{-\frac{1}{2}z^2}.\left(z-\frac{d}{dz}\right)^{m'} e^{-\frac{1}{2}z^2}\,dz, \tag{31}$$

where the dot has the meaning that operators to the left of it do not operate on functions to the right of it. For $m > 0$, (31) goes over by partial integration into

$$\int_{-\infty}^{\infty}\left(z-\frac{d}{dz}\right)^{m-1} e^{-\frac{1}{2}z^2}.\left(z+\frac{d}{dz}\right)\left(z-\frac{d}{dz}\right)^{m'} e^{-\frac{1}{2}z^2}\,dz$$

$$=\int_{-\infty}^{\infty}\left(z-\frac{d}{dz}\right)^{m-1} e^{-\frac{1}{2}z^2}.\left\{\left(z-\frac{d}{dz}\right)^{m'}\left(z+\frac{d}{dz}\right)+2m'\left(z-\frac{d}{dz}\right)^{m'-1}\right\} e^{-\frac{1}{2}z^2}\,dz$$

$$=2m'\int_{-\infty}^{\infty}\left(z-\frac{d}{dz}\right)^{m-1} e^{-\frac{1}{2}z^2}.\left(z-\frac{d}{dz}\right)^{m'-1} e^{-\frac{1}{2}z^2}\,dz.$$

Applying this procedure m times, one gets zero if $m' < m$ and $2^m m!\,\pi^{\frac{1}{2}}$ if $m' = m$. Since there is symmetry between m and m', the result is

$$\int_{-\infty}^{\infty}\left(z-\frac{d}{dz}\right)^m e^{-\frac{1}{2}z^2}.\left(z-\frac{d}{dz}\right)^{m'} e^{-\frac{1}{2}z^2}\,dz = 2^m m!\,\pi^{\frac{1}{2}}\delta_{mm'}.$$

Now substitute for z each of the variables x_0, x_1, x_2, x_3 in turn, with m equal to n, r, s, t and m' equal to n', r', s', t' respectively, and multiply the four equations so obtained. The result, after dividing through by $\pi^2 n!^2\, 2^{n+r+s+t}$, is

$$\iiiint F_{nrst}F_{n'r's't'}\,dx_0\,dx_1\,dx_2\,dx_3 = n!^{-1}r!\,s!\,t!\,\delta_{nn'}\delta_{rr'}\delta_{ss'}\delta_{tt'}.$$

This proves the theorem for the case when the expansors each have only one non-vanishing component, which is sufficient to prove it generally.

One can obviously get a unitary representation of the Lorentz group by considering the transformations of a set of vectors in a space of an infinite number of dimensions, where each vector corresponds to a function of four Lorentz variables x_0, x_1, x_2, x_3, and where the squared length of a vector is defined as the integral of the square of the function over all x_0, x_1, x_2, x_3. The above work shows how this obvious unitary representation is connected with the expansor theory.

6. Applications to quantum mechanics

The four x's of the preceding section may be looked upon as the co-ordinates of a four-dimensional harmonic oscillator, the four operators $i\partial/\partial x_\mu$ being the conjugate momenta p^μ, and the energy of the oscillator may be taken to be the Lorentz invariant

$$\tfrac{1}{2}[x_1^2+x_2^2+x_3^2-x_0^2+(p^1)^2+(p^2)^2+(p^3)^2-(p^0)^2]. \tag{32}$$

The 1, 2, 3 components of the oscillator thus have positive energies and the 0 component negative energy.

The state of the oscillator for which the 0, 1, 2, 3 components are in the nth, rth, sth, tth quantum states respectively is then represented by the function F_{nrst} defined by (29), with a suitable normalizing factor. This representative may be transformed to the ξ-representation and becomes $\xi_0^{-n-1}\xi_1^r\xi_2^s\xi_3^t$. Thus a state of the oscillator for which each of its components is in a quantum state corresponds to an expansor with one non-vanishing component. A general state of the oscillator therefore corresponds to a general expansor with integral n-values. A stationary state of the oscillator corresponds to a homogeneous expansor, the degree of the expansor giving the energy of the state with neglect of zero-point energy.

Four-dimensional harmonic oscillators of the above type occur in the theory of the electromagnetic field. Each Fourier component of the field, specified by a particular frequency and a particular direction of motion of the waves, provides one such oscillator, its four components coming from the four electromagnetic potentials. Thus a state of the electromagnetic field in quantum mechanics is described by a number of expansors, one for each Fourier component. By using the electromagnetic equation which gives the value of the divergence of the potentials, one can eliminate in a non-relativistic way the 0 component and one other component of each of the four-dimensional oscillators, so that only two-dimensional oscillators are left. This circumstance has made it possible for people to develop quantum electrodynamics without using expansors.

Another possible application of expansors is to the spins of particles. The wave function describing a particle may be a function of the four co-ordinates x_μ of the particle in space-time and also of the four variables ξ_μ whose coefficients are the components of an expansor. As a simple example of relativistic wave equations for such a particle in the absence of external forces, one may consider

$$\left(\hbar^2\frac{\partial^2}{\partial x_\mu \partial x^\mu}+m^2\right)\psi = 0, \quad \xi_\mu\frac{\partial}{\partial \xi_\mu}\psi = -\psi, \quad \xi_\mu\frac{\partial}{\partial x_\mu}\psi = 0. \tag{33}$$

The first of these is the usual equation for the motion of the particle as a whole. The second shows that at each point in space-time the wave function ψ is homogeneous in the ξ's of degree -1. The third shows that the state for which the momentum-energy four-vector of the particle has the value p_μ is represented by the wave function

$$\psi = \frac{m}{\xi_0 p_0 - (\xi \mathbf{p})} e^{-i[p_0x_0-(\mathbf{p}\mathbf{x})]/\hbar}$$

in three-dimensional vector notation. This may be expanded as

$$\psi = m e^{-i[p_0x_0-(\mathbf{p}\mathbf{x})]/\hbar}\sum_{n=0}^{\infty}\frac{(\xi\mathbf{p})^n}{\xi_0^{n+1}p_0^{n+1}}. \tag{34}$$

For the state for which the particle is at rest, $p_1 = p_2 = p_3 = 0$, $p_0 = m$, and ψ reduces to

$$\psi = e^{-imx_0/\hbar}\xi_0^{-1}.$$

This ψ is spherically symmetrical, showing that when the particle is at rest it has no spin. But when the particle is moving, it is represented by the general ψ (34) and has a finite probability of a non-zero spin. In fact, taking for simplicity $p_2 = p_3 = 0$, the particle has a probability $(mp_1^n/p_0^{n+1})^2$ of being in a state of spin corresponding to the transformations of ξ_1^n under three-dimensional rotations.

This example shows there is a possibility of a particle having no spin when at rest but acquiring a spin when moving, a state of affairs which was not allowed by previous theory. It is desirable that the new spin possibilities opened up by the present theory should be investigated to see whether they could in some cases give an improved description of Nature. The present theory of expansors applies, of course, only to integral spins, but probably it will be possible to set up a corresponding theory of two-valued representations of the Lorentz group, which will apply to half odd integral spins.

Appendix

The rules (5), (11) for forming scalar products are not always convenient for direct use. There are various ways of transforming them and making them more suitable for practical application. One such way has been given (Dirac 1942, equation 3·22) for the case of a single ξ with ascending power series. Another way, applicable to the case of homogeneous functions of ξ_1, ξ_2, ξ_3, is provided by the following.

By partial integration with respect to ξ', one gets, for $m > 0$,

$$\iint_{-\infty}^{\infty} \xi^m \xi'^n e^{i\xi\xi'} d\xi d\xi' = \int_{-\infty}^{\infty} \xi^{m-1} d\xi \left\{ \left[-i\xi'^n e^{i\xi\xi'} \right]_{\xi'=-\infty}^{\xi'=\infty} + in \int_{-\infty}^{\infty} \xi'^{n-1} e^{i\xi\xi'} d\xi' \right\}.$$

If the integrals are made precise in the sense of Cesaro, which means neglecting oscillating terms like $\xi'^n e^{i\xi\xi'}$ for ξ' infinite, this gives

$$\iint_{-\infty}^{\infty} \xi^m \xi'^n e^{i\xi\xi'} d\xi d\xi' = in \iint_{-\infty}^{\infty} \xi^{m-1} \xi'^{n-1} e^{i\xi\xi'} d\xi d\xi'.$$

Taking $m \geqslant n$ and applying the partial integration process n times, one gets

$$\begin{aligned} \iint_{-\infty}^{\infty} \xi^m \xi'^n e^{i\xi\xi'} d\xi d\xi' &= i^n n! \iint_{-\infty}^{\infty} \xi^{m-n} e^{i\xi\xi'} d\xi d\xi' \\ &= 2\pi i^n n! \int_{-\infty}^{\infty} \xi^{m-n} \delta(\xi) d\xi \\ &= 2\pi i^n n! \delta_{mn}. \end{aligned} \tag{35}$$

It follows that if A and B are homogeneous functions of ξ_1, ξ_2, ξ_3 of degree u, their scalar product according to (5) is

$$(AB) = (2\pi)^{-3} i^{-u} \iint_{\cdots\cdots} A(\xi_1 \xi_2 \xi_3) B(\xi_1' \xi_2' \xi_3') e^{i(\xi_1\xi_1' + \xi_2\xi_2' + \xi_3\xi_3')} d\xi_1 d\xi_1' d\xi_2 d\xi_2' d\xi_3 d\xi_3'. \tag{36}$$

As an application of this rule, take

$$A = (\xi_1^2+\xi_2^2+\xi_3^2)^r S_{u-2r}, \quad B = (\xi_1^2+\xi_2^2+\xi_3^2)^s S_{u-2s},$$

where the S's are solid harmonic functions. Then, using three-dimensional vector notation, (36) gives

$$(AB) = (2\pi)^{-3} i^{-u} \iint_{\cdots\cdots} \boldsymbol{\xi}^{2r}\boldsymbol{\xi}'^{2s} S_{u-2r}(\xi)\, S_{u-2s}(\xi')\, e^{i(\xi\xi')}\, d\xi_1 \ldots d\xi_3'. \tag{37}$$

From Green's theorem

$$\iiint\left[e^{i(\xi\xi')}\left(\frac{\partial^2}{\partial\xi_1^2}+\frac{\partial^2}{\partial\xi_2^2}+\frac{\partial^2}{\partial\xi_3^2}\right)\left\{\boldsymbol{\xi}^{2r}S_{u-2r}(\xi)\right\} - \boldsymbol{\xi}^{2r}S_{u-2r}(\xi)\left(\frac{\partial^2}{\partial\xi_1^2}+\frac{\partial^2}{\partial\xi_2^2}+\frac{\partial^2}{\partial\xi_3^2}\right)e^{i(\xi\xi')}\right]d\xi_1 d\xi_2 d\xi_3$$

equals a surface integral of an oscillating kind which is to be counted as vanishing at infinity. This result reduces to

$$4r(u-r+\tfrac{1}{2})\iiint_{-\infty}^{\infty} \boldsymbol{\xi}^{2(r-1)}S_{u-2r}(\xi)\, e^{i(\xi\xi')}\, d\xi_1 d\xi_2 d\xi_3 + \boldsymbol{\xi}'^2\iiint_{-\infty}^{\infty} \boldsymbol{\xi}^{2r}S_{u-2r}(\xi)\, e^{i(\xi\xi')}\, d\xi_1 d\xi_2 d\xi_3 = 0.$$

so (37) becomes, for $r, s > 0$,

$$(AB) = (2\pi)^{-3} i^{-u+2}\, 4r(u-r+\tfrac{1}{2}) \iint_{\cdots\cdots} \boldsymbol{\xi}^{2(r-1)}\boldsymbol{\xi}'^{2(s-1)} S_{u-2r}(\xi)\, S_{u-2s}(\xi')\, e^{i(\xi\xi')}\, d\xi_1 \ldots d\xi_3'. \tag{38}$$

Now suppose $s \geqslant r$ and apply the procedure by which (37) was changed to (38) r times. The result is

$$(AB) = (2\pi)^{-3} i^{-u+2r}\, 4^r r!(u-r+\tfrac{1}{2})!\,(u-2r+\tfrac{1}{2})!^{-1}$$
$$\times \iint_{\cdots\cdots} \boldsymbol{\xi}'^{2(s-r)} S_{u-2r}(\xi)\, S_{u-2s}(\xi')\, e^{i(\xi\xi')}\, d\xi_1 \ldots d\xi_3'. \tag{39}$$

where $n!$ means $\Gamma(n+1)$ for n not an integer. If $s > r$, the procedure can be applied once more, and then shows that

$$(AB) = 0 \quad \text{for} \quad r \neq s. \tag{40}$$

If $s = r$, (39) shows that $\quad (AB) = c4^r r!(u-r+\tfrac{1}{2})!,$ (41)

where c depends only on $u-2r$ and on the two S functions.

Reference

Dirac, P. A. M. 1942 *Proc. Roy. Soc.* A, **180**, 1–40.

PHYSICAL REVIEW VOLUME 133, NUMBER 5B 9 MARCH 1964

Feynman Rules for Any Spin*

STEVEN WEINBERG†
Department of Physics, University of California, Berkeley, California
(Received 21 October 1963)

The explicit Feynman rules are given for massive particles of any spin j, in both a $2j+1$-component and a $2(2j+1)$-component formalism. The propagators involve matrices which transform like symmetric traceless tensors of rank $2j$; they are the natural generalizations of the 2×2 four-vector σ^μ and 4×4 four-vector γ^μ for $j=\frac{1}{2}$. Our calculation uses field theory, but only as a convenient instrument for the construction of a Lorentz-invariant S matrix. This approach is also used to prove the spin-statistics theorem, crossing symmetry, and to discuss T, C, and P.

I. INTRODUCTION

THIS article will develop the relativistic theory of higher spin, from a point of view midway between that of the classic Lagrangian field theories and the more recent S-matrix approach. Our chief aim is to present the explicit Feynman rules for perturbation calculations, in a formalism that varies as little as possible from one spin to another. Such a formalism should be useful if we are to treat particles like the 3–3 resonance as if they were elementary, and is perhaps inindispensable if we are ever to construct a relativistic perturbation theory of Regge poles.

Our treatment[1] is based on three chief assumptions.

(*1*) *Perturbation theory.* We assume that the S matrix can be calculated from Dyson's formula:

$$S=\sum_{n=0}^{\infty}\frac{(-i)^n}{n!}\int_{-\infty}^{\infty}dt_1\cdots dt_n T\{H'(t_1)\cdots H'(t_n)\}. \quad (1.1)$$

Here we have split the Hamiltonian H into a free-particle part H_0 and an interaction H', and define $H'(t)$ as the interaction in the interaction representation:

$$H'(t)\equiv\exp(iH_0t)H'\exp(-iH_0t). \quad (1.2)$$

(*2*) *Lorentz invariance of the S matrix.* We require that S be invariant under proper orthochronous Lorentz transformations. This certainly imposes a much stronger restriction on H_0 and H' than that they just transform like energies. A sufficient and probably necessary condition for the invariance of S is:

$$H'(t)=\int d^3x\mathcal{H}(\mathbf{x},t), \quad (1.3)$$

where:

(a) $\mathcal{H}(x)$ is a scalar. That is, to every inhomogeneous Lorentz transformation $x^\mu\to\Lambda^\mu{}_\nu x^\nu+a^\mu$ there corresponds a unitary operator $U[\Lambda,a]$ such that

$$U[\Lambda,a]\mathcal{H}(x)U^{-1}[\Lambda,a]=\mathcal{H}(\Lambda x+a). \quad (1.4)$$

(b) For $(x-y)$ spacelike,

$$[\mathcal{H}(x),\mathcal{H}(y)]=0. \quad (1.5)$$

The necessity of (a) is rather obvious if we use (1.3) to rewrite (1.1) as

$$S=\sum_{n=0}^{\infty}\frac{(-i)^n}{n!}\int d^4x_1\cdots d^4x_n T\{\mathcal{H}(x_1)\cdots\mathcal{H}(x_n)\}. \quad (1.6)$$

But (a) is certainly not sufficient, because the θ functions $\theta(x_i-x_j)$ implicit in the definition of the time-ordered product are not scalars unless their argument is timelike or lightlike. Condition (b) guarantees that no θ ever appears with a spacelike argument.

(*3*) *Particle interpretation.* We require that $\mathcal{H}(x)$ be constructed out of the creation and annihilation operators for the free particles described by H_0. The only known way of making sure that such an $\mathcal{H}(x)$ will satisfy the restrictions 2(a) and 2(b), is to form it as a function of one or more fields $\psi_n(x)$, which are linear combinations of the creation and annihilation operators, and which have the properties:

(a) The fields transform according to

$$U[\Lambda,a]\psi_n(x)U^{-1}[\Lambda,a]=\sum_m D_{nm}[\Lambda^{-1}]\psi_m(\Lambda x+a), \quad (1.7)$$

where $D_{nm}[\Lambda]$ is some representation of Λ.

(b) For $(x-y)$ spacelike

$$[\psi_n(x),\psi_m(y)]_\pm=0, \quad (1.8)$$

where $[\ ]_\pm$ may be either a commutator or anticommutator. Condition 3(a) enables us to satisfy 2(a) by coupling the $\psi_n(x)$ in various invariant combinations, while 3(b) guarantees the validity of 2(b), provided that $\mathcal{H}(x)$ contains an even number of fermion field factors. (There are some fine points about the case $x=y$ which will be discussed in Sec. V.)

Equations (1.7) and (1.8) will dictate how the fields are to be constructed. We have not pretended to derive these equations as inescapable consequences of assumptions (1)–(3), but our discussion suggests strongly that they can be understood as necessary to the Lorentz

* Research supported in part by the U. S. Air Force Office of Scientific Research, Grant No. AF–AFOSR-232–63.
† Alfred P. Sloan Foundation Fellow.

[1] I have recently learned that a similar approach is used by E. H. Wichmann in the manuscript of his forthcoming book in quantum field theory.

Reprinted from *Phys. Rev.* **133**, B1318 (1964).

invariance of the S matrix, without any recourse to separate postulates of causality or analyticity.[2]

Nowhere have we mentioned field equations or Lagrangians, for they will not be needed. In fact, our refusal to get enmeshed in the canonical formalism has a number of important physical (and pedagogical) advantages:

(1) We are able to use a $2j+1$-component field for a massive particle of spin j. This is often thought to be impossible, because such fields do not satisfy any free-field equations (besides the Klein-Gordon equation). The absence of field equations is irrelevant in our approach, because the fields do satisfy (1.7) and (1.8); a free-field equation is nothing but an invariant record of which components are superfluous.

The $2j+1$-component fields are ideally suited to weak interaction theory, because they transform simply under **T** and **CP** but not under **C** or **P**. In order to discuss theories with parity conservation it is convenient to use $2(2j+1)$-component fields, like the Dirac field. These do obey field equations, which can be *derived* as incidental consequences of (1.7) and (1.8).

(2) Schwinger[3] has noticed a serious difficulty in the quantization of theories of spin $j\geqq\frac{3}{2}$ by the canonical method. This can be taken to imply either that particles with $j\geqq\frac{3}{2}$ cannot be elementary, or it might be interpreted as a shortcoming of the Lagrangian approach.

(3) Pauli's proof[4] of the connection between spin and statistics is straightforward for integer j, but rather indirect for half-integer j. We take the particle interpretation of $\psi_n(x)$ as an assumption, and are able to show almost trivially that (1.8) makes sense only with the usual choice between commutation and anticommutation relations. Crossing symmetry arises in the same way.

(4) By avoiding the principle of least action, we are able to remain somewhat closer throughout our development of field theory to ideas of obvious physical significance.

At any rate the ambiguity in choosing $\mathcal{H}(x)$ is no worse than for $\mathcal{L}(x)$. The one place where the Lagrangian approach does suggest a specific interaction is in the theory of massless particles like the photon and graviton. Our work in this paper will be restricted to massive particles, but we shall come back to this point in a later article.

The transformation properties of states, creation and annihilation operators, and fields are reviewed in Sec. II. The $2j+1$-component field is constructed in Sec. III so that it satisfies the transformation rule (1.7). The "causality" requirement (1.8) is invoked in Sec. IV, yielding the spin-statistics connection and crossing symmetry. Section V is devoted to a statement of the Feynman rules. The inversions **T**, **C**, and **P** are studied in Sec. VI. They suggest the use of a $2(2j+1)$-component field whose propagator is calculated in Sec. VII. More general fields are considered briefly in Sec. VIII. The propagator for $2j+1$- and $2(2j+1)$-component fields involves a set of matrices which transform like symmetric traceless tensors of rank $2j$, and which form the natural generalizations of the 2×2 vector $\{\boldsymbol{\sigma},1\}$ and the 4×4 vector γ_μ, respectively. These matrices are discussed in two appendices, where we also derive the general formulas for a spin j propagator. The $2j+1\times 2j+1$ propagators for spin $j\leqq3$ are listed in Table I, and the $2(2j+1)\times2(2j+1)$ propagators for $j\leqq2$ are listed in Table II.

This article treats a quantum field as a mere artifice to be used in the construction of an invariant S matrix. It is therefore not unlikely that most of the work presented here could be translated into the language of pure S-matrix theory, with unitarity replacing our assumptions (1) and (3).

TABLE I. The scalar matrix $\Pi(q)=(-)^{2j}t^{\mu_1\mu_2\cdots}q_{\mu_1}q_{\mu_2}\cdots$ for spins $j\leqq3$. In each case $\mathbf{J}$ is the usual $2j+1$-dimensional matrix representation of the angular momentum. The propagator for a particle of spin j is $S(q)=-i(-im)^{-2j}\Pi(q)/q^2+m^2-i\epsilon$.

$$\Pi^{(0)}(q)=1$$
$$\Pi^{(1/2)}(q)=q^0-2(\mathbf{q}\cdot\mathbf{J})$$
$$\Pi^{(1)}(q)=-q^2+2(\mathbf{q}\cdot\mathbf{J})(\mathbf{q}\cdot\mathbf{J}-q^0)$$
$$\Pi^{(3/2)}(q)=-q^2(q^0-2\mathbf{q}\cdot\mathbf{J})+\tfrac{1}{6}[(2\mathbf{q}\cdot\mathbf{J})^2-\mathbf{q}^2][3q^0-2\mathbf{q}\cdot\mathbf{J}]$$
$$\Pi^{(2)}(q)=(-q^2)^2-2q^2(\mathbf{q}\cdot\mathbf{J})(\mathbf{q}\cdot\mathbf{J}-q^0)+\tfrac{2}{3}(\mathbf{q}\cdot\mathbf{J})[(\mathbf{q}\cdot\mathbf{J})^2-\mathbf{q}^2][\mathbf{q}\cdot\mathbf{J}-2q^0]$$
$$\Pi^{(5/2)}(q)=(-q^2)^2(q^0-2\mathbf{q}\cdot\mathbf{J})-\tfrac{1}{6}q^2[(2\mathbf{q}\cdot\mathbf{J})^2-\mathbf{q}^2][3q^0-2\mathbf{q}\cdot\mathbf{J}]+\frac{1}{120}[(2\mathbf{q}\cdot\mathbf{J})^2-\mathbf{q}^2][(2\mathbf{q}\cdot\mathbf{J})^2-9\mathbf{q}^2][5q^0-2\mathbf{q}\cdot\mathbf{J}]$$
$$\Pi^{(3)}(q)=(-q^2)^3+2(-q^2)(\mathbf{q}\cdot\mathbf{J})(\mathbf{q}\cdot\mathbf{J}-q^0)-\tfrac{2}{3}q^2(\mathbf{q}\cdot\mathbf{J})[(\mathbf{q}\cdot\mathbf{J})^2-\mathbf{q}^2][\mathbf{q}\cdot\mathbf{J}-2q^0]+\frac{4}{45}(\mathbf{q}\cdot\mathbf{J})[(\mathbf{q}\cdot\mathbf{J})^2-\mathbf{q}^2][(\mathbf{q}\cdot\mathbf{J})^2-4\mathbf{q}^2][\mathbf{q}\cdot\mathbf{J}-3q^0]$$

TABLE II. The scalar matrix $\mathcal{P}(q)=-i^{2j}\gamma^{\mu_1\mu_2\cdots\mu_{2j}}q_{\mu_1}q_{\mu_2}\cdots q_{\mu_{2j}}$ for spins $j\leqq2$. In each case

$$\mathcal{J}=\begin{bmatrix}\mathbf{J}^{(j)} & 0\\ 0 & \mathbf{J}^{(j)}\end{bmatrix}.$$

The propagator for a particle of spin j is

$$S(q)=-im^{-2j}[\mathcal{P}(q)+m^{2j}]/q^2+m^2-i\epsilon.$$

$$\mathcal{P}^{(0)}(q)=1$$
$$\mathcal{P}^{(1/2)}(q)=q^0\beta-2(\mathbf{q}\cdot\mathcal{J})\gamma_5\beta$$
$$\mathcal{P}^{(1)}(q)=-q^2\beta+2(\mathbf{q}\cdot\mathcal{J})(\mathbf{q}\cdot\mathcal{J}\beta-q^0\gamma_5\beta)$$
$$\mathcal{P}^{(3/2)}(q)=-q^2(q^0\beta-2\mathbf{q}\cdot\mathcal{J}\gamma_5\beta)+\tfrac{1}{6}[(2\mathbf{q}\cdot\mathcal{J})^2-\mathbf{q}^2][3q^0\beta-2\mathbf{q}\cdot\mathcal{J}\gamma_5\beta]$$
$$\mathcal{P}^{(2)}(q)=(-q^2)^2\beta-2q^2(\mathbf{q}\cdot\mathcal{J})[\mathbf{q}\cdot\mathcal{J}\beta-q^0\gamma_5\beta]+\tfrac{2}{3}(\mathbf{q}\cdot\mathcal{J})[(\mathbf{q}\cdot\mathcal{J})^2-\mathbf{q}^2][\mathbf{q}\cdot\mathcal{J}\beta-2q^0\gamma_5\beta]$$

[2] In this connection, it is very interesting that a Hamiltonian without particle creation and annihilation can yield a Lorentz-invariant S matrix, but *not* if we use perturbation theory. See R. Fong and J. Sucher, University of Maryland (to be published).

[3] J. Schwinger, Phys. Rev. **130**, 800 (1963).

[4] W. Pauli, Phys. Rev. **58**, 716 (1940).

II. LORENTZ TRANSFORMATIONS

In our noncanonical approach it is essential to begin with a description of the Lorentz transformation properties of free-particle states, or equivalently, of creation and annihilation operators. The transformation rules are simple and unambiguous, and have been well understood for many years,[5] but it will be useful to review them once again here.

The proper homogeneous orthochronous Lorentz transformations are defined by

$$x^\mu \to \Lambda^\mu{}_\nu x^\nu,$$
$$g_{\mu\nu}\Lambda^\mu{}_\lambda\Lambda^\nu{}_\rho = g_{\lambda\rho}, \tag{2.1}$$
$$\det\Lambda = 1; \quad \Lambda^0{}_0 > 0.$$

These will be referred to simply as "Lorentz transformations" from now on. Our metric is

$$g_{ij} = \delta_{ij}; \quad g_{00} = -1; \quad g_{i0} = g_{0i} = 0. \tag{2.2}$$

To each Λ there corresponds a unitary operator $U[\Lambda]$, which acts on the Hilbert space of physical states, and has the group property

$$U[\Lambda_2]U[\Lambda_1] = U[\Lambda_2\Lambda_1]. \tag{2.3}$$

Of particular importance for us is the "boost" $L(\mathbf{p})$, which takes a particle of mass m from rest to momentum $\mathbf{p}$:

$$L^i{}_j(\mathbf{p}) = \delta_{ij} + \hat{p}_i\hat{p}_j[\cosh\theta - 1],$$
$$L^i{}_0(\mathbf{p}) = L^0{}_i(\mathbf{p}) = \hat{p}_i \sinh\theta, \tag{2.4}$$
$$L^0{}_0(\mathbf{p}) = \cosh\theta.$$

Here $\hat{p}$ is the unit vector $\mathbf{p}/|\mathbf{p}|$, and

$$\sinh\theta = |\mathbf{p}|/m, \quad \cosh\theta = \omega/m = [\mathbf{p}^2 + m^2]^{1/2}/m. \tag{2.5}$$

Strictly speaking, this should be called $L(\mathbf{p}/m)$.

We can use $L(\mathbf{p})$ to define the one-particle state of momentum $\mathbf{p}$, mass m, spin j, and z-component of spin σ $(\sigma = j, j-1, \cdots, -j)$ by

$$|\mathbf{p},\sigma\rangle = [m/\omega(\mathbf{p})]^{1/2} U[L(\mathbf{p})]|\sigma\rangle, \tag{2.6}$$

where $|\sigma\rangle$ is the state of the particle at rest with $J_z = \sigma$. Our normalization is conventional,

$$\langle \mathbf{p},\sigma|\mathbf{p}',\sigma'\rangle = \delta^3(\mathbf{p}-\mathbf{p}')\delta_{\sigma\sigma'}. \tag{2.7}$$

The effect of an arbitrary Lorentz transformation $\Lambda^\mu{}_\nu$ on these one-particle states is

$$\begin{aligned} U[\Lambda]|\mathbf{p},\sigma\rangle &= [m/\omega(\mathbf{p})]^{1/2} U[\Lambda]U[L(\mathbf{p})]|\sigma\rangle \\ &= [m/\omega(\mathbf{p})]^{1/2} U[L(\Lambda\mathbf{p})]U[L^{-1}(\Lambda\mathbf{p})\Lambda L(\mathbf{p})]|\sigma\rangle \\ &= [m/\omega(\mathbf{p})]^{1/2} \textstyle\sum_{\sigma'} U[L(\Lambda\mathbf{p})]|\sigma'\rangle \\ &\qquad \times \langle\sigma'|U[L^{-1}(\Lambda\mathbf{p})\Lambda L(\mathbf{p})]|\sigma\rangle, \end{aligned}$$

and finally

$$U[\Lambda]|\mathbf{p},\sigma\rangle = [\omega(\Lambda\mathbf{p})/\omega(\mathbf{p})]^{1/2} \textstyle\sum_{\sigma'} |\Lambda\mathbf{p},\sigma'\rangle \times D_{\sigma'\sigma}^{(j)}[L^{-1}(\Lambda\mathbf{p})\Lambda L(\mathbf{p})]. \tag{2.8}$$

The coefficients $D_{\sigma'\sigma}^{(j)}$ are

$$D_{\sigma'\sigma}^{(j)}[R] = \langle\sigma'|U[R]|\sigma\rangle. \tag{2.9}$$

In (2.8), R is the pure rotation $L^{-1}(\Lambda\mathbf{p})\Lambda L(\mathbf{p})$ (the so-called "Wigner rotation") so that $D^{(j)}[R]$ here is nothing but the familiar $2j+1$-dimensional unitary matrix representation[6] of the rotation group.

A general state containing several free particles will transform like (2.8), with a factor $[\omega'/\omega]^{1/2}D$ for each particle. These states can be built up by acting on the bare vacuum with creation operators $a^*(\mathbf{p},\sigma)$ which satisfy either the usual Bose or Fermi rules[7]:

$$[a(\mathbf{p},\sigma), a^*(\mathbf{p}',\sigma')]_\pm = \delta_{\sigma\sigma'}\delta^3(\mathbf{p}-\mathbf{p}'), \tag{2.10}$$

so the general transformation law can be summarized by replacing (2.8) with

$$\begin{aligned} &U[\Lambda]a^*(\mathbf{p},\sigma)U^{-1}[\Lambda] \\ &= [\omega(\Lambda\mathbf{p})/\omega(\mathbf{p})]^{1/2} \textstyle\sum_{\sigma'} D_{\sigma'\sigma}^{(j)}[L^{-1}(\Lambda\mathbf{p})\Lambda L(\mathbf{p})]a^*(\Lambda\mathbf{p},\sigma'). \end{aligned} \tag{2.11}$$

Taking the adjoint and using the unitarity of $D^{(j)}[R]$ gives

$$\begin{aligned} &U[\Lambda]a(\mathbf{p},\sigma)U^{-1}[\Lambda] \\ &= [\omega(\Lambda\mathbf{p})/\omega(\mathbf{p})]^{1/2} \textstyle\sum_{\sigma'} D_{\sigma\sigma'}^{(j)}[L^{-1}(\mathbf{p})\Lambda^{-1}L(\Lambda\mathbf{p})]a(\Lambda\mathbf{p},\sigma'). \end{aligned} \tag{2.12}$$

It will be convenient to rewrite (2.11) in a form closer to that of (2.12). Note that the ordinary complex conjugate of the rotation-representation D is given by a unitary transformation[8]

$$D^{(j)}[R]^* = CD^{(j)}[R]C^{-1}, \tag{2.13}$$

where C is a $2j+1 \times 2j+1$ matrix with

$$C^*C = (-)^{2j}; \quad C^\dagger C = 1. \tag{2.14}$$

[With the usual phase conventions, C can be taken as the matrix

$$C_{\sigma\sigma'} = (-)^{j+\sigma}\delta_{\sigma',-\sigma},$$

but we won't need this here.] Since $D^{(j)}[R]$ is unitary, (2.13) can be written

$$D_{\sigma'\sigma}^{(j)}[R] = \{CD^{(j)}[R^{-1}]C^{-1}\}_{\sigma\sigma'} \tag{2.15}$$

so (2.11) becomes

$$\begin{aligned} &U[\Lambda]a^*(\mathbf{p},\sigma)U^{-1}[\Lambda] \\ &= [\omega(\Lambda\mathbf{p})/\omega(\mathbf{p})]^{1/2} \textstyle\sum_{\sigma'} \{CD^{(j)}[L^{-1}(\mathbf{p})\Lambda^{-1}L(\Lambda\mathbf{p})]C^{-1}\}_{\sigma\sigma'} \\ &\qquad \times a^*(\Lambda\mathbf{p},\sigma'). \end{aligned} \tag{2.16}$$

[5] E. P. Wigner, Ann. Math. **40**, 149 (1939).

[6] See, for example, M. E. Rose, *Elementary Theory of Angular Momentum* (John Wiley and Sons, Inc., New York, 1957), p. 48 ff.

[7] We use an asterisk to denote the adjoint of an operator on the physical Hilbert space, or the ordinary complex conjugate of a c number or a c-number matrix. A dagger is used to indicate the adjoint of a c-number matrix. Other possible statistics than allowed by (2.10) will not be considered here.

[8] Reference 6, Eq. (4.22).

We speak of one particle as being the antiparticle of another if their masses and spins are equal, and all their charges, baryon numbers, etc., are opposite. We won't assume that every particle has an antiparticle, since this is a well-known consequence of field theory, which will be proved from our standpoint in Sec. IV. But if an antiparticle exists then its states will transform like those of the corresponding particle. In particular, the operator $b^*(\mathbf{p},\sigma)$ which creates the antiparticle of the particle destroyed by $a(\mathbf{p},\sigma)$ transforms by the same rule (2.16) as $a^*(\mathbf{p},\sigma)$:

$$U[\Lambda]b^*(\mathbf{p},\sigma)U^{-1}[\Lambda] = [\omega(\Lambda\mathbf{p})/\omega(\mathbf{p})]^{1/2}\sum_{\sigma'}\{CD^{(j)}[L^{-1}(\mathbf{p})\Lambda^{-1}L(\Lambda\mathbf{p})]C^{-1}\}_{\sigma\sigma'} \times b^*(\Lambda\mathbf{p},\sigma'). \quad (2.17)$$

To some extent this is a convention, but it has the advantage of not forcing us to use different notation for purely neutral particles and for particles with distinct antiparticles.

It cannot be stressed too strongly that the transformation rules (2.12) and (2.17) have nothing to do with representations of the homogeneous Lorentz group, but only involve the familiar representations of the ordinary rotation group. If a stranger asks how the spin states of a moving particle with $j=1$ transform under some Lorentz transformation, it is not necessary to ask him whether he is thinking of a four-vector, a skew symmetric tensor, a self-dual skew symmetric tensor, or something else. One need only refer him to (2.16) or (2.8), and hope that he knows the $j=1$ rotation matrices.

The complexities of higher spin enter only when we try to use $a(\mathbf{p},\sigma)$ and $b^*(\mathbf{p},\sigma)$ to construct a *field* which transforms simply under the homogeneous Lorentz group. We will need to use only a little of the classic theory of the representations of this group, but it will be convenient to recall its vocabulary. Any representation is specified by a representation of the infinitesimal Lorentz transformations. These are of the form

$$\Lambda^\mu{}_\nu=\delta^\mu{}_\nu+\omega^\mu{}_\nu, \quad (2.18)$$

where the ω's form an infinitesimal "six-vector"

$$\omega_{\mu\nu}=-\omega_{\nu\mu}. \quad (2.19)$$

The corresponding unitary operators are of the form

$$U[1+\omega]=1+(i/2)J_{\mu\nu}\omega^{\mu\nu}, \quad (2.20)$$

$$J_{\mu\nu}{}^\dagger=J_{\mu\nu}=-J_{\nu\mu}. \quad (2.21)$$

It is very convenient to group the six operators $J_{\mu\nu}$ into two Hermitian three-vectors

$$J_i=\tfrac{1}{2}\epsilon_{ijk}J_{jk}, \quad (2.22)$$

$$K_i=J_{i0}=-J_{0i}. \quad (2.23)$$

It follows from (2.3) that

$$[J_i,J_j]=i\epsilon_{ijk}J_k, \quad (2.24)$$

$$[J_i,K_j]=i\epsilon_{ijk}K_k, \quad (2.25)$$

$$[K_i,K_j]=-i\epsilon_{ijk}J_k. \quad (2.26)$$

The $\mathbf{J}$ generate rotations and the $\mathbf{K}$ generate boosts. In particular, the unitary operator for the finite boost (2.4) is

$$U[L(\mathbf{p})]=\exp(-i\hat{p}\cdot\mathbf{K}\theta). \quad (2.27)$$

The commutation rules (2.24)–(2.26) can be decoupled by defining a new pair of non-Hermitian generators:

$$\mathbf{A}=\tfrac{1}{2}[\mathbf{J}+i\mathbf{K}], \quad (2.28)$$

$$\mathbf{B}=\tfrac{1}{2}[\mathbf{J}-i\mathbf{K}], \quad (2.29)$$

with commutation rules

$$\mathbf{A}\times\mathbf{A}=i\mathbf{A}, \quad (2.30)$$

$$\mathbf{B}\times\mathbf{B}=i\mathbf{B}, \quad (2.31)$$

$$[A_i,B_j]=0. \quad (2.32)$$

The $(2A+1)(2B+1)$-dimensional irreducible representation (A,B) is defined for any integer values of $2A$ and $2B$ by

$$\langle a,b|\mathbf{A}|a',b'\rangle=\delta_{bb'}\mathbf{J}_{aa'}{}^{(A)}, \quad (2.33)$$

$$\langle a,b|\mathbf{B}|a',b'\rangle=\delta_{aa'}\mathbf{J}_{bb'}{}^{(B)}, \quad (2.34)$$

where a and b run by unit steps from $-A$ to $+A$ and from $-B$ to $+B$, respectively, and $\mathbf{J}^{(j)}$ is the usual $2j+1$-dimensional representation of the rotation group:

$$(J_x{}^{(j)}\pm iJ_y{}^{(j)})_{\sigma'\sigma}=\delta_{\sigma',\sigma\pm1}[(j\mp\sigma)(j\pm\sigma+1)]^{1/2}, \quad (J_z{}^{(j)})_{\sigma'\sigma}=\delta_{\sigma'\sigma}\sigma. \quad (2.35)$$

The representations (A,B) exhaust all finite dimensional irreducible representations of the homogeneous Lorentz group. None of them are unitary, except for (0,0).

We will be particularly concerned with the simplest irreducible representations $(j,0)$ and $(0,j)$. These are respectively characterized by

$$\mathbf{J}\to\mathbf{J}^{(j)},\quad \mathbf{K}\to -i\mathbf{J}^{(j)},\quad \text{for } (j,0) \quad (2.36)$$

and

$$\mathbf{J}\to\mathbf{J}^{(j)},\quad \mathbf{K}\to +i\mathbf{J}^{(j)},\quad \text{for } (0,j), \quad (2.37)$$

where $J^{(j)}$ is given as always by (2.35). We denote the $2j+1$-dimensional matrix representing a finite Lorentz transformation Λ by $D^{(j)}[\Lambda]$ and $\bar{D}^{(j)}[\Lambda]$ in the $(j,0)$ and $(0,j)$ representations, respectively. The two representations are related by

$$D^{(j)}[\Lambda]=\bar{D}^{(j)}[\Lambda^{-1}]^\dagger. \quad (2.38)$$

In particular the boost $L(\mathbf{p})$ is represented according to

(2.27) and (2.36) or (2.37) by

$$D^{(j)}[L(\mathbf{p})]=\exp(-\hat{p}\cdot\mathbf{J}^{(j)}\theta), \quad (2.39)$$

$$\bar{D}^{(j)}[L(\mathbf{p})]=\exp(+\hat{p}\cdot\mathbf{J}^{(j)}\theta), \quad (2.40)$$

with $\sinh\theta\equiv|\mathbf{p}|/m$. For pure rotations both $D^{(j)}[R]$ and $\bar{D}^{(j)}[R]$ reduce to the usual rotation matrices.

III. $2j+1$-COMPONENT FIELDS

We want to form the free field by taking linear combinations of creation and annihilation operators. The transformation property under translations required by (1.7) forces us to do this by setting the field equal to some sort of Fourier transform of these operators. But (2.12) and (2.17) show that each $a(\mathbf{p},\sigma)$ and $b^*(\mathbf{p},\sigma)$ behaves under Lorentz transformations in a way that depends on the individual momentum $\mathbf{p}$, so that the ordinary Fourier transform would not have a covariant character. In order to construct fields with simple transformation properties, it is necessary to extend $D^{(j)}[R]$ to a representation of the homogeneous Lorentz group, so that the $\mathbf{p}$-dependent factors in (2.12) and (2.17) can be grouped[9] with the $a(\mathbf{p},\sigma)$ and $b^*(\mathbf{p},\sigma)$. There are as many ways of doing this as there are representations of the Lorentz group, but for the present we shall use the $(j,0)$ representation defined by (2.36) and (2.35). [The $(0,j)$ representation will be considered in Sec. VI, the $(j,0)\oplus(0,j)$ in Sec. VII, and the general case in Sec. VIII.]

Having extended the definition of the $2j+1\times2j+1$ matrix $D^{(j)}$ in this way, we can split the rotation matrix appearing in (2.12) and (2.17) into three factors

$$D^{(j)}[L^{-1}(\mathbf{p})\Lambda^{-1}L(\Lambda\mathbf{p})] =D^{(j)-1}[L(\mathbf{p})]D^{(j)}[\Lambda^{-1}]D^{(j)}[L(\Lambda\mathbf{p})]. \quad (3.1)$$

This allows us to write (2.12) and (2.17) as[9]

$$U[\Lambda]\alpha(\mathbf{p},\sigma)U^{-1}[\Lambda]=\sum_{\sigma'}D_{\sigma\sigma'}{}^{(j)}[\Lambda^{-1}]\alpha(\Lambda\mathbf{p},\sigma'), \quad (3.2)$$

$$U[\Lambda]\beta(\mathbf{p},\sigma)U^{-1}[\Lambda]=\sum_{\sigma'}D_{\sigma\sigma'}{}^{(j)}[\Lambda^{-1}]\beta(\Lambda\mathbf{p},\sigma'), \quad (3.3)$$

with

$$\alpha(\mathbf{p},\sigma)\equiv[2\omega(\mathbf{p})]^{1/2}\sum_{\sigma'}D_{\sigma\sigma'}{}^{(j)}[L(\mathbf{p})]a(\mathbf{p},\sigma'), \quad (3.4)$$

$$\beta(\mathbf{p},\sigma)\equiv[2\omega(\mathbf{p})]^{1/2}\sum_{\sigma'}\{D^{(j)}[L(\mathbf{p})]C^{-1}\}_{\sigma\sigma'}b^*(\mathbf{p},\sigma'). \quad (3.5)$$

The operators α and β transform simply, so the field can be constructed now by a Lorentz invariant Fourier transform

$$\varphi_\sigma(x)=(2\pi)^{-3/2}\int\frac{d^3\mathbf{p}}{2\omega(\mathbf{p})} \times[\xi\alpha(\mathbf{p},\sigma)e^{ip\cdot x}+\eta\beta(\mathbf{p},\sigma)e^{-ip\cdot x}], \quad (3.6)$$

with constants ξ and η to be determined in the next section. It is clear that this is the most general linear combination of the a's and the b^*'s which has the simple Lorentz transformation property

$$U[\Lambda,a]\varphi_\sigma(x)U^{-1}[\Lambda,a] =\sum_{\sigma'}D_{\sigma\sigma'}{}^{(j)}[\Lambda^{-1}]\varphi_{\sigma'}(\Lambda x+a). \quad (3.7)$$

[We choose to combine a and b^*, so that $\varphi_\sigma(x)$ also behaves simply under gauge transformations.]

In terms of the original creation and annihilation operators, the field is

$$\varphi_\sigma(x)=(2\pi)^{-3/2}\int\frac{d^3\mathbf{p}}{[2\omega(\mathbf{p})]^{1/2}} \times\sum_{\sigma'}[\xi D_{\sigma\sigma'}{}^{(j)}[L(\mathbf{p})]a(\mathbf{p},\sigma')e^{ip\cdot x} +\eta\{D^{(j)}[L(\mathbf{p})]C^{-1}\}_{\sigma\sigma'}b^*(\mathbf{p},\sigma')e^{-ip\cdot x}]. \quad (3.8)$$

We have already derived a formula [Eq. (2.39)] for the wave function appearing in (3.8):

$$D_{\sigma\sigma'}{}^{(j)}[L(\mathbf{p})]=\{\exp(-\hat{p}\cdot\mathbf{J}^{(j)}\theta)\}_{\sigma\sigma'}.$$

The field obeys the Klein-Gordon equation

$$(\Box^2-m^2)\varphi_\sigma(x)=0, \quad (3.9)$$

but it does not obey any other field equations. As discussed in the introduction, we consider this to be a distinct advantage of the $(j,0)$ representation, because any field equation [except (3.9)] is nothing but a confession that the field contains superfluous components.

If a particle has no antiparticle (including itself) then we have to set $\eta=0$ in (3.6) and (3.8). In the other extreme, a theory with full crossing symmetry would have $|\eta|=|\xi|$. We will now show that the choice of ξ and η is dictated by requirement (1.8), and hence essentially by the Lorentz invariance of the S matrix.

IV. CROSSING AND STATISTICS

We are assuming, on the basis of their particle interpretation, that the a's and b's satisfy either the usual Bose commutation *or* Fermi anticommutation rules:

$$\begin{aligned}[a(\mathbf{p},\sigma),a^*(\mathbf{p},\sigma')]_\pm&=\delta(\mathbf{p}-\mathbf{p}')\delta_{\sigma\sigma'},\\ [b(\mathbf{p},\sigma),b^*(\mathbf{p},\sigma')]_\pm&=\delta(\mathbf{p}-\mathbf{p}')\delta_{\sigma\sigma'},\end{aligned} \quad (4.1)$$

with all others vanishing. It is then easy to work out the commutation or anticommutation rule for the field defined by (3.8):

$$[\varphi_\sigma(x),\varphi_{\sigma'}{}^\dagger(y)]_\pm =\frac{m^{-2j}}{(2\pi)^3}\int\frac{d^3\mathbf{p}}{2\omega(\mathbf{p})}\Pi_{\sigma\sigma'}{}^{(j)}(\mathbf{p},\omega(\mathbf{p})) \times\{|\xi|^2\exp[ip\cdot(x-y)]\pm|\eta|^2\exp[-ip\cdot(x-y)]\}, \quad (4.2)$$

[9] This step corresponds to Stapp's replacement of the S matrix by the "M-functions." See H. Stapp, Phys. Rev. **125**, 2139 (1962) for $j=\frac{1}{2}$; and A. O. Barut, I. Muzinich, D. N. Williams, Phys. Rev. **130**, 442 (1963) for general j.

where the matrix $\Pi(p)$ is given by

$$m^{-2j}\Pi(\mathbf{p},\omega)=D^{(j)}[L(\mathbf{p})]D^{(j)}[L(\mathbf{p})]^{\dagger} \quad (4.3)$$

$$=\exp(-2\hat{p}\cdot\mathbf{J}\theta) \quad (4.4)$$

$$[\cosh\theta=p^0/m=\omega(\mathbf{p})/m].$$

This matrix is evaluated explicitly in the Appendix and given for $j\leqq 3$ in Table I. For our present purposes, the important point is that

$$\Pi_{\sigma\sigma'}(p)=(-)^{2j}t_{\sigma\sigma'}{}^{\mu_1\mu_2\cdots\mu_{2j}}p_{\mu_1}p_{\mu_2}\cdots p_{\mu_{2j}}, \quad (4.5)$$

where t is a constant symmetric traceless tensor. It follows then from (4.2) that

$$[\varphi_\sigma(x),\varphi_{\sigma'}{}^\dagger(y)]_\pm$$
$$=(2\pi)^{-3}(-im)^{-2j}t_{\sigma\sigma'}{}^{\mu_1\mu_2\cdots\mu_{2j}}\partial_{\mu_1}\partial_{\mu_2}\cdots\partial_{\mu_{2j}}$$
$$\times\int\frac{d^3\mathbf{p}}{2\omega(\mathbf{p})}\{|\xi|^2\exp[ip\cdot(x-y)]$$
$$\pm(-)^{2j}|\eta|^2\exp[-ip\cdot(x-y)]\}. \quad (4.6)$$

It is well known[4] that such an integral will vanish outside the light-cone if, and only if, the coefficients of $\exp[ip\cdot(x-y)]$ and $\exp[-ip\cdot(x-y)]$ are equal and opposite, i.e.,

$$|\xi|^2=\mp(-)^{2j}|\eta|^2. \quad (4.7)$$

Thus the requirement of causality leads immediately to the two most important consequences of field theory:

(a) *Statistics:* Eq. (4.7) makes sense only if

$$\mp(-)^{2j}=1, \quad (4.8)$$

so a particle with integer spin must be a boson, with a (−) sign in (4.1), while a particle with half-integer spin must be a fermion, with a (+) sign in (4.1).[10]

(b) *Crossing:* Eq. (4.7) also requires that

$$|\xi|=|\eta|. \quad (4.9)$$

Thus every particle must have an antiparticle (perhaps itself) which enters into interactions with equal coupling strength. There is no reason why we cannot redefine the *phase* of $a(\mathbf{p},\sigma)$ and $b^*(\mathbf{p},\sigma)$ and the phase and normalization of $\varphi_\sigma(x)$ as we like, so Eq. (4.9) allows us to take

$$\xi=\eta=1 \quad (4.10)$$

without any loss of generality.

[10] As a demonstration that the causality requirement cannot be satisfied with the wrong statistics, this is certainly inferior to the more modern proof of P. N. Burgoyne, Nuovo Cimento **8**, 607 (1958). Our purpose in this section is to show that causality can be satisfied, but only with the right statistics and with crossing symmetry.

The field is now in its final form:

$$\varphi_\sigma(x)=(2\pi)^{-3/2}\int\frac{d^3\mathbf{p}}{[2\omega(\mathbf{p})]^{1/2}}$$
$$\times\sum_{\sigma'}[D_{\sigma\sigma'}{}^{(j)}[L(\mathbf{p})]a(\mathbf{p},\sigma')e^{ip\cdot x}$$
$$+\{D^{(j)}[L(\mathbf{p})]C^{-1}\}_{\sigma\sigma'}b^*(\mathbf{p},\sigma')e^{-ip\cdot x}]. \quad (4.11)$$

The commutator or anticommutator is

$$[\varphi_\sigma(x),\varphi_{\sigma'}{}^\dagger(y)]_\pm$$
$$=i(-im)^{-2j}t_{\sigma\sigma'}{}^{\mu_1\mu_2\cdots\mu_{2j}}\partial_{\mu_1}\partial_{\mu_2}\cdots\partial_{\mu_{2j}}\Delta(x-y), \quad (4.12)$$

where Δ is the usual causal function

$$\Delta(x)=\frac{-i}{(2\pi)^3}\int\frac{d^3\mathbf{p}}{2\omega(\mathbf{p})}[e^{ip\cdot(x-y)}-e^{-ip\cdot(x-y)}]. \quad (4.13)$$

V. THE FEYNMAN RULES

Suppose now that the interaction Hamiltonian is given as some invariant polynomial in the $\varphi_\sigma(x)$ and their adjoints. For example, the only possible nonderivative interaction among three particles of spin j_1, j_2, and j_3 would be

$$\mathcal{H}(x)=g\sum_{\sigma_1\sigma_2\sigma_3}\begin{pmatrix}j_1 & j_2 & j_3\\ \sigma_1 & \sigma_2 & \sigma_3\end{pmatrix}$$
$$\times\varphi_{\sigma_1}{}^{(1)}(x)\varphi_{\sigma_2}{}^{(2)}(x)\varphi_{\sigma_3}{}^{(3)}(x)+\text{H.c.}, \quad (5.1)$$

the "vertex function" being given here by the usual $3j$ symbol.

The S matrix can be calculated from (1.1) by using Wick's theorem as usual to derive the Feynman rules:

(a) For each vertex include a factor $(-i)$ times whatever coefficients appear with the fields in $\mathcal{H}(x)$. For example, each vertex arising from (5.1) will contribute a factor

$$-ig\begin{pmatrix}j_1 & j_2 & j_3\\ \sigma_1 & \sigma_2 & \sigma_3\end{pmatrix}. \quad (5.2)$$

(b) For each internal line running from a vertex at x to a vertex at y include a propagator

$$\langle T\{\varphi_\sigma(x)\varphi_{\sigma'}{}^\dagger(y)\}\rangle_0=\theta(x-y)\langle\varphi_\sigma(x)\varphi_{\sigma'}{}^\dagger(y)\rangle_0$$
$$+(-)^{2j}\theta(y-x)\langle\varphi_{\sigma'}{}^\dagger(y)\varphi_\sigma(x)\rangle_0 \quad (5.3)$$

(c) For an external line corresponding to a particle of spin j, $J_z=\mu$, and momentum $\mathbf{p}$, include a wave function

$$\frac{1}{[2\omega(\mathbf{p})]^{1/2}(2\pi)^{3/2}}D_{\sigma\mu}{}^{(j)}[L(\mathbf{p})]\exp(ip\cdot x) \qquad \text{[particle destroyed]},$$
$$\frac{1}{[2\omega(\mathbf{p})]^{1/2}(2\pi)^{3/2}}D_{\sigma\mu}{}^{(j)*}[L(\mathbf{p})]\exp(-ip\cdot x) \qquad \text{[particle created]},$$
$$\frac{1}{[2\omega(\mathbf{p})]^{1/2}(2\pi)^{3/2}}[D^{(j)}[L(\mathbf{p})]C^{-1}]_{\sigma\mu}\exp(-ip\cdot x) \quad \text{[antiparticle created]}, \tag{5.4}$$
$$\frac{1}{[2\omega(\mathbf{p})]^{1/2}(2\pi)^{3/2}}[D^{(j)}[L(\mathbf{p})]C^{-1}]_{\sigma\mu}{}^{*}\exp(ip\cdot x) \quad \text{[antiparticle destroyed]}.$$

These wave functions can be calculated from Eq. (2.39). In conjunction with (4.4), this tells us that

$$D^{(j)}[L(\mathbf{p})]=m^{-2j}\Pi^{(j)}(p'), \tag{5.5}$$

where the 4-vector p' is defined to have $\theta'=\theta/2$, i.e.,

$$p'=\{\hat{p}[\tfrac{1}{2}m(\omega-m)]^{1/2},[\tfrac{1}{2}m(\omega+m)]^{1/2}\}. \tag{5.6}$$

The matrix $\Pi^{(j)}$ is calculated in the Appendix; see also Table I.

(d) Integrate over all vertex positions x, y, etc. and sum over all dummy indices σ, σ', etc.

(e) Supply a $(-)$ sign for each fermion loop.

The problem still remaining is to calculate the propagator (5.3). An elementary calculation using (4.11) and (4.3) gives

$$\langle\varphi_\sigma(x)\varphi_{\sigma'}{}^\dagger(y)\rangle_0 = (2\pi)^{-3}m^{-2j}\int\frac{d^3p}{2\omega(\mathbf{p})}\Pi_{\sigma\sigma'}(p)\exp[ip\cdot(x-y)]$$
$$\langle\varphi_{\sigma'}{}^\dagger(y)\varphi_\sigma(x)\rangle_0 = (2\pi)^{-3}m^{-2j}\int\frac{d^3p}{2\omega(\mathbf{p})}\Pi_{\sigma\sigma'}(p)\exp[-ip\cdot(x-y)].$$

Formula (4.5) for $\Pi(p)$ lets us write this as

$$\langle\varphi_\sigma(x)\varphi_{\sigma'}{}^\dagger(y)\rangle_0 =i(-im)^{-2j}t_{\sigma\sigma'}{}^{\mu_1\mu_2\cdots\mu_{2j}}\partial_{\mu_1}\partial_{\mu_2}\cdots\partial_{\mu_{2j}}\Delta_+(x-y), \tag{5.7}$$
$$(-)^{2j}\langle\varphi_{\sigma'}{}^\dagger(y)\varphi_\sigma(x)\rangle_0 =i(-im)^{-2j}t_{\sigma\sigma'}{}^{\mu_1\mu_2\cdots\mu_{2j}}\partial_{\mu_1}\partial_{\mu_2}\cdots\partial_{\mu_{2j}}\Delta_+(y-x),$$

where

$$i\Delta_+(x)\equiv\frac{1}{(2\pi)^3}\int\frac{d^3p}{2\omega(\mathbf{p})}\exp(ip\cdot x).$$

At this point we encounter an infamous difficulty. If the θ function in (5.3) could be commuted past the derivatives in (5.7), then the propagator (5.3) would be

$$S_{\sigma\sigma'}(x-y)=-i(-im)^{-2j}t_{\sigma\sigma'}{}^{\mu_1\mu_2\cdots\mu_{2j}} \times\partial_{\mu_1}\partial_{\mu_2}\cdots\partial_{\mu_{2j}}\Delta^C(x-y), \tag{5.8}$$

where $-i\Delta^C(x-y)$ is the usual spin-zero propagator:

$$-i\Delta^C(x)=i\theta(x)\Delta_+(x)+i\theta(-x)\Delta_+(-x) =\tfrac{1}{2}[\Delta_1(x)+i\epsilon(x)\Delta(x)] \tag{5.9}$$

and, as usual,

$$\epsilon(x)\equiv\theta(x)-\theta(-x),$$
$$\Delta_1(x)\equiv i[\Delta_+(x)+\Delta_+(-x)], \tag{5.10}$$
$$\Delta(x)\equiv\Delta_+(x)-\Delta_+(-x).$$

It is well known that $\Delta^C(x)$ is scalar, because $\epsilon(x)$ is scalar unless x is spacelike, in which case $\Delta(x)=0$. Using the tensor transformation rule (A.5) for the $t^{\mu\nu}\cdots$ we find that

$$D^{(j)}[\Lambda]S(x)D^{(j)}[\Lambda]^\dagger=S(\Lambda x). \tag{5.11}$$

This is just the right behavior to guarantee a Lorentz-invariant S matrix.

But unfortunately the propagator (5.3) arising from Wick's theorem is *not* equal to the covariant propagator $S(x)$ defined by (5.8), except for $j=0$ and $j=\frac{1}{2}$. The trouble is that the derivatives in (5.8) act on the ϵ function in $\Delta^C(x)$ as well as on the functions Δ and Δ_1. This gives rise to extra terms proportional to equal-time δ functions and their derivatives. These extra terms are not covariant by themselves, but are needed to make $S(x)$ covariant; we must conclude then that (5.3) is not covariant.

For example, for spin 1 Eq. (5.3) gives

$$\langle T\{\varphi_\sigma(x)\varphi_{\sigma'}{}^\dagger(y)\}\rangle_0=\tfrac{1}{2}im^{-2}t_{\sigma\sigma'}{}^{\mu\nu} \times[\partial_\mu\partial_\nu\Delta_1(x-y)+i\epsilon(x-y)\partial_\mu\partial_\nu\Delta(x-y)],$$

while (5.8) gives

$$S_{\sigma\sigma'}(x-y)=\tfrac{1}{2}im^{-2}t_{\sigma\sigma'}{}^{\mu\nu}\partial_\mu\partial_\nu \times[\Delta_1(x-y)+i\epsilon(x-y)\Delta(x-y)].$$

The difference can be readily calculated by using the familiar properties of $\Delta(x)$. We find that

$$\langle T\{\varphi_\sigma(x)\varphi_{\sigma'}{}^\dagger(y)\}\rangle_0 =S_{\sigma\sigma'}(x-y)-2m^{-2}t_{\sigma\sigma'}{}^{00}\delta^4(x-y), \tag{5.12}$$

and the second term is definitely not covariant in the

sense of Eq. (5.11). [This problem does not arise for spin 0, where there are no derivatives, nor for spin $\frac{1}{2}$, where there is just one derivative and the extra term is proportional to

$$t^\mu\Delta(x-y)\partial_\mu\epsilon(x-y)=2t^0\Delta(x-y)\delta(x^0-y^0)=0.$$

But it does occur for any $j\geqq 1$.]

This problem has nothing to do with our noncanonical approach or our use of $2j+1$-component fields. For example, in the conventional theory of spin 1 (using the four-component $(\frac{1}{2},\frac{1}{2})$ representation) the propagator is

$$\begin{aligned}\langle T\{A_\mu(x)A_\nu(y)\}\rangle_0 &= -(i/2)[(g_{\mu\nu}-m^{-2}\partial_\mu\partial_\nu)\Delta_1(x-y) \\ &\quad +i\epsilon(x-y)(g_{\mu\nu}-m^{-2}\partial_\mu\partial_\nu)\Delta(x-y)] \\ &= -i(g_{\mu\nu}-m^{-2}\partial_\mu\partial_\nu)\Delta^C(x-y)-2m^{-2}\delta_\mu{}^0\delta_\nu{}^0\delta^4(x-y);\end{aligned}$$

so here also there appears a noncovariant term like that in (5.12). The general reason why the S matrix turns out to be noncovariant is that condition (1.5) is not really satisfied by an interaction like (5.1) if any of the spins are higher than $\frac{1}{2}$, because the commutators (4.12) of such fields are too singular at the apex of the light cone.

The cure is well known. We must add noncovariant "contact" terms to $\mathcal{H}(x)$ in such a way as to cancel out the noncovariant terms in the propagator. If we used a Lagrangian formalism, then such noncovariant contact terms would be generated automatically in the transition from $\mathcal{L}(x)$ to $\mathcal{H}(x)$, although the proof[11] of this general Matthews theorem is very complicated. For our purposes it is only necessary to remark that we take the invariance of the S matrix as a postulate and not a theorem, so that we have no choice but to add contact terms to $\mathcal{H}(x)$ which will just cancel the noncovariant parts of the propagator, such as the second term in (5.12).

In summary, we are to construct the S matrix according to the Feynman rules (a)–(e), but with the slight modifications:

(a′) Pay no attention to the noncovariant contact interactions; compute the vertex factors using only the original covariant part of $\mathcal{H}(x)$.

(b′) Do not use (5.3) for internal lines; instead use the covariant propagator

$$S_{\sigma\sigma'}(x-y)=-i(-im)^{-2j}t_{\sigma\sigma'}{}^{\mu_1\mu_2\cdots\mu_{2j}} \times\partial_{\mu_1}\partial_{\mu_2}\cdots\partial_{\mu_{2j}}\Delta^C(x-y). \quad (5.8)$$

Similar modifications are required when $\mathcal{H}(x)$ includes derivative interactions.

The Feynman rules could also be stated in momentum space. The propagator (5.8) would then become

$$\begin{aligned}S_{\sigma\sigma'}(q)&=\int d^4x e^{-iq\cdot x}S_{\sigma\sigma'}(x)\\ &=-i(-m)^{-2j}\Pi_{\sigma\sigma'}(q)/q^2+m^2-i\epsilon. \quad (5.13)\end{aligned}$$

The monomials $\Pi(q)$ are calculated in the Appendix, and presented explicitly for $j\leqq 3$ in Table I.

VI. T, C, AND P

The effect of time-reversal (**T**), charge-conjunction (**C**), and space-inversion (**P**) on the free-particle states is well known. It can be summarized by specifying the transformation properties of the annihilation operators:

$$\mathbf{T}a(\mathbf{p},\sigma)\mathbf{T}^{-1}=\eta_T\textstyle\sum_{\sigma'}C_{\sigma\sigma'}a(-\mathbf{p},\sigma'), \quad (6.1)$$

$$\mathbf{T}b(\mathbf{p},\sigma)\mathbf{T}^{-1}=\bar\eta_T\textstyle\sum_{\sigma'}C_{\sigma\sigma'}b(-\mathbf{p},\sigma'), \quad (6.2)$$

$$\mathbf{C}a(\mathbf{p},\sigma)\mathbf{C}^{-1}=\eta_C b(\mathbf{p},\sigma), \quad (6.3)$$

$$\mathbf{C}b(\mathbf{p},\sigma)\mathbf{C}^{-1}=\bar\eta_C a(\mathbf{p},\sigma), \quad (6.4)$$

$$\mathbf{P}a(\mathbf{p},\sigma)\mathbf{P}^{-1}=\eta_P a(-\mathbf{p},\sigma), \quad (6.5)$$

$$\mathbf{P}b(\mathbf{p},\sigma)\mathbf{P}^{-1}=\bar\eta_P b(-\mathbf{p},\sigma). \quad (6.6)$$

The η's and $\bar\eta$'s are phase factors[12] representing a degree of freedom in the definition of these inversions. The operator **T** is antiunitary, while **C** and **P** are unitary. The matrix $C_{\sigma\sigma'}$ was defined in Sec. II, and has the properties

$$C\mathbf{J}^{(j)}C^{-1}=-\mathbf{J}^{(j)*}, \quad (6.7)$$

$$C^*C=(-)^{2j};\quad C^\dagger C=1 \quad (6.8)$$

where $\mathbf{J}^{(j)}$ are the usual $2j+1$- dimensional angular-momentum matrices.

In order to describe the effect that **C** and **P** have on the field $\varphi_\sigma(x)$, it will be necessary to introduce a second $2j+1$-component field:

$$\chi_\sigma(x)\equiv(2\pi)^{-3/2}\int\frac{d^3\mathbf{p}}{[2\omega(\mathbf{p})]^{1/2}}\sum_{\sigma'}\Big[D_{\sigma\sigma'}{}^{(j)}[L(-\mathbf{p})]a(\mathbf{p},\sigma')e^{ip\cdot x}+(-)^{2j}\sum_{\sigma'}\{D^{(j)}[L(-\mathbf{p})]C^{-1}\}_{\sigma\sigma'}b^*(\mathbf{p},\sigma')e^{-ip\cdot x}\Big]. \quad (6.9)$$

This is the field that we would have constructed instead of $\varphi_\sigma(x)$ had we chosen to represent the "boost" generators by

$$\mathbf{K}^{(j)}=+i\mathbf{J}^{(j)} \quad (2.37)$$

instead of Eq. (2.36). The field $\chi_\sigma(x)$ transforms under the $(0,j)$ representation of the Lorentz group:

$$U[\Lambda]\chi_\sigma(x)U^{-1}[\Lambda]=\textstyle\sum_{\sigma'}\bar D_{\sigma\sigma'}{}^{(j)}[\Lambda^{-1}]\chi_{\sigma'}(\Lambda x), \quad (6.10)$$

$$\bar D^{(j)}[\Lambda]\equiv D^{(j)\dagger}[\Lambda^{-1}], \quad (6.11)$$

[11] See, for example, H. Umezawa, *Quantum Field Theory* (North-Holland Publishing Company, Amsterdam, 1956), Chap. X.

[12] For a general discussion of these phases, see G. Feinberg and S. Weinberg, Nuovo Cimento **14**, 571 (1959). The discussion there was limited to (0,0), $(\frac{1}{2},\frac{1}{2})$, and $(\frac{1}{2},0)\oplus(0,\frac{1}{2})$ fields, but can be easily adapted to the general case.

the matrix $\bar{D}$ appearing instead of D because we use (2.40) instead of (2.39). Like $\varphi_\sigma(x)$, the field $\chi_\sigma(x)$ obeys the Klein-Gordon equation (and no other field equation) and commutes with its adjoint outside the light-cone. It also has causal commutation relations with $\varphi_\sigma(x)$, *but only because of our choice of the sign* $(-)^{2j}$ *in Eq. (6.9).*

The effect of **T**, **C**, and **P** on $\varphi_\sigma(x)$ and $\chi_\sigma(x)$ can be readily calculated by use of the formula:

$$D^{(j)}[L(\mathbf{p})]^* = CD^{(j)}[L(-\mathbf{p})]C^{-1}. \tag{6.12}$$

We find that:

$$\mathbf{T}\varphi_\sigma(x)\mathbf{T}^{-1} = \eta_T \textstyle\sum_{\sigma'} C_{\sigma\sigma'}\varphi_{\sigma'}(\mathbf{x}, -x^0), \tag{6.13}$$

$$\mathbf{T}\chi_\sigma(x)\mathbf{T}^{-1} = \eta_T \textstyle\sum_{\sigma'} C_{\sigma\sigma'}\chi_{\sigma'}(\mathbf{x}, -x^0), \tag{6.14}$$

$$\mathbf{C}\varphi_\sigma(x)\mathbf{C}^{-1} = \eta_C \textstyle\sum_{\sigma'} C_{\sigma\sigma'}^{-1}\chi_{\sigma'}^\dagger(x), \tag{6.15}$$

$$\mathbf{C}\chi_\sigma(x)\mathbf{C}^{-1} = \eta_C(-)^{2j} \textstyle\sum_{\sigma'} C_{\sigma\sigma'}^{-1}\varphi_{\sigma'}^\dagger(x), \tag{6.16}$$

$$\mathbf{P}\varphi_\sigma(x)\mathbf{P}^{-1} = \eta_P\chi_\sigma(-\mathbf{x}, x^0), \tag{6.17}$$

$$\mathbf{P}\chi_\sigma(x)\mathbf{P}^{-1} = \eta_P\varphi_\sigma(-\mathbf{x}, x^0), \tag{6.18}$$

provided that the antiparticle inversion phases are chosen as

$$\bar{\eta}_T = \eta_T^*; \quad \bar{\eta}_C = \eta_C^*; \quad \bar{\eta}_P = \eta_P^*(-)^{2j}. \tag{6.19}$$

Any other choice of the $\bar{\eta}$ would result in the creation and annihilation parts of φ_σ and χ_σ transforming with different phases, destroying the possibility of simple transformation laws.[13]

If a particle is its own antiparticle then we call it "purely neutral," and set

$$a(\mathbf{p},\sigma) = b(\mathbf{p},\sigma). \tag{6.20}$$

In this special case the (j 0) and (0 j) fields are related by

$$\chi_\sigma^\dagger(x) = \textstyle\sum_{\sigma'} C_{\sigma\sigma'}\varphi_{\sigma'}(x), \tag{6.21}$$

$$\varphi_\sigma^\dagger(x) = (-)^{2j} \textstyle\sum_{\sigma'} C_{\sigma\sigma'}\chi_{\sigma'}(x). \tag{6.22}$$

The fields are *not* Hermitian, except of course for $j=0$. Nevertheless, Eq. (6.20) requires the phases $\bar{\eta}_I$ to be equal to the corresponding η_I, and (6.19) then implies that these phases can only take the real values ± 1, except that η_P must be $\pm i$ for purely neutral fermions.

We see that the fields $\varphi_\sigma(x)$ and $\chi_\sigma(x)$ transform separately under **T**, and also under the combined operation **CP**:

$$\mathbf{CP}\varphi_\sigma(x)\mathbf{P}^{-1}\mathbf{C}^{-1} = \eta_C\eta_P \textstyle\sum_{\sigma'} C_{\sigma\sigma'}^{-1}\varphi_{\sigma'}^\dagger(-\mathbf{x}, x^0), \tag{6.23}$$

$$\mathbf{CP}\chi_\sigma(x)\mathbf{P}^{-1}\mathbf{C}^{-1} = \eta_C\eta_P(-)^{2j} \textstyle\sum_{\sigma'} C_{\sigma\sigma'}^{-1}\chi_{\sigma'}^\dagger(-\mathbf{x}, x^0). \tag{6.24}$$

[Under **CPT** the transformation law is just that of a spinless field:

$$\mathbf{CPT}\varphi_\sigma(x)\mathbf{T}^{-1}\mathbf{P}^{-1}\mathbf{C}^{-1} = \eta_C\eta_P\eta_T\varphi_\sigma^\dagger(-x), \tag{6.25}$$

$$\mathbf{CPT}\chi_\sigma(x)\mathbf{T}^{-1}\mathbf{P}^{-1}\mathbf{C}^{-1} = \eta_C\eta_P\eta_T(-)^{2j}\chi_\sigma^\dagger(-x), \tag{6.26}$$

permitting a great simplification in the proof of the **CPT** theorem.] The use of $2j+1$-component fields (either φ_σ or χ_σ) for massive particles as well as for neutrinos would seem very appropriate in theories of the weak interactions, where **CP** and **T** are conserved but **C** and **P** are not.

[13] An important consequence is that a particle-antiparticle pair has intrinsic parity

$$\eta_P\bar{\eta}_P = (-)^{2j},$$

a well-known result that would be inexplicable on the basis of nonrelativistic quantum mechanics.

VII. $2(2j+1)$-COMPONENT FIELDS

Any parity-conserving interaction must involve both the $(j,0)$ field $\varphi_\sigma(x)$ and the $(0,j)$ field $\chi_\sigma(x)$. It is therefore convenient to unite these two $(2j+1)$-component fields into a single $2(2j+1)$-component field:

$$\psi(x) = \begin{bmatrix} \varphi(x) \\ \chi(x) \end{bmatrix}. \tag{7.1}$$

This field transforms according to the $(j,0)\oplus(0,j)$ representation, i.e.,

$$U[\Lambda]\psi_\alpha(x)U^{-1}[\Lambda] = \textstyle\sum_\beta \mathfrak{D}_{\alpha\beta}^{(j)}[\Lambda^{-1}]\psi_\beta(\Lambda x), \tag{7.2}$$

where

$$\mathfrak{D}^{(j)}[\Lambda] = \begin{bmatrix} D^{(j)}[\Lambda] & 0 \\ 0 & \bar{D}^{(j)}[\Lambda] \end{bmatrix}, \tag{7.3}$$

the representations $D^{(j)}$ and $\bar{D}^{(j)}$ being defined by (2.36) and (2.37) respectively. The representation $\mathfrak{D}^{(j)}$ can be defined also by specifying that the generators of rotations are to be represented by

$$\mathfrak{J}^{(j)} = \begin{bmatrix} 0 & \mathbf{J}^{(j)} \\ 0 & \mathbf{J}^{(j)} \end{bmatrix}, \tag{7.4}$$

and that the generators of boosts are represented by

$$\mathfrak{K}^{(j)} = -i\gamma_5\mathfrak{J}^{(j)}, \tag{7.5}$$

where γ_5 is the $2(2j+1)$-dimensional matrix:

$$\gamma_5 \equiv \begin{bmatrix} 1 & 0 \\ 0 & -1 \end{bmatrix}. \tag{7.6}$$

This satisfies (2.24)–(2.26) because $\gamma_5^2=1$.

The $(j,0)\oplus(0,j)$ representation (7.3) differs from the $(j,0)$ and $(0,j)$ representations in the important respect that $\mathfrak{D}^\dagger$ is equivalent to $\mathfrak{D}^{-1}$:

$$\mathfrak{D}^{(j)}[\Lambda]^\dagger = \beta\mathfrak{D}^{(j)}[\Lambda^{-1}]\beta, \tag{7.7}$$

where

$$\beta = \begin{bmatrix} 0 & 1 \\ 1 & 0 \end{bmatrix}; \quad \beta^2 = 1. \tag{7.8}$$

[See Eq. (6.11).] This has the consequence that

$$U[\Lambda]\bar{\psi}_\alpha(x)U^{-1}[\Lambda] = \textstyle\sum_\beta \bar{\psi}_\beta(\Lambda x)\mathfrak{D}_{\beta\alpha}^{(j)}[\Lambda], \tag{7.9}$$

where $\bar{\psi}$ is the covariant adjoint

$$\bar{\psi}(x) \equiv \psi^\dagger(x)\beta. \tag{7.10}$$

The **T**, **C**, and **P** transformation properties of $\psi_\alpha(x)$ can be read off immediately from (6.13)–(6.18):

$$\mathbf{T}\psi(x)\mathbf{T}^{-1} = \eta_T \mathcal{C}\psi(\mathbf{x}, -x^0) \tag{7.11}$$

$$\mathbf{C}\psi(x)\mathbf{C}^{-1} = \begin{matrix} \eta_C \mathcal{C}^{-1}\beta\psi^*(x) \text{ (bosons)}, \\ \eta_C \mathcal{C}^{-1}\gamma_5\beta\psi^*(x) \text{ (fermions)}, \end{matrix} \tag{7.12}$$

$$\mathbf{P}\psi(x)\mathbf{P}^{-1} = \eta_P \beta\psi(-\mathbf{x}, x^0), \tag{7.13}$$

with

$$\mathcal{C} = \begin{bmatrix} C & 0 \\ 0 & C \end{bmatrix}. \tag{7.14}$$

A purely neutral particle will have a field which satisfies the reality condition

$$\psi(x) = \begin{matrix} \mathcal{C}^{-1}\beta\psi^*(x) \text{ (bosons)} \\ \mathcal{C}^{-1}\gamma_5\beta\psi^*(x) \text{ (fermions)}. \end{matrix} \tag{7.15}$$

Its inversion phases η_T, η_C, η_P must be real, except that $\eta_P = \pm i$ for purely neutral fermions.

The field $\psi(x)$ of course satisfies the Klein-Gordon equation

$$(\Box^2 - m^2)\psi_\alpha(x) = 0. \tag{7.16}$$

But $\psi(x)$ has twice as many components as the operators $a(\mathbf{p},\sigma)$ and $b^*(\mathbf{p},\sigma)$, so it has a chance of also satisfying some other homogeneous field equation. In fact, it does. Using (A.12) and (A.40), we can easily show that the $(j,0)$ and $(0,j)$ fields are related by

$$\bar{\Pi}(-i\partial)\varphi(x) = m^{2j}\chi(x), \tag{7.17}$$

$$\Pi(-i\partial)\chi(x) = m^{2j}\varphi(x), \tag{7.18}$$

where $\Pi(q)$ and $\bar{\Pi}(q)$ are defined by (A.10) and (A.41). In the $2(2j+1)$-dimensional matrix notation this reads

$$[\gamma^{\mu_1\mu_2\cdots\mu_{2j}}\partial_{\mu_1}\partial_{\mu_2}\cdots\partial_{\mu_{2j}} + m^{2j}]\psi(x) = 0, \tag{7.19}$$

where the generalized γ matrices, $\gamma^{\mu_1\mu_2\cdots}$, are defined by

$$\gamma^{\mu_1\mu_2\cdots\mu_{2j}} = -i^{2j}\begin{bmatrix} 0 & t^{\mu_1\mu_2\cdots\mu_{2j}} \\ \bar{t}^{\mu_1\mu_2\cdots\mu_{2j}} & 0 \end{bmatrix}, \tag{7.20}$$

and are discussed and evaluated in Appendix B.

The field ψ obviously obeys causal commutation relations, since φ and χ commute with both $\varphi^\dagger$ and $\chi^\dagger$ at spacelike separations. Its homogeneous Green's functions are

$$\langle\psi_\alpha(x)\bar{\psi}_\beta(y)\rangle_0 = (2\pi)^{-3}m^{-2j}\int\frac{d^3\mathbf{p}}{2\omega(\mathbf{p})}M_{\alpha\beta}(p) \times\exp\{ip\cdot(x-y)\}, \tag{7.21}$$

$$\langle\bar{\psi}_\beta(y)\psi_\alpha(x)\rangle_0 = (2\pi)^{-3}m^{-2j}\int\frac{d^3\mathbf{p}}{2\omega(\mathbf{p})}N_{\alpha\beta}(p) \times\exp\{ip\cdot(y-x)\}, \tag{7.22}$$

where

$$M(p) = \begin{bmatrix} m^{2j} & \Pi(p) \\ \bar{\Pi}(p) & m^{2j} \end{bmatrix}, \tag{7.23}$$

$$N(p) = \begin{bmatrix} (-m)^{2j} & \Pi(p) \\ \bar{\Pi}(p) & (-m)^{2j} \end{bmatrix} = (-)^{2j}M(-p). \tag{7.24}$$

The "raw" propagator is then

$$\langle T\{\psi_\alpha(x)\bar{\psi}_\beta(y)\}\rangle_0 \equiv \theta(x-y)\langle\psi_\alpha(x)\bar{\psi}_\beta(y)\rangle_0 + (-)^{2j}\theta(y-x)\langle\bar{\psi}_\beta(y)\psi_\alpha(x)\rangle_0$$
$$= (2\pi)^{-3}m^{-2j}\int\frac{d^3\mathbf{p}}{2\omega(\mathbf{p})}[\theta(x-y)M_{\alpha\beta}(-i\partial)\exp\{ip\cdot(x-y)\} + \theta(y-x)M_{\alpha\beta}(-i\partial)\exp\{ip\cdot(y-x)\}]. \tag{7.25}$$

As discussed in Sec. V, this is *not* the covariant propagator to be used in conjunction with the Feynman rules. We must add certain noncovariant contact terms to (7.25) which allow us to move the derivatives in $M(-i\partial)$ to the left of the θ functions. The true propagator is

$$S_{\alpha\beta}(x-y) = (2\pi)^{-3}m^{-2j}M_{\alpha\beta}(-i\partial)\int\frac{d^3\mathbf{p}}{2\omega(\mathbf{p})}[\theta(x-y)\exp\{ip\cdot(x-y)\} + \theta(y-x)\exp\{ip\cdot(y-x)\}]$$
$$= -im^{-2j}M_{\alpha\beta}(-i\partial)\Delta^C(x-y), \tag{7.26}$$

where $\Delta^C(x)$ is the invariant $j=0$ propagator (5.9). This can be written in a more familiar form by using (B.13); we find that

$$S(x) = im^{-2j}[\gamma^{\mu_1\mu_2\cdots\mu_{2j}}\partial_{\mu_1}\partial_{\mu_2}\cdots\partial_{\mu_{2j}} - m^{2j}]\Delta^C(x). \tag{7.33}$$

It is easy to see from (B.4) that this has the correct transformation property:

$$\mathfrak{D}^{(j)}[\Lambda]S(x)\mathfrak{D}^{(j)-1}[\Lambda] = S(\Lambda x).$$

In momentum space we replace ∂_μ by iq_μ, so that

$$S(q) = -im^{-2j}[\mathcal{P}(q) + m^{2j}]/q^2 + m^2 - i\epsilon,$$

where

$$\mathcal{P}(q) = -i^{2j}\gamma^{\mu_1\mu_2\cdots\mu_{2j}}q_{\mu_1}q_{\mu_2}\cdots q_{\mu_{2j}}.$$

General formulas for $\mathcal{P}(q)$ are given in Appendix B; the results for $j \leqq 2$ are in Table 2. The wave functions for creation and annihilation of particles and antiparticles can be read off from (7.1), (4.11), and (6.9), or alternatively found from the solutions of (7.19). This whole formalism reduces to the Dirac theory for $j=\frac{1}{2}$.

VIII. GENERAL FIELDS

We started in Sec. III by introducing a field $\varphi(x)$ which transforms according to the $(j,0)$ representation. Then, in order to discuss parity conserving theories, we introduced the $(0,j)$ field $\chi(x)$ in Sec. VI and used it in Sec. VII to construct a field $\psi(x)$ which transforms under the (reducible) representation $(j,0)\oplus(0,j)$. These particular fields have the advantage of depending very simply and explicitly on the particular value of j, but φ, χ, and ψ are certainly not otherwise unique. In fact, the usual tensor representation of a field with integer j is $(j/2,j/2)$, while the Rarita-Schwinger representation for half-integer j is based on the $(2j+1)^2$-dimensional reducible representation:

$$[(\tfrac{1}{2},0)\oplus(0,\tfrac{1}{2})]\otimes\left(\frac{2j-1}{4},\frac{2j-1}{4}\right).$$

Our simpler fields agree with these conventional representations only for the case $j=\frac{1}{2}$.

We now consider the general case. Let $D_{nm}[\Lambda]$ be any representation (perhaps reducible) of the Lorentz group. Assume that when Λ is restricted to be a rotation R, the representation $D[R]$ contains a particular component $D^{(j)}[R]$. By this we mean that there must be a rotation basis of vectors $u_n(\sigma)$, such that

$$\textstyle\sum_m D_{nm}[R]u_m(\sigma)=\sum_{\sigma'} u_n(\sigma')D_{\sigma'\sigma}{}^{(j)}[R]. \quad (8.1)$$

We can form an operator $\alpha_n(\mathbf{p})$ analogous to (3.4):

$$\alpha_n(\mathbf{p})=[2\omega(p)]^{1/2}\textstyle\sum_{\sigma m} D_{nm}[L(\mathbf{p})]u_m(\sigma)a(\mathbf{p},\sigma) \quad (8.2)$$

which transforms simply:

$$U[\Lambda]\alpha_n(\mathbf{p})U^{-1}[\Lambda]=\textstyle\sum_m D_{nm}[\Lambda^{-1}]\alpha_m(\Lambda\mathbf{p}). \quad (8.3)$$

[Use (8.1) and (2.12).] For the antiparticles we can use another basis $v_m(\sigma)$, which in general may or may not be the same as the $u_m(\sigma)$, but which must also satisfy

$$\textstyle\sum_m D_{nm}[R]v_m(\sigma)=\sum_{\sigma'} v_n(\sigma')D_{\sigma'\sigma}{}^{(j)}[R]. \quad (8.4)$$

The operator $\beta_n(\mathbf{p})$ analogous to (3.5) is now formed as

$$\beta_n(\mathbf{p})=[2\omega(\mathbf{p})]^{1/2}\sum_{\sigma,\sigma',m} D_{nm}[L(\mathbf{p})]\times v_m(\sigma')C_{\sigma'\sigma}{}^{-1}b^*(\mathbf{p},\sigma). \quad (8.5)$$

Using (8.4) and (2.17), we see that it transforms just like $\alpha_n(\mathbf{p})$:

$$U[\Lambda]\beta_n(\mathbf{p})U^{-1}[\Lambda]=\textstyle\sum_m D_{nm}[\Lambda^{-1}]\beta_m(\Lambda\mathbf{p}). \quad (8.6)$$

The field is constructed as the invariant Fourier transform

$$\psi_n(x)=(2\pi)^{-3/2}\int\frac{d^3\mathbf{p}}{2\omega(\mathbf{p})}[\alpha_n(\mathbf{p})e^{ip\cdot x}+\beta_n(\mathbf{p})e^{-ip\cdot x}], \quad (8.7)$$

or going back to a and b^*

$$\psi_n(x)=\int d^3p\sum_\sigma [u_n(\mathbf{p},\sigma)a(\mathbf{p},\sigma)e^{ip\cdot x}+v_n(\mathbf{p},\sigma)b^*(\mathbf{p},\sigma)e^{-ip\cdot x}], \quad (8.8)$$

where the "wave functions" in (8.8) are

$$u_n(\mathbf{p},\sigma)=(2\pi)^{-3/2}[2\omega(\mathbf{p})]^{-1/2}\sum_m D_{nm}[L(\mathbf{p})]u_m(\sigma), \quad (8.9)$$

$$v_n(\mathbf{p},\sigma)=(2\pi)^{-3/2}[2\omega(\mathbf{p})]^{-1/2}\sum_{m,\sigma'} D_{nm}[L(\mathbf{p})]\times v_m(\sigma')C_{\sigma'\sigma}{}^{-1}. \quad (8.10)$$

This field transforms correctly

$$U[\Lambda,a]\psi_n(x)U^{-1}[\Lambda,a]=\textstyle\sum_m D_{nm}[\Lambda^{-1}]\psi_m(\Lambda x+a). \quad (8.11)$$

It obeys the Klein-Gordon equation, and may or may not obey other field equations as well. The causality condition (1.8) can be satisfied if we choose

$$\textstyle\sum_\sigma u_n(\sigma)u_m{}^*(\sigma)=\sum_\sigma v_n(\sigma)v_m{}^*(\sigma), \quad (8.12)$$

and if we use the usual connection between spin and statistics. We will not pursue these matters further here.

The chief point to be learned from this general construction is that the wave functions (8.9), (8.10) which enter into the Feynman rules are always determined by the matrices $D_{nm}[L(\mathbf{p})]$ representing a boost.

ACKNOWLEDGMENTS

I am very grateful to P. N. Burgoyne, K. M. Watson, and E. H. Wichmann, for conversations on this subject.

APPENDIX A: SPINOR CALCULUS FOR ANY SPIN

Everyone knows that the three Pauli matrices together with the 2×2 unit matrix make up a four vector t^μ:

$$\mathbf{t}\equiv\boldsymbol{\sigma};\quad t^0\equiv1 \quad (A1)$$

in the sense that

$$D^{(1/2)}[\Lambda]t^\mu D^{(1/2)}[\Lambda]^\dagger=\Lambda_\nu{}^\mu t^\nu. \quad (A2)$$

Here Λ is a general proper homogeneous Lorentz transformation, and $D^{(1/2)}[\Lambda]$ is the corresponding 2×2 matrix in the $(\frac{1}{2},0)$ representation, defined by representing the generators of infinitesimal transformations as

$$\mathbf{J}=\frac{1}{2}\boldsymbol{\sigma},\quad \mathbf{K}=-\frac{i}{2}\boldsymbol{\sigma}. \quad (A3)$$

This famous construction of the vector t^μ is the basis of the familiar spinor calculus, which can also be employed in a rather cumbrous fashion to discuss spins higher than $\frac{1}{2}$.

We shall instead show here that this construction of a vector out of two-dimensional matrices can be *directly* generalized to the construction of a tensor of rank $2j$ out of $2j+1$-dimensional matrices.[14] We shall also show that the commutator and propagator of a $(2j+1)$-component field of spin j are proportional to this tensor.

We first prove that for any integral or half-integral j there exists a set of quantities

$$t_{\sigma\sigma'}{}^{\mu_1\mu_2\cdots\mu_{2j}}\begin{pmatrix}\sigma, \sigma'=j, j-1, \cdots, -j \\ \mu_1, \mu_2, \cdots, \mu_{2j}=0, 1, 2, 3\end{pmatrix}$$

with the properties:

(a) t is symmetric in all μ's.
(b) t is traceless in all μ's, i.e.,

$$g_{\mu_1\mu_2}t_{\sigma\sigma'}{}^{\mu_1\mu_2\cdots\mu_{2j}}=0. \quad (A4)$$

(c) t is a tensor, in the sense that

$$D^{(j)}[\Lambda]t^{\mu_1\mu_2\cdots\mu_{2j}}D^{(j)}[\Lambda]^\dagger = \Lambda_{\nu_1}{}^{\mu_1}\Lambda_{\nu_2}{}^{\mu_2}\cdots\Lambda_{\nu_{2j}}{}^{\mu_{2j}}t^{\nu_1\nu_2\cdots\nu_{2j}}, \quad (A5)$$

where $D^{(j)}[\Lambda]$ is the $2j+1$-dimensional matrix corresponding to Λ in the $(j,0)$ representation. [These $D^{(j)}[\Lambda]$ are the same as used in the text, and are defined by Eqs. (2.36) and (2.35). Ordinary matrix multiplication is understood on the left-hand side of (A5). Eq. (A5) reduces to (A2) for $j=\frac{1}{2}$.]

Existence Proof:

Let u_σ be a $2j+1$-dimensional basis transforming according to the $(j,0)$ representation of the Lorentz group, i.e.,

$$u_\sigma \to \sum_{\sigma'} D_{\sigma'\sigma}{}^{(j)}[\Lambda]u_{\sigma'}. \quad (A6)$$

The quantities $u_\sigma u_\tau^*$ evidently furnish a $(2j+1)^2$-dimensional representation, the direct product of the $(j\,0)$ representation with its complex conjugate. But this is

$$(j,0)\otimes(0,j)=(j,j) \quad (A7)$$

so the quantities $u_\sigma u_\tau^*$ transform under the (j,j) representation. The (j,j) representation consists of all symmetric traceless tensors of rank $2j$, so it must be possible to form such a tensor basis by taking linear combinations of the $u_\sigma u_\tau^*$, i.e.,

$$T^{\mu_1\mu_2\cdots\mu_{2j}}(u)=\sum_{\sigma\tau} t_{\sigma\tau}{}^{\mu_1\mu_2\cdots\mu_{2j}}u_\sigma u_\tau^* \quad (A8)$$

in such a way that the transformation (A6) gives

$$T^{\mu_1\mu_2\cdots\mu_{2j}}(u)\to\Lambda_{\nu_1}{}^{\mu_1}\Lambda_{\nu_2}{}^{\mu_2}\cdots\Lambda_{\nu_{2j}}{}^{\mu_{2j}}T^{\nu_1\nu_2\cdots\nu_{2j}}(u). \quad (A9)$$

But this requires that the t coefficients must satisfy Eq. (A5). They must also be symmetric and traceless with respect to the μ_i, because $T(u)$ is symmetric and traceless for all u. Q.E.D.

Having proved the existence of the t's, we must now establish a formula which will allow us to calculate them, and which will also provide a connection with the Green's functions of field theory. For any four-vector q, we define a scalar matrix

$$\Pi_{\sigma'\sigma}{}^{(j)}(q)\equiv(-)^{2j}t_{\sigma'\sigma}{}^{\mu_1\mu_2\cdots\mu_{2j}}q_{\mu_1}q_{\mu_2}\cdots q_{\mu_{2j}}. \quad (A10)$$

We will prove that if q is in the forward light-cone,

$$q^2=-m^2;\quad q^0>0, \quad (A11)$$

then

$$\Pi^{(j)}(q)=m^{2j}D^{(j)}[L(\mathbf{q})]^2=m^{2j}\exp(-2\theta\hat{q}\cdot\mathbf{J}^{(j)}) \quad (A12)$$

where

$$\hat{q}=\mathbf{q}/|\mathbf{q}|,\qquad \sinh\theta=|\mathbf{q}|/m, \quad (A13)$$

and $\mathbf{J}^{(j)}$ is the usual $2j+1$-dimensional representation of the angular momentum. [The constant factor in (A12) is of course arbitrary, but is chosen here so that the normalization of t will be as simple as possible.]

Proof of (A12):

The transformation law (A5) implies that

$$D^{(j)}[\Lambda]\Pi^{(j)}(q)D^{(j)\dagger}[\Lambda]=\Pi^{(j)}(\Lambda q). \quad (A14)$$

Let us fix q to have the rest-value $q=q(m)$, where

$$q^0(m)=m;\quad \mathbf{q}(m)=0. \quad (A15)$$

(a) If Λ is a rotation then $D^{(j)}[\Lambda]$ is the unitary matrix

$$D^{(j)}[\Lambda]=\exp(i\mathbf{e}\cdot\mathbf{J}^{(j)}), \quad (A16)$$

where $\mathbf{J}^{(j)}$ is the usual $2j+1$-dimensional representation of the angular momentum vector $\mathbf{J}$. The vector (A15) is rotation-invariant, so (A14) gives

$$[\mathbf{J}^{(j)},\Pi^{(j)}(q(m))]=0. \quad (A17)$$

But the three matrices $\mathbf{J}^{(j)}$ are irreducible, so Schur's Lemma tells us that $\Pi^{(j)}(q(m))$ must be proportional to the unit matrix. We will fix its normalization so that

$$\Pi_{\sigma\sigma'}{}^{(j)}(q(m))=m^{2j}\delta_{\sigma\sigma'}, \quad (A18)$$

and therefore

$$t_{\sigma\sigma'}{}^{00\cdots0}=\delta_{\sigma\sigma'}. \quad (A19)$$

Equation (A14) therefore gives

$$\Pi^{(j)}(\Lambda q(m))=m^{2j}D^{(j)}[\Lambda]D^{(j)\dagger}[\Lambda]. \quad (A20)$$

(b) If Λ is the "boost" $L(\mathbf{p})$ defined by Eq. (2.4), then $D^{(j)}[\Lambda]$ is the Hermitian matrix

$$D^{(j)}[L(\mathbf{p})]=\exp(-\theta\hat{p}\cdot\mathbf{J}^{(j)}) \quad (A21)$$

and

$$L(\mathbf{p})q(m)=p. \quad (A22)$$

Formula (A12) now follows immediately.

The exponential in (A12) may be calculated as a polynomial of order $2j$ in the matrix

$$z\equiv2(\hat{q}\cdot\mathbf{J}^{(j)}). \quad (A23)$$

Recall that z is an Hermitian matrix with integer eigenvalues $2j, 2j-2, \cdots, -2j$, and that therefore

$$(z-2j)(z-2j+2)\cdots(z+2j)=0. \quad (A24)$$

[14] These are a special case of the matrices constructed by Barut, Muzinich, and Williams, Ref. 4, by induction from the $j=\frac{1}{2}$ case.

This can be rewritten to give z^{2j+1} as a polynomial of order $2j$ in z. It follows then that $\Pi^{(j)}(q)$ must itself be such a polynomial, since all powers of z beyond the $2j$th in the Taylor series for the exponential can be reduced to polynomials in z of order $2j$.

For example, in the case of spin $j=\frac{1}{2}$, Eq. (A24) gives, $z^2=1$, so that

$$\exp(-z\theta)=1-z\theta+\tfrac{1}{2}\theta^2-\tfrac{1}{6}z\theta^3+\cdots=\cosh\theta-z\sinh\theta.$$

Then (A12) gives

$$\Pi^{(1/2)}(q)=m[\cosh\theta-2(\hat{q}\cdot\mathbf{J}^{(1/2)})\sinh\theta]=q^0-2(\mathbf{q}\cdot\mathbf{J}^{(1/2)}). \quad \text{(A25)}$$

Setting this equal to $-t^\mu q_\mu$ then gives (A1).

To go through this sort of calculation for general j would be tedious and difficult. We shall approach the problem of representing $\exp(-2\theta z)$ as a polynomial in z more directly. First split it into even and odd parts,

$$\exp(-\theta z)=\cosh\theta z-\sinh\theta z. \quad \text{(A26)}$$

We consider separately the cases of j integer and half-integer.

1. Integer Spin

The eigenvalues $2j, 2j-2$, etc., of the Hermitian matrix $z=2(\hat{q}\cdot\mathbf{J})$ are even integers. If follows that[15]

$$\cosh z\theta=1+\sum_{n=0}^{j-1}\frac{z^2(z^2-2^2)(z^2-4^2)\cdots(z^2-(2n)^2)}{(2n+2)!}\sinh^{2n+2}\theta, \quad \text{(A27)}$$

$$\sinh z\theta=z\cosh\theta\sum_{n=0}^{j-1}\frac{(z^2-2^2)(z^2-4^2)\cdots(z^2-(2n)^2)}{(2n+1)!}\sinh^{2n+1}\theta. \quad \text{(A28)}$$

Using (A26) and (A12) gives for all q:

$$\Pi^{(j)}(q)=(-q^2)^j+\sum_{n=0}^{j-1}\frac{(-q^2)^{j-1-n}}{(2n+2)!}(2\mathbf{q}\cdot\mathbf{J})[(2\mathbf{q}\cdot\mathbf{J})^2-(2\mathbf{q})^2][(2\mathbf{q}\cdot\mathbf{J})^2-(4\mathbf{q})^2]\cdots$$
$$\times[(2\mathbf{q}\cdot\mathbf{J})^2-(2n\mathbf{q})^2][2\mathbf{q}\cdot\mathbf{J}-(2n+2)q^0] \quad \text{(A29)}$$

or

$$\Pi^{(j)}(q)=(-q^2)^j+\frac{(-q^2)^{j-1}}{2!}(2\mathbf{q}\cdot\mathbf{J})[2\mathbf{q}\cdot\mathbf{J}-2q^0]+\frac{(-q^2)^{j-2}}{4!}(2\mathbf{q}\cdot\mathbf{J})[(2\mathbf{q}\cdot\mathbf{J})^2-(2\mathbf{q})^2][2\mathbf{q}\cdot\mathbf{J}-4q^0]$$
$$+\frac{(-q^2)^{j-3}}{6!}(2\mathbf{q}\cdot\mathbf{J})[(2\mathbf{q}\cdot\mathbf{J})^2-(2\mathbf{q})^2][(2\mathbf{q}\cdot\mathbf{J})^2-(4\mathbf{q})^2][2\mathbf{q}\cdot\mathbf{J}-6q^0]+\cdots. \quad \text{(A30)}$$

The series (A30) cuts itself off automatically after $j+1$ terms. The terms we have listed are sufficient to calculate Π for $j=0, 1, 2, 3$; the results are in Table I.

2. Half-Integer Spin

The eigenvalues $2j, 2j-2$, etc., of $z=2(\hat{q}\cdot\mathbf{J})$ are now odd integers. It follows that[15]

$$\cosh z\theta=\cosh\theta\left[1+\sum_{n=1}^{j-1/2}\frac{(z^2-1^2)(z^2-3^2)\cdots(z^2-(2n-1)^2)}{(2n)!}\sinh^{2n}\theta\right], \quad \text{(A31)}$$

$$\sinh z\theta=z\sinh\theta\left[1+\sum_{n=1}^{j-1/2}\frac{(z^2-1^2)(z^2-3^2)\cdots(z^2-(2n-1)^2)}{(2n+1)!}\sinh^{2n}\theta\right]. \quad \text{(A32)}$$

Using (A26) and (A12) now gives:

$$\Pi^{(j)}(q)=(-q^2)^{j-1/2}[q^0-2\mathbf{q}\cdot\mathbf{J}]+\sum_{n=1}^{j-1/2}\frac{(-q^2)^{j-n-1/2}}{(2n+1)!}$$
$$\times[(2\mathbf{q}\cdot\mathbf{J})^2-\mathbf{q}^2][(2\mathbf{q}\cdot\mathbf{J})^2-(3\mathbf{q})^2]\cdots[(2\mathbf{q}\cdot\mathbf{J})^2-([2n-1]\mathbf{q})^2][(2n+1)q^0-2\mathbf{q}\cdot\mathbf{J}], \quad \text{(A33)}$$

[15] For (A27) and (A31) see, for example, H. B. Dwight, *Table of Integrals and Other Mathematical Data* (The Macmillan Company, New York, 1961), fourth edition, formulas 403.11 and 403.13, respectively. Equations (A28) and (A32) can be checked by differentiating with respect to θ; we get (A27) and (A31). I would like to thank C. Zemach for suggesting the existence of such expressions and a method of deriving them.

or

$$\Pi^{(j)}(q)=(-q^2)^{j-1/2}[q^0-2\mathbf{q}\cdot\mathbf{J}]+\frac{1}{3!}(-q^2)^{j-3/2}[(2\mathbf{q}\cdot\mathbf{J})^2-\mathbf{q}^2][3q^0-2\mathbf{q}\cdot\mathbf{J}]$$

$$+\frac{1}{5!}(-q^2)^{j-5/2}[(2\mathbf{q}\cdot\mathbf{J})^2-\mathbf{q}^2][(2\mathbf{q}\cdot\mathbf{J})^2-(3\mathbf{q})^2][5q^0-2\mathbf{q}\cdot\mathbf{J}]+\cdots. \quad \text{(A34)}$$

The series (A34) cuts itself off after $j+\frac{1}{2}$ terms. The terms we have listed suffice to calculate Π for $j=\frac{1}{2},\frac{3}{2},\frac{5}{2}$; the results are in Table I.

Having calculated $\Pi(q)$, the coefficients $t^{\mu_1\mu_2\cdots}$ may be determined by inspection. For example, in the case $j=1$, Eq. (A30) gives

$$\Pi^{(1)}(q)=-q^2+2(\mathbf{q}\cdot\mathbf{J})(\mathbf{q}\cdot\mathbf{J}-q^0). \quad \text{(A35)}$$

Setting this equal to $t^{\mu\nu}q_\mu q_\nu$ gives

$$t^{00}=1$$
$$t^{0i}=t^{i0}=+J_i \quad \text{(A36)}$$
$$t^{ij}=\{J_j,J_i\}-\delta_{ij}.$$

Observe that this *is* traceless, because

$$t_\mu{}^\mu=[2\mathbf{J}^2-3]-1=2(\mathbf{J}^2-2)=0. \quad \text{(A37)}$$

We won't bother extracting the $t^{\mu\nu\cdots}$ for $j>1$, because it is $\Pi(q)$ that we really need to know.

We could have gone through this whole analysis using the $(0,j)$ instead of the $(j,0)$ representation in (A5). In that case we should have defined a symmetric traceless object $\bar{t}^{\mu_1\mu_2\cdots\mu_{2j}}$ which is a tensor in the sense that

$$\bar{D}^{(j)}[\Lambda]\bar{t}^{\mu_1\mu_2\cdots\mu_{2j}}\bar{D}^{(j)}[\Lambda]^\dagger$$
$$=\Lambda_{\nu_1}{}^{\mu_1}\Lambda_{\nu_2}{}^{\mu_2}\cdots\Lambda_{\nu_{2j}}{}^{\mu_{2j}}\bar{t}^{\nu_1\nu_2\cdots\nu_{2j}}, \quad \text{(A38)}$$

where $\bar{D}^{(j)}[\Lambda]$ is the matrix corresponding to Λ in the $(0,j)$ representation:

$$\bar{D}^{(j)}[\Lambda]=D^{(j)}[\Lambda^{-1}]^\dagger. \quad \text{(A39)}$$

The fundamental formula (A12) would then read

$$\bar{\Pi}^{(j)}(q)=m^{2j}\bar{D}^{(j)}[L(\mathbf{q})]^2=m^{2j}D^{(j)}[L(-\mathbf{q})]^2$$
$$=m^{2j}\exp(2\theta\hat{q}\cdot\mathbf{J}^{(j)}), \quad \text{(A40)}$$

where

$$\bar{\Pi}^{(j)}(q)\equiv(-)^{2j}\bar{t}^{\mu_1\mu_2\cdots\mu_{2j}}q_{\mu_1}q_{\mu_2}\cdots q_{\mu_{2j}}. \quad \text{(A41)}$$

Hence

$$\bar{t}^{\mu_1\mu_2\cdots\mu_{2j}}=(\pm)t^{\mu_1\mu_2\cdots\mu_{2j}}, \quad \text{(A42)}$$

the sign being $+1$ or -1 according to whether the μ's contain altogether an even or an odd number of spacelike indices. There is another relation between barred and unbarred matrices which follows from (6.7):

$$\bar{\Pi}^{(j)}(q)^*=C\Pi^{(j)}(q)C^{-1}, \quad \text{(A43)}$$

and so

$$\bar{t}^{\mu_1\mu_2\cdots\mu_{2j}*}=Ct^{\mu_1\mu_2\cdots\mu_{2j}}C^{-1}. \quad \text{(A44)}$$

Equation (A44) in conjunction with (A42) yields the reality condition

$$t^{\mu_1\mu_2\cdots\mu_{2j}*}=(\pm)Ct^{\mu_1\mu_2\cdots\mu_{2j}}C^{-1}. \quad \text{(A45)}$$

It follows immediately from (A12) and (A40) that

$$\Pi^{(j)}(q)\bar{\Pi}^{(j)}(q)=\bar{\Pi}^{(j)}(q)\Pi^{(j)}(q)=(-q^2)^{2j}. \quad \text{(A46)}$$

Substitution of (A10) and (A41) into (A46) gives

$$t^{\mu_1\mu_2\cdots\mu_{2j}}\bar{t}^{\nu_1\nu_2\cdots\nu_{2j}}q_{\mu_1}q_{\mu_2}\cdots q_{\mu_{2j}}q_{\nu_1}q_{\nu_2}\cdots q_{\nu_{2j}}$$
$$=\bar{t}^{\mu_1\mu_2\cdots\mu_{2j}}t^{\nu_1\nu_2\cdots\nu_{2j}}q_{\mu_1}q_{\mu_2}\cdots q_{\mu_{2j}}q_{\nu_1}q_{\nu_2}\cdots q_{\nu_{2j}}$$
$$=(-q^2)^{2j}. \quad \text{(A47)}$$

Since this holds true for any q, we can use it to derive formulas for any symmetrized product of t and $\bar{t}$. For $j=\frac{1}{2}$:

$$\tfrac{1}{2}[t^\mu\bar{t}^\nu+t^\nu\bar{t}^\mu]=\tfrac{1}{2}[\bar{t}^\mu t^\nu+\bar{t}^\nu t^\mu]=-g^{\mu\nu}. \quad \text{(A48)}$$

APPENDIX B: DIRAC MATRICES FOR ANY SPIN

We will use the $2j+1$-dimensional matrices $t^{\mu\nu\cdots}$, $\bar{t}^{\mu\nu\cdots}$ discussed in Appendix A to construct a set of $2(2j+1)$-dimensional matrices:

$$\gamma^{\mu_1\mu_2\cdots\mu_{2j}}\equiv-i^{2j}\begin{bmatrix}0 & t^{\mu_1\mu_2\cdots\mu_{2j}}\\ \bar{t}^{\mu_1\mu_2\cdots\mu_{2j}} & 0\end{bmatrix}, \quad \text{(B1)}$$

$$\gamma_5\equiv\begin{bmatrix}1 & 0\\ 0 & -1\end{bmatrix}, \quad \text{(B2)}$$

$$\beta\equiv\begin{bmatrix}0 & 1\\ 1 & 0\end{bmatrix}. \quad \text{(B3)}$$

Their properties follow immediately from the work of Appendix A.

1. Lorentz Transformations

It follows from (A5), (A38), and (A39) that the γ's are tensors, in the sense that

$$\mathfrak{D}^{(j)}[\Lambda]\gamma^{\mu_1\mu_2\cdots\mu_{2j}}\mathfrak{D}^{(j)-1}[\Lambda]$$
$$=\Lambda_{\nu_1}{}^{\mu_1}\Lambda_{\nu_2}{}^{\mu_2}\cdots\Lambda_{\nu_{2j}}{}^{\mu_{2j}}\gamma^{\nu_1\nu_2\cdots\nu_{2j}}, \quad \text{(B4)}$$

where $\mathfrak{D}^{(j)}$ is the $(j,0)\oplus(0,j)$ representation

$$\mathfrak{D}^{(j)}[\Lambda]\equiv\begin{bmatrix}D^{(j)}[\Lambda] & 0\\ 0 & \bar{D}^{(j)}[\Lambda]\end{bmatrix}. \quad \text{(B5)}$$

Obviously γ_5 is a scalar

$$\mathfrak{D}^{(j)}[\Lambda]\gamma_5\mathfrak{D}^{(j)-1}[\Lambda]=\gamma_5, \quad \text{(B6)}$$

but β is not, because

$$\beta=-i^{-2j}\gamma^{00\cdots0}. \quad \text{(B7)}$$

2. Symmetry and Tracelessness

The t and $\bar{t}$ are symmetric and traceless in the μ indices, so γ is also:

$$\gamma^{\mu_1\mu_2\cdots\mu_{2j}}=\gamma^{\mu_1'\mu_2'\cdots\mu_{2j}'} \quad \text{(any permutation)}, \tag{B8}$$

$$g_{\mu_1\mu_2}\gamma^{\mu_1\mu_2\cdots\mu_{2j}}=0. \tag{B9}$$

3. Algebra

I have not studied the algebra generated by these γ matrices in detail, but there is one simple relation that can be derived very easily. It follows from (A47) that for any q:

$$\gamma^{\mu_1\mu_2\cdots\mu_{2j}}\gamma^{\nu_1\nu_2\cdots\nu_{2j}}q_{\mu_1}q_{\mu_2}\cdots q_{\mu_{2j}}q_{\nu_1}q_{\nu_2}\cdots q_{\nu_{2j}}=(q^2)^{2j}. \tag{B10}$$

Cancellation of the q's gives the symmetrized product of two γ's as a symmetrized product of $g^{\mu\nu}$. For example, it follows from (B10) that

$$j=\tfrac{1}{2}:\quad \{\gamma^\mu,\gamma^\nu\}=2g^{\mu\nu}, \tag{B11}$$

$$j=1:\quad \{\gamma^{\mu\rho},\gamma^{\nu\lambda}\}+\{\gamma^{\mu\nu},\gamma^{\rho\lambda}\}+\{\gamma^{\mu\lambda},\gamma^{\rho\nu}\} =2[g^{\mu\nu}g^{\rho\lambda}+g^{\mu\rho}g^{\nu\lambda}+g^{\mu\lambda}g^{\nu\rho}], \tag{B12}$$

and so on.

4. Evaluation

Comparison with (A10) and (A41) shows that

$$\mathcal{P}(q)\equiv -i^{2j}\gamma^{\mu_1\mu_2\cdots\mu_{2j}}q_{\mu_1}q_{\mu_2}\cdots q_{\mu_{2j}} =\begin{bmatrix} 0 & \Pi(q) \\ \bar{\Pi}(q) & 0 \end{bmatrix}. \tag{B13}$$

The matrix $\Pi(q)$ was evaluated in Appendix A, and $\bar{\Pi}(q)$ is just

$$\bar{\Pi}(q)=\Pi(-\mathbf{q},q^0). \tag{B14}$$

It follows that we can calculate $\mathcal{P}(q)$ from the formulae (A29) and (A33) for $\Pi(q)$, by making the substitution

$$\mathbf{J}^{(j)}\to\mathfrak{J}^{(j)}\gamma_5,\quad \text{where}\quad \mathfrak{J}^{(j)}\equiv\begin{bmatrix} \mathbf{J}^{(j)} & 0 \\ 0 & \mathbf{J}^{(j)} \end{bmatrix} \tag{B15}$$

and then multiplying the whole resulting formula on the right by β. We find that for integer j:

$$\mathcal{P}^{(j)}(q)=(-q^2)^j\beta +\sum_{n=0}^{j-1}\frac{(-q^2)^{j-1-n}}{(2n+2)!}(2\mathbf{q}\cdot\mathfrak{J})[(2\mathbf{q}\cdot\mathfrak{J})^2-(2\mathbf{q})^2] \times[(2\mathbf{q}\cdot\mathfrak{J})^2-(4\mathbf{q})^2]\cdots[(2\mathbf{q}\cdot\mathfrak{J})^2-(2n\mathbf{q})^2] \times[2\mathbf{q}\cdot\mathfrak{J}\beta-(2n+2)q^0\gamma_5\beta], \tag{B16}$$

and for half-integer j:

$$\mathcal{P}^{(j)}(q)=(-q^2)^{j-1/2}[q^0\beta-2\mathbf{q}\cdot\mathfrak{J}\gamma_5\beta] +\sum_{n=0}^{j-1/2}\frac{(-q^2)^{j-n-1/2}}{(2n+1)!}[(2\mathbf{q}\cdot\mathfrak{J})^2-\mathbf{q}^2] \times[(2\mathbf{q}\cdot\mathfrak{J})^2-(3\mathbf{q})^2]\cdots[(2\mathbf{q}\cdot\mathfrak{J})^2-([2n-1]\mathbf{q})^2] \times[(2n+1)q^0\beta-2\mathbf{q}\cdot\mathfrak{J}\gamma_5\beta]. \tag{B17}$$

The results for $j\leqq 2$ are presented in Table II.

5. Spin $\frac{1}{2}$ and 1

Table II gives

$$\mathcal{P}^{(1/2)}(q)\equiv -i\gamma^\mu q_\mu=q^0\beta-2(\mathbf{q}\cdot\mathfrak{J})\gamma_5\beta,$$

so that

$$\gamma^0=-i\beta=\begin{bmatrix} 0 & -i \\ -i & 0 \end{bmatrix}, \qquad \boldsymbol{\gamma}=-2i\mathfrak{J}\gamma_5\beta=\begin{bmatrix} 0 & -i\boldsymbol{\sigma} \\ i\boldsymbol{\sigma} & 0 \end{bmatrix}. \tag{B18}$$

This is just the standard representation of the Dirac matrices with γ_5 diagonal.

For spin 1, Table II gives

$$\mathcal{P}^{(1)}(q)\equiv\gamma^{\mu\nu}q_\mu q_\nu=-q^2\beta+2(\mathbf{q}\cdot\mathfrak{J})(\mathbf{q}\cdot\mathfrak{J}\beta-q^0\gamma_5\beta),$$

so that

$$\gamma^{00}=\beta,\qquad \gamma^{i0}=\gamma^{0i}=\mathfrak{J}_i\gamma_5\beta,\qquad \gamma^{ij}=\{\mathfrak{J}_i,\mathfrak{J}_j\}\beta-\delta_{ij}\beta. \tag{B19}$$

Notes added in proof. (1) The external-line wave functions are much simpler in the Jacob-Wick helicity formalism. They are given for both massive and massless particles in a second article on the Feynman rules for any spin (Phys. Rev., to be published). (We also give general rules for constructing Lorentz-invariant interactions involving derivatives, field adjoints, etc.) (2) It is not strictly necessary to introduce $2(2j+1)$-component fields in order to satisfy **P** and **C** conservation, because the χ_σ fields in (6.15) and (6.17) may be expressed in terms of φ_σ by using (7.17). I would like to thank H. Stapp for a discussion on this point.

Representations of the Poincaré group for relativistic extended hadrons

Y. S. Kim

Center for Theoretical Physics, Department of Physics and Astronomy, University of Maryland, College Park, Maryland 20742

Marilyn E. Noz

Department of Radiology, New York University, New York, New York 10016

S. H. Oh

Laboratory of Nuclear Science and Department of Physics, Massachusetts Institute of Technology, Cambridge, Massachusetts 02139

(Received 7 June 1978; revised manuscript received 23 October 1978)

Representations of the Poincaré group are constructed from the relativistic harmonic oscillator wave functions which have been effective in describing the physics of internal quark motions in the relativistic quark model. These wave functions are solutions of the Lorentz-invariant harmonic oscillator differential equation in the "cylindrical" coordinate system moving with the hadronic velocity in which the time-separation variable is treated separately. This result enables us to assert that the hadronic mass spectrum is generated by the internal quark level excitation, and that the hadronic spin is due to the internal orbital angular momentum. An addendum relegated to PAPS contains discussions of detailed calculational aspects of the Lorentz transformation, and of solutions of the oscillator equation which are diagonal in the Casimir operators of the homogeneous Lorentz group. It is shown there that the representation of the homogeneous Lorentz group consists of solutions of the oscillator partial differential equation in a "spherical" coordinate system in which the Lorentz-invariant Minkowskian distance between the constituent quarks is the radial variable.

I. INTRODUCTION

In building models of relativistic extended hadrons, we have to keep in mind the fundamental fact that the overall space–time symmetry structure is that of the Poincaré group.[1] In our previous papers on physical applications of the relativistic harmonic oscillator,[2] our primary purpose was to devise a calculational scheme for explaining experimental observations. As was pointed out by Biedenharn *et al.*,[3] the question of the Poincaré symmetry has not been systematically discussed.

The purpose of the present paper is to address this symmetry problem. We are considering a model hadron consisting of two spinless quarks bound together by a harmonic oscillator potential. In this case, we are led to consider the center-of-mass coordinate which specifies the space–time location of the hadron, and the relative coordinate which specifies the internal space–time separation between the quarks.

Both the hadronic and internal coordinates are subject to Poincaré transformations consisting of translations and Lorentz transformations. The hadronic coordinate undergoes Poincaré transformation in the usual manner. However, the internal coordinate is invariant under translations. This coordinate should, nonetheless, satisfy the Poincaré symmetry as a whole. We discuss in this paper the role of this internal coordinate, and show that internal excitations generate the hadronic mass spectrum, and that the internal angular momentum corresponds to the spin of the hadron.

In Sec. II, we formulate the problem using a model hadron consisting of two spinless quarks bound together by a harmonic oscillator potential of unit strength, and then discuss the generators of the Poincaré group applicable to the entire system. In Sec. III, we present the oscillator wave functions which are diagonal in the invariant Casimir operators of the Poincaré group.

II. FORMULATION OF THE PROBLEM

In our previous papers on physical applications of the relativistic harmonic oscillators, we started with the following Lorentz-invariant differential equation:

$$\{2[\Box_1 + \Box_2] - \tfrac{1}{16}(x_1 - x_2)^2 + m_0^2\}\phi(x_1,x_2) = 0, \quad (1)$$

where x_1 and x_2 are the space–time coordinates for the two spinless quarks bound together by a harmonic oscillator potential with unit spring constant. In order to simplify the above equation, let us define new coordinate variables

$$X = \tfrac{1}{2}(x_1 + x_2), \quad x = (1/2\sqrt{2})(x_1 - x_2). \quad (2)$$

The X coordinate represents the space–time specification of the hadron as a whole, while the x variable measures the relative space–time separation between the quarks. In terms of these variables, Eq. (1) can be written as

$$\left[\frac{\partial^2}{\partial X_\mu^{\ 2}} + m_0^2 + \frac{1}{2}\left(\frac{\partial^2}{\partial x_\mu^{\ 2}} - x_\mu^{\ 2}\right)\right]\phi(X,x) = 0. \quad (3)$$

The above equation is separable in the X and x variables. Thus we write

$$\phi(X,x) = f(X)\psi(x), \quad (4)$$

where $f(X)$ and $\psi(x)$ satisfy the following differential equations respectively:

 0022-2488/79/071341-04$01.00

Reprinted from *J. Math. Phys.* 20, 1341 (1979).

$$\left[\frac{\partial^2}{\partial X_\mu{}^2} + m_0{}^2 + (\lambda + 1)\right] f(X) = 0, \tag{5}$$

$$\frac{1}{2}\left(\frac{\partial^2}{\partial x_\mu{}^2} - x_\mu{}^2\right)\psi(x) = (\lambda + 1)\psi(x). \tag{6}$$

The differential equation of Eq. (5) is a Klein–Gordon equation, and its solutions are well known. $f(X)$ takes the form

$$f(X) = \exp(\pm ip \cdot X), \tag{7}$$

with

$$p^2 = m_0{}^2 + (\lambda + 1), \tag{8}$$

where p is the 4-momentum of the hadron. p^2 is, of course, the mass of the hadron and is numerically constrained to take the values allowed by Eq. (8). The separation constant λ is determined from the solutions of the harmonic oscillator differential equation of Eq. (6). The physical solutions of the oscillator equation satisfy the subsidiary condition

$$p^\mu a_\mu^\dagger \psi_\beta(x) = 0, \tag{9}$$

where

$$a_\mu^\dagger = x_\mu + \frac{\partial}{\partial x^\mu}$$

The physics of this subsidiary condition has been extensively discussed in the literature.[2,4]

The space–time transformation of the total wave function of Eq. (4) is generated by the following ten generators of the Poincaré group. The operators

$$P_\mu = i\frac{\partial}{\partial X^\mu} \tag{10}$$

generate space–time translations. Lorentz transformations, which include boosts and rotations, are generated by

$$M_{\mu\nu} = L^{\bullet}_{\mu\nu} + L_{\mu\nu}, \tag{11}$$

where

$$L^{\bullet}_{\mu\nu} = i\left(X_\mu \frac{\partial}{\partial X^\nu} - X_\nu \frac{\partial}{\partial X^\mu}\right),$$

$$L_{\mu\nu} = i\left(x_\mu \frac{\partial}{\partial x^\nu} - x_\nu \frac{\partial}{\partial x^\mu}\right).$$

The translation operators P_μ act only on the hadronic coordinate, and do not affect the internal coordinate. The operators $L^{\bullet}_{\mu\nu}$ and $L_{\mu\nu}$ Lorentz-transform the hadronic and internal coordinates respectively. The above ten generators satisfy the commutation relations for the Poincaré group.

In order to consider irreducible representations of the Poincaré group, we have to construct wave functions which are diagonal in the invariant Casimir operators of the group, which commute with all the generators of Eqs. (10) and (11). The Casimir operators in this case are

$$P^\mu P_\mu \quad \text{and} \quad W^\mu W_\mu, \tag{12}$$

where

$$W_\mu = \epsilon_{\mu\nu\alpha\beta} P^\nu M^{\alpha\beta}.$$

The eigenvalues of the above P^2 and W^2 represent respectively the mass and spin of the hadron.

III. PHYSICAL WAVE FUNCTIONS AND REPRESENTATIONS OF THE POINCARÉ GROUP

In constructing wave functions diagonal in the Casimir operators of the Poincaré group, we note first that the operator which acts on the wave function in the subsidiary condition of Eq. (9) commutes with these invariant operators:

$$[P^2, p^\mu a_\mu^\dagger] = 0, \tag{13}$$

$$[W^2, p^\mu a_\mu^\dagger] = 0. \tag{14}$$

Therefore, the wave functions satisfying the condition of Eq. (9) can be diagonal in the Casimir operators.

In order to obtain the solutions explicitly, let us assume without loss of generality that the hadron moves along the z direction with the velocity parameter β. Then we are led to consider the Lorentz frame where the hadron is at rest, and the coordinate variables are given by

$$\begin{aligned} x' &= x, \quad y' = y, \\ z' &= (z - \beta t)/(1 - \beta^2)^{1/2}, \\ t' &= (t - \beta z)/(1 - \beta^2)^{1/2}. \end{aligned} \tag{15}$$

The Lorentz-invariant oscillator equation of Eq. (6) is separable in the above variables. In terms of these primed variables, we can construct a complete set of wave functions

$$\psi_\beta(x) = f_b(x') f_s(y') f_n(z') f_k(t'), \tag{16}$$

where

$$f_n(z') = (\sqrt{\pi}\, 2^n n!)^{-1/2} H_n(z') \exp(-z'^2/2),$$

$$f_k(t') = (\sqrt{\pi}\, 2^k k!)^{-1/2} H_k(t') \exp(-t'^2/2).$$

If the excitation numbers, $b,\ldots,k$ are allowed to take all possible nonnegative integer values, the solutions in Eq. (16) form a complete set. However, the eigenvalues λ takes the form

$$\lambda = b + s + n - k. \tag{17}$$

Because the coefficient of k is negative in the above expression, λ has no lower bound, and there is an infinite degeneracy for a given value of λ.

In terms of the primed coordinates, the subsidiary condition of Eq. (9) takes the simple form

$$\left(\frac{\partial}{\partial t'} + t'\right)\psi_\beta(x) = 0. \tag{18}$$

This limits $f_k(t')$ to $f_0(t')$, and the eigenvalue λ becomes

$$\lambda = b + s + n, \tag{19}$$

The physical wave functions satisfying the subsidiary condition of Eq. (9) or (18) have nonnegative values of λ.

As far as the x', y', z' coordinates are concerned, they form an orthogonal Euclidean space, and $f_b(x')$, $f_s(y')$, $f_n(z')$ form a complete set in this three-dimensional space. The Hermite polynomials in these Cartesian wave functions can then be combined to form the eigenfunctions of W^2 which, in terms of the primed coordinate variables, takes the form

$$W^2 = M^2(\mathbf{L}')^2, \tag{20}$$

where

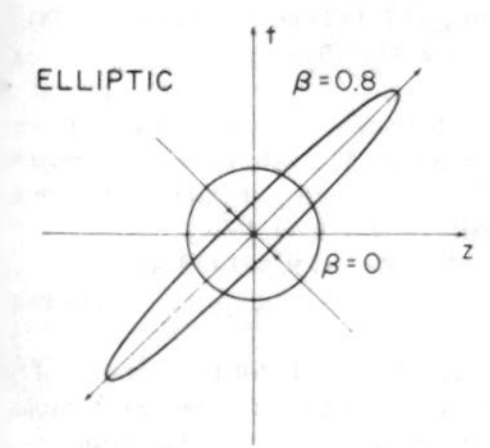

FIG. 1. Elliptic and hyperbolic localizations in space–time. The wave functions in the present paper are elliptically localized, and undergo Lorentz deformation as the hadron moves. The Lorentz invariant form $x^\mu x_\mu$, to which we are accustomed, is hyperbolically localized, and is basically different from the form used in the present paper.

$$L'_i = -i\epsilon_{ijk}x'_j \frac{\partial}{\partial x'_k},$$

and M is the hadronic mass.

The physical wave functions now take the form

$$\psi_\beta^{\lambda lm}(x) = (1/\pi)^{1/4}[\exp(-t'^2)]R_{\lambda l}(r')Y_{lm}(\theta',\phi'), \quad (21)$$

where r', θ', ϕ' are the radial and spherical variables in the three-dimensional space spanned by x', y', z'. $R_{\lambda l}(r')$ is the normalized radial wave function for the three-dimensional isotropic harmonic oscillator, and its form is well known. The above wave function is diagonal in W^2 for which the eigenvalue is $l(l+1)M^2$, and l represents the total spin of the hadron in the present case. The quantum number m corresponds to the helicity.

Since the eigenvalue p^2 of the Casimir operator P^2 is constrained to take the numerical values allowed by Eq. (8), the hadronic mass is given by

$$M^2 = m_0^2 + (\lambda + 1). \quad (22)$$

If we relax the subsidiary condition of Eq. (18), we indeed obtain a complete set. In this case, λ of Eq. (17) can become negative for sufficiently large values of k. For $\lambda > 0$, the solutions become

$$\psi_\beta^{\lambda lmk}(x) = [\sqrt{\pi}\, 2^k k!]^{-1/2} H_k(t')[\exp(-t'^2/2)] \times R_{\lambda+k,l}(r')Y_{lm}(\theta',\phi'). \quad (23)$$

For $\lambda < 0$, the solutions take the form

$$\psi^{\lambda lmk}(x) = [\sqrt{\pi}\, 2^{(k-\lambda)}(k-\lambda)!]^{-1/2} H_{k-\lambda}(t') \times [\exp(-t'^2)]R_{k,l}(r')Y_{lm}(\theta',\phi'). \quad (24)$$

The eigenvalues of P^2 and W^2 are again $m_0^2 + (\lambda + 1)$ and $l(l+1)M^2$ respectively. In both of the above cases, k is allowed to take all possible integer values.

The functional forms of Eqs. (23) and (24) are relatively simple, and they suggest that this representation of the Poincaré group corresponds to the solution of the Lorentz-invariant oscillator differential equation in a "cylindrical" coordinate system moving with the hadronic velocity where the t' variable is treated separately. We are then led to the question of why this fact was not known.

Even though the above representations take simple forms, the wave functions contain the following nonconventional features. The first point to note is that they are written as functions of the x', y', z', t' variables. The transverse variables x', y' are simply x and y respectively. However, z' and t' are linear combinations of z and t. Because the physical meaning of the time-separation variable was not clearly understood, the t dependence discouraged us in the past from using it explicitly in representation theory. The explicit use of this variable in the present paper is based on the progress that has been made in our physical understanding of this time-separation variable in terms of measurable quantities, and in terms of the relativistic wave functions carrying a covariant probability interpretation.[2]

Another factor which used to discourage the use of the t variable was that we are accustomed to its appearance through the form

$$x^\mu x_\mu = t^2 - r^2, \quad (25)$$

where

$$r^2 = x^2 + y^2 + z^2.$$

In terms of this form, it is very inconvenient, if not impossible, to describe functions which are localized in a finite space–time region.

In contrast to the above hyperbolic case, the wave functions which we constructed in this paper are well localized within the region

$$(z'^2 + t'^2) < 2, \quad (26)$$

due to the Gaussian factor appearing in the wave functions. This elliptic form was obtained from the covariant expression

$$-x^\mu x_\mu + 2(x\cdot p/M)^2 = x'^2 + y'^2 + z'^2 + t'^2. \quad (27)$$

The x' and y' variables have been omitted in Eq. (26) because they are trivial. In terms of z and t, the above inequality takes the form

$$\left[\frac{1-\beta}{1+\beta}(z+t)^2 + \frac{1+\beta}{1-\beta}(z-t)^2\right] < 2. \quad (28)$$

We are therefore dealing with the function localized within an elliptic region defined by this inequality, and can control the t variable in the same manner as we do in the case of the spatial variables appearing in nonrelativistic quantum mechanics. This localization property together with the hyperbolic case is illustrated in Fig. 1.

IV. CONCLUDING REMARKS

We have shown in this paper that the wave functions used in our previous papers are diagonal in the Casimir operators of the Poincaré group, which specify covariantly the mass and total spin of the hadron. These wave functions are well localized in a space–time region, and undergoes elliptic Lorentz deformation.

An addendum to this paper containing a discussion of Lorentz transformation of the physical wave function and a construction of the representation of the homogeneous Lorentz group is relegated to PAPS.[5] It is shown there that solutions of the oscillator equation diagonal in the Casimir operators of the homogeneous Lorentz group are localized within the Lorentz-invariant hyperbolic region illustrated in Fig. 1.

[1]E.P. Wigner, Ann. Math. **40**, 149 (1939).
[2]Y.S. Kim and M.E. Noz, Phys. Rev. D **8**, 3521 (1973), **12**, 129 (1975); **15**, 335 (1977); Y.S. Kim, J. Korean Phys. Soc. **9**, 54 (1976); **11**, 1 (1978); Y.S. Kim and M.E. Noz, Prog. Theor. Phys. **57**, 1373 (1977); **60**, 801 (1978); Y.S. Kim and M.E. Noz, Found. Phys. **9**, 375 (1979); Y.S. Kim, M.E. Noz, and S.H. Oh, "Lorentz Deformation and the Jet Phenomenon," Found. Phys. (to be published). For review articles written for teaching purposes, see Y.S. Kim and M.E. Noz, Am. J. Phys. **46**, 480, 486 (1978). For a review written for the purpose of formulating a field theory of extended hadrons, see T.J. Karr, Ph.D. thesis (University of Maryland, 1976).
[3]L.C. Biedenharn and H. van Dam, Phys. Rev. D **9**, 471 (1974).
[4]T. Takabayashi, Phys. Rev. **139**, B1381 (1965); S. Ishida and J. Otokozawa, Prog. Theor. Phys. **47**, 2117 (1972).
[5]See AIP document no. PAPS JMAPA–20–1336–12 for twelve pages of discussions of the Lorentz transformation of the physical wave functions, and of the representations of the homogeneous Lorentz group. Order by PAPS number and journal reference from American Institute of Physics, Physics Auxiliary Publication Service, 335 East 45th Street, New York, N.Y. 10017. The price is $1.50 for each microfiche (98 pages), or $5 for photocopies of up to 30 pages with $0.15 for each additional page over 30 pages. Airmail additional. Make checks payable to the American Institute of Physics. This material also appears in *Current Physics Microfilm*, the monthly microfilm edition of the complete set of journals published by AIP, on the frames immediately following this journal article.

A simple method for illustrating the difference between the homogeneous and inhomogeneous Lorentz groups

Y. S. Kim
Center for Theoretical Physics, Department of Physics and Astronomy, University of Maryland, College Park, Maryland 20742

Marilyn E. Noz
Department of Radiology, New York University, New York, New York 10016

S. H. Oh
Laboratory for Nuclear Science and Department of Physics, Massachusetts Institute of Technology, Cambridge, Massachusetts 02139
(Received 5 January 1979; accepted 12 June 1979)

A method is proposed for illustrating the difference between the Poincaré (inhomogeneous Lorentz) and homogeneous Lorentz groups. Representations of the Poincaré group are constructed from solutions of the relativistic harmonic oscillator equation whose physical wave functions have been effective in describing basic high-energy hadronic features in the relativistic quark model. It is shown that the Poincaré group can be represented by solutions of the relativistic oscillator equation in a "moving cylindrical" coordinate system in which the time-separation variable in the hadronic rest frame is treated separately. Representations which are diagonal in the Casimir operators of the homogeneous Lorentz group are also constructed from solutions of the same oscillator differential equation in a hyperbolic coordinate system. It is pointed out that the difference between the Poincaré and homogeneous Lorentz groups mainfests itself in the coordinate systems in which the oscillator differential equation in separable.

I. INTRODUCTION

In our previous papers in this journal, we discussed the possibility of using the relativistic harmonic oscillator formalism in teaching group theory,[1] and also in discussing high-energy physics in the first-year course in quantum mechanics.[2] It was emphasized in Ref. 1 that the oscillator formalism with its mathematical simplicity can serve as an illustrative example in teaching one of the difficult theorems in group theory. As in the case of Ref. 1, we are interested here in a group theory course for graduate students who have some background in quantum mechanics and special relativity. In such a group theory course, the inhomogeneous Lorentz group, which is often called the Poincaré group, occupies an important place because its representations play the fundamental role of specifying covariantly the mass and spin of a relativistic particle.[3]

The purpose of the present paper is to give a concrete example for illustrating how the Poincaré group is different from a direct product of a translation and the homogeneous Lorentz group which we call here simply the Lorentz group. There are at present two different ways to tell students how the Poincaré group is different from the Lorentz group. One way is to point out that the Casimir operators for the former are different from those of the latter.[4] Another approach is to show that the homogeneous Lorentz group is only a little group of the Poincaré group and tell students that this little group has no physical significance.[5] However, it would be better if we could use concrete examples to illustrate this difference.

In this paper, we use the covariant harmonic oscillator formalism as an illustrative example. We show that the Poincaré and Lorentz groups can be represented respectively by two different sets of solutions of the same Lorentz-invariant partial differential equation. It is shown that the (homogeneous) Lorentz group can be represented by the solutions of the harmonic oscillator equation in a hyperbolic coordinate system in which the Minkowskian space-time distance between the quarks is a "radial" variable, while the Poincaré group is represented by the solutions of the same differential equation in a "cylindrical" coordinate system in which the time-separation variable in the hadronic rest frame is treated separately.

In Sec. II, we formulate the problem in terms of the group generators applicable to solutions of the Lorentz-invariant differential equation for the hadron consisting of two quarks bound together by a harmonic oscillator force. It is pointed out that the Poincaré group is not a direct product but is instead a semidirect product of a translation and a Lorentz transformation, and therefore that the Casimir operators of the Poincaré group are quite different from those of the Lorentz group.

In Sec. III, we construct the wave functions which are diagonal in the Casimir operators of the Poincaré group. It is shown that these wave functions are solutions in a "moving cylindrical" coordinate system in which the time-separation variable is treated separately. It is shown also that the physical solutions form a subset of a complete basis for the Poincaré group. Transformation properties of the physical solutions are also discussed.

Section IV deals with solutions of the oscillator equation which are diagonal in the Casimir operators of the homogeneous Lorentz group. Unlike the Poincaré case, the solutions forming a representation of the homogeneous Lorentz group are obtained in the hyperbolic coordinate system.

 0002-9505/79/100892-06$00.50

Reprinted from *Am. J. Phys.* **47**, 892 (1979).

II. STATEMENT OF THE PROBLEM

Because of its mathematical simplicity, the harmonic oscillator has served as a concrete solution to many physical theories. It played an important role also in the development of relativistic theories of particle physics where hadrons are regarded as bound states of quarks.[6,7] Let us consider in this paper a hadron consisting of two quarks held together by a harmonic oscillator potential of unit strength, and start with the differential equation[7]

$$\{2[\Box_1 + \Box_2] - (1/16)(x_1 - x_2)^2 + m_0^2\}\phi(x_1,x_2) = 0. \tag{1}$$

This partial differential equation has many different solutions depending on the coordinate system in which the equation is separated, and on the boundary conditions. The purpose of this paper is to discuss two different sets of solutions which are useful for teaching group theory.

In order to simplify the above equation, let us define new coordinate variables

$$X = (x_1 + x_2)/2,$$
$$x = (x_1 - x_2)/2\sqrt{2}. \tag{2}$$

The X coordinate represents the space-time specification of the hadron as a whole, while the x variable measures the relative space-time separation between the quarks. In terms of these variables, Eq. (1) can be written

$$\left[\frac{\partial^2}{\partial X_\mu^2} + m_0^2 + \frac{1}{2}\left(\frac{\partial^2}{\partial x_\mu^2} - x_\mu^2\right)\right]\phi(X,x) = 0. \tag{3}$$

This equation is separable in the X and x variables. Thus

$$\phi(X,x) = f(X)\psi(x), \tag{4}$$

and $f(X)$ and $\psi(x)$ satisfy the following differential equations, respectively:

$$\left(\frac{\partial^2}{\partial X_\mu^2} + m_0^2 + (\lambda + 1)\right)f(X) = 0, \tag{5}$$

$$\frac{1}{2}\left(\frac{\partial^2}{\partial x_\mu^2} - x_\mu^2\right)\psi(x) = (\lambda + 1)\psi(x). \tag{6}$$

Equation (5) is a Klein-Gordon equation, and its solution takes the form

$$f(X) = \exp(\pm ip^\mu X_\mu), \tag{7}$$

with

$$p^2 = M^2 = m_0^2 + (\lambda + 1),$$

where M and p are, respectively, the mass and four momentum of the hadron. The eigenvalue λ is determined from the solutions of Eq. (6).

The wave function $\psi(x)$ of Eq. (6) is a solution of the harmonic oscillator differential equation whose physical solutions have been discussed extensively in the literature. Because it is separable in many different coordinate systems, the oscillator differential equation given in Eq. (6) generates many solutions of mathematical interest, in addition to those which carry a physical interpretation.

We shall use these properties of the oscillator equation to illustrate the following important point to be covered in group theory courses for physicists. The most general space-time transformation consists of a translation T and a Lorentz transformation Λ. We call this a Poincaré transformation and write it as

$$A = (T,\Lambda). \tag{8}$$

For two successive Poincaré transformations A_1 and A_2,

$$A_2A_1 = (T_2(\Lambda_2T_1),\Lambda_2\Lambda_1),$$
$$\neq (T_2T_1,\Lambda_2\Lambda_1). \tag{9}$$

For this reason, the Poincaré group is not a direct product but is only a semidirect product of T and Λ.

In terms of the operators applicable to the wave function of Eq. (4), space-time translations are generated by

$$P_\mu = i\frac{\partial}{\partial X^\mu}, \tag{10}$$

and the Lorentz transformation generators take the form

$$M_{\mu\nu} = L_{\mu\nu}^* + L_{\mu\nu}, \tag{11}$$

where

$$L_{\mu\nu}^* = i\left(X_\mu\frac{\partial}{\partial X^\nu} - X_\nu\frac{\partial}{\partial X^\mu}\right), \tag{12}$$

$$L_{\mu\nu} = i\left(x_\mu\frac{\partial}{\partial x^\nu} - x_\nu\frac{\partial}{\partial x^\mu}\right). \tag{13}$$

The above four translation and six Lorentz-transformation operators form the ten generators of the Poincaré group. The invariant Casimir operators which commute with all of the Poincaré group generators are

$$P^2 = P^\mu P_\mu,$$

and

$$W^2 = W^\mu W_\mu, \tag{14}$$

where $W_\mu = \epsilon_{\mu\nu\alpha\beta}P^\nu M^{\alpha\beta}$.

The solutions of the oscillator equation which are diagonal in the above Casimir operators have been discussed in the literature.[1,2,8] In Sec. III of this paper, we discuss explicitly the transformation properties of the wave functions representing the Poincaré group.

Since the relative coordinate x defined in Eq. (2) is not affected by translation operations, it is difficult to resist the temptation to construct as well solutions which are diagonal in the operators which commute only with the six $L_{\mu\nu}$ operators of Eq. (13). The Casimir operators for this homogeneous Lorentz group are

$$C_1 = (1/2)L^{\mu\nu}L_{\mu\nu}, \tag{15}$$

$$C_2 = (1/4)\epsilon_{\mu\nu\alpha\beta}L^{\mu\nu}L^{\alpha\beta}. \tag{16}$$

These operators are basically different from those of the Poincaré group.[4] We shall see this difference more explicitly by constructing representations of the homogeneous Lorentz group in Sec. IV.

III. REPRESENTATIONS OF THE POINCARÉ GROUP

In this section we discuss wave functions which are diagonal in the Casimir operators of Eq. (14). Since the hadron in this case has a definite four momentum p specified in Eq. (7), we are naturally interested in the Lorentz frame

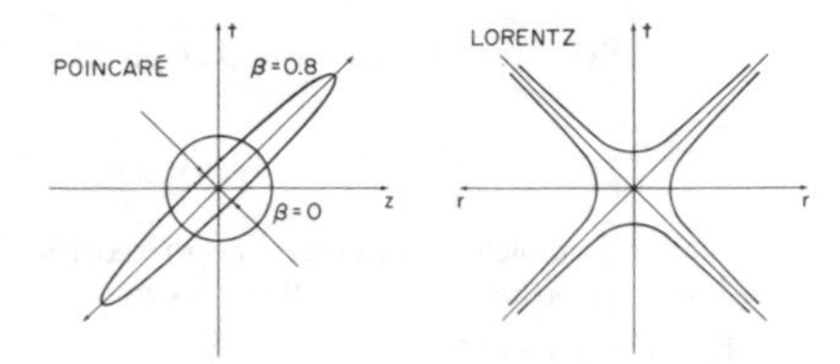

Fig. 1. Concentration regions for the wave functions representing the Poincaré and homogeneous Lorentz groups. The concentration region in both cases is dictated by the Gaussian factor in the wave function. In the case the Poincaré group, the wave function is concentrated within an ellipse specified in the figure. If $\beta = 0$, the ellipse becomes a circle. As β increases, the ellipse becomes more eccentric. Unlike the Poincaré case, the wave functions representing the homogeneous Lorentz group are concentrated within the hyperbolic region around the light cones. This concentration region does not depend on β, and is therefore Lorentz invariant.

in which the hadron is at rest.[9] The coordinate variables for this frame are

$$x' = x, \; y' = y,$$
$$z' = (z - \beta t)/(1 - \beta^2)^{1/2}, \tag{17}$$
$$t' = (t - \beta z)/(1 - \beta^2)^{1/2},$$

where $\beta = p_z/p_0$. We assume here that the hadron moves along the z axis.

In terms of these rest-frame variables, we can write the Casimir operator W^2 of Eq. (14) simply as

$$W^2 = M^2(\mathbf{L}')^2, \tag{18}$$

where

$$L_i' = i\epsilon_{ijk}x_j' \frac{\partial}{\partial x_k'} .$$

The Casimir operator P^2 is independent of the variable x, but is constrained to take the eigenvalue of the oscillator equation given in Eq. (6). In terms of the rest-frame coordinate variables, we can write the P^2 operator applicable to the wave function $\psi(x)$ as

$$P^2 = m_0^2 + \frac{1}{2}\left(\frac{\partial^2}{\partial x_\mu'^2} - x_\mu'^2\right). \tag{19}$$

With the above forms of the Casimir operators, it is not difficult to construct normalizable wave functions which are diagonal in W^2 and P^2. First, we observe that the oscillator equation of Eq. (6) is Lorentz invariant and is separable also in the x', y', z', t' variables. We can therefore choose the cylindrical coordinate system in which the t' variable is treated separately, and write the wave function as

$$\psi(x) = f(t')g(\mathbf{x}'), \tag{20}$$

where $f(t')$ is a solution of the one-dimensional harmonic oscillator equation and takes the form

$$f_k(t') = (\sqrt{\pi}2^k k!)^{-1/2}H_k(t')\exp(-t'^2/2). \tag{21}$$

This portion of the wave function is not affected by the Casimir operator W^2. $g(\mathbf{x}')$ satisfies the three-dimensional isotropic oscillator equation, and the solutions diagonal in W^2 can be written

$$g(\mathbf{x}') = R_{\mu l}(r')Y_l^m(\theta',\phi'), \tag{22}$$

where

$$r' = (x'^2 + y'^2 + z'^2)^{1/2}.$$

$R_{\mu l}(r')$ is the normalized radial wave function for the oscillator, and its form is well known. The total wave function now takes the form

$$\psi_\beta^{\lambda lmk}(x) = R_{\mu l}(r')Y_l^m(\theta',\phi')f_k(t'), \tag{23}$$

with $\lambda = \mu - k$. This function generates the eigenvalues M^2 and $l(l+1)M^2$ for the Casimir operators P^2 and W^2, respectively. The physical wave functions discussed in the literature are those of the above functions with $k = 0$. Due to the Gaussian factors, the above wave functions decrease rapidly outside the region:

$$(r'^2 + t'^2) < 2. \tag{24}$$

If we ignore the trivial transverse coordinates, the above inequality becomes

$$z'^2 + t'^2 = \frac{1}{2}\left(\frac{1-\beta}{1+\beta}(z+t)^2 + \frac{1+\beta}{1-\beta}(z-t)^2\right) < 2. \tag{25}$$

This β-dependent elliptic region is illustrated in Fig. 1.

The wave functions given in Eq. (23) can serve a useful purpose in illustrating the group theoretical identities given in the classic paper of Naimark which is standard reference material in group theory courses for physicists.[10,11] Reference 10 gives a detailed discussion of the recurrence relations for the generators of the Lorentz transformation as well as those for the well-known case of the rotation group. The $L_{\mu\nu}$ of Eq. (13) applicable to the x coordinate represents three rotation and three boost generators. The rotation operators are L_{ij}, where $i,j = 1,2,3$. The boost generators consist of $K_i = L_{io}$. The rotation generators take the form

$$L_i = i\epsilon_{ijk}x_j \frac{\partial}{\partial x_k}, \tag{26}$$

and the boost generators take the form

$$K_i = -i\left(x_i \frac{\partial}{\partial t} + t\frac{\partial}{\partial x_i}\right), \tag{27}$$

where $i = 1,2,3$.

The rotation around the $\mathbf{n}$ axis through an angle ξ is represented by

$$R(\mathbf{n},\xi) = \exp[-i\xi\mathbf{n}\cdot\mathbf{L}], \tag{28}$$

and its mathematics is well known. The Lorentz boost along the direction $\mathbf{n}$ by η is

$$T(\mathbf{n},\eta) = \exp[-i\eta\mathbf{n}\cdot\mathbf{K}], \tag{29}$$

where

$$\eta = \sinh^{-1}[\beta/(1-\beta^2)^{1/2}].$$

As in the case of rotation, let us discuss first the effect of the generators K_i on the wave functions. It is more convenient to work with K_3 and $K_\pm$,[10] where

$$K_\pm = K_1 \pm iK_2. \tag{30}$$

If we apply these operators to the wave functions of Eq (23) with $\beta = 0$, we obtain

$$iK_3\psi_{lm}^{\lambda k} = A_3\left(\frac{(l+m+1)(l-m+1)}{(2l+1)(2l+3)}\right)^{1/2} Y_{l+1}^m(\theta,\phi) + B_3\left(\frac{(l+1)(l-m)}{(2l+1)(2l-1)}\right)^{1/2} Y_{l-1}^m(\theta,\phi), \quad (31)$$

$$iK_\pm\psi_{lm}^{\lambda k} = A_\pm\left(\frac{(l\pm m+1)(l\pm m+2)}{(2l+1)(2l+3)}\right)^{1/2} Y_{l+1}^{m\pm1}(\theta,\phi) + B_\pm\left(\frac{(l\mp m)(l\mp m-1)}{(2l+1)(2l-1)}\right)^{1/2} Y_{l-1}^{m\mp1}(\theta,\phi).$$

The notation for the wave function in the above expression is slightly different from that used in Eq. (23), but this difference should not cause any confusion. The coefficients of the spherical harmonics given above have been calculated before by Naimark.[10] What is new in the present work is that the coefficients A and B can be calculated in terms explicit functions. These coefficients take the form

$$A_3 = Q_{-l}F_l^{\lambda k}(r,t), \quad B_3 = Q_{l+1}F_l^{\lambda k}(r,t),$$
$$A_\pm = \mp Q_{-l}F_l^{\lambda k}(r,t), \; B_\pm = \pm Q_{l+1}F_l^{\lambda k}(r,t), \quad (32)$$

where

$$Q_l = t\frac{\partial}{\partial r} + r\frac{\partial}{\partial t} + l\frac{t}{r},$$

$$F_l^{\lambda k}(r,t) = \frac{\psi_{lm}^{\lambda k}(x)}{Y_l^m(\theta,\phi)}.$$

In order to understand the nature of the Lorentz transformation $T(\mathbf{n},\eta)$ more clearly, let us concentrate our efforts on the case where the boost is along the z direction. The form of K_3 applied to $\psi_{\lambda k}^{lm}$ in Eq. (31) indicates that the helicity quantum number m is conserved under this boost transformation. However, the transformed wave function contains all possible values of l. This is a reflection of the following nonvanishing commutator:

$$[(\mathbf{L})^2,(\mathbf{L}')^2] \neq 0. \quad (33)$$

The forms of A_3 and B_3 in Eq. (32) indicate that the transformed wave function will contain radial and timelike wave functions which are different from those given initially. In order to see this and other points more clearly, let us consider a finite boost along the z axis. For this purpose, we go back to Eq. (23), and consider a Lorentz transformation of a wave function without timelike oscillations for which the calculation is relatively simple.

Because only the z and t components are affected by the boost along the z direction, we have to rewrite the wave function in terms of the Cartesian variables and their Hermite polynomials. The portion of the wave function which is affected by this transformation is

$$\psi_0^{n,o}(z,t) = \left(\frac{1}{\pi 2^n n!}\right)^{1/2} H_n(z)\exp\left(-\frac{1}{2}(z^2+t^2)\right). \quad (34)$$

The superscript o indicates that there are no timelike excitations: $k = 0$. We now consider the transformation

$$\psi_\beta^{n,o}(z,t) = T(\beta)\psi_0^{n,o}(z,t), \quad (35)$$

$$\psi_\beta^{n,o}(z,t) = \psi_0^{n,o}(z',t'), \quad (36)$$

and ask what $T(\beta)$ does on $\psi_0^{n,o}(z,t)$. In order to answer this question, we write Eq. (35) as

$$\psi_\beta^{n,o}(z,t) = \sum_{n',k'} A_{n',k'}^{n,o}(\beta)\psi_0^{n',k'}(z,t). \quad (37)$$

It is shown in Ref. 1 that this expression can be simplified to

$$\psi_\beta^{n,o}(z,t) = \sum_k^\infty A_k^n(\beta)\psi_0^{n+k,k}(z,t). \quad (38)$$

The remaining problem is to determine the coefficient $A_k^n(\beta)$. Using the orthogonality relation, we can write

$$A_k^n(\beta) = \int dzdt\psi_\beta^{n,o}(z,t)\psi_0^{n+k,k}(z,t)$$
$$= \frac{1}{\pi}\left(\frac{1}{2}\right)^n\left(\frac{1}{2^k n!(n+k)!}\right)^2 \quad (39)$$
$$\times \int dzdtH_{n+k}(z)H_k(t)H_n(z')$$
$$\times \exp\left(-\frac{1}{2}(z^2+z'^2+t^2+t'^2)\right).$$

In the above integral, the Hermite polynomials and the Gaussian form are mixed with the kinematics of Lorentz transformation. However, if we use the generating function for the Hermite polynomial as Ruiz did in his paper,[12] this integral can be calculated easily, and

$$A_k^n(\beta) = (1-\beta^2)^{(n+1)/2}\beta^k\left(\frac{(n+k)!}{n!k!}\right)^{1/2}. \quad (40)$$

Thus Eq. (37) can be rewritten

$$\psi_\beta^{n,o}(z,t) = \left(\frac{1}{\pi}\right)^{1/2}\left(\frac{1}{2}\right)^{n/2}(1-\beta^2)^{(n+1)/2}$$
$$\times\left(\sum_{k=0}^\infty \frac{\beta^k}{2^k k!}H_{n+k}(z)H_k(t)\right)$$
$$\times\exp\left(-\frac{1}{2}(z^2+t^2)\right). \quad (41)$$

Let us now examine the implications of the above result. Since the expression in Eq. (41) requires a sum over the longitudinal excitations equal to or higher than n, the Lorentz transformed wave function with a given l value in the moving frame is a sum over all corresponding l values of the wave function at rest. This result is not inconsistent with $K_3\psi$ of Eq. (31) and the nonvanishing commutator of Eq. (33).

In the hadronic rest frame, the wave function with $k > 0$ is not a physical solution. Therefore the wave function of Eq. (41) is a sum of nonphysical solutions in the rest frame. However, after the summation, these wave functions in the rest frame form a physical wave function corresponding to a hadron moving with velocity parameter β. The wave function of Eq. (41) satisfies the subsidiary condition

$$\left(\frac{\partial}{\partial t'} + t'\right)\psi_\beta(x) = 0, \quad (42)$$

whose physical interpretation has been extensively discussed in the literature.[1,2]

For the Lorentz transformation of more general wave functions with $k > 0$, the calculation becomes more complicated. However, the mathematics is essentially the same.

IV. REPRESENTATIONS OF THE HOMOGENEOUS LORENTZ GROUP

The purpose of this paper is to show that the representations of the Poincaré group discussed in Sec. III are different from those of the Lorentz group. The best way to teach this point is to construct solutions of the same differential equation, Eq. (6), which are diagonal in the Casimir operators of the Lorentz group given in Eqs. (15) and (16).

In terms of the rotation and boost generators, the Casimir operators take the form

$$C_1 = \mathbf{L}^2 - \mathbf{K}^2, \quad C_2 = \mathbf{L}\cdot\mathbf{K}. \tag{43}$$

If we evaluate C_2 using the explicit expression for $\mathbf{L}$ and $\mathbf{K}$, this operator vanishes for the present spinless case. In order to construct solutions diagonal in C_1, we use a hyperbolic coordinate with the Lorentz invariant distance

$$\rho = \{|t^2 - r^2|\}^{1/2}, \tag{44}$$

where

$$r = (x^2 + y^2 + z^2)^{1/2}, \quad t = \pm\,\rho\cosh\alpha, \quad r = \rho\sinh\alpha, \tag{45}$$

for $|t| > r$, and

$$t = \rho\sinh\alpha, \quad r = \rho\cosh\alpha, \tag{46}$$

for $|t| < r$. For both cases, we use the usual three-dimensional spherical coordinate for x,y,z:

$$\begin{aligned} x &= r\sin\theta\cos\phi,\\ y &= r\sin\theta\sin\phi,\\ z &= r\cos\theta. \end{aligned} \tag{47}$$

In terms of ρ, α, θ, ϕ, the differential equation of Eq. (6) takes the form

$$\frac{1}{\rho^3}\frac{\partial}{\partial\rho}\left(\rho^3\frac{\partial\psi}{\partial\rho}\right) + \left(\frac{1}{\rho^2}(\mathbf{K}^2 - \mathbf{L}^2) - \rho^2\right)\psi = \epsilon\psi, \tag{48}$$

where $\epsilon = \pm 2(\lambda + 1)$ for the timelike and spacelike cases respectively. The form of $\mathbf{L}$ is well known. The operator $(\mathbf{L}^2 - \mathbf{K}^2)$ takes the form

$$(\mathbf{L}^2 - \mathbf{K}^2) = \frac{1}{\sinh^2\alpha}\frac{\partial}{\partial\alpha}\left(\sinh^2\alpha\frac{\partial}{\partial\alpha}\right) - \frac{1}{\sinh^2\alpha}\mathbf{L}^2, \tag{49}$$

for $|t| > r$, and

$$(\mathbf{L}^2 - \mathbf{K}^2) = \frac{1}{\cosh^2\alpha}\left(\frac{\partial}{\partial\alpha}\cosh^2\alpha\frac{\partial}{\partial\alpha}\right) + \frac{1}{\cosh^2\alpha}\mathbf{L}^2, \tag{50}$$

for $|t| < r$. We are interested in representations which are diagonal in the above operators.

In order to construct the desired representation, we solve the partial differential equation given in Eq. (48) by separating the variables

$$\psi(x) = R(\rho)B(\alpha,\theta,\phi). \tag{51}$$

In terms of $R(\rho)$ and $B(\alpha,\theta,\phi)$, Eq. (48) is separated into

$$\left[\frac{1}{\rho^3}\frac{\partial}{\partial\rho}\left(\rho^3\frac{\partial}{\partial\rho}\right) - \frac{\eta}{\rho^2} + \rho^2 - \epsilon\right]R(\rho) = 0, \tag{52}$$

and

$$(\mathbf{L}^2 - \mathbf{K}^2)B(\alpha,\theta,\phi) = \eta B(\alpha,\theta,\phi). \tag{53}$$

In order that the radial equation have regular solutions,

$$\eta = n(n+1), \quad n = 0,1,2,\ldots. \tag{54}$$

The radial wave function in this case takes the form

$$R^l_{\mu,n}(\rho) = \rho^n L^{(n+1)}_\mu(\rho^2)\exp(-\rho^2/2). \tag{55}$$

with $\epsilon = 2(2\mu + n)$, $\mu = 0,1,2,\ldots$ $L^{(n+1)}_\mu(\rho^2)$ is the generalized Laguerre function.[13]

With this preparation, we now write the "angular" function B as

$$B^l_n(\alpha,\theta,\phi) = A^l_n(\alpha)Y^m_l(\theta,\phi). \tag{56}$$

For the timelike region where $|t| > r$, we use the notation

$$A^l_n(\alpha) = T^l_n(\alpha), \tag{57}$$

and for the spacelike region,

$$A^l_n(\alpha) = S^l_n(\alpha). \tag{58}$$

Then $T^l_n(\alpha)$ and $S^l_n(\alpha)$ satisfy the following differential equations, respectively:

$$\frac{\partial}{\partial\alpha}(\sinh^2\alpha T^l_n) - [n(n+2) + l(l+1)]T^l_n = 0, \tag{59}$$

$$\frac{\partial}{\partial\alpha}(\cosh^2\alpha S^l_n) - [n(n+2) - l(l+1)]S^l_n = 0. \tag{60}$$

If $l = 0$, the solutions to the above equations take the form

$$\begin{aligned} T^0_n(\alpha) &= \sinh(n+1)\alpha/\sinh\alpha,\\ S^0_n(\alpha) &= \cosh(n+1)\alpha/\cosh\alpha. \end{aligned} \tag{61}$$

For nonvanishing values of l,

$$\begin{aligned} T^l_n(\alpha) &= (\sinh\alpha)^l\left(\frac{1}{\sinh\alpha}\frac{d}{d\alpha}\right)^l T^0_n(\alpha),\\ S^l_n(\alpha) &= (\cosh\alpha)^l\left(\frac{1}{\cosh\alpha}\frac{d}{d\alpha}\right)^l S^0_n(\alpha). \end{aligned} \tag{62}$$

The solutions given in Eqs. (61) and (62) become infinite when $\alpha \to \infty$. This means that the Lorentz harmonics are singular along the light cones. At this point, we are tempted to make n imaginary in order to make $T^l_n(\alpha)$ and $S^l_n(\alpha)$ normalizable. In fact, this and other interesting possibilities have been extensively discussed in the literature.[14] However, if n takes noninteger values, the radial wave function becomes singular along the light cones. In either case, the light-cone singularity is unavoidable.

The wave functions which are diagonal in the Casimir operators C_1 are now

$$\psi^{\mu,n}_{l,m}(x) = R^n_\mu(\rho)A^l_n(\alpha)Y^m_l(\theta,\phi), \tag{63}$$

where R^n_μ, A^l_n are given in Eqs. (55), and (57), and (58), respectively.

The localization property of the above solution is dictated by the Gaussian factor in the radial function $R^n_\mu(\rho)$, and is illustrated in Fig. 1. Unlike the case of wave functions representing the Poincaré group, the hyperbolic localization region is independent of the hadronic velocity and is thus Lorentz invariant. We can of course carry out the mathematics of the operators L_i and K_i applied to the wave functions given in Eq. (63). However, it is not yet clear whether this wave function carries any physical interpretation.[5]

V. CONCLUDING REMARKS

We have discussed in this paper the difference between the representations of the Poincaré group and those of the Lorentz group starting from the same differential equation. The Poincaré group is represented by solutions in a moving "cylindrical" coordinate system, while the representations of the Lorentz group are formed from solutions in a hyperbolic coordinate system. This result may serve a useful purpose in a group theory course for physicists.

As was clearly stated by Michel,[4] confusion concerning this difference is not uncommon in the literature. This fact was not recognized in Refs. 1 and 2. The present paper clarifies this important point.

ACKNOWLEDGMENT

In carrying out the calculations for Sec. IV, we have been guided by M. Rubin's unpublished note. We thank Professor Rubin for making his note available to us.

[1]Y. S. Kim and M. E. Noz, Am. J. Phys. **46,** 480 (1978).
[2]Y. S. Kim and M. E. Noz, Am. J. Phys. **46,** 484 (1978).
[3]E. P. Wigner, Ann. Math. **40,** 149 (1939).
[4]L. Michel, in *Group Theoretical Concepts and Methods in Elementary Particle Physics*, edited by F. Gürsey (Gordon and Breach, New York, 1963).
[5]F. R. Halpern, *Special Relativity and Quantum Mechanics* (Prentice-Hall, Englewood Cliff, NJ, 1968).
[6]H. Yukawa, Phys Rev. **91,** 416 (1953); K. Fujimura, T. Kobayashi, and M. Namiki, Prog. Theor. Phys. **43,** 73 (1979); Y. S. Kim and M. E. Noz, Phys. Rev. D **15,** 335 (1977).
[7]R. P. Feynman, M. Kislinger, and F. Ravndal, Phys. Rev. D **3,** 2706 (1971).
[8]Y. S. Kim, M. E. Noz, and S. H. Oh, J. Math. Phys. **20,** 1341 (1979).
[9]For more detailed discussions of this coordinate system, see Y. S. Kim and M. E. Noz, Phys. Rev. D **8,** 3521 (1973); Found. Phys. **9,** 375 (1979). See also, Y. S. Kim, Phys. Rev. D **14,** 273 (1976).
[10]M. A. Naimark, Uspehi Mat. Nauk **9,** 19 (1954); Am. Math. Soc. Trans. **6,** 379 (1957).
[11]W. Miller, *Symmetry Groups and Their Applications* (Academic, New York and London, 1972).
[12]M. J. Ruiz, Phys. Rev. D **10,** 4306 (1974).
[13]W. Magnus and F. Oberhettinger, *Formulas and Theorems for the Functions in Mathematical Physics* (Chelsea, New York, 1949).
[14]E. G. Kalnins and W. Miller, J. Math. Phys. **18,** 1 (1977), and the references contained in this paper.

Chapter III

The Time-Energy Uncertainty Relation

According to Heisenberg's uncertainty principle, the position and momentum are noncommuting q-numbers. However, the time variable is a c-number. This causes a problem when we make a Lorentz transformation. If the time variable is purely a c-number in one Lorentz frame, it is no longer a c-number in different Lorentz frames. This was first pointed out by Dirac in 1927. Indeed, the question of where the time-energy uncertainty relation stands in quantum mechanics is constantly debated in the literature.

As Wigner pointed out in 1972, the reason why the time energy uncertainty relation is so peculiar is that there are not many physical phenomena which can be regarded as direct consequences of the time energy uncertainty relation. In 1985, Hussar, Kim, and Noz showed how the time-energy uncertainty relation can be combined covariantly with the position-momentum uncertainty relation in the relativistic quark model, and pointed out there are direct consequences of this uncertainty relation in high-energy physics.

The Quantum Theory of the Emission and Absorption of Radiation.

By P. A. M. DIRAC, St. John's College, Cambridge, and Institute for Theoretical Physics, Copenhagen.

(Communicated by N. Bohr, For. Mem. R.S.—Received February 2, 1927.)

§ 1. *Introduction and Summary.*

The new quantum theory, based on the assumption that the dynamical variables do not obey the commutative law of multiplication, has by now been developed sufficiently to form a fairly complete theory of dynamics. One can treat mathematically the problem of any dynamical system composed of a number of particles with instantaneous forces acting between them, provided it is describable by a Hamiltonian function, and one can interpret the mathematics physically by a quite definite general method. On the other hand, hardly anything has been done up to the present on quantum electrodynamics. The questions of the correct treatment of a system in which the forces are propagated with the velocity of light instead of instantaneously, of the production of an electromagnetic field by a moving electron, and of the reaction of this field on the electron have not yet been touched. In addition, there is a serious difficulty in making the theory satisfy all the requirements of the restricted

Reprinted from *Proc. Roy. Soc. (London)* **A114**, 243 (1927).

principle of relativity, since a Hamiltonian function can no longer be used. This relativity question is, of course, connected with the previous ones, and it will be impossible to answer any one question completely without at the same time answering them all. However, it appears to be possible to build up a fairly satisfactory theory of the emission of radiation and of the reaction of the radiation field on the emitting system on the basis of a kinematics and dynamics which are not strictly relativistic. This is the main object of the present paper. The theory is non-relativistic only on account of the time being counted throughout as a c-number, instead of being treated symmetrically with the space co-ordinates. The relativity variation of mass with velocity is taken into account without difficulty.

The underlying ideas of the theory are very simple. Consider an atom interacting with a field of radiation, which we may suppose for definiteness to be confined in an enclosure so as to have only a discrete set of degrees of freedom. Resolving the radiation into its Fourier components, we can consider the energy and phase of each of the components to be dynamical variables describing the radiation field. Thus if E_r is the energy of a component labelled r and θ_r is the corresponding phase (defined as the time since the wave was in a standard phase), we can suppose each E_r and θ_r to form a pair of canonically conjugate variables. In the absence of any interaction between the field and the atom, the whole system of field plus atom will be describable by the Hamiltonian

$$H = \Sigma_r E_r + H_0 \tag{1}$$

equal to the total energy, H_0 being the Hamiltonian for the atom alone, since the variables E_r, θ_r obviously satisfy their canonical equations of motion

$$\dot{E}_r = -\frac{\partial H}{\partial \theta_r} = 0, \quad \dot{\theta}_r = \frac{\partial H}{\partial E_r} = 1.$$

When there is interaction between the field and the atom, it could be taken into account on the classical theory by the addition of an interaction term to the Hamiltonian (1), which would be a function of the variables of the atom and of the variables E_r, θ_r that describe the field. This interaction term would give the effect of the radiation on the atom, and also the reaction of the atom on the radiation field.

In order that an analogous method may be used on the quantum theory, it is necessary to assume that the variables E_r, θ_r are q-numbers satisfying the standard quantum conditions $\theta_r E_r - E_r \theta_r = ih$, etc., where h is $(2\pi)^{-1}$ times the usual Planck's constant, like the other dynamical variables of the problem. This assumption immediately gives light-quantum properties to the radiation.* For if ν_r is the frequency of the component r, $2\pi\nu_r\theta_r$ is an angle variable, so that its canonical conjugate $E_r/2\pi\nu_r$ can only assume a discrete set of values differing by multiples of h, which means that E_r can change only by integral multiples of the quantum $(2\pi h)\,\nu_r$. If we now add an interaction term (taken over from the clasical theory) to the Hamiltonian (1), the problem can be solved according to the rules of quantum mechanics, and we would expect to obtain the correct results for the action of the radiation and the atom on one another. It will be shown that we actually get the correct laws for the emission and absorption of radiation, and the correct values for Einstein's A's and B's. In the author's previous theory,† where the energies and phases of the components of radiation were c-numbers, only the B's could be obtained, and the reaction of the atom on the radiation could not be taken into account.

It will also be shown that the Hamiltonian which describes the interaction of the atom and the electromagnetic waves can be made identical with the Hamiltonian for the problem of the interaction of the atom with an assembly of particles moving with the velocity of light and satisfying the Einstein-Bose statistics, by a suitable choice of the interaction energy for the particles. The number of particles having any specified direction of motion and energy, which can be used as a dynamical variable in the Hamiltonian for the particles, is equal to the number of quanta of energy in the corresponding wave in the Hamiltonian for the waves. There is thus a complete harmony between the wave and light-quantum descriptions of the interaction. We shall actually build up the theory from the light-quantum point of view, and show that the Hamiltonian transforms naturally into a form which resembles that for the waves.

The mathematical development of the theory has been made possible by the author's general transformation theory of the quantum matrices.‡ Owing to the fact that we count the time as a c-number, we are allowed to use the notion of the value of any dynamical variable at any instant of time. This value is

* Similar assumptions have been used by Born and Jordan [' Z. f. Physik,' vol. 34, p. 886 (1925)] for the purpose of taking over the classical formula for the emission of radiation by a dipole into the quantum theory, and by Born, Heisenberg and Jordan [' Z. f. Physik,' vol. 35, p. 606 (1925)] for calculating the energy fluctuations in a field of black-body radiation.

† ' Roy. Soc. Proc.,' A, vol. 112, p. 661, § 5 (1926). This is quoted later by, *loc. cit.*, I.

‡ ' Roy. Soc. Proc.,' A, vol. 113, p. 621 (1927). This is quoted later by *loc. cit.*, II. An essentially equivalent theory has been obtained independently by Jordan [' Z. f. Physik,' vol. 40, p. 809 (1927)]. See also, F. London, ' Z. f. Physik,' vol. 40, p. 193 (1926).

a q-number, capable of being represented by a generalised " matrix " according to many different matrix schemes, some of which may have continuous ranges of rows and columns, and may require the matrix elements to involve certain kinds of infinities (of the type given by the δ functions*). A matrix scheme can be found in which any desired set of constants of integration of the dynamical system that commute are represented by diagonal matrices, or in which a set of variables that commute are represented by matrices that are diagonal at a specified time.† The values of the diagonal elements of a diagonal matrix representing any q-number are the characteristic values of that q-number. A Cartesian co-ordinate or momentum will in general have all characteristic values from $-\infty$ to $+\infty$, while an action variable has only a discrete set of characteristic values. (We shall make it a rule to use unprimed letters to denote the dynamical variables or q-numbers, and the same letters primed or multiply primed to denote their characteristic values. Transformation functions or eigenfunctions are functions of the characteristic values and not of the q-numbers themselves, so they should always be written in terms of primed variables.)

If $f(\xi, \eta)$ is any function of the canonical variables ξ_k, η_k, the matrix representing f at any time t in the matrix scheme in which the ξ_k at time t are diagonal matrices may be written down without any trouble, since the matrices representing the ξ_k and η_k themselves at time t are known, namely,

$$\left.\begin{aligned} \xi_k(\xi'\xi'') &= \xi_k'\,\delta(\xi'\xi''), \\ \eta_k(\xi'\xi'') &= -i\hbar\,\delta(\xi_1'-\xi_1'')\ldots\delta(\xi_{k-1}'-\xi_{k-1}'')\,\delta'(\xi_k'-\xi_k'')\,\delta(\xi_{k+1}'-\xi_{k+1}'')\ldots \end{aligned}\right\} . \quad (2)$$

Thus if the Hamiltonian H is given as a function of the ξ_k and η_k, we can at once write down the matrix $H(\xi'\ \xi'')$. We can then obtain the transformation function, (ξ'/α') say, which transforms to a matrix scheme (α) in which the Hamiltonian is a diagonal matrix, as (ξ'/α') must satisfy the integral equation

$$\int H(\xi'\xi'')\,d\xi''\,(\xi''/\alpha') = W(\alpha')\,.\,(\xi'/\alpha'), \quad (3)$$

of which the characteristic values $W(\alpha')$ are the energy levels. This equation is just Schrödinger's wave equation for the eigenfunctions (ξ'/α'), which becomes an ordinary differential equation when H is a simple algebraic function of the

* *Loc. cit.* II, § 2.

† One can have a matrix scheme in which a set of variables that commute are at all times represented by diagonal matrices if one will sacrifice the condition that the matrices must satisfy the equations of motion. The transformation function from such a scheme to one in which the equations of motion are satisfied will involve the time explicitly. See p. 628 in *loc. cit.*, II.

ξ_k and η_k on account of the special equations (2) for the matrices representing ξ_k and η_k. Equation (3) may be written in the more general form

$$\int H(\xi'\xi'')\, d\xi''\, (\xi''/\alpha') = ih\, \partial(\xi'/\alpha')/\partial t, \tag{3'}$$

in which it can be applied to systems for which the Hamiltonian involves the time explicitly.

One may have a dynamical system specified by a Hamiltonian H which cannot be expressed as an algebraic function of any set of canonical variables, but which can all the same be represented by a matrix $H(\xi'\xi'')$. Such a problem can still be solved by the present method, since one can still use equation (3) to obtain the energy levels and eigenfunctions. We shall find that the Hamiltonian which describes the interaction of a light-quantum and an atomic system is of this more general type, so that the interaction can be treated mathematically, although one cannot talk about an interaction potential energy in the usual sense.

It should be observed that there is a difference between a light-wave and the de Broglie or Schrödinger wave associated with the light-quanta. Firstly, the light-wave is always real, while the de Broglie wave associated with a light-quantum moving in a definite direction must be taken to involve an imaginary exponential. A more important difference is that their intensities are to be interpreted in different ways. The number of light-quanta per unit volume associated with a monochromatic light-wave equals the energy per unit volume of the wave divided by the energy $(2\pi h)\nu$ of a single light-quantum. On the other hand a monochromatic de Broglie wave of amplitude a (multiplied into the imaginary exponential factor) must be interpreted as representing a^2 light-quanta per unit volume for all frequencies. This is a special case of the general rule for interpreting the matrix analysis,* according to which, if (ξ'/α') or $\psi_{\alpha'}(\xi_k')$ is the eigenfunction in the variables ξ_k of the state α' of an atomic system (or simple particle), $|\psi_{\alpha'}(\xi_k')|^2$ is the probability of each ξ_k having the value ξ_k', [or $|\psi_{\alpha'}(\xi_k')|^2\, d\xi_1'\, d\xi_2' \ldots$ is the probability of each ξ_k lying between the values ξ_k' and $\xi_k' + d\xi_k'$, when the ξ_k have continuous ranges of characteristic values] on the assumption that all phases of the system are equally probable. The wave whose intensity is to be interpreted in the first of these two ways appears in the theory only when one is dealing with an assembly of the associated particles satisfying the Einstein-Bose statistics. There is thus no such wave associated with electrons.

* *Loc. cit.*, II, §§ 6, 7.

§ 2. *The Perturbation of an Assembly of Independent Systems.*

We shall now consider the transitions produced in an atomic system by an arbitrary perturbation. The method we shall adopt will be that previously given by the author,† which leads in a simple way to equations which determine the probability of the system being in any stationary state of the unperturbed system at any time.‡ This, of course, gives immediately the probable number of systems in that state at that time for an assembly of the systems that are independent of one another and are all perturbed in the same way. The object of the present section is to show that the equations for the rates of change of these probable numbers can be put in the Hamiltonian form in a simple manner, which will enable further developments in the theory to be made.

Let H_0 be the Hamiltonian for the unperturbed system and V the perturbing energy, which can be an arbitrary function of the dynamical variables and may or may not involve the time explicitly, so that the Hamiltonian for the perturbed system is $H = H_0 + V$. The eigenfunctions for the perturbed system must satisfy the wave equation

$$ih\,\partial\psi/\partial t = (H_0 + V)\,\psi,$$

where $(H_0 + V)$ is an operator. If $\psi = \Sigma_r a_r \psi_r$ is the solution of this equation that satisfies the proper initial conditions, where the ψ_r's are the eigenfunctions for the unperturbed system, each associated with one stationary state labelled by the suffix r, and the a_r's are functions of the time only, then $|a_r|^2$ is the probability of the system being in the state r at any time. The a_r's must be normalised initially, and will then always remain normalised. The theory will apply directly to an assembly of N similar independent systems if we multiply each of these a_r's by $N^{\frac{1}{2}}$ so as to make $\Sigma_r |a_r|^2 = N$. We shall now have that $|a_r|^2$ is the probable number of systems in the state r.

The equation that determines the rate of change of the a_r's is§

$$ih\dot{a}_r = \Sigma_s V_{rs} a_s, \tag{4}$$

where the V_{rs}'s are the elements of the matrix representing V. The conjugate imaginary equation is

$$-ih\dot{a}_r^* = \Sigma_s V_{rs}^* a_s^* = \Sigma_s a_s^* V_{sr}. \tag{4'}$$

† *Loc. cit.* I.

‡ The theory has recently been extended by Born ['Z. f. Physik,' vol. 40, p. 167 (1926)] so as to take into account the adiabatic changes in the stationary states that may be produced by the perturbation as well as the transitions. This extension is not used in the present paper.

§ *Loc. cit.*, I, equation (25).

If we regard a_r and $ih\, a_r^*$ as canonical conjugates, equations (4) and (4′) take the Hamiltonian form with the Hamiltonian function $F_1 = \Sigma_{rs} a_r^* V_{rs} a_s$, namely,

$$\frac{da_r}{dt} = \frac{1}{ih}\frac{\partial F_1}{\partial a_r^*}, \quad ih\frac{da_r^*}{dt} = -\frac{\partial F_1}{\partial a_r}.$$

We can transform to the canonical variables N_r, ϕ_r by the contact transformation

$$a_r = N_r^{\frac{1}{2}} e^{-i\phi_r/h}, \quad a_r^* = N_r^{\frac{1}{2}} e^{i\phi_r/h}.$$

This transformation makes the new variables N_r and ϕ_r real, N_r being equal to $a_r a_r^* = |a_r|^2$, the probable number of systems in the state r, and ϕ_r/h being the phase of the eigenfunction that represents them. The Hamiltonian F_1 now becomes

$$F_1 = \Sigma_{rs} V_{rs} N_r^{\frac{1}{2}} N_s^{\frac{1}{2}} e^{i(\phi_r - \phi_s)/h},$$

and the equations that determine the rate at which transitions occur have the canonical form

$$\dot{N}_r = -\frac{\partial F_1}{\partial \phi_r}, \quad \dot{\phi}_r = \frac{\partial F_1}{\partial N_r}.$$

A more convenient way of putting the transition equations in the Hamiltonian form may be obtained with the help of the quantities

$$b_r = a_r\, e^{-iW_r t/h}, \quad b_r^* = a_r^*\, e^{iW_r t/h},$$

W_r being the energy of the state r. We have $|b_r|^2$ equal to $|a_r|^2$, the probable number of systems in the state r. For $\dot{b}_r$ we find

$$\begin{aligned} ih\,\dot{b}_r &= W_r b_r + ih\,\dot{a}_r\, e^{-iW_r t/h} \\ &= W_r b_r + \Sigma_s V_{rs} b_s e^{i(W_s - W_r)t/h} \end{aligned}$$

with the help of (4). If we put $V_{rs} = v_{rs} e^{i(W_r - W_s)t/h}$, so that v_{rs} is a constant when V does not involve the time explicitly, this reduces to

$$\begin{aligned} ih\,\dot{b}_r &= W_r b_r + \Sigma_s v_{rs} b_s \\ &= \Sigma_s H_{rs} b_s, \end{aligned} \tag{5}$$

where $H_{rs} = W_r \delta_{rs} + v_{rs}$, which is a matrix element of the total Hamiltonian $H = H_0 + V$ with the time factor $e^{i(W_r - W_s)t/h}$ removed, so that H_{rs} is a constant when H does not involve the time explicitly. Equation (5) is of the same form as equation (4), and may be put in the Hamiltonian form in the same way.

It should be noticed that equation (5) is obtained directly if one writes down the Schrödinger equation in a set of variables that specify the stationary states of the unperturbed system. If these variables are ξ_h, and if $H(\xi'\xi'')$ denotes a matrix element of the total Hamiltonian H in the (ξ) scheme, this Schrödinger equation would be

$$ih\, \partial\psi(\xi')/\partial t = \Sigma_{\xi''} \mathrm{H}(\xi'\xi'')\, \psi(\xi''), \tag{6}$$

like equation (3′). This differs from the previous equation (5) only in the notation, a single suffix r being there used to denote a stationary state instead of a set of numerical values ξ_k' for the variables ξ_k, and b_r being used instead of $\psi(\xi')$. Equation (6), and therefore also equation (5), can still be used when the Hamiltonian is of the more general type which cannot be expressed as an algebraic function of a set of canonial variables, but can still be represented by a matrix $\mathrm{H}(\xi'\xi'')$ or H_{rs}.

We now take b_r and ihb_r^* to be canonically conjugate variables instead of a_r and iha_r^*. The equation (5) and its conjugate imaginary equation will now take the Hamiltonian form with the Hamiltonian function

$$\mathrm{F} = \Sigma_{rs} b_r^* \mathrm{H}_{rs} b_s. \tag{7}$$

Proceeding as before, we make the contact transformation

$$b_r = \mathrm{N}_r^{\frac{1}{2}} e^{-i\theta_r/h}, \qquad b_r^* = \mathrm{N}_r^{\frac{1}{2}} e^{i\theta_r/h}, \tag{8}$$

to the new canonical variables N_r, θ_r, where N_r is, as before, the probable number of systems in the state r, and θ_r is a new phase. The Hamiltonian F will now become

$$\mathrm{F} = \Sigma_{rs} \mathrm{H}_{rs} \mathrm{N}_r^{\frac{1}{2}} \mathrm{N}_s^{\frac{1}{2}} e^{i(\theta_r - \theta_s)/h},$$

and the equations for the rates of change of N_r and θ_r will take the canonical form

$$\dot{\mathrm{N}}_r = -\frac{\partial \mathrm{F}}{\partial \theta_r}, \qquad \dot{\theta}_r = \frac{\partial \mathrm{F}}{\partial \mathrm{N}_r}.$$

The Hamiltonian may be written

$$\mathrm{F} = \Sigma_r \mathrm{W}_r \mathrm{N}_r + \Sigma_{rs} v_{rs} \mathrm{N}_r^{\frac{1}{2}} \mathrm{N}_s^{\frac{1}{2}} e^{i(\theta_r - \theta_s)/h}. \tag{9}$$

The first term $\Sigma_r \mathrm{W}_r \mathrm{N}_r$ is the total proper energy of the assembly, and the second may be regarded as the additional energy due to the perturbation. If the perturbation is zero, the phases θ_r would increase linearly with the time, while the previous phases ϕ_r would in this case be constants.

§ 3. *The Perturbation of an Assembly satisfying the Einstein-Bose Statistics.*

According to the preceding section we can describe the effect of a perturbation on an assembly of independent systems by means of canonical variables and Hamiltonian equations of motion. The development of the theory which naturally suggests itself is to make these canonical variables q-numbers satisfying the usual quantum conditions instead of c-numbers, so that their Hamiltonian equations of motion become true quantum equations. The Hamiltonian function will now provide a Schrödinger wave equation, which must be solved and interpreted in the usual manner. The interpretation will give not merely the probable number of systems in any state, but the probability of any given distribution of the systems among the various states, this probability being, in fact, equal to the square of the modulus of the normalised solution of the wave equation that satisfies the appropriate initial conditions. We could, of course, calculate directly from elementary considerations the probability of any given distribution when the systems are independent, as we know the probability of each system being in any particular state. We shall find that the probability calculated directly in this way does not agree with that obtained from the wave equation except in the special case when there is only one system in the assembly. In the general case it will be shown that the wave equation leads to the correct value for the probability of any given distribution when the systems obey the Einstein-Bose statistics instead of being independent.

We assume the variables b_r, ihb_r^* of § 2 to be canonical q-numbers satisfying the quantum conditions

$$b_r \,.\, ih\, b_r^* - ih\, b_r^* \,.\, b_r = ih$$

or

$$b_r b_r^* - b_r^* b_r = 1,$$

and

$$b_r b_s - b_s b_r = 0, \qquad b_r^* b_s^* - b_s^* b_r^* = 0,$$

$$b_r b_s^* - b_s^* b_r = 0 \qquad (s \neq r).$$

The transformation equations (8) must now be written in the quantum form

$$\left.\begin{aligned} b_r &= (N_r + 1)^{\frac{1}{2}}\, e^{-i\theta_r/h} = e^{-i\theta_r/h} N_r^{\frac{1}{2}} \\ b_r^* &= N_r^{\frac{1}{2}} e^{i\theta_r/h} = e^{i\theta_r/h}(N_r + 1)^{\frac{1}{2}}, \end{aligned}\right\} \tag{10}$$

in order that the N_r, θ_r may also be canonical variables. These equations show that the N_r can have only integral characteristic values not less than zero,† which provides us with a justification for the assumption that the variables are q-numbers in the way we have chosen. The numbers of systems in the different states are now ordinary quantum numbers.

† See § 8 of the author's paper 'Roy. Soc. Proc.,' A, vol. 111, p. 281 (1926). What are there called the c-number values that a q-number can take are here given the more precise name of the characteristic values of that q-number.

The Hamiltonian (7) now becomes

$$\begin{aligned} F &= \Sigma_{rs} b_r^* H_{rs} b_s = \Sigma_{rs} N_r^{\frac{1}{2}} e^{i\theta_r/h} H_{rs} (N_s + 1)^{\frac{1}{2}} e^{-i\theta_s/h} \\ &= \Sigma_{rs} H_{rs} N_r^{\frac{1}{2}} (N_s + 1 - \delta_{rs})^{\frac{1}{2}} e^{i(\theta_r - \theta_s)/h} \end{aligned} \quad (11)$$

in which the H_{rs} are still c-numbers. We may write this F in the form corresponding to (9)

$$F = \Sigma_r W_r N_r + \Sigma_{rs} v_{rs} N_r^{\frac{1}{2}} (N_s + 1 - \delta_{rs})^{\frac{1}{2}} e^{i(\theta_r - \theta_s)/h} \quad (11')$$

in which it is again composed of a proper energy term $\Sigma_r W_r N_r$ and an interaction energy term.

The wave equation written in terms of the variables N_r is†

$$ih \frac{\partial}{\partial t} \psi (N_1', N_2', N_s' \ldots) = F\psi (N_1', N_2', N_3' \ldots), \quad (12)$$

where F is an operator, each θ_r occurring in F being interpreted to mean $ih\, \partial/\partial N_r'$. If we apply the operator $e^{\pm i\theta_r/h}$ to any function $f(N_1', N_2', \ldots N_r', \ldots)$ of the variables $N_1', N_2', \ldots$ the result is

$$\begin{aligned} e^{\pm i\theta_r/h} f(N_1', N_2', \ldots N_r', \ldots) &= e^{\mp \partial/\partial N_r'} f(N_1', N_2', \ldots N_r' \ldots) \\ &= f(N_1', N_2', \ldots N_r' \mp 1, \ldots). \end{aligned}$$

If we use this rule in equation (12) and use the expression (11) for F we obtain‡

$$\begin{aligned} & ih \frac{\partial}{\partial t} \psi(N_1', N_2', N_3' \ldots) \\ &= \Sigma_{rs} H_{rs} N_r'^{\frac{1}{2}} (N_s' + 1 - \delta_{rs})^{\frac{1}{2}} \psi (N_1', N_2' \ldots N_r' - 1, \ldots N_s' + 1, \ldots). \end{aligned} \quad (13)$$

We see from the right-hand side of this equation that in the matrix representing F, the term in F involving $e^{i(\theta_r - \theta_s)/h}$ will contribute only to those matrix elements that refer to transitions in which N_r decreases by unity and N_s increases by unity, *i.e.*, to matrix elements of the type $F(N_1', N_2' \ldots N_r' \ldots N_s';\ N_1', N_2' \ldots N_r' - 1 \ldots N_s' + 1 \ldots)$. If we find a solution $\psi(N_1', N_2' \ldots)$ of equation (13) that is normalised [*i.e.*, one for which $\Sigma_{N_1', N_2' \ldots} |\psi(N_1', N_2' \ldots)|^2 = 1$] and that satisfies the proper initial conditions, then $|\psi(N_1', N_2' \ldots)|^2$ will be the probability of that distribution in which N_1' systems are in state 1, N_2' in state 2, ... at any time.

Consider first the case when there is only one system in the assembly. The probability of its being in the state q is determined by the eigenfunction

† We are supposing for definiteness that the label r of the stationary states takes the values 1, 2, 3, ...

‡ When $s = r$, $\psi(N_1', N_2' \ldots N_r' - 1 \ldots N_s' + 1)$ is to be taken to mean $\psi(N_1' N_2' \ldots N_r' \ldots)$.

$\psi(N_1', N_2', \ldots)$ in which all the N′'s are put equal to zero except N_q', which is put equal to unity. This eigenfunction we shall denote by $\psi\{q\}$. When it is substituted in the left-hand side of (13), all the terms in the summation on the right-hand side vanish except those for which $r = q$, and we are left with

$$ih \frac{\partial}{\partial t} \psi\{q\} = \Sigma_r H_{qs} \psi\{s\},$$

which is the same equation as (5) with $\psi\{q\}$ playing the part of b_q. This establishes the fact that the present theory is equivalent to that of the preceding section when there is only one system in the assembly.

Now take the general case of an arbitrary number of systems in the assembly, and assume that they obey the Einstein-Bose statistical mechanics. This requires that, in the ordinary treatment of the problem, only those eigenfunctions that are symmetrical between all the systems must be taken into account, these eigenfunctions being by themselves sufficient to give a complete quantum solution of the problem.† We shall now obtain the equation for the rate of change of one of these symmetrical eigenfunctions, and show that it is identical with equation (13).

If we label each system with a number n, then the Hamiltonian for the assembly will be $H_A = \Sigma_n H(n)$, where $H(n)$ is the H of § 2 (equal to $H_0 + V$) expressed in terms of the variables of the nth system. A stationary state of the assembly is defined by the numbers $r_1, r_2 \ldots r_n \ldots$ which are the labels of the stationary states in which the separate systems lie. The Schrödinger equation for the assembly in a set of variables that specify the stationary states will be of the form (6) [with H_A instead of H], and we can write it in the notation of equation (5) thus :—

$$ih \dot{b}(r_1 r_2 \ldots) = \Sigma_{s_1, s_2 \ldots} H_A(r_1 r_2 \ldots ; s_1 s_2 \ldots) b(s_1 s_2 \ldots), \tag{14}$$

where $H_A(r_1 r_2 \ldots ; s_1 s_2 \ldots)$ is the general matrix element of H_A [with the time factor removed]. This matrix element vanishes when more than one s_n differs from the corresponding r_n; equals $H_{r_m s_m}$ when s_m differs from r_m and every other s_n equals r_n; and equals $\Sigma_n H_{r_n r_n}$ when every s_n equals r_n. Substituting these values in (14), we obtain

$$ih \dot{b}(r_1 r_2 \ldots) = \Sigma_m \Sigma_{s_m \neq r_m} H_{r_m s_m} b(r_1 r_2 \ldots r_{m-1} s_m r_{m+1} \ldots) + \Sigma_n H_{r_n r_n} b(r_1 r_2 \ldots). \tag{15}$$

We must now restrict $b(r_1 r_2 \ldots)$ to be a symmetrical function of the variables $r_1, r_2 \ldots$ in order to obtain the Einstein-Bose statistics. This is permissible since if $b(r_1 r_2 \ldots)$ is symmetrical at any time, then equation (15) shows that

† *Loc. cit.*, I, § 3.

$\dot{b}(r_1 r_2 \ldots)$ is also symmetrical at that time, so that $b\ (r_1 r_2 \ldots)$ will remain symmetrical.

Let N_r denote the number of systems in the state r. Then a stationary state of the assembly describable by a symmetrical eigenfunction may be specified by the numbers $N_1, N_2 \ldots N_r \ldots$ just as well as by the numbers $r_1, r_2 \ldots r_n \ldots$, and we shall be able to transform equation (15) to the variables $N_1, N_2 \ldots$. We cannot actually take the new eigenfunction $b\ (N_1, N_2 \ldots)$ equal to the previous one $b\ (r_1 r_2 \ldots)$, but must take one to be a numerical multiple of the other in order that each may be correctly normalised with respect to its respective variables. We must have, in fact,

$$\Sigma_{r_1, r_2 \ldots} |b(r_1\, r_2 \ldots)|^2 = 1 = \Sigma_{N_1, N_2 \ldots} |b(N_1, N_2 \ldots)|^2,$$

and hence we must take $|b(N_1, N_2 \ldots)|^2$ equal to the sum of $|b(r_1 r_2 \ldots)|^2$ for all values of the numbers $r_1, r_2 \ldots$ such that there are N_1 of them equal to 1, N_2 equal to 2, etc. There are $N!/N_1!\,N_2!\ldots$ terms in this sum, where $N = \Sigma_r N_r$ is the total number of systems, and they are all equal, since $b(r_1 r_2 \ldots)$ is a symmetrical function of its variables $r_1, r_2 \ldots$. Hence we must have

$$b(N_1, N_2 \ldots) = (N!/N_1!\,N_2!\ldots)^{\frac{1}{2}}\, b(r_1 r_2 \ldots).$$

If we make this substitution in equation (15), the left-hand side will become $ih\,(N_1!\,N_2!\ldots/N!)^{\frac{1}{2}}\,\dot{b}\,(N_1, N_2 \ldots)$. The term $H_{r_m s_m} b\,(r_1 r_2 \ldots r_{m-1} s_m r_{m+1} \ldots)$ in the first summation on the right-hand side will become

$$[N_1!\,N_2!\ldots(N_r-1)!\ldots(N_s+1)!\ldots/N!]^{\frac{1}{2}}\,H_{rs} b\,(N_1, N_2 \ldots N_r - 1 \ldots N_s + 1 \ldots), \quad (16)$$

where we have written r for r_m and s for s_m. This term must be summed for all values of s except r, and must then be summed for r taking each of the values $r_1, r_2 \ldots$. Thus each term (16) gets repeated by the summation process until it occurs a total of N_r times, so that it contributes

$$\begin{aligned} N_r [N_1!\,N_2!\ldots(N_r-1)!\ldots(N_s+1)!\ldots/N!]^{\frac{1}{2}}\,H_{rs} b\,(N_1, N_2 \ldots N_r - 1 \ldots N_s + 1 \ldots) \\ = N_r^{\frac{1}{2}} (N_s+1)^{\frac{1}{2}} (N_1!\,N_2!\ldots/N!)^{\frac{1}{2}}\,H_{rs} b\,(N_1, N_2 \ldots N_r - 1 \ldots N_s + 1 \ldots) \end{aligned}$$

to the right-hand side of (15). Finally, the term $\Sigma_n H_{r_n r_n} b\,(r_1, r_2 \ldots)$ becomes

$$\Sigma_r N_r H_{rr} . b\,(r_1 r_2 \ldots) = \Sigma_r N_r H_{rr} . (N_1!\,N_2!\ldots/N!)^{\frac{1}{2}}\, b\,(N_1, N_2 \ldots).$$

Hence equation (15) becomes, with the removal of the factor $(N_1!\,N_2!\ldots/N!)^{\frac{1}{2}}$,

$$\begin{aligned} ih\dot{b}\,(N_1, N_2 \ldots) = \Sigma_r \Sigma_{s \neq r} N_r^{\frac{1}{2}} (N_s+1)^{\frac{1}{2}}\,H_{rs} b\,(N_1, N_2 \ldots N_r - 1 \ldots N_s + 1 \ldots) \\ + \Sigma_r N_r H_{rr} b\,(N_1, N_2 \ldots), \quad (17) \end{aligned}$$

which is identical with (13) [except for the fact that in (17) the primes have been omitted from the N's, which is permissible when we do not require to refer to the N's as q-numbers]. We have thus established that the Hamiltonian (11) describes the effect of a perturbation on an assembly satisfying the Einstein-Bose statistics.

§ 4. *The Reaction of the Assembly on the Perturbing System.*

Up to the present we have considered only perturbations that can be represented by a perturbing energy V added to the Hamiltonian of the perturbed system, V being a function only of the dynamical variables of that system and perhaps of the time. The theory may readily be extended to the case when the perturbation consists of interaction with a perturbing dynamical system, the reaction of the perturbed system on the perturbing system being taken into account. (The distinction between the perturbing system and the perturbed system is, of course, not real, but it will be kept up for convenience.)

We now consider a perturbing system, described, say, by the canonical variables J_k, ω_k, the J's being its first integrals when it is alone, interacting with an assembly of perturbed systems with no mutual interaction, that satisfy the Einstein-Bose statistics. The total Hamiltonian will be of the form

$$H_T = H_P(J) + \Sigma_n H(n),$$

where H_P is the Hamiltonian of the perturbing system (a function of the J's only) and $H(n)$ is equal to the proper energy $H_0(n)$ plus the perturbation energy $V(n)$ of the nth system of the assembly. $H(n)$ is a function only of the variables of the nth system of the assembly and of the J's and w's, and does not involve the time explicitly.

The Schrödinger equation corresponding to equation (14) is now

$$ih\dot{b}(J', r_1r_2 \ldots) = \Sigma_{J''} \Sigma_{s_1, s_2} \ldots H_T(J', r_1r_2 \ldots; J'', s_1s_2 \ldots)\, b(J'', s_1s_2 \ldots),$$

in which the eigenfunction b involves the additional variables J_k'. The matrix element $H_T(J', r_1r_2 \ldots; J'', s_1s_2 \ldots)$ is now always a constant. As before, it vanishes when more than one s_n differs from the corresponding r_n. When s_m differs from r_m and every other s_n equals r_n, it reduces to $H(J'r_m; J''s_m)$, which is the $(J'r_m; J''s_m)$ matrix element (with the time factor removed) of $H = H_0 + V$, the proper energy plus the perturbation energy of a single system of the assembly; while when every s_n equals r_n, it has the value $H_P(J')\delta_{J'J''} + \Sigma_n H(J'r_n; J''r_n)$. If, as before, we restrict the eigenfunctions to be symmetrical in the variables $r_1, r_2 \ldots$, we can again transform to the variables $N_1, N_2 \ldots$, which will lead, as before, to the result

$$i\hbar \dot{b}(J', N_1', N_2' \ldots) = H_P(J')\, b(J', N'_1, N_2' \ldots)$$
$$+ \Sigma_{J''} \Sigma_{r,s} N_r'^{\frac{1}{2}} (N_s' + 1 - \delta_{rs})^{\frac{1}{2}} H(J'r; J''s)\, b(J'', N_1', N_2' \ldots N_r' - 1 \ldots N_s' + 1 \ldots) \quad (18)$$

This is the Schrödinger equation corresponding to the Hamiltonian function

$$F = H_P(J) + \Sigma_{r,s} H_{rs} N_r^{\frac{1}{2}} (N_s + 1 - \delta_{rs})^{\frac{1}{2}} e^{i(\theta_r - \theta_s)/h}, \quad (19)$$

in which H_{rs} is now a function of the J's and w's, being such that when represented by a matrix in the (J) scheme its (J′ J″) element is $H(J'r; J''s)$. (It should be noticed that H_{rs} still commutes with the N's and θ's.)

Thus the interaction of a perturbing system and an assembly satisfying the Einstein-Bose statistics can be described by a Hamiltonian of the form (19). We can put it in the form corresponding to (11′) by observing that the matrix element $H(J'r; J''s)$ is composed of the sum of two parts, a part that comes from the proper energy H_0, which equals W_r when $J_k'' = J_k'$ and $s = r$ and vanishes otherwise, and a part that comes from the interaction energy V, which may be denoted by $v(J'r; J''s)$. Thus we shall have

$$H_{rs} = W_r \delta_{rs} + v_{rs},$$

where v_{rs} is that function of the J's and w's which is represented by the matrix whose (J′ J″) element is $v(J'r; J''s)$, and so (19) becomes

$$F = H_P(J) + \Sigma_r W_r N_r + \Sigma_{r,s} v_{rs} N_r^{\frac{1}{2}} (N_s + 1 - \delta_{rs})^{\frac{1}{2}} e^{i(\theta_r - \theta_s)/h}. \quad (20)$$

The Hamiltonian is thus the sum of the proper energy of the perturbing system $H_P(J)$, the proper energy of the perturbed systems $\Sigma_r W_r N_r$ and the perturbation energy $\Sigma_{r,s} v_{rs} N_r^{\frac{1}{2}} (N_s + 1 - \delta_{rs})^{\frac{1}{2}} e^{i(\theta_r - \theta_s)/h}$.

§5. *Theory of Transitions in a System from One State to Others of the Same Energy.*

Before applying the results of the preceding sections to light-quanta, we shall consider the solution of the problem presented by a Hamiltonian of the type (19). The essential feature of the problem is that it refers to a dynamical system which can, under the influence of a perturbation energy which does not involve the time explicitly, make transitions from one state to others of the same energy. The problem of collisions between an atomic system and an electron, which has been treated by Born,* is a special case of this type. Born's method is to find a *periodic* solution of the wave equation which consists, in so far as it involves the co-ordinates of the colliding electron, of plane waves,

* Born, 'Z. f. Physik,' vol. 38, p. 803 (1926).

representing the incident electron, approaching the atomic system, which are scattered or diffracted in all directions. The square of the amplitude of the waves scattered in any direction with any frequency is then assumed by Born to be the probability of the electron being scattered in that direction with the corresponding energy.

This method does not appear to be capable of extension in any simple manner to the general problem of systems that make transitions from one state to others of the same energy. Also there is at present no very direct and certain way of interpreting a periodic solution of a wave equation to apply to a non-periodic physical phenomenon such as a collision. (The more definite method that will now be given shows that Born's assumption is not quite right, it being necessary to multiply the square of the amplitude by a certain factor.)

An alternative method of solving a collision problem is to find a *non-periodic* solution of the wave equation which consists initially simply of plane waves moving over the whole of space in the necessary direction with the necessary frequency to represent the incident electron. In course of time waves moving in other directions must appear in order that the wave equation may remain satisfied. The probability of the electron being scattered in any direction with any energy will then be determined by the rate of growth of the corresponding harmonic component of these waves. The way the mathematics is to be interpreted is by this method quite definite, being the same as that of the beginning of § 2.

We shall apply this method to the general problem of a system which makes transitions from one state to others of the same energy under the action of a perturbation. Let H_0 be the Hamiltonian of the unperturbed system and V the perturbing energy, which must not involve the time explicitly. If we take the case of a continuous range of stationary states, specified by the first integrals, α_k say, of the unperturbed motion, then, following the method of § 2, we obtain

$$ih\,\dot{a}(\alpha') = \int \mathrm{V}(\alpha'\alpha'')\,d\alpha'' \,.\, a(\alpha''), \tag{21}$$

corresponding to equation (4). The probability of the system being in a state for which each α_k lies between α_k' and $\alpha_k' + d\alpha_k'$ at any time is $|a(\alpha')|^2 d\alpha_1' \,.\, d\alpha_2' \ldots$ when $a(\alpha')$ is properly normalised and satisfies the proper initial conditions. If initially the system is in the state α^0, we must take the initial value of $a(\alpha')$ to be of the form $a^0 \,.\, \delta(\alpha' - \alpha^0)$. We shall keep a^0 arbitrary, as it would be inconvenient to normalise $a(\alpha')$ in the present case. For a first approximation we may substitute for $a(\alpha'')$ in the right-hand side of (21) its initial value. This gives

$$ih\,\dot{a}(\alpha') = a^0 V(\alpha'\alpha^0) = \alpha^0 v(\alpha'\alpha^0)\, e^{i[W(\alpha')-W(\alpha^0)]t/h},$$

where $v(\alpha'\alpha^0)$ is a constant and $W(\alpha')$ is the energy of the state α'. Hence

$$ih\,a(\alpha') = a^0\,\delta(\alpha'-\alpha^0) + a^0 v(\alpha'\alpha^0)\,\frac{e^{i[W(\alpha')-W(\alpha^0)]t/h}-1}{i[W(\alpha')-W(\alpha^0)]/h}. \tag{22}$$

For values of the α_k' such that $W(\alpha')$ differs appreciably from $W(\alpha^0)$, $a(\alpha')$ is a periodic function of the time whose amplitude is small when the perturbing energy V is small, so that the eigenfunctions corresponding to these stationary states are not excited to any appreciable extent. On the other hand, for values of the α_k' such that $W(\alpha') = W(\alpha^0)$ and $\alpha_k' \neq \alpha_k^0$ for some k, $a(\alpha')$ increases uniformly with respect to the time, so that the probability of the system being in the state α' at any time increases proportionally with the square of the time. Physically, the probability of the system being in a state with exactly the same proper energy as the initial proper energy $W(\alpha^0)$ is of no importance, being infinitesimal. We are interested only in the integral of the probability through a small range of proper energy values about the initial proper energy, which, as we shall find, increases linearly with the time, in agreement with the ordinary probability laws.

We transform from the variables $\alpha_1, \alpha_2 \ldots \alpha_u$ to a set of variables that are arbitrary independent functions of the α's such that one of them is the proper energy W, say, the variables $W, \gamma_1, \gamma_2, \ldots \gamma_{u-1}$. The probability at any time of the system lying in a stationary state for which each γ_k lies between γ_k' and $\gamma_k' + d\gamma_k'$ is now (apart from the normalising factor) equal to

$$d\gamma_1' \,.\, d\gamma_2' \ldots d\gamma_{u-1}' \int |a(\alpha')|^2\, \frac{\partial(\alpha_1', \alpha_2' \ldots \alpha_u')}{\partial(W', \gamma_1' \ldots \gamma_{u-1}')}\, dW'. \tag{23}$$

For a time that is large compared with the periods of the system we shall find that practically the whole of the integral in (23) is contributed by values of W' very close to $W^0 = W(\alpha^0)$. Put

$$a(\alpha') = a(W', \gamma') \quad\text{and}\quad \partial(\alpha_1', \alpha_2' \ldots \alpha_u')/\partial(W', \gamma_1' \ldots \gamma_{u-1}') = J(W', \gamma').$$

Then for the integral in (23) we find, with the help of (22) (provided $\gamma_k' \neq \gamma_k^0$ for some k)

$$\int |a(W', \gamma')|^2\, J(W', \gamma')\, dW'$$

$$= |a^0|^2 \int |v(W', \gamma'; W^0, \gamma^0)|^2\, J(W', \gamma')\, \frac{[e^{i(W'-W^0)t/h}-1]\,[e^{-i(W'-W^0)t/h}-1]}{(W'-W^0)^2}\, dW'$$

$$= 2\,|a^0|^2 \int |v(W', \gamma'; W^0, \gamma^0)|^2\, J(W', \gamma')[1-\cos(W'-W^0)t/h]/(W'-W^0)^2 \,.\, dW'$$

$$= 2\,|a^0|^2\, t/h \,.\, \int |v(W^0 + hx/t, \gamma'; W^0, \gamma^0)|^2\, J(W^0 + hx/t, \gamma')\,(1-\cos x)/x^2 \,.\, dx,$$

if one makes the substitution $(W'-W^0)t/h=x$. For large values of t this reduces to

$$2\,|a^0|^2 t/h\,.\,|v(W^0,\gamma'\,;\,W^0,\gamma^0)|^2\,J(W^0,\gamma')\int_{-\infty}^{\infty}(1-\cos x)/x^2\,.\,dx$$

$$=2\pi\,|a^0|^2 t/h\,.\,|v(W^0,\gamma'\,;\,W^0,\gamma^0)|^2\,J(W^0,\gamma').$$

The probability per unit time of a transition to a state for which each γ_k lies between γ_k' and $\gamma_k'+d\gamma_k'$ is thus (apart from the normalising factor)

$$2\pi\,|a^0|^2/h\,.\,|v(W^0,\gamma'\,;\,W^0,\gamma^0)|^2\,J(W^0,\gamma')\,d\gamma_1'\,.\,d\gamma_2'\,\ldots\,d\gamma_{u-1}', \qquad (24)$$

which is proportional to the square of the matrix element associated with that transition of the perturbing energy.

To apply this result to a simple collision problem, we take the α's to be the components of momentum p_x, p_y, p_z of the colliding electron and the γ's to be θ and ϕ, the angles which determine its direction of motion. If, taking the relativity change of mass with velocity into account, we let P denote the resultant momentum, equal to $(p_x^2+p_y^2+p_z^2)^{\frac{1}{2}}$, and E the energy, equal to $(m^2c^4+P^2c^2)^{\frac{1}{2}}$, of the electron, m being its rest-mass, we find for the Jacobian

$$J=\frac{\partial(p_x,p_y,p_z)}{\partial(E,\theta,\phi)}=\frac{EP}{c^2}\sin\theta.$$

Thus the $J(W^0,\gamma')$ of the expression (24) has the value

$$J(W^0,\gamma')=E'P'\sin\theta'/c^2, \qquad (25)$$

where E′ and P′ refer to that value for the energy of the scattered electron which makes the total energy equal the initial energy W^0 (*i.e.*, to that value required by the conservation of energy).

We must now interpret the initial value of $a(\alpha')$, namely, $a^0\,\delta(\alpha'-\alpha^0)$, which we did not normalise. According to § 2 the wave function in terms of the variables α_k is $b(\alpha')=a(\alpha')\,e^{-iW't/h}$, so that its initial value is

$$a^0\,\delta(\alpha'-\alpha^0)\,e^{-iW't/h}=a^0\,\delta(p_x'-p_x^0)\,\delta(p_y'-p_y^0)\,\delta(p_z'-p_z^0)\,e^{-iW't/h}.$$

If we use the transformation function*

$$(x'/p')=(2\pi h)^{-3/2}e^{i\Sigma_{xyz}p_x'x'/h},$$

and the transformation rule

$$\psi(x')=\int(x'/p')\,\psi(p')\,dp_x'\,dp_y'\,dp_z',$$

we obtain for the initial wave function in the co-ordinates x, y, z the value

$$a^0\,(2\pi h)^{-3/2}\,e^{i\Sigma_{xyz}p_x^0x'/h}\,e^{-iW't/h}.$$

* The symbol x is used for brevity to denote x, y, z.

This corresponds to an initial distribution of $|a^0|^2(2\pi h)^{-3}$ electrons per unit volume. Since their velocity is P^0c^2/E^0, the number per unit time striking a unit surface at right-angles to their direction of motion is $|a^0|^2 P^0c^2/(2\pi h)^3 E^0$. Dividing this into the expression (24) we obtain, with the help of (25),

$$4\pi^2 (2\pi h)^2 \frac{E'E^0}{c^4} |v(p'\,;\; p^0)|^2 \frac{P'}{P^0} \sin\theta'\, d\theta'\, d\phi'. \tag{26}$$

This is the effective area that must be hit by an electron in order that it shall be scattered in the solid angle $\sin\theta'\, d\theta'\, d\phi'$ with the energy E'. This result differs by the factor $(2\pi h)^2/2mE'\,.\, P'/P^0$ from Born's.* The necessity for the factor P'/P^0 in (26) could have been predicted from the principle of detailed balancing, as the factor $|v(p'\,;\, p^0)|^2$ is symmetrical between the direct and reverse processes.†

§ 6. *Application to Light-Quanta.*

We shall now apply the theory of § 4 to the case when the systems of the assembly are light-quanta, the theory being applicable to this case since light-quanta obey the Einstein-Bose statistics and have no mutual interaction. A light-quantum is in a stationary state when it is moving with constant momentum in a straight line. Thus a stationary state r is fixed by the three components of momentum of the light-quantum and a variable that specifies its state of polarisation. We shall work on the assumption that there are a finite number of these stationary states, lying very close to one another, as it would be inconvenient to use continuous ranges. The interaction of the light-quanta with an atomic system will be described by a Hamiltonian of the form (20), in which $H_P(J)$ is the Hamiltonian for the atomic system alone, and the coefficients v_{rs} are for the present unknown. We shall show that this form for the Hamiltonian, with the v_{rs} arbitrary, leads to Einstein's laws for the emission and absorption of radiation.

The light-quantum has the peculiarity that it apparently ceases to exist when it is in one of its stationary states, namely, the zero state, in which its momentum, and therefore also its energy, are zero. When a light-quantum is absorbed it can be considered to jump into this zero state, and when one is emitted it can be considered to jump from the zero state to one in which it is

* In a more recent paper ('Nachr. Gesell. d. Wiss.,' Gottingen, p. 146 (1926)) Born has obtained a result in agreement with that of the present paper for non-relativity mechanics, by using an interpretation of the analysis based on the conservation theorems. I am indebted to Prof. N. Bohr for seeing an advance copy of this work.

† See Klein and Rosseland, 'Z. f. Physik,' vol. 4, p. 46, equation (4) (1921).

physically in evidence, so that it appears to have been created. Since there is no limit to the number of light-quanta that may be created in this way, we must suppose that there are an infinite number of light-quanta in the zero state, so that the N_0 of the Hamiltonian (20) is infinite. We must now have θ_0, the variable canonically conjugate to N_0, a constant, since

$$\dot{\theta}_0 = \partial F/\partial N_0 = W_0 + \text{terms involving } N_0^{-\frac{1}{2}} \text{ or } (N_0 + 1)^{-\frac{1}{2}}$$

and W_0 is zero. In order that the Hamiltonian (20) may remain finite it is necessary for the coefficients v_{r0}, v_{0r} to be infinitely small. We shall suppose that they are infinitely small in such a way as to make $v_{r0}N_0^{\frac{1}{2}}$ and $v_{0r}N_0^{\frac{1}{2}}$ finite, in order that the transition probability coefficients may be finite. Thus we put

$$v_{r0}(N_0 + 1)^{\frac{1}{2}} e^{-i\theta_0/h} = v_r, \quad v_{0r}N_0^{\frac{1}{2}}e^{i\theta_0/h} = v_r^*,$$

where v_r and v_r^* are finite and conjugate imaginaries. We may consider the v_r and v_r^* to be functions only of the J's and w's of the atomic system, since their factors $(N_0 + 1)^{\frac{1}{2}} e^{-i\theta_0/h}$ and $N_0^{\frac{1}{2}}e^{i\theta_0/h}$ are practically constants, the rate of change of N_0 being very small compared with N_0. The Hamiltonian (20) now becomes

$$F = H_P(J) + \Sigma_r W_r N_r + \Sigma_{r \neq 0}[v_r N_r^{\frac{1}{2}} e^{i\theta_r/h} + v_r^*(N_r + 1)^{\frac{1}{2}} e^{-i\theta_r/h}]$$
$$+ \Sigma_{r \neq 0}\Sigma_{s \neq 0} v_{rs} N_r^{\frac{1}{2}}(N_s + 1 - \delta_{rs})^{\frac{1}{2}} e^{i(\theta_r - \theta_s)/h}. \qquad (27)$$

The probability of a transition in which a light-quantum in the state r is absorbed is proportional to the square of the modulus of that matrix element of the Hamiltonian which refers to this transition. This matrix element must come from the term $v_r N_r^{\frac{1}{2}} e^{i\theta_r/h}$ in the Hamiltonian, and must therefore be proportional to $N_r'^{\frac{1}{2}}$ where N_r' is the number of light-quanta in state r before the process. The probability of the absorption process is thus proportional to N_r'. In the same way the probability of a light-quantum in state r being emitted is proportional to $(N_r' + 1)$, and the probability of a light-quantum in state r being scattered into state s is proportional to $N_r'(N_s' + 1)$. Radiative processes of the more general type considered by Einstein and Ehrenfest,† in which more than one light-quantum take part simultaneously, are not allowed on the present theory.

To establish a connection between the number of light-quanta per stationary state and the intensity of the radiation, we consider an enclosure of finite volume, A say, containing the radiation. The number of stationary states for light-quanta of a given type of polarisation whose frequency lies in the

† 'Z. f. Physik,' vol. 19, p. 301 (1923).

range ν_r to $\nu_r + d\nu_r$ and whose direction of motion lies in the solid angle $d\omega_r$ about the direction of motion for state r will now be $A\nu_r^2 d\nu_r d\omega_r/c^3$. The energy of the light-quanta in these stationary states is thus $N_r' . 2\pi h\nu_r . A\nu_r^2 d\nu_r d\omega_r/c^3$. This must equal $Ac^{-1}I_r d\nu_r d\omega_r$, where I_r is the intensity per unit frequency range of the radiation about the state r. Hence

$$I_r = N_r' (2\pi h)\nu_r^3/c^2, \tag{28}$$

so that N_r' is proportional to I_r and $(N_r' + 1)$ is proportional to $I_r + (2\pi h)\nu_r^3/c^2$. We thus obtain that the probability of an absorption process is proportional to I_r, the incident intensity per unit frequency range, and that of an emission process is proportional to $I_r + (2\pi h)\nu_r^3/c^2$, which are just Einstein's laws.* In the same way the probability of a process in which a light-quantum is scattered from a state r to a state s is proportional to $I_r [I_s + (2\pi h)\nu_r^3/c^2]$, which is Pauli's law for the scattering of radiation by an electron.†

§7. *The Probability Coefficients for Emission and Absorption.*

We shall now consider the interaction of an atom and radiation from the wave point of view. We resolve the radiation into its Fourier components, and suppose that their number is very large but finite. Let each component be labelled by a suffix r, and suppose there are σ_r components associated with the radiation of a definite type of polarisation per unit solid angle per unit frequency range about the component r. Each component r can be described by a vector potential κ_r chosen so as to make the scalar potential zero. The perturbation term to be added to the Hamiltonian will now be, according to the classical theory with neglect of relativity mechanics, $c^{-1} \Sigma_r \kappa_r \dot{X}_r$, where X_r is the component of the total polarisation of the atom in the direction of κ_r, which is the direction of the electric vector of the component r.

We can, as explained in §1, suppose the field to be described by the canonical variables N_r, θ_r, of which N_r is the number of quanta of energy of the component r, and θ_r is its canonically conjugate phase, equal to $2\pi h\nu_r$ times the θ_r of §1. We shall now have $\kappa_r = a_r \cos \theta_r/h$, where a_r is the amplitude of κ_r, which can be connected with N_r as follows:—The flow of energy per unit area per unit time for the component r is $\frac{1}{2}\pi c^{-1} a_r^2 \nu_r^2$. Hence the intensity

* The ratio of stimulated to spontaneous emission in the present theory is just twice its value in Einstein's. This is because in the present theory either polarised component of the incident radiation can stimulate only radiation polarised in the same way, while in Einstein's the two polarised components are treated together. This remark applies also to the scattering process.

† Pauli, 'Z. f. Physik,' vol. 18, p. 272 (1923).

per unit frequency range of the radiation in the neighbourhood of the component r is $I_r = \frac{1}{2}\pi c^{-1} a_r^2 \nu_r^2 \sigma_r$. Comparing this with equation (28), we obtain $a_r = 2(h\nu_r/c\sigma_r)^{\frac{1}{2}} N_r^{\frac{1}{2}}$, and hence

$$\kappa_r = 2(h\nu_r/c\sigma_r)^{\frac{1}{2}} N_r^{\frac{1}{2}} \cos\theta_r/h.$$

The Hamiltonian for the whole system of atom plus radiation would now be, according to the classical theory,

$$F = H_P(J) + \Sigma_r (2\pi h\nu_r) N_r + 2c^{-1} \Sigma_r (h\nu_r/c\sigma_r)^{\frac{1}{2}} \dot{X}_r N_r^{\frac{1}{2}} \cos\theta_r/h, \qquad (29)$$

where $H_P(J)$ is the Hamiltonian for the atom alone. On the quantum theory we must make the variables N_r and θ_r canonical q-numbers like the variables J_k, w_k that describe the atom. We must now replace the $N_r^{\frac{1}{2}} \cos\theta_r/h$ in (29) by the real q-number

$$\tfrac{1}{2}\{N_r^{\frac{1}{2}} e^{i\theta_r/h} + e^{-i\theta_r/h} N_r^{\frac{1}{2}}\} = \tfrac{1}{2}\{N_r^{\frac{1}{2}} e^{i\theta_r/h} + (N_r+1)^{\frac{1}{2}} e^{-i\theta_r/h}\}$$

so that the Hamiltonian (29) becomes

$$F = H_P(J) + \Sigma_r (2\pi h\nu_r) N_r + h^{\frac{1}{2}} c^{-\frac{3}{2}} \Sigma_r (\nu_r/\sigma_r)^{\frac{1}{2}} \dot{X}_r \{N_r^{\frac{1}{2}} e^{i\theta_r/h} + (N_r+1)^{\frac{1}{2}} e^{-i\theta_r/h}\}. \qquad (30)$$

This is of the form (27), with

$$v_r = v_r^* = h^{\frac{1}{2}} c^{-\frac{3}{2}} (\nu_r/\sigma_r)^{\frac{1}{2}} \dot{X}_r \qquad (31)$$

and

$$v_{rs} = 0 \qquad (r, s \neq 0).$$

The wave point of view is thus consistent with the light-quantum point of view and gives values for the unknown interaction coefficient v_{rs} in the light-quantum theory. These values are not such as would enable one to express the interaction energy as an algebraic function of canonical variables. Since the wave theory gives $v_{rs} = 0$ for $r, s \neq 0$, it would seem to show that there are no direct scattering processes, but this may be due to an incompleteness in the present wave theory.

We shall now show that the Hamiltonian (30) leads to the correct expressions for Einstein's A's and B's. We must first modify slightly the analysis of § 5 so as to apply to the case when the system has a large number of discrete stationary states instead of a continuous range. Instead of equation (21) we shall now have

$$ih\,\dot{a}(\alpha') = \Sigma_{\alpha''} V(\alpha'\alpha'')\, a(\alpha'').$$

If the system is initially in the state α^0, we must take the initial value of $a(\alpha')$ to be $\delta_{\alpha'\alpha^0}$, which is now correctly normalised. This gives for a first approximation

$$ih\,\dot{a}(\alpha') = V(\alpha'\alpha^0) = v(\alpha'\alpha^0)\, e^{i[W(\alpha')-W(\alpha^0)]t/h},$$

which leads to

$$ih\, a(\alpha') = \delta_{\alpha'\alpha^0} + v(\alpha'\alpha^0)\, \frac{e^{i[W(\alpha')-W(\alpha^0)]t/h} - 1}{i[W(\alpha')-W(\alpha^0)]/h},$$

corresponding to (22). If, as before, we transform to the variables W, γ_1, $\gamma_2 \dots \gamma_{u-1}$, we obtain (when $\gamma' \neq \gamma^0$)

$$a(W'\gamma') = v(W', \gamma'; W^0, \gamma^0)\,[1 - e^{i(W'-W^0)t/h}]/(W'-W^0).$$

The probability of the system being in a state for which each γ_k equals γ_k' is $\Sigma_{W'} |a(W'\gamma')|^2$. If the stationary states lie close together and if the time t is not too great, we can replace this sum by the integral $(\Delta W)^{-1}\int |a(W'\gamma')|^2 dW'$, where ΔW is the separation between the energy levels. Evaluating this integral as before, we obtain for the probability per unit time of a transition to a state for which each $\gamma_k = \gamma_k'$

$$2\pi/h\Delta W \,.\, |v(W^0, \gamma'; W^0, \gamma^0)|^2. \qquad (32)$$

In applying this result we can take the γ's to be any set of variables that are independent of the total proper energy W and that together with W define a stationary state.

We now return to the problem defined by the Hamiltonian (30) and consider an absorption process in which the atom jumps from the state J^0 to the state J' with the absorption of a light-quantum from state r. We take the variables γ' to be the variables J' of the atom together with variables that define the direction of motion and state of polarisation of the absorbed quantum, but not its energy. The matrix element $v(W^0, \gamma'; W^0, \gamma^0)$ is now

$$h^{1/2}c^{-3/2}(\nu_r/\sigma_r)^{1/2}\dot{X}_r(J^0J')N_r^0,$$

where $\dot{X}_r(J^0J')$ is the ordinary (J^0J') matrix element of $\dot{X}_r$. Hence from (32) the probability per unit time of the absorption process is

$$\frac{2\pi}{h\Delta W}\cdot\frac{h\nu_r}{c^3\sigma_r}|\dot{X}_r(J^0J')|^2 N_r^0.$$

To obtain the probability for the process when the light-quantum comes from any direction in a solid angle $d\omega$, we must multiply this expression by the number of possible directions for the light-quantum in the solid angle $d\omega$, which is $d\omega\,\sigma_r\Delta W/2\pi h$. This gives

$$d\omega\frac{\nu_r}{hc^3}|\dot{X}_r(J_0J')|^2 N_r^0 = d\omega\frac{1}{2\pi h^2 c\nu_r^2}|\dot{X}_r(J^0J')|^2 I_r$$

with the help of (28). Hence the probability coefficient for the absorption process is $1/2\pi h^2 c\nu_r^2 \,.\, |\dot{X}_r(J^0J')|^2$, in agreement with the usual value for Einstein's absorption coefficient in the matrix mechanics. The agreement for the emission coefficients may be verified in the same manner.

The present theory, since it gives a proper account of spontaneous emission, must presumably give the effect of radiation reaction on the emitting system, and enable one to calculate the natural breadths of spectral lines, if one can overcome the mathematical difficulties involved in the general solution of the wave problem corresponding to the Hamiltonian (30). Also the theory enables one to understand how it comes about that there is no violation of the law of the conservation of energy when, say, a photo-electron is emitted from an atom under the action of extremely weak incident radiation. The energy of interaction of the atom and the radiation is a q-number that does not commute with the first integrals of the motion of the atom alone or with the intensity of the radiation. Thus one cannot specify this energy by a c-number at the same time that one specifies the stationary state of the atom and the intensity of the radiation by c-numbers. In particular, one cannot say that the interaction energy tends to zero as the intensity of the incident radiation tends to zero. There is thus always an unspecifiable amount of interaction energy which can supply the energy for the photo-electron.

I would like to express my thanks to Prof. Niels Bohr for his interest in this work and for much friendly discussion about it.

Summary.

The problem is treated of an assembly of similar systems satisfying the Einstein-Bose statistical mechanics, which interact with another different system, a Hamiltonian function being obtained to describe the motion. The theory is applied to the interaction of an assembly of light quanta with an ordinary atom, and it is shown that it gives Einstein's laws for the emission and absorption of radiation.

The interaction of an atom with electromagnetic waves is then considered, and it is shown that if one takes the energies and phases of the waves to be q-numbers satisfying the proper quantum conditions instead of c-numbers, the Hamiltonian function takes the same form as in the light-quantum treatment. The theory leads to the correct expressions for Einstein's A's and B's.

The Quantum Theory of Dispersion.

By P. A. M. Dirac, St. John's College, Cambridge; Institute for Theoretical Physics, Göttingen.

(Communicated by R. H. Fowler, F.R.S.—Received April 4, 1927.)

§ 1. *Introduction and Summary.*

The new quantum mechanics could at first be used to answer questions concerning radiation only through analogies with the classical theory. In Heisenberg's original matrix theory, for instance, it is assumed that the matrix elements of the polarisation of an atom determine the emission and absorption of radiation analogously to the Fourier components in the classical theory. In more recent theories* a certain expression for the electric density obtained from the quantum mechanics is used to determine the emitted radiation by the same formulæ as in the classical theory. These methods give satisfactory results in many cases, but cannot even be applied to problems where the classical analogies are obscure or non-existent, such as resonance radiation and the breadths of spectral lines.

A theory of radiation has been given by the author which rests on a more definite basis.† It appears that one can treat a field of radiation as a dynamical system, whose interaction with an ordinary atomic system may be described by a Hamiltonian function. The dynamical variables specifying the field are the energies and phases of its various harmonic components, each of which

* E. Schrödinger, 'Ann. d. Physik,' vol. 81, p. 109 (1926); W. Gordon, 'Z. f. Physik,' vol. 40, p. 117 (1926); O. Klein, 'Z. f. Physik,' vol. 41, p. 407 (1927).

† 'Roy. Soc. Proc.,' A, vol. 114, p. 243 (1927). This is referred to later by *loc. cit.*

Reprinted from *Proc. Roy. Soc. (London)* **A114**, 710 (1927).

is effectively a simple harmonic oscillator. One must, of course, in the quantum theory take these variables to be q-numbers satisfying the proper quantum conditions. One finds then that the Hamiltonian for the interaction of the field with an atom is of the same form as that for the interaction of an assembly of light-quanta with the atom. There is thus a complete formal reconciliation between the wave and light-quantum points of view.

In applying the theory to the practical working out of radiation problems one must use a perturbation method, as one cannot solve the Schrödinger equation directly. One can assume that the term (V say) in the Hamiltonian due to the interaction of the radiation and the atom is small compared with that representing their proper energy, and then use V as the perturbing energy. Physically the assumption is that the mean life time of the atom in any state is large compared with its periods of vibration. In the present paper we shall apply the theory to determine the radiation scattered by the atom, considering also the case when the frequency of the incident radiation coincides with that of a spectral line of the atom. The method used will be that in which one finds a solution of the Schrödinger equation that satisfies certain initial conditions, corresponding to a given initial state for the atom and field. In general terms it may be described as follows :—

If V_{mn} are the matrix elements of the perturbing energy V, where each suffix m or n refers to a stationary state of the whole system of atom plus field the stationary state of the atom being specified by its action variables, J say, and that of the field by a given distribution of energy among its harmonic components, or by a given distribution of light-quanta), then each V_{mn} gives rise to transitions from state n to state m*; more accurately, it causes the eigenfunction representing state m to grow if that representing state n is already excited, the general formula for the rate of change of the amplitude a_m of an eigenfunction being†

$$ih/2\pi \, . \, \dot{a}_m = \Sigma_n V_{mn} a_n = \Sigma_n v_{mn} a_n e^{2\pi i (W_m - W_n) t/h}, \tag{1}$$

where v_{mn} is the constant amplitude of the matrix element V_{mn}, and W_m is the

* In *loc. cit.*, § 6, it was in error assumed that V_{mn} caused transitions from state m to state n, and consequently the information there obtained about an absorption (or emission) process in terms of the number of light-quanta existing before the process should really apply to an emission (or absorption) process in terms of the number of light-quanta in existence after the process. This change, of course, does not affect the results (namely the proof of Einstein's laws) which can depend on $|V_{mn}|^2 = |V_{nm}|^2$.

† *Loc. cit.*, equation (4). In the present paper h is taken to mean just Planck's constant [instead of $(2\pi)^{-1}$ times this quantity as in *loc. cit.*] which is preferable when one has to deal much with quanta $h\nu$ of radiation.

total proper energy of the state m. To solve these equations one obtains a first approximation by substituting for the a's on the right-hand side their initial values, a second approximation by substituting for these a's their values given by the first approximation, and so on. One or two such approximations will usually be sufficient to give a solution that is fairly accurate for times that are small compared with the life time, but may all the same be large compared with the periods of the atom. From the first approximation, namely,

$$a_m = a_{mo} + \Sigma_n v_{mn} a_{no} \left(1 - e^{2\pi i (W_m - W_n) t/h}\right)/(W_m - W_n), \tag{2}$$

where a_{no} denotes the initial value of a_n, one sees readily that when two states m and n have appreciably different proper energies, the amplitude a_m gets changed only by a small extent, varying periodically with the time, on account of transitions from state n. Only when two states, m and m' say, have the same energy does the amplitude a_m of one of them grow continually at the expense of that of the other, as is necessary for physically recognisable transitions to occur, and the rate of growth is then proportional to $v_{mm'}$.

The interaction term of the Hamiltonian function obtained in *loc. cit.* [equation (30)] does not give rise to any direct scattering processes, in which a light-quantum jumps from one state to another of the same frequency but different direction of motion (*i.e.*, the corresponding matrix element $v_{mm'} = 0$). All the same, radiation that has apparently been scattered can appear by a double process in which a third state, n say, with different proper energy from m and m', plays a part. If initially all the a's vanish except $a_{m'}$, then a_n gets excited on account of transitions from state m' by an amount proportional to $v_{nm'}$, and although it must itself always remain small, a calculation shows that it will cause a_m to grow continually with the time at a rate proportional to $v_{mn} v_{nm'}$. The scattered radiation thus appears as the result of the two processes $m' \rightarrow n$ and $n \rightarrow m$, one of which must be an absorption and the other an emission, in neither of which is the total proper energy even approximately conserved.

The more accurate expression for the interaction energy obtained in § 3 of the present paper does give rise to direct scattering processes, whose effect is of the same order of magnitude as that of the double processes, and must be added to it. The sum of the two will be found to give just Kramers' and Heisenberg's dispersion formula* when the incident frequency does not coincide with that of an absorption or emission line of the atom. If, however, the incident frequency coincides with that of, say, an absorption line, one of the

* Kramers and Heisenberg, 'Z. f. Physik,' vol. 31, p. 681 (1925).

terms in the Kramers-Heisenberg formula becomes infinite. The present theory shows that in this case the scattered radiation consists of two parts, of which the amount of one increases proportionally to the time since the interaction commenced, and that of the other proportionally to the square of this time. The first part arises from those terms in the Kramers-Heisenberg formula that remain finite, with perhaps a contribution from the infinite term, while the second, which is much larger, is just what one would get from transitions of the atom to the upper state and down again governed by Einstein's laws of absorption and emission.

A difficulty that appears in the present treatment of radiation problems should be here pointed out. If one tries to calculate, for instance, the total probability of a light-quantum having been emitted by a given time, one obtains as result a sum or integral with respect to the frequency of the emitted light-quantum that does not converge in the high frequencies. This difficulty is not due to any fundamental mistake in the theory, but comes from the fact that the atom has, for the purpose of its interaction with the field, been counted simply as a varying electric dipole, and the field produced by a dipole, when resolved into its Fourier components, has an infinite amount of energy in the short wave-lengths, owing to the infinite field in its immediate neighbourhood. If one does not make the approximation of regarding the atom as a dipole, but uses the exact expression for the interaction energy, then the fact that the singularity in the field is of a lower order of magnitude and remains constant is sufficient to make the series or integral converge. The exact interaction energy is too complicated to be used as a basis for radiation theory at present, and we shall here use only the dipole energy, which will mean that divergent series are always liable to appear in the calculation. The best method to adopt under such circumstances is first to work out the general theory of any effect using arbitrary coefficients v_{mn}, and then to substitute for these coefficients in the final result their values given by the dipole interaction energy. If one then finds that the series all converge, one can assume that the result is a correct first approximation; if, however, any of them do not converge, one must conclude that a dipole theory is inadequate for the treatment of that particular effect. We shall find that for the phenomena of dispersion and resonance radiation dealt with in the present paper, there are no divergent series in the first approximation, so that the dipole theory is sufficient. If, however, one tries to calculate the breadth of a spectral line, one meets with a divergent series, so that a dipole theory of the atom is presumably inadequate for the correct treatment of this question.

§ 2. *Preliminary Formulæ.*

We consider the electromagnetic field to be resolved into its components of plane, plane-polarised, progressive waves, each component r having a definite frequency, direction of motion and state of polarisation, and being associated with a certain type of light-quanta. (To save writing we shall in future suppose the words " direction of motion " applied to a light-quantum or a component of the field to imply also its state of polarisation, and a sum or integral taken over all directions of motion to imply also the summation over both states of polarisation for each direction of motion. This is convenient because the two variables, direction of motion and state of polarisation, are always treated mathematically in the same way.) For an electromagnetic field of infinite extent there will be a continuous three-dimensional range of these components. As this would be inconvenient to deal with mathematically, we suppose it to be replaced by a large number of discrete components. If there are σ_r components per unit solid angle of direction of motion per unit frequency range, we can keep σ_r an arbitrary function of the frequency and direction of motion of the component r, provided it is large and reasonably continuous, and shall find that it always cancels from the final results of a calculation, which fact appears to justify our replacement of the continuous range by the discrete set.

We can express σ_r in the form $\sigma_r = (\Delta\nu_r \, \Delta\omega_r)^{-1}$, where $\Delta\nu_r$ can be regarded as the frequency interval between successive components in the neighbourhood of the component r, and $\Delta\omega_r$ is in the same way the solid angle of direction of motion to be associated with this component. The quantities $\Delta\nu_r$, $\Delta\omega_r$ enable one to pass directly from sums to integrals. Thus if f_r is any function of the frequency and direction of motion of the component r that varies only slightly from one component to a neighbouring one, the sum of $f_r \, \Delta\nu_r$ for all components having a specified direction of motion is

$$\Sigma_\nu f_r \, \Delta\nu_r = \int f_r \, d\nu_r, \tag{3}$$

and the sum of $f_r \, \Delta\omega_r$ for all components having a specified frequency is

$$\Sigma_\omega f_r \, \Delta\omega_r = \int f_r \, d\omega_r. \tag{3'}$$

Also the sum of $f_r \, (\sigma_r)^{-1}$ for all components is

$$\Sigma f_r (\sigma_r)^{-1} = \Sigma f_r \, \Delta\nu_r \, \Delta\omega_r = \int f_r d\nu_r \, d\omega_r. \tag{3''}$$

If the number* N_s of quanta of energy of the component s varies only slightly from one component to a neighbouring one, one can give a meaning to the intensity of the radiation per unit frequency range. By supposing the discreteness in the number of components to arise from the radiation being confined in an enclosure (which would imply stationary waves and a special function σ_s) one obtains† for the rate of flow of energy per unit area per unit solid angle per unit frequency range

$$I_{\nu\omega} = N_s h\nu_s^3/c^2, \tag{4}$$

a result which may be taken to hold generally for arbitrary σ_s and progressive waves.‡ If only those components with a specified direction of motion are excited, we have instead that the rate of flow of energy per unit area per unit frequency range is

$$I_\nu = N_s h\nu_s^3/c^2 \,.\, \Delta\omega_s; \tag{5}$$

while if only a single component s is excited, we have that the rate of flow of energy per unit area is

$$I = N_s h\nu_s^3/c^2 \,.\, \Delta\omega_s \, \Delta\nu_s = N_s h\nu_s^3/c^2\sigma_s. \tag{6}$$

In this last case the amplitude of the electric force has the value E given by

$$E^2 = 8\pi I/c = 8\pi N_s h\nu_s^3/c^3\sigma_s, \tag{7}$$

and the amplitude a of the magnetic vector potential, when chosen so that the electric potential is zero, is

$$a = cE/2\pi\nu_s = 2\,(h\nu_s/2\pi c\sigma_s)^{\frac{1}{2}} N_s^{\frac{1}{2}}. \tag{8}$$

§ 3. *The Hamiltonian Function.*

We shall now determine the Hamiltonian function that describes the interaction of the field with an atom more accurately than in *loc. cit.* We consider the atom to consist of a single electron moving in an electrostatic field of potential ϕ. According to the classical theory its relativity Hamiltonian equation when undisturbed is

$$p_x^2 + p_y^2 + p_z^2 - (W + e\phi)^2/c^2 + m^2c^2 = 0,$$

so that its Hamiltonian function is

$$H = W = c\,\{m^2c^2 + p_x^2 + p_y^2 + p_z^2\}^{\frac{1}{2}} - e\phi. \tag{9}$$

* The rule given in *loc. cit.* that symbols representing c-number values for q-number variables should be primed need not always be observed if no confusion thus arises, as in the present case.

† *Loc. cit.*, § 6, equation (28).

‡ This is justified by the fact that one can obtain the result by an alternative method that does not rquire a finite enclosure, namely by using a quantum-mechanical argument similar to that of *loc. cit.* (lower part of p. 259), applied to the case of discrete momentum values.

If now there is a perturbing field of radiation, given by the magnetic vector potential κ_x, κ_y, κ_z chosen so that the electric scalar potential is zero, the Hamiltonian equation for the perturbed system will be

$$\left(p_x + \frac{e}{c}\kappa_x\right)^2 + \left(p_y + \frac{e}{c}\kappa_y\right)^2 + \left(p_z + \frac{e}{c}\kappa_z\right)^2 - \frac{(\mathrm{W} + e\phi)^2}{c^2} + m^2c^2 = 0,$$

which gives for the Hamiltonian function

$$\mathrm{H} = \mathrm{W} = c\left\{m^2c^2 + \left(p_x + \frac{e}{c}\kappa_x\right)^2 + \left(p_y + \frac{e}{c}\kappa_y\right)^2\left(p_z + \frac{e}{c}\kappa_z\right)^2\right\}^{\frac{1}{2}} - e\phi$$

$$= c\{[m^2c^2 + p_x^{\ 2} + p_y^{\ 2} + p_z^{\ 2}] + [2e/c\,.\,(p_x\kappa_x + p_y\kappa_y + p_z\kappa_z) + e^2/c^2\,.\,(\kappa_x^{\ 2} + \kappa_y^{\ 2} + \kappa_z^{\ 2})]\}^{\frac{1}{2}} - e\phi.$$

By expanding the square root, counting the second term in square brackets [ ] as small, and then neglecting relativity corrections for this term, one finds approximately

$$\mathrm{H} = c\,[m^2c^2 + p_x^{\ 2} + p_y^{\ 2} + p_z^{\ 2}]^{\frac{1}{2}} - e\phi + e/c\,.\,(\dot{x}\kappa_x + \dot{y}\kappa_y + \dot{z}\kappa_z) + e^2/2mc^2\,.\,(\kappa_x^{\ 2} + \kappa_y^{\ 2} + \kappa_z^{\ 2})$$

$$= \mathrm{H}_0 + e/c\,.\,(\dot{x}\kappa_x + \dot{y}\kappa_y + \dot{z}\kappa_z) + e^2/2mc^2\,.\,(\kappa_x^{\ 2} + \kappa_y^{\ 2} + \kappa_z^{\ 2}), \qquad (10)$$

where H_0 is the Hamiltonian for the unperturbed system given by (9). When one counts the radiation field as a dynamical system, one must add on its proper energy $\Sigma \mathrm{N}_r h\nu_r$ to the Hamiltonian (10).

According to the classical theory, the magnetic vector potential for any component r of the radiation is

$$\kappa_r = a_r \cos 2\pi\theta_r/h = 2\,(h\nu_r/2\pi c\sigma_r)^{\frac{1}{2}}\,\mathrm{N}_r^{\frac{1}{2}} \cos 2\pi\theta_r/h \qquad (11)$$

from (8), where θ_r increases uniformly with the time such that $\dot{\theta}_r = h\nu_r$, and is the variable that must be taken to be the canonical conjugate of N_r when the radiation field is treated as a dynamical system. The direction of this vector potential is that of the electric vector of the component of radiation. Hence the total value of the component of the vector potential in any direction, say that of the x-axis, is

$$\kappa_x = \Sigma_r \kappa_r \cos\alpha_{xr} = 2\,(h/2\pi c)^{\frac{1}{2}}\,\Sigma_r \cos\alpha_{xr}\,(\nu_r/\sigma_r)^{\frac{1}{2}}\,\mathrm{N}_r^{\frac{1}{2}} \cos 2\pi\theta_r/h, \qquad (12)$$

where α_{xr} is the angle between the electric vector of the component r and the x-axis. In the quantum theory, where the variables N_r, θ_r are q-numbers, the expression $2\mathrm{N}_r^{\frac{1}{2}} \cos 2\pi\theta_r/h$ must be replaced by the real q-number $\mathrm{N}_r^{\frac{1}{2}} e^{2\pi i\theta_r/h} + (\mathrm{N}_r + 1)^{\frac{1}{2}}\, e^{-2\pi i\theta_r/h}$. With this change one can take over the

Hamiltonian (10) into the quantum theory, which gives, when one includes the term $\Sigma N_r h\nu_r$,

$$H = H_0 + \Sigma N_r h\nu_r + eh^{\frac{1}{2}}/(2\pi)^{\frac{1}{2}} c^{\frac{1}{2}} \,.\, \Sigma_r \dot{x}_r (\nu_r/\sigma_r)^{\frac{1}{2}} [N_r^{\frac{1}{2}} e^{2\pi i\theta_r/h} + (N_r+1)^{\frac{1}{2}} e^{-2\pi i\theta_r/h}]$$
$$+ e^2h/4\pi mc^3 \,.\, \Sigma_{r,s} \cos\alpha_{rs} (\nu_r\nu_s/\sigma_r\sigma_s)^{\frac{1}{2}} [N_r^{\frac{1}{2}} e^{2\pi i\theta_r/h} + (N_r+1)^{\frac{1}{2}} e^{-2\pi i\theta_r/h}]$$
$$\times [N_s^{\frac{1}{2}} e^{2\pi i\theta_s/h} + (N_s+1)^{\frac{1}{2}} e^{-2\pi i\theta_s/h}] \quad (13)$$

where x_r denotes the component of the vector (x, y, z) in the direction of the electric vector of the component r, *i.e.*,

$$x_r = x\cos\alpha_{xr} + y\cos\alpha_{yr} + z\cos\alpha_{zr},$$

and α_{rs} denotes the angle between the electric vectors of the components r and s, *i.e.*

$$\cos\alpha_{rs} = \cos\alpha_{xr}\cos\alpha_{xs} + \cos\alpha_{yr}\cos\alpha_{ys} + \cos\alpha_{zr}\cos\alpha_{zs}.$$

The terms in the first line of (13) are just those obtained in *loc. cit.*, equation (30), and give rise only to emission and absorption processes. The remaining terms (*i.e.*, those in the double summation) were neglected in *loc. cit.* These terms may be divided into three sets :—

(i) Those terms that are independent of the θ's, which can be added to the proper energy $H_0 + \Sigma N_r h\nu_r$. The sum of all such terms, which can arise only when $r = s$, is

$$e^2h/4\pi mc^3 \,.\, \Sigma_r \nu_r/\sigma_r \,.\, [N_r^{\frac{1}{2}} e^{2\pi i\theta_r/h} (N_r+1)^{\frac{1}{2}} e^{-2\pi i\theta_r/h}$$
$$+ (N_r+1)^{\frac{1}{2}} e^{-2\pi i\theta_r/h} N_r^{\frac{1}{2}} e^{2\pi i\theta_r/h}]$$
$$= e^2h/4\pi mc^3 \,.\, \Sigma_r \nu_r/\sigma_r \,.\, (2N_r+1).$$

The terms $e^2h/4\pi mc^3 \,.\, \Sigma\nu_r/\sigma_r \,.\, 2N_r$ are negligible compared with $\Sigma N_r h\nu_r$, owing to the very large quantity σ_r in the denominator, while the terms $e^2h/4\pi mc^3 \,.\, \Sigma\nu_r/\sigma_r$ may be ignored since they do not involve any of the dynamical variables, in spite of the fact that the sum $\Sigma\nu_r/\sigma_r$, equal to $\int \nu_r\, d\nu_r\, d\omega_r$ from (3″), does not converge for the high frequencies.

(ii) The terms containing a factor of the form $e^{2\pi i(\theta_r-\theta_s)/h}$ $(r \neq s)$, whose sum is

$$e^2h/4\pi mc^3 \,\Sigma_r \Sigma_{s\neq r} \cos\alpha_{rs} (\nu_r\nu_s/\sigma_r\sigma_s)^{\frac{1}{2}} [N_r^{\frac{1}{2}} (N_s+1)^{\frac{1}{2}} e^{2\pi i(\theta_r-\theta_s)/h}$$
$$+ (N_r+1)^{\frac{1}{2}} N_s^{\frac{1}{2}} e^{-2\pi i(\theta_r-\theta_s)/h}]$$
$$= e^2h/2\pi mc^3 \,\Sigma_r \Sigma_{s\neq r} \cos\alpha_{rs} (\nu_r\nu_s/\sigma_r\sigma_s)^{\frac{1}{2}} N_r^{\frac{1}{2}} (N_s+1)^{\frac{1}{2}} e^{2\pi i(\theta_r-\theta_s)/h}. \quad (14)$$

These terms, which are the only important ones in the three sets, give rise to transitions in which a light-quantum jumps directly from a state s to a state r.

Such transitions may be called true scattering processes, to distinguish them from the double scattering processes described in § 1.

(iii) The remaining terms, each of which involves a factor of one or other of the forms $e^{\pm 4\pi i\theta_r/h}$, $e^{\pm 2\pi i(\theta_r+\theta_s)/h}$. These terms correspond to processes in which two light-quanta are emitted or absorbed simultaneously, and cannot arise in a light-quantum theory in which there are no forces between the light quanta. The effects of these terms will be found to be negligible, so that the disagreement with the light-quantum theory is not serious.

§ 4. *Discussion of the Emission and True Scattering Processes.*

We shall consider now the simple emission processes, in order to discuss the divergent integral that arises in this question. Suppose a light-quantum to be emitted in state r, with a simultaneous jump of the atom from the state $J = J'$ to the state $J = J''$. If we label the final state of the whole system of atom plus field m and the initial state k, the value at time t of the amplitude a_m of the eigenfunction of the final state will be in the first approximation

$$a_m = v_{mk}\,(1 - e^{2\pi i(W_m - W_k)t/h})\,/\,(W_m - W_k), \tag{15}$$

obtained by putting $a_{k_0} = 1$, $a_{n_0} = 0$ $(n \neq k)$ in equation (2). The only term in the Hamiltonian (13) that can contribute anything to the matrix element v_{mk} is the one involving $e^{2\pi i\theta_r/h}$, whose $(J'', N_1', N_2' \ldots N_r' + 1 \ldots;\ J', N_1', N_2' \ldots N_r' \ldots)$ matrix element is $eh^{\frac{1}{2}}/(2\pi)^{\frac{1}{2}}c^{\frac{3}{2}} \,.\, \dot{x}_r(J''J')\,(\nu_r/\sigma_r)^{\frac{1}{2}}\,(N_r' + 1)^{\frac{1}{2}}$, where $\dot{x}_r(J''J')$ is the ordinary $(J''J')$ matrix element of $\dot{x}_r$. If there is no incident radiation we must take all the N''s zero, which gives

$$v_{mk} = eh^{\frac{1}{2}}\,/\,(2\pi)^{\frac{1}{2}}c^{\frac{3}{2}} \,.\, \dot{x}_r(J''J')\,(\nu_r/\sigma_r)^{\frac{1}{2}},$$

and also

$$W_k = H_0(J') \qquad W_m = H_0(J'') + h\nu_r.$$

Thus

$$W_m - W_k = H_0(J'') + h\nu_r - H_0(J') = h\,[\nu_r - \nu(J'\,J'')]$$

where $\nu(J'\,J'') = [H_0(J') - H_0(J'')]/h$ is the transition frequency between states J' and J'', if one assumes J' to be the higher one. Hence from (15)

$$|a_m|^2 = \frac{e^2}{\pi hc^3}\,|\dot{x}_r(J'J'')|^2\,\frac{\nu_r}{\sigma_r}\,\frac{1 - \cos 2\pi[\nu_r - \nu(J'J'')]t}{[\nu_r - \nu(J'J'')]^2}.$$

To obtain the total probability of any light-quantum being emitted within the solid angle $\delta\omega$ about the direction of motion of a given light-quantum r with this jump of the atom, we must multiply $|a_m|^2$ by $\delta\omega/\Delta\omega_r$ and sum for all frequencies. This gives, with the help of (3)

$$\delta\omega\,\Sigma_\nu\,\frac{|a_m|^2}{\Delta\omega_r} = \delta\omega\,\frac{e^2}{\pi hc^3}\,|\dot{x}_r(J'J'')|^2\int_0^\infty \nu_r\,d\nu_r\,\frac{1 - \cos 2\pi[\nu_r - \nu(J'J'')]t}{[\nu_r - \nu(J'J'')]^2}. \tag{16}$$

The integral does not converge for the high frequencies. This is due, as mentioned in § 1, to the non-legitimacy of taking only the dipole action of the atom into account, which is what one does when one substitutes for the magnetic potential in (10) its value given by (12), which is its value at some fixed point such as the nucleus instead of its value where the electron is momentarily situated. To obtain the interaction energy exactly, one should put $\cos 2\pi [\theta_r/h - \nu_r \xi_r/c]$ instead of $\cos 2\pi\theta r/h$ in (11), where ξ_r is the component of the vector (x, y, z) in the direction of motion of the component r of radiation. This will make no appreciable change for low frequencies ν_r, but will cause a new factor $\cos 2\pi\nu_r\xi_r/c$ or $\sin 2\pi\nu_r\xi_r/c$, whose matrix elements tend to zero as ν_r tends to infinity, to appear in the coefficients of (13). This will presumably cause the integral in (16) to converge when corrected, as its divergence when uncorrected is only logarithmic.

Assuming that the integrand in (16) has been suitably modified in the high frequencies, one sees that for values of t large compared with the periods of the atom (but small compared with the life time in order that the approximations may be valid) practically the whole of the integral is contributed by values of ν_r close to ν (J′ J″), which means physically that only radiation close to a transition frequency can be spontaneously emitted. One finds readily for the total probability of the emission, by performing the integration,

$$\delta\omega e^2/\pi hc^3 . |\dot{x}_r(\mathrm{J'J''})|^2 . 2\pi^2 t\nu(\mathrm{J'J''}),$$

which leads to the correct value for Einstein's A coefficient per unit solid angle, namely,

$$2\pi e^2/hc^3 . |\dot{x}_r(\mathrm{J'J''})|^2\nu(\mathrm{J'J''}) = 8\pi^3 e^2/hc^3 . |x_r(\mathrm{J'J''})|^2\nu^3(\mathrm{J'J''}).$$

We shall now determine the rate at which true scattering processes occur, caused by the terms (14) in the Hamiltonian. We see at once that the frequency of occurrence of these processes is independent of the nature of the atom, and is thus the same for a bound as for a free electron. The true scattering is the only kind of scattering that can occur for a free electron, so that we should expect the terms (14) to lead to the correct formula for the scattering of radiation by a free electron, with neglect of relativity mechanics and thus of the Compton effect.

Suppose that initially the atom is in the state J′ and all the N's vanish except one of them, N_s say, which has the value N_s'. We label this state for the whole system by k, and the state for which $\mathrm{J} = \mathrm{J}'$ and $\mathrm{N}_s = \mathrm{N}_s' - 1$, $\mathrm{N}_r = 1$ with all the other N's zero by m. In the first approximation a_m is again given by (15), where we now have

$$v_{mk} = e^2 h/2\pi mc^3 . \cos \alpha_{rs} (\nu_r \nu_s/\sigma_r \sigma_s)^{\frac{1}{2}} N_s'^{\frac{1}{2}}, \tag{17}$$

$$W_k = H_0(J') + N_s' h\nu_s, \qquad W_m = H_0(J') + (N_s' - 1) h\nu_s + h\nu_r. \tag{18}$$

Thus

$$W_m - W_k = h(\nu_r - \nu_s), \tag{19}$$

and hence

$$|a_m|^2 = \frac{e^4}{2\pi^2 m^2 c^6} \cos^2 \alpha_{rs} \frac{\nu_r \nu_s}{\sigma_r \sigma_s} N_s' \frac{1 - \cos 2\pi(\nu_r - \nu_s)t}{(\nu_r - \nu_s)^2}.$$

To obtain the total probability of a scattered light-quantum being in the solid angle $\delta\omega$ we must, as before, multiply $|a_m|^2$ by $\delta\omega/\Delta\omega_r$ and sum for all frequencies ν_r, which gives*

$$\delta\omega \Sigma_\nu \frac{|a_m|^2}{\Delta\omega_r} = \delta\omega \frac{e^4}{2\pi^2 m^2 c^6} \cos^2 \alpha_{rs} \frac{\nu_s}{\sigma_s} N_s' \int \nu_r d\nu_r \frac{1 - \cos 2\pi(\nu_r - \nu_s)t}{(\nu_r - \nu_s)^2}. \tag{20}$$

We again obtain a divergent integral, of the same form as before, which we may assume becomes convergent in the more exact theory. We now have that practically the whole of the integral is contributed by values of ν_r close to ν_s and the total probability for the scattering process is

$$\delta\omega \frac{e^4}{2\pi^2 m^2 c^6} \cos^2 \alpha_{rs} \frac{\nu_s}{\sigma_s} N_s' . 2\pi^2 t\nu_s = \delta\omega \frac{e^4}{hm^2 c^4 \nu_s} \cos^2 \alpha_{rs} . tI$$

from (6), where I is the rate of flow of incident energy per unit area. The rate of emission of scattered energy per unit solid angle is thus

$$e^4/m^2c^4 . \cos^2 \alpha_{rs} I,$$

where α_{rs} is the angle between the electric vectors of the incident and scattered radiation, which is the correct classical formula.

* The reason why there is a small probability for the scattered frequency ν_r differing by a finite amount from the incident frequency ν_s is because we are considering the scattered radiation, after the scattering process has been acting for only a finite time t, resolved into its Fourier components. One sees from the formula (20) that as the time t gets greater, the scattered radiation gets more and more nearly monochromatic with the frequency ν_s. If one obtained a periodic solution of the Schrödinger equation corresponding to permanent physical conditions, one would then find that the scattered frequency was exactly equal to the incident frequency.

§ 5. *Theory of Dispersion.*

We shall now work out the second approximation to the solution of equations (1), taking the case when the system is initially in the state k, so that the first approximation, given by (2) with $a_{n0} = \delta_{nk}$, reduces to

$$a_m = \delta_{mk} + v_{mk}(1 - e^{2\pi i(W_m - W_k)t/h})/(W_m - W_k).$$

When one substitutes these values for the a_n's in the right-hand side of (1), one obtains

$$\begin{aligned} ih/2\pi \, . \, \dot{a}_m &= v_{mk}\, e^{2\pi i(W_m - W_k)t/h} \\ &\quad + \Sigma_n v_{mn} v_{nk} (1 - e^{2\pi i(W_n - W_k)t/h})\, e^{2\pi i(W_m - W_n)t/h} / (W_n - W_k) \\ &= \left(v_{mk} - \Sigma_n \frac{v_{mn} v_{nk}}{W_n - W_k}\right) e^{2\pi i(W_m - W_k)t/h} + \Sigma_n \frac{v_{mn} v_{nk}}{W_n - W_k}\, e^{2\pi i(W_m - W_n)t/h}, \end{aligned}$$

and hence when $m \neq k$

$$\begin{aligned} a_m &= \left(v_{mk} - \Sigma_n \frac{v_{mn} v_{nk}}{W_n - W_k}\right) \frac{1 - e^{2\pi i(W_m - W_k)t/h}}{W_m - W_k} \\ &\quad + \Sigma_n \frac{v_{mn} v_{nk}}{W_n - W_k} \frac{1 - e^{2\pi i(W_m - W_n)t/h}}{W_m - W_n}. \end{aligned} \tag{21}$$

We may suppose the diagonal elements v_{nn} of the perturbing energy to be zero, since if they were not zero they could be included with the proper energy W_n. There will then be no terms in (21) with vanishing denominators, provided all the energy levels are different.

Suppose now that the proper energy of tho state m is equal to that of the initial state k. Then the first term on the right-hand side of (21) ceases to be periodic in the time, and becomes

$$\{v_{mk} - \Sigma_n v_{mn} v_{nk} / (W_n - W_k)\}\, 2\pi t/ih,$$

which increases linearly with the time. The rate of increase consists of a part, proportional to v_{mk}, that is due to direct transitions from state k, together with a sum of parts, each of which is proportional to a $v_{mn}v_{nk}$, and is due to transitions first from k to n and then from n to m, although the amplitude a_n of the eigenfunction of the intermediate state always remains small.

When one applies the theory to the scattering of radiation one must consider not a single final state with exactly the same proper energy as the initial state, but a set of final states with proper energies lying close together in a range that contains the initial proper energy, corresponding to all the possible scattered light-quanta with different frequencies but the same direction of motion that may appear. One must now determine the total probability of the system lying in any one of these final states, which is

$$\Sigma |a_m|^2 = \int (\Delta W_m)^{-1} |a_m|^2 dW_m,$$

where ΔW_m is the interval between the energy levels. The second term in the expression (21) for a_m may be neglected since it always remains small (except in the case of resonance which will be considered later) and hence

$$\Sigma |a_m|^2 = \int \left| v_{mk} - \Sigma_n \frac{v_{mn} v_{nk}}{W_n - W_k} \right|^2 \frac{2[1 - \cos 2\pi (W_m - W_k) t/h]}{\Delta W_m . (W_m - W_k)^2} dW_m.$$

If one assumes that the integral converges, so that for large values of t practically the whole of it is contributed by values of W_m close to W_k, one obtains

$$\Sigma |a_m|^2 = \frac{4\pi^2 t}{h \Delta W_m} \left| v_{mk} - \Sigma_n \frac{v_{mn} v_{nk}}{W_n - W_k} \right|^2, \tag{22}$$

where the quantities on the right refer to that final state that has exactly the initial proper energy.

We take the states k and m to be the same as for the true scattering process considered in the preceding section, so that equations (17), (18) and (19) still hold, and $\Delta W_m = h \Delta \nu_r = h/\sigma_r \, \Delta \omega_r$. We can now take the state n to be either the state $J = J''$, $N_s = N_s' - 1$, $N_t = 0$ $(t \neq s)$ for any J'', which would make the process $k \rightarrow n$ an absorption of an s-quantum and $n \rightarrow m$ an emission of an r-quantum, or the state $J = J''$, $N_s = N_s'$, $N_r = 1$, $N_t = 0$ $(t \neq s, r)$, which would make $k \rightarrow n$ the emission and $n \rightarrow m$ the absorption. In the first case we should have

$$v_{nk} = \frac{e}{c} \left(\frac{h\nu_s}{2\pi c \sigma_s} \right)^{\frac{1}{2}} \dot{x}_s (J''J') \, N_s'^{\frac{1}{2}} \qquad v_{mn} = \frac{e}{c} \left(\frac{h\nu_r}{2\pi c \sigma_r} \right)^{\frac{1}{2}} \dot{x}_r (J'J''),$$

and

$$W_n = H_0 (J'') + (N_s' - 1) h\nu_s \qquad W_n - W_k = h [\nu (J''J') - \nu_s]^*,$$

and in the second

$$v_{nk} = \frac{e}{c} \left(\frac{h\nu_r}{2\pi c \sigma_r} \right)^{\frac{1}{2}} \dot{x}_r (J''J') \qquad v_{mn} = \frac{e}{c} \left(\frac{h\nu_s}{2\pi c \sigma_s} \right)^{\frac{1}{2}} \dot{x}_s (J'J'') \, N_s'^{\frac{1}{2}},$$

and

$$W_n = H_0 (J'') + N_s' h\nu_s + h\nu_r \qquad W_n - W_k = h [\nu (J''J') + \nu_r].$$

We shall neglect the other possible states n, namely those for which the matrix elements v_{mn}, v_{nk} come from terms in the double summation in the Hamiltonian

* The frequency ν (J″J′) is not necessarily positive.

(13), as we are working only to the first order in these terms. (We are working to the second order only in the emission and absorption terms, which, as we shall find, is the same as the first order in the terms of the double summation.) We now obtain for the right-hand side of (22) in which we must take $\nu_r = \nu_s$,

$$N_s' t \Delta\omega_r \frac{e^4 \nu_s^2}{h^2 c^6 \sigma_s} \left| \frac{h}{m} \cos\alpha_{rs} - \Sigma_{J''} \left\{ \frac{\dot{x}_r(J'J'')\dot{x}_s(J''J')}{\nu(J''J') - \nu_s} + \frac{\dot{x}_s(J'J'')\dot{x}_r(J''J')}{\nu(J''J') + \nu_s} \right\} \right|^2 \quad (23).$$

The most convenient way of expressing this result is to find the amplitude P (a vector) of the electric moment of that vibrating dipole of frequency ν_s that would, according to the classical theory, emit the same distribution of radiation as that actually scattered by the atom. The number of light-quanta of the type r (with $\nu_r = \nu_s$) emitted by the dipole P in time t per unit solid angle is

$$2\pi^3 \nu_s^3 / hc^3 \, . \, P_r^2 t,$$

where P_r is the component of P in the direction of the electric vector of the light-quanta r. Comparing this with (23) (which must first be divided by $\Delta\omega_r$ to change it to the probability of a light quantum being scattered per unit solid angle) one finds for P_r

$$P_r = \left(\frac{8\pi N_s'}{hc^3 \nu_s \sigma_s}\right)^{\frac{1}{2}} \frac{e^2}{4\pi^2} \left| \frac{h}{m} \cos\alpha_{rs} - \Sigma_{J''} \left\{ \frac{\dot{x}_r(J'J'')\dot{x}_s(J''J')}{\nu(J''J') - \nu_s} + \frac{\dot{x}_s(J'J'')\dot{x}_r(J''J')}{\nu(J''J') + \nu_s} \right\} \right|$$

$$= E \frac{e^2}{h} \frac{1}{\nu_s^2} \left| \frac{h}{4\pi^2 m} \cos\alpha_{rs} - \Sigma_{J''} [\nu(J''J')]^2 \left\{ \frac{x_r(J'J'')\, x_s(J''J')}{\nu(J''J') - \nu_s} + \frac{x_s(J'J'')\, x_r(J''J')}{\nu(J''J') + \nu_s} \right\} \right|, \quad (24)$$

using (7), where E is the amplitude of the electric vector of the incident radiation.

We can put this result in a different form by using the following relations, which follow from the quantum conditions,

$$\Sigma_{J''} [x_r(J'J'')\, x_s(J''J') - x_s(J'J'')\, x_r(J''J')] = [x_r x_s - x_s x_r](J'J') = 0 \quad (25)$$

and

$$\Sigma_{J''} [x_r(J'J'')\, \dot{x}_s(J''J') - \dot{x}_s(J'J'')\, x_r(J''J')] = [x_r \dot{x}_s - \dot{x}_s x_r](J'J') = ih/2\pi m \, . \, \cos\alpha_{rs}, \quad (26)$$

which gives

$$\Sigma_{J''} [x_r(J'J'')\, x_s(J''J')\, \nu(J''J') + x_s(J'J'')\, x_r(J''J')\, \nu(J''J')] = h/4\pi^2 m \, . \, \cos\alpha_{rs}. \quad (27)$$

Multiplying (25) by ν_s and adding to (27), we obtain

$$\Sigma_{J''} [x_r (J' J'') x_s (J'' J') \{\nu (J'' J') + \nu_s\} + x_s (J' J'') x_r (J'' J') \{\nu (J'' J') - \nu_s\}] = h/4\pi^2 m \,.\, \cos \alpha_{rs}. \qquad (28)$$

With the help of this equation, (24) reduces to

$$P_r = E \frac{e^2}{h} \left| \Sigma_{J''} \left\{ \frac{x_r (J'J'') x_s (J''J')}{\nu (J''J') - \nu_s} + \frac{x_s (J'J'') x_r (J''J')}{\nu (J''J') + \nu_s} \right\} \right|,$$

so that the vector P is equal to

$$P = E \frac{e^2}{h} \left| \Sigma_{J''} \left\{ \frac{x (J'J'') x_s (J''J')}{\nu (J''J') - \nu_s} + \frac{x_s (J'J'') x (J''J')}{\nu (J''J') + \nu_s} \right\} \right|, \qquad (29)$$

where x without a suffix means the vector (x, y, z). This is identical with Kramers' and Heisenberg's result.*

In applying the formula (22), instead of taking the final state m of the system to be one for which the atom is again in its initial state $J = J'$, we can take a new final state for the atom, $J = J'''$ say. The frequency ν_r for the scattered radiation that gives no change of total proper energy is now

$$\nu_r = \nu_s - \nu (J'''J') = \nu_s + \nu (J''J''') - \nu (J''J'), \qquad (30)$$

which differs from the incident frequency ν_s, so that we obtain in this way the non-coherent scattered radiation. (We assume that this ν_r is positive as otherwise there would be no non-coherent scattered radiation associated with the final state $J = J'''$ of the atom.) In the present case we have $v_{mk} = 0$, corresponding to the fact that the true scattering process does not contribute to the non-coherent radiation. We now obtain for P_r, after a similar and almost identical calculation to that leading to equation (24),

$$P_r = E \frac{e^2}{h} \frac{1}{\nu_r \nu_s} \left| \Sigma_{J''} \nu (J''J') \nu (J''J''') \left\{ \frac{x_r (J'''J'') x_s (J''J')}{\nu (J''J') - \nu_s} + \frac{x_s (J''' J'') x_r (J'' J')}{\nu (J''J') + \nu_r} \right\} \right| \qquad (31)$$

This result can be put in the form corresponding to (29) with the help of equations analogous to (25) and (26) referring to the non-diagonal $(J'''J')$ matrix elements of $[x_r x_s - x_s x_r]$ and $[x_r \dot{x}_s - \dot{x}_s x_r]$. These equations give, corresponding to (28),

$$\Sigma_{J''} [x_r (J'''J'') x_s (J''J') \{\nu (J''J') + \nu_r\} + x_s (J'''J'') x_r (J''J') \{\nu (J''J''') - \nu_r\}] = 0.$$

* Kramers and Heisenberg, *loc. cit.*, equation (18). For previous quantum-theoretical deductions of the dispersion formula see Born, Heisenberg and Jordan, 'Z. f. Physik,' vol. 35, p. 557, Kap. 1, equation (40) (1926); Schrödinger, *loc. cit.*, § 2, equation (23); and Klein, *loc. cit.*, § 5, equation (82).

When the left-hand side of this equation is subtracted from the summation in (31) one obtains, on account of the relations

$$\nu(\mathrm{J''J'})\,\nu(\mathrm{J''J'''}) = \nu(\mathrm{J''J'})\,[\nu(\mathrm{J''J'}) + \nu_r - \nu_s]$$
$$= [\nu(\mathrm{J''J'}) - \nu_s]\,[\nu(\mathrm{J''J'}) + \nu_r] + \nu_r\nu_s,$$

and

$$\nu(\mathrm{J''J'})\,\nu(\mathrm{J''J'''}) = [\nu(\mathrm{J''J'''}) - \nu_r]\,[\nu(\mathrm{J''J'}) + \nu_r] + \nu_r\nu_s$$

which follow from (30), the result

$$\mathrm{P}_r = \mathrm{E}\frac{e^2}{h}\left|\Sigma_{\mathrm{J''}}\left\{\frac{x_r(\mathrm{J'''J''})\,x_s(\mathrm{J''J'})}{\nu(\mathrm{J''J'}) - \nu_s} + \frac{x_s(\mathrm{J'''J''})\,x_r(\mathrm{J''J'})}{\nu(\mathrm{J''J'}) + \nu_r}\right\}\right|,$$

again in agreement with Kramers and Heisenberg.

§ 6. *The Case of Resonance.*

The dispersion formulæ obtained in the preceding section can no longer hold when the frequency of the incident radiation coincides with that of an absorption or emission line of the atom, on account of a vanishing denominator. One easily sees where a modification must be made in the deduction of the formulæ. Since one of the intermediate states n now has the same energy as the initial state k, the term in the second summation in (21) referring to this n becomes large and can no longer be neglected.

In investigating this case of resonance one must, for generality, suppose the incident radiation to consist of a distribution of light-quanta over a range of frequencies including the resonance frequency, instead of entirely of light-quanta of a single frequency, as the results will depend very considerably on how nearly monochromatic the incident radiation is. Thus one must take the initial state k of the system to be given by $\mathrm{J} = \mathrm{J'}$ and $\mathrm{N}_s = \mathrm{N}_s'$, where N_s' is zero except for light-quanta of a specified direction, and is for these light-quanta (roughly speaking) a continuous function of the frequency, so that the rate of flow of incident energy per unit area per unit frequency range is given by (5). The final state m for a process of coherent scattering is one for which $\mathrm{J} = \mathrm{J'}$ again, and a light-quantum s has been absorbed and one r of approximately the same frequency emitted. Thus we have

$$\mathrm{W}_m - \mathrm{W}_k = h(\nu_r - \nu_s). \tag{32}$$

As before, the intermediate states n will be those for which $\mathrm{J} = \mathrm{J''}$ (arbitrary) and either the s-quantum has already been absorbed or the r-quantum has already been emitted. If we take for definiteness the case when the range of incident frequencies includes only one resonance frequency, and this is an absorption frequency to the state of the atom $J = J'$, say, then that intermediate state of the system for which $J = J^l$ and for which the s-quantum has already been absorbed will have very nearly the same proper energy as the initial state. Calling this intermediate state l we have

$$W_l - W_k = h(\nu_0 - \nu_s) \qquad W_m - W_l = h(\nu_r - \nu_0) \tag{33}$$

where ν_0 is the resonance frequency, equal to $[H(J^l) - H(J')]/h$.

In equation (21) we can now neglect only those terms of the second summation for which $n \neq l$. This gives

$$a_m = \left(v_{mk} - \Sigma_{n \neq l} \frac{v_{mn} v_{nk}}{W_n - W_k}\right) \frac{1 - e^{2\pi i (W_m - W_k) t/h}}{W_m - W_k}$$
$$+ \frac{v_{ml} v_{lk}}{W_l - W_k} \left\{ \frac{1 - e^{2\pi i (W_m - W_l) t/h}}{W_m - W_l} - \frac{1 - e^{2\pi i (W_m - W_k) t/h}}{W_m - W_k} \right\},$$

which, with the help of (32) and (33), may be written

$$a_m = \left(v_{mk} - \Sigma_{n \neq l} \frac{v_{mn} v_{nk}}{W_n - W_k}\right) \frac{1 - e^{2\pi i (\nu_r - \nu_s) t}}{h(\nu_r - \nu_s)}$$
$$+ \frac{v_{ml} v_{lk}}{h^2(\nu_0 - \nu_s)} \left\{ \frac{1 - e^{2\pi i (\nu_r - \nu_0) t}}{\nu_r - \nu_0} - \frac{1 - e^{2\pi i (\nu_r - \nu_s) t}}{\nu_r - \nu_s} \right\}.$$

We must now determine the total probability of a specified light-quantum r being emitted with the absorption of any one of the incident light-quanta s, which is given by $\Sigma_{\nu_s} |a_m|^2$, equal to $\int (\Delta\nu_s)^{-1} |a_m|^2 d\nu_s$. To evaluate this we require the following integrals

$$\int_0^\infty \frac{\left|1 - e^{2\pi i (\nu_r - \nu_s) t}\right|^2}{(\nu_r - \nu_s)^2} d\nu_s = 4\pi^2 t$$

$$\int_0^\infty \frac{1}{(\nu_0 - \nu_s)^2} \left| \frac{1 - e^{2\pi i (\nu_r - \nu_0) t}}{\nu_r - \nu_0} - \frac{1 - e^{2\pi i (\nu_r - \nu_s) t}}{\nu_r - \nu_s} \right|^2 d\nu_s$$
$$= 4\pi \frac{2\pi(\nu_r - \nu_0) t - \sin 2\pi(\nu_r - \nu_0) t}{(\nu_r - \nu_0)^3}$$

$$\int_0^\infty \frac{1 - e^{2\pi i (\nu_r - \nu_s) t}}{(\nu_r - \nu_s)(\nu_0 - \nu_s)} \left\{ \frac{1 - e^{-2\pi i (\nu_r - \nu_0) t}}{\nu_r - \nu_0} - \frac{1 - e^{-2\pi i (\nu_r - \nu_s) t}}{\nu_r - \nu_s} \right\} d\nu_s$$
$$= 2\pi \left\{ \frac{2\pi(\nu_r - \nu_0) t - \sin 2\pi(\nu_r - \nu_0) t}{(\nu_r - \nu_0)^2} + i \frac{1 - \cos 2\pi(\nu_r - \nu_0) t}{(\nu_r - \nu_0)^2} \right\},$$

for large t, and with their help obtain,

$$\Sigma_{\nu_s} |a_m|^2 = \left| v_{mk} - \Sigma_{n \neq l} \frac{v_{mn} v_{nk}}{W_n - W_k} \right|^2 \frac{4\pi^2 t}{h^2 \Delta\nu_s}$$

$$+ \frac{|v_{ml} v_{lk}|^2}{h^4} \frac{4\pi}{\Delta\nu_s} \frac{2\pi(\nu_r - \nu_0)t - \sin 2\pi(\nu_r - \nu_0)t}{(\nu_r - \nu_0)^3}$$

$$+ 2R\left(v_{mk} - \Sigma_{n \neq l} \frac{v_{mn} v_{nk}}{W_n - W_k}\right) \frac{v_{kl} v_{lm}}{h^3 \Delta\nu_s} 2\pi \left\{ \frac{2\pi(\nu_r - \nu_0)t - \sin 2\pi(\nu_r - \nu_0)t}{(\nu_r - \nu_0)^2} + i \frac{1 - \cos 2\pi(\nu_r - \nu_0)t}{(\nu_r - \nu_0)^2} \right\} \quad (34)$$

where the quantities on the right now refer to that incident light-quantum s for which $\nu_s = \nu_r$, and R means the real part of all that occurs in the term after it.

The first of these three terms is just the contribution of those terms of the dispersion formula (22) that remain finite, the second is that which replaces the contribution of the infinite term,* and the third gives the interference between the first two, and replaces the cross terms obtained when one squares the dispersion electric moment. One can see the meaning of the second term more clearly if one sums it for all frequences ν_r of the scattered radiation in a small frequency range $\nu_0 - \alpha'$ to $\nu_0 + \alpha''$ about the resonance frequency ν_0 (which frequency range must be large compared with the theoretical breadth of the spectral line in order that the approximations may be valid). This is equivalent to multiplying the term by $(\Delta\nu_r)^{-1}$ and integrating through the frequency range. If, for brevity, one denotes the quantity $4\pi |v_{ml} v_{lk}|^2 / h^4 \, \Delta\nu_r \, \Delta\nu_s$ by $f(\nu_r)$, the result is, neglecting terms that do not increase indefinitely with t or that tend to zero as the α's tend to zero,

$$\int_{\nu_0 - \alpha'}^{\nu_0 + \alpha''} f(\nu_r) \frac{2\pi(\nu_r - \nu_0)t - \sin 2\pi(\nu_r - \nu_0)t}{(\nu_r - \nu_0)^3} d\nu_r$$

$$= f(\nu_0) \int_{\nu_0 - \alpha'}^{\nu_0 + \alpha''} \frac{2\pi(\nu_r - \nu_0)t - \sin 2\pi(\nu_r - \nu_0)t}{(\nu_r - \nu_0)^3} d\nu_r + f'(\nu_0) \int_{\nu_0 - \alpha'}^{\nu_0 + \alpha''} \frac{2\pi(\nu_r - \nu_0)t - \sin 2\pi(\nu_r - \nu_0)t}{(\nu_r - \nu_0)^2} d\nu_r$$

$$= f(\nu_0)(2\pi t)^2 \left[\tfrac{1}{2}\pi - \frac{1}{2\pi t \alpha''} - \frac{1}{2\pi t \alpha'}\right] + f'(\nu_0)\, 2\pi t \log \alpha''/\alpha'.$$

* It should be noticed that this second term does not reduce to the square of the l term in the summation (22) when ν_r is not a resonance frequency, but to double this amount. This difference is due to the fact that processes involving a change of proper energy are not entirely negligible for the initial conditions used in the present paper, and one such scattering process, which was neglected in § 5, becomes in the resonance case a process with no change of proper energy and is included in the calculation.

Thus the contribution of the second term in (34) to the small frequency range $\nu_0 - \alpha'$ to $\nu_0 + \alpha''$ consists of two parts, one of which increases proportionally to t^2 and the other proportionally to t. The part that increases proportionally to t^2, namely,

$$\tfrac{1}{2}\pi f(\nu_0)(2\pi t)^2 = \tfrac{1}{2}(2\pi)^4 \mid v_{ml} v_{lk} \mid^2 / h^4 \, \Delta\nu_r \, \Delta\nu_s \, . \, t^2,$$

is just that which would arise from actual transitions to the higher state of the atom and down again governed by Einstein's laws, since the probability that the atom has been raised to the higher state by the time τ is* $(2\pi)^2 \mid \nu_{lk} \mid^2 / h^2 \, \Delta\nu_s \, . \, \tau$, and when it is in the higher state the probability per unit time of its jumping down again with emission of a light-quantum in the required direction is $(2\pi)^2 \mid v_{ml} \mid^2 / h^2 \, \Delta\nu_r$, so that the total probability of the two transitions taking place within a time t is

$$\frac{(2\pi)^2 \mid v_{lk} \mid^2}{h^2 \Delta\nu_s} \cdot \frac{(2\pi)^2 \mid v_{ml} \mid^2}{h^2 \, \Delta\nu_r} \int_0^t \tau \, d\tau = \frac{(2\pi)^4 \mid v_{ml} v_{lk} \mid^2}{h^4 \, \Delta\nu_r \, \Delta\nu_s} \tfrac{1}{2} t^2.$$

The part that increases linearly with the time may be added to the contributions of the first and third terms, which also increase according to this law. For values of t large compared with the periods of the atom, the terms proportional to t will be negligible compared with those proportional to t^2, and hence the resonance scattered radiation is due practically entirely to absorptions and emissions according to Einstein's laws.

* This result and the one for the emission follow at once from formula (32) of *loc. cit.*

14

On the Time–Energy Uncertainty Relation

Eugene P. Wigner

1. INTRODUCTION AND SUMMARY

There is only one well-known application for the time–energy uncertainty relation: the connection between the life-time and the energy-width of resonance states. The relation in question was commonly known even before quantum mechanics was established; its first quantum mechanical derivation was based on Dirac's original theory of the interaction between matter and radiation.[1] The point which should be noted is that the uncertainty relation does not apply to time and energy *in abstracto* but to the life-time of a definite state of a system. In the example referred to, this is the state in which an atom is in an excited state but there is no radiation present.

The preceding formulation of the time–energy uncertainty relation appears to be different from Heisenberg's well-known position–momentum uncertainty relation.[2] This postulates that the state of any quantum mechanical system is, at every instant of time, such that the product of the position and momentum spreads s_x and s_p

$$s_x s_p \geqslant \tfrac{1}{2}\hbar. \tag{1}$$

The spreads in question are defined as the positive square roots of

$$s_x^2 = \langle\psi|(x-x_0)^2|\psi\rangle/\langle\psi|\psi\rangle$$
$$s_p^2 = \langle\psi|(p-p_0)^2|\psi\rangle/\langle\psi|\psi\rangle \tag{1a}$$

and holds for every ψ and all x_0, p_0. In (1a) ψ is an arbitrary state vector, x and p position and momentum operators, respectively. The clause 'at every instant of time' in the preceding formulation of the uncertainty principle is, evidently, important since the state vector ψ in (1) represents the state of the system under consideration only for a definite instant of time and this must be the same for both expressions in (1a).

Reprinted from *Aspects in Quantum Theory* (1972).

It follows that, if one wishes to formulate the time–energy analogue of the usual position–momentum uncertainty relation, one will have to restrict one's attention to the situation along a single value of a space coordinate (which will be chosen as the x coordinate) just as a single time coordinate entered (1). The simplest form of the relation then becomes the statement that the product of the energy spread and the spread in the probability that the definite value of the coordinate x be assumed at time t, is at least $\hbar/2$. For a single non-relativistic particle these spreads τ and ϵ are defined as the positive square roots of

$$\tau^2 = \frac{\iiint |\psi(x,y,z,t)|^2 (t-t_0)^2 \, dy \, dz \, dt}{\iiint |\psi(x,y,z,t)|^2 \, dy \, dz \, dt} \tag{2}$$

$$\epsilon^2 = \langle\psi|(H-E_0)^2|\psi\rangle / \langle\psi|\psi\rangle$$

or

$$\epsilon^2 = \frac{\iiint |\phi(x,y,z,E)|^2 (E-E_0)^2 \, dy \, dz \, dE}{\iiint |\phi(x,y,z,E)|^2 \, dy \, dz \, dE}, \tag{2a}$$

where

$$\phi(x,y,z,E) = \int \psi(x,y,z,t) e^{iEt/\hbar} \, dt \tag{2b}$$

is the Fourier transform of ψ from time into energy. It will be seen (as is pretty evident) that the uncertainty relation

$$\tau\epsilon > \tfrac{1}{2}\hbar \tag{3}$$

holds again for all values of t_0 and E_0. However, whereas there are, for any x_0 and p_0 in (1), state vectors for which the equality sign is valid in (1), namely those for which the x dependence of ψ is a Gaussian of $x-x_0$ multiplied with exp ip_0x, the inequality sign always holds in (3). This is a consequence of the fact that the energy is a positive definite operator (or has, in the non-relativistic case, a lower bound). The lower limit of $\tau\epsilon$ is, naturally, independent of t_0 but does depend on E_0 and increases substantially as E_0 approaches the lower bound of H from above. Naturally, it increases even further as E_0 crosses that bound and decreases further. All this, as well as equation (2a), will be further discussed below.

2. A GENERAL OBSERVATION ON THE INTERCHANGE OF THE SPACE AND TIME COORDINATES

Actually, the point just mentioned does not constitute the most interesting difference between the position–momentum and the time–energy uncertainty relations. The difference which appears most interesting to this writer will be first formulated in a universe with only one spatial dimension. In such a universe, it is natural to ask[3] for the probability that, at a definite time, the spatial coordinate of the particle have the value x. It is less natural to ask for the probability that the particle be, at a definite landmark in space, just at the time t. It would be more natural to ask, instead, for the probability that the particle crosses the aforementioned landmark at the time t from the left, and also that it crosses that landmark, at a given time, from the right. The sum of these probabilities, when integrated over all times t, is 1 for a free particle. The difference between the two cases arises from the fact that a particle's world line can cross the $t=$constant line only in one direction (in the direction of increasing t); it can cross the $x=$constant line in both directions. If we replace 'line' in the last sentence by 'plane', we have the generalization of the distinction to the actual four-dimensional universe. In Dirac's theory of the electron,[4] the probability that the particle be, at the definite time t, at the point x, y, z is given by $\psi^{\dagger}\alpha_0\psi$, the space–time variables of ψ being x, y, z, and t. The probabilities for crossing the $x=$constant plane, at y, z and at time t, in the two different directions, are given, essentially, by $\frac{1}{2}\psi^{\dagger}(\alpha_0+\alpha_x)\psi$ and $\frac{1}{2}\psi^{\dagger}(\alpha_0-\alpha_x)\psi$. This point of this paragraph, interesting though it may be, will not be elaborated further.

Even though the idea of space–time, and hence the similarity between space and time coordinates, appears natural from the point of view of relativity theory, initial conditions characterizing the state of the system at a definite instant of time appear more natural to us than initial conditions giving the state of the system for all times but only for a single value of one of the spatial coordinates. There are, indeed, valid reasons for this preference. The transition from the position–momentum uncertainty relation to the time–energy uncertainty relation is, however, based on the second type of description of the state of the system. It would be interesting to develop equations of motion giving the spatial derivative of the second type of characterization of a state and to explore the properties of the resulting equations.

3. A GENERALIZATION OF HEISENBERG'S UNCERTAINTY RELATION

There are uncertainty relations for practically any two non-commuting operators, but the generalization of Heisenberg's relation to be pointed out here is a very special one and bears a close resemblance to the original form of the relation. We denote, first, the variables of ψ by

x and r, the latter one standing for all variables different from x. The relation (1) then remains valid if one replaces ψ in (1a) by

$$\psi(x, r) \rightarrow \int \phi(r)^* \psi(x, r) dr \tag{4}$$

ϕ being any function of r and the integration is over all values of the continuous coordinates implied by r and summation over the discrete ones. This is, of course, a well-known fact, most commonly used with a ϕ which is a delta function of all coordinates different from x. The derivation of the relation (1), given by Robertson,[2] remains equally valid, however, if r is assumed to involve also the time, with $\psi(x, r)$ satisfying Schrödinger's equation and ϕ depending on time in an arbitrary fashion. The right side of (4) is then a generalized transition amplitude which can be thought of as corresponding to a measurement lasting a finite length of time, but not affecting x.

Accepting the generalization of Heisenberg's relation just outlined, one sees that the time–energy uncertainty relation referring to the life-time of resonance states, which was mentioned in the first paragraph of this article, is not as different from (3) as it first appeared. It is in fact included in the generalization of (3) which is the analogue of the generalization of (1) just pointed to. The generalization replaces in this case $\psi(x, y, z, t)$ by $\langle u|\psi(t)\rangle = \chi(t)$ where u is any state vector. The time spread τ is then the positive square root of

$$\tau^2 = \frac{\int |\langle u|\psi(t)\rangle|^2 (t-t_0)^2 \, dt}{\int |\langle u|\psi(t)\rangle|^2 \, dt} = \frac{\int |\chi(t)|^2 (t-t_0)^2 \, dt}{\int |\chi(t)|^2 \, dt}. \tag{5}$$

In order to define the energy spread, we calculate, in analogy to (2b)

$$\phi(E) = (2\pi\hbar)^{-\frac{1}{2}} \int \psi(t) e^{iEt/\hbar} dt. \tag{5a}$$

This is a stationary state of energy E. Hence, ϵ^2 will be

$$\epsilon^2 = \frac{\int |\langle u|\phi(E)\rangle|^2 (E-E_0)^2 \, dE}{\int |\langle u|\phi(E)\rangle|^2 \, dE} = \frac{\int |\eta(E)|^2 (E-E_0)^2 \, dE}{\int |\eta(E)|^2 \, dE} \tag{5b}$$

where

$$\begin{aligned} \eta(E) = \langle u|\phi(E)\rangle &= (2\pi\hbar)^{-\frac{1}{2}} \int \langle u|\psi(t)\rangle e^{iEt/\hbar} dt \\ &= (2\pi\hbar)^{-\frac{1}{2}} \int \chi(t) e^{iEt/\hbar} dt \end{aligned} \tag{5c}$$

is the Fourier transform of χ. The factor $(2\pi\hbar)^{-\frac{1}{2}}$ renders the denominators of (5b) and (5) equal. With these definitions, and identifying u with the state vector of the resonance, (3) will represent the uncertainty relation giving the minimum energy spread of a resonance with given life-time. This is the time–energy uncertainty relation mentioned in the first paragraph of the present article. On the other hand, if one inserts a delta function of x for u, the close analogue of the original Heisenberg uncertainty relations given by (2), (2*a*), and (3) results.

We will go over next to the proof of (3) with the τ and ϵ given by (5) and (5*b*) and will give some estimates for the lowest possible value of $\tau\epsilon$ as function of E_0. The lower bound of $\tau\epsilon$ will turn out, naturally, to be independent of t_0 and, perhaps somewhat less obviously, depends only on the ratio E_0/ϵ, not on E_0 and ϵ separately.

4. MINIMAL TIME–ENERGY UNCERTAINTY

PRELIMINARY REMARKS

As was mentioned already in the first section, the product of the spread in the time of presence at a definite plane and the spread in energy, $\tau\epsilon$, has to exceed $\hbar/2$. (The time of presence at a definite plane was, actually, generalized in the preceding section to the time of being in any definite quantum mechanical state.) The rest of this article will be concerned with the lower bound of $\tau\epsilon$, for a given ϵ, by characterizing the ψ which renders $\tau\epsilon$ to a minimum. This ψ, and the corresponding $\tau\epsilon$, will in this case depend on the energy E_0 around which the spread of the energy of ψ is defined as ϵ. The minimum of the quantity $\tau\epsilon$ will be independent of t_0 around which the spread of the time of presence, that is τ, will be calculated, ψ will depend on t_0 only in a trivial way. The last two statements follow, of course, from time displacement invariance. The minimum of $\tau\epsilon$ (and the corresponding ψ) will, on the other hand, depend on the choice of ϵ: clearly, if ϵ can be chosen to be very small as compared with the excess of E_0 over the lower bound of the energy, the existence of the lower bound will have very little significance. The lower bound $\hbar/2$ of $s_x s_p$ of the usual Heisenberg uncertainty relation was independent of s_p because the momentum p has no lower bound. In our case, the lower bound of $\tau\epsilon$ can be expected to increase with increasing ϵ (or, rather, with the increase of the ratio $\epsilon/(E_0 - E_b)$ where E_b is the lower bound of the energy).

Actually, what the subsequent calculations are aimed at is the determination of

$$\chi(t) = \langle u|\psi(t)\rangle \tag{6}$$

or of its Fourier transform

$$\eta(E) = \langle u|\phi(E)\rangle = (2\pi\hbar)^{-\frac{1}{2}} \int \chi(t) e^{iEt/\hbar} dt \tag{6a}$$

which render

$$\tau^2\epsilon^2 = \frac{\int|\chi(t)|^2 (t-t_0)^2\, dt \int|\eta(E)|^2 (E-E_0)^2\, dE}{(\int|\eta(E)|^2\, dE)^2} \tag{7}$$

s

to a minimum. In the calculation which follows $\eta(E)$ will be the dependent variable because the essential limitation of the problem, the existence of a lower bound for the energy, is most easily expressed in terms of $\eta(E)$. This must vanish for all E below the lower bound. It will be convenient to choose an energy scale in which the lower bound is 0; the transformation to any other lower bound is trivial. As a result, the integrations over E will extend from 0 to ∞ and will not be specified explicitly.

We wish to establish, next, that for $0 \leq E$ the function $\eta(E)$ can be chosen arbitrarily, subject only to the condition that it approach 0 fast enough as E increases toward infinity. This requires three assumptions. First, the Hamiltonian will be assumed to have a continuous spectrum, extending from its lower bound to infinity. This is surely true for any isolated system. Second, it will be assumed that, if $|u\rangle$ is expanded in terms of the characteristic functions $|E\rangle$ of the Hamiltonian

$$|u\rangle = \int b(E)|E\rangle dE \tag{8}$$

$b(E)$ does not vanish for any positive E. This condition is surely fulfilled for all states $|u\rangle$ which restrict the system to a spatial plane, i.e., for the original form (3) of the time–energy uncertainty relation. It is fulfilled also for the resonance states discussed later but is, of course, not true for all u. The second assumption, therefore, restricts our considerations to a certain degree. The third assumption is that the $b(E)$ of (8) does not go to 0 at very large E as fast as $\exp(-\alpha E^2)$ with any α. This is again fulfilled in the two aforementioned cases but restricts u somewhat further.

If the preceding conditions are satisfied, we expand $\psi(t)$ also in terms of the characteristic functions of the Hamiltonian

$$\psi(t) = \int a(E) e^{-iEt/\hbar} |E\rangle dE. \tag{9}$$

If the spectrum of the Hamiltonian is degenerate, the $|E\rangle$ in (9) shall be the same which appear in the expansion of $|u\rangle$ in (8). We then have, by (6)

$$\begin{aligned} \chi(t) &= \int\int b(E)^* \langle E|E'\rangle a(E') e^{-iEt/\hbar} dE dE' \\ &= \int b(E)^* a(E) e^{-iEt/\hbar} dE \end{aligned} \tag{10}$$

and hence

$$\eta(E) = (2\pi\hbar)^{\frac{1}{2}} b(E)^* a(E). \tag{10a}$$

Hence, any $\eta(E)$ can be obtained by a proper choice of $a(E)$ as long as $b(E)$ remains different from zero for all E. The square integrability of $a(E)$, does, though, impose a condition on the way η must go to 0 as

$E \to \infty$. It will turn out, however, that the η which will be obtained under the sole condition that it be square integrable goes to 0 as, $\exp(-\frac{1}{2}cE^2)$. Hence, if u satisfies the last condition imposed above, the $a(E)$ needed to furnish this η will be automatically square integrable. We can, therefore, proceed to the determination of the $\eta(E)$ which gives the smallest $\tau\epsilon$, permitting η to be, for positive E, an arbitrary function of E.

5. MINIMAL TIME–ENERGY UNCERTAINTY. EQUATION FOR η WHICH MINIMIZES UNCERTAINTY

If $\eta(E)$ were finite at the lower bound of the energy (at $E=0$), or if it had a discontinuity somewhere, its Fourier transform

$$\chi(t) = (2\pi\hbar)^{-\frac{1}{2}} \int \eta(E) e^{-iEt/\hbar} dE$$

would go to 0 at $t = \infty$ only as $1/t$. The integral in (5) then would become infinite. This surely would not give the minimum value of $\tau\epsilon$. It follows that $\eta(0)=0$. Hence

$$\tau^2 = \frac{\int_{-\infty}^{\infty} dt(t-t_0)^2 \int \eta(E) e^{-iEt/\hbar} dE \int \eta(E')^* e^{iE't/\hbar} dE'}{2\pi\hbar \int |\eta(E)|^2 dE}. \tag{11}$$

Introducing $\xi = (t-t_0)/\hbar$ as variable instead of t, and writing

$$\eta_0(E) = \eta(E) e^{-iEt_0/\hbar} \tag{12}$$

one obtains

$$\tau^2 = \frac{\hbar^2 \int_{-\infty}^{\infty} d\xi\, \xi^2 \int \eta_0(E) e^{-iE\xi} dE \int \eta_0(E')^* e^{iE'\xi} dE'}{2\pi \int |\eta_0(E)|^2 dE}$$

$$= \frac{\hbar^2}{2\pi} \frac{\int_{-\infty}^{\infty} d\xi \iint \eta_0(E)\, \eta_0(E')^* (\partial^2/\partial E \partial E') e^{i(E'-E)\xi} dE dE'}{\int |\eta_0(E)|^2 dE}. \tag{13}$$

Since $\eta_0(E)$ and $\eta_0(E')^*$ both vanish at both ends of the integration, at 0 and ∞, partial integration with respect to E and E' gives

$$\tau^2 = \frac{\hbar^2}{2\pi} \frac{\int_{-\infty}^{\infty} d\xi \iint (\partial\eta_0(E)/\partial E)(\partial\eta_0(E')^*/\partial E') e^{i(E'-E)\xi} dE dE'}{\int |\eta_0(E)|^2 dE}$$

$$= \hbar^2 \int |\partial\eta_0(E)/\partial E|^2 dE / \int |\eta_0(E)|^2 dE. \tag{14}$$

The last line follows from Fourier's theorem and is the expression which had to be expected. In fact, the calculation was carried out in such detail principally to show the necessity to assume $\eta(0)=0$ in order to obtain (14).

We now have

$$\tau^2=\hbar^2 \int|\partial\eta_0(E)/\partial E|^2 dE/ \int|\eta_0(E)|^2 dE \tag{15}$$

and

$$\epsilon^2= \int|\eta_0(E)|^2(E-E_0)^2 dE/ \int|\eta_0(E)|^2 dE. \tag{15a}$$

If one writes

$$\eta_0(E)=\alpha(E)e^{i\beta(E)} \tag{16}$$

with both α and β real, β drops out from the expression for ϵ^2 and the denominator of (15). The integral in the numerator becomes

$$\int|\partial\alpha/\partial E+i\alpha\partial\beta/\partial E|^2 dE= \int(\partial\alpha/\partial E)^2+\alpha^2(\partial\beta/\partial E)^2 dE \tag{17}$$

and this will be decreased if one sets $\partial\beta/\partial E=0$. It follows that the minimum of $\tau\epsilon$ will be assumed for a real η_0 and such an η_0 will be assumed henceforth. The real nature of η_0 could have been inferred also from time inversion invariance.

We are ready to obtain the minimum of $\tau^2\epsilon^2$ for given ϵ^2 (and E_0). To obtain it, we set the variation of

$$\tau^2\epsilon^2+\lambda'\epsilon^2=\hbar^2\epsilon^2\frac{\int(\partial\eta_0/\partial E)^2\, dE}{\int\eta_0^2\, dE}+\lambda'\frac{\int\eta_0^2(E-E_0)^2\, dE}{\int\eta_0^2\, dE} \tag{18}$$

equal to 0; the λ' is a Lagrange multiplier. This gives us the equation

$$\int\eta_0^2\, dE\left[-2\hbar^2\epsilon^2\frac{\partial^2\eta_0}{\partial E^2}+2\lambda'(E-E_0)^2\eta_0\right]$$
$$=[\hbar^2\epsilon^2 \int(\partial\eta_0/\partial E)^2\, dE+\lambda' \int\eta_0^2(E-E_0)^2\, dE]2\eta_0. \tag{19}$$

Elimination of the integrals by means of (15) and (15*a*) and division by $2\tau^2\epsilon^2$ gives then

$$-\frac{\hbar^2}{\tau^2}\frac{\partial^2\eta_0}{\partial E^2}+\frac{\lambda}{\epsilon^2}(E-E_0)^2\eta_0=(1+\lambda)\eta_0 \tag{20}$$

where

$$\lambda=\lambda'/\tau^2 \tag{20a}$$

has been introduced to make (20) somewhat more symmetric; λ must be so determined that (15*a*) become valid.

Since η_0 must vanish for both $E=0$ and $E=\infty$, (20) is essentially a characteristic value – characteristic function equation. It is well known, and can be easily verified, that its solution which approaches 0 for very large E approaches 0 as exp $(-\frac{1}{2}cE^2)$ with $c=\lambda^{\frac{1}{2}}\tau/\hbar\epsilon$. This verifies the statement made about η_0 at the end of the last section.

The solution of (20) is easily obtained for very large E_0/ϵ and for $E_0=0$. In the former case, one can set $\lambda=1$ and obtains $\tau=\hbar/2\epsilon$ (that is $\tau\epsilon=\hbar/2$ as expected) and

$$\eta_0=\exp(-(E-E_0)^2/4\epsilon^2). \tag{21}$$

This does not quite satisfy the boundary condition at $E=0$ but for large E_0 it satisfies it quite closely. Similarly, (15*a*) is satisfied closely enough. In the present case, actually, the minimum of $\tau\epsilon$ is independent of ϵ as long as this remains very much smaller than E_0.

The other case in which (20) can be easily solved is $E_0=0$. In this case the only solutions of (20) which satisfy the boundary conditions have the form

$$\eta_0=E\exp(-\tfrac{1}{2}cE^2). \tag{22}$$

One can determine c by (15*a*) to be

$$c=3/2\epsilon^2. \tag{22a}$$

This then solves the differential equation (20) if one sets again $\lambda=1$ and $\tau\epsilon$ becomes

$$\tau\epsilon=3\hbar/2. \tag{22b}$$

Because of the low value of E_0, the uncertainty is much larger than in Heisenberg's relation or for the large E_0 of (21). In this case again, the minimum of $\tau\epsilon$ is independent of ϵ – it is, as we shall see next, independent in no other case but the two just considered. Let us then proceed to the determination of λ and the discussion of η_0 in the general case.

6. MINIMAL TIME–ENERGY UNCERTAINTY. DISCUSSION

Let us observe, first, that the lower limit of $\tau\epsilon$ depends on ϵ and E_0 only in terms of E_0/ϵ. The two examples considered in the last section correspond to the values of ∞ and 0, respectively, of this quantity. It is for this reason that we found the minimum of $\tau\epsilon$ to be independent of ϵ and E_0 separately.

In order to see that the minimum of $\tau\epsilon$ depends only on E_0/ϵ, let us denote the solution of (20) for $\epsilon=1$ and the value e_0 of E_0 by $g(E, e_0)$. This solves (20) with $\epsilon=1$, $E_0=e_0$ and a $\lambda(e_0)$ such that the solution g satisfies the condition (15)

$$\int g(E,e_0)^2(E-e_0)^2dE=\int g(E,e_0)^2dE \tag{23}$$

and the differential equation

$$-\frac{\hbar^2}{\tau^2}g''(E, e_0)+\lambda(E-e_0)^2 g(E, e_0)=(1+\lambda)g(E, e_0), \tag{23a}$$

the primes denoting differentiation with respect to the first variable and τ the value for which $g(0, e_0)=0$, and g tending to 0 for large E.

We can then set

$$\eta_0(E)=g(E/\epsilon, E_0/\epsilon) \tag{24}$$

i.e., choose $e_0=E_0/\epsilon$ and expand the first variable of g by the factor ϵ. If one then substitutes (24) into (23a), one finds that η_0 satisfies (20). Similarly,

$$\int\eta_0(E)^2(E-E_0)^2dE= \int g(E/\epsilon, E_0/\epsilon)^2(E-E_0)^2dE$$
$$\epsilon^3 \int g(E', E_0/\epsilon)^2(E'-E_0/\epsilon)^2dE' = \epsilon^2 \int g(E/\epsilon, E_0/\epsilon)^2dE=\epsilon^2 \int\eta_0(E)^2dE.$$

The before-last member is a consequence of (23). This establishes the theorem which was plausible anyway. It follows that the wave functions η_0 of the minimal $\tau\epsilon$, that is time–energy uncertainty, can all be obtained by means of (24) by solving (23a) with the subsidiary condition (23).

In order to solve (23a) with this subsidiary condition, one will choose a $\lambda\tau^2$, multiply (23a) with τ^2, whereupon this will be a simple characteristic value equation for τ^2. One obtains its lowest characteristic value and the corresponding characteristic function g by Ritz' or some other method and compare then the two sides of (23). If the left side is larger than the right, one will choose a larger $\lambda\tau^2$; if the left side is smaller, one will try a smaller $\lambda\tau^2$. It should be possible to satisfy (23) after not too many trials.

It may be of some interest to deduce a few identities between λ, τ and the function η_0 which renders $\tau\epsilon$ to a minimum. For this purpose, one multiplies (20) with certain factors and integrates the resulting equation with respect to E. Multiplication with η_0 and partial integration of the first term gives

$$(\hbar^2/\tau^2) \int(\partial\eta_0/\partial E)^2dE+(\lambda/\epsilon^2) \int(E-E_0)^2\eta_0{}^2dE=(1+\lambda) \int\eta_0{}^2dE. \tag{25}$$

The integrated terms are 0 in this case because $\eta_0(0)=0$. Because of (15) and (15a), this is an identity, λ dropping out.

Multiplication of (20) with $\partial\eta_0/\partial E$ gives in a similar way

$$\frac{\hbar^2}{2\tau^2}\eta'_0(0)^2=\frac{\lambda}{\epsilon^2}\int\eta_0(E)^2(E-E_0)\, dE \tag{25a}$$

showing that the mean value of the energy is always larger than E_0.

Finally, multiplication of (20) with $E\partial\eta_0/\partial E$ gives with (15) and (15*a*)

$$\tfrac{1}{2} - \tfrac{3}{2}\lambda - \frac{\lambda E_0 \int \eta_0(E)^2 (E - E_0)\, dE}{\epsilon^2 \int \eta_0(E)^2\, dE} = -\tfrac{1}{2} - \tfrac{1}{2}\lambda. \tag{25b}$$

This last equation does not contain τ, the former one gives an expression for $\lambda\tau^2$ in terms of η_0. Naturally, (25*a*) and (25*b*) can be combined in various ways to eliminate λ or the integral in (25*a*). These equations play the role of the virial theorem which applies for the wave function giving minimal position-energy spread and Heisenberg's uncertainty relation. The last term on the left of (25*b*) vanishes in the simple cases considered in the last section: $E_0 = 0$ in the second case and the integral in the numerator vanishes for $E_0 = 0$, giving $\lambda = 1$ in both cases. In all other cases, $\lambda < 1$.

REFERENCES

1. P. A. M. Dirac, *Proc. Roy. Soc.* (*London*) **A114**, 243, 710 (1927). The calculation was carried out by V. Weisskopf and E. Wigner, *Z. Physik* **63,** 54 (1930). See also the article of the same authors, *ibid.* **65,** 18 (1930) and many subsequent discussions of the same subject and of resonance states decaying by the emission of particles rather than radiation.
2. W. Heisenberg, *Z. Physik* **43,** 172 (1927). For a rigorous derivation, see H. P. Robertson, *Phys. Rev.* **34,** 163 (1929). The derivation of section 5 is patterned on that of this article.
3. G. R. Allcock, *Ann. Phys.* (*N.Y.*) **53,** 253, 286, 311 (1969). Section II of the first of these articles gives a very illuminating discussion of the ideas which underlie also the present section. It also contains a review of the literature of the time–energy uncertainty principle, making it unnecessary to give such a review here. The review also gives a criticism of some of the unprecise interpretations of the time–energy uncertainty relation which are widely spread in the literature. The later parts of the aforementioned articles arrive at a pessimistic view on the possibility of incorporating into the present framework of quantum mechanics time measurements as described by Allcock in section II or in the present section. This pessimism, which is not shared by the present writer, is expressed, however, quite cautiously and is mitigated by the various assumptions on which it is based.
4. The notation of chapter XI of P. A. M. Dirac's *The Principles of Quantum Mechanics* (Oxford University Press, various editions) is used.

Time-energy uncertainty relation and Lorentz covariance

P. E. Hussar[a)] and Y. S. Kim
Center for Theoretical Physics, Department of Physics and Astronomy, University of Maryland, College Park, Maryland 20742

Marilyn E. Noz
Department of Radiology, New York University, New York, New York 10016

(Received 2 December 1983; accepted for publication 14 March 1984)

The uncertainty relations applicable to space and time separations between the quarks in a hadron are discussed. It is pointed out that the time-energy uncertainty relation between the time and energy separations is the same as the relationship between the widths and lifetimes of unstable states. It is then shown that this relation can be combined with Heisenberg's position-momentum uncertainty relation to give the uncertainty principle in a covariant form. It is pointed out that this effect manifests itself in Feynman's parton picture.

I. INTRODUCTION

The time-energy uncertainty relation in the form $(\Delta t)(\Delta E)\simeq\hbar$ was known to exist even before the present form of quantum mechanics was formulated.[1,2] However, the treatment of this subject in existing quantum-mechanics textbooks is not adequate. Students and teachers alike are frequently confused on the following three issues:

(a) Is the time-energy uncertainty relation a consequence of the time-dependent Schrödinger equation, or is this relation expected to hold even in systems which cannot be described by the Schrödinger equation?[3,4]

(b) While there exists the time-energy uncertainty relation in the real world, possibly with the form $[t, H] = i\hbar$,[5] this commutator is zero in the case of Schrödinger quantum mechanics. As was noted by Dirac in 1927,[2] the time variable is a c number. Then, is the c number time–energy uncertainty relation universal, or true only in nonrelativistic quantum mechanics?

(c) If the time variable is a c number and the position variables are q numbers, then the coordinate variables in a different Lorentz frame are mixtures of c and q numbers. This cannot be consistent with special relativity, as was also pointed out by Dirac.[2]

One of the reasons why we are not able to get satisfactory answers to these questions from the existing literature is that, while the uncertainty relation is to be formulated from experimental observations,[6] there are not many experimental phenomena which can be regarded as direct manifestations of the time-energy uncertainty relation.

The uncertainty relation between the lifetime and the energy width of unstable states was known to exist before 1927.[1,2] It is now believed that this phenomenon is independent of the Schrödinger equation.[1,7]

The Schrödinger equation leads to a form of the time-energy uncertainty relation when we calculate transition rates in the first-order time-dependent perturbation theory. However, the second-order time-dependent perturbation theory requires a separate time-energy uncertainty relation in the initial condition.

It is now widely believed that the time-energy uncertainty relation is responsible for off-mass-shell intermediate particles in quantum field theory. In this case, however, we are talking about particles which are not observable. Indeed, the role of the time-energy uncertainty relation in the present form of quantum field theory requires further investigation.[8]

The purpose of the present paper is to study the relativistic quark model as a physical example in which the time-energy uncertainty relation leads to a directly observable effect. For a hadron consisting of two quarks whose space-

Reprinted from *Am. J. Phys.* **53**, 142 (1985).

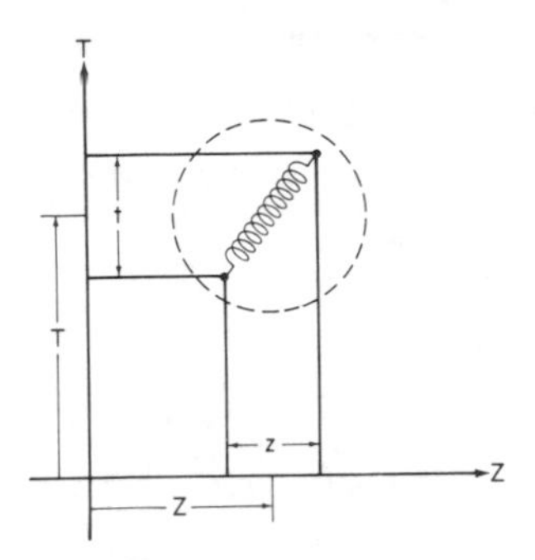

Fig. 1. Hadronic and internal coordinate systems. Each of the two quarks in the hadron has its own space-time coordinate. These two coordinates can be translated into the hadronic coordinate (Z,T) and the separation coordinate (z,t).

time coordinates are x_1 and x_2, we can define new variables:

$$X=(x_1+x_2)/2,$$
$$x=(x_1-x_2)/2\sqrt{2}. \tag{1}$$

Then X and x correspond, respectively, to the overall hadronic coordinate and space-time separation between the quarks. As is specified in Fig. 1, the coordinate variable X specifies the space-time position of the hadron, and the x coordinate is for the space-time separation between the quarks. The spatial component of the four-vector X specifies where the hadron is, and its time component tells how old the hadron and the quarks become. The spatial components of the four-vector x specify the relative spatial separation between the quarks. Its time component is the time interval or separation between the quarks.

Because the time-separation variable is not contained in nonrelativistic quantum mechanics,[9] the quark model provides an excellent testing ground to examine whether there exists a time-energy uncertainty relation which does not come from the Schrödinger equation. If there exists an uncertainty relation along the time separation coordinate, we should see whether this time-energy uncertainty relation is the same as the currently accepted form largely based on nonrelativistic quantum mechanics.

According to the currently accepted version,[10,11] the time variable is a c number or

$$[t,H]=0. \tag{2}$$

In quantum mechanics, the above commutator means that there is no Hilbert space in which t and $i\partial/\partial t$ act as operators.[12] This means that the Robertson procedure[13] applicable to Heisenberg's position-momentum uncertainty relation does not work here. Classically, this corresponds to the fact that t and H are not canonically conjugate variables. However, it is important to note that there still exists a "Fourier" relation between time and energy which limits the precision to[14]

$$(\Delta t)(\Delta E)\simeq 1. \tag{3}$$

We shall use for simplicity the convention $\hbar=c=1$.

In Sec. II, we discuss in detail the uncertainty relations applicable to the internal space-time separation variable in the quark model. In Sec. III, the covariant harmonic oscillator formalism is used for giving a wave-function interpretation to the commutator form of the uncertainty relations. In Sec. IV, Dirac's light-cone coordinate system is shown to play a decisive role in defining the uncertainty relations in a Lorentz-invariant manner. In Sec. V, we discuss briefly measurable consequences of the time-energy uncertainty relation combined covariantly with the position-momentum uncertainty relation.

II. TIME-ENERGY UNCERTAINTY RELATION APPLICABLE TO INTERNAL SPACE-TIME SEPARATION COORDINATES IN THE QUARK MODEL

In order to discuss the uncertainty relations, we need momentum-energy variables in addition to the space-time coordinates. Let us define the four-momenta:

$$P=p_1+p_2,$$
$$q=\sqrt{2}(p_1-p_2), \tag{4}$$

where p_1 and p_2 are the four-momenta of the first and second quarks, respectively. P is the sum of the momenta of the two quarks and is therefore the four-momentum of the hadron. q measures the difference between the quark four-momenta.

Without loss of generality, we assume that the hadron has a definite four-momentum and moves along the z or Z direction with velocity parameter β. Then it is possible to find the Lorentz frame in which the hadron is at rest. We shall use x^*, y^*, z^*, and t^* to denote the space-time separations in this Lorentz frame, and $q_x^*, q_y^*, q_z^*, q_0^*$ for momentum-energy separations.

In this Lorentz frame, the uncertainty principle applicable to the space-time separation of quarks is expected to be the same as the presently accepted form largely based on nonrelativistic quantum mechanics. The usual Heisenberg uncertainty relation holds for each of the three spatial coordinates:

$$[x^*,q_x^*]=i,$$
$$[y^*,q_y^*]=i, \tag{5}$$
$$[z^*,q_z^*]=i.$$

On the other hand, the time-separation variable is a c number and therefore does not cause quantum excitations. The commutator of Eq. (2) in this case takes the form

$$[t^*,q_0^*]=0, \tag{6}$$

with

$$(\Delta t^*)(\Delta q_0^*)\simeq 1. \tag{7}$$

This means that, according to the presently accepted form of the time-energy uncertainty relation, the c-number commutation relation of Eq. (6) should be accompanied by the "Fourier relation"[14] of Eq. (7). The crucial question is how these uncertainty relations appear to an observer in the laboratory frame with the space-time-separation variables x, y, z, and t.

Since the hadron moves along the z axis, the x and y coordinates are not affected by boosts, and the first two commutation relations of Eq. (5) remain invariant:

$$[x,q_x]=i,$$
$$[y,q_y]=i. \tag{8}$$

As for the third commutation relation of Eq. (5) for the longitudinal direction, we have to consider the Lorentz transformation of the coordinate system:

$$z = (z^* + \beta t^*)/(1-\beta^2)^{1/2},$$
$$t = (t^* + \beta z^*)/(1-\beta^2)^{1/2}. \tag{9}$$

Likewise, the transformation equations for the momentum-energy variables can be written as

$$q_z = (q_z^* + \beta q_0^*)/(1-\beta^2)^{1/2},$$
$$q_0 = (q_0^* + \beta q_z^*)/(1-\beta^2)^{1/2}. \tag{10}$$

In terms of the lab-frame variables, the uncertainty relation $[z^*, q_z^*] = i$ of Eq. (5) and the TE relation of Eq. (6) can be written as

$$[z, q_z] = i/(1-\beta^2),$$
$$[t, q_0] = i\beta^2/(1-\beta^2). \tag{11}$$

In addition, because the Lorentz transformation is not an orthogonal transformation, the commutation relations between z and q_0 and between t and q_z do not vanish:

$$[z, q_0] = i\beta/(1-\beta^2),$$
$$[t, q_z] = i\beta/(1-\beta^2). \tag{12}$$

The commutation relations of Eqs. (8), (11), and (12) can now be combined into a covariant form[14]:

$$[x_\mu, q_\nu] = -g_{\mu\nu} + P_\mu P_\nu/M^2, \tag{13}$$

with the covariant form of Eq. (7):

$$[\Delta(P\cdot x/M)][\Delta(P\cdot q/M)] \simeq 1, \tag{14}$$

where M is the hadronic mass. We use the convention $g_{00} = -g_{11} = -g_{22} = -g_{33} = 1$.

Although the uncertainty relations can be brought to the above convariant form, it is very difficult to give physical interpretations to them. First, in Eq. (11), the right-hand side is dependent on the velocity parameter. Does this mean that Planck's constant becomes dependent on the velocity? Second, the commutators of Eq. (12) do not vanish while there are no conjugate relations between the variables involved.[15] In order to resolve these puzzles, we have to resort to an interpretation based on wave functions.

III. USE OF THE HARMONIC OSCILLATOR FORMALISM

Traditonally, in nonrelativistic quantum mechanics, the harmonic oscillator has been very useful in giving interpretations to the uncertainty relations. It is therefore not unreasonable to expect that a relativistic harmonic oscillator model may prove useful in clarifying the questions raised at the end of Sec. II. Is there then a model which can be used for this purpose?

It has been shown that the covariant harmonic oscillator formalism introduced to this journal by the present authors in 1978 serves many useful purposes.[16] It can explain the basic hadronic features in the quark model, including the mass spectrum,[17] form factors,[18] parton picture,[19,20] and the jet phenomenon.[21] In addition, the relativistic bound-state wave functions in the oscillator formalism form a vector space for the representations of the Poincaré group diagonal in the Casimir operators corresponding to (mass)^2 and (intrinsic spin)^2 of the hadron.[22,23] As a consequence, the harmonic oscillator model constitutes a solution of Dirac's "Poisson bracket" equations for his "instant form" quantum mechanics.[24,25]

While the exact form for the hadronic wave function is somewhat complicated, the essential element of the wave function takes the form[16]

$$\psi_{nk}(x) = (1/\pi 2^{n+k} n! k!)^{1/2} H_n(z^*) H_k(t^*) \times \exp\{-[(z^*)^2 + (t^*)^2]/2\}. \tag{15}$$

For simplicity, we assume here that the harmonic oscillator has a unit strength. In the above expression, we have suppressed all the factors which are not affected by the Lorentz transformation along the z axis. This is possible because the oscillator wave functions are separable in both the Cartesian and spherical coordinate systems.

In terms of the standard step-up and step-down operators,

$$a_\mu = \frac{1}{\sqrt{2}}\left(x_\mu - \frac{\partial}{\partial x^\mu}\right),$$
$$a_\mu^\dagger = \frac{1}{\sqrt{2}}\left(x_\mu + \frac{\partial}{\partial x^\mu}\right), \tag{16}$$

and the oscillator wave function of Eq. (15) satisfies the differential equation

$$a_\mu^\dagger a^\mu \psi(x) = (\lambda + 1)\psi(x), \tag{17}$$

where the eigenvalue λ, together with transverse excitations, determines the (mass)^2 of the hadron.[23]

The operators given in Eq. (16) satisfy the algebraic relation

$$[a_\mu, a_\nu^\dagger] = -g_{\mu\nu}. \tag{18}$$

This commutation relation is Lorentz invariant. The timelike component of the above commutator is -1 in every Lorentz frame. This allows timelike excitations. Indeed, it is possible to construct a covariant Hilbert space of harmonic oscillator wave functions in which timelike excitations are allowed in all Lorentz frames.[26]

On the other hand, as we shall see in Sec. V, there is no evidence to indicate the existence of such timelike excitations in the real world.[17] This is perfectly consistent with the fact that the basic space-time symmetry of confined quarks is that of the $O(3)$-like little group of the Poincaré group.[22,27] We can suppress timelike excitations in the hadronic rest frame by imposing the subsidiary condition[16,23].

$$P^\mu a_\mu^\dagger \psi_{nk}(x) = 0. \tag{19}$$

Then only the solutions with $k = 0$ are allowed, and the commutator given in Eq. (18) is not consistent with the above subsidiary condition.

How can we then construct a covariant commutator consistent with the condition of Eq. (17)? In order to attack this problem, let us divide the four-dimensional Minkowskian space-time into the one-dimensional timelike space t^* parallel to the hadronic four-momentum and to the three-dimensional spacelike hyperplane perpendicular to the four-momentum spanned by x^*, y^*, and z^*.[15] This leads us to consider the operator

$$b_\mu = a_\mu - (P_\mu P^\nu/M^2)a_\nu. \tag{20}$$

Then b_μ is perpendicular to P^μ, and satisfies the constraint condition

$$P^\mu b_\mu = P^\mu b_\mu^\dagger = 0. \quad (21)$$

b_μ and $b_\mu^\dagger$ satisfy the covariant commutation relation[28]

$$[b_\mu, b_\nu^\dagger] = -g_{\mu\nu} + P_\mu P_\nu / M^2. \quad (22)$$

The right-hand side of the above expression is symmteric in μ and ν, and satisfies the relation

$$P^\mu(-g_{\mu\nu} + P_\mu P_\nu/M^2) = 0. \quad (23)$$

Therefore the covariant commutation relation given in Eq. (21) is consistent with the subsidiary condition of Eq. (19).

The covariant form given in Eq. (22) represents the usual Heisenberg uncertainty relations on the three-dimensional spacelike hypersurface perpendicular to the hadronic four-momentum. This form enables us to treat separately the uncertainty relation applicable to the timelike direction, without destroying covariance. The existence of the t^* distribution due to the ground-state wave function of Eq. (15) restricted to $k=0$ by Eq. (18) allows us to write the time-energy uncertainty relation in the form

$$(\Delta t^*)(\Delta q_0^*) \simeq 1, \quad (24)$$

without postulating the commutation relation.

Now let us go back to the commutators given in Sec. II. It is not difficult to see that Eq. (22) is the harmonic oscillator realization of the covariant commutator given in Eq. (13). Since we forbid timelike excitations in the hadronic rest frame by imposing the subsidiary condition of Eq. (19), the time variable is a c number. However, we still have the time-energy uncertainty relation due to the ground-state oscillator wave function. This is the harmonic oscillator realization of the relation given in Eq. (14). Indeed, the oscillator wave function of Eq. (15), together with the subsidiary condition of Eq. (19), constitutes a covariant realization of the Eqs. (13) and (14). We shall study in Sec. IV the covariance property of the oscillator wave function.

IV. LORENTZ-INVARIANT FORM OF THE UNCERTAINTY RELATIONS

In this section, we shall complete the study of the uncertainty relations by studying the localization properties of the wave functions and see whether there is a manner in which the uncertainty relations can be stated in a Lorentz-invariant manner.

One of the difficulties noted in Sec. II was that the orthogonality of the coordinate system is not preserved under Lorentz boosts. In order to rectify the situation, we can consider Dirac's light-cone coordinate system which preserves the orthogonality relations.[24,30] The coordinate variables in the light-cone coordinate system are

$$u = (t+z)/\sqrt{2}, \quad v = (t-z)/\sqrt{2}, \quad (25)$$

and the Lorentz transformation of Eq. (19) takes a simpler form:

$$u = [(1+\beta)/(1-\beta)]^{1/2} u^*, \quad v = [(1-\beta)/(1+\beta)]^{1/2} v^*. \quad (26)$$

Under the Lorentz transformation, one axis becomes elongated while the other goes through a contraction so that the product uv will stay constant:

$$uv = u^* v^* = (t^2 - z^2)/2 = [(t^*)^2 - (z^*)^2]/2. \quad (27)$$

This transformation property is explained in detail in Figs.

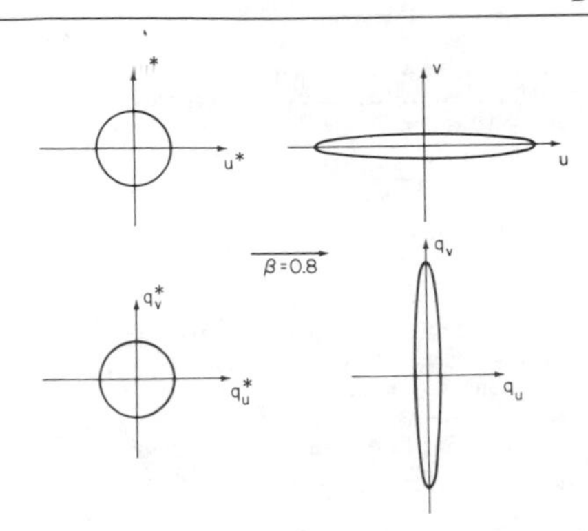

Fig. 2. The Lorentz deformation in the light-cone coordinate system. The major (minor) axis in the uv coordinate system is conjugate to the minor (major) axis in the $q_u q_v$ coordinate system. If we rotate these figures by 45°, they become Fig. 3 of Ref. 3 which explains the peculiarities observed in Feynman's parton picture.

1 and 2 of Ref. 30. If the stationary space-time region is a square in the uv or zt plane, then the moving region will appear like a rectangle with the same area.

The Lorentz deformation property in the $q_z q_0$ plane is the same as that for the zt plane, and it is possible to use the light-cone coordinate system for these variables:

$$q_+ = (q_z + q_0)/\sqrt{2}, \quad q_- = (q_z - q_0)/\sqrt{2}, \quad (28)$$

with the transformation property

$$q_+ = [1+\beta)/(1-\beta)]^{1/2} q_+^*, \quad q = [(1-\beta)/(1+\beta)]^{1/2} q_-^*. \quad (29)$$

The basic advantage of using the light-cone variables is that the coordinate system remains orthogonal. It is not difficult to visualize the deformation of the regions given in Fig. 2 due to the boost. The circular region in the hadronic rest frame will appear as an ellipse to the lab-frame observer.

With this understanding, let us use the Gaussian wave function which has the circular uncertainty distribution in the hadronic rest frame:

$$\psi(z,t) = (1/\sqrt{\pi}) \exp\{-(1/2)[(z^*)^2 + (t^*)^2]\}. \quad (30)$$

This is the ground-state space-time wave function in the covariant oscillator formalism, which has been discussed extensively in Sec. III. Since

$$(z^*)^2 + (t^*)^2 = (u^*)^2 + (v^*)^2, \quad (31)$$

the wave function of Eq. (30) can be written as

$$\psi(z,t) = (\Omega/\pi)^{1/2} \exp\left[-\frac{1}{4}\left(\frac{1-\beta}{1+\beta}(z+t)^2 + \frac{1+\beta}{1-\beta}(z-t)^2\right)\right]. \quad (32)$$

The momentum-energy wave function becomes

$$\phi(q_z, q_0) = \left(\frac{1}{2\pi}\right) \int \exp[i(q_z z - q_0 t)] \psi(z,t)\, dt\, dz = \left(\frac{1}{\sqrt{\pi}}\right) \exp\{-[(q_z^*)^2 + (q_0^*)^2]/2\}, \quad (33)$$

which can also be written in a form similar to Eq. (32).

The Fourier relations[14] between the space-time and momentum-energy coordinates are

$$q_u = q_- = -i\left(\frac{\partial}{\partial u}\right),$$
$$q_v = q_+ = -i\left(\frac{\partial}{\partial v}\right). \tag{34}$$

This means that the major and minor axes of the momentum-energy coordinates are the "Fourier conjugates" of the minor and major axes of the space-time coordinates, respectively. This aspect is illustrated in Fig. 2. Thus we have the following Lorentz-invariant relations.[31]

$$(\Delta u)(\Delta q_-) = (\Delta u^*)(\Delta q^*_-) \simeq 1,$$
$$(\Delta v)(\Delta q_+) = (\Delta v^*)(\Delta q^*_+) \simeq 1. \tag{35}$$

This is indeed a Lorentz-invariant statement of the c number time-energy uncertainty relation combined with Heisenberg's position-momentum uncertainty relations.

V. OBSERVABLE CONSEQUENCES

In order that the time-separation variable be a c number, it is essential that there be no timelike oscillations. We implement this concept by imposing the subsidiary condition of Eq. (19), which restricts the wave functions of Eq. (15) to those with $k = 0$.

On the other hand, if we allow timelike oscillations, the eigenvalue of the oscillator wave function which corresponds to the (mass)^2 spectrum will be[23]

$$\lambda = (n - k). \tag{36}$$

For a given value n of the longitudinal excitation, k can take an arbitrarily large number. Thus (mass)^2 can take negative numbers with no lower bound. This does not happen in nature. Furthermore, for a given value of λ, there are infinite possible combinations of n and k, resulting in infinite degeneracy. There is no evidence to indicate the existence of such a degeneracy in hadronic mass spectra.[17]

We noted in Sec. IV that the Lorentz deformation property of the harmonic oscillator wave function enables us to define the uncertainty products in a Lorentz-invariant manner. Then the question is whether the deformation property described in Fig. 2 manifests itself in the real world. This question has been addressed in Refs. 19 and 20. The point is that Lorentz deformation given in Fig. 2 can be described in the zt and $q_z q_0$ planes, which are simply 45° rotations of the figures in Fig. 2. Figure 2 of the present paper and Fig. 3 of Ref. 20 are only two different forms of the same figure. Figure 2 is designed to explain the Lorentz invariance of the uncertainty relation, while Fig. 3 of Ref. 20 is designed to explain Feynman's parton picture.

It is by now a widely accepted view that hadrons such as nucleons and mesons are bound states of quarks, if they do not move rapidly. However, Feynman observed in 1969 that, if a hadron moves with a velocity close to that of light, it appears as a collection of partons,[32] as is illustrated in Fig. 3. It is also believed that partons are quarks. The parton picture, which has been a primary vehicle toward our present understanding of high-energy hadronic interactions, has the following peculiarities.

(a) The picture is valid only for hadrons moving with velocity close to that of light.

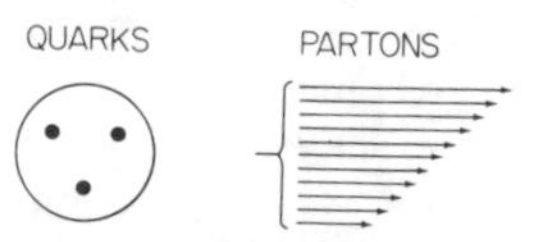

Fig. 3. Pictures of the proton in the quark and parton models. Suppose that a proton is sitting quietly on the desk. According to the quark model, it appears like a bound state of three quarks to an observer who is sitting on the chair. However, to an observer who is on a jet plane with its speed close to that of light, the proton would look like a collection of free particles with a wide momentum distribution. This is called Feynman's parton picture.

(b) The interaction time between the quarks becomes dilated, and the partons become free to allow an incoherent sum of cross sections of the constituent particles.

(c) The momentum distribution of partons becomes widespread as the hadron moves very fast.

(d) The number of partons appears to be much larger than that of quarks.

Because the hadron is believed to be a bound state of two or three quarks, each of the above phenomena appears as a paradox, particularly (b) and (c) together. How can bound-state quarks become free while the momentum distribution becomes widespread? This question has been discussed in Ref. 20. Peculiarities (a) and (b) have been addressed in Ref. 19. According to Hussar's calculation,[33] the ground-state harmonic oscillator wave function leads to a parton distribution in good agreement with the experimental data. The parton phenomenon is therefore a direct manifestation of the time-energy uncertainty relation combined covariantly with Heisenberg's position-momentum uncertainty relation.

[a] Present address: Illinois Institute of Technology Research Institute, Annapolis, Maryland 21401.

[1] E. P. Wigner, in *Aspects of Quantum Theory, in Honour of P. A. M. Dirac's 70th Birthday*, edited by A. Salam and E. P. Wigner (Cambridge University, London, 1972).
[2] P. A. M. Dirac, Proc. R. Soc. London Ser. A **114**, 234, 710 (1927).
[3] E. P. Wigner and V. Weisskopf, Z. Phys. **63**, 54 (1930); **65**, 18 (1930).
[4] E. P. Wigner, Phys. Rev. **70**, 15, 609 (1946); E. P. Wigner and L. Eisenbud, *ibid.* **70**, 29 (1947); M. Moshinsky, Rev. Mex. Fis. **1**, 28 (1952); Phys. Rev. **81**, 347 (1951); **84**, 525, 533 (1951); **88**, 625 (1952). P. T. Mathews and A. Salam, *ibid.* **115**, 1979 (1959); F. T. Smith, *ibid.* **118**, 349 (1960); Y. Aharonov and D. Bohm, *ibid.* **122**, 1649 (1961); V. A. Fock, J. Exp. Theor. Phys. (U.S.S.R.) **42**, 1135 (1962); Sov. Phys. JETP **15**, 784 (1962); Y. Aharonov and D. Bohm, Phys. Rev. **134**, 1417 (1964); B. A. Lippmann, *ibid.* **151**, 1023 (1966); G. R. Allcock, Ann. Phys. **53**, 253 (1969); **53**, 286 (1969); **53**, 311 (1969); J. H. Eberly and L. P. S. Singh, Phys. Rev. D7, 359 (1973). See also articles by J. Rayski and J. M. Rayski, Jr.; by E. Recami; and by E. W. R. Papp, in *The Uncertainty Principles and Quantum Mechanics*, edited by W. C. Price and S. S. Chissick (Wiley, New York, 1977); M. Bauer and P. A. Mello, Ann. Phys. (NY) **111**, 38 (1978); M. Bauer, *ibid.* **150**, 1 (1983).
[5] W. Heisenberg, Z. Phys. **43**, 172 (1927); **45**, 172 (1927).
[6] W. Heisenberg, Am. J. Phys. **43**, 389 (1975).
[7] The relation between the size of wave train and the linewidth was known in classical optics long before quantum mechanics was formulated. However, this is only the relation between the lifetime and linewidth of an unstable system mentioned in Ref. 1. See W. Heitler, *The Quantum Theory of Radiation* (Oxford University, London, 1954), 3rd ed.
[8] D. Han, Y. S. Kim, and M. E. Noz, Found. Phys. **11**, 895 (1981).

[9]For papers dealing with the time-separation variable in bound systems, see H. Yukawa, Phys. Rev. **79**, 416 (1953); G. C. Wick, *ibid*. **96**, 1124 (1954); M. Markov, Suppl. Nuovo Cimento **3**, 760 (1956); T. Takabayasi, Nuovo Cimento **33**, 668 (1964); S. Ishida, Prog. Theor. Phys. **46**, 1570, 1905 (1971); R. P. Feynman, M. Kislinger, and F. Ravndal, Phys. Rev. D **3**, 2706 (1972); G. Preparata and N. S. Craigie, Nucl. Phys. **B102**, 478 (1976); Y. S. Kim, Phys. Rev. D **14**, 273 (1976); J. Lukierski and M. Oziewics, Phys. Lett. **69B**, 339 (1977); D. Dominici and G. Longhi, Nuovo Cimento A **42**, 235 (1977); T. Goto, Prog. Theor. Phys. **58**, 1635 (1977); H. Leutwyler and J. Stern, Phys. Lett. **73B**, 75 (1978); and Nucl. Phys. **B157**, 327 (1979); I. Fujiwara, K. Wakita, and H. Yoro, Prog. Theor. Phys. **64**, 363 (1980); J. Jersak and D. Rein, Z. Phys. C **3**, 339 (1980); I. Sogami and H. Yabuki, Phys. Lett. **94B**, 157 (1980); M. Pauri, in *Group Theoretical Methods in Physics*, Proceedings of the 9th International Colloquium, Cocoyoc, Mexico, edited by K. B. Wolf (Springer-Verlag, Berlin, 1980); G. Marchesini and E. Onofri, Nuovo Cimento **A65**, 298 (1981); E. C. G. Sudarshan, N. Mukunda, and C. C. Chiang, Phys. Rev. D **25**, 3237 (1982).

[10]For a recent pedagogical paper on this problem, see C. H. Blanchard, Am. J. Phys. **50**, 642 (1982).

[11]For a possible departure from the accepted view, see E. Prugovecki, Found. Phys. **12**, 555 (1982); Phys. Rev. Lett. **49**, 1065 (1982).

[12]E. Prugovecki, *Quantum Mechanics in Hilbert Space* (Academic, New York, 1981), 2nd ed.

[13]H. P. Robertson, Phys. Rev. **34**, 163 (1929).

[14]The word "Fourier relation" was used earlier by Blanchard in Ref. 10. This word is necessary because the energy is not a variable canonically conjugate to the time variable.

[15]G. N. Fleming, Phys. Rev. **137**, B188 (1965); G. N. Fleming, J. Math. Phys. **11**, 1959 (1966).

[16]Y. S. Kim and M. E. Noz, Am. J. Phys. **46**, 484 (1978).

[17]For some of the latest papers on hadronic mass spectra, see N. Isgur and G. Karl, Phys. Rev. D **19**, 2653 (1978); D. P. Stanley and D. Robson, Phys. Rev. Lett. **45**, 235 (1980). For review articles written for teaching purposes, see P. E. Hussar, Y. S. Kim, and M. E. Noz, Am. J. Phys. **48**, 1038 (1980); **48**, 1043 (1980).

[18]For papers dealing with form factor behavior, see K. Fujimura, T. Kobayashi, and M. Namiki, Prog. Theor. Phys. **43**, 73 (1970); R. G. Lipes, Phys. Rev. D **5**, 2849 (1972); Y. S. Kim and M. E. Noz, *ibid*. **8**, 3521 (1973).

[19]Y. S. Kim and M. E. Noz, Phys. Rev. D **15**, 335 (1977).

[20]Y. S. Kim and M. E. Noz, Am. J. Phys. **51**, 368 (1983).

[21]For papers dealing with the jet phenomenon, see T. Kitazoe and S. Hama, Phys. Rev. D **19**, 2006 (1979); Y. S. Kim, M. E. Noz, and S. H. Oh, Found. Phys. **9**, 947 (1979); T. Kitazoe and T. Morii, Phys. Rev. D **21**, 685 (1980); Nucl. Phys. **B164**, 76 (1980).

[22]E. P. Wigner, Ann. Math. **40**, 149 (1939).

[23]Y. S. Kim, M. E. Noz, and S. H. Oh, J. Math. Phys. **10**, 1341 (1979); Am. J. Phys. **47**, 892 (1979); J. Math Phys. **21**, 1224 (1980).

[24]P. A. M. Dirac, Rev. Mod. Phys. **21**, 392 (1949).

[25]D. Han and Y. S. Kim, Am. J. Phys. **49**, 1157 (1981).

[26]F. C. Rotbart, Phys. Rev. D **23**, 3078 (1981). For a physical basis for Rotbart's calculation, see L. P. Horwitz and C. Piron, Helv. Phys. Acta **46**, 316 (1973). See also Ref. 16.

[27]D. Han, M. E. Noz, and Y. S. Kim, Phys. Rev. D **25**, 1740 (1982).

[28]D. Han, M. E. Noz, Y. S. Kim, and D. Son, Phys. Rev. D **27**, 3032 (1983).

[29]This commutator can be translated into wave-function formalism. For a wave-function description of this commutation relation, see M. J. Ruiz, Phys. Rev. D **10**, 4306 (1974).

[30]For a pedagogical treatment of the light-cone coordinate system, see Y. S. Kim and M. E. Noz, Am. J. Phys. **50**, 721 (1982).

[31]Y. S. Kim and M. E. Noz, Found Phys. **9**, 375 (1979); J. Math Phys. **22**, 2289 (1981).

[32]R. P. Feynman, in *High Energy Collisions*, Proceedings of the Third International Conference, Stony Brook, NY, edited by C. N. Yang *et al.* (Gordon and Breach, New York, 1969); *Photon Hadron Interactions* (Benjamin, New York, 1972). See also J. D. Bjorken and E. A. Paschos, Phys. Rev. **185**, 1975 (1969).

[33]P. E. Hussar, Phys. Rev. D **23**, 2781 (1981).

Chapter IV

Covariant Picture of Quantum Bound States

The success of quantum field theory in the 1950's led a large number of physicists to believe that field theory will solve all dynamical problems. Naturally, they attempted to solve the hydrogen atom problem within the framework of field theory. The Bethe-Salpeter equation is the most suitable field theoretic equation for quantum bound-state problems. Although this equation offers us mathematical challenge and generates some useful solutions, it is plagued with fundamental difficulties, as Wick pointed out clearly in 1954. During the 1970's, field theoretic bound state models have been proven to be ineffective in dealing with hadrons in the quark model. We are thus fully justified to construct a relativistic model of bound states consistent with special relativity and quantum mechanics, but not necessarily within the framework of the mathematical framework of field theory, as is indicated in Figure 2.

The question is then whether we can construct a relativistic bound-state model without the fundamental difficulties contained in the Bether-Salpeter wave function. In 1973, Kim and Noz investigated this possibility, and showed that the harmonic oscillator model of Yukawa (1953) can satisfy this condition. Yukawa's 1953 papers are based on his earlier effort made in 1950 to formulate a field theory of particles with space-time extension. Yukawa's aim was not different from that of the present day string models.

The covariant harmonic oscillator model has been studied extensively by the present authors and their associates. The orthogonality relation and Lorentz transformation properties have also been studied by Ruiz (1974) and Rotbart (1981). In 1981, Han and Kim showed that the covariant oscillator formalism can serve as a solution of the Poisson-bracket equations for relativistic bound state formulated by Dirac in 1949.

The paper of Kim, Noz, and Oh in Chapter II shows that the same oscillator formalism constitutes a representation of the Poincaré group for relativistic extended hadrons.

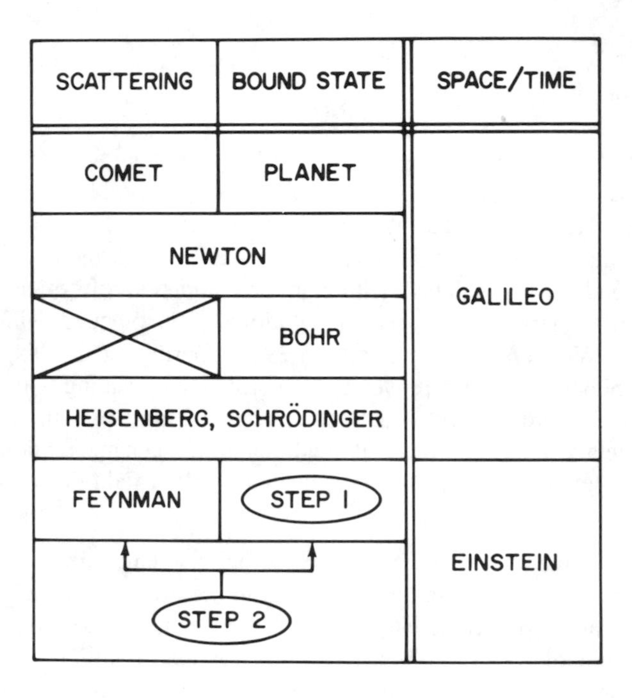

FIG. 2. History of dynamical and kinematical developments. Mankind's unified understanding of scattering and bound states has been very brief. It is therefore not unusual to expect that separate theoretical models be developed for scattering and bound states, before a completely satisfactory theory can be developed. The successes and limitations of the Feynman diagram approach are well known. There is an urgent need for a relativistic bound state model. This figure is from D. Han, Y. S. Kim, and M. E. Noz, Found. of Physics, 11, 895 (1981).

REVIEWS OF MODERN PHYSICS VOLUME 21, NUMBER 3 JULY, 1949

Forms of Relativistic Dynamics

P. A. M. DIRAC
St. John's College, Cambridge, England

For the purposes of atomic theory it is necessary to combine the restricted principle of relativity with the Hamiltonian formulation of dynamics. This combination leads to the appearance of ten fundamental quantities for each dynamical system, namely the total energy, the total momentum and the 6-vector which has three components equal to the total angular momentum. The usual form of dynamics expresses everything in terms of dynamical variables at one instant of time, which results in specially simple expressions for six of these ten, namely the components of momentum and of angular momentum. There are other forms for relativistic dynamics in which others of the ten are specially simple, corresponding to various sub-groups of the inhomogeneous Lorentz group. These forms are investigated and applied to a system of particles in interaction and to the electromagnetic field.

1. INTRODUCTION

EINSTEIN'S great achievement, the principle of relativity, imposes conditions which all physical laws have to satisfy. It profoundly influences the whole of physical science, from cosmology, which deals with the very large, to the study of the atom, which deals with the very small. General relativity requires that physical laws, expressed in terms of a system of curvilinear coordinates in space-time, shall be invariant under all transformations of the coordinates. It brings gravitational fields automatically into physical theory and describes correctly the influence of these fields on physical phenomena.

Gravitational fields are specially important when one is dealing with large-scale phenomena, as in cosmology, but are quite negligible at the other extreme, the study

Reprinted from *Rev. Mod. Phys.* **21**, 392 (1949).

of the atom. In the atomic world the departure of space-time from flatness is so excessively small that there would be no point in taking it into account at the present time, when many large effects are still unexplained. Thus one naturally works with the simplest kind of coordinate system, for which the tensor $g^{\mu\nu}$ that defines the metric has the components

$$g^{00}=-g^{11}=-g^{22}=-g^{33}=1$$
$$g^{\mu\nu}=0 \quad \text{for} \quad \mu\neq\nu.$$

Einstein's restricted principle of relativity is now of paramount importance, requiring that physical laws shall be invariant under transformations from one such coordinate system to another. A transformation of this kind is called an inhomogeneous Lorentz transformation. The coordinates u_μ transform linearly according to the equations

$$\left.\begin{aligned} &u_\mu{}^*=\alpha_\mu+\beta_\mu{}^\nu u_\nu \\ \text{with}\quad &\beta_\mu{}^\nu\beta^{\mu\rho}=g^{\nu\rho}, \end{aligned}\right\}\tag{1}$$

the α's and β's being constants.

A transformation of the type (1) may involve a reflection of the coordinate system in the three spacial dimensions and it may involve a time reflection, the direction du_0 in space-time changing from the future to the past. I do not believe there is any need for physical laws to be invariant under these reflections, although all the exact laws of nature so far known do have this invariance. The restricted principle of relativity arose from the requirement that the laws of nature should be independent of the position and velocity of the observer, and any change the observer may make in his position and velocity, taking his coordinate system with him, will lead to a transformation (1) of a kind that can be built up from infinitesimal transformations and cannot involve a reflection. Thus it appears that restricted relativity will be satisfied by the requirement that physical laws shall be invariant under infinitesimal transformations of the coordinate system of the type (1). Such an infinitesimal transformation is given by

$$\left.\begin{aligned} &u_\mu{}^*=u_\mu+a_\mu+b_\mu{}^\nu u_\nu \\ \text{with}\quad &b_{\mu\nu}=-b_{\nu\mu}, \end{aligned}\right\}\tag{2}$$

the a's and b's being infinitesimal constants.

A second general requirement for dynamical theory has been brought to light through the discovery of quantum mechanics by Heisenberg and Schrödinger, namely the requirement that the equations of motion shall be expressible in the Hamiltonian form. This is necessary for a transition to the quantum theory to be possible. In atomic theory one thus has two over-riding requirements. The problem of fitting them together forms the subject of the present paper.

The existing theories of the interaction of elementary particles and fields are all unsatisfactory in one way or another. The imperfections may well arise from the use of wrong dynamical systems to represent atomic phenomena, i.e., wrong Hamiltonians and wrong interaction energies. *It thus becomes a matter of great importance to set up new dynamical systems and see if they will better describe the atomic world.* In setting up such a new dynamical system one is faced at the outset by the two requirements of special relativity and of Hamiltonian equations of motion. The present paper is intended to make a beginning on this work by providing the simplest methods for satisfying the two requirements simultaneously.

2. THE TEN FUNDAMENTAL QUANTITIES

The theory of a dynamical system is built up in terms of a number of algebraic quantities, called dynamical variables, each of which is defined with respect to a system of coordinates in space-time. The usual dynamical variables are the coordinates and momenta of particles at particular times and field quantities at particular points in space-time, but other kinds of quantities are permissible, as will appear later.

In order that the dynamical theory may be expressible in the Hamiltonian form, it is necessary that any two dynamical variables, ξ and η, shall have a P.b. (Poisson bracket) $[\xi, \eta]$, subject to the following laws,

$$\left.\begin{aligned} &[\xi,\eta]=-[\eta,\xi] \\ &[\xi,\eta+\zeta]=[\xi,\eta]+[\xi,\zeta] \\ &[\xi,\eta\zeta]=[\xi,\eta]\zeta+\eta[\xi,\zeta] \\ &[[\xi,\eta],\zeta]+[[\eta,\zeta],\xi]+[[\zeta,\xi],\eta]=0. \end{aligned}\right\}\tag{3}$$

A number or physical constant may be counted as a special case of a dynamical variable, and has the property that its P.b. with anything vanishes.

Dynamical variables change when the system of coordinates with respect to which they are defined changes, and must do so in such a way that P.b. relations between them remain invariant. This requires that with an infinitesimal change in the coordinate system (2) each dynamical variable ξ shall change according to the law

$$\xi^*=\xi+[\xi, F],\tag{4}$$

where F is some infinitesimal dynamical variable independent of ξ, depending only on the dynamical system involved and the change in the coordinate system. We are thus led to associate one F with each infinitesimal transformation of coordinates.

Let us apply two infinitesimal transformations of coordinates in succession. Suppose the first one changes the dynamical variable ξ to ξ^* according to

$$\xi^*=\xi+[\xi, F_1],$$

and the second one changes ξ^* to $\xi^\dagger$ according to

$$\xi^\dagger=\xi^*+[\xi^*, F_2{}^*]=\xi^*+[\xi, F_2]^*.$$

The two transformations together change ξ to $\xi^\dagger$

according to

$$\xi^\dagger = \xi + [\xi, F_1] + [\xi, F_2] + [[\xi, F_2], F_1],$$

to the accuracy of the order F_1F_2 (with neglect of terms of order F_1^2 or F_2^2). If these two transformations are applied in the reverse order, they change ξ to $\xi^{\dagger\dagger}$ according to

$$\xi^{\dagger\dagger} = \xi + [\xi, F_2] + [\xi, F_1] + [[\xi, F_1], F_2].$$

Thus

$$\xi^{\dagger\dagger} = \xi^\dagger + [[\xi, F_1], F_2] - [[\xi, F_2], F_1] = \xi^\dagger + [\xi, [F_1, F_2]],$$

with the help of the first and last of Eqs. (3). This gives the change in a dynamical variable associated with that change of the coordinate system which is the commutator of the two previous changes. It is of the standard form

$$\xi^{\dagger\dagger} = \xi^\dagger + [\xi^\dagger, F],$$

with an F that is the P. b. of the F's associated with the two previous changes of coordinates. Thus the commutation relations between the various infinitesimal changes of coordinates correspond to the P.b. relations between the associated F's.

The F associated with the transformation (2) must depend linearly on the infinitesimal numbers a_μ, $b_{\mu\nu}$ that fix this transformation. Thus we can put

$$\left.\begin{aligned} F &= -P^\mu a_\mu + \tfrac{1}{2}M^{\mu\nu}b_{\mu\nu} \\ M^{\mu\nu} &= -M^{\nu\mu}, \end{aligned}\right\} \quad (5)$$

where P^μ, $M^{\mu\nu}$ are finite dynamical variables, independent of the transformation of coordinates.

The ten quantities P_μ, $M_{\mu\nu}$ are characteristic for the dynamical system. They will be called the ten *fundamental quantities*. They determine how all dynamical variables are affected by a change in the coordinate system of the kind that occurs in special relativity. Each of them is associated with a type of infinitesimal transformation of the inhomogeneous Lorentz group. Seven of them have simple physical interpretations, namely, P_0 is the total energy of the system, P_r ($r=1$, 2, 3) is the total momentum, and M_{rs} is the total angular momentum about the origin. The remaining three M_{r0} do not correspond to any such well-known physical quantities, but are equally important in the general dynamical scheme.

From the commutation relations between particular infinitesimal transformations of the coordinate system we get at once the P.b. relations between the ten fundamental quantities,

$$\left.\begin{aligned} &[P_\mu, P_\nu] = 0 \\ &[M_{\mu\nu}, P_\rho] = -g_{\mu\rho}P_\nu + g_{\nu\rho}P_\mu \\ &[M_{\mu\nu}, M_{\rho\sigma}] = -g_{\mu\rho}M_{\nu\sigma} + g_{\nu\rho}M_{\mu\sigma} - g_{\mu\sigma}M_{\rho\nu} + g_{\nu\sigma}M_{\rho\mu}. \end{aligned}\right\} \quad (6)$$

To construct a theory of a dynamical system one must obtain expressions for the ten fundamental quantities that satisfy these P.b. relations. *The problem of finding a new dynamical system reduces to the problem of finding a new solution of these equations.*

An elementary solution is provided by the following scheme. Take the four coordinates q_μ of a point in space-time as dynamical coordinates and let their conjugate momenta be p_μ, so that

$$[q_\mu, q_\nu] = 0, \quad [p_\mu, p_\nu] = 0$$
$$[p_\mu, q_\nu] = g_{\mu\nu}.$$

The q's will transform under an infinitesimal transformation of the coordinate system in the same way as the u's in (2). This leads to

$$P_\mu = p_\mu, \quad M_{\mu\nu} = q_\mu p_\nu - q_\nu p_\mu, \quad (7)$$

and provides a solution of the P.b. relations (6). The solution (7) does not seem to be of any practical importance, but it may be used as a basis for obtaining other solutions that are of practical importance, as the next three sections will show.

The foregoing discussion of the requirements for a relativistic dynamical theory may be generalized somewhat. We may work with dynamical variables that are connected by one or more relations for all states of motion that occur physically. Such relations are called *subsidiary equations*. They will be written

$$A \approx 0 \quad (8)$$

to distinguish them from dynamical equations. They are less strong than dynamical equations, because with a dynamical equation one can take the P.b. of both sides with any dynamical variable and get another equation, while with a subsidiary equation one cannot do this in general. The lesser assumption is made, however, that from two subsidiary equations $A \approx 0$, $B \approx 0$ one can infer a third

$$[A, B] \approx 0. \quad (9)$$

A subsidiary equation must remain a subsidiary equation under any change of coordinate system. This enables one to infer from (8)

$$[P_\mu, A] \approx 0, \quad [M_{\mu\nu}, A] \approx 0. \quad (10)$$

A dynamical variable is of physical importance only if its P.b. with any subsidiary equation gives another subsidiary equation, i.e., its P.b. with A in (8) must vanish in the subsidiary sense. Such a dynamical variable will be called a *physical variable*. The P.b. of two physical variables is a physical variable. Equations (10) show that the ten fundamental quantities are physical variables.

The physical variables are the only ones that are really important. One could eliminate the non-physical variables from the theory altogether and one could then make the subsidiary equations into dynamical equations. However, the elimination may be awkward and may spoil some symmetry feature in the scheme of equations, so it is desirable to retain the possibility of subsidiary equations in the general theory.

3. THE INSTANT FORM

The ten fundamental quantities for dynamical systems that occur in practice are usually such that some of them are specially simple and the others are complicated. The complicated ones will be called the Hamiltonians. They play jointly the rôle of the single Hamiltonian in non-relativistic dynamics. Since the P.b. of two simple quantities is a simple quantity, the simple ones of the ten fundamental quantities must be those associated with some sub-group of the inhomogeneous Lorentz group.

In the usual form of dynamics one works with dynamical variables referring to physical conditions at some instant of time, e.g., the coordinates and momenta of particles at that instant. An instant in the four-dimensional relativistic picture is a flat three-dimensional surface containing only directions which lie outside the light-cone. The simplest instant referred to the u coordinate system is given by the equation

$$u_0=0. \tag{11}$$

The effect of working with dynamical variables referring to physical conditions at this instant will be to make specially simple those of the fundamental quantities associated with transformations of coordinates that leave the instant invariant, namely P_1, P_2, P_3, M_{23}, M_{31}, M_{12}. The remaining ones, P_0, M_{10}, M_{20}, M_{30}, will be complicated in general and will be the Hamiltonians. We get in this way a form of dynamics which is associated with the sub-group of the inhomogeneous Lorentz group that leaves the instant invariant, and which may appropriately be called the *instant form*.

Let us take as an example a single particle by itself. The ten fundamental quantities in this case are well known, but they will be worked out again here to illustrate a method that can be used also with the other forms of dynamics.

We take as dynamical coordinates the three coordinates of the particle at the instant (11). Calling these coordinates q_r, we can base our work on the scheme (7), with the additional equation

$$q_0=0. \tag{12}$$

With this equation p_0 no longer has a meaning. We must therefore modify the expressions for the ten fundamental quantities given by (7) so as to eliminate p_0 from them, without invalidating the P.b. relations (6).

Let us change the expressions for the ten fundamental quantities by multiples of $p^\sigma p_\sigma - m^2$, where m is a constant, i.e., let us put

$$\left.\begin{aligned} P_\mu &= p_\mu + \lambda_\mu(p^\sigma p_\sigma - m^2) \\ M_{\mu\nu} &= q_\mu p_\nu - q_\nu p_\mu + \lambda_{\mu\nu}(p^\sigma p_\sigma - m^2), \end{aligned}\right\} \tag{13}$$

where

$$\lambda_{\mu\nu} = -\lambda_{\nu\mu}$$

and the coefficients λ are functions of the q's and p's that do not become infinitely great when one puts $p^\sigma p_\sigma = m^2$ with $p_0 > 0$. Since $p^\sigma p_\sigma - m^2$ has zero P.b. with all the expressions (7), the modified expressions (13) must still satisfy the P.b. relations (6), apart from multiples of $p^\sigma p_\sigma - m^2$, with any choice of the λ's. If we now choose the λ's so as to make the P_μ, $M_{\mu\nu}$ given by (13) independent of p_0, the P.b. relations (6) must be satisfied apart from terms that are independent of p_0 as well as being multiples of $p^\sigma p_\sigma - m^2$. Such terms must vanish, so we get in this way a solution of our problem.

The λ's have the values

$$\left.\begin{aligned} \lambda_r &= 0, \quad \lambda_0 = -\{p_0 + (p_s p_s + m^2)^{\frac{1}{2}}\}^{-1}, \\ \lambda_{sr} &= 0, \quad \lambda_{r0} = -q_r\{p_0 + (p_s p_s + m^2)^{\frac{1}{2}}\}^{-1}, \end{aligned}\right\} \tag{14}$$

and Eqs. (13) become

$$P_r = p_r, \qquad M_{rs} = q_r p_s - q_s p_r, \tag{15}$$

$$P_0 = (p_s p_s + m^2)^{\frac{1}{2}}, \quad M_{r0} = q_r(p_s p_s + m^2)^{\frac{1}{2}}, \tag{16}$$

with the help of (12). Equations (15) and (16) give all the ten fundamental quantities for a particle with rest-mass m. Those given by (15) are the simple ones: those given by (16) are the Hamiltonians.

For a dynamical system composed of several particles, P_r and M_{rs} will be just the sum of their values for the particles separately,

$$P_r = \sum p_r, \quad M_{rs} = \sum(q_r p_s - q_s p_r). \tag{17}$$

The Hamiltonians P_r, M_{r0} will be the sum of their values for the particles separately plus interaction terms,

$$\left.\begin{aligned} P_0 &= \sum(p_s p_s + m^2)^{\frac{1}{2}} + V, \\ M_{r0} &= \sum q_r(p_s p_s + m^2)^{\frac{1}{2}} + V_r. \end{aligned}\right\} \tag{18}$$

The V's here must be chosen to make P_0, M_{r0} satisfy all the P.b. relations (6) in which they appear.

Some of these relations are linear in the V's and are easily fulfilled. The P.b. relations for $[M_{rs}, P_0]$ and $[M_{rs}, M_{t0}]$ are fulfilled provided V is a three-dimensional scalar (in the space u_1, u_2, u_3) and V_r a three-dimensional vector. The P.b. relation for $[P_r, P_0]$ will be fulfilled provided V is independent of the position of the origin in the three-dimensional space u_1, u_2, u_3. The P.b. relation for $[M_{r0}, P_s]$ will be fulfilled provided

$$V_r = q_r V + V_r', \tag{19}$$

where the q_r are the coordinates of any one of the particles and the V_r' are independent of the position of the origin in three-dimensional space.

The remaining conditions for the V's are quadratic, involving $[V, V_r]$ or $[V_r, V_s]$. These conditions are not easily fulfilled and provide the real difficulty in the problem of constructing a theory of a relativistic dynamical system in the instant form.

4. THE POINT FORM

One can build up a dynamical theory in terms of dynamical variables that refer to physical conditions on some three-dimensional surface other than an instant. The surface must satisfy the condition that the world-line of every particle must meet it, otherwise the particle could not be described by variables on the surface, and preferably the world-line should meet it only once, for the sake of uniqueness.

To get a simple form of theory one should take the surface to be such that it is left invariant by some sub-group of the group of inhomogeneous Lorentz transformations. A possible sub-group is the group of rotations about some point, say the origin $u_\mu=0$. The surface may then be taken to be a branch of a hyperboloid

$$u^\rho u_\rho=\kappa^2, \quad u_0>0, \tag{20}$$

with κ a constant. The fundamental quantities associated with the infinitesimal transformations of the sub-group, namely the $M_{\mu\nu}$, will then be specially simple, while the others, namely the P_μ, will be complicated in general and will be the Hamiltonians. A new form of dynamics is thus obtained, which may be called the *point form*, as it is characterized by being associated with the sub-group that leaves a point invariant.

To illustrate the new form, let us take again the example of a single particle. The dynamical coordinates must determine the point where the world-line of the particle meets the hyperboloid (20). Let the four coordinates of this point in the u system of coordinates be q_μ. Only three of these are independent, but instead of eliminating one of them, it is more convenient to work with all four and introduce the subsidiary equation

$$q^\rho q_\rho\approx\kappa^2. \tag{21}$$

It is then necessary that the ten fundamental quantities, and indeed all physical variables, shall have zero P.b. with $q^\rho q_\rho$. The condition for this is that they should involve the p's only through the combinations $q_\mu p_\nu - q_\nu p_\mu$.

The ten fundamental quantities may be obtained by a method parallel to that of the preceding section, with the subsidiary equation (21) taking the place of Eq. (12). We again assume Eqs. (13), and now choose the λ's so as to make their right-hand sides have zero P.b. with $q^\rho q_\rho$. The resulting expressions for the ten fundamental quantities will again satisfy the P.b. relations (6), as may be inferred by a similar argument to the one given in the preceding section.

We find at once

$$\lambda_{\mu\nu}=0.$$

To obtain λ_μ, instead of arranging directly for the P_μ to have zero P.b. with $q^\rho q_\rho$, it is easier to make $q_\mu P_\nu - q_\nu P_\mu$ and $P^\mu P_\mu$ have zero P.b. with $q^\rho q_\rho$. Now

$$q_\mu P_\nu - q_\nu P_\mu = q_\mu p_\nu - q_\nu p_\mu + (q_\mu\lambda_\nu - q_\nu\lambda_\mu)(p^\sigma p_\sigma - m^2),$$

so we must have

$$q_\mu\lambda_\nu - q_\nu\lambda_\mu=0,$$

and hence

$$\lambda_\mu=q_\mu B,$$

where B is some dynamical variable independent of μ. Further

$$\begin{aligned}P^\mu P_\mu &= \{p^\mu+q^\mu B(p^\sigma p_\sigma-m^2)\}\{p_\mu+q_\mu B(p^\rho p_\rho-m^2)\}\\ &= m^2+\{1+2p^\mu q_\mu B+q^\mu q_\mu B^2(p^\sigma p_\sigma-m^2)\}(p^\rho p_\rho-m^2).\end{aligned}$$

In order that $P^\mu P_\mu$ shall have zero P.b. with $q^\rho q_\rho$, we must take

$$1+2p^\mu q_\mu B+q^\mu q_\mu B^2(p^\sigma p_\sigma-m^2)=0,$$

so that

$$B(p^\sigma p_\sigma-m^2)=(q^\rho q_\rho)^{-1}\{[(p^\nu q_\nu)^2 - q^\lambda q_\lambda(p^\sigma p_\sigma-m^2)]^{\frac{1}{2}}-p^\nu q_\nu\}$$

The right-hand side here tends to zero as $p^\sigma p_\sigma - m^2\to 0$, so it is a multiple of $p^\sigma p_\sigma - m^2$, as it ought to be. We now get finally

$$\left.\begin{aligned}P_\mu &= p_\mu+q_\mu\kappa^{-2}\{[(p^\nu q_\nu)^2-\kappa^2(p^\sigma p_\sigma-m^2)]^{\frac{1}{2}}-p^\nu q_\nu\}\\ M_{\mu\nu} &= q_\mu p_\nu - q_\nu p_\mu,\end{aligned}\right\} \tag{22}$$

in which the expression for P_μ has been simplified with the help of (21).

It is permissible to take $\kappa=0$ and so to have a light-cone instead of a hyperboloid. The expression for B then becomes much simpler and gives

$$P_\mu=p_\mu-\tfrac{1}{2}q_\mu(p^\nu q_\nu)^{-1}(p^\sigma p_\sigma-m^2). \tag{23}$$

instead of the first of Eqs. (22).

For a dynamical system composed of several particles, the $M_{\mu\nu}$ will be just the sum of their values for the particles separately,

$$M_{\mu\nu}=\sum(q_\mu p_\nu - q_\nu p_\mu). \tag{24}$$

The Hamiltonians P_μ will be the sum of their values for the particles separately plus interaction terms,

$$P_\mu=\sum\{p_\mu+q_\mu B(p^\sigma p_\sigma-m^2)\}+V_\mu. \tag{25}$$

The V_μ must be chosen so as to make the P_μ satisfy the correct P.b. relations. The relations for $[M_{\mu\nu}, P_\rho]$ are satisfied provided the V_μ are the components of a 4-vector. The remaining relations, which require the P_μ to have zero P.b. with one another, lead to quadratic conditions for the V_μ. These cause the real difficulty in the problem of constructing a theory of a relativistic dynamical system in the point form.

5. THE FRONT FORM

Consider the three-dimensional surface in space-time formed by a plane wave front advancing with the velocity of light. Such a surface will be called a *front* for brevity. An example of a front is given by the equation

$$u_0-u_3=0. \tag{26}$$

We may set up a dynamical theory in which the dynamical variables refer to physical conditions on a front. This will make specially simple those of the fundamental quantities associated with infinitesimal transformations of coordinates that leave the front invariant, and will give a third form of dynamics, which may be called the *front form*.

If A_μ is any 4-vector, put

$$A_0+A_3=A_+,\quad A_0-A_3=A_-.$$

We get a convenient notation by using the + and − suffixes freely as tensor suffixes, together with 1 and 2. They may be raised with the help of

$$g^{++}=g^{--}=0,\quad g^{+-}=\tfrac{1}{2},$$
$$g^{i+}=g^{i-}=0,\quad \text{for } i=1, 2,$$

as one can verify by noting that these g values lead to the correct value for $g^{\mu\nu}A_\mu A_\nu$ when μ and ν are summed over 1, 2, +, −.

The equation of the front (26) becomes in this notation

$$u_-=0.$$

The fundamental quantities P_1, P_2, P_-, M_{12}, M_{+-}, M_{1-}, M_{2-} are associated with transformations of coordinates that leave this front invariant and will be specially simple. The remaining ones P_+, M_{1+}, M_{2+} will be complicated in general and will be the Hamiltonians.

Let us again work out the example of a single particle. The dynamical coordinates are now q_1, q_2, q_+. We again assume Eqs. (13), and add to them the further equation $q_-=0$. We must now choose the λ's so as to make the right-hand sides of (13) independent of p_+. The resulting expressions for the ten fundamental quantities will then again satisfy the required P.b. relations.

We find

$$\lambda_+=-1/p_-,\quad \lambda_{i+}=-q_i/p_-,$$

the other λ's vanishing. Thus

$$\left.\begin{aligned} &P_i=p_i,\quad P_-=p_-,\\ &M_{12}=q_1p_2-q_2p_1,\quad M_{i-}=q_ip_-,\quad M_{+-}=q_+p_-,\end{aligned}\right\}\quad(27)$$

$$\left.\begin{aligned} &P_+=(p_1^2+p_2^2+m^2)/p_-,\\ &M_{i+}=q_i(p_1^2+p_2^2+m^2)/p_--q_+p_i.\end{aligned}\right\}\quad(28)$$

Equations (27) give the simple fundamental quantities. Equations (28) give the Hamiltonians.

For a dynamical system composed of several particles, P_i, P_-, M_{12}, M_{+-}, M_{i-} will be just the sum of their values for the particles separately. The Hamiltonians P_+, M_{i+} will be the sum of their values for the particles separately plus interaction terms,

$$\left.\begin{aligned} &P_+=\sum(p_1^2+p_2^2+m^2)/p_-+V\\ &M_{i+}=\sum\{q_i(p_1^2+p_2^2+m^2)/p_--q_+p_i\}+V_i.\end{aligned}\right\}\quad(29)$$

The V's must satisfy certain conditions to make the Hamiltonians satisfy the correct P.b. relations.

As before, some of these conditions are linear and some are quadratic. The linear conditions for V require that it shall be invariant under all transformations of the three coordinates u_1, u_2, u_+ of the front except those for which du_+ gets multiplied by a factor, and for the latter transformations V must get multiplied by the same factor. The linear conditions for V_i require it to be of the form

$$V_i=q_iV+V_i',\quad(30)$$

where q_i are the coordinates 1, 2 of any one of the particles, and V_i' has the same properties as V with regard to all transformations of the three coordinates of the front except rotations associated with M_{12}, and under these rotations it behaves like a two-dimensional vector. The quadratic conditions for the V's are not easily fulfilled and give rise to the real difficulty in the construction of a theory of a relativistic dynamical system in the front form.

6. THE ELECTROMAGNETIC FIELD

To set up the dynamical theory of fields on the lines discussed in the three preceding sections, one may take as dynamical variables the three-fold infinity of field quantities at all points on the instant, hyperboloid, or front, and use them in place of the discrete set of variables of particle theory. The ten fundamental quantities P_μ, $M_{\mu\nu}$ are to be constructed out of them, satisfying the same P.b. relations as before.

For a field which allows waves moving with the velocity of light, a difficulty arises with the point form of theory, because one may have a wave packet that does not meet the hyperboloid (20) at all. Thus physical conditions on the hyperboloid cannot completely describe the state of the field. One must introduce some extra dynamical variables besides the field quantities on the hyperboloid. A similar difficulty arises, in a less serious way, with the front form of theory. Waves moving with the velocity of light in exactly the direction of the front cannot be described by physical conditions on the front, and some extra variables must be introduced for dealing with them.

An alternative method of setting up the dynamical theory of fields is obtained by working with dynamical variables that describe the Fourier components of the field. This method has various advantages. It disposes of the above difficulty of extra variables, and it usually lends itself more directly to physical interpretation. It leads to expressions for the ten fundamental quantities that can be used with all three forms. For a field by itself, there is then no difference between the three forms. A difference occurs, of course, if the field is in interaction with something. The dynamical variables of the field are then to be understood as the Fourier components that the field would have, if the interaction were suddenly cut off at the instant, hyperboloid or front, after the cutting off.

Let us take as an example the electromagnetic field, first without any interaction. We may work with the four potentials $A_\lambda(u)$ satisfying the subsidiary equation

$$\partial A_\lambda/\partial u_\lambda \approx 0. \tag{31}$$

Their Fourier resolution is

$$A_\lambda(u)=\int\{A_{k\lambda}\exp(ik^\mu u_\mu)+A^\dagger_{k\lambda}\exp(-ik^\mu u_\mu)\}k_0^{-1}d^3k, \tag{32}$$

with

$$k_0=(k_1^2+k_2^2+k_3^2)^{\frac{1}{2}},\quad d^3k=dk_1dk_2dk_3.$$

The factor k_0^{-1} inserted in (32) leads to simpler transformation laws for the Fourier coefficients $A_{k\lambda}$, since the differential element $k_0^{-1}d^3k$ is Lorentz invariant. We now take the $A_{k\lambda}$, $A^\dagger_{k\lambda}$ as dynamical variables.

Under the transformation of coordinates (2) the potential $A_\lambda(u)$ at a particular point u changes to a potential at the point with the same u-values in the new coordinate system, i.e., the point with the coordinates $u_\mu - a_\mu - b_\mu{}^\nu u_\nu$ in the original coordinate system. This causes a change in $A_\lambda(u)$ of amount

$$-(a_\mu+b_\mu{}^\nu u_\nu)\partial A_\lambda/\partial u_\mu.$$

There is a further change, of amount $b_\lambda{}^\nu A_\nu$, owing to the change in the direction of the axes. Thus, from (4) and (5)

$$\begin{aligned}[A_\lambda(u),\ -P^\mu a_\mu+\tfrac{1}{2}M^{\mu\nu}b_{\mu\nu}]&=A_\lambda(u)^*-A_\lambda(u)\\&=-(a_\mu+b_\mu{}^\nu u_\nu)\partial A_\lambda/\partial u_\mu+b_\lambda{}^\nu A_\nu,\end{aligned}$$

and hence

$$\left.\begin{aligned}&[A_\lambda(u),\ P^\mu]=\partial A_\lambda/\partial u_\mu,\\&[A_\lambda(u),\ M_{\mu\nu}]=u_\mu\partial A_\lambda/\partial u^\nu-u_\nu\partial A_\lambda/\partial u^\mu\\&\qquad+g_{\lambda\mu}A_\nu-g_{\lambda\nu}A_\mu.\end{aligned}\right\}\tag{33}$$

Taking Fourier components according to (32), we get

$$\left.\begin{aligned}&[A_{k\lambda},\ P_\mu]=ik_\mu A_{k\lambda},\\&[A_{k\lambda},\ M_{\mu\nu}]=(k_\mu\partial/\partial k^\nu-k_\nu\partial/\partial k^\mu)A_{k\lambda}\\&\qquad+g_{\lambda\mu}A_{k\nu}-g_{\lambda\nu}A_{k\mu},\end{aligned}\right\}\tag{34}$$

in which $A_{k\lambda}$ may be considered as a function of four independent k's for the purpose of applying the differential operator $k_\mu\partial/\partial k^\nu-k_\nu\partial/\partial k^\mu$ to it.

The Maxwell theory gives for the energy and momentum of the electromagnetic field

$$P_\mu=-4\pi^2\int k_\mu A_k{}^\lambda A^\dagger_{k\lambda}k_0^{-1}d^3k, \tag{35}$$

the $-$ sign being needed to make the transverse components contribute a positive energy. In order that this may agree with the first of Eqs. (34), we must have the P.b. relations

$$\left.\begin{aligned}&[A_{k\lambda},\ A_{k'\mu}]=0\\&[A_{k\lambda},\ A^\dagger_{k'\mu}]=-ig_{\lambda\mu}/4\pi^2\\&\qquad\cdot k_0\delta(k_1-k_1')\delta(k_2-k_2')\delta(k_3-k_3').\end{aligned}\right\}\tag{36}$$

The second of Eqs. (34) then leads to

$$M_{\mu\nu}=4\pi^2 i\int\{A^\dagger_{k\lambda}(k_\mu\partial/\partial k^\nu-k_\nu\partial/\partial k^\mu)A_k{}^\lambda+A^\dagger_{k\mu}A_{k\nu}-A^\dagger_{k\nu}A_{k\mu}\}k_0^{-1}d^3k. \tag{37}$$

Equations (35) and (37) give the ten fundamental quantities.

For the electromagnetic field in interaction with charged particles, the ten fundamental quantities will be the sum of their values for the field alone, given by (35) and (37), and their values for the particles, given in one of the three preceding sections, with interaction terms involving the field variables $A_{k\lambda}$, $A^\dagger_{k\lambda}$ as well as the particle variables. One usually assumes that there is no direct interaction between the particles, only interaction between each particle and the field. The ten fundamental quantities then take the form

$$\left.\begin{aligned}P_\mu&=P_\mu{}^F+\textstyle\sum_a P_\mu{}^a\\M_{\mu\nu}&=M_{\mu\nu}{}^F+\textstyle\sum_a M_{\mu\nu}{}^a,\end{aligned}\right\}\tag{38}$$

where $P_\mu{}^F$, $M_{\mu\nu}{}^F$ are the contributions of the field alone, given by (35) and (37), and $P_\mu{}^a$, $M_{\mu\nu}{}^a$ are the contributions of the a-th particle, consisting of terms for the particle by itself and interaction terms. For point charges, the interaction terms will involve the field variables only through the $A_\lambda(q)$ and their derivatives at the point q where the world-line of the particle meets the instant, hyperboloid or front. The expressions for $P_\mu{}^a$, $M_{\mu\nu}{}^a$ may easily be worked out for this case by a generalization of the method of the three preceding sections, as follows.

Suppose there is only one particle, for simplicity. We must replace Eqs. (13) by

$$\left.\begin{aligned}P_\mu&=P_\mu{}^F+p_\mu+\lambda_\mu(\pi^\sigma\pi_\sigma-m^2)\\M_{\mu\nu}&=M_{\mu\nu}{}^F+q_\mu p_\nu-q_\nu p_\mu+\lambda_{\mu\nu}(\pi^\sigma\pi_\sigma-m^2),\end{aligned}\right\}\tag{39}$$

where

$$\pi_\sigma=p_\sigma-eA_\sigma(q),$$

and $P_\mu{}^F$, $M_{\mu\nu}{}^F$ are the right-hand sides of (35) and (37). From (33),

$$\begin{aligned}&[A_\lambda(q),\ P_\mu{}^F+p_\mu]=0\\&[A_\lambda(q),\ M_{\mu\nu}{}^F+q_\mu p_\nu-q_\nu p_\mu]=g_{\lambda\mu}A_\nu(q)-g_{\lambda\nu}A_\mu(q),\end{aligned}$$

and hence

$$\begin{aligned}&[\pi_\lambda,\ P_\mu{}^F+p_\mu]=0\\&[\pi_\lambda,\ M_{\mu\nu}{}^F+q_\mu p_\nu-q_\nu p_\mu]=g_{\lambda\mu}\pi_\nu-g_{\lambda\nu}\pi_\mu.\end{aligned}$$

It follows that $\pi^\sigma\pi_\sigma$ has zero P.b. with each of the quantities $P_\mu{}^F+p_\mu$, $M_{\mu\nu}{}^F+q_\mu p_\nu-q_\nu p_\mu$. One can now infer, by the same argument as in the case of no field, that if the λ's in (39) are chosen to make P_μ, $M_{\mu\nu}$ have zero P.b. with q_0, $q^\rho q_\rho$ or q_-, the P.b. relations (6) will all be satisfied. Such a choice of λ's, in conjunction with one of the equations $q_0=0$, $q^\rho q_\rho\approx\kappa^2$, $q_-=0$, will provide the ten fundamental quantities for a charged particle in interaction with the field in the instant, point and front forms, respectively. The subsidiary Eq. (31) must be modified when a charge is present.

The point form will be worked out as an illustration. In this case we have at once $\lambda_{\mu\nu}=0$. We can get λ_μ conveniently by arranging that $q_\mu(P_\nu-P_\nu{}^F)-q_\nu(P_\mu-P_\mu{}^F)$ and $\{P^\mu-P^{\mu F}-eA^\mu(q)\}\{P_\mu-P_\mu{}^F-eA_\mu(q)\}$ shall have zero P.b. with $q^\rho q_\rho$. The first condition gives $\lambda_\mu=q_\mu B$. The second then gives

$$1+2\pi^\mu q_\mu B+q^\mu q_\mu B^2(\pi^\sigma\pi_\sigma-m^2)=0.$$

Thus we get finally

$$\left.\begin{aligned} P_\mu&=P_\mu{}^F+p_\mu+q_\mu\kappa^{-2}\{[(\pi^\nu q_\nu)^2 \\ &\qquad -\kappa^2(\pi^\sigma\pi_\sigma-m^2)]^{\frac{1}{2}}-\pi^\nu q_\nu\} \\ M_{\mu\nu}&=M_{\mu\nu}{}^F+q_\mu p_\nu-q_\nu p_\mu. \end{aligned}\right\}(40)$$

The above theory of point charges is subject to the usual difficulty that infinities will arise in the solution of the equations of motion, on account of the infinite electromagnetic energy of a point charge. The present treatment has the advantage over the usual treatment of the electromagnetic equations that it offers simpler opportunities for departure from the point-charge model for elementary particles.

8. DISCUSSION

Three forms have been given in which relativistic dynamical theory may be put. For particles with no interaction, any one of the three is possible. For particles with interaction, it may be that all three are still possible, or it may be that only one is possible, depending on the kind of interaction. If one wants to set up a new kind of interaction between particles in order to improve atomic theory, the way to proceed would be to take one of the three forms and try to find the interaction terms V, or to find directly the Hamiltonians, satisfying the required P.b. relations. The question arises, which is the best form to take for this purpose.

The instant form has the advantage of being the one people are most familiar with, but I do not believe it is intrinsically any better for this reason. The four Hamiltonians P_0, M_{r0} form a rather clumsy combination.

The point form has the advantage that it makes a clean separation between those of the fundamental quantities that are simple and those that are the Hamiltonians. The former are the components of a 6-vector, the latter are the components of a 4-vector. Thus the four Hamiltonians can easily be treated as a single entity. All the equations with this form can be expressed neatly and concisely in four-dimensional tensor notation.

The front form has the advantage that it requires only three Hamiltonians, instead of the four of the other forms. This makes it mathematically the most interesting form, and makes any problem of finding Hamiltonians substantially easier. The front form has the further advantage that there is no square root in the Hamiltonians (28), which means that one can avoid negative energies for particles by suitably choosing the values of the dynamical variables in the front, without having to make a special convention about the sign of a square root. It may then be easier to eliminate negative energies from the quantum theory. This advantage also occurs with the point form with $\kappa=0$, there being no square root in (23).

There is no conclusive argument in favor of one or other of the forms. Even if it could be decided that one of them is the most convenient, this would not necessarily be the one chosen by nature, in the event that only one of them is possible for atomic systems. Thus all three forms should be studied further.

The conditions discussed in this paper for a relativistic dynamical system are necessary but not sufficient. Some further condition is needed to ensure that the interaction between two physical objects becomes small when the objects become far apart. It is not clear how this condition can be formulated mathematically. Present-day atomic theories involve the assumption of localizability, which is sufficient but is very likely too stringent. The assumption requires that the theory shall be built up in terms of dynamical variables that are each localized at some point in space-time, two variables localized at two points lying outside each other's light-cones being assumed to have zero P.b. A less drastic assumption may be adequate, e.g., that there is a fundamental length λ such that the P.b. of two dynamical variables must vanish if they are localized at two points whose separation is space-like and greater than λ, but need not vanish if it is less than λ.

I hope to come back elsewhere to the transition to the quantum theory.

PHYSICAL REVIEW VOLUME 77, NUMBER 2 JANUARY 15, 1950

Quantum Theory of Non-Local Fields. Part I. Free Fields

HIDEKI YUKAWA*
Columbia University, New York, New York
(Received September 27, 1949)

The possibility of a theory of non-local fields, which is free from the restriction that field quantities are always point functions in the ordinary space, is investigated. Certain types of non-local fields, each satisfying a set of mutually compatible commutation relations, which can be obtained by extending familiar field equations for local fields in conformity with the principle of reciprocity, are considered in detail. Thus a scalar non-local field is obtained, which represents an assembly of particles with the mass, radius and spin 0, provided that the field is quantized according to the procedure similar to the method of second quantization in the usual field theory. Non-local vector and spinor fields corresponding to assemblies of particles with the finite radius and the spins 1 and ½ respectively are obtained in the similar way.

I. INTRODUCTION

IT has been generally believed for years that well-known divergence difficulties in quantum theory of wave fields could be solved only by taking into account the finite size of the elementary particles consistently. Recent success of quantum electrodynamics, which took advantage of the relativistic covariance to the utmost,[1] however, seemed to have weakened to some extent the necessity of introducing so-called universal length or any substitute for it into field theory. In fact, all infinities which had been familiar in previous formulations of quantum electrodynamics were reduced to unobservable renormalization factors for the mass and the electric charge in the newer formalism. Furthermore, in order to get rid of the remaining difficulties that these renormalization factors were still either infinite or indefinite, main efforts were concentrated in the direction of introducing various kinds of auxiliary fields, either real or only formal, rather than in the direction of introducing explicitly the universal length or the finite radius of the elementary particles. So far as the results of the investigations in the former direction are concerned, however, the prospect is not so encouraging. Namely, an ingenious method of regulators, which was investigated by Pauli extensively,[2] can be regarded as a formalistic generalization of the theory of mixed fields,[3] but cannot be replaced by a combination of neutral vector fields and charged spinor fields with different masses, unless we admit the introduction of bosons with negative energies and fermions with imaginary charges as pointed out by Feldman.[4] More generally, according to recent investigations by Umezawa and others[5] and by Feldman,[4] no combination of quantized fields with spins 0, ½, and 1 can be free from all of the divergence difficulties, as long as only *positive* energy states for bosons and *real* coupling constants for the interactions between fermions and bosons are taken into account.

Nevertheless, the difficulties remaining in quantum electrodynamics are not so serious as those which appear in meson theory. In the latter case, we know that straightforward calculations very often lead to divergent results for directly observable quantities such as the probabilities of certain types of meson decay.[6] Although the application of Pauli's regulators to meson theory was found useful for obtaining finite results, it can hardly be considered as a satisfactory solution of the problem for reasons mentioned above. It seems to the present author that, at least, a part of the defect of the present meson theory is due to the lack of a consistent method of dealing with the finite extension of the elementary particle such as the nucleon, whereas the effect of the finite extension is usually very small so far as electrodynamical phenomena in the narrowest sense are concerned, except for its decisive effect on the renormalizations of the mass and the electric charge.

Under these circumstances, it seems worth while to investigate again the possibility of extension of the present field theory in the direction of introducing the finite radius of the elementary particle. In this paper, as the continuation of the preceding papers,[7] the possibility of a theory of quantized *non-local* fields, which is free from the restriction that field quantities are always point functions in the ordinary space, will be discussed in detail. One may be very sceptical about the necessity of such a drastic change in field theory, because

* On leave of absence from Kyoto University, Kyoto, Japan.

[1] As to the list of recent works by Tomonaga, Schwinger and others, see V. Weisskopf, Rev. Mod. Phys. 21, 305 (1949).

[2] W. Pauli and F. Villars, Rev. Mod. Phys. 21, 433 (1949). The method of regulators is an extension of cut-off procedures by R. P. Feynman, Phys. Rev. 74, 1430 (1948) and by D. Rivier and E. C. G. Stueckelberg, Phys. Rev. 74, 218 (1948).

[3] Field theories by Bopp, Podolsky, Dirac, and others are more formalistic in that negative energy bosons are taken into account, whereas those by Pais, Sakata, and Hara are more realistic.

[4] D. Feldman, Phys. Rev. 76, 1369 (1949). The author is indebted to Dr. Feldman for discussing the subject before publication of his paper.

[5] Umezawa, Yukawa, and Yamada, Prog. Theor. Phys. 4, 25, 113 (1949). See also R. Jost and J. Rayski, Helv. Phys. Acta 22, 457 (1949).

[6] H. Fukuda, and Y. Miyamoto, Prog. Theor. Phys. 4, 235 (1949); Sasaki, Oneda, and Ozaki, Prog. Theor. Phys. (to be published); J. Steinberger, Phys. Rev. 76, 1180 (1949). See further a comprehensive survey of recent works on meson theory by H. Yukawa, Rev. Mod. Phys. 21, 474 (1949).

[7] A preliminary account of the content of this paper was published in H. Yukawa, Phys. Rev. 76, 300 (1949), which will be cited as I.

Reprinted from *Phys. Rev.* 77, 219 (1950).

other possibilities such as the introduction of *local* fields corresponding to particles with spins higher than 1 are not yet fully investigated. However, present theory of elementary particles with spins higher than 1 suffers from the difficulty associated with the necessity of of auxiliary conditions, and even if this is overcome by some revision of the formalism as proposed by Bhabha,[8] we can hardly expect a satisfactory solution of the whole problem, because the admixture of higher spin fields may well give rise to newer types of divergence in return for the elimination of more familiar ones. Moreover, it does not seem to the present author that the theory of non-local fields is necessarily contradictory to the theory of mixed local fields. They can rather be complementary to each other in that a non-local field may well happen to be approximately equivalent to some mixture of local fields. The most essential point, which is in favor of the non-local field, is that the convergence of field theory might be guaranteed by introducing a new type of *irreducible* field instead of a mixture, which is *reducible*.

In this paper, as in the preceding papers, we confine our attention to certain types of non-local field, each satisfying a set of mutually compatible commutation relations, which can be obtained by extending familiar field equations for local fields in conformity with the principle of reciprocity. The solutions of these operator equations can be interpreted as a field-theoretical representation of assemblies of elementary particles, each having a definite mass and a definite radius. In this connection, recent attempt by Born and Green[9] is interesting particularly in that they made use of the principle of reciprocity as a postulate for determining possible masses of elementary particles of various types. However, it is not yet clear whether their method of density operators contains something essentially different from the usual theory of mixture of local fields. The most important question of the interaction of two or more non-local fields will be discussed in Part II of this paper.

II. AN EXAMPLE OF THE NON-LOCAL SCALAR FIELD

In order to see what comes out by generalizing a field theory so as to include non-local fields, we start from a particular case of the non-local scalar field. A scalar operator U, which is supposed to describe a non-local scalar field, can be represented, in general, by a matrix with rows and columns, each characterized by a set of values of space and time coordinates. Alternatively, we can regard this operator U as a certain function of four space-time operators x^μ ($x'=x_1\equiv x$, $x^2=x_2\equiv y$, $x^3=x_3\equiv z$, $x^4=-x_4\equiv ct$) as well as of four space-time displacement operators p_μ, which satisfy well-known commutation relations

$$[x^\mu, p_\nu]=i\hbar\delta_{\mu\nu}, \tag{1}$$

where

$$[A, B]\equiv AB-BA \tag{2}$$

for any two operators A and B. Usual local fields are included as the particular case, in which the field operator U is a function of x^μ alone, so that it can be represented by a diagonal matrix in the representation, in which the operators x^μ themselves are diagonal. In this particular case, it is customary to start from the second-order wave equation

$$\left(\frac{\partial^2}{\partial x_\mu \partial x^\mu}-\kappa^2\right)U(x^\mu)=0, \quad \kappa=mc/\hbar \tag{3}$$

for the local field $U(x^\mu)$, in order that it can reproduce, when quantized, an assembly of identical particles with a definite mass m and the spin 0. Equation (3) is equivalent to the relation between the operator U and the operators p_μ

$$[p_\mu[p^\mu, U]]+m^2c^2U=0 \tag{4}$$

for this case. We assume that the non-local scalar field U in question satisfies the commutation relation of the same form as (4). However, in our case, we need further the commutation relation between U and x^μ, in contrast to the case of local field, in which U and x^μ are simply commutative with each other. In order to guess the correct form for it, some heuristic idea is needed. The principle of reciprocity seems to be very useful for this purpose. Namely, we assume that the commutation relation between U and x^μ has a form

$$[x_\mu[x^\mu, U]]-\lambda^2U=0, \tag{5}$$

where λ is a constant with the dimension of length and can be interpreted as the radius of the elementary particle in question, as will be shown below. The relations (4) and (5) are not exactly the same in form, but differ from each other by plus and minus signs of the last terms on the left-hand sides of (4) and (5). Thus, the two operator equations (4) and (5) can be said to be mutually reciprocal rather than perfectly symmetrical, indicating that the radius of the elementary particle λ must be introduced as something reciprocal to the mass m.

Now the operator U can be represented by a matrix $(x_\mu'|U|x_\mu'')$ in the representation, in which x_μ are diagonal matrices. The matrix elements, in turn, can be considered as a function $U(X_\mu, r_\mu)$ of two sets of real variables

$$X_\mu=\tfrac{1}{2}(x_\mu'+x_\mu''), \quad r_\mu=x_\mu'-x_\mu''. \tag{6}$$

Accordingly, the relations (4), (5) can be replaced by

$$(\partial^2/\partial X_\mu\partial X^\mu-\kappa^2)U(X_\mu, r_\mu)=0, \tag{7}$$

$$(r_\mu r^\mu-\lambda^2)U(X_\mu, r_\mu)=0, \tag{8}$$

[8] H. J. Bhabha, Proc. Ind. Acad. Sci. **A21**, 241 (1945); Rev. Mod. Phys. **17**, 200 (1945).

[9] M. Born, Nature **163**, 207 (1949); H. S. Green, Nature **163**, 208 (1949); M. Born and H. S. Green, Proc. Roy. Soc. Edinburgh **A92**, 470 (1949).

respectively. Equations (7) and (8) are obviously compatible with each other and the former implies that $U(X_\mu, r_\mu)$ is, in general, a superposition of plane waves of the form $\exp ik_\mu X^\mu$ with k_μ satisfying the condition

$$k_\mu k^\mu+\kappa^2=0, \tag{9}$$

whereas the latter implies that $U(X_\mu, r_\mu)$ can be different from zero only for those values of r_μ, which satisfy the condition

$$r_\mu r^\mu-\lambda^2=0. \tag{10}$$

Thus the most general solution of the simultaneous Eqs. (7) and (8) has the form

$$U(X_\mu, r_\mu)=\int\cdots\int(dk^4)u(k_\mu, r_\mu)\delta(\gamma_\mu r^\mu-\lambda^2)\times\delta(k_\mu k^\mu+\kappa^2)\exp(ik_\mu X^\mu), \tag{11}$$

where $u(k_\mu, r_\mu)$ is an arbitrary function of two sets of variables k_μ and r_μ.

The above considerations suggest us that one set X_μ of the real variables could be identified with the conventional space and time coordinates of the elementary particle regarded as a material point in the limit of $\lambda\rightarrow 0$, whereas the other set r_μ could be interpreted as variables describing the internal motion in general case, in which the finite extension of the elementary particle in question could not be ignored. Thus, we might expect that the field U of the above type is equivalent to an assembly of elementary particles with the mass m, the radius λ and the spin 0, if it is further quantized according to the familiar method of second quantization. However, we can easily anticipate that the equivalence is incomplete, because $U(X_\mu, r_\mu)$ is different from zero for arbitrary large values of r_μ, so far as they satisfy the condition (10), even when only one term of the right-hand side of (11) corresponding to a definite set of values of k_μ is taken into account. In other words, we need another condition for restricting the possible form of $U(X_\mu, r_\mu)$ or $u(k_\mu, r_\mu)$ in order to complete the equivalence above mentioned. For this purpose, we introduce an auxiliary condition

$$[p_\mu[x^\mu U]]=0, \tag{12}$$

which can be said to be self-reciprocal in that the relation

$$[x^\mu[p_\mu, U]]=0 \tag{13}$$

can be deduced from (12) immediately on account of the commutation relation (1). Both of (12) and (13) are equivalent to the condition

$$r_\mu\frac{\partial U(X_\mu, r_\mu)}{\partial X^\mu}=0 \tag{14}$$

for $U(X_\mu, r_\mu)$, or the restriction that $u(k_\mu, r_\mu)$ should be zero unless k_μ and r_μ satisfy the condition

$$k_\mu r^\mu=0. \tag{15}$$

Thus the most general form of $U(X_\mu, r_\mu)$, which satisfies all the relations (7), (8), and (14), is

$$U(X_\mu, r_\mu)=\int\cdots\int(dk)^4 u(k_\mu, r_\mu)\delta(k_\mu k^\mu+\kappa^2)\times\delta(r_\mu r^\mu-\lambda^2)\delta(k_\mu r^\mu)\exp(ik_\mu X^\mu), \tag{16}$$

where $u(k_\mu, r_\mu)$ is again an arbitrary function of k_μ and r_μ.[10]

Now a simple physical interpretation can be given to the non-local field of the form (16) by considering the corresponding particle picture: Suppose that the particle is at rest with respect to a certain reference system. In this particular case, the motion of the particle as a whole, or the motion of its center of mass, can be represented presumably by a plane wave in X-space with the wave vector $k_1=k_2=k_3=0$, $k_4=-\kappa$. The corresponding form of $U(X_\mu, r_\mu)$ is, apart from the factor independent of X_μ, r_μ,

$$u(0, 0, 0, -\kappa; r_\mu)\delta(r_\mu r^\mu-\lambda^2)\delta(\kappa r_4)\exp(-i\kappa X^4) \tag{17}$$

which is different from zero only for those values of r_μ, which satisfy the conditions

$$r_1^2+r_2^2+r_3^2=\lambda^2,\quad r_4=0. \tag{18}$$

Thus, the form of $U(X_\mu, r_\mu)$ in this case is determined completely by giving $u(0, 0, 0, -\kappa; r_\mu)$ as defined on the surface of the sphere with the radius λ in r-space. In other words, the internal motion can be described by the wave function $u(\theta, \varphi)$ depending only on the polar angles θ, φ, which are defined by

$$r_1=r\sin\theta\cos\varphi,\quad r_2=r\sin\theta\sin\varphi,\quad r_3=r\cos\theta. \tag{19}$$

In general, $u(\theta, \varphi)$ can be expanded into series of spherical harmonics:

$$u(\theta, \varphi)=\sum_{l, m} c(0, 0, 0, -\kappa; l, m)P_l^m(\theta, \varphi), \tag{20}$$

which is equivalent to decomposing the internal rotation into various states characterized by the azimuthal quantum number l and the magnetic quantum number m.

In the case when the center of mass of the particle is moving with the velocity v_x, v_y, v_z, it can be described by a plane wave in X-space with the wave vector k_μ, which is connected with the velocity by the relations

$$v_x=-k_1c/k_4,\quad v_y=-k_2c/k_4,\quad v_z=-k_3c/k_4,\quad k_4=-(k_1^2+k_2^2+k_3^2+\kappa^2)^{\frac{1}{2}}. \tag{21}$$

In this case, $U(X_\mu, r_\mu)$ has the form

$$u(k_\mu, r_\mu)\delta(r_\mu r^\mu-\lambda^2)\delta(k_\mu r^\mu)\exp(ik_\mu X^\mu), \tag{22}$$

which is different from zero only on the surface of the sphere with the radius λ in r-space, the sphere itself

[10] $U(X_\mu, r_\mu)$ as given by the expression (6) in I was not the most general form in that the coefficients $b(k_\mu)$ were independent of l_μ, which corresponded to ignore the internal rotation. The author is indebted to Professor R. Serber for calling attention to this point.

moving with the velocity v_x, v_y, v_z. Accordingly, we perform first the Lorentz transformation

$$x_\mu' = a_{\mu\nu} x_\nu \tag{23}$$

with the transformation matrix

$$(a_{\mu\nu}) \equiv \begin{pmatrix} 1+(k_1/K)^2 & k_1k_2/K^2 & k_1k_3/K^2 & k_1/\kappa \\ k_1k_2/K^2 & 1+(k_2/K)^2 & k_2k_3/K^2 & k_2/\kappa \\ k_1k_3/K^2 & k_2k_3/K^2 & 1+(k_3/K)^2 & k_3/\kappa \\ k_1/\kappa & k_2/\kappa & k_3/\kappa & -k_4/\kappa \end{pmatrix}, \tag{24}$$

where $K=(\kappa(\kappa-k_4))^{\frac{1}{2}}$. Then the wave function for the internal motion can be described by a function $u'(\theta', \varphi')$ of the polar angle θ', φ' defined by

$$\left.\begin{aligned} r_1' &= a_{1\nu}r_\nu = r'\sin\theta'\cos\varphi', & r_2' &= a_{2\nu}r_\nu = r'\sin\theta'\sin\varphi', \\ r_3' &= a_{3\nu}r_\nu = r'\cos\theta', & r_4' &= a_{4\nu}r_\nu = k_\mu r^\mu/\kappa. \end{aligned}\right\} \tag{25}$$

Incidentally, r_4' as defined by the last expression in (25) is nothing but the proper time multiplied by $-c$ for the particle, which is moving with the velocity v_x, v_y, v_z. Again, $u'(\theta', \varphi')$ can be expanded into series of spherical harmonics:

$$u'(\theta', \varphi') = \sum_{l,m} c(k_\mu, l, m) P_l^m(\theta', \varphi'). \tag{26}$$

Since the above arguments are in conformity with the principle of relativity perfectly, the non-local field in question can be regarded as a field-theoretical representation of a system of identical particles, each with the mass m, the radius and the spin 0, which can rotate as the relativistic rigid sphere without any change in shape other than the Lorentz contraction associated with the change of the proper time axis.

The non-local field U given by (16) reduces to the ordinary local scalar field in the limit $\lambda \to 0$, as it should be, provided that the rest mass m is different from zero. Namely, $(x_\mu'|U|x_\mu'')$ is different from zero only for $x_\mu'=x_\mu''$, because the only possible solution of the simultaneous Eqs. (9), (11), and (15) with $m\neq 0$ and $\lambda=0$ is $r_1=r_2=r_3=r_4=0$. On the contrary, the case of the zero rest mass $m=0$ is exceptional in that the non-local field U does not necessarily reduce to the local field in the limit $\lambda=0$. This is because the simultaneous Eqs. (9), (11), and (15) with $m=0$ and $\lambda=0$ have solutions of the form

$$r_\mu = \pm(\lambda')^2 k_\mu, \quad k_4 = \pm(k_1^2+k_2^2+k_3^2)^{\frac{1}{2}}, \tag{27}$$

where λ' is an arbitrary constant with the dimension of length. More generally, the simultaneous equations with $m=0$ and $\lambda\neq 0$ has the general solution of the form

$$r_\mu = r_\mu' \pm (\lambda')^2 k_\mu, \quad k_4 = \pm(k_1^2+k_2^2+k_3^2)^{\frac{1}{2}}, \tag{28}$$

where r_μ' is any particular solution of the same equations. Thus the radius of the particle without the rest mass cannot be defined so naturally as in the case of the particle with the rest mass, corresponding to the circumstance that there is no rest system in the former case. Detailed discussions of this particular case will be made elsewhere.

III. QUANTIZATION OF NON-LOCAL SCALAR FIELD

In order to show that the non-local field above considered represents exactly the assembly of identical particles with the finite radius, we have to quantize the field on the same lines as the method of second quantization in ordinary field theory. For this purpose, it is convenient to write (16) in another form

$$U(X_\mu, r_\mu) = \int \cdots \int (dk)^4 (dl)^4 u(k_\mu, l_\mu) \times \delta(k_\mu k^\mu + \kappa^2)\delta(l_\mu l^\mu - \lambda^2)\delta(k_\mu l^\mu) \times \exp(ik_\mu X^\mu) \prod_\mu \delta(r_\mu + l_\mu), \tag{29}$$

where l_μ is a four vector. The integrand is different from zero only for those values of k_μ, l_μ, which satisfy the relations

$$k_\mu k^\mu + \kappa^2 = 0, \quad l_\mu l^\mu - \lambda^2 = 0, \quad k_\mu l^\mu = 0. \tag{30}$$

Accordingly, the matrix elements for the operator U are

$$(x_\mu'|U|x_\mu'') = \int \cdots \int (dk)^4 (dl)^4 u(k_\mu, l_\mu) \times \delta(k_\mu k^\mu + \kappa^2)\delta(l_\mu l^\mu - \lambda^2) \exp(ik^\mu x_\mu'/2) \times \prod_\mu \delta(x_\mu' - x_\mu'' + l_\mu) \exp(ik^\mu x_\mu''/2), \tag{31}$$

which is equivalent to the relation

$$U = \int \cdots \int (dk)^4 (dl)^4 \bar{u}(k_\mu, l_\mu) \exp(ik_\mu x^\mu/2) \times \exp(il^\mu p_\mu/\hbar) \exp(ik_\mu x^\mu/2), \tag{32}$$

between the operators x^μ, p_μ and U, where

$$\bar{u}(k_\mu, l_\mu) = u(k_\mu, l_\mu)\delta(k_\mu k^\mu + \kappa^2)\delta(l_\mu l^\mu - \lambda^2)\delta(k_\mu l^\mu). \tag{33}$$

As the operators $k_\mu x^\mu$ and $l^\mu p_\mu$ in the same term on the right-hand side of (32) are commutative with each other on account of the relations (1) and (30), (32) can also be written in the form

$$U = \int \cdots \int (dk)^4 (dl)^4 \bar{u}(k_\mu, l_\mu) \exp(ik_\mu x^\mu) \times \exp(il^\mu p_\mu/\hbar). \tag{32'}$$

Similarly the operator U^*, which is the Hermitian

conjugate of U, can be written in the form

$$U^*=\int\cdots\int(dk^4)(dl)^4\bar{u}^*(k_\mu,l_\mu)\times\exp(-ik_\mu x^\mu)\exp(-il^\mu p_\mu/\hbar). \quad (34)$$

Now the method of second quantization can be applied to our case in the following way: $\bar{u}(k_\mu, l_\mu)$ and $\bar{u}^*(k_\mu, l_\mu)$ in Eqs. (32′) and (34) are regarded as operators, which are Hermitian conjugate to each other and are non-commutative in general. The fact that the operators defined by

$$U(k_\mu,l_\mu)\equiv\exp(ik_\mu x^\mu)\exp(il^\mu p_\mu/\hbar);\quad U^*(k_\mu,l_\mu)\equiv\exp(-ik_\mu x^\mu)\exp(-il^\mu p_\mu/\hbar) \quad (35)$$

are unitary, i.e., satisfy the relation

$$U(k_\mu,l_\mu)U^*(k_\mu,l_\mu)=U^*(k_\mu,l_\mu)U(k_\mu,l_\mu)=1 \quad (36)$$

suggests us the commutation relations

$$\begin{aligned}&[\bar{u}(k_\mu,l_\mu),\bar{u}^*(k_\mu',l_\mu')]\\&=-\frac{k_4}{|k_4|}\Pi_\mu\,\delta(k_\mu-k_\mu')\delta(l_\mu-l_\mu')\cdot\delta(k_\mu k^\mu+\kappa^2)\times\delta(l_\mu l^\mu-\lambda^2)\delta(k_\mu l^\mu), \quad (37)\\&[\bar{u}(k_\mu,l_\mu),\bar{u}(k_\mu',l_\mu')]=0,\\&[\bar{u}^*(k_\mu,l_\mu),\bar{u}^*(k_\mu',l_\mu')]=0,\end{aligned}$$

which are obviously invariant with respect to the whole group of Lorentz transformations. In order to make the physical meaning of the relations (37) clear, we suppose the field in a cube with the edges of the length L, which is very large compared with λ. Then the effects of non-localizability of the field are negligible, because they are confined to small regions very near the surface of the cube.[11] In this case, the integrations with respect to k_μ on the right-hand side of Eqs. (32′) and (34) are replaced by the summations with respect to k_μ, which take the values

$$k_1=(2\pi/L)n_1,\quad k_2=(2\pi/L)n_2,\quad k_3=(2\pi/L)n_3,\quad k_4=\pm(k_1^2+k_2^2+k_3^2+\kappa^2)^{\frac{1}{2}}, \quad (38)$$

where n_1, n_2, n_3 are integers, either positive or negative, including zero. The integrations with respect to l_μ with fixed k_μ are replaced by those with respect to l_μ' defined by

$$l_\mu'=a_{\mu\nu}l_\nu, \quad (39)$$

where the coefficients $a_{\mu\nu}$ are given by (24). Further, we introduce the polar angle Θ, Φ, which are connected with l_1', l_2', l_3' just as θ', φ' are connected with r_1', r_2', r_3' by the relations (25). Thus we obtain

$$\begin{aligned}U=\sum_{k_1k_2k_3}\int\int\left(\frac{2\pi}{L}\right)^3\frac{\lambda\sin\Theta d\Theta d\Phi}{4\kappa(k^2+\kappa^2)^{\frac{1}{2}}}&\times\{u(k,\Theta,\Phi)U(k,\Theta,\Phi)\\&+v^*(k,\Theta,\Phi)U^*(k,\Theta,\Phi)\}, \quad (40)\end{aligned}$$

[11] More precisely, L must be large compared with $\lambda/(1-\beta^2)^{\frac{1}{2}}$, where βc is the maximum velocity of particles in consideration.

where

$$\left.\begin{aligned}&u(k,\Theta,\Phi)\equiv u(k_1,k_2,k_3,-(k^2+\kappa^2)^{\frac{1}{2}};l_\mu),\\&v^*(k,\Theta,\Phi)\equiv u(-k_1,-k_2,-k_3,(k^2+\kappa^2)^{\frac{1}{2}};-l_\mu),\\&U(k,\Theta,\Phi)\equiv\exp(ikx+i(k^2+\kappa^2)^{\frac{1}{2}}x_4)\times\exp(i\lambda l^\mu p_\mu/\hbar),\\&U^*(k,\Theta,\Phi)\equiv\exp(-ikx-i(k^2+\kappa^2)^{\frac{1}{2}}x_4)\times\exp(-i\lambda l^\mu p_\mu/\hbar).\end{aligned}\right\} \quad (41)$$

Finally, by expanding u and v^* into series of spherical harmonics, we obtain

$$U=\sum_{k_1k_2k_3}\sum_{l,m}\left(\frac{2\pi}{L}\right)^3\frac{\lambda}{4\kappa(k^2+\kappa^2)^{\frac{1}{2}}}\{u(k,l,m)\times U(k,l,m)+v^*(k,l,m)U^*(k,l,m)\}, \quad (42)$$

where

$$\left.\begin{aligned}&u(k,l,m)\equiv\int\int u(k,\Theta,\Phi)\times\bar{P}_l^m(\Theta,\Phi)\sin\Theta d\Theta d\Phi,\\&v^*(k,l,m)\equiv\int\int v^*(k,\Theta,\Phi)\times P_l^m(\Theta,\Phi)\sin\Theta d\Theta d\Phi,\end{aligned}\right\} \quad (43)$$

$$\left.\begin{aligned}&U(k,l,m)\equiv\int\int U(k,\Theta,\Phi)\times P_l^m(\Theta,\Phi)\sin\Theta\, d\Theta d\Phi\\&U^*(k,l,m)\equiv\int\int U^*(k,\Theta,\Phi)\times\bar{P}_l^m(\Theta,\Phi)\sin\Theta d\Theta d\Phi\end{aligned}\right\} \quad (44)$$

assuming that the spherical harmonics $P_l^m(\Theta, \Phi)$ and their complex conjugate $\bar{P}_l^m(\Theta, \Phi)$ are normalized according to the rule:

$$\int\int\bar{P}_l^m(\Theta,\Phi)P_l^m(\Theta,\Phi)\sin\Theta d\Theta d\Phi_m=1. \quad (45)$$

Similarly, U^* is transformed into the form

$$U^*=\sum_{k_1k_2k_3}\sum_{l,m}\left(\frac{2\pi}{L}\right)^3\frac{\lambda}{4\kappa(k^2+\kappa^2)^{\frac{1}{2}}}\{v(k,l,m)\times U(k,l,m)+u^*(k,l,m)U^*(k,l,m)\}, \quad (46)$$

where

$$\left.\begin{aligned}&u^*(k,l,m)\equiv\int\int u^*(k,\Theta,\Phi)P_l^m(\Theta,\Phi)\times\sin\Theta d\Theta d\Phi,\\&v(k,l,m)\equiv\int\int v(k,\Theta,\Phi)\bar{P}_l^m(\Theta,\Phi)\times\sin\Theta d\Theta d\Phi.\end{aligned}\right\} \quad (47)$$

By the same transformation, we obtain from Eq. (37) the commutation relations

$$\left.\begin{aligned}&[a(k,l,m),a^*[k',l',m')]=\delta(k,k')\delta(l,l')\delta(m,m'),\\&[b(k,l,m),b^*(k',l',m')]=\delta(k,k')\delta(l,l')\delta(m,m'),\\&[a(k,l,m),b(k',l',m')]=0,\ \text{etc.}\end{aligned}\right\}\quad(48)$$

for the operators defined by

$$\left.\begin{aligned}a(k,l,m)&\equiv\left(\left(\frac{2\pi}{L}\right)^3\frac{\lambda}{4\kappa(k^2+\kappa^2)^{\frac{1}{2}}}\right)^{\frac{1}{2}}\cdot u(k,l,m),\\a^*(k,l,m)&\equiv\left(\left(\frac{2\pi}{L}\right)^3\frac{\lambda}{4\kappa(k^2+\kappa^2)^{\frac{1}{2}}}\right)^{\frac{1}{2}}\cdot u^*(k,l,m),\\b(k,l,m)&\equiv\left(\left(\frac{2\pi}{L}\right)^3\frac{\lambda}{4\kappa(k^2+\kappa^2)^{\frac{1}{2}}}\right)^{\frac{1}{2}}\cdot v(k,l,m),\\b^*(k,l,m)&\equiv\left(\left(\frac{2\pi}{L}\right)^3\frac{\lambda}{4\kappa(k^2+\kappa^2)^{\frac{1}{2}}}\right)^{\frac{1}{2}}\cdot v^*(k,l,m).\end{aligned}\right\}\quad(49)$$

Hence, each of the operators defined by

$$\begin{aligned}n^+(k,l,m)&\equiv a^*(k,l,m)a(k,l,m);\\n^-(k,l,m)&\equiv b^*(k,l,m)b(k,l,m)\end{aligned}\quad(50)$$

has eigenvalues 0, 1, 2, $\cdots$ and can be interpreted as the number of particles in the state characterized by the quantum numbers k, l, m with either positive or negative charge. Thus the non-local field above considered corresponds to the assembly of charged bosons with the mass m, the radius λ and the spin 0. It can easily be shown that in the limit $\lambda\to 0$, U reduces to the familiar quantized local field for bosons apart from the extra factor

$$\delta(x_1'-x_1'')\delta(x_2'-x_2'')\delta(x_3'-x_3'')\delta(x_4'-x_4'')\quad(51)$$

which must be omitted, whenever we go over from non-local to local field.

The non-local neutral field can be obtained, if we assume that the field operator U is Hermitian, i.e., $U=U^*$. In this case, we cannot discriminate between u and v, or a and b, so that we have instead of Eqs. (42) and (46) the relation

$$U=\sum_{k_1k_2k_3}\sum_{l,m}\left(\frac{2\pi}{L}\right)^3\frac{\lambda}{4\kappa(k^2+\kappa^2)^{\frac{1}{2}}}\{u(k,l,m)\times U(k,l,m)+u^*(k,l,m)U^*(k,l,m)\}.\quad(52)$$

It should be noticed, further, that we could start from the commutation relations

$$\left.\begin{aligned}&[\bar{u}(k_\mu,l_\mu),\bar{u}^*(k_\mu',l_\mu')]_+\\&\quad=\prod_\mu\delta(k_\mu-k_\mu')\delta(l_\mu-l_\mu')\cdot\delta(k_\mu k^\mu+\kappa^2)\\&\qquad\times\delta(l_\mu l^\mu-\lambda^2)\delta(k_\mu l^\mu),\\&[\bar{u}(k_\mu,l_\mu),\bar{u}(k_\mu',l_\mu')]_+=0,\\&[\bar{u}^*(k_\mu,l_\mu),\bar{u}^*(k_\mu',l_\mu')]_+=0.\end{aligned}\right\}\quad(37)$$

instead of Eq. (37), where

$$[A,B]_+\equiv AB+BA\quad(53)$$

for any two operators A and B. However, in this case, we arrive at the well-known contradiction in the limit of $\lambda\to 0$, which prohibits the elementary particles with spin 0 from obeying Fermi statistics.

IV. NON-LOCAL SPINOR FIELD

The above considerations can easily be extended to the non-local vector field without introducing anything essentially new which needs detailed discussions. On the contrary, the case of the non-local spinor field must be investigated from the beginning. We start from the spinor operator ψ with four components, which transform as the components of Dirac wave function. Each of these components can be considered as a non-local operator just like the operator U in the case of the scalar field. As an extension of Dirac's wave equations for the local spinor field, we assume the relations between the operators x^μ, p_μ and ψ:

$$\gamma^\mu[p_\mu,\psi]+mc\psi=0,\quad(54)$$

$$\beta_\mu[x^\mu,\psi]+\lambda\psi=0,\quad(55)$$

where γ^μ are well-known Dirac matrices forming a four vector, which satisfy the commutation relations among themselves:

$$[\gamma^\mu,\gamma_\nu]_+=-2\delta_{\mu\nu}.\quad(56)$$

We assume similar commutation relations for matrices β_μ:

$$[\beta^\mu,\beta_\mu]_+=2\delta_{\mu\nu}.\quad(57)$$

Then, we obtain by iteration the relations

$$[p^\mu[p_\mu,\psi]]+m^2c^2\psi=0,\quad(58)$$

$$[x_\mu[x^\mu,\psi]]-\lambda^2\psi=0,\quad(59)$$

which have the same form as the relations (4) and (5) for the scalar field. However, the matrices β_μ have to be so chosen as to satisfy the demand that the relations (54) and (55) are compatible with each other. Namely, from the relations

$$\beta_\mu\gamma^\mu[x^\mu[p_\nu,\psi]]=\lambda mc\psi,\quad(60)$$

$$\gamma^\nu\beta_\mu[p_\nu[x^\mu,\psi]]=\lambda mc\psi,\quad(61)$$

which can be readily obtained by considering Eqs. (54) and (55), must have the same form, so that β_μ must satisfy an additional condition:

$$[\beta_\mu,\gamma^\nu][x^\mu[p_\nu,\psi]]=0.\quad(62)$$

This condition reduces to the form

$$[x^\mu[p_\mu,\psi]]=0,\quad(63)$$

which is the same as the condition (12) or (13) for the scalar field, if β_μ are so chosen as to satisfy the commutation relations

$$[\beta_\mu,\gamma^\nu]=C\delta_{\mu\nu},\quad(64)$$

where C is a matrix with the determinant different from zero. Equation (64) can be satisfied by matrices γ^μ, β_μ which are expressed in the form

$$\gamma^1=i\rho_2\sigma_1,\quad \gamma^2=i\rho_2\sigma_2,\quad \gamma^3=i\rho_2\sigma_3,\quad \gamma^4=\rho_3, \tag{65}$$

$$\beta_1=\rho_3\sigma_1,\quad \beta_2=\rho_3\sigma_2,\quad \beta_3=\rho_3\sigma_3,\quad \beta_4=-i\rho_2, \tag{66}$$

in terms of sets of mutually independent Pauli matrices $\sigma_1, \sigma_2, \sigma_3$ and ρ_1, ρ_2, ρ_3. It is well known that the matrices as given by (66) do not form an ordinary vector, but a pseudovector. Thus, if we confine our attention to the proper Lorentz transformation, the relations (54) and (55) are both invariant. However, if we perform the improper Lorentz transformation, for which the determinant of the transformation matrix has the value -1 instead of $+1$, the form of the relation (55) changes into

$$\beta_\mu[x^\mu, \psi]-\lambda\psi=0, \tag{67}$$

whereas the relation (54) is invariant. In other words, the fundamental equations for the non-local spinor field, which has similar properties as the non-local scalar field considered in the preceding sections, can be constructed so as to be invariant with respect to the whole group of Lorentz transformations including reflections, only if both forms (55) and (67) are put together into one relation for one spinor field with the components twice as many as the four components for the usual spinor field. This is equivalent to introduce one more independent set of Pauli matrices ω_1, ω_2, ω_3 and to assume that all of the matrices γ^μ, β_μ have each eight rows and columns characterized by eight combinations of eigenvalues of σ_3, ρ_3, ω_3. Therewith the spinor must have eight components, first four components and the remaining four corresponding respectively to the eigenvalues $+1$ and -1 of ω_3.

In order to establish the invariance of fundamental laws for the non-local spinor field with respect to the whole group of Lorentz transformations, we assume further that ω_2 and ω_3 change sign under improper Lorentz transformation, whereas ω_1 does not. We can now adopt the relation

$$\beta_\mu[x^\mu, \psi]+\omega_3\lambda\psi=0 \tag{68}$$

in place of Eq. (55). It is clear from the above arguments that the fundamental Eqs. (54) and (68) are invariant with respect to the whole group of Lorentz transformations. However, for the purpose of proving it more explicitly, we consider the transformation properties of ψ with respect to the Lorentz transformation, whereby we assume that the matrices γ^μ, β_μ have prescribed forms as defined by Eqs. (65), (66) independent of the coordinate system. In the usual theory, in which the spinor field ψ has four components, we have the linear transformation

$$\psi'=S\psi \tag{69}$$

associated with each of the Lorentz transformations for the coordinates:

$$x_\mu'=a_{\mu\nu}x_\nu, \tag{70}$$

where S is a matrix with four rows and four columns.[12] In our case, in which the spinor ψ has eight components, we assume the same form for S in Eq. (69) except that the numbers of rows and columns are doubled, when Eq. (70) is a proper Lorentz transformation with the determinant $+1$, whereas we have to replace Eq. (69) by

$$\psi'=\omega_1 S\psi, \tag{71}$$

when Eq. (70) is an improper Lorentz transformation with the determinant -1. This guarantees the invariance of the relation (68) with respect to improper as well as proper Lorentz transformations.

However, the above procedure is unsatisfactory, particularly because it is difficult to give a simple physical meaning to the new degree of freedom. As will be shown in the additional remark at the end of this paper, there is an alternative way, in which we have no need to increase the number of components of ψ from 4 to 8.

Now, each component ψ_i $(i=1, 2, 3, 4)$ of the spinor ψ can be represented as a matrix $(x_\mu'|\psi_i|x_\mu'')$ in the representation, in which x_μ are diagonal. $(x_\mu'|\psi_i|x_\mu'')$ can be regarded, in turn, as a function $\psi_i(X_\mu, r_\mu)$ of X_μ, r_μ, where X_μ, r_μ are defined by Eq. (6). Therewith the relations (54) and (68) can be represented by

$$\gamma^\mu(\partial\psi(X_\mu, r_\mu)/\partial X^\mu)+i\kappa\psi(X_\mu, r_\mu)=0, \tag{72}$$

$$\beta_\mu r^\mu\psi(X_\mu, r_\mu)+\lambda\psi(X_\mu, r_\mu)=0, \tag{73}$$

respectively, where $\psi(X_\mu, r_\mu)$ is a spinor with four components $\gamma_i(X_\mu, r_\mu)$ $(i=1, 2, 3, 4)$. The simultaneous Eqs. (72), (73) for $\psi(X_\mu, r_\mu)$ have a particular solution of the form

$$\psi(X_\mu, r_\mu)=\bar{u}(k_\mu, r_\mu)\exp(ik_\mu X^\mu), \tag{74}$$

where $\bar{u}(k_\mu, r_\mu)$ is a spinor with four components satisfying

$$\gamma^\mu k_\mu\bar{u}+\kappa\bar{u}=0,\quad \beta_\mu r^\mu\bar{u}+\lambda\bar{u}=0. \tag{75}$$

It follows immediately from (75) that $\bar{u}$ must satisfy

$$(k_\mu k^\mu+\kappa^2)\bar{u}=0,\quad (r_\mu r^\mu-\lambda^2)\bar{u}=0,\quad k_\mu r^\mu\bar{u}=0 \tag{76}$$

so that $\bar{u}$ can be written in the form

$$\bar{u}=u(k_\mu, r_\mu)\delta(k_\mu k^\mu+\kappa^2)\delta(r_\mu r^\mu-\lambda^2)\delta(k_\mu r^\mu). \tag{77}$$

Each of four components of u can be expanded in the same way as the scalar operator u in the preceding sections. The second quantization can be performed by assuming commutation relations of the type (37) between field quantities, so that the non local field represents an assembly of fermions with the mass m, the radius λ and the spin $\frac{1}{2}$. Further analysis of the non-local spinor field will be made in Part II of this paper. At any rate it is now clear that there exist non-local scalar, vector, and spinor fields, each corresponding to the assembly of particles with the mass, radius, and the spin 0, 1, and $\frac{1}{2}$.

[12] See, for example, W. Pauli, *Handbuch der Physik* 24, Part 1, 83 (1933).

Now the question, with which we are met first, when we go over to the case of two or more non-local fields interacting with each other, is whether we can start from Schrödinger equation for the total system (or any substitute for it), thus retaining the most essential feature of quantum mechanics. We know that Schrödinger equation in its simplest form is not obviously relativistic in that it is a differential equation with the time variable as independent variable, space coordinates being regarded merely as parameters. It can be extended to a relativistic form as in Dirac's many-time formalism or, more satisfactorily, in Tomonaga-Schwinger's super-many-time formalism, as long as we are dealing with local fields satisfying the infinitesimal commutation relations. However, if we introduce the non-local fields or the non-localizability in the interaction between local fields, the clean-cut distinction between space-like and time-like directions is impossible in general. This is because the interaction term in the Lagrangian or Hamiltonian for the system of non-local fields contains the displacement operators in the time-like directions as well as those in the space-like directions. Thus, even if there exists an equation of Schrödinger type, it cannot be solved, in general, by giving the initial condition at a certain time in the past. Under these circumstances, we must have recourse to more general formalism such as the S-matrix scheme, which was proposed by Heisenberg.[13] In other words, we had better start from the integral formalism rather than the differential formalism. In local field theory, the integral formalism such as that, which was developed by Feynman, can be deduced from the ordinary differential formalism.[14,15] In non-local field theory, however, it may well happen that we are left only with some kind of integral formalism. In fact it will be shown in Part II that the non-local fields above considered can be fitted into the S-matrix scheme.

[13] W. Heisenberg, Zeits. f. Physik **120**, 513, 673 (1943); Zeits. f. Naturforsch. **1**, 608 (1946); C. Møller, Kgl. Danske Vid. Sels. Math. Fys. Medd. **23**, Nr. 1 (1945); **22**, Nr. 19 (1946).
[14] R. P. Feynman, Phys. Rev. **76**, 749, 769 (1949).
[15] F. J. Dyson, Phys. Rev. **75**, 486, 1736 (1949). See also many papers by E. C. G. Stueckelberg, which appeared mainly in Helv. Phys. Acta.

This work was done during the author's stay at The Institute for Advanced Study, Princeton. The author is grateful to Professor J. R. Oppenheimer for giving him the opportunity of staying there and also for stimulating discussions. He is also indebted to Dr. A. Pais and Professor G. Uhlenbeck for fruitful conversation.

ADDITIONAL REMARKS ON NON-LOCAL SPINOR FIELD

The problem of invariance of the relation (55) with respect to improper Lorentz transformation can be solved without introducing extra components to the spinor field. Namely, we take advantage of the anti-symmetric tensor of the fourth rank with the components $\epsilon_{\kappa\lambda\mu\nu}$, which are $+1$ or -1 according as $(\kappa, \lambda, \mu, \nu)$ are even or odd permutations of (1, 2, 3, 4) and 0 otherwise. Further we take into account the relations

$$i\beta_\nu = \gamma^\kappa\gamma^\lambda\gamma^\mu, \tag{78}$$

where $(\kappa, \lambda, \mu, \nu)$ are even permutations of (1, 2, 3, 4). Then (55) can be written in the form

$$\tfrac{1}{6}\sum_{\kappa\lambda\mu\nu} \epsilon_{\kappa\lambda\mu\nu}\gamma^\kappa\gamma^\lambda\gamma^\mu[x^\nu, \psi]+i\lambda\psi=0, \tag{79}$$

which is obviously invariant with respect to the whole group of Lorentz transformation. The invariance can be proved more explicitly by associating a linear transformation

$$\psi' = S\psi, \tag{80}$$

with each of the Lorentz transformation (70), where S is a matrix with four rows and columns satisfying the relations

$$S\gamma^\mu S^{-1} = a_{\nu\mu}\gamma^\nu. \tag{81}$$

It should be noticed, however, that the relation (79) is a unification of the relations (55) and (67) rather than the simple reproduction of (55), because (79) must be identified with (67) in the coordinate system, which is connected with the original coordinate system by an improper Lorentz transformation with the determinant -1.

PHYSICAL REVIEW VOLUME 80, NUMBER 6 DECEMBER 15, 1950

Quantum Theory of Non-Local Fields. Part II. Irreducible Fields and their Interaction*

HIDEKI YUKAWA†
Columbia University, New York, New York
(Received August 7, 1950)

General properties of non-local operators are considered in connection with the problem of invariance with respect to the group of inhomogeneous Lorentz transformations. It is shown that irreducible fields can be classified by the eigenvalues of four invariant quantities. Three of these quantities can be interpreted, respectively, as the mass, radius, and magnitude of the internal angular momentum of the particles associated with the quantized non-local field in question. Further, space-time displacement operators are introduced as a particular kind of non-local operator. As a tentative method of dealing with the interaction of non-local fields, an invariant matrix is defined by the space-time integral of a certain invariant operator, which is a sum of products of non-local field operators and displacement operators. It is shown that the matrix thus constructed satisfies the requirements that it be unitary and invariant and that the matrix elements are different from zero only if the initial and final states had the same energy and momentum. However, the remaining conditions of correspondence and convergence cannot be fulfilled simultaneously, in general, by the S-matrix for the non-local fields. It is yet to be investigated whether all of these requirements are satisfied by an appropriate change in the definition of the S-matrix.

I. ELEMENTARY NON-LOCAL SYSTEMS

THE notion of an elementary particle has been intimately connected with the procedure of decomposing a quantized field into its irreducible parts. Accordingly, if the concept of the field itself is so extended as to include the non-local field, the definition of the elementary particle will be altered in its turn. In Part I,[1] we confined our attention to certain types of non-local fields which satisfied a set of operator equations and were supposed to represent assemblies of elementary particles with finite radii. Our problem is now to decompose more general non-local fields into irreducible parts. Again we start from an arbitrary unquantized non-local scalar field U, which can be represented by an arbitrary matrix $(x'|U|x'')$, where x' and x'' stand for x_μ' and x_μ'' $(\mu=1, 2, 3, 4)$, respectively. The matrix $(x'|U|x'')$ can be regarded as a function $U(X, r)$ of two sets of real variables,

$$X_\mu=\tfrac{1}{2}(x_\mu'+x_\mu''),\quad r_\mu=x_\mu'-x_\mu'' \tag{1}$$

as in Part I. Then an arbitrary function $U(X, r)$ can be expanded in the form

$$U(X, r)=\int\cdots\int u(k, r)\exp(ik_\mu X^\mu)(dk_\mu)^4 \tag{2}$$

and further in the form

$$U(X, r)=\int\cdots\int u(k, l)\exp(ik_\mu X^\mu)\times\prod_\mu\delta(r_\mu-l_\mu)(dk_\mu)^4(dl_\mu)^4, \tag{3}$$

where $u(k, r)$ and $u(k, l)$ are arbitrary functions of parameters k, r and k, l, respectively.

Now, if we perform an arbitrary homogeneous Lorentz transformation,

$$x_\mu'=a_{\mu\nu}x_\nu, \tag{4}$$

where x_μ' $(\mu=1, 2, 3, 4)$ denote this time the space-time operators in the new coordinate system. Therewith, two sets of parameters, X and r, are transformed into

$$X_\mu'=a_{\mu\nu}X_\nu,\quad r_\mu'=a_{\mu\nu}r_\nu \tag{5}$$

and $U(X, r)$ becomes

$$U(X', r')=\int\cdots\int u'(k', l')\exp(ik_\mu'X'^\mu)\times\prod_\mu\delta(r_\mu'-l_\mu')(dk_\mu')^4(dl_\mu')^4, \tag{6}$$

where $u'(k', l')=u(k, l)$. k', l' are connected with k, l just as X', r' are connected with X, r. In order that Eq. (6) retain the same form as Eq. (3) for an arbitrary Lorentz transformation (4), either one of the following two requirements must be satisfied:

(i) $u(k, l)$ is a function of k and l, which retains its form under an arbitrary Lorentz transformation;

(ii) $u(k, l)$ is not a mere function of k and l, but is an ensemble of quantities, which are distinguished by the parameters k and l and which are to be subject to second quantization.

In the first case, it is required that

$$u(k', l')=u(k, l) \tag{7}$$

for an arbitrary transformation

$$k_\mu'=a_{\mu\nu}k_\nu,\quad l_\mu'=a_{\mu\nu}l_\nu, \tag{8}$$

so that $u(k, l)$ must be the function of invariant quantities such as $k_\mu k^\mu$, $l_\mu l^\mu$ and $k_\mu l^\mu$ alone. In many cases, however, we can confine our attention to the subgroup of the homogeneous Lorentz group which does not include the reversal of the time, so that $u(k, l)$ may depend also on $k_4/|k_4|$, provided that k_μ is a time-like vector, and similarly for l_μ. Thus $U(X, r)$ can be

* Publication assisted by the Ernest Kempton Adams Fund.
† On leave of absence from Kyoto University, Kyoto, Japan.
[1] H. Yukawa, Phys. Rev. 77, 219 (1950). See also B. Kwal, J. de phys. et rad. 11, 213 (1950).

Reprinted from *Phys. Rev.* 80, 1047 (1950).

written, in general, in the form

$$U(X,r)=\int\cdots\int w(K,L,M)\delta(k_\mu k^\mu-K)\delta(l_\mu l^\mu-L)$$
$$\times\delta(k_\mu l^\mu-M)\exp(ik_\mu X^\mu)\prod_{\mu=1}^{4}\delta(r_\mu-l_\mu)$$
$$\times(dk_\mu)^4(dl_\mu)^4dKdLdM, \quad (9)$$

where $w(K, L, M)$ is an arbitrary function of the real parameters K, L, and M, which can be positive, negative, or zero. If we restrict the transformations to those belonging to the subgroup mentioned above, w may depend also on $k_4/|k_4|$ (and on $l_4/|l_4|$). In any case, the operator U of this type has nothing to do with the quantized non-local field, because there is no room for the application of the method of second quantization. However, an important family of space-time displacement operators belongs to this category as will be shown later on.

The case (ii) is the more important, because the field operator U can be quantized as in the usual theory. Namely, the coefficients $u(k, l)$ can be regarded as an ensemble of creation and annihilation operators for the quanta associated with the field U. The requirement of invariance is fulfilled simply by identifying $u(k, l)$ $[=u'(k', l')]$ with the creation or annihilation operator for a particle in the quantum state characterized by k and l according as k_4 is positive or negative. The only effect of a transformation of the type (4) or (8) is to give a new notation $u'(k', l')$ to the operator $u(k, l)$, owing to the change in name for the same quantum states caused by the change of the reference system. Thus, there is the one-to-one correspondence between $u(k, l)$ and $u'(k', l')$ in two representations, (3) and (6), of the same operator U. Since $k_\mu k^\mu$, $l_\mu l^\mu$, and $k_\mu l^\mu$ are invariant with respect to any Lorentz transformation, the one-to-one correspondence remains, even if the domain of integrations on the right-hand side of Eq. (3) is restricted to definite values K, L, M of these invariant quantities $k_\mu k^\mu$, $l_\mu l^\mu$, and $k_\mu l^\mu$, respectively. In such a case, $U(X, r)$ reduces to

$$U(X,r)=\int\cdots\int u(k,l)\exp(ik_\mu X^\mu)\prod_\mu\delta(r_\mu-l_\mu)$$
$$\times\delta(k_\mu k^\mu-K)\delta(l_\mu l^\mu-L)\delta(k_\mu l^\mu-M)(dk_\mu)^4(dl_\mu)^4. \quad (10)$$

It is now clear that the scalar non-local field, which was dealt with in detail in Part I, is a particular example with

$$K\equiv-\kappa^2,\quad L\equiv+\lambda^2,\quad M\equiv0. \quad (11)$$

More generally, L and M can be either positive or negative including zero, but K can only be negative or zero, because a positive K has no correspondence with the classical model of particles with real mass. Positive values of L correspond to the assembly of elementary particles with a finite dimension which is extended to space-like directions, whereas negative values of L corresponds to that which is extended to time-like directions.

It should be noticed, however, that the field characterized by a set of values of K, L, and M can be decomposed further into irreducible parts, each of which corresponds to a definite value for the absolute magnitude of the internal angular momentum. Namely, as shown in Part I, the non-local scalar field U with a given set of values K, L, M can be expanded in the form

$$U=\sum_{k_1k_2k_3}\sum_{l,m}(2\pi/L)^3[\lambda/4\kappa(k^2+\kappa^2)^{\frac{1}{2}}]$$
$$\times[u(k,l,m)U(k,l,m)+v^*(k,l,m)U^*(k,l,m)], \quad (12)$$

provided that K is negative, L is positive, and M is zero, where u, v^*, U, U^* are defined by the expressions (43) and (44) in Part I. The parameter l in (12) is the quantum number which characterizes the magnitude of the internal angular momentum in the coordinate system moving with the particle with a given wave vector k_1, k_2, k_3. Since l thus defined is invariant with respect to Lorentz transformations, each part of U with a definite value of l transforms into itself and constitutes an irreducible representation of the non-local scalar field. Thus, it is possible that the elementary particles with the integer spins are classified by the value of four constants K, L, M, and l, provided that they are represented by irreducible representations of the non-local scalar field. Among these four constants, $-K$, L, and l can be interpreted, apart from the numerical factors depending on $\hbar$ and c alone, as the mass, radius, and magnitude of the internal angular momentum of the particle, whereas M has no immediate physical meaning.[2]

As was pointed out recently by Fierz,[3] each of these irreducible representations of the non-local scalar field finds its counterpart in the usual field theory of elementary particles with arbitrary integer spin, so far as the behavior with respect to Lorentz transformations is concerned. The essential difference between local and non-local fields will be clear only when the interaction between fields is taken into account. In the case of the non-local spinor field, however, the situation is somewhat different. Namely, a non-local spinor operator ψ_i ($i=1, 2, 3, 4$) is equivalent to a set of four functions $\psi_i(X, r)$, which can be expanded in the form

$$\psi_i(X,r)=\int\cdots\int u_i(k_\mu,l_\mu)\exp(ik_\mu X^\mu)$$
$$\times\prod_\mu\delta(r_\mu-l_\mu)(dk_\mu)^4(dl^\mu)^4. \quad (13)$$

This can be decomposed into parts in an invariant manner by giving each of $k_\mu k^\mu$, $l_\mu l^\mu$, $k_\mu l^\mu$ a definite value. Each part can further be regarded as a sum of operators, which differ from one another by their behaviors with respect to space rotations in the rest

[2] D. Yennie, Phys. Rev., following paper.
[3] M. Fierz, Phys. Rev. 78, 184 (1950); Helv. Phys. Acta 23, 412 (1950).

system. Each of these operators thus obtained is not yet irreducible in general, because it is a mixture of two types of fields belonging to the same resultant (half-integral) spin. For instance, the operator corresponding to the resultant spin 1/2 may have an internal orbital angular momentum of either zero or unity. In the usual local field theory, however, a spinor field with the spin 1/2, for example, is already irreducible. Thus the difference between the non-local spinor field and the local spinor fields with arbitrary half-integral spins is apparent without taking into account the interaction between the fields.[4]

So far we have considered the problem of invariance of non-local operators with respect to homogeneous Lorentz transformations. Now we go over to the more general inhomogeneous Lorentz transformation of the type

$$x_\mu' = a_{\mu\nu}(x_\nu + b_\nu) \tag{14}$$

or

$$x_\mu' = a_{\mu\nu}x_\nu + b_\mu', \tag{15}$$

with $b_\mu' = a_{\mu\nu}b_\nu$, X and r are transformed thereby into

$$X_\mu' = a_{\mu\nu}(X_\nu + b_\nu), \quad r_\mu' = a_{\mu\nu}r_\nu. \tag{16}$$

Accordingly, we have

$$k_\mu' = a_{\mu\nu}k_\nu, \quad l_\mu' = a_{\mu\nu}l_\nu, \tag{17}$$

and

$$u'(k', l') = \exp(-ik_\mu b^\mu)u(k, l), \tag{18}$$

in order that U be invariant with respect to the transformation (14). The implication of the relation (18) must be considered for the cases (i) and (ii) separately.

In case (i), relation (18) is compatible with the assumption that $u(k, l)$ is an invariant function of k and l, only if $u(k, l)$ is zero for all values of k_μ except $k_\mu = 0$ ($\mu = 1, 2, 3, 4$). This is equivalent to the following statement:

(i)′ A non-local operator U which satisfies requirement (i) is invariant with respect to the whole group of inhomogeneous Lorentz transformations only if $U(X, r)$ is an invariant function of r alone.

It will be shown in the next section that some of the invariant operators satisfying the requirement (i)′ will be of importance in constructing the S-matrix for the interacting non-local fields.

In case (ii), relation (18) reflects the situation that the creation or annihilation operator $u(k, l)$ or $u^*(k, l)$ is defined unambiguously except for an arbitrary phase factor. In spite of this ambiguity or complication, the operator $u^*(k, l)u(k, l)/|k_4|$, which is to be identified with the occupation operator for particles in the quantum state characterized by k_μ and l_μ apart from the purely numerical factor, is defined uniquely and is invariant with respect to the whole group of inhomogeneous Lorentz transformations. So are the commutation relations for $u(k, l)$ and $u^*(k, l)$ as given by Eq. (37) of Part I, for example. Of course, it must always be kept in mind that the time-reversal is associated with the interchange of the annihilation operator $u(k, l)$ and the creation operator $u^*(k, l)$.

These arguments can be applied to non-local spinor fields without essential change. In this way we arrive at the following suggestion: according to the non-local field theory it is possible that there are only two kinds of elementary particles, Bose-Einstein particles and Fermi-Dirac particles, which are described by a scalar field and a spinor field, respectively. The customary discrimination of particles with spins 0, 1, 2, etc., among Bose-Einstein particles, for instance, may well be reduced to the difference in the quantum number l for the internal motion of the same kind of particles.

II. S-MATRIX IN NON-LOCAL FIELD THEORY[5]

Now we must undertake the problem of interaction between non-local fields. In the usual field theory we could always start from the Schrödinger equation for the total system. The Hamiltonian in the Schrödinger equation is derived from the Lagrangian which, in turn, is so chosen as to give the correct field equations for unquantized fields, when the classical variation principle was applied to the system consisting of unquantized fields. In the non-local field theory, however, it is difficult to follow the same procedure as in local field theories for two reasons. Firstly, even in the case of the free field, it is difficult to deduce all of the field equations, (4), (5), and (12), for example, for the scalar non-local field from an invariant operator which is supposed to correspond to the Lagrangian in the usual theory. Moreover, the procedure of variation itself is ambiguous.[6] Secondly, it is rather dubious whether the differentiation of the Schrödinger function with respect to time will play an important role in non-local field theory because other operators, in general, are related to two time instants, which differ from each other by a finite amount. Even the existence of the Schrödinger function in the same sense as in the local field theory is not at all certain.

Although it is not yet clear whether these difficulties could be overcome without renouncing the fundamental principles of quantum mechanics, there seems to exist a tentative solution which retains many of the characteristics of the present field theory. Namely, we can start from the so-called interaction representation in the usual theory, laying aside for the moment the question of whether the free field equations in non-local field theory can be deduced from the Lagrangian formalism or not. Furthermore, we can adopt the integral formalism of the usual theory, which has been

[4] Detailed discussions of non-local spinor field will be made elsewhere.

[5] A preliminary account of the subject was published by H. Yukawa, Phys. Rev. 77, 849 (1950).

[6] Variation principles in the non-local field theory were discussed by C. Bloch, Kgl. Danske Vid. Sel. Math.-Fys. Medd. See also C. Gregory, Phys. Rev. 78, 67, 479 (1950).

proved to be equivalent to the differential formalism and in which the S-matrix, instead of the Schrödinger wave function, came in the foreground. Then the S-matrix for local fields can be transformed in the following manner so as to be easily extended to the case of non-local fields. We consider a system of local fields, for which the interaction Hamiltonian density $H'(x, y, z, t)$ is invariant and is equal to $-L'(x, y, z, t)$, where L' is the interaction part in the Lagrangian density for the system. In the usual one-time formalism, the Schrödinger equation has the form

$$i\hbar\partial\Psi(n', t)/\partial t = \sum_{n''}(n'|\bar{H}'(t)|n'')\Psi(n'', t), \quad (19)$$

where each of n' and n'' stands for a set of eigenvalues of occupation operators of various types of particles in the system in various quantum states. $\bar{H}'(t) = -\bar{L}'(t)$ is the space integral of the Hamiltonian density $H'(x, y, z, t)$ or $-L'(x, y, z, t)$. The differential equation (19) can be integrated with respect to time, at least formally, by the method of successive approximation and we obtain

$$\Psi(n', +\infty) = \Psi(n', -\infty) + (i/\hbar)\int_{-\infty}^{+\infty}\sum_{n''}(n'|\bar{L}'(t)|n'')dt\cdot\Psi(n'', -\infty) + (i/\hbar)^2\int\int_{t>t'}\sum_{n'', n'''}(n'|\bar{L}'(t)|n'')(n''|\bar{L}'(t')|n''') \times dtdt'\Psi(n''', -\infty) + \cdots, \quad (20)$$

where $\Psi(n', +\infty)$ and $\Psi(n', -\infty)$ are Schrödinger wave functions in the infinite future and infinite past, respectively. Thus the S-matrix for this case is given by

$$(n'|S|n'') = (n'|1|n'') + (i/\hbar)\int(n'|\bar{L}'(t)|n'')dt + (i/\hbar)^2\int\int_{t>t'}\sum_{n'''}(n'|\bar{L}'(t)|n''')(n'''|\bar{L}'(t')|n'') \times dtdt' + \cdots. \quad (21)$$

In order to generalize this expression for the S-matrix to the case of the system of non-local fields, we introduce an invariant Hermitian non-local operator L' which is represented by a matrix $(n', x'|L'|n'', x'')$ and which reduces to

$$(n', x'|L'|n'', x'') = (n'|L'(x')|n'')\prod_{\mu}\delta(x'^{\mu} - x''^{\mu}) \quad (22)$$

in the limiting case of the system of local fields, where each of x' and x'' stands for a set of eigenvalues of space-time operators $x^1 = x$, $x^2 = y$, $x^3 = z$, $x^4 = ct$. With the help of Eq. (22) and of another non-local operator ϵ which is represented by the matrix

$$(n', x'|\epsilon|n'', x'') = \tfrac{1}{2}\{[(x'^4 - x''^4)/|x'^4 - x''^4|] + 1\}(n'|1|n^4). \quad (23)$$

Equation (21) can be written in the form

$$(n'|S|n'') = (n'|1|n'') + (i/\hbar c)\int\int(n', x'|L'|n'', x'')(dx')^4(dx'')^4 + (i/\hbar c)^2\int\cdots\int\sum_{n''', n^{\mathrm{IV}}}(n', x'|L'|n''', x''') \times(n''', x'''|\epsilon|n^{\mathrm{IV}}, x^{\mathrm{IV}})(n^{\mathrm{IV}}, x^{\mathrm{IV}}|L'|n'', x'') \times(dx')^4(dx'')^4(dx''')^4(dx^{\mathrm{IV}})^4 + \cdots. \quad (24)$$

If we define $\{A\}$ for an arbitrary non-local operator A by

$$(n'|\{A\}|n'') = \int\cdots\int(n', x'|A|n'', x'')(dx')^4(dx'')^4, \quad (25)$$

the S-matrix with matrix elements as given by Eq. (24) can be written symbolically in the form

$$S = 1 + (i/\hbar c)\{L'\} + (i/\hbar c)^2\{L'\epsilon L'\} + (i/\hbar c)^3\{L'\epsilon L'\epsilon L'\} + \cdots. \quad (26)$$

This could be used as the definition of the S-matrix in non-local field theory as well as in local field theory. Alternatively, we can define S or $S-1$ by

$$S - 1 = \{R\}, \quad (27)$$

where

$$R = (i/\hbar c)L' + (i/\hbar c)^2 L'\epsilon L' + (i/\hbar c)^3 L'\epsilon L'\epsilon L' + \cdots. \quad (28)$$

Incidentally, the non-local operator R satisfies a linear operator equation

$$R = (i/\hbar c)L' + (i/\hbar c)L'\epsilon R. \quad (29)$$

The physical interpretation of the S-matrix remains the same as in the usual theory in spite of the fact that the S-matrix is non-local field theory is defined directly by Eq. (26) or by Eqs. (27) and (28) without recourse to the Schrödinger equation of type (19). Thus, $|(n'|S|n'')|^2$ is the probability that the system will be in the state characterized by n' in the infinite future provided that it was in the state characterized by n'' in the infinite past. In fact, the S-matrix as defined by (26) satisfies two conditions:

(i) S is a unitary matrix which satisfies the relation

$$S^*S = SS^* = 1. \quad (30)$$

(ii) The matrix element $(n'|S|n'')$ is different from zero only if the states characterized by n' and n'', respectively, have the same total energy and momentum.

Before going into the proof of these statements, we have to take into account the third condition:

(iii) S must be an invariant matrix.

In local field theories, the S-matrix defined above is

invariant, in spite of the fact that the operator ϵ as defined by Eq. (23) is not invariant with respect to Lorentz transformations. This is due to the fact that the Hamiltonian density $H'(x')$ at a point x' is commutative with the density $H'(x'')$ at any other point x'', which is located in a space-like direction with respect to x'. It is not so, in general, in non-local field theory. An obvious way of guaranteeing the invariance of the S-matrix in such a case is to replace the operator ϵ in Eq. (26) by a suitable invariant non-local operator D_+ such that conditions (i) and (ii) are still fulfilled. Thus the S-matrix for the system of non-local fields takes the form

$$S=1+(i/\hbar c)\{L'\}+(i/\hbar c)^2\{L'D_+L'\}+(i/\hbar c)^3\{L'D_+L'D_+L'\}+\cdots. \quad (31)$$

The actual form of the operator D_+ can be determined in the following manner. If we assume that the invariant operator L' is a sum of products of non-local field operators, condition (ii) is satisfied for any displacement operator D_+ whose matrix element $(x'|D_+|x'')\equiv D_+(X, r)$ is an invariant function of r_μ alone. The proof is simple. Any non-local operator A can be represented by a matrix $(x'|A|x'')$ or a function $A(X, r)$ and Eq. (25) can be written alternatively in the form

$$(n'|\{A\}|n'')=\int\int(n'|A(X,r)|n'')(dX)^4(dr)^4. \quad (32)$$

If the operator A consists of a sum of products of L' and D_+, $(n'|A(X,r)|n'')$ can be expanded into a series with the typical term

$$(n'|a(k_\mu^{(i)}, r_\mu)|n'')\exp(iK_\mu X^\mu), \quad (33)$$

where

$$K_\mu=\sum_i n''^{(i)}k_\mu^{(i)}-\sum_i n'^{(i)}k_\mu^{(i)}. \quad (34)$$

Evidently $\hbar K$ is the difference in momenta between the initial state n'' and the final state n' and $-\hbar cK_4$ is the difference in energies of the states n'' and n'. If we insert Eq. (33) into Eq. (32) and integrate with respect to X, we find that each term of $(n'|\{A\}|n'')$ contains a factor $\prod_\mu \delta(K_\mu'-K_\mu'')$, so that $(n'|\{A\}|n'')$ is different from zero only if the states n' and n'' have the same energy and momentum. It should be noticed, however, that we mean by the energy and momentum of a particle the energy and momentum of its center of mass. Thus the energy of internal motion is supposed to be included already in the mass $\hbar\kappa/c$. In other words, κ must be, in general, a function of other constants such as λ and l. The problem of determining the form of such a function is still completely open.

The condition (i) is also fulfilled, if we further imply the condition

$$D_++D_+^*=E \quad (35)$$

on D_+, where D_+^* is the Hermitian conjugate of D_+ and E is an invariant displacement operator with the matrix element

$$(n', x'|E|n'', x'')=(n'|1|n'') \quad (36)$$

for any values of x' and x''. In order to prove this, we have only to multiply S as given by Eq. (31) by

$$S^*=1-(i/\hbar c)\{L'\}+(i/\hbar c)^2\{L'D_+^*L'\}-(i/\hbar c)^3\{L'D_+^*L'D_+^*L'\}+\cdots. \quad (37)$$

Then the condition of unitarity

$$\sum_{n'''}(n'|S^*|n''')(n'''|S|n'')=\sum_{n'''}(n'|S|n''')(n'''|S^*|n'')=(n'|1|n'') \quad (38)$$

comes out by the help of Eq. (32) and the relation

$$\{AEB\}=\{A\}\{B\}, \quad (39)$$

which holds for any two non-local operators A and B.

The operators D_+^* and D_+ which satisfy all of these conditions are given by matrices

$$(n', x'|D_+|n'', x'')=(n'|1|n''),\ \tfrac{1}{2}(n'|1|n''),\text{ or } 0;$$
$$(n', x'|D_+^*|n'', x'')=0,\ \tfrac{1}{2}(n'|1|n''),\text{ or } (n'|1|n''), \quad (40)$$

according as $x'-x''$ is future-like, space-like, or past-like.

This modification of the definition of S-matrix gives rise to the new question: does it reduce to the usual definition (21) in the limit of local fields? This question is very intimately connected with another, and probably the most important, question: is the S-matrix for non-local fields convergent? In order to answer these questions, we begin with the investigation of the particular matrix element $(0|S|0)$ of $(n'|S|n'')$, where both the initial state n'' and the final state n' are complete vacua; i.e., all eigenvalues n' and n'' are zero. Now $(0|S|0)$ has the general form

$$(0|S|0)=1+(i/\hbar c)(0|\{L'\}|0)+(i/\hbar c)^2(0|\{L'D_+L'\}|0)+\cdots. \quad (41)$$

Let us consider a very simple case of a system consisting of a complex non-local scalar field V, V^* and a real non-local scalar field U with the interaction of the form

$$L'=gV^*UV. \quad (42)$$

We have first

$$(0|\{L'\}|0)=0 \quad (43)$$

because L' is linear in U and hence has no term which connects the state 0 with itself. As for the third term in Eq. (38), we have the relation

$$(0|\{L'D_+L'\}|0)=\tfrac{1}{2}\sum_{n'}(0|\{L'\}|n')(n'|\{L'\}|0)+\tfrac{1}{2}(0|\{L'DL'\}|0) \quad (44)$$

on account of relations (32) and (36), where the operator D is defined by

$$D=D_+-D_+^* \quad (45)$$

with the matrix element

$$(n', x'|D|n'', x'')=(n'|1|n''),\quad 0,\quad -(n'|1|n''), \quad (46)$$

according as $x'-x''$ is future-like, space-like, or past-like.[7] The first term on the right-hand side of (44)

[7] This operator was introduced by Koba independently. See Z. Koba, Prog. Theor. Phys. 5, 139 (1950).

vanishes on account of the fact that $(n'|\{L'\}|0)$ is zero provided that $\kappa_U < 2\kappa_V$ and the second term also vanishes for the following reason: first we expand U, V, V^*, and D in Fourier series and integrate each of the terms of $(0|\{L'DL'\}|0)$ with respect to all of the space-time parameters. Actually we have eight sets of such parameters. Then we are left with the expression of the form

$$\int\int\int f(k_\mu^{(1)}, k_\mu^{(2)}, k_\mu^{(3)})\delta'(K_\mu K^\mu) \times (dk_\mu^{(1)})^4(dk_\mu^{(2)})^4(dk_\mu^{(3)})^4, \quad (47)$$

where $K_\mu = \sum_i k_\mu^{(i)}$ and $k_\mu^{(1)}$, $k_\mu^{(2)}$, $k_\mu^{(3)}$ are the wave vectors of the three particles created in the intermediate state. The first of them is a particle of U-type and the other two are particles of $V-V^*$-type. $f(\cdots)$ is a function of $k_\mu^{(1)}$, $k_\mu^{(2)}$, $k_\mu^{(3)}$, which could be determined by elementary calculations, but it is not necessary for our purpose to write it explicitly. δ' denotes the derivative of the δ-function with respect to the argument, which comes from the Fourier transform of the operator D, as discussed in detail by Yennie.[2] Thus, $(0|\{L'DL'\}|0)$ must be zero, unless the condition

$$K_\mu K^\mu = 0 \quad (48)$$

is fulfilled. The condition (48) can be satisfied by certain sets of $k_\mu^{(1)}$, $k_\mu^{(2)}$, $k_\mu^{(3)}$ only if both types of particles have the rest mass zero.

The above arguments can be applied to local fields as well as to non-local fields. According to the usual theory of local fields, the third term on the right-hand side of Eq. (41) must be the divergent self-energy of the vacuum, whereas it is actually zero according to our formalism, except for the very particular case of particles both with the rest mass zero. The same argument can be applied to the case of charged particles interacting with the electromagnetic field, and according to our formalism the self-energy of the vacuum is zero, at least up to the second order, if we assume that there is no charged particle with the rest mass zero. Thus, the discrepancy between our formalism and the usual theory is already clear; they give different answers to the same problem for local fields.

Next we consider the matrix element $(1|S|1)$ of S, where only one particle of the same type in the same state exists in the initial and final states. The second-order term of $(1|S|1)$ corresponds to the divergent self-energy of the particle in local field theory. As discussed by Yennie in detail,[2] if we start from a system of two non-local scalar fields of U-type and $V-V^*$-type with the interaction operator L' as given by Eq. (42), the self-energy term is again divergent. However, the fields U and $V-V^*$ can be decomposed further into positive frequency and negative frequency parts without destroying the invariance with respect to the subgroup of Lorentz transformation, in which the direction of time is not reversed. Namely, we can write

$$U = U_+ + U_-, \quad V = V_+ + V_-, \quad V^* = V_+^* + V_-^*, \quad (49)$$

where U_+, U_- are positive and negative frequency parts of U, while V_+, V_+^* and V_-, V_-^* are corresponding parts of V and V^*. If we take the new interaction operator

$$L' = g\{V_+^*UV_+ + V_-UV_-^* + V_+^*UV_- + V_-^*UV_+\} \quad (50)$$

instead of Eq. (39), the self-energy terms for the U-type particle as well as the $V-V^*$-type particles are convergent, although there still remains an undesirable feature, as discussed by Yennie.[2]

Now, in order to remove the discrepancy between the present formalism and the usual formalism in the limit of local fields, we may imagine that D-operator above defined is a limit of the operator with the matrix element, which is a function of r_μ and is different from zero in a narrow region outside the light cone in r-space. Then the correspondence between the present formalism and the usual formalism in the limit of local fields is restored up to the second order, but the essential difference between ϵ- and D-operators remains in the third- and higher order terms. Moreover, the divergences reappear in the case of non-local fields. It is very difficult to construct an S-matrix which is convergent and which reduces to the usual S-matrix in the limit of local fields. It is not yet clear whether the S-matrix formalism itself is not adequate for dealing with the problem of interaction of non-local fields. It might be possible that the S-matrix as defined by Eq. (24) is invariant, if the interaction operator L' has an appropriate form, even in the case of non-local fields. However, it is more probable that the clean-cut separation of the free fields from their interaction is justified only if we are dealing with the weak coupling between local fields. If so, we must go back in search of the Lagrangian formalism for the whole system of non-local fields interacting with one another. In any case, the compatibility conditions for the field equations or the integrability conditions for any substitute for the Schrödinger equation will be of fundamental importance.

In this connection it should be noticed that so far we have not been able to find any relation between the mass and other constants. It is clear that a relation which connects the mass of an elementary particle with other constants such as the radius, the internal angular momentum, and the constants of coupling with other particles will be of vital importance in any future theory of elementary particles. Again this is closely related to the problem of finding the Lagrangian operator for the whole system or any substitute for it.

The author wishes to express his appreciation to Columbia University, where this work was done, for the hospitality shown to him, and to the Rockefeller Foundation for financial support. He is indebted also to Mr. Yennie for his useful criticism and elucidation of some of the important consequences of the formalism.

Structure and Mass Spectrum of Elementary particles. I. General Considerations

Hideki Yukawa*
Columbia University, New York, New York
(Received May 25, 1953)

As discussed in previous papers,[1] the nonlocal field was introduced in order to describe relativistically a system which was elementary in the sense that it could no longer be decomposed into more elementary constituents, but was so substantial, nevertheless, as to be able to contain implicitly a great variety of particles with different masses, spins, and other intrinsic properties. However, the conclusions reached so far were very unsatisfactory in many respects.[2] Among other things, the masses of the particles associated with the irreducible nonlocal fields remained completely arbitrary and simple and plausible assumptions concerning the interaction between fields did not result in the expected convergence of self-energies. It seems to the author that these disappointing consequences are not inherent in nonlocal field theory, in general, but are rather related to the particular type of field to which the author restricted himself. Instead, if we start anew from less restricted nonlocal fields, a more promising aspect of possible nonlocal theories is revealed, as shown in the following.

Let us take a scaler (or pseudoscalar) nonlocal field,

$$(x_\mu'|\phi|x_\mu') \equiv \phi(X_\mu, r_\mu),$$

where x_μ', x_μ'' ($\mu = 1, 2, 3, 4$) stand for two sets of space-time parameters and

$$X_\mu = (x_\mu' + x_\mu'')/2, \qquad r_\mu = x_\mu' - x_\mu''.$$

The free field equation is supposed to have a general form

$$F(\partial/\partial X_\mu, r_\mu, \partial/\partial r_\mu)\phi(X_\mu, r_\mu), \tag{1}$$

where the operator F is a certain invariant function of $\partial/\partial X_\mu$, r_μ, and $\partial/\partial r_\mu$ and is independent of X_μ so that it is invariant under any inhomogeneous Lorentz transformation. In particular, if we assume that F is linear in $\partial^2/\partial X_\mu \partial\lambda_\mu$ and separable, i.e.,

$$F \equiv \frac{\partial^2}{\partial X_\mu \partial X_\mu} + F^{(r)}\left[r_\mu r_\mu, \frac{\partial^2}{\partial r_\mu \partial r_\mu}, r_\mu \frac{\partial}{\partial r_\mu}\right], \tag{2}$$

we have eigensolutions of the form $\phi \equiv u(X)\,\chi(r)$, where u and χ satisfy

Reprinted from *Phys. Rev.* **91**, 415 (1953).

$$(\partial^2 / \partial X_\mu \partial X_\mu - \mu)\, u\,(X) = 0, \tag{3}$$

$$(F^{(r)} - \mu)\, X\,(r) = 0, \tag{4}$$

μ being the separation constant. Thus, the masses of the free particles associated with the nonlocal field ϕ are given as the eigenvalues of $\mu^{1/2}$ in Eq. (4) for the internal eigenfunction χ. If one chooses the operator $F^{(r)}$ such that the eigenvalues $\mu_n \equiv m_n^2$ are all positive and discrete, one can expand an arbitrary nonlocal field ϕ into a series of internal eigenfunctions, $\chi_n(r)$:

$$\phi(X\,,r) = \sum_n u_n(X)\chi_n(r). \tag{5}$$

Now, the field equations for a scalar nonlocal field $(x'|\phi|x'')$ interacting with a local spinor field $\psi(x')$, for instance, can be deduced from an appropriate Lagrangian and are

$$\left\{-\frac{\partial^2}{\partial X_\mu \partial X_\mu} + F^{(r)}\right\}\phi(X,r) = -g\sum_\alpha \overline{\psi}_\alpha(X + {}^1\!/\!_2 r)\psi_\alpha(X - {}^1\!/\!_2 r), \tag{6}$$

$$\gamma_\mu \frac{\partial \psi(x')}{\partial x_\mu{}'} + M\psi(x') = -g\int \psi(x'')(x''|\phi|x')dx''. \tag{7}$$

We insert (5) in (6), multiply both sides by the complex conjugate $\chi_n^*(r)$, and integrate over the four-dimensional space of $r_1, r_2, r_3, r_0 = -ir_4$. The result is

$$\left[\frac{\partial^2}{\partial x_m{}'' \partial x_\mu{}''} - m_n^2\right] u_n(x'') = \int \phi_n(x', x'', x''') \sum{}_\alpha \overline{\psi}_\alpha(x')\psi_\alpha(x''')dx'dx''', \tag{8}$$

where

$$\phi_n(x', x'', x''') \equiv g\chi_n^*(x' - x''')\delta({}^1\!/\!_2(x' + x''') - x''). \tag{9}$$

Similarly, we obtain from (7) the equation

$$\gamma_\mu \frac{\partial \psi(x')}{\partial x_\mu{}'} + M\psi\,(x') = -\sum_n \int \phi_n(x', x'', x''')\, u_n\,(x'')\psi(x''')dx''dx'''. \tag{10}$$

If we compare these equations with the corresponding equations (19) of Møller and Kristensen[3] in the theory of nonlocal interaction between a local scalar (or pseudoscalar) field and a local spinor field, we notice that the internal eigenfunction $\chi_n(r)$ plays the role of a convergence factor. There is, however, an essential difference between their equations and ours. Namely, in our theory, we are obliged to take into account simultaneously all the particles with different masses m_n which were derived from an eigenvalue problem. Furthermore, the form function for each of these particles is uniquely determined by the same eigenvalue problem.

In the following letter, the above general considerations will be illustrated and further details will be examined.

[*]Now at Kyoto University, Kyoto, Japan, on leave of absence from Columbia University (July, 1953).

[1]H. Yukawa, Phys. Rev. 77, 219 (1950); 80, 1047 (1950).

[2]D. R. Yennie, Phys. Rev. 80, 1053 (1950); J. Rayski, Acta. Phys. Polonica 10, 103 (1950); Proc.

Phys. Soc. (London) A64, 957 (1951); M. Fierz, Helv. Phys. Acta 23, 412 (1950); Z. Tokuoka and Y. Katayama, Progr. Theoret. Phys. 6, 132 (1951); C. Bloch, Kgl. Danske Videnskab. Selskab, Mat.-fys. Medd. 24, No. 1 (1950); Progr. Theoret, Phys. 5, 606 (1950); O. Hara and H. Shimazu, Prog. Theoret. Phys. 5, 1055 (1950); 7, 255 (1952); 9, 137 (1953).

[3]P. Kristensen and C. Møller, Kgl. Danske Videnskab. Selkab, Mat.-fys. Medd. 27, No. 7 (1952); C. Bloch, Kgl. Danske Videnskab. Seiskab, Mat.-fys. Medd. 27, No. 8 (1952). Y. Katayama, Progr. Theoret. Phys. 8, 381 (1952).

Structure and Mass Spectrum of Elementary Particles II. Oscillator Model

Hideki Yukawa*
Columbia University, New York, New York
(Received May 25, 1953)

As an illustration of the general considerations on nonlocal fields in the preceding letter, let us assume that the operator F has a very simple form

$$F \equiv -\frac{\partial^2}{\partial X_\mu \partial X_\mu} + \frac{\lambda^2}{2}\left[-\frac{\partial^2}{\partial r_\mu \partial r_\mu} + \frac{1}{\lambda^4} r_\mu r_\mu\right]^2, \quad (1)$$

where λ is a small constant with the dimension of length. One may call this the four-dimensional oscillator model for the elementary particle, which was considered first by Born[1] in connection with his idea of a self-reciprocity. However, our model differs from his model in that we have introduced internal degrees of freedom of the particles which are related to the nonlocalizability of the field itself. The internal eigenfunctions in our case are

$$X n_1 n_2 n_3 n_0 (r) = H_{n_1}(r_1/\lambda) H_{n_2}(r_2/\lambda) H_{n_3}(r_3/\lambda) H_{n_0}(r_0/\lambda) \times \exp\{-(r_1^2 + r_2^2 + r_3^2 + r_0^2)/2\lambda^2\} \quad (2)$$

and the corresponding eigenvalues for the mass become

$$m n_1 n_2 n_3 n_0 = (\sqrt{2}/\lambda \mid n_1 + n_2 + n_3 - n_0 + 1 \mid, \quad (3)$$

where $r_0 = ir_4$ is a real variable and n_1, n_2, n_3, n_0 are quantum numbers which can take only zero or positive integer values. $H_n(x)$ denotes the Hermite polynomial of x of degree n. All these eigenfunctions (2) decrease rapidly in any direction whatsoever in the four-dimensional r space. Furthermore, the Fourier transform of each of these eigenfunctions has exactly the same form as the original function due to the self-reciprocity. Thus, the form function (9) in the preceding letter seems to be sufficient to cut off high energy-momentum intermediate states in such a way that each term corresponding to each Feynman diagram in the expansion of the nonlocal S-matrix according to the Bloch-Kristensen-Møller formulation is convergent. However, since we have to take into account all of infinitely many of different mass states of the nonlocal system, the number of terms in the S matrix increases very rapidly with the increasing power of the coupling constant, so that we can claim nothing for the moment concerning the convergence or divergence of the S matrix as

Reprinted from *Phys. Rev.* **91**, 416 (1953).

a whole.

The totality of the internal eigenfunctions (2) constitutes a complete set of orthogonal and quadratically integrable functions in the four-dimensional r space and can be regarded as the eigenvectors for an infinite-dimensional unitary representation of the Lorentz group. The eigenvalues (3) for the mass are all infinitely degenerate. For instance, all those values of n's which satisfy $n_1 + n_2 + n_3 - n_0 = 0$ give the same mass, $m_0 = \sqrt{2}/\lambda$. This is not a peculiar feature of the oscillator model; it is common to all those models for which the operator F is separable, because there can be no unitary representation of finite dimensions for the Lorentz group. Presumably, such an undesired degeneracy could be removed either by introducing interaction with other fields or by first introducing the coupling between the external and internal degrees of freedom. The latter possibility can be illustrated by the addition of the coupling term,

$$-\beta^2\lambda^2\left\{-\left[\frac{\partial^2}{\partial X_\mu \partial X_\mu}\right]^2 + \frac{1}{\lambda^4}\left[r_\mu \frac{\partial}{\partial X_\mu}\right]^2\right\}, \tag{4}$$

to the expression (1) for F, where β is a dimensionless real constant. The free field equation becomes

$$\left[k_\mu k_\mu + \frac{\lambda^2}{2}\left[-\frac{\partial^2}{\partial r_\mu \partial r_\mu} + \frac{1}{\lambda^4} r_\mu r_\mu\right]^2 + \beta^2\lambda^2\left\{-\left[k_\mu \frac{\partial}{\partial r_\mu}\right]^2 + \frac{1}{\lambda^4}\left[k_\mu r_\mu\right]^2\right\}\right]\chi(k_\mu, r_\mu) = 0, \tag{5}$$

in the eight-dimensional space k_μ and r_μ, where $x\ (k_\mu, r_\mu)$ is the Fourier transform of $\phi\ (X_\mu, r_\mu)$ as defined by

$$\phi\ (X_\mu, r_\mu) = \int \exp\ (ik_\mu X_\mu)\ \chi\ (k_\mu, r_\mu)\ (dk_\mu)^4. \tag{6}$$

One can solve Eq. (5) in the coordinate system in which only one component of the wave vector is different from zero.[2] Thus, one obtains the mass spectrum

$$m_{n_1 n_2 n_3 n_0} = \frac{\sqrt{2}\ |n_1 + n_2 + n_3 - n_0 + 1|}{\lambda\ \left[1 - 2\beta^2(n_0 + ½)\right]^{½}}, \tag{7}$$

where n_0 is restricted by the condition

$$n_0 < ½(1/\beta^2 - 1). \tag{8}$$

If we take, for instance, $\beta = 1/\sqrt{2}$, only $n_0 = 0$ is allowed and the mass spectrum reduces to

$$m_{n_1 n_2 n_3 n_0} = (2/\lambda)\ n_1 + n_2 + n_3 + 1), \tag{9}$$

and the degree of degeneracy of the mass eigenvalues is now finite. In particular, the lowest mass, $m_0 = 2/\lambda$, is free from degeneracy and the corresponding solution of (5) is given by

$$\chi_{0000}(k_\mu, r_\mu) = \exp\left\{ -\frac{1}{2\lambda^2}\left[r_\mu r_\mu + \frac{2(k_\mu r_\mu)^2}{m_0^2} \right] \right\} \tag{10}$$

in an arbitrary coordinate system, where $k_\mu k_\mu = -m_0^2$.

The above advantage of introducing the coupling between external and internal degrees of freedom is offset, however, by a complication which is almost prohibitive if we further take into account the interaction with other fields, because the general method of reducing the theory of nonlocal fields to that of the nonlocal interaction between local fields as discussed in the preceding letter can no longer be applied straightforwardly to our case. On the other hand, it may well be that one could arrive at the desired removal of the infinite degeneracy as a consequence of the interaction between nonlocal fields without assuming the coupling between external and internal degrees of freedom for each of the nonlocal fields. This is plausible, because the submatrix of the S matrix corresponding to one-particle states can always be represented by an equivalent coupling between the external and internal variables for the particle in question, so that one can hope that a reasonable mass spectrum which is free from the infinite degeneracy may come out even without assuming the coupling between external and internal degrees of freedom at the outset.

A detailed account of all these points, including the quantization of nonlocal fields, will be given in a forthcoming paper.

[*] Now at Kyoto University, Kyoto, Japan, on leave of absence from Columbia University (July, 1953).

[1] M. Born and H. S. Green, Proc. Roy. Soc. Edinburgh 62, 470 (1949); M. Born, Revs. Modern Phys. 21, 463 (1949).

[2] Equation (5) has no solution which is quadratically integral unless k is time-like, i.e., $k_\mu k_m < 0$.

PHYSICAL REVIEW VOLUME 96, NUMBER 4 NOVEMBER 15, 1954

Properties of Bethe-Salpeter Wave Functions

G. C. WICK
Carnegie Institute of Technology, Pittsburgh, Pennsylvania
(Received June 30, 1954)

A boundary condition at $t=\pm\infty$ (t being the "relative" time variable) is obtained for the four-dimensional wave function of a two-body system in a bound state. It is shown that this condition implies that the wave function can be continued analytically to complex values of the "relative time" variable; similarly the wave function in momentum space can be continued analytically to complex values of the "relative energy" variable p_0. In particular one is allowed to consider the wave function for purely imaginary values of t, or respectively p_0, i.e., for real values of $x_4=ict$ and $p_4=ip_0$. A wave equation satisfied by this function is obtained by rotation of the integration path in the complex plane of the variable p_0, and it is further shown that the formulation of the eigenvalue problem in terms of this equation presents several advantages in that many of the ordinary mathematical methods become available.

In an especially simple case ("ladder approximation" equation for two spinless particles bound by a scalar field of zero rest mass) an integral representation method is presented which allows one to reduce the problem exactly (and for arbitrary values of the total energy of the bound state) to an eigenvalue problem of the Sturm-Liouville type. A complete set of solutions for this problem is obtained in the subsequent paper by Cutkosky.

1. INTRODUCTION

THE formulation of a completely relativistic wave equation for two-body systems[1] has, in a certain sense, solved a long-standing problem of quantum mechanics. The natural and simple way in which relativistic invariance is achieved is, of course, very real progress, which may lead one to hope that the main features of the equation are more permanent than the solidity of its present field theoretic foundation might suggest. Furthermore, it is hardly necessary to recall that the usefulness of the equation has been amply demonstrated in several high-precision calculations of energy levels.[2]

Nevertheless, it is generally recognized that several serious and valid doubts remain about the significance and the self-consistency of the equation. Some of these doubts, of course, stem from the remaining unresolved convergence questions of renormalized quantum electrodynamics (and other similar theories). It goes without saying, however, that these deeper questions lie entirely beyond the scope of the present investigation[3] The questions and doubts we shall be concerned with[4] arise at a less formidable level; they have to do with the several unfamiliar features of the equation itself.

These are (and the list is probably incomplete):

(a) The appearance of a relative time (or respectively a relative energy) variable, the physical role of which is not entirely clear; in particular, it is admitted that the boundary conditions on the wave function for infinite values of the relative time have not been adequately formulated.

(b) The presence of strong singularities in the interaction kernel, to be avoided by special prescriptions. Standard mathematics has practically nothing to say about integral equations of this type. In particular, the prescriptions referred to imply properties of analyticity, about which one would like to know a lot more.

(c) The absence of a positive-definite norm for the wave function and of any orthogonality theorem.

(d) The fact that when the coupling constant λ is set equal to zero, the equation admits obviously improper solutions. Notwithstanding all that can be said about it, this feature is a little disturbing. It is connected to the other feature that the "order" of the differential operator in the equation is higher than that of the corresponding one-body equation. This leads to the expectation that the equation may have "too many" solutions. On the other hand, circumstance (b) has led some authors to suspect that there are no solutions at all!

(e) Finally, as explained by Goldstein,[4] we are faced with the paradoxical circumstance that, owing to the nonrelativistic perturbation approach employed, the highly successful numerical results obtained do not really offer any direct clue as to the actual properties of the relativistic equation.

The investigation described in the following pages was aimed at throwing some light on these questions. It really consists of two quite different lines of attack. The first of these starts from the remark (Sec. 2) that an additional condition for the Bethe-Salpeter (B-S) wave function follows from its definition[5] supplemented by simple stability requirements. From this, then, some unexpected consequences can be derived about

[1] E. E. Salpeter and H. A. Bethe, Phys. Rev. **84**, 1232 (1951); J. Schwinger, Proc. Natl. Acad. Sci. **37**, 455 (1951). Other closely related but more general relativistic schemes recently developed by various authors will not be discussed here.

[2] E. E. Salpeter, Phys. Rev. **87**, 328 (1952); R. Karplus and A. Klein, Phys. Rev. **87**, 848 (1952).

[3] In particular, expressions such as "the general structure" of the equation, "the analytic properties" of the interaction kernel, etc., will be used on the assumption that such properties may be inferred correctly from truncated expressions of finite order in the coupling constant, for example, from the lowest-order ("ladder") approximation.

[4] See, especially, J. S. Goldstein, Phys. Rev. **91**, 1516 (1953).

[5] M. Gell-Mann and F. Low, Phys. Rev. **84**, 350 (1951).

Reprinted from *Phys. Rev.* **96**, 1124 (1954).

the analytic continuation of the wave function to complex values of the relative time (or relative energy) variable. As far as we can tell these properties cannot be obtained from the B-S equation itself. Vice versa, they can be used (Sec. 3) to transform the equation, by rotation of the integration path in the complex plane, to an equation in which $x_4=ix_0$ (respectively $p_4=ip_0$) is real. While the concept of an imaginary relative time variable does not help physical intuition, it has mathematically several advantages. A discussion of the eigenvalue problem in terms of the transformed equation will be given (Sec. 4), and the existence of solutions will be shown to follow, under fairly general assumptions, from considerations similar to those commonly employed in the nonrelativistic case. No claim of completeness or rigor is made for this "proof." Finally in Sec. 5 we shall merely itemize various approximation methods that have been studied, but will be reserved for another publication.

The second line of attack (Sec. 6), which is the subject of a more extensive investigation in the subsequent paper by Cutkosky,[6] is rather different in nature. It is an attempt to make much more specific statements about the exact solutions of the equation, by restricting the character of the equation to an especially simple type. It has not been possible so far to extend this approach to any case of real practical interest. But the fact that in one case, which is not entirely artificial, one can get a complete picture of all the solution (as is shown more completely in the following paper[6]) is not perhaps devoid of general interest. In particular the presence of "abnormal" solutions, which do not possess a nonrelativistic limit, and the circumstances under which they occur may well give a qualitative indication as to properties that will occur also in the cases of real physical interest.

2. THE STABILITY CONDITIONS

The relativistic wave function $\chi(x)$ for a system of two particles, a and b, bound together in a state $|\alpha\rangle$ is defined[5] as the matrix element, between α and the "true" vacuum state $|0\rangle$, of the time ordered product of the Heisenberg field operators ψ_a and ψ_b describing the two kinds of particles. If, for example, the relative time $t=t_a=t_b$ is positive,

$$\chi(x)=e^{-iP\cdot X}\langle 0|\Psi_a(x_a)\Psi_b(x_b)|\alpha\rangle, \tag{1}$$

where $x=x_a-x_b$, $X=(m_ax_a+m_bx_b)/(m_a+m_b)$, and $P\cdot X$ is the four-dimensional scalar product of X with the total momentum P of the system in state α. If for simplicity we assume that the compound system is at rest, then $P=(\mathbf{0}, iE)$, E being the total energy. For a bound state,

$$E=m_a+m_b-B<m_a+m_b. \tag{2}$$

Now the matrix element in (1) can be written

$$\sum_n\langle 0|\Psi_a(x_a)|n\rangle\langle n|\Psi_b(x_b)|\alpha\rangle. \tag{3}$$

[6] R. Cutkosky, following paper [Phys. Rev. 96, 1135 (1954)].

The sum extends in principle over all states, but in fact the states n giving a nonzero contribution will belong to a rather special class. Consider for example the case where a and b are an electron and proton, respectively. If Ψ_a and Ψ_b were noninteracting fields, it is obvious that only one-electron states would have to be considered in the sum (3). In the presence of interaction, the states n may also contain photons, electron-positron pairs and proton-antiproton pairs. But at any rate the fundamental integrals of the motion N_a (number of electrons—number of positrons) and N_b (number of protons—number of antiprotons) must have the same values,

$$N_a=1,\quad N_b=0, \tag{4}$$

as the one-electron states. This may be rigorously shown from the commutation properties of N_a and N_b with the field operators, $(N_a+1)\Psi_a=\Psi_aN_a$, etc.

In a similar manner, one can show that the total angular momentum quantum number J for a state n, when measured in a system of reference in which the total momentum $\mathbf{p}$ is zero, must be equal to $\frac{1}{2}$.

Now all states known to us in nature, and satisfying condition (4), also satisfy the inequality,

$$E_n^2-\mathbf{p}^2\geqslant m_a, \tag{5}$$

E_n and $\mathbf{p}$ being the total energy and momentum in the state n. Furthermore, the equality sign holds true only for one-electron states.

The inequality (5) means that among all the states having the same values of the fundamental constants of the motion $\mathbf{p}$, N_a, N_b, etc., as a one-electron state, the latter is the state of lowest energy. We shall refer to (5), therefore, as the stability condition for an electron.

In a similar way, when the relative time t is negative, the wave function χ may be shown to depend on the sum

$$\sum_{n'}\langle 0|\Psi_b(x_b)|n'\rangle\langle n'|\Psi_a(x_a)|\alpha\rangle, \tag{3'}$$

in which the contributing states n' must satisfy the condition,

$$N_a=0;\quad N_b=1, \tag{4'}$$

and hence the inequality,

$$E_{n'2}-\mathbf{p}^2\geqslant m_b, \tag{5'}$$

which shall be called the stability condition for a proton.

Summing up, we have three inequalities (2), (5), and (5'), which will form the basis of the following discussion. It should be pointed out that the above considerations can be extended to other systems. If a and b were a neutron and proton, bound together in the ground state α of the deuteron by a meson field, with the customary assumptions, one would then have, as integrals of the motion, the number of nucleons minus antinucleons N and the total electric charge Q. The states n could be shown to have values $N=1$, $Q=0$ and the states n' the values $N=1$, $Q=1$. In a theory which neglects the β-decay interaction, one has the right to regard both neutron and proton as essentially stable particles. If there were states $n(n')$ not satisfying conditions (5) (5′) the neutron (proton) could decay into those states by emission of photons, without violating any of the known conservation theorems. Thus it is extremely reasonable to postulate that these conditions must again be satisfied.

Now going back to (1) and using (3) with the conditions (2) and (5), we see that for $t>0$, and assuming $P=(0,iE)$, $\chi(x)$ is of the form

$$\chi(x)=\int d\mathbf{p}\int_{\omega_{\min}}^{+\infty} d\omega f(\mathbf{p},\omega)\exp(i\mathbf{p}\cdot\mathbf{x}-i\omega t), \quad (6)$$

where

$$\omega_{\min}=B\mu_a+(m_a^2+\mathbf{p}^2)^{\frac{1}{2}}-m_a>B\mu_a>0, \quad (7)$$

with $\mu_a=m_a/(m_a+m_b)$. Thus, when $t>0$, $\chi(x)$ is a superposition of positive frequency terms only.

Similarly, from (2) and (5′) it follows that, when $t<0$, $\chi(x)$ contains negative frequencies only. Thus we find that $\chi(x)$ has properties with which we are familiar in the case of Feynman propagation kernels. There is, of course, an analogy between the definition of these kernels and Eq. (1).

Let us now consider t as a complex variable. Equation (6) shows that $\chi(x)$ can be continued analytically in the lower half-plane, in the region $0\geqslant\arg t>-\pi$. Similarly starting from the negative real axis, $\chi(x)$ can be continued in the upper half-plane, in the region $\pi\geqslant\arg t>0$. There is, of course, no analytic continuation from one half-plane to the other; the two regions touch one another at one point only, $t=0$.

It should be pointed out that the statements just made are not dependent on the assumption that the state α is bound; they follow from well-known properties of the Laplace transform from the mere fact that ω is finite. If, however, $B>0$ and hence $\omega_{\min}>0$, we can further assert that $\chi(x)\rightarrow0$ when t tends to ∞ in any direction in the lower or upper half-plane different from the real axis. This suggests that the eigenvalue problem may take a more familiar and a simpler form if the wave function and the wave equation are considered on the imaginary t axis (i.e., for $x_4=it$ real).

In order to examine this possibility carefully, it is desirable to go over to momentum space. We write $\chi(x)=\chi_1+\chi_2$, where $\chi_1=0$ for $t<0$ and $\chi_2=0$ for $t>0$. Let us calculate the Fourier transform of χ_1.

$$\phi_1(\mathbf{p},p_0)=(2\pi)^{-4}\int d_3x e^{-i\mathbf{p}\cdot\mathbf{x}}\int_0^{+\infty} dt e^{ip_0t}\chi(x). \quad (8)$$

From (6) one easily finds

$$\phi_1(\mathbf{p},p_0)=\frac{1}{2\pi i}\int_{\omega_{\min}}^{+\infty} f(\mathbf{p},\omega)(\omega-p_0-i\epsilon)^{-1}d\omega, \quad (9)$$

where ϵ is an infinitesimal positive constant. We must assume, of course, that the wave function exists for real values of p_0 [i.e., that the integral (9) converges]. From the theory of Stieltjes transforms, we then infer that (9) defines an analytic function of p_0 in the whole complex plane, in the region

$$2\pi>\arg(p_0-\omega_{\min})\geqslant0. \quad (10)$$

Similarly ϕ_2 is defined in the region

$$-\pi<\arg(p_0-\omega_{\max})<\pi, \quad (11)$$

where

$$-\omega_{\max}=B\mu_b+(m_b^2+\mathbf{p})^{\frac{1}{2}}-m_b>B\mu_b>0.$$

Thus $\phi(p)=\phi(\mathbf{p},p_0)=\phi_1+\phi_2$ is defined in the complex p_0 plane with two cuts from $\omega_{\min}$ to $+\infty$ and from $-\infty$ to $\omega_{\max}$ (Fig. 1). In this case analytic continuation from the lower to the upper half-plane is ensured through the gap between the two cuts. ($B>0$ is essential for the existence of the gap.) Notice also that the sense of rotation implied by (10) and (11) is the opposite of that in the t plane. From the real p_0 axis one goes continuously into the upper half-plane if $p_0>\omega_{\min}>0$, into the lower half-plane if $p_0<\omega_{\max}<0$.

3. TRANSFORMATION OF THE B–S EQUATION

We shall now use the analytic properties of the wave function to transform the B-S equation by a rotation of the axis of integration in the complex p_0 (respectively x_0) plane.

The equation[1] may be written

$$F_aF_b\phi=I_{ab}\phi, \quad (12)$$

where ϕ is the wave function in momentum space, i.e., the Fourier transform of $\chi(x)$; it is a function of the relative momentum p defined by

$$p_a=\mu_aP+p, \quad p_b=\mu_bP-p, \quad (13)$$

P, μ_a and μ_b being the total momentum and the mass ratios previously defined. F_a and F_b are one-particle propagators, which, if one neglects radiative corrections

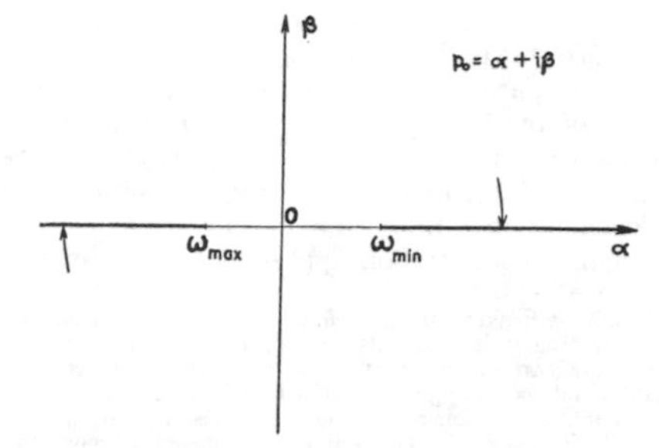

FIG. 1. The complex plane of the variable p_0. The wave function is analytic everywhere, excluding the cuts (heavy lines) on real axis.

reduce to

$$F_a=\gamma_a p_a - im_a, \quad F_b=\gamma_b p_b - im_b, \quad \text{(Dirac particles)}$$
$$F_a=p_a^2+m_a^2, \quad F_b=p_b^2+m_b^2. \quad \text{(Klein-Gordon)} \tag{14}$$

Finally, I_{ab} is the interaction operator, which has different forms, depending on the kind of theory. The following form[4] covers several cases, for the lowest order ("ladder") approximation:

$$I_{ab}\phi(p)=(\lambda/\pi^2)\int \frac{[dk]}{(p-k)^2+\kappa^2}\rho_a\rho_b\phi(k), \tag{15}$$

where $[dk]=idk_0d\mathbf{k}$. The various cases are obtained from the various possible assumptions about the "photon mass" κ, and the factors $\rho_a\rho_b(\rho_a\rho_b=1$, scalar interaction, etc.).

For simplicity we shall carry out the transformation under the assumptions (14), (15), but the proof can be easily generalized to include radiative corrections to any desired order.[7]

Let us consider the right-hand side of Eq. (12), as given by (15). The poles of the interaction kernel are at

$$k_0=p_0\pm[(\mathbf{p}-\mathbf{k})^2+\kappa^2]^{\frac{1}{2}}. \tag{16}$$

Let us carry out the integration over k_0 first. The integration is along the real axis in the plane of the complex k_0 variable, passing just under the cut on the negative axis and above the cut on the positive axis. It is also important to remember that κ in (16) is assumed to have an infinitesimal negative imaginary part, so that the pole with the larger real part lies under the integration path and the pole with the smaller real part above the path. Suppose for instance $p_0>0$, then depending on the relative magnitude of the two terms in (16) the poles will lie as in Fig. 2(a) or 2(b). For simplicity, the cuts of Fig. 1 are not indicated in Fig. 2, but they do not interfere with the following operations. First the integral path may be deformed along the dashed line [there is an *assumption* here, that $\phi(k)$ tends to zero at least like k_0^{-2} when $k_0\to\infty$ in any direction]. Now we move p_0 upwards along a circle so as to end on the positive imaginary axis. In Fig. 2(b) the path need not be changed. In Fig. 2(a) the left pole, around which the path is bent, moves to the left of the imaginary axis, and the path can be straightened. In both cases we end up with p_0 on the positive imaginary axis, and the integral over k_0 along the imaginary axis, from $-i\infty$ to $+i\infty$.

A similar consideration applies when p_0 is on the

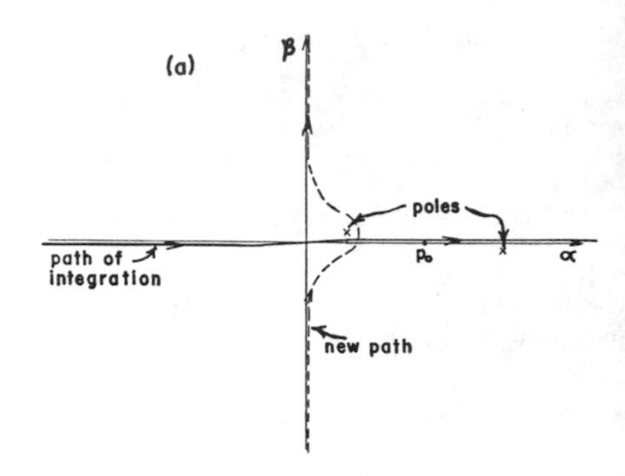

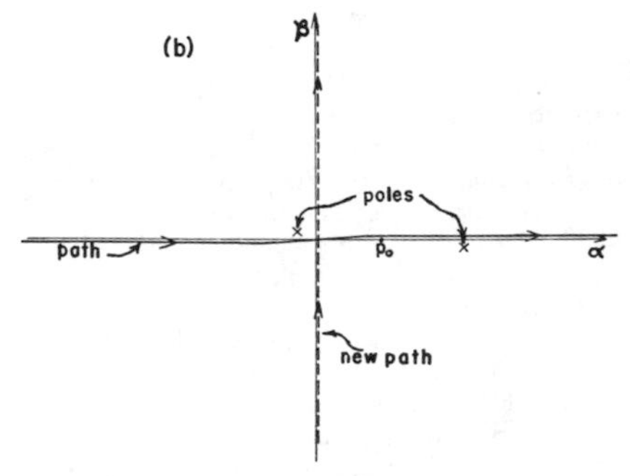

FIG. 2. Integration paths for the variable k_0 in Eq. (15).

negative real axis; it then moves to the negative imaginary axis. The net result is a counter-clockwise rotation of the axis on which the wave function is used, on both sides of the equation.

Equation (12) is thus reduced to an integral equation in a Euclidean vector space, with the metric

$$p^2=p_1^2+p_2^2+p_3^2+p_4^2. \tag{17}$$

One does not really have to change anything to the equation, except for the understanding that a real vector now has a real component p_4 and that, in Eq. (15)

$$[dk]=dk_1dk_2dk_3dk_4 \tag{18}$$

the integral over k_4 being from $-\infty$ to $+\infty$. The fixed vector $P=(\mathbf{0}, iE)$ is now, of course, regarded as pure imaginary.

One sees at once several advantages of this transformation. The singularities of the interaction kernel [and with them the difficulties mentioned under (b) in the Introduction] are eliminated, and what is equally important, the zeros of the Klein-Gordon factors (14), i.e., the singularities of the inverses F_a^{-1}, F_b^{-1}, have similarly disappeared from the space of real p vectors. Furthermore the symmetry group of the equation is no longer the Lorentz group, but the group

[7] A higher-order term includes, in general, a number of integrations over fourth components k_0, k_0', k_0'', . . . The proof is most easily carried out if all these are regarded as complex variables and their integration paths are rotated simultaneously. An examination of higher-order corrections also requires a closer look at the factors F_a, F_b. The analytical nature of the propagators F_a^{-1}, F_b^{-1} (i.e., of the S_F', Δ_F' functions) is well understood (see reference 14), and it is easy to show that they have no singularities that stand in the way of our transformation.

of real rotations in four dimensions.[8] This is important in the first place, because the group determines the polar variables, which may be used with advantage. In the Lorentz case integrals over a surface $p^2=\text{const}$, or $x^2=\text{const}$ are usually divergent; there are no orthogonality theorems for spherical harmonics, no completeness theorems, etc. Here instead we have the whole familiar machinery at our disposal.

Other advantages appear in the configuration space formulation of the equation, as we shall presently see.

4. DISCUSSION OF THE EIGENVALUE PROBLEM

We shall now examine several cases and show that the transformed equation presents us with an eigenvalue problem, to which many of the ordinary methods and conclusions can be applied.

We shall begin, like Goldstein,[9] with the extreme case $E=0$, where the equation acquires full four-dimensional symmetry in relative momentum space. Unlike Goldstein, however, and for reasons to appear later, we shall choose in Eq. (14) the K.G. (Klein-Gordon) form of the factors F_a and F_b. That is, we assume that a and b have zero spin. The equation for $E=0$ thus has the form

$$(p^2+m_a^2)(p^2+m_b^2)\phi(p)=\lambda\pi^{-2}\int\frac{[dk]}{(p-k)^2+\kappa^2}\phi(k). \quad (19)$$

We shall often use, in the following, the abbreviation $\lambda I_\kappa\phi$ for the right-hand side of (19). In particular I_0 shall designate the interaction operator when the "photon" mass κ is zero.

We can now, of course, separate ϕ, using polar variables, and reduce the problem to a one-dimensional integral equation. If for example ϕ is a function of p^2 only, the integration over angular variables on the right-hand side of (19) is quite elementary. For simplicity we shall write the one-dimensional integral equation for this case only. Let $p^2=s$, $\phi(p)=u(s)$; then

$$(s+m_a^2)(s+m_b^2)u(s)=2\lambda\int_0^\infty tu(t)dt/\{s+t+\kappa^2+[(s+t+\kappa^2)^2-4st]^{\frac{1}{2}}\}. \quad (20)$$

With the further change of variables

$$x=f(s),\quad y=f(t),$$

$$f(s)=\int_0^s ds'/(s'+m_a^2)(s'+m_b^2), \quad (21)$$

$$s^{\frac{1}{2}}(s+m_a^2)(s+m_b^2)u(s)=v(x),$$

Eq. (20) becomes a symmetric integral equation,

$$v(x)=\lambda\int_0^a K(x,y)v(y)dy, \quad (22)$$

with the finite kernel,

$$K(x,y)=2(st)^{\frac{1}{2}}/\{s+t+\kappa^2+[(s+t+\kappa^2)^2-4st]^{\frac{1}{2}}\}, \quad (23)$$

and the finite interval $a=f(\infty)$. Fredholm's theory can then be applied, to conclude that (22) has a discrete eigenvalue spectrum. The case where ϕ is proportional to a four-dimensional spherical harmonic can be similarly handled.

It may be pointed out that if $\kappa=0$, Eq. (20) can be reduced to a second order differential equation either by differentiating twice, or by a parametric representation of the solution. Both methods will be used later, and especially in the subsequent paper by Cutkosky,[6] to obtain more precise information about this case.

Let us now consider briefly Goldstein's Eq. (10), which applies to the case of two Dirac particles. When written in our notation, the equation is quite similar to (19) except that it contains only one quadratic factor in p on the left. Goldstein manages to reduce the equation to the one dimensional form, his Eq. (14), in exact analogy to our Eq. (20); the transformation in the usual frame, however, is far from trivial.[10] Unlike Eq. (20), however, Goldstein's (14) is not reducible to the Fredholm type. The difference in behavior is not an effect of our transformation, but is really due to the different power of p^2 on the left-hand side. The difficulties which Goldstein encounters in defining the eigenvalue spectrum, and which he surmounts by a special cut-off procedure, are thus not a general property of the B-S equation, but rather of the special case considered by him.

For the purpose of obtaining a more general viewpoint, let us now examine the problem in configuration space, i.e., in terms of the function $\chi(x)$. Consider first again the case $E=0$. The Fourier transform of Eq. (19) is

$$[(-\Box+m_a^2)(-\Box+m_b^2)-\lambda V(R)]\chi(x)=0, \quad (24)$$

where the "potential" $V(R)$ is

$$V(R)=4\kappa R^{-1}K_1(\kappa R),\quad R=(x_\mu x_\mu)^{\frac{1}{2}}, \quad (25)$$

K_1 being a modified Hankel function. The expression for $V(R)$ in the case $\kappa=0$,

$$V(R)=4R^{-2}, \quad (25a)$$

also gives the singularity of V at the origin in the general case.

Goldstein's Eq. (10) becomes similarly

$$[\Box-m^2+\lambda V(R)]\chi(x)=0, \quad (26)$$

[8] Four-dimensional rotations must be applied simultaneously, of course, to the relative momentum p and to the total P. If one uses the c.m. system to begin with, so that P is pure imaginary, it will stay pure imaginary after a real rotation. For Dirac particles, a linear transformation of the χ or ϕ function must accompany the rotation; this can be established in the usual way. Contrary to the Lorentz case, however, the transformation here is always *unitary*.

[9] See reference 4. Like Goldstein, we find it convenient, in general, to regard E as given, λ as the eigenvalue to be found.

[10] The author is indebted to Dr. Goldstein for various interesting conversations, and in particular for pointing out to him the peculiar "Euclidean" nature of his Eq. (14). This remark was one of the early motivations for the present study.

which presents a striking analogy to the ordinary three-dimensional Schrödinger equation. With x_4 real, (26) is, of course, an elliptic differential equation. This, together with the boundary condition $\chi(x)\rightarrow 0$ at infinity, allows a discussion of the eigenvalue problem along familiar lines.

A special difficulty, also encountered by Goldstein, is presented by the boundary condition at the origin $R=0$, about which we have unfortunately no definite indication from general field-theoretic considerations. The difficulty arises because of the Fuchsian singularity (25a); if the potential were regular everywhere, there would be little doubt that $\chi(x)$ must be regular too.

One can see at once, however, that the singularity of the potential affects (24) and (26) in a very different manner. Consider, for example, spherically symmetric solutions. The radial equation corresponding to (26), or

$$[d^2/dR^2+(3/R)(d/dR)-m^2+\lambda V(R)]\chi=0, \quad (26a)$$

has two solutions near the origin, of the type $\chi=R^\alpha \times(1+c_1R+\cdots)$ with

$$\alpha=-1\pm(1-4\lambda)^{\frac{1}{2}}. \quad (27)$$

Thus, if $\lambda<\frac{1}{4}$, it is possible to make a distinction between the "regular" (less singular) and the "irregular" solution. If $\lambda>\frac{1}{4}$, it seems highly unlikely that a plausible condition to determine the right solution can be found. In the case $\kappa=0$, moreover, the equation can be solved explicitly,[4] the "regular" solution being $R^{-1}J_n(iR)$, where $n=+(1-4\lambda)^{\frac{1}{2}}$. This solution, however, never satisfies the condition at infinity. We thus reach the conclusion that no value $\lambda<\frac{1}{4}$ is an eigenvalue. In our opinion, for $\lambda>\frac{1}{4}$ the eigenvalue problem becomes ill-defined. We shall not try to discuss further here[11] whether the limiting case $\lambda=\frac{1}{4}$ can actually be regarded as an eigenvalue.[4]

In Eq. (24), on the other hand, the singularity (25a) does not affect the indicial equation. The radial equation for a spherically symmetric solution, for example, has four independent solutions near origin, say $\chi_1, \chi_2, \chi_3, \chi_4$, behaving respectively like R^2, R^0, $\ln R$, and R^{-2}. If there were no potential, we would clearly say that the acceptable solution is a linear combination $c_1\chi_1+c_2\chi_2$ of the two "regular" solutions. We shall make the same assumption when there is a potential.[12] Likewise we can define, for large R values, four solutions behaving respectively like $R^{-\frac{3}{2}}\exp(\pm\mu_aR)$ and $R^{-\frac{3}{2}}\exp(\pm\mu_bR)$. The solution $c_1\chi_1+c_2\chi_2$ will be a linear combination of these four. In order to satisfy the condition $\chi\rightarrow 0$ at infinity, two coefficients must be zero; that is, we have two conditions. One of these may be satisfied by a suitable choice of c_1/c_2; the remaining one gives a condition on λ. This will, in general, determine a discrete spectrum of eigenvalues.

We shall see later that for $\kappa=0$ the analysis can be carried much further. Let us now turn to the more interesting general case $E\neq 0$. Let us write (in the c.m. system)

$$P=(0,iE)=i(m_a+m_b)\eta, \quad (28)$$

where η is the four vector

$$\eta=(0,\epsilon), \quad \epsilon=E/(m_a+m_b).$$

Notice that

$$\eta^2=\epsilon^2<1. \quad (29)$$

The factor on the left of Eq. (19) now becomes, remembering (13):

$$(m_a^2+p_a^2)(m_b^2+p_b^2)=p^4+(m_a^2+m_b^2)(1-\eta^2)p^2 +4m_am_b(p\eta)^2+m_a^2m_b^2(1-\eta^2)^2 +2i(m_a-m_b)(p^2-m_am_b)(p\eta). \quad (30)$$

It is at first sight rather puzzling that the equation now contains an imaginary term whose presence depends on m_a being $\neq m_b$. In configuration space this means that the operator corresponding to (30) is self-adjoint only when $m_a=m_b$. One can show that this feature is connected with the time-reversal properties of the equation.

We shall point out, when the occasion arises, the differences produced by the term in m_a-m_b. For the moment, we shall consider only the case $m_a=m_b$ ($=m$, say). The analog of Eq. (24) then is

$$\{[-\Box+m^2(1-\eta^2)]^2-4m^2\eta^2\partial^2/\partial x_4^2\}\chi(x) =\lambda V(R)\chi(x). \quad (31)$$

Since complete separation of variables is impossible, a solution must now be a superposition $\chi=\sum_n f_n(R)Y_n$ of four-dimensional spherical harmonics Y_n of different orders. The radial functions f_n satisfy a system of coupled fourth-order differential equations, and it is no longer possible to discuss the eigenvalue problem in terms of a single radial function. This is a considerable complication, but one may notice, nevertheless, that the term in (31) which produces the coupling is of second order only, so that the indicial equation for each radial function f_n is the same as in Eq. (24). If one writes $f_n(R)=R^\alpha(1+c_1R+\ \ldots\ )$ the possible values for α are $\pm n$, $\pm(n+2)$; we may assume that only the positive values are allowed in a "regular" solution, just as in Eq. (24). Thus there is no qualitative difference between the two equations, with regard to the behavior of solutions near $R=0$.

The asymptotic behavior of $\chi(x)$ at infinity, on the other hand, is more interesting. It will be shown below that when x tends to infinity, χ behaves asymptotically like $\exp[-R\varphi(\theta_4)]$, i.e., it tends to zero exponentially but with a coefficient depending on the direction, more specifically on the angle θ_4 with the "4" axis. For our present purpose, however, it is only interest-

[11] It may be remarked that in reference 6 Goldstein's eigenvalue is also obtained from Eq. (19) in the limit $m_b/m_a\rightarrow 0$ (and $\kappa=0$).

[12] One can argue that $\chi\sim\ln R$ is not really a solution of (24) since it gives an additional term $\sim\delta_4(x)$. $\chi\sim R^{-2}$ gives a term $\Box\delta_4(x)$.

ing to notice that $\varphi(\theta_4)$ has a positive lower limit $\varphi \geqq 1-\epsilon$ so that, in a certain sense, there is again no fundamental difference in behavior between the solutions of (31) and those of (24), and we may expect that in both cases the boundary conditions at $R=0$ and $R=\infty$ will determine a discrete λ spectrum.

The elementary considerations developed previously seemed of interest, because of the analogy with considerations often made with regard to the ordinary Schrödinger equation. In this sense we may say that (31) presents an analogy to the Schrödinger equation for a particle in an asymmetric field, where again the reduction of the eigenvalue problem to a simple one-dimensional Sturm-Liouville problem is not feasible.

In either case, a rigorous discussion of the eigenvalue problem can only be achieved by less elementary means, such as the reduction of the problem to an integral equation. We do not wish to carry out such a study here, but we may point out along what lines it could be carried out.

We already have, of course, in Eq. (19) and its generalization for $E\neq 0$, an integral formulation of the problem. In the case $m_a=m_b$ corresponding to Eq. (31), the equation can be reduced to the real symmetric form

$$\Phi(p)=\lambda\int H(p,k)\Phi(k)[dk], \tag{32}$$

where

$$\begin{aligned}\Phi(p)&=f^{\frac{1}{2}}(p)\phi(p)\\ f(p)&=[p^2+m^2(1-\eta)]^2+4m^2\eta^2p_4^2\\ H(p,k)&=\pi^{-2}[f(p)]^{-\frac{1}{2}}[(p-k)^2+\kappa^2]^{-1}[f(k)]^{-\frac{1}{2}}.\end{aligned} \tag{33}$$

Now by counting powers of p and k it is easy to see that

$$\int H^2(p,k)[dp][dk]<\infty, \tag{34}$$

which together with other similar inequalities, which the mathematically inclined reader can readily discover, may be used to show that (32) is "nonsingular" and thus possesses a discrete λ spectrum. Furthermore all eigenvalues are real. Finally, one can see that the kernel is positive-definite[13] so that $\lambda>0$.

An alternative integral formulation can be obtained as usual in configuration space. In fact, Eq. (31) together with the regularity condition at the origin and the boundary condition $\chi(x)\rightarrow 0$ at infinity, can be replaced by an integral equation,

$$\chi(x)=\lambda\int G(x-x')V(R')\chi(x')[dx'], \tag{35}$$

or $\chi=\lambda GV\chi$, where G is the inverse of the differential operator on the left-hand side of (31). When $m_a=m_b$, the function $G(x)$ is even: $G(x)=G(-x)$, so that (35) can be easily symmetrized. The function $G(x)$ is constructed in the Appendix, and it may be seen from Eqs. (A7) and (A8) there that $G(x)$ has a very weak singularity at the origin (it is in fact finite at $x=0$) and tends to zero at infinity like

$$G(x)\sim ge^{-R\varphi(\theta_4)}, \tag{36}$$

where g is a factor which varies slowly compared to the exponential and

$$\begin{aligned}\varphi(\theta_4)&=m(1-\epsilon\cos\theta_4) \qquad |\cos\theta_4|>\epsilon\\ &=m(1-\epsilon^2)^{\frac{1}{2}}\sin\theta_4 \qquad |\cos\theta_4|<\epsilon.\end{aligned} \tag{37}$$

If $V(R)\rightarrow 0$ sufficiently rapidly when $R\rightarrow\infty$, the asymptotic behavior of $\chi(x)$ as given by the integral in Eq. (35) will reflect that of $G(x)$, from which the conclusions previously mentioned may be obtained. Incidentally it may be noticed that in the nonrelativistic limit, $\epsilon\approx 1$, the lower form in Eq. (37) covers almost the whole solid angle, and furthermore $\varphi\approx(mB)^{\frac{1}{2}}\sin\theta_4$, $R\varphi(\theta_4)=(mB)^{\frac{1}{2}}r$, where $r^2=x_1^2+x_2^2+x_3^2$. We thus find the typical exponential of the three-dimensional Schrödinger function. It is indeed rather remarkable that in this region, i.e., with the exception of a narrow cone around the x_4 axis, the asymptotic form of $\chi(x)$ is not time-dependent.

In the foregoing discussion we have, perhaps, laid too much stress on the special case of two spinless particles with the special interaction I_κ of Eq. (19). It is clear that none of the conclusions we have reached as to discreteness of the λ-spectrum, etc., must necessarily remain true if we change the propagators F_a, F_b or the interaction kernel.

If, for example, we write the analog of (32) with Dirac propagators, the conclusion that the equation is nonsingular no longer holds true. As pointed out above, Goldstein[4] already met this situation for the special case $E=0$. It is, of course, also possible to formulate the problem in a form similar to (35), namely,

$$\chi(x)=\lambda\int G_D(x-x')V(R')\chi(x')[dx'], \tag{38}$$

where

$$\begin{aligned}G_D(x)=[\gamma_a(\partial/\partial x)-m_a(1+\gamma_a\eta)]\\ \times[\gamma_b(\partial/\partial x)+m_b(1+\gamma\eta)]G(x).\end{aligned} \tag{39}$$

In this case the singular character of the equation comes about because $G_D(x)$ has a much stronger singularity than $G(x)$, near $x=0$. When this is combined with the $1/R^2$ singularity of $V(R)$ [Eqs. (25) and (25a)], Eq. (38) becomes singular. This does not mean that discrete eigenvalues of λ will not exist, but only that a much more detailed study of the equation will be necessary. One could, of course, also consider the possibility of less singular potentials $V(R)$, in which case the general theory of integral equations might again be applicable.

It seems pointless at present to investigate in detail such possibilities. One will bear in mind, however, that within the framework of our transformed system of

[13] G. C. Wick, Nuovo cimento (to be published).

coordinates, such questions can be attacked by ordinary mathematical methods.

5. APPROXIMATION METHODS

It is also possible to show that our transformed equation has several advantages if one wants to employ approximate methods of solution. We have in mind, in particular: (a) a perturbation expansion in the neighborhood of $E=0$ (see also reference 4), (b) variational principles, (c) nonrelativistic approximations, without special restrictions as to the form of $V(R)$. These questions will be discussed in a paper which the author hopes to present shortly in another periodical.[13]

6. EXACT SOLUTIONS FOR $\kappa=0$

A comparison of Eqs. (25) and (25a) suggests that the problem of solving the B-S equation exactly may be far more elementary in the latter ($\kappa=0$) case. This is borne out by Goldstein's solution[4] for Eq. (26), and we shall see in a moment that also Eq. (24) has quite simple solutions if $\kappa=0$ and $m_a=m_b$. And, of course, one will remember that the ordinary nonrelativistic Schrödinger problem is far more elementary with a Coulomb than with a Yukawa potential.

At first, however, one would regard this analogy as encouraging only for the special case $E=0$, when the B-S equation is separable. We were, therefore, quite surprised when we first realized that for $\kappa=0$ even the nonseparable Eq. (31) can be reduced to a one-dimentional integral equation, or alternatively to a one-dimensional eigenvalue problem of the Sturm-Liouville type. We shall explain the basic idea for the simplest type of solution and for $m_a=m_b$ only. The extension to other cases was carried out by Cutkosky and is described in the accompanying paper.

Choosing $m_a=m_b$ ($=m$, say), let us first examine the separable case, Eq. (24). In momentum space, the equation has the form

$$(p^2+m^2)^2\phi(p)=\lambda I_0\phi(p), \tag{40}$$

which is very similar to the nonrelativistic hydrogen equation in momentum space. The latter, of course, is a three-dimensional equation and does not have the square power on the left, but it will appear that the analogy is closest when the two changes are made simultaneously.

In particular, the ground-state wave function of hydrogen: $\phi(p)=(p^2+p_0^2)^{-2}$, is duplicated here by the solution

$$\phi(p)=(p^2+m^2)^{-3} \tag{41}$$

corresponding to the eigenvalue $\lambda=2m^2$. That (41) satisfies Eq. (40) can be verified most easily if one first writes, à la Feynman:

$$(k^2-2k\cdot p+p^2)^{-1}(k^2+m^2)^{-3}=\int_0^1 3(1-x)^2dx \times[(k-xp)^2+(1-x)(m^2+xp^2)]^{-4}. \tag{42}$$

One then finds easily that

$$I_0\phi=(1/2m^2)(p^2+m^2)^{-1}, \tag{43}$$

showing that Eq. (40) is satisfied.

More generally, one can see that I_0 applied to $(p^2+2p\cdot q+M^2)^{-3}$, where M^2 and the vector q are constants, gives $(p^2+2p\cdot q+M^2)^{-1}$, apart from a proportionality factor. This peculiar self-reproducing property of a quadratic form in p, under the operation I, is characteristic of the case $\kappa=0$.

Consider now the equation for $E\neq 0$. For simplicity let $m=1$ from now on. The equation is

$$[p^2+2ip\cdot\eta+1-\eta^2][p^2-2ip\cdot\eta+1-\eta^2]\phi=I_0\phi. \tag{44}$$

Clearly ϕ cannot be a function of p^2 alone; it must be at least a function of p^2 and $p\cdot\eta$ (for an S state). The above considerations suggest that we may be able to generalize solution (41) by writing ϕ as a superposition of terms[14] of the type $(p^2+2p\cdot q+M^2)^{-3}$ where q is parallel to η, say, $q=iz\eta$. That is

$$\phi(p)=\int dzdM^2g(z,M^2)[p^2+2izp\cdot\eta+M^2]^{-3}. \tag{45}$$

One then sees immediately that

$$I_0\phi=\tfrac{1}{2}\int dzdM^2g_1(z,M^2)[p^2+2izp\cdot\eta+M^2]^{-1}, \quad g_1(z,M^2)=g(z,M^2)/(M^2+z^2\eta^2). \tag{46}$$

Inserting on the right of (44) and dividing by the two quadratic factors on the left, one then tries to reduce the result again to the form (45) by reassembling the three quadratic denominators into a cube [in a similar way as in Eqs. (42) and (A3) in the Appendix]. One sees at once that if $M^2=1-\eta^2$ the "mass term" reproduces itself. Thus we set

$$g(z,M^2)=g(z)\delta(M^2-1+\eta^2). \tag{47}$$

Carrying out the transformations indicated above and writing

$$Q(z)=1-\eta^2+z^2\eta^2, \tag{48}$$

we find

$$\phi(p)=\tfrac{1}{2}\lambda\int Q^{-1}(z)g(z)dz\int_{-1}^{+1}dy\int_0^1 \times xdx[p^2+2i\zeta p\cdot\eta+1-\eta^2]^{-3}, \quad \zeta=xy+(1-x)z. \tag{49}$$

[14] An expression of this type has a certain resemblance to the parametric representations for S_F' and Δ_F' developed by M. Gell-Mann and F. E. Low, Phys. Rev. **95**, 1300 (1954). G. Källén [Helv. Phys. Acta **25**, 417 (1952)] has previously used similar representations for other quantities that are a little less closely related to the B-S wave function, Eq. (1). In the case of these quantities, and of the functions $S_F'\Delta_F'$, it is possible as the above-mentioned authors have shown, to derive the general form of the parametric representation from the definition of the quantities, and from considerations of relativistic invariance. The author has not been able to do the same for Eq. (1). Nevertheless the analogy with S_F' and Δ_F' was used to "guess" the form of Eq. (45).

Eliminating y in favor of ζ, and carrying out the integrations over x and z first, (49) acquires indeed the general form required by (45) and (47). Writing that the two expressions are identical gives an integral equation for $g(z)$.

To this end notice that if z in (45) is allowed to vary between -1 and $+1$, ζ will also vary between the same limits. Writing $d\zeta = xdy$ and noting that for given z and ζ,

$$\int dx = R(\zeta,z) = \begin{cases} (1+\zeta)/(1+z) & \text{if } z>\zeta \\ (1-\zeta)/(1-z) & \text{if } z<\zeta, \end{cases} \quad (50)$$

one finds

$$\phi(p) = \int_{-1}^{+1} \gamma(\zeta)[p^2+2i\zeta p\cdot\eta+1-\eta^2]^{-3}d\zeta, \quad (51)$$

where $\gamma(\zeta)$ is given by the right-hand side of Eq. (52) below. The condition $g=\gamma$ thus gives the integral equation

$$g(\zeta) = \tfrac{1}{2}\lambda \int_{-1}^{+1} R(\zeta,z)Q^{-1}(z)g(z)dz. \quad (52)$$

This is, of course, an integral equation of Fredholm's type, and has a discrete λ spectrum. We thus have achieved the surprising result that the B-S equation (44), although nonseparable (as far as we can tell), can be reduced to a one-dimensional problem.

Further Reduction of the Problem

Equations (45), (47), and (52), of course, do not give all the solutions; they do not even give all the S states. The necessary generalizations, however, are natural and will be described in the accompanying paper. Let us instead study (52) a little further. From (50) and (52) one can see that

$$g(+1) = g(-1) = 0. \quad (53)$$

Furthermore, differentiating (52) twice, we get

$$g''(z)+\lambda(1-z^2)^{-1}Q^{-1}(z)g(z) = 0. \quad (54)$$

These equations formulate the problem as a Sturm-Liouville eigenvalue problem. Thus it is easy to predict qualitatively the dependence of λ on η^2.

Thus consider first $\eta^2=0$; then $g(z)=(1-z^2)$ is a solution, and clearly it corresponds to the lowest eigenvalue since it has no nodes. The higher solutions are also polynomials.[15] The lowest eigenvalue is $\lambda=2$, as we know already. The "potential" $Q^{-1}(z)$ is an increasing function of the parameter η^2. Hence every eigenvalue λ must decrease as η^2 increases.

When $\eta^2\to 1$, $Q^{-1}(z)$ develops a singularity at $z=0$, in fact,

$$Q^{-1}(z) \approx (1-\eta^2+z^2)^{-1} \approx \pi(1-\eta^2)^{-\frac{1}{2}}\delta(z). \quad (55)$$

[15] See the general discussion in reference 6.

The lowest eigenfunction simply develops a kink at $z=0$, while the behavior of the higher states is more complicated; if one inserts the approximation (55) into (52), one finds

$$g(\zeta) = \tfrac{1}{2}\pi(1-\eta^2)^{-\frac{1}{2}}g(0)\lambda(1-|\zeta|), \quad (56)$$

which requires

$$\lambda = (2/\pi)(1-\eta^2)^{\frac{1}{2}}. \quad (57)$$

This is, of course, just what one expects from the nonrelativistic Balmer formula for the lowest eigenvalue.

Clearly the limit $\eta^2\to 1$ requires a more careful treatment for the higher eigenvalues. The reason is that all the nodes of the eigenfunction tend to concentrate near $z=0$ so that the approximation (55) is not adequate.

It is easy to see that λ does not tend to zero for the higher eigenvalues. Thus, none of the higher eigenvalues of Eq. (54) has anything to do with the states known from the nonrelativistic case. It will be shown by Cutkosky[6] that the other known states are contained in other families of solutions of the B-S Equation; each of these families, however, contains in addition "abnormal" solutions that have no nonrelativistic limit.

We shall now examine the behavior of the "abnormal" eigenvalues of Eq. (54) when $\eta\to 1$ and show that all these eigenvalues converge to a common limit $\lambda\to\frac{1}{4}$. First we can see that $\lambda<\frac{1}{4}$ cannot be an eigenvalue other than (57). Consider in fact the second eigenvalue; the corresponding eigenfunction must be odd and have a node at $z=0$. Hence we need only examine a solution of (54) with the boundary conditions $g(0)=g(1)=0$. Assuming

$$1-\eta^2 \ll 1, \quad (58)$$

we divide the interval $0-1$ into two parts,

$$0<z<z_0 \quad \text{and} \quad z_0<z<1, \quad (59)$$

choosing z_0 to satisfy

$$(1-\eta^2)^{\frac{1}{2}} \ll z_0 \ll 1. \quad (60)$$

In the first interval we write the equation with a slight change of variables,

$$\begin{gathered} x = \eta(1-\eta^2)^{-\frac{1}{2}}z, \\ d^2g/dx^2+\lambda(1+x^2)^{-1}g = 0, \end{gathered} \quad (61)$$

neglecting terms of order $\leq(I-\eta^2)^{\frac{1}{2}}$. (One can see *a posteriori* that this approximation is justified for our purposes.) Equation (61) is of Riemann's type, and the solution we want is

$$\begin{gathered} g = g^+ - g^-, \\ g^\pm = (1+x^2)\,{}_2F_1(\tfrac{3}{2}+\rho, \tfrac{3}{2}-\rho; 2; \tfrac{1}{2}(1\pm ix)), \\ \rho = (\tfrac{1}{4}-\lambda)^{\frac{1}{2}}. \end{gathered} \quad (62)$$

In the second interval we write $Q(z)=z^2$, again neglecting terms of order $(I-\eta^2)^{\frac{1}{2}}$ at most, and write

$$s=z^2,\quad d^2g/ds^2+\tfrac{1}{2}s^{-1}dg/ds+\tfrac{1}{4}\lambda g/(1-s)s^2=0, \tag{63}$$

which again is of Riemann's type. The solution satisfying $g=0$ at $z=1$ is

$$g=(1-z^2)z^{\frac{1}{2}+\rho}{}_2F_1(1\tfrac{1}{4}+\tfrac{1}{2}\rho,\tfrac{3}{4}+\tfrac{1}{2}\rho;2;1-z^2). \tag{64}$$

We will first show that if $\lambda<\frac{1}{4}$, the "internal" and "external" solutions (62) and (64) cannot join smoothly at $z=z_0$, i.e., $x=x_0=\eta z_0(1-\eta^2)^{-\frac{1}{2}}$. In fact, since $x_0\gg1$, we may evaluate (62) by means of the asymptotic formula for the hypergeometric function. One finds, omitting a proportionality factor,

$$g_{\text{int}}\sim x^{\rho+\frac{1}{2}}(1+\cdots)+A(\rho)x^{-\rho+\frac{1}{2}}(1+\cdots). \tag{62'}$$

The dots indicate expansions in powers of x^{-1}, and since $\rho<\frac{1}{2}$ it is consistent to keep the first term of the second expansion, while neglecting the higher terms of the first expansion. Furthermore,

$$A(\rho)=2^{2\rho}\tan(\tfrac{1}{4}\pi-\tfrac{1}{2}\pi\rho)\Gamma(2\rho)/\Gamma(-2\rho) \tag{65}$$

is a negative quantity which varies from 0 to -1 as λ varies from 0 to $\frac{1}{4}$.

Similarly, (64) may be evaluated for small values of z by means of the known transformation of $F(a, b, c, 1-s)$ to hypergeometric functions of the variable s. One finds

$$g_{\text{ext}}\sim z^{\rho+\frac{1}{2}}(1+\cdots)+B(\rho)z^{\rho+\frac{1}{2}}(1+\cdots), \tag{64'}$$

where the dots now indicate expansions in powers of z^2, and

$$B(\rho)=2^{2\rho}\Gamma(\rho)\Gamma(\tfrac{3}{2}-\rho)/\Gamma(-\rho)\Gamma(\tfrac{3}{2}+\rho) \tag{66}$$

is a quantity which on the whole interval $0<\lambda<\frac{1}{4}$ $(0<\rho<\frac{1}{2})$ stays quite close to -1 (and is in fact <-1).

Rewriting (62′) in terms of the variable z and omitting again a proportionality factor, we find

$$g_{\text{int}}\sim z^{\rho+\frac{1}{2}}+A(\rho)(1-\eta^2)^{\rho}z^{-\rho+\frac{1}{2}}, \tag{67}$$

which is of the same form as (64′), but with a coefficient for the second term which is smaller than $B(\rho)$ in absolute value, for all values of ρ in the stated interval. Hence (64′) and (67) can never join smoothly. In addition it is easy to verify that the slope g'/g is larger for (67) than for (64′), as one expects if λ is *too low* to be an eigenvalue.

Let us now turn to the case $\lambda>\frac{1}{4}$. One can see that essentially the same formulae will hold, except that ρ will be a pure imaginary, say $\rho=i\sigma$, $\sigma=(\lambda-\frac{1}{4})^{\frac{1}{2}}$. One sees, then, that (64′) and (67) take the respective forms

$$g_{\text{ext}}\sim z^{\frac{1}{2}}\sin(\sigma\ln z+\beta) \tag{64''}$$

and

$$g_{\text{int}}\sim z^{\frac{1}{2}}\sin(\sigma\ln z-\tfrac{1}{2}\sigma\ln(1-\eta^2)+\alpha), \tag{67'}$$

where α and β are phases depending on σ, which for small values of σ are of the form

$$\alpha=a\sigma,\quad \beta=b\sigma, \tag{68}$$

a and b being constants, whose precise value we shall not determine.

Obviously (64″) and (67′) can be joined smoothly if $\alpha-\beta-\frac{1}{2}\sigma\ln(1-\eta^2)=n\pi$, where n is an integer. If $1-\eta^2$ is so small that $-\ln(1-\eta^2)\gg1$, the above equation will have small roots σ so that by using (68),

$$\sigma\equiv(\lambda-\tfrac{1}{4})^{\frac{1}{2}}=n\pi/[a-b-\tfrac{1}{2}\ln(1-\eta^2)]. \tag{69}$$

To an even cruder approximation, one has

$$\sigma\sim-2\pi n/\ln(1-\eta^2);\quad \lambda=\tfrac{1}{4}+[2\pi n/\ln(1-\eta^2)]^2. \tag{70}$$

Equation (70), for $n=0, 1, 2,\cdots$ gives an infinity of eigenvalues all tending to $\lambda=\frac{1}{4}$ when $\eta^2\to1$. It should be pointed out that these correspond to odd eigenfunctions. In a similar way one can show, however, that the same formula, with $2n$ replaced by $2n+1$, gives the eigenvalues for the even eigenfunctions.

About the possible significance of these "abnormal" solutions we shall not try to speculate here. Since they occur only for finite values of λ $(\lambda\geqslant\frac{1}{4})$, it would be unwise to assume that they are a property of the complete B-S Equation. Certainly the ladder approximation cannot be trusted to that extent. If the theory is used only for small values of the coupling constant, the abnormal solutions do not exist, in the case we have studied, and no contradiction with known facts can be established. Nevertheless it would seem that these solutions deserve further study.

ACKNOWLEDGMENTS

The present work was begun while the author was a guest of the Institute for Advanced Study, Princeton, New Jersey. The author is happy to acknowledge his indebtedness to the Director of the Institute, Professor J. R. Oppenheimer, for the stimulation and encouragement he derived from a year's stay at the Institute. Various members of the Institute, in particular, Professor F. Dyson, Professor G. Källén, Professor A. Pais, Professor W. Pauli, and Professor R. Jost gave kind encouragement and invaluable criticism. Special thanks are due to Dr. Murray Gell-Mann for suggesting that the analogy discussed in reference 14 might be of help.

APPENDIX

We shall construct here the Green's function $G(x)$, which is a solution of $(p_a^2+m_a^2)(p_b^2+m_b^2)G(x)=\delta(x)$, p_a and p_b being defined by Eqs. (13) and (28), with $p=-i$ Grad. We shall calculate G for the general case $m_a\neq m_b$, since this involves no additional difficulty. Using Fourier transforms, one sees at once that

$$G(x)=(2\pi)^{-4}\int[(p_a^2+m_a^2)(p_b^2+m_b^2)]^{-1}e^{ipx}[dp]. \tag{A1}$$

In the following we use $\frac{1}{2}(m_a+m_b)$ as the unit of mass, setting

$$m_a=1+\Delta, \quad m_b=1-\Delta. \tag{A2}$$

Furthermore we transform, *à la* Feynman,

$$[(p_a^2+m_a^2)(p_b^2+m_b^2)]^{-1}=\tfrac{1}{2}\int_{-1}^{+1}[p,y,\Delta]^{-2}dy, \tag{A3}$$

where

$$[p,y,\Delta]\equiv p^2+2i(y+\Delta)(p\cdot\eta) +(1-\eta^2)(1+2y\Delta+\Delta^2). \tag{A4}$$

Furthermore, applying to $Q\equiv[p,y,\Delta]$ the formula

$$Q^{-2}=\int_0^\infty e^{-\alpha Q}\alpha d\alpha$$

and inserting into (A1), the integration over p may be performed, with the result

$$G(x)=(32\pi^2)^{-1}\int_{-1}^{+1}dy\int_0^\infty \alpha^{-1}d\alpha \times\exp[-\alpha U-\tfrac{1}{4}R^2\alpha^{-1}+(y+\Delta)(x\eta)], \tag{A5}$$

with

$$U=(1+2y\Delta+\Delta^2)(1-\eta^2)+\eta^2(y+\Delta)^2. \tag{A6}$$

Owing to (29), U is positive for $|y|\leq 1$; hence the integral over α in (A5) is always meaningful.

We then find that

$$G(x)=(4\pi)^{-2}e^{\epsilon\Delta x_4}\int_{-1}^{+1}dy \times e^{y\epsilon x_4}K_0(R(1-\eta^2+\Delta^2+2y\Delta+y^2\eta^2)^{\frac{1}{2}}), \tag{A7}$$

where $K_0(z)$ is the modified Hankel function, $i(\pi/2)\times H_0^1(iz)$. The asymptotic behavior of (A7) when $R\to\infty$ in a specified direction (i.e., keeping x_4/R constant) is found noting that $K_0(z)\sim(\pi/2z)^{\frac{1}{2}}e^{-z}$. The exponential part of (A7) is then

$$G(x)\sim\cdots\int dy\exp[-Rf(y)],$$

where

$$f(y)=[(1-\eta^2)(1-y^2)+(y+\Delta)^2]^{\frac{1}{2}}-\epsilon(y+\Delta)x_4R^{-1}.$$

It is easy to see that $f(y)>0$ in the whole interval $-1\leq y\leq+1$. Hence $G(x)$ satisfies the boundary condition $G\to 0$ as $R\to\infty$ in any direction. If y_m is the point in the interval where $f(y)$ is a minimum, then the strongest factor in the asymptotic dependence of $G(x)$ is

$$G(x)\sim\exp[-Rf(y_m)]. \tag{A8}$$

Notice that y_m depends on the direction. Consider, for example, the simplest case $\Delta=0$. Then if $|x_4|<\epsilon R$, y_m is defined by the minimum condition

$$y_m\epsilon R=x_4(1-\epsilon^2+y^2\epsilon^2)^{\frac{1}{2}}; \tag{A9}$$

that is, writing $x_4/R=\cos\theta_4$, $y_m\epsilon=(1-\epsilon^2)^{\frac{1}{2}}\cot\theta_4$. If $|\cos\theta_4|>\epsilon$ the root (A9) is not inside the interval, so the minimum of $f(y)$ occurs at $y=\pm1$, according as $\cos\theta_4\gtrless 0$; summarizing, one has

$$\begin{aligned} &|\cos\theta_4|>\epsilon \quad f(y_m)=1-\epsilon|\cos\theta_4| \\ &|\cos\theta_4|<\epsilon \quad f(y_m)=(1-\epsilon^2)^{\frac{1}{2}}\sin\theta_4. \end{aligned} \tag{A10}$$

Notice that in the latter case,

$$G(x)\sim\exp[-(1-\epsilon^2)^{\frac{1}{2}}r\sin\theta_4]=\exp[-(1-\epsilon^2)^{\frac{1}{2}}r],$$

if r is the length of the space component of x. In the former case, instead, $G(x)\sim e^{-R+\epsilon|x_4|}$; in particular, in the time direction $G(x)$ tends to zero like $\exp[-(1-\epsilon)\times|x_4|]$.

PHYSICAL REVIEW D VOLUME 8, NUMBER 10 15 NOVEMBER 1973

Covariant Harmonic Oscillators and the Quark Model*

Y. S. Kim

Center for Theoretical Physics, Department of Physics and Astronomy, University of Maryland, College Park, Maryland 20742

Marilyn E. Noz

Department of Physics, Indiana University of Pennsylvania, Indiana, Pennsylvania 15701

(Received 22 March 1973; revised manuscript received 20 July 1973)

An attempt is made to give a physical interpretation to the phenomenological wave function of Yukawa, which gives a correct nucleon form factor in the symmetric quark model. This wave function is first compared with the Bethe-Salpeter wave function. It is shown that they have similar Lorentz-contraction properties in the high-momentum limit. A hyperplane harmonic oscillator is then introduced. It is shown that the Yukawa wave function, which is defined over the entire four-dimensional Euclidean space, can be interpreted in terms of the three-dimensional hyperplane oscillators. It is shown further that this wave function satisfies a Lorentz-invariant differential equation from which excited harmonic-oscillator states can be constructed, and from which a gauge-invariant electromagnetic interaction can be generated.

I. INTRODUCTION

The quark,[1,2] which was originally introduced to explain SU(3) symmetry and its consequences, has gained considerable ground as a fundamental constituent particle in hadrons. The inventors of the quark did not make any commitment to its existence.[1] In spite of this and other well-known difficulties, model calculations based on this constituent particle are producing increasingly encouraging results.[3]

In both the successful and the disappointing features of the quark model, there seems to be one crucial question: What "forces" are responsible for making quarks stay together in hadrons?[4] In the early days of the quark model, quarks were put into an infinite potential purely for convenience,[2,5] and no attempts were made to assert that these simplified forces were of fundamental importance. In the symmetric quark model, for instance, Greenberg used the harmonic-oscillator potential in order to borrow the well-known formalism from the nuclear shell model.

As the resonance spectrum became richer, the search for quantum numbers that correspond to binding forces continued.[6] It was Kim and Noz[7] who established the existence of harmonic-oscillator-like radial modes for nonstrange baryon resonances for which there is barely enough experimental data to test the linearity of the three lowest energy levels.

There are numerous calculations of decay rates in the harmonic-oscillator model.[8] A more important analysis seems to be that of elastic form factors. The first objection to the use of the harmonic oscillator, that is the Gaussian, wave function is that the form factor decreases exponentially for large t values. This discrepancy with the real world together with our field-theoretic common sense once led us to conclude that the harmonic-oscillator potential, which is manifestly analytic at the origin, cannot be the fundamental force between the quarks.[9] However, an encouraging development was that the relativistic effect suitably applied on the Gaussian wave function eliminates this unwanted exponential decrease and gives the desired dipole effect.[10-13]

The above-mentioned relativistic models are essentially one or another form of the Gaussian wave function multiplied by a Lorentz contraction factor, and they do not necessarily represent a completely consistent picture of the relativistic bound state. The important fact, however, is that all those "wrong" models give a correct form factor. We are thus led to believe that there is some truth in the Lorentz contraction of quantum-mechanical wave functions.[14]

We realize that there are no completely consistent relativistic measurement theories and that we are not going to solve this difficult problem in this paper. For this reason, we can give relativistic interpretations only in terms of the existing languages that have been developed to answer this ultimate question. One commonly used language is the Bethe-Salpeter equation.[14,15] This equation is well known to us and has been used extensively in both low- and high-energy physics.[16]

Another important language developed for the same purpose is the hyperplane generalization of the Schrödinger equation. The concept of a spacelike hyperplane played a crucial role in the early days of quantum field theory.[17] This hyperplane technique was used recently by Fleming to understand the Newton-Wigner localization problem.[18,19] We shall use this hyperplane language in order to understand Lorentz-contracted Gaussian wave

Reprinted from *Phys. Rev. D* **8**, 3521 (1973).

functions.

We are specifically interested in the covariant oscillator wave function first introduced by Yukawa[20] and used by Fujimura *et al.*[10,11] in their successful calculation of the nucleon form factor in the symmetric quark model. In spite of their numerical success, there does not seem to be any physical basis for the covariant differential equation from which the wave function is derived. Thus it is fair to say that the Yukawa wave function has been a purely phenomenological entity. The purpose of this paper is to give a physical meaning to this wave function in terms of the accepted relativistic languages.

In Sec. II, we compare the Yukawa wave function with the Bethe-Salpeter wave function. It is pointed out that both wave functions are to be integrated over the four-dimensional Euclidean space in the low-momentum region. We note that both the Yukawa and the Bethe-Salpeter wave functions have the same Lorentz contraction properties in the large-momentum limit. Since the Bethe-Salpeter equation is a field-theoretic model, we believe that this is the point where Yukawa's nonlocal theory makes contact with local field theory.

In Sec. III, we introduce the hyperplane technique. The nonrelativistic harmonic oscillator can be generalized to covariant hyperplanes. We present a hyperplane interpretation of the Yukawa wave function which is consistent with the Lorentz-invariant probability and the observed nucleon form factor.

In Sec. IV, we discuss a Lorentz-invariant differential equation which the Yukawa wave function satisfies. This equation can generate a gauge-invariant electromagnetic interaction. It is shown that this harmonic-oscillator differential equation can be separated in the normal coordinate variables which are Lorentz transformations of the space-time variables, and that the excited states can be constructed in this normal coordinate system. A Lorentz-invariant mass eigenvalue is given.

In Sec. V, we discuss briefly the experimental basis upon which the harmonic-oscillator quark model is built.

II. PROPERTIES OF THE YUKAWA AND THE BETHE-SALPETER WAVE FUNCTIONS

In this section, we compare the covariance properties of the Bethe-Salpeter and the Yukawa wave functions. In the early days of nonrelativistic quantum mechanics, the standing-wave properties for the square well, the harmonic oscillator, and the other bound-state potentials were described by different mathematical techniques. However, the inherent similarities enabled the creators of quantum mechanics to formulate a new concept of bound states in terms of the quantum superposition principle. By studying the properties that are common to the Bethe-Salpeter and the Yukawa wave functions, which have different mathematical forms, we expect to work toward finding a possible new form of relativistic dynamics.

Since the Bethe-Salpeter equation and its wave functions are well known,[15] we will only describe here how Yukawa arrived at his covariant harmonic-oscillator model. Yukawa noticed that Born's reciprocity relation[21] gives an oscillator-like Hamiltonian and attempted to write down a normalizable wave function in terms of the relative internal coordinates. The covariance requirement, however, forced him to introduce time-like excitations with negative energies. As a consequence, the energy levels were infinitely degenerate. In order to eliminate this undesirable feature, Yukawa introduced a coupling with an external momentum. His wave function takes the form

$$\Psi(x,p)=\exp\{-\tfrac{1}{2}\omega[\,x^2+2(p\cdot x)^2/m^2]\}\,, \tag{1}$$

where x is the relative space-time four-vector and p is the total four-momentum of the bound state. Throughout this paper we use the space-favored metric where $x^2=(\vec{x})^2-x_0^2$.

The bound-state Bethe-Salpeter Green's function takes the form[14,15]

$$G(x,p)=\left(\frac{1}{4\pi}\right)^2\int_0^1 d\alpha\cos(\tfrac{1}{2}\alpha p\cdot x)\times K_0(\tfrac{1}{2}(x^2)^{1/2}[\,4M^2-(1-\alpha^2)m^2\,]^{1/2})\,. \tag{2}$$

This Green's function is seen to be a function of x and p as in Yukawa's function above [Eq. (1)]. The mass of the bound state is given by m. We consider here the bound state of two equal-mass particles whose individual mass is M. This Green's function is not the solution of the equation but contains most of the features of the exact wave function.[15]

We are now ready to discuss the covariance properties that are common to Eq. (1) and Eq. (2). We start from the rest frame where $\vec{p}=0$. In this system, Eq. (1) becomes a harmonic-oscillator wave function in the four-dimensional Euclidean space of $\vec{x}$ and t, and is manifestly normalizable. We can make the Bethe-Salpeter Green's function of Eq. (2) normalizable in the four-dimensional Euclidean space of $(\vec{x},t)$ by making the Wick rotation.[15]

As we increase $|\vec{p}|$, this property holds for Eq. (2) until the kinetic energy becomes larger than the binding energy.[14] For $|\vec{p}|$ larger than the binding energy, the Bethe-Salpeter wave function is no longer normalizable in the above-mentioned four-dimensional Euclidean space. The harmonic-oscillator wave function of Eq. (1) does not suffer from this effect and remains normalizable for large values of $|\vec{p}|$. This is expected because particles bound by an oscillator potential have infinite binding energy.

Let us rewrite the oscillator wave function assuming that $\vec{p}$ is in the z direction. We use E for p_0 and p for p_z. Then

$$\Psi(x,p) = \exp[-\tfrac{1}{2}\omega(x^2+y^2)] \times \exp\{(-\omega/4m^2)[(E-p)^2(t+z)^2 + (E+p)^2(t-z)^2]\} . \quad (3)$$

For large p,

$$\frac{\omega(E-p)^2}{4m^2} \to \frac{\omega}{16}\left(\frac{m}{p}\right)^2, \qquad \frac{\omega(E+p)^2}{4m^2} \to \omega\left(\frac{p}{m}\right)^2. \quad (4)$$

Thus

$$\Psi(x,p) \to \exp[-\tfrac{1}{2}\omega(x^2+y^2)] \times \exp[-\tfrac{1}{16}\omega(m/p)^2(t+z)^2] \times \exp[-\omega(p/m)^2(t-z)^2] . \quad (5)$$

The last factor becomes $(\sqrt{\pi}/\omega)(m/p)\delta(t-z)$ for large p, and the dependence on the variable $(t+z)$ becomes insensitive by the factor $(m/p)^2$. This contraction behavior is strikingly similar to that of the Bethe-Salpeter equation.[14] The Bethe-Salpeter wave function is a model derivable from field theory. The oscillator function is a phenomenological wave function giving correct form factors. It is interesting to note that these two wave functions have the same Lorentz contraction properties in the large-p limit.

We now restrict ourselves to the Yukawa wave function. Let us analyze the form factor calculation of Fujimura *et al.*[10] in the Breit system. We can sketch the initial and final "Lorentz-contracted" wave functions as in Fig. 1. The form-factor integral

$$F(q^2) = \int d^4x\, \psi_f^*(x)\psi_i(x)\exp(iq\cdot x) , \quad (6)$$

where q is the momentum transfer, receives contributions only from the small overlapping region indicated in Fig. 1. This region shrinks as the momentum transfer increases, and this coherent

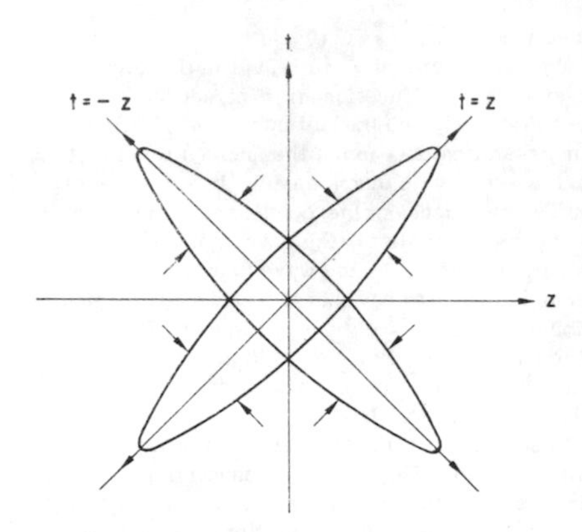

FIG. 1. Lorentz-contracted wave functions with two equal and opposite momenta. The form-factor integral of Fujimura *et al.* receives contributions primarily from the small overlapping region.

shrinkage is responsible for the nonexponential decrease of the form factor.

In Eq. (6), the integral is performed over Euclidean space-time. We know clearly the physical meaning of the probability distribution over the three-dimensional space, but we do not know what physics, if any, the timelike probability distribution corresponds to. We shall discuss this problem in Sec. III.

III. HYPERPLANE FORMALISM OF HARMONIC OSCILLATOR

Here we study Yukawa's phenomenological wave function from the point of view of the nonrelativistic-harmonic-oscillator wave function, generalized to covariant hyperplanes.

Let us start with the nonrelativistic harmonic oscillator. The Hamiltonian is separable and the wave function is Gaussian multiplied by the appropriate polynomials corresponding to excited energy levels. Because the ground-state wave function depends only on $(\vec{x})^2$ in the exponent, we can Lorentz-generalize $\vec{x}$ to the three-vector on the hyperplane which is perpendicular to the total four-momentum of the system. We follow the standard method of constructing this three-vector and

$$\tilde{x}_\mu = \left(\delta_\mu^\nu + \frac{p_\mu p^\nu}{m^2}\right)x_\nu . \quad (7)$$

When the momentum $\vec{p}$ is zero, $\tilde{x}_\mu$ becomes $\vec{x}$. For nonzero $\vec{p}$,

$$p^\mu \tilde{x}_\mu = 0$$

and

$$\tilde{x}^\mu \tilde{x}_\mu = x^\mu x_\mu + \left(\frac{p\cdot x}{m}\right)^2 . \tag{8}$$

Assume now that $\vec{p}$ is in the z direction. Using p for p_z and β for v/c, Eq. (8) becomes

$$\tilde{x}^\mu \tilde{x}_\mu = x^\mu x_\mu + (1-\beta^2)^{-1}(t-\beta z)^2 . \tag{9}$$

The three independent hyperplane coordinate variables are

$$x,\ y,\ \text{and}\ (1-\beta^2)^{-1/2}(z-\beta t) . \tag{10}$$

The hyperplane ground-state oscillator wave function then takes the form

$$\phi(x,\beta) = \exp\{-\tfrac{1}{2}\omega[x^2 + (1-\beta^2)^{-1}(t-\beta z)^2]\} . \tag{11}$$

There are two important differences between the above wave function and that of Eq. (1). First, the coefficients of $(p\cdot x/m)^2$ are different. In Eq. (1), it is 2, while it is 1 in Eq. (11). Next, Eq. (1) is integrated over the entire four-space while Eq. (11) is integrated only over the three-dimensional hyperplane. The purpose of this section is to point out that we can indeed give a hyperplane interpretation to the Yukawa wave function of Eq. (1).

The wave function given in Eq. (11), which depends explicitly on β, is the ground-state wave function. We can excite the harmonic oscillator just as in the nonrelativistic case. If we multiply ϕ or its excited form by $\exp[-\frac{1}{2}\omega(1-\beta^2)^{-1}(t-\beta z)^2]$, it does not change the hyperplane oscillator because the variable $-(1-\beta^2)^{-1/2}(t-\beta z)$ is perpendicular to the three hyperplane variables given in Eq. (9). If we perform the integration over the variable $-(1-\beta^2)^{-1/2}(t-\beta z)$ after this multiplication, this certainly leaves the hyperplane oscillator intact. Therefore we can write the inner product of two wave functions belonging to the same hyperplane as

$$(\phi_n, \phi_m) = \int \exp[-\omega(1-\beta^2)^{-1}(t-\beta z)^2] \times \phi_n^*(x,\beta)\phi_m(x,\beta)d^4x . \tag{12}$$

The integration measure d^4x is invariant under Lorentz transformation. Because of this the above quantity does not depend on the hyperplane parameter β. Hence, we have introduced a multiplication factor, $\exp\{-\frac{1}{2}\omega[\frac{1}{2}(p\cdot x)]^2\}$, and an inner product of the form of Eq. (12), while leaving the hyperplane oscillator intact. By doing this we have been able to show that the hyperplane probability is Lorentz-invariant.

Let us consider next the inner product between wave functions belonging to two different hyperplanes.[22] Since nonrelativistic quantum mechanics does not say anything about Lorentz transformation, it cannot give the transition probability between two such wave functions. We believe that this is one of the most pressing problems of our time and that we can solve this problem only by building models that can produce the observed experimental data.

In order to build such a model, we go back to our original rule that $\exp[-\frac{1}{2}\omega(p\cdot x/m)^2]$ multiply each wave function and that the integral be performed over the entire four-space. Then the inner product becomes

$$(\phi_1, \phi_2) = \int \exp\left\{-\frac{\omega}{2}\left[\left(\frac{p_1\cdot x}{m_1}\right)^2 + \left(\frac{p_2\cdot x}{m_2}\right)^2\right]\right\} \times \phi_1(x,\beta_1)\phi_2(x,\beta_2)d^4x . \tag{13}$$

Here again the integration measure d^4x is hyperplane-independent and is good for both the β_1 and the β_2 plane. The above expression becomes Eq. (12) when β_1 and β_2 are equal.

The next and most crucial question is whether the above inner product produces experimentally measurable effects. The answer is contained in the fact that because of the additional exponential factor, the form factor calculation with this inner product becomes exactly the phenomenological form of Fujimura *et al.* which we discussed in Sec. II. The single-oscillator ground-state form factor becomes in the Breit system

$$F(q^2) = \int d^4x \exp[i\vec{q}\cdot\vec{x}] \exp[-\omega(x^2+y^2)] \times \exp\left[-\frac{\omega}{m^2}(m^2+2q^2)(t^2+z^2)\right] . \tag{14}$$

For large q^2, the time integral is like a δ-function integral, and hence this form becomes that of Licht and Pagnamenta[12] who proposed the instant $(t=0)$ probability integral.

We have thus generalized the time-independent harmonic oscillator to covariant hyperplanes, and then introduced a covariant inner product. This operation leaves the hyperplane oscillator intact, produces the Lorentz-invariant probability for the states belonging to the same hyperplane and gives the correct nucleon form factor.

At this point, we may mention that the mathematics of the covariant harmonic oscillator is very similar to that of the quantization of the electromagnetic field. There are two well-known approaches to the electromagnetic field quantization. One uses the Lorentz gauge, and the other uses the Coulomb gauge. The Coulomb-gauge method is not manifestly covariant, but its main advantage is that we do not introduce unphysical photons and thus we can make quick references to the real world.

There have been many previous attempts to

understand the covariant harmonic oscillator.[23] In this paper, we used the hyperplane coordinates to avoid timelike excitations. The advantages are similar to those in the Coulomb gauge case. By eliminating completely the burden of handling those unphysical excitations, we have been able to separate clearly what can be done and what cannot be done in the framework of nonrelativistic quantum mechanics. We emphasize here that a relativistic measurement theory has yet to be constructed.[24]

IV. COVARIANT DIFFERENTIAL EQUATION AND EXCITED STATES

In the preceding sections, we studied a possible physical interpretation of the Gaussian factor which corresponds to a ground-state harmonic oscillator. In order to construct excited states, we use the Lorentz-invariant differential equation which is needed in generating a gauge-invariant electromagnetic interaction of the harmonic-oscillator quarks.[11]

We rewrite here the ground-state solution

$$\psi_0(x,p)=\exp\left\{-\frac{\omega}{2}\left[x^2+2\left(\frac{x\cdot p}{m}\right)^2\right]\right\}$$

as

$$\psi_0(x,p)=\psi_0(y)=\exp[-\tfrac{1}{2}\omega(y_1^2+y_2^2+y_3^2+y_0^2)]\,, \qquad (15)$$

where

$$\begin{aligned} y_1&=x_1\,,\quad y_2=x_2\,,\\ y_3&=(1-\beta^2)^{-1/2}(x_3-\beta t)\,,\\ y_0&=(1-\beta^2)^{-1/2}(t-\beta x_3)\,. \end{aligned} \qquad (16)$$

The above linear transformation is a homogeneous Lorentz transformation of the original coordinate variables. Thus $\psi_0(x,p)$ satisfies the equation

$$\left\{-\nabla_y^2+\frac{\partial^2}{\partial y_0^2}+\omega^2[(\vec{y})^2-y_0^2]\right\}\psi_0(y)=\lambda\psi_0(y)\,. \qquad (17)$$

Since the transformation of Eq. (16) is a Lorentz transformation, we also have

$$\left\{-\nabla^2+\frac{\partial^2}{\partial t^2}+\omega^2[(\vec{x})^2-t^2]\right\}\psi_0(x,p)=\lambda\psi_0(x,p)\,. \qquad (18)$$

Eq. (17) and Eq. (18) represent the same equation. The form of Eq. (18) has been discussed in the literature and is used in constructing a gauge-invariant electromagnetic interaction.

The Gaussian form of Eq. (15) is not separable in the x-coordinate variables. It is separable in the y variables which contain the p dependence. Thus we have to use Eq. (17) to construct excited states. Because of the Lorentz invariance of the harmonic-oscillator operator, the excited-state wave functions also satisfy the differential equation of Eq. (18).

We now write the excited-state solution as

$$\psi_\lambda(y)=H_{n_1}(y_1)H_{n_2}(y_2)H_{n_3}(y_3)H_{n_0}(y_0)\times\exp[-\tfrac{1}{2}\omega(\vec{y}^2+y_0^2)]\,, \qquad (19)$$

where

$$\lambda=\omega(n_1+n_2+n_3-n_0+1)\,. \qquad (20)$$

The above solution is possible because the starting differential equation of Eq. (17) is separable and remains separable as we change the value of the total four-momentum p. The quantum numbers n_i are separation constants. Our Lorentz transformation therefore preserves this separability. Because of the minus sign in front of n_0, the eigenvalues of Eq. (20) are infinitely degenerate. In order to remove this ambiguity, we set $n_0=0$; the physics of this procedure has been discussed in Sec. III. Thus

$$\lambda=\omega(N+1)\,,$$

where (21)

$$N=n_1+n_2+n_3\,.$$

Since the separability is preserved, the $n_0=0$ condition is invariant under a Lorentz transformation. The covariant harmonic oscillator now has three normal excitation variables, namely, y_1, y_2, and y_3, and they are precisely the hyperplane variables mentioned in Sec. III. They are O(3)-invariant within the hyperplane and generate covariant excited-state wave functions in exactly the same way as in the nonrelativistic oscillator.

The eigenvalue λ can serve as the mass of the covariant harmonic oscillator or as its mass squared. There have been many previous attempts to express the bound-state mass as an eigenvalue of a differential equation.[8,10,20,23] In all these attempts, except possibly that of Lipes,[25] the x variables are used to excite the harmonic oscillator. Since the Gaussian factor is separable in the x variables only in the rest frame, the mass quantum numbers are good only in that frame, and an attempt to boost will bring in an infinite number of unphysical wave functions.

It has been twenty years since Yukawa introduced the Gaussian factor corresponding to the Lorentz contraction.[20] The concept of quark did not exist at that time. It was stated in Yukawa's paper that the differential equation representing the coupling of the total momentum to the internal oscillation is so complicated that the study of the interaction of the internal mode with the external field is very difficult. We have shown in this paper that the differential equation is similar to the Klein-Gordon equation, and that the interaction can be manufactured in the usual way.

V. CONCLUDING REMARKS

In this paper, we discussed, first, Lorentz contraction properties of the covarient Gaussian factor. We then proposed the use of the hyperplane technique to study possible relativistic ingredients in quantum mechanics. Finally, we introduced the normal-coordinate method in solving the covariant harmonic-oscillator equation, and showed that this method is technically equivalent to the hyperplane method.

The normal-coordinate method is the most powerful weapon in attacking harmonic-oscillator problems. It is a convenient way of describing covariantly the orbital and radial quantum numbers. Therefore we have studied in this paper a possible theoretical tool which can link the basic concepts of quantum mechanics to quantities that can be measured experimentally.

The most widely available numbers that can be both calculated and measured are decay rates.[8] Since the decay rate calculations are not sensitive to the exact shape of the wave function, the decay rate alone does not force us to accept the harmonic-oscillator model.

The form factor study such as the one discussed in this paper strengthens our assertion on the harmonic oscillator and enables us to relate the observed curve to Lorentz contractions.[26]

The most important characteristic of the harmonic oscillator is, of course, the linearity of its eigenvalues. In order to study the linearity in the observed mass spectra, we need at least three radial modes. For nonstrange baryons, we barely have these three levels, and the present authors studied this linearity.[7]

Radial quantum numbers ↓	Baryons		Mesons	
	non-strange	strange	non-strange	strange
n = 0	A	A	A	A
n = 1	A	A^-	B^-	C
n = 2	A^-	C	D	D
n = 3				

FIG. 2. Summary of the present status of the multiplet scheme in the symmetric quark model. A means "excellent", B means "good", etc.

In this paper, we have restricted ourselves to nonstrange baryons. We realize that there are some difficulties in pionic form factors.[10] As we see in the experimental summary of Fig. 2, we do not yet have enough experimental information from which a linear mass spectrum can be derived for the mesons. Therefore we cannot and do not insist on the simple harmonic oscillator for the mesons. Consequently, we do not have to explain the above-mentioned difficulty at this time.

ACKNOWLEDGMENTS

This work was started when one of us (M. E. N.) was visiting the University of Maryland during the summer of 1972. This visit was supported by the National Science Foundation. She would like to thank her colleauges at Maryland for the hospitality extended to her during the summer.

*Work supported in part by the National Science Foundation Grant No. NSF GP 8748.

[1]M. Gell-Mann, Phys. Lett. 8, 214 (1964).

[2]G. Zweig, CERN Report Nos. TH401 and TH412, 1964 (unpublished).

[3]J. J. J. Kokkedee, *The Quark Model* (Benjamin, New York, 1969).

[4]For the latest attempt to keep quarks inside the hadron, see K. Johnson, Phys. Rev. D 6, 1101 (1972).

[5]O. W. Greenberg, Phys. Rev. Lett. 13, 598 (1964).

[6]O. W. Greenberg and M. Resnikoff, Phys. Rev. 163, 1844 (1967); D. R. Divgi and O. W. Greenberg, *ibid.* 175, 2024 (1968). For the latest numerical analysis of the $N = 1$ and $N = 2$ multiplets, see C. T. Chen-Tsai and T. Y. Lee, Phys. Rev. D 6, 2459 (1972).

[7]Y. S. Kim and M. E. Noz, Nuovo Cimento 11A, 513 (1972). See also T. De, Y. S. Kim, and M. E. Noz, *ibid.* 13A, 1089 (1973).

[8]R. P. Feynman, M. Kislinger, and F. Ravndal, Phys. Rev. D 3, 2706 (1971), and the references contained therein.

[9]S. D. Drell, A. Finn, and M. Goldhaber, Phys. Rev. 157, 1402 (1967).

[10]K. Fujimura, T. Kobayashi, and M. Namiki, Prog.

Theor. Phys. 43, 73 (1970).

[11]For a complete and through treatment of the form-factor calculation, see R. Lipes, Phys. Rev. D 5, 2849 (1972).

[12]A. L. Licht and A. Pagnamenta, Phys. Rev. D 2, 1150 (1970); *ibid.* 2, 1156 (1970).

[13]G. Cocho, C. Fronsdal, I. T. Grodsky, and R. White, Phys. Rev. 162, 1662 (1967).

[14]For a discussion of the Lorentz contraction of the Bethe-Salpeter wave function, see Y. S. Kim and R. Zaoui, Phys. Rev. D 4, 1764 (1971).

[15]G. C. Wick, Phys. Rev. 96, 1124 (1954).

[16]For the latest high-energy applications, see S. D. Drell and T. D. Lee, Phys. Rev. D 5, 1738 (1972); C. H. Woo, *ibid.* 6, 1127 (1972).

[17]S. Tomonaga, Prog. Theor. Phys. 1, 27 (1946); J. Schwinger, Phys. Rev. 82, 914 (1951).

[18]G. N. Fleming, Phys. Rev. 137, B188 (1965); G. N. Fleming, J. Math. Phys. 11, 1959 (1966).

[19]T. D. Newton and E. P. Wigner, Rev. Mod. Phys. 21, 400 (1949).

[20]H. Yukawa, Phys. Rev. 91, 416 (1953).

[21]M. Born, Rev. Mod. Phys. 21, 463 (1949).

[22]It was pointed out by Kogut and Susskind that the problem of physical systems belonging to two different hyperplanes is a dynamical question. It is of course a relativistic dynamical question. See J. Kogut and L. Susskind, Phys. Rep. 8, 75 (1973).

[23]For the latest discussion of the covariant oscillators, see S. Ishida and J. Otokozawa, Prog. Theor. Phys. 47, 2117 (1972).

[24]G. F. Chew, Phys. Rev. D 4, 2330 (1971). In this paper, Chew states that the construction of a theoretical model which is demonstrably compatible both with the quantum superposition principle and with relativistic space-time is one of the most pressing problems. We agree with him. See also Y. S. Kim and K. V. Vasavada, Phys. Rev. D 5, 1002 (1972).

[25]Lipes[11] uses the $\bar{x}_\mu$ of Eq. (7) as his independent variables. They are not linearly independent, and they do not form a set of variables in which the four-dimensional oscillator equation is completely separable. They are not the y variables we use in this paper.

[26]For the latest experimental indication of the harmonic-oscillator characteristic, see P. S. Kummer, E. Ashburn, F. Foster, G. Hughes, R. Siddle, J. Allison, B. Dickinson, E. Evangelides, M. Ibboton, R. S. Lawson, R. S. Meaburn, H. E. Montgomery, and W. J. Shuttleworth, Phys. Rev. Lett. 30, 873 (1973).

Orthogonality relation for covariant harmonic-oscillator wave functions

Michael J. Ruiz
Center or Theoretical Physics, Department of Physics and Astronomy,
University of Maryland, College Park, Maryland 20742
(Received 20 May 1974)

Orthogonality relations for the Kim-Noz covariant harmonic-oscillator wave functions are discussed. It is shown that the wave functions belonging to different Lorentz frames satisfy an orthogonality relation. Furthermore, it is shown that for $n = m$ transitions there is a contraction factor of $(1 - \alpha^2)^{(n+1)/2}$, where α is the velocity difference between the two Lorentz frames.

The covariant harmonic-oscillator wave functions recently proposed by Kim and Noz[1] can be applied to a wide range of hadronic processes. The harmonic-oscillator characteristics prominently show up in hadronic mass spectra. The Lorentz-contraction properties of the oscillator wave functions can be seen in the nucleon elastic form factors and other electromagnetic transition amplitudes.[2]

In their paper, Kim and Noz are primarily concerned with the probability interpretation of their covariant wave functions. Their oscillator wave functions satisfy all the requirements of nonrelativistic quantum mechanics and enable us to extend the probability concept to the relativistic region. Kim and Noz, however, did not explicitly calculate the overlap integral of harmonic-oscillator wave functions belonging to two different Lorentz frames. The purpose of this note is to perform the overlap integral.

Kim and Noz start with the following differential equation:

$$\frac{1}{2}\left\{ -\nabla^2 + \frac{\partial^2}{\partial t^2} + \omega^2[(\vec{x})^2 - t^2]\right\} \psi(x) = \lambda\psi(x) . \tag{1}$$

This harmonic-oscillator equation is separable in the $\vec{x}$ and t variables. Kim and Noz then observe that the above equation can also be written as

Reprinted from *Phys. Rev. D* **10**, 4306 (1974).

$$\frac{1}{2}\left\{-\nabla_y^2 + \frac{\partial^2}{\partial y_0^2} + \omega^2[(\vec{y})^2 - y_0^2]\right\} \psi(y) = \lambda\psi(y), \tag{2}$$

where the y variables are the Lorentz transforms of the x variables:

$$\begin{aligned} y_1 = x_1, \quad y_2 &= x_2, \\ y_3 &= (1 - \beta^2)^{-1/2}(x_3 - \beta t), \\ y_0 &= (1 - \beta^2)^{-1/2}(t - \beta x_3). \end{aligned} \tag{3}$$

Equation (2) is also separable in the y variables. The normalizable solutions in the y variables are the Kim-Noz wave functions. Their wave function has the form

$$\psi_\lambda(y) = NH_{n_1}(y_1)H_{n_2}(y_2)H_{n_3}(y_3)\exp[-\frac{1}{2}\omega(\vec{y}^2 + y_0^2)], \tag{4}$$

where $\lambda = \omega(n_1 + n_2 + n_3 + 1)$, and N is the normalization constant. The elimination of timelike oscillations can be done covariantly.[1] Since the transverse oscillations do not undergo Lorentz transformations, we shall assume that $n_1 = n_2 = 0$ in the following discussion.

The purpose of this note is to evaluate the following integral:

$$T_{nm}(\beta, \beta') = \int \psi_n(y)\psi_m(y')d^4x, \tag{5}$$

where the y' variables are the y variables of Eq. (3) with β'. We can evaluate the above integral using the generating function for Hermite polynomials. This generating function has the form

$$\exp(-s^2 + 2sy_3) = \sum_{n=o}^{\infty} \frac{s^n}{n!}H_n(y_3). \tag{6}$$

We can now use the integral

$$\begin{aligned} I(s,r) = \left[\frac{\omega}{\pi}\right]^2 &\int d^4x \exp(-s^2 + 2sy_3)\exp(-r^2 + 2ry'_3) \\ &\times \exp[-\frac{1}{2}\omega(\vec{y}^2 + y_0^2 + \vec{y}'^2 + y'^2_0)] \end{aligned} \tag{7}$$

to evaluate the integral in Eq. (5). Both y and y' are functions of $\vec{x}$ and t. The transverse integrals can be performed trivially. For the t and z integrals we can use the variables ξ and η defined as

$$z = \frac{1}{\sqrt{2}}(\xi + \eta), \quad t = \frac{1}{\sqrt{2}}(\xi - \eta) \tag{8}$$

to complete the square of the exponent of the Gaussian factor. We obtain

$$I(s, r) = (1 - \alpha^2)^{1/2}\exp[2rs(1 - \alpha^2)^{1/2}], \tag{9}$$

where $\alpha = (\beta - \beta')(1 - \beta\beta')^{-1}$. Since the power-series expansion of Eq. (9) contains only equal powers of r and s, the integral of Eq. (5) vanishes unless $m = n$. When $m = n$, we find from the coefficient of $(rs)^n$ in the expansion of Eq. (9)

$$T_{nn}\ (\beta, \beta') = (1 - \alpha^2)^{(n+1)/2} . \tag{10}$$

The above results can be expressed in the following orthogonality relation:[3]

$$T_{nm}\ (\beta, \beta') = (1 - \alpha^2)^{(n+1)/2}\delta_{nm} . \tag{11}$$

The above result suggests that the covariant harmonic oscillators behave like nonrelativistic oscillators if they are in the same Lorentz frame. If two oscillators are in different frames, the orthogonality is preserved. The ground-state oscillator wave function, consisting of one half-wave, is contracted by $(1 - \alpha^2)^{1/2.}$ The nth excited state, consisting of (n + 1) half-waves, is contracted by $(1 - \alpha^2)^{(n+1)/2}$.

The author would like to thank Professor Y. S. Kim for suggesting this problem.

[1]Y. S. Kim and M. E. Noz, Phys. Rev. D *8*, 3521 (1973).

[2]R. Lipes, Phys. Rev. D *5*, 2849 (1972). Lipes calculates his transition amplitudes in the Lorentz frame where the excited-state resonance is at rest, and his wave functions coincide with those of Kim and Noz in this particular frame. For this reason, the use of the Kim-Noz wave function will lead to Lipe's calculation.

[3]The first attempt to get the Lorentz contraction factor was made by Markov, who obtained a similar result for n = 0 and β = 0. See M. Markov, Nuovo Cimento Suppl *3*, 760 (1956).

Complete orthogonality relations for the covariant harmonic oscillator

F.C. Rotbart
Department of Physics and Astronomy, Tel Aviv University, Ramat Aviv, Israel
(Received 12 June 1980)

Within relativistic quantum mechanics the complete orthogonality relations for the covariant harmonic oscillator are derived. These relations include time-axis excitations and are valid for wave functions belonging to different Lorentz frames.

In relativistic quantum mechanics (RQM),[1] the covariant harmonic oscillator appears as a natural extension of the nonrelativistic case.[2] As pointed out by Kim and Noz[3] in an earlier development of a covariant oscillator wave function, these functions can be applied to a wide range of hadronic processes. The RQM formalism in general, and the covariant oscillator wave functions in particular, allow the probability interpretation to be extended to the relativistic domain. This raises the important question of orthogonality between states in different Lorentz frames.

Reprinted from *Phys. Rev. D* **23**, 3078 (1981).

For spatial excitations, this was investigated by Ruiz.[4] However, the RQM oscillator functions differ from those of Kim and Noz, not only in that they derive from a more general covariant formalism, but also in that time-axis excitation states are not excluded from the ground state as they are in the formalism by Kim and Noz. Of course, in both formalisms these time excitation states are necessary for completeness of the Hilbert space. The purpose of this note is to calculate the complete orthogonality relations for states in differing Lorentz frames, including the time-axis excitations. We shall, to some extent, be following the paper of Ruiz.

The covariant harmonic oscillator in a stationary state is described by[2]

$$\left[\frac{1}{2m}\left(-\nabla_x^2+\frac{\partial^2}{\partial x_0^2}\right)+\frac{1}{2}k(\vec{x}^2-x_0^2)\right]\phi(x)=\kappa\psi(x)\,, \tag{1}$$

where m and k are positive constants and the x in $\psi(x)$ denotes $x^\mu=(x^0,x^1,x^2,x^3)$. Under the Lorentz transformation

$$y^1=x^1\,,\quad y^2=x^2\,,$$

$$y^3=\gamma\,(x^3-\beta x^0)\,, \tag{2}$$

$$y^0=\gamma(x^0-\beta x^3)\,,\quad \gamma=(1-\beta^2)^{-1/2}\,,$$

the above equation becomes

$$\left[\frac{1}{2m}\left(-\nabla_y^2+\frac{\partial^2}{\partial y_0^2}\right)+\frac{1}{2}k(\vec{y}^2-y_0^2)\right]\psi(y)=\kappa\psi(y)\,, \tag{3}$$

which has the solution

$$\psi_k(y)=N_{n_1n_2n_3n_0}H_{n_1}(\sigma y_1)H_{n_2}(\sigma y_2)H_{n_3}(\sigma y_3)H_{n_0}(\sigma y_0)\exp[-\frac{1}{2}\sigma^2(\vec{y}^2+y_0^2)]\,, \tag{4}$$

where κ is the invariant $(n_1+n_2+n_3-n_0+1)\omega$, $\sigma=\sqrt{m\omega}$, $\omega=\sqrt{k/m}$, and $N_{n_1n_2n_3n_0}$ is the normalization constant. Since the transverse directions y_1 and y_2 are invariant under transformation (2), we set, for convenience, $n_1=n_2=0$ in what follows.

Since the $\psi_n(y)$ [which we shall now denote as $\psi_{n_3n_0}(y)$], span a closed subspace, we can express a state in another Lorentz frame, that is, a frame characterized by β' instead of β in the transformation (2), as a linear sum of $\psi_{n'_3n_0}(y)$; namely

$$\psi_{n'_3n'_0}(y')=\sum_{n_3,n_0}A_{n_3n_0n'_3n'_0}(\beta,\beta')\psi_{n_3n_0}(y)\,, \tag{5}$$

where for orthonormal $\psi_{n_3n_0}$

$$A_{n_3n_0n'_3n'_0}(\beta,\beta')=\int_{-\infty}^{\infty}\psi^\dagger_{n_3n_0}(y)\psi_{n'_3n'_0}(y')d^4x\,. \tag{6}$$

These are the orthogonality relations we wish to evaluate.

Using the generating functions for the Hermite polynomials,

$$e^{-s^2+2s\xi}=\sum_{n=o}^{\infty}\frac{s^n}{n!}H_n(\xi)\,, \tag{7}$$

the calculation of expression (6) entails the integral

$$I(s,r,u,v)=\int_{-\infty}^{\infty} d^4x\ \exp[-s^2+2s\sigma y_3-r^2+2r\sigma y'_3-u^2+2u\sigma y'_0-v^2$$
$$+2v\sigma y'_0-\frac{1}{2}\sigma^2\,(\vec{y}^2+y_0^2+\vec{y'}^2+y'^2_0)]\,, \tag{8}$$

where y and y' are, by (2), functions of x. Introducing the variables ξ and η, defined by

$$x_3=\frac{1}{\sqrt{2}}\,(\xi+\eta)\,,\quad x_0=\frac{1}{\sqrt{2}}\,(\xi-\eta)\,, \tag{9}$$

this integral (8) yields

$$I(s,r,u,v)$$
$$=\frac{\pi^2}{\sigma^4}(1-\alpha^2)^{½}\exp[2sr(1-\alpha^2)^{½}-2su\alpha+2rv\alpha+2uv(1-\alpha^2)^{½}]\,, \tag{10}$$

where $\alpha=(\beta-\beta')(1-\beta\beta')^{-1}$.

This is expanded as a power series and the problem is to compare coefficients of powers in s, r, u, and v with those of

$$I(s,r,u,v)=\frac{\pi}{\sigma^2}\sum_{n_3 n_0 n'_3 n'_0}\frac{s^{n_3}r^{n'_3}u^{n_0}v^{n'_0}}{n_3!n'_3!n_0!n'_0!}\int_{-\infty}^{\infty}dx_3\,dx_0 H_{n_3}(\sigma y_3)H_{n'_3}(\sigma y'_3)$$
$$\times H_{n_0}(\sigma y_0)H_{n'_0}(\sigma y'_0)\exp[-\frac{1}{2}\sigma^2\,(y_3^2+y_0^2+y'^2_3+y'^2_0)]\,. \tag{11}$$

A typical term in the expansion of (10) is

$$\frac{\pi^2}{\sigma^4}\,\frac{[2sr(1-\alpha^2)^{1/2}]^a}{a!}\,\frac{[-2su\alpha]^b}{b!}\,\frac{[2rv\alpha]^c}{c!}\,\frac{[2uv(1-\alpha^2)^{1/2}]^d}{d!}(1-\alpha)^{1/2}$$
$$=\frac{\pi^2}{\sigma^4}\,\frac{2^{a+b+c+d}}{a!b!c!d!}\,(1-\alpha^2)^{(a+d+1)/2}(-1)^b\alpha^{b+c}s^{a+b}r^{a+c}u^{b+d}v^{c+d}\,, \tag{12}$$

with a, b, c, and d integers. Comparison with (11) shows that

$$a+b=n_3\,,\quad a+c=n'_3\,,$$
$$b+d=n_0,\quad c+d=n'_0. \tag{13}$$

These are not independent since we see that

$$n_3+n'_0=a+b+c+d=n'_3+n_0$$

or

$$n_0 - n_3 = n'_0 - n'_3. \tag{14}$$

This is to be expected from the invariance of k. Hence for a given three of n_3, n'_3, n_0 and n'_0 the fourth is defined, and we are left with three independent relations between the four variables a, b, c, and d. For a given n_1, n_0, and n'_0, say, we can take b as equal to some integer λ, a free parameter, and we find that, for $n'_0 > n_0$,

$$a = n_3 - \lambda, \quad b = \lambda, \quad c = n'_0 - n_0 + \lambda, \quad d = n_0 - \lambda,$$

as λ goes from zero to the smaller of n_3 or n_0. In this way, for the given n_3, n_0, and n'_0, (12) shows the coefficient to be a sum over λ, and so we arrive at the final result

$$\int_{-\infty}^{\infty} \psi^{\dagger}{}_{n_3 n_0}(y)\psi_{n'_3 n'_0}(y')d^4x = \delta^{n_0 - n_3}_{n'_0 - n'_3}(1-\alpha^2)^{(n_0+n_3+1)/2}\,\alpha^{n'_0 - n_0}$$
$$\times \sum_{\lambda=0}^{n_0} \frac{(i\alpha)^{2\lambda}(1-\alpha^2)^{-\lambda}(n_3!n'_3!n_0!n'_0!)^{1/2}}{(n_3-\lambda)!(n_0-\lambda)!(n'_0-n_0+\lambda)!\lambda!} \tag{15}$$

for $n'_0 \geq n_0$. Note that the requirement that λ be summed over the smaller of n_0 or n_3 is automatically taken care of by the factorials in the denominators. In the case of $n_0 \geq n'_0$ we simply interchange the primed quantities on the right-hand side with the unprimed ones and replace α with $(-\alpha)$.

For $n_0 = n'_0 = 0$, these orthogonality relations reduce to the relations obtained by Ruiz. In addition, it can be seen from (15) that Lorentz transformations contract the wave function not only along the z axis, but also along the time axis. This is due to the fact that time in the wave function is on an equal footing with the space coordinates. A hypothetical observer will measure the extent of the wave function in time, in his frame, at equal z in essentially the same way that he would measure the length at equal time. This extent in time is not to be confused with the time interval between two events which occur at the same z only in one particular frame.

It is amusing to note that in evaluating the sum in expression (15), the exponent of $(1 - \alpha)$ ranges from $\frac{1}{2}(n_0 + n_3 + 1)$ to $\frac{1}{2}(|n_0 - n_3| + 1)$. This behavior is reminiscent of angular momentum representations and suggests that a "spherical" coordinate representation for the covariant harmonic oscillator might be useful.

This research was partially supported by the Binational Science Foundation (BSF), Jerusalem, Israel.

[1]L. P. Horwitz and C. Piron, Helv. Phys. *46*, 316 (1973).
[2]L. P. Horwitz and F. C. Rotbart (unpublished).
[3]Y. S. Kim and M. E. Noz, Phys. Rev. D *8*, 3521 (1973); Found. Phys. *9* 375 (1979); Y. S. Kim, M. E. Noz, and S. H. Oh, ibid *9*, 947 (1979).
[4]M. J. Ruiz, Phys. Rev. D *10* 4306 (1974).

Dirac's form of relativistic quantum mechanics

D. Han
Systems and Applied Sciences Corporation, Riverdale, Maryland 20840

Y. S. Kim
Center for Theoretical Physics, Department of Physics and Astronomy, University of Maryland, College Park, Maryland 20742

(Received 29 November 1979; accepted 22 January 1981)

It is shown that Dirac's "instant form" dynamics provides a theoretical framework in which models of relativistic quantum mechanics can be constructed. The convariant harmonic oscillator formalism discussed in previous papers is shown to be such a model. Dirac's "point" and "front" forms are shown to generate a space-time geometry convenient for describing Lorentz deformation properties of relativistic extended hadrons.

I. INTRODUCTION

Quantum mechanics and special relativity were formulated before most of us were born and are likely to remain as the two most important scientific languages for many years to come. They are, in fact, the most important and exciting subjects for the students who plan to become physicists. For these reasons, it is not uncommon for students in the first-year quantum mechanics class to ask the question of how these two physical theories can be combined.

The usual answer to this question is that quantum field theory takes care of the job. However, this answer cannot prevent the students from reasoning in the following manner. The basic tool for quantum mechanics is Schrödinger's wave function, and the mathematical apparatus for special relativity is the coordinate transformation known as the Lorentz transformation. Then, why can we not construct wave functions that can be Lorentz transformed?

This is indeed a good homework problem not only for students but also for "grown-up" physicists. With this point in mind, Kim and Noz discussed the possibility of including the relativistic harmonic oscillator formalism in the first-year quantum mechanics curriculum.[1] The oscillator wave functions discussed in Ref. 1 are covariant, carry a probabilistic interpretation, and are mathematically simple enough to be included in the first-year quantum mechanics curriculum. This oscillator model can explain basic high-energy hadronic features associated with the Lorentz deformation of extended objects including the hadronic form factors, mass spectra, and the peculiarities in Feynman's patron picture. It was emphasized[1] that it is now possible to understand some space-time symmetry properties of high-energy hadrons without resorting to the techniques of quantum field theory.

In Ref. 1, Kim and Noz explain why this line of teaching physics does not do injustice to students who will eventually learn quantum field theory, by reviewing the successes and limitations of the present form of field theory. We are not disputing here the numerical successes and promises of the present form of quantum field theory. However, we should realize also that field theory has not yet provided a physical interpretation to the space-time boundary condition that is essential for formulating a bound-state picture in terms of localized probability distribution.[2,3] For this reason, we are allowed to construct relativistic bound-state wave functions from the first principles of quantum mechanics and relativity, in a manner consistent with what we observe in the real world, particularly in high-energy physics.[2]

In Ref. 1, it was pointed out that the relativistic harmonic oscillator wave function indeed satisfies the above-mentioned requirement. Students are then likely to ask the following question. There are two ways of doing quantum mechanics. One way is to construct wave functions, and the other approach is to write down a system of commutators that is equivalent to the algorithm based on wave functions. If the relativistic oscillator wave function discussed in Ref. 1 is to represent a form of relativistic quantum mechanics, then where is its broadly based commutator form?

The purpose of the present paper is to show that the wave function approach presented in Ref. 1 is equivalent to Dirac's "instant form" quantum mechanics.[4] Dirac's dynamical system consists of the ten generators of the Poincaré group and one constraint condition that reduces the four-dimensional space-time into a three-dimensional Euclidian space in which nonrelativistic quantum mechanics is expected to be valid. We show in this paper that the oscillator wave functions discussed in Ref. 1 represent a space-time solution of Dirac's "Poisson bracket" equations.[4]

In Sec. II, we formulate Dirac's instant form quantum mechanics for a system of two bound-state quarks. In Sec. III, the covariant harmonic oscillator formalism discussed in Refs. 1 and 5 is shown to form a space-time solution of Dirac's "Poisson bracket" equations. In Sec. IV, the con-

Reprinted from *Am. J. Phys.* **49**, 1157 (1981).

straint condition of Dirac and subsidiary condition of the oscillator formalism are discussed in detail, and they are shown to serve the same purpose.

In Sec. V, some geometrical consequences are derived from Dirac's "point" and "front" forms. It is shown that the Lorentz deformation property of the harmonic oscillator wave functions is derivable from this space-time geometry. It is shown further that this deformation property is consistent with the representation of the Poincaré group discussed in Secs. II and III.

II. DIRAC'S FORMULATION OF RELATIVISTIC THEORY OF "ATOM"

Dirac's atom in modern language is a hadron that is a bound state of quarks and/or antiquarks. Let us consider here a hadron consisting of two quarks whose space-time coordinates are x_1 and x_2. Then the standard procedure is to introduce the variables

$$X = (x_1 + x_2)/2, \quad x = (x_1 - x_2)/2\sqrt{2}. \tag{1}$$

Both X and x have their respective spatial and time components. The spatial variable appearing in the Schrödinger equation for the hydrogen atom is the distance between the proton and electron, and thus corresponds to the spatial component of the x coordinate. The Schrödinger equation does not contain the time component of x that corresponds to the time separation between the constituent particles.

In order to control this time-separation variable, Dirac considered in his "instant form" the condition

$$x_0 \simeq 0, \tag{2}$$

whose covariant form is

$$xP \simeq 0, \tag{3}$$

where P is the total four-momentum of the hadron. Equation (3) becomes Eq. (2) when the hadron is at rest. Dirac did not use the exact numerical equality in writing down the above constraint condition in order to allow further physical interpretations consistent with quantum mechanics and relativity. In particular, Dirac had in mind the possibility of the left-hand side becoming an operator acting on state vectors.

Dirac then points out that the relativistic dynamical system should consist of transformation operators that generate space-time translations, rotations, and Lorentz transformations. He points out further that the generators of the Poincaré group form the desired dynamical system. The generators in the present case take the form

$$P_\mu = i(\partial/\partial X^\mu) \quad \text{and} \quad M_{\mu\nu} = L^*_{\mu\nu} + L_{\mu\nu}, \tag{4}$$

where

$$L^*_{\mu\nu} = i[X_\mu\, \partial/\partial X^\nu - X_\nu\, \partial/\partial X^\mu], \quad L_{\mu\nu} = i[x_\mu\, \partial/\partial x^\nu - x_\nu\, \partial/\partial x^\mu]. \tag{5}$$

The operators P_μ generate space-time translations. $M_{\mu\nu}$ is antisymmetric under the interchange of μ and ν. Three M_{ij}, with $i, j = 1, 2, 3$, generate rotations, and three M_{0i} are the generators of Lorentz transformations.

These generators satisfy the following commutation relations:

$$[P_\mu, P_\nu] = 0,$$
$$[M_{\mu\nu}, P_\rho] = -g_{\mu\rho} P_\nu + g_{\nu\rho} P_\mu, \tag{6}$$
$$[M_{\mu\nu}, M_{\rho\sigma}] = -g_{\mu\rho} M_{\nu\sigma} + g_{\nu\rho} M_{\mu\sigma} - g_{\mu\sigma} M_{\rho\nu} + g_{\nu\sigma} M_{\rho\mu}.$$

Dirac emphasizes in his paper that *the problem of finding a new dynamical system reduces to the problem of finding a new solution of the above "Poisson brackets."*

The unusual feature in the above generators is that $L_{\mu\nu}$ operating on the x coordinate is not affected by translations and acts as the spin operator for the hadron, while operating basically on the space-time coordinates. Therefore, the problem of finding a solution of the commutator system is to find wave functions $\phi(X, x)$, which form a representation of the Poincaré group. The procedure of finding such a representation is to construct the wave functions that are diagonal in the Casimir operators[6]

$$P^2 = P^\mu P_\mu \quad \text{and} \quad W^2 = W^\mu W_\mu, \tag{7}$$

where

$$W_\mu = \epsilon_{\mu\nu\alpha\beta} P^\nu M^{\alpha\beta}.$$

These Casimir operators commute with each other and with the ten generators given in Eq. (4). Physically, P^2 and W^2 specify the mass and total spin of the hadron, respectively.

The final step in constructing Dirac's dynamical system is to make the constraint condition consistent with the generators of the Poincaré group. Dirac noted in particular that the constraint condition of Eq. (3) can be an operator equation, and that its "Poisson brackets" with other dynamical variables should be zero or become zero in the manner in which the right-hand side of Eq. (3) vanishes. We shall examine this point in detail in Sec. IV.

III. COVARIANT HARMONIC OSCILLATOR FORMALISM

In terms of the coordinate variables X and x defined in Sec. II, the starting partial differential equation for the oscillator formalism can be written as[2]

$$\tfrac{1}{2}[(\partial/\partial X_\mu)^2 + (\partial/\partial x_\mu)^2 - x_\mu^2 + m_0^2]\phi(X, x) = 0, \tag{8}$$

with the subsidiary condition

$$(\partial/\partial X_\mu) a_\mu^+ \phi(X, x) = 0$$

with

$$a_\mu^+ = x_\mu + \partial/\partial x^\mu.$$

In Eq. (1), the X and x variables are completely separable, and solution $\phi(X, x)$ takes the form

$$\phi(X, x) = \psi(x, P)\exp(\pm iPX), \tag{9}$$

where the internal wave function $\psi(x, P)$ satisfies the harmonic oscillator differential equation

$$H(x)\psi(x, P) = \lambda\psi(x, P), \tag{10}$$

with

$$H(x) = \tfrac{1}{2}[(\partial/\partial x_\mu)^2 - x_\mu^2].$$

The subsidiary condition in Eq. (8) becomes

$$P^\mu a_\mu^+ \psi(x, P) = 0. \tag{11}$$

The (mass)^2 of the hadron P^2 is constrained by the eigenvalues of the above oscillator equation:

$$P^2 = \lambda + m_0^2. \tag{12}$$

The harmonic oscillator equation given in Eq. (10) is separable in many different coordinate systems. We are interested here in the coordinate that is most convenient for constructing solutions that are diagonal in the Casimir operators of Eq. (7). These operators take the simplest form in the Lorentz coordinate system in which the hadron is at rest:

$$\begin{aligned} x' &= x, \quad y' = y, \\ z' &= (z - \beta t)/(1-\beta^2)^{1/2}, \\ t' &= (t - \beta z)/(1-\beta^2)^{1/2}. \end{aligned} \tag{13}$$

We assume here that the hadron moves along the z direction with velocity parameter β.

In terms of the above coordinate variables, we can write the Casimir operators as

$$P^2 = -(\partial/\partial X'_\mu)^2 \tag{14}$$

and

$$W^2 = M^2(\mathbf{L}')^2, \tag{15}$$

where

$$L'_i = -i\epsilon_{ijk}x'_j(\partial/\partial x'_k).$$

We also have to take into account the fact that the hadron (mass)^2 operators is constrained to take the eigenvalues determined by the oscillator equation for the internal wave function and that we have to consider P^2 of the form

$$P^2 = m_0^2 + H(x'). \tag{16}$$

In addition, we have to consider the form of the subsidiary condition given in Eq. (11). In terms of this moving coordinate system, the subsidiary condition takes the form

$$(t' + \partial/\partial t')\psi(x, P) = 0. \tag{17}$$

By using the moving coordinate system, we have achieved considerable simplification in the expressions for the Casimir operator W^2 and the subsidiary condition, without complicating the forms for P^2 given in Eqs. (14) and (16). This subsidiary condition restricts the t' dependence to that of the ground state and forbids excitations along the time-separation variable in the Lorentz frame where the hadron is at rest.

As was discussed in Ref. 1, the subsidiary condition of Eq. (17) forbids timelike excitations that contribute negatively to the total eigenvalue. This condition therefore guarantees the existence of the lower limit in the (mass)^2 spectrum. The absence of such timelike excitations are perfectly consistent with what we observe in the real world.

After these preparations, it is a simple matter to write down the solutions which form the desired representation of the Poincaré group[5]:

$$\phi(X, x) = \psi(x, P)\exp(\pm iPX), \tag{18}$$

with

$$\psi(x, P) = (1/\pi)^{1/2}[\exp(-t'^2/2)]R_n(r')Y^m(\theta', \phi'),$$

where r', θ', ϕ' are the spherical coordinate variables in a three-dimensional Euclidian space spanned by x, y, and z'.

IV. FURTHER CONSIDERATIONS OF THE CONSTRAINT CONDITION

As was pointed out in Sec. II, Dirac was interested in a possible quantum-mechanical form of his "instant form" constraint of Eq. (3). The key question at this point is whether the subsidiary condition given in Eq. (11) meets this requirement.

What Dirac wanted from his "conditional" equality was to freeze the motion along the time separation variable in a manner consistent with quantum mechanics and relativity. This means that we can allow a time–energy uncertainty along this timelike axis without excitations, in accordance with Dirac's own "C-number" time–energy uncertainty relation.[7] This time–energy uncertainty without excitation is widely observed in the relation between the decay lifetime and energy width of unstable systems.[8] The C number in the matrix language is one-by-one matrix, and is the ground state with no excitations in the harmonic oscillator system.

We have observed in Sec. III that the subsidiary condition of Eq. (11) becomes Eq. (17) in terms of the coordinate variables in which the hadron is at rest. Eq. (17) restricts the t' dependence to that of the ground state. Equation (11) therefore eliminates all timelike excitations in the Lorentz frame where the hadron is at rest, and makes the uncertainty associated with the t' direction a C-number uncertainty relation. We can therefore conclude that the subsidiary condition of Eq. (11) is a quantum-mechanical form of Dirac's "instant form" constraint given in Eq. (3).

In order that the dynamical system be completely consistent, the subsidiary condition should commute with the generators of the Poincaré group:

$$\begin{aligned} [P_\alpha, P^\mu a_\mu^+] &= 0, \\ [M_{\alpha\beta}, P^\mu a_\mu^+] &= 0. \end{aligned} \tag{19}$$

The above equations follow immediately from the fact that the operator $P^\mu a_\mu^+$ is invariant under translations and Lorentz transformations.

Since the Casimir operators are constructed from the generators of the Poincaré group, we are tempted to conclude that the constraint operator commutes also with the invariant Casimir operators. However, we have to note that the operator P^2 also takes the form of Eq. (16). Therefore, it should commute with $H(x)$ given in Eq. (10). However, a simple calculation gives

$$[H(x), P^\mu a_\mu^+] = P^\mu a_\mu^+. \tag{20}$$

This means that the right-hand side is not identically zero, but vanishes when applied to the wave functions satisfying the subsidiary condition of Eq. (11).

In his paper, Dirac considered also the commutation relations between dynamical quantities and the constraint condition that is only "approximately" zero. He asserted that the resulting "Poisson bracket" should also vanish in the same "approximate" sense. The commutator of Eq. (20) indeed vanishes in the manner prescribed by Dirac.[4]

It is well known that Dirac concludes his paper by noting some difficulties associated with the potential term in making his system of "Poisson brackets" completely consistent. The crucial question is whether the harmonic oscillator formalism can resolve Dirac's "real difficulty." We shall attack this problem here by carrying out some explicit calculations.

In formulating his scheme to solve the commutator equations for the generators of the Poincaré group, Dirac chose to adopt the view that each constituent particle in "atom" (bound or confined state) is on its mass shell, and that the total energy is the sum of all the free-particle energies and the potential energy. This potential term indeed causes the real difficulty in making the commutator system self-consistent.

In the covariant oscillator formalism, we observe that the Casimir operators of the Poincaré group clearly indicate that the mass of the hadron is a Poincaré-invariant constant, but they do not tell anything about the masses of constituent particles. Let us write down the four-momentum operators for the constituents in terms of the X and x variables:

$$p_{1\mu} = (i/2)\partial/\partial X^{\mu} + (i/2\sqrt{2})\partial/\partial x^{\mu},$$
$$p_{2\mu} = (i/2)\partial/\partial X^{\mu} - (i/2\sqrt{2})\partial/\partial x^{\mu}. \qquad (21)$$

In order that the constituent mass be a Poincaré-invariant constant, p_1^2 and p_2^2 have to commute with the Casimir operators of Eq. (7) and with the harmonic oscillator operator $H(x)$ of Eq. (10). The constituent (mass)^2 operators derivable from the above forms do not commute with $H(x)$ due to its "potential" term. We have therefore translated Dirac's "real difficulty" into

$$[p_1^2, H(x)] \neq 0,$$
$$[p_2^2, H(x)] \neq 0. \qquad (22)$$

The above commutators indicate that the (mass)^2 of the constituent quark is not a Poincaré-invariant quantity. In 1949, when Dirac's paper was written, the idea of an off-mass-shell particle was not accepted, and therefore the nonvanishing of the commutators was regarded as a problem. Because of our experience with, among other things, quantum field theory, we no longer regard the necessity of a bound particle being off-mass shell as a problem. Only the full bound system must be on the mass shell as is specified by the condition given in Eq. (16).

We note further that the concept of virtual off-mass-shell particle is derivable from the violation of causality allowed by the time–energy uncertainty relation.[9] It is interesting to see that Dirac's own work on this subject resolves the difficulty he mentioned in his 1949 paper.[4]

V. FURTHER GEOMETRICAL CONSIDERATIONS

In addition to the instant form, Dirac considered two other kinematical constraints that generate three-dimensional Euclidian subspaces of the four-dimensional Minkowskian space-time. They are "point" and "front" forms. We do not know how to take advantage of these two forms in constructing representations of the Poincaré group,[10] but they seem to provide a convenient space-time geometry for discussing Lorentz deformation properties of the harmonic oscillator wave functions, and thus of relativistic extended hadrons.

In his point form, Dirac imposes the constraint

$$x^{\mu}x_{\mu} = C = \text{const.} \qquad (23)$$

This condition is not unlike the mathematics of the mass-shell condition for free particles, and the resulting Euclidian space consists of x, y, and z, with the time separation variable constraint to take the values

$$t = \pm[C + x^2 + y^2 + z^2]^{1/2}. \qquad (24)$$

In his front form, Dirac introduced the light-cone variables

$$u = (t+z)/\sqrt{2},$$
$$v = (t-z)/\sqrt{2}, \qquad (25)$$

and considered the constraint

$$v = 0. \qquad (26)$$

The basic advantage of using the light-cone variables is that the Lorentz transformation takes a very simple form. For instance, the transformation given in Eq. (13) can be written as

$$u' = [(1+\beta)/(1-\beta)]^{1/2}u,$$
$$v' = [(1-\beta)/(1+\beta)]^{1/2}v. \qquad (27)$$

In discussing detailed geometrical properties, we can ignore the transverse coordinates that are not affected by Lorentz transformations. Let us first look at the point form constraint of Eq. (23) using the coordinate variables for the front form. Eq. (23) then becomes

$$uv = u'v' = (t^2 - z^2)/2 = A/4, \qquad (28)$$

where A is a constant. We are accustomed to associate the above form with a hyperbola in the zt plane. It is interesting to note that this equation also represents a rectangle in the uv plane as is specified in Fig. 1. The area of this rectangle remains invariant under Lorentz transformations.

Let us next look at the space-time geometry of the covariant oscillator wave function. Because of the oscillator equation in Eq. (10) is separable also in the Cartesian variables x, y, z', and t', the solution given in Eq. (18) can be written as a linear combination of the Cartesian solutions. The portion of the wave function that is affected by the Lorentz transformation is[5,10]

$$\psi(x, P) = H_n(z')\exp[-(z'^2 + t'^2)/2]. \qquad (29)$$

The localization property of this wave function is dictated by its Gaussian form. If we write this exponential function in terms of the light-cone variables u and v,[11]

$$\psi(x, P) \sim \exp\left[-\frac{1}{2}\left(\frac{1-\beta}{1+\beta}u^2 + \frac{1+\beta}{1-\beta}v^2\right)\right]. \qquad (30)$$

When the hadron is at rest and $\beta = 0$, the above wave function is concentrated within a circular region around the origin. As the hadron moves and β increases, the circle becomes an ellipse whose area remains constant. This space-time geometry is basically the same as that of the square and rectangle given in Fig. 1. The harmonic oscillator model is therefore a specific example of the more general space-time geometry derivable from Dirac's "point" and "front" forms.

It is important to note here that this form of hadronic Lorentz deformation is consistent with what we observe in high-energy laboratories.[12–15] In particular, it is possible, in

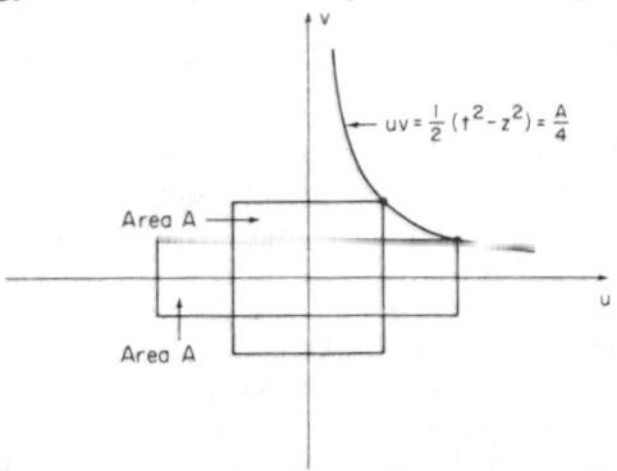

Fig. 1. Space-time geometry derivable from Dirac's point and front forms. The hyperbola represents the point-form geometry where $x^{\mu}x_{\mu}$ = const. The area of the rectangle is also Lorentz invariant. This "front form" geometry is very convenient in describing Lorentz deformation properties of relativistic extended hadrons.

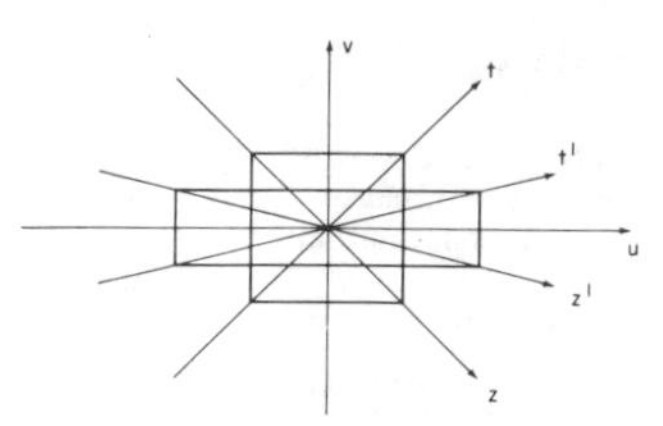

Fig. 2. Space-time geometry for the instant form in the light-cone coordinate system. The usual quantum excitations take place along the z' direction. The "C-number" uncertainty relation holds along the t' axis, and there are no excitations along this timelike direction.

terms of this Lorentz deformation picture, to explain Feynman's parton phenomenon[16] both qualitatively[17] and quantitatively.[18]

Finally, let us describe the "instant form" quantum mechanics discussed in Secs. II, III, and IV using the light-cone coordinate system. As is seen in Eq. (29), the Gaussian form is diagonal also in the instant-form coordinate variables. The z' and t' axes in the uv coordinate system are shown in Fig. 2. There are usual quantum excitations along the z' axis. Along the t' direction, Dirac's C-number time–energy uncertainty relation holds.

VI. CONCLUDING REMARKS

As we stated in Sec. I, quantum mechanics and relativity are two of the most important subjects in the physics curriculum. Because of its mathematical simplicity, the harmonic oscillator is one of the most effective teaching instruments. In their recent papers,[1,5] Kim *et al.* emphasized that the oscillator model can serve an effective purpose in teaching relativistic quantum mechanics and high-energy physics at the level of the first- and second-year graduate curriculum.

The wave functions in the oscillator formalism are compatible with the known principles of quantum mechanics and relativity. However, what was missing in the past has been a broader theoretical base from which this specific model is derivable. In this paper, we have shown that this theoretical base had already been given by Dirac in his "instant" form quantum mechanics.

Since the appearance of Driac's original paper in 1949,[4] many authors have made and are still making attempts to construct solutions of the "Poisson brackets" given in Eq. (6).[19] The point is that the commutator equations are basically differential equations without any specific form for potentials. The final form of solutions therefore depends on boundary conditions and/or forms of potentials. It is thus possible to end up with solutions that are not covariant.[20] In the present paper, we discussed a solution with a covariant form for potential satisfying a covariant space-time boundary condition.

[1] Y. S. Kim and M. E. Noz, Am. J. Phys. **46**, 484 (1978).

[2] R. P. Feynman, M. Kislinger, and F. Ravndal, Phys. Rev. D **3**, 2706 (1971). The point of this paper is that the inventor of Feynman diagrams stated that it is not practical, if not impossible, to use Feynman diagrams for relativistic bound-state problems. Feynman *et al.* suggested that the relativistic harmonic oscillator model, even if it is not totally consistent, can serve useful purposes. The point of Ref. 1 is that the oscillator model does not have to be imperfect, and therefore that it can be made consistent with the known rules of quantum mechanics and special relativity.

[3] The most successful bound-state model in field theory is of course the Bethe–Salpeter equation. However, the Bethe–Salpeter wave function does not yet have proper quantum-mechanical interpretation. See Sec. I of G. C. Wick, Phys. Rev. **96**, 1124 (1954). The difficulty in giving a physical interpretation to the relative time-separation variable between two bound-state particles was mentioned earlier by Karplus and Klein. See R. Karplus and A. Klein, Phys. Rev. **87**, 848 (1952).

[4] P. A. M. Dirac, Rev. Mod. Phys. **21**, 392 (1949).

[5] Y. S. Kim, M. E. Noz, and S. H. Oh, Am. J. Phys. **47**, 892 (1979); J. Math. Phys. **20**, 1341 (1979).

[6] E. P. Wigner, Ann. Math. **40**, 149 (1939).

[7] P. A. M. Dirac, Proc. R. Soc. London A **114**, 243, 710 (1927).

[8] E. P. Wigner, in *Aspects of Quantum Field Theory, in Honour of P. A. M. Dirac's 70th Birthday*, edited by A. Salam and E. P. Wigner (Cambridge University, London, 1972).

[9] W. Heitler, *The Quantum Theory of Radiation*, 3rd ed. (Oxford University, London, 1954). See also D. Han, Y. S. Kim, and M. E. Noz, Found. Phys. (to be published).

[10] Y. S. Kim, M. E. Noz, and S. H. Oh, J. Math. Phys. **21**, 1224 (1980).

[11] This exponential form is also derivable from Yukawa's work. See Eq. (10) of H. Yukawa, Phys. Rev. **91**, 416 (1953). For an interpretation of this original paper, see D. Han and Y. S. Kim, Prog. Theor. Phys. **64**, 1852 (1980).

[12] The fact that the proton (one of hadrons) is not a point particle and has a space-time extension was discovered by Hofstadter. See R. Hofstadter, Rev. Mod. Phys. **28**, 214 (1956).

[13] Since Hofstadter's discovery, there have been many attempts to construct theoretical models for relativistic extended hadrons. See, for instance, V. N. Gribov, B. L. Ioffe, and I. Ya. Pomeranchuk, J. Nucl. Phys. (USSR) **2**, 768 (1965) or Sov. J. Nucl. Phys. **2**, 549 (1966); N. Byers and C. N. Yang, Phys. Rev. **142**, 796 (1966); J. D. Bjorken and E. A. Paschos, *ibid.* **185**, 1975 (1969); B. L. Ioffee, Phys. Lett. B **30**, 123 (1969); K. Fujimura, T. Kobayashi, and M. Namiki, Prog. Theor. Phys. **43**, 73 (1970); A. L. Licht and A. Pagnamenta, Phys. Rev. D **2**, 1150, 1156 (1970); S. D. Drell and T. M. Yan, Ann. Phys. (NY) **60**, 578 (1971). Y. S. Kim and R. Zaoui, Phys. Rev. D **4**, 1764 (1971); R. G. Lipes, *ibid.* **5**, 2849 (1972); S. Ishida and J. Otokozawa, Prog. Theor. Phys. **47**, 2117 (1972); T. D. Lee, Phys. Rev. D **5**, 1738 (1972); G. Feldman, T. Fulton, and J. Townsend, *ibid.* **7**, 1814 (1973). Y. S. Kim and M. E. Noz, *ibid.* **8**, 3521 (1973). See also Refs. 1, 2, and the references contained therein.

[14] Perhaps one of the current models of extended hadrons is the "MIT bag model," as is explained by K. Johnson in Sci. Am. **241** (1), 112 (July 1979). One interesting question in this model is how "bags" would look to moving observers.

[15] The quark confinement problem is regarded as one of the most important current problems in the particle theory front. The ultimate goal of this program is to find a potential that confines the quarks inside hadrons within the field theoretic framework of QCD (quantum chromodynamics). The basic question is then this. What are we going to do with this confining potential? The next step is naturally to construct bound-state wave functions, which eventually leads to the question of their Lorentz transformation properties. As was noted in QED (quantum electrodynamics),[3] this does not as yet appear to be an easy problem. For an introductory review article on QCD, see W. Marciano and H. Pagels, Phys. Rep. **36** C, 138 (1978).

[16] We have to say that the most important observation made on Lorentz-deformed hadrons was Feynman's parton model. See R. P. Feynman, in *High Energy Collisions*, Proceedings of the 3rd International Conference, Stony Brook, New York, edited by C. N. Yang *et al.* (Gordon and Breach, New York, 1969); *Photon-Hadron Interactions* (Benjamin, Reading, MA, 1972).

[17] For an explanation of the peculiarities in Feynman's parton picture, see Y. S. Kim and M. E. Noz, Phys. Rev. D **15**, 335 (1977). For a graphical interpretation of the formulas in this paper, see Y. S. Kim and M. E. Noz, Found. Phys. **9**, 375 (1979).

[18] For a calculation of the proton structure function, see P. E. Hussar, Phys. Rev. D **23**, 2781 (1981).

[19] For one of the most recent papers on this subject, see A. Kihlberg, R. Marnelius, and N. Mukunda, Phys. Rev. D **23**, 2201 (1981).

[20] See, for instance, R. Fong and J. Sucher, J. Math. Phys. **5**, 456 (1964).

Chapter V

Lorentz-Dirac Deformation in High Energy Physics

In 1955, Hofstadter and McAllister observed that the proton is not a point particle. Although several field theoretic approaches had been made immediately after this discovery to explain the space-time extension of the proton, a satisfactory answer to this question can be found in the quark model, in which the proton is a bound state of three quarks.

In order to explain the high momentum-transfer behavior in the Hofstadter experiment, we need a wave function for the proton which can be Lorentz boosted. The covariant harmonic wave function discussed in the papers of Chapter IV is a suitable wave function for this purpose. In 1970, Fujimura, Kobayashi, and Namiki calculated the form factor of the proton, and showed that the asymptotic behavior of the form factor is due to the Lorentz deformation of the wave function.

The most peculiar behavior in high-energy physics is Feynman's parton picture. In 1969, Feynman observed that a rapidly moving proton can be regarded as a collection of an infinite number of partons whose properties appear quite different from those of quarks. This model is clearly spelled out in the paper of Bjorken and Paschos (1969). In 1977, using the covariant oscillator formalism, Kim and Noz showed that the static quark model and Feynman's parton picture are two different limiting cases of one covariant physics. Hussar in 1981 calculated the parton distribution for the rapidly moving proton using the covariant harmonic oscillator wave function.

Reprinted from: *Physical Review, Volume 98, 1955, pp. 183-184*

Electron Scattering from the Proton[*†‡]

Robert Hofstadter and Robert W. McAllister

Department of Physics and High-Energy Physics Laboratory,
Stanford University, Stanford, California
(Received January 24, 1955)

With apparatus previously described[1,2] we have studied the elastic scattering of electrons of energies 100, 188, and 236 MeV from protons initially at rest. At 100 MeV and 188 MeV, the angular distributions of scattered electrons have been examined in the ranges 60°-138° and 35°-138°, respectively, in the laboratory frame. At 236 MeV, because of an inability of the analyzing magnet to bend electrons of energies larger than 192 MeV, we have studied the angular distribution between 90° and 138° in the laboratory frame. In all cases a gaseous hydrogen target was used.

We have found that deviations in excess of Mott scattering are readily apparent at large scattering angles. The early results (reported at the Seattle meeting, July, 1954) at smaller angles showed the expected agreement with the Mott formula within experimental error. Deviations from the Mott formula such as we have found may be anticipated at large angles because of additional scattering from the magnetic moment of the proton.[3] We have observed this additional scattering but in an amount smaller than predicted by theory.

The experimental curve at 188 MeV is given in Fig 1. It may be observed that the experimental points do not fit either the Mott curve or the theoretical curve of Rosenbluth,[3] computed for a point charge and point (anomalous) magnetic moment of the proton. Furthermore, the experimental curve does not fit a Rosenbluth curve with the Dirac magnetic moment and a point charge. The latter curve would lie close to the Mott curve and slightly above it. Similar behavior is observed at 236 MeV.

The correct interpretation of these results will require a more elaborate explanation (probably involving a good meson theory) than can be given at the moment, although Rosenbluth already has made weak-coupling calculations in meson theory which predict an effect of the kind we have observed.[4]

Reprinted from *Phys. Rev.* **98**, 183 (1953).

Nevertheless, if we make the naive assumption that the proton charge cloud and its magnetic moment are both spread out in the same proportions we can calculate simple form factors for various values of the proton "size". When these calculations are carried out we find that the experimental curves can be represented very well by the following choices of size. At 188 MeV, the data are fitted accurately by an rms radius of $(7.0\pm2.4)\times10^{-14}$ cm. At 236 MeV, the data are well fitted by an rms radius of $(7.8\pm2.4)\times10^{-14}$ cm. At 100 MeV the data are relatively insensitive to the radius but the experimental results are fitted by both choices given above. The 100-MeV data serve therefore as a valuable check of the apparatus. A compromise value fitting all the experimental results is $(7.4\pm2.4)\times10^{-14}$ cm. If the proton were a spherical ball of charge, this rms radius would indicate a true radius of 9.5×10^{-14} cm, or in round numbers 1.0×10^{-13} cm. It is to be noted that if our interpretation is correct the Coulomb law of force has not been violated at distances as small as 7×10^{-14} cm.

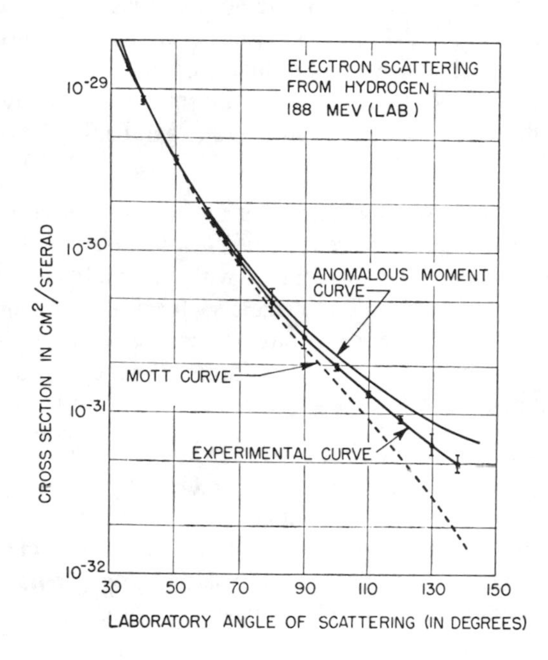

Fig. 1. The figure shows the experimental curve, the Mott curve, and the point-charge, point-magnetic-moment curve. The experimental curve passes through the points with the attached margins of error. The margins of error are not statistical; statistical error would be much smaller than the errors shown. The limits of error are, rather, the largest deviations observed in the many complete and partial runs taken over a period of several months. Absolute cross sections given in the ordinate scale were not measured experimentally but were taken from theory. The radiative corrections of Schwinger have been ignored since they affect the angular distribution hardly at all. The radiative corrections do influence the absolute cross sections. Experimental points in the figure refer to areas under the elastic peaks taken over an

energy interval of ± 1.5 MeV centering about the peak. The data at the various points are unchanged in relation to each other when the energy interval is increased to ± 2.5 MeV about the peak; the latter widths include essentially all the area under the peak.

These results will be reported in more detail in a paper now in preparation.

We wish to thank Dr. D. R. Yennie for his generous aid in discussions of the theory. We wish to thank Mr. E. E. Chambers for assistance with several phases of the work. In the early phases of this research, the late Miss Eva Wiener made important contributions.

*The research reported in this document was supported jointly by the U. S. Navy (Office of Naval Research) and the U. S. Atomic Energy Commission, and the U. S. Air Force through the Office of Scientific Research of the Air Research and Development Command.

†Aided by a grant from the Research Corporation.

‡Early results were presented at the Seattle Meeting of the American Physical Society [Phys. Rev. *96*, 854(A) (1954)]. More recent results were presented at the Berkeley meeting of the American Physical Society [Bull Am. Phys. Soc. *29*, No. 8, 29 (1954)].

[1] Hofstadter, Fechter, and McIntyre, Phys. Rev. *92*, 978 (1953).

[2] Hofstadter, Hahn, Knudsen, and McIntyre, Phys. Rev. *95*, 512 (1954).

[3] M.N. Rosenbluth, Phys. Rev. *79*, 615 (1950).

[4] See also the classical calculation of L. I. Schiff reported in Rosenbluth's paper.

Progress of Theoretical Physics, Vol. 43, No. 1, January 1970

Nucleon Electromagnetic Form Factors at High Momentum Transfers in an Extended Particle Model Based on the Quark Model

Kimio FUJIMURA, Tsunehiro KOBAYASHI and Mikio NAMIKI

Department of Physics, Waseda University, Tokyo

(Received August 21, 1969)

Taking account of the Lorentz contraction effect of the extended nucleon core as a nucleon but not as a quark, it is shown that the Gaussian inner orbital wave function can produce the form factor very close to the dipole formula.

Recent experiments show that the nucleon electromagnetic form factor are empirically described by the "scaling law" $e^{-1}G_E{}^P=\mu_P{}^{-1}G_M{}^P=\mu_n{}^{-1}G_M{}^n$ ($\equiv F$) and $G_E{}^n=0$ and by the "dipole formula" $F=(1+K^{-2}|t|)^{-2}$, where we have followed the usual notations and $K^2=0.71$ $(\text{GeV}/c)^2$. The scaling law was already discussed on the theoretical basis of the nonrelativistic urbaryon (quark) model.[1],[2] Ishida et al.[2] and Drell et al.[3] attempted to extract information about the inner orbital wave function at short distances from the $|t|$-dependence of F in a wide region of $|t|$ over M^2 (M being the nucleon mass), using nonrelativistic formulas. In this note we show that if possible relativistic effects as a nucleon (not as a quark), especially the Lorentz contraction of the nucleon core, are taken into account in a proper way, their conclusions become never true but the simple Gaussian inner orbital wave function can produce the form factor very close to the dipole formula.

Those who are working with the nonrelativistic quark model have believed that if $|t|\ll M_q{}^2$ (M_q being the quark mass), nonrelativistic formulas can be used for everything. As for the form factor, therefore, they have used

$$F=W_{\rm NR}\quad \text{(Drell et al. and others)}, \tag{1a}$$

$$F=(1+|t|/m_{\rm V}^2)^{-1}W_{\rm NR}\quad \text{(Ishida et al.)}, \tag{1b}$$

with the nonrelativistic formula

$$W_{\rm NR}=\iint e^{iqx}|\phi(x,\cdots)|^2dxd(\cdots), \tag{2}$$

where $m_{\rm V}^2$ is the mean square mass of ρ and ω mesons and q the momentum transfer. $\phi(x,\cdots)$ stands for the inner orbital wave function, where independent inner coordinates are denoted by x and $\cdots$. Assuming the simple Gaussian function for ϕ, we have got

Reprinted from *Prog. Theor. Phys.* **43**, 73 (1970).

$$W_0 = \exp(-\tfrac{1}{6}\langle r^2\rangle_c |t|), \tag{3}$$

where $\langle r^2\rangle_c$ is the mean square radius of the nucleon core. It is evident that the simple Gaussian function never gives us the form factor consistent with the dipole formula for $|t| \gtrsim M^2$. This is the reason why Ishida et al. introduced a singular wave function and Drell et al. discussed singular potentials among constituent particles. It is, however, to be noted that Eq. (2) is a nonrelativistic formula to be verified not only for $|t| \ll M_q{}^2$ but also for $|t| \ll M^2$. Here we want to emphasize that relativistic effects as a nucleon (but not as a quark) become very important for $|t| \gtrsim M^2$. Indeed, we can see that the Lorents contraction effect as a nucleon for $|t| > M^2$ should reduce $\langle r^2\rangle_c$ in Eq. (3) by the Lorentz factor, Γ^{-1}, approximately proportional to $M^2|t|^{-1}$. Hence Eq. (3) must be modified essentially in its $|t|$-dependence in the following way:

$$\exp\ (-\tfrac{1}{6}\Gamma^{-1}\langle r^2\rangle_c |t|)\, O_{\mathrm{FI}}(q^2), \tag{4}$$

where $O_{\mathrm{FI}}(q^2)$ is the overlap integral defined by $(\phi_{\mathrm{F}}, \phi_{\mathrm{I}})$.

Hence we cannot exclude the Gaussian inner wave function. Furthermore note that the region $M_q{}^2 \gg |t| \gtrsim M^2$ covers a wide range from one $(\mathrm{GeV}/c)^2$ to several ten $(\mathrm{GeV}/c)^2$ if $M_q \simeq (5 \text{ to } 10) \times M$.

In order to take properly the relativistic effect into account we must inevitably use the four-dimensional inner orbital wave function. In the quark model, the nucleon is assumed to be a composite particle of three quarks. Suppose that the three quarks have, respectively, four-position coordinates x_1, x_2 aed x_3. After separating the center-of-mass coordinate $X = (x_1 + x_2 + x_3)/3$, we keep two independent relative coordinates $r = (x_2 - x_3)/\sqrt{6}$ and $s = (-2x_1 + x_2 + x_3)/3\sqrt{2}$. Now, as the simplest example, we can choose the four-dimensional Gaussian function

$$\psi(r, s; P) = (\alpha/\pi)^2 \exp\left[\frac{\alpha}{2}\left\{r^2 + s^2 - \frac{2}{M^2}(P\cdot r)^2 - \frac{2}{M^2}(P\cdot s)^2\right\}\right] \tag{5}$$

for the inner orbital wave function,*) where P stands for the center-of-mass momentum of the composite system, i.e. the nucleon momentum, and $(\alpha/\pi)^2$ is the normalization costant determined by $\int\int |\psi|^2 d^4r d^4s = 1$. The constant α is related to the mean square radius of the nucleon core through $\alpha^{-1} = \langle r^2\rangle_c/3$. It may be worth while to emphasize another reason why Eq. (5) is used here: Equation (5) represents the ground state eigenfunction of the Hamiltonian of a four-dimensional harmonic oscillator consistent with the famous linearly raising trajectory in an extended particle model.[4] Our procedure should be regarded as one theoretical attempt in an extended particle model represented by a trilocal field based on the quark model, rather than one in the naive relativistic quark model.

---

*) Note that $r^2 = r_0{}^2 - \boldsymbol{r}^2$.

Nucleon Electromagnetic Form Factors at High Momentum Transfers 75

The relativistic form factor is given by the formula*)

$$W_{\mathrm{R}}=\iint\psi^*(r,\ s;\ P_{\mathrm{F}})\,e^{iq\cdot(ar+bs)}\psi(r,\ s;\ P_{\mathrm{I}})\,d^4r d^4s\,, \tag{6}$$

or symbolically

$$W_{\mathrm{R}}=(\psi_{\mathrm{F}},\ e^{iqx}\psi_{\mathrm{I}}), \tag{6'}$$

where P_{I} and P_{F} are, respectively, the initial and final momenta of the nucleon and $q=P_{\mathrm{F}}-P_{\mathrm{I}}$. a and b is one of the following pairs; $(0,\ -\sqrt{2})$, $(\sqrt{3/2},\ 1/\sqrt{2})$ and $(-\sqrt{3/2},\ 1/\sqrt{2})$. Note here that $a^2+b^2=2$ for every pair. Inserting Eq. (5) into Eq. (6), one obtains

$$W_{\mathrm{R}}=\left(\frac{1}{\sqrt{\Gamma}}\right)^B\exp\left(-\frac{1}{6}\langle r^2\rangle_c\Gamma^{-1}|t|\right). \tag{7}$$

Here the first factor is not other than the overlap integral

$$O_{\mathrm{FI}}(q^2)=(\psi_{\mathrm{F}},\ \psi_{\mathrm{I}})=\left(\frac{1}{\sqrt{\Gamma}}\right)^B$$

with (8)

$$\Gamma=1+\frac{|t|}{2M^2}\quad\text{and}\quad B=4\,,$$

where $B=4$ means the number of dimensions giving the Lorentz contraction, namely, the longitudinal space-like inner coordinates and two time-like inner coordinates. As mentioned above we can see in Eq. (7) that $\sqrt{\Gamma}$ behaves just like the effective Lorentz contraction factor and the exponential function in Eq. (7) goes to a constant as $|t|$ increases over M^2. The $|t|$-dependence of the form factor for $M_q^2\gg|t|\gtrsim M^2$ is, therefore, governed mainly by the overlapping-effect factor $(1/\sqrt{\Gamma})^B$ which goes to $(|t|/2M^2)^{-B/2}$. We can remark that power B is nothing other than the number of independent relative coordinates. Thus the dipole-like behavior of W_{R} can be obtained from $B=4$, namely, the inner freedom of motion equivalent to four indepencent relative coordinates. Thus the dipole-like behavior of W_{R} can be obtained from $B=4$, namely, the inner freedom of motion equivalent to four independent relative coordinates.

Equations (1a) and (1b) should, respectively, be replaced with

$$F=W_{\mathrm{R}}\,, \tag{9a}$$

$$F=(1+\tfrac{1}{6}\langle r^2\rangle_{\mathrm{V}}\Gamma^{-1}|t|)^{-1}W_{\mathrm{R}}\,, \tag{9b}$$

where $\langle r^2\rangle_{\mathrm{V}}=6m_{\mathrm{V}}^{-2}$ is the mean square radius of the vector meson cloud. Both formulas (9a) and (9b) together with Eq. (7) behave like the dipole formula

*) It is to be noted that most of the infinite component field theories have identified the form factor with the overlap function $(\psi_F,\ \psi_I)$ but never with $(\psi_F,\ e^{iqx}\psi_I)$ itself. In fact, some authors have derived our later result, $(\psi_F,\ \psi_I)=(1+|t|/2M^2)^{-2}$ as the form factor, using the infinite component field theory. See A. O. Barut's lecture given at the Colorado Summer School in 1967. We must emphasize here that the form factor should not be given by $(\psi_F,\ \psi_I)$.

modified by a constant factor for $M_q^2 \gg |t| \gtrsim M^2$. Note that they contain only one free parameter $\langle r^2 \rangle_c$ to be adjusted. Let us first compare the theoretical form factor given by Eq. (9a) together with Eq. (7)——call it Case (i)——with experiment[5] in Fig. 1, in which we have used $\langle r^2 \rangle_c = 7.50$ (GeV/c)$^{-2}$. From them one can see that the theoretical curve given by Eqs. (9a) and (7) is not inconsistent with the experimental plot but quite different from the nonrelativistic Gaussian form factor W_0. Next we examine Case (ii) in which Eq. (9b) is combined with Eq. (7), namely, each quark has the vector meson cloud. Choosing $\langle r^2 \rangle_c = 1.82$ (GeV/c)$^{-2}$, we see in Fig. 2 that the theoretical curve can reproduce the experimental plot in a wide range of $|t|$ from zero to about 25 (GeV/c)^2. It is repeatedly noted that this fit has been obtained by adjusting only one parameter $\langle r^2 \rangle_c$, and that the nonrelativistic Gaussian form factor is strongly modified in its essence.

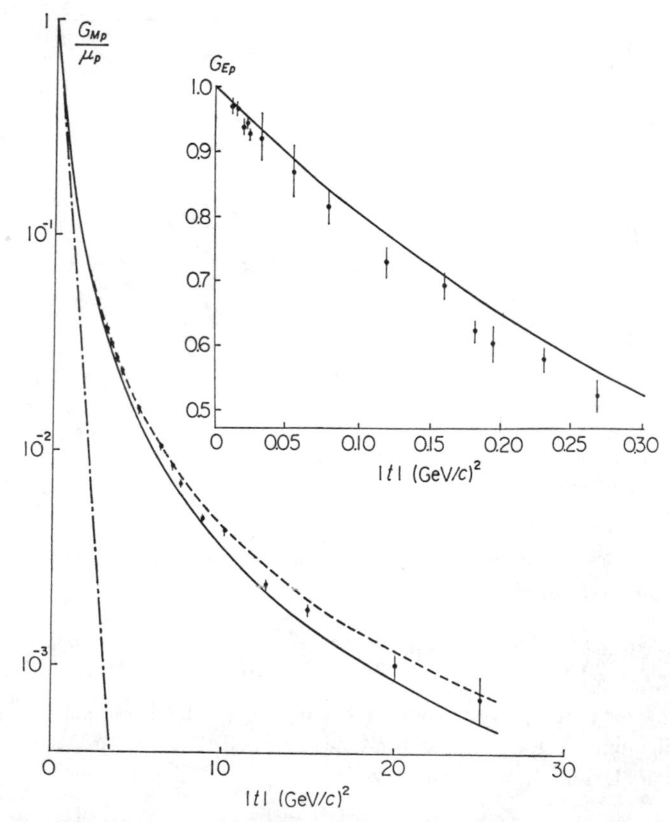

Fig. 1. Comparison of the theoretical form factors with experiments in Case (i) with $\langle r^2 \rangle_c = 7.50$ (GeV/c)$^{-2}$. The dipole formula and the nonrelativistic Gaussian form factor are, respectively, shown by the broken and chain lines.

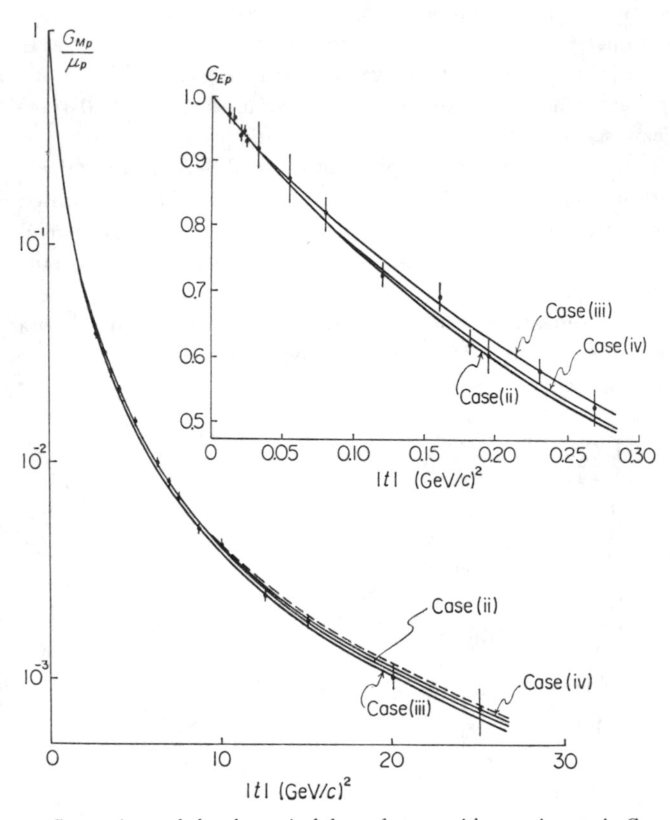

Fig. 2. Comparison of the theoretical form factors with experiments in Case (ii) with $\langle r^2\rangle_c=1.82$ $(\text{GeV}/c)^{-2}$, in Case (iii) with $\langle r^2\rangle_c=8.81$ $(\text{GeV}/c)^{-2}$ and $\lambda=1.4$, and in Case (iv) with $\langle r^2\rangle_c=1.20$ $(\text{GeV}/c)^{-2}$ and $\lambda=0.9$. The broken line shows the dipole formula.

Here we want to introduce a new parameter, say λ, into the inner orbital wave function as follows:

$$\psi_\lambda(r,\ s;\ P)=\left(\frac{\alpha}{\pi}\right)^2\sqrt{2\lambda-1}\ \exp\left[\frac{\alpha}{2}\left\{r^2+s^2-\frac{2\lambda}{M^2}(P\cdot r)^2-\frac{2\lambda}{M^2}(P\cdot s)^2\right\}\right]. \qquad (10)$$

It is easy to see that the parameter λ distinguishes the time-like extension from the space-like one of the inner orbital motion, and that $\lambda=1$ gives us the original one Eq. (5). Using ψ_λ, we have got

$$W_R^{(\lambda)}=\Gamma_\lambda^{-1}\left(1+\frac{\lambda}{2\lambda-1}\frac{|t|}{2M^2}\right)^{-1}\exp\left[-\frac{1}{6}\langle r^2\rangle_c\Gamma_\lambda^{-1}|t|\right], \qquad (11)$$

where

$$\Gamma_\lambda = 1 + \lambda \frac{|t|}{2M^2}. \tag{12}$$

The form factor F_λ is obtained by Eq. (11) together with the modified formulas

$$F_\lambda = W_R^{(\lambda)}, \tag{13a}$$

$$F_\lambda = (1 + \tfrac{1}{6}\langle r^2\rangle_V \Gamma_\lambda^{-1}|t|)^{-1} W_R^{(\lambda)}. \tag{13b}$$

In case (iii) Eq. (13a) is combined with Eq. (11), and Case (iv) is given by Eq. (13b) together with Eq. (11). In both cases the form factor goes to one proportional to $(2M^2|t|^{-1})^2$ like the dipole formula. Figure 2 shows us that the experimental plot can be fitted by the theoretical curves with $\langle r^2\rangle_c = 8.81$ (GeV/c)$^{-2}$ and $\lambda = 1.4$ in Case (iii) and with $\langle r^2\rangle_c = 1.20$ (GeV/c)$^{-2}$ and $\lambda = 0.9$ in Case (iv). The theoretical curves are in good agreement with experiment. Needless to say, Cases (iii) and (iv) include Cases (i) and (ii), respectively, as their special cases with $\lambda = 1$.

From the above arguments we have inferred that the Lorentz contraction of the extended nucleon core can be a possible origin of the "dipole formula". The same effects will appear also in inelastic electron proton collisions leading to the isobar excitation. In the nonrelativistic quark model we have got the differential cross section for the inelastic collision in the following form:[6)]

$$\frac{d^2\sigma}{d\Omega dE} \propto (\langle r^2\rangle_c q^2)^L \exp[-A\langle r^2\rangle_c q^2] \tag{14}$$

using the simple Gaussian wave function, where L is an integral number determined by the type of transition, and A a numerical factor. If we take the Lorentz contraction factor into account, then we can infer that Eq. (14) should be replaced with

$$(\Gamma^{-1}\langle r^2\rangle_c q^2)^L \exp[-A\Gamma^{-1}\langle r^2\rangle_c q^2] |O_{FI}(q^2)|^2, \tag{15}$$

where $\sqrt{\Gamma}$ and $O_{FI}(q^2)$ are, respectively, the effective Lorentz contraction factor and the overlap integral in the inelastic collision. The similar structure of the wave function suggests us that $\Gamma \to q^2$ and $O_{FI}(q^2) \to (q^2)^{-2}$ as q^2 goes over M^2, and then that the inelastic cross section would behave like the dipole formula squared for $q^2 \gtrsim M^2$. Indeed, it seems to us that recent experiments indicate such a behavior for the cross section.[7)] Detailed discussions will be given in a forthcoming paper in which the full nucleon and isobar wave functions and the quark current to be valid for $M_q^2 \gg |t| \gtrsim M^2$ will be formulated.

The earlier form of this work was done when one of the authors (M. N.) was working in the Niels Bohr Institute in Copenhagen. He would like to express his sincere gratitude to Professor A. Bohr for his kind hospitality and to Professor Z. Koba for many discussions. He is also much indepted to Professor T. Takabayashi for helpful discussions.

Nucleon Electromagnetic Form Factors at High Momentum Transfers 79

References

1) Y. Kinoshita, T. Kobayashi, S. Machida and M. Namiki, Prog. Theor. Phys. **36** (1966), 107. In the first several sections they derived the scaling law, assuming that a virtual photon couples directly to one point quark in the nucleon core subject to the 56-dimensional representation of the *SU*(6) symmetry.

2) S. Ishida, K. Konno and H. Shimodaira, Prog. Theor. Phys. **36** (1966), 1243. They first derived the scaling law in a semiphenomenological way, assuming that a virtual photon couples to the nucleon only through ρ and ω mesons. They have also got the quark-theoretical form factor, Eq. (1b), together with Eq. (2), assuming each quark to have a vector meson cloud.

3) S. D. Drell, A. C. Finn and M. H. Goldhaber, Phys. Rev. **157** (1967), B157.

4) T. Takabayasi, Phys. Rev. **139** (1965), B1381.

5) D. H. Coward, H. DeStalbler, R. A. Early, J. Litt, A. Minten, L. W. Mo, W. K. H. Panofsky, R. E. Taylor, M. Breidenbach, J. I. Friedman, H. W. Kendall, P. N. Kirk, B. C. Barish, J. Mar and J. Pine, Phys. Rev. Letters **20** (1968), 292.
L. N. Hand, D. G. Miller and R. Wilson, Rev. Mod. Phys. **35** (1963), 335.
T. Jansens, R. Hofstadter, E. B. Hughes and M. R. Yearian, Phys. Rev. **142** (1966), 922.
W. Albrecht, H. J. Behrend, W. Flauger, H. Hultshig and K. G. Steffen, Phys. Rev. Letters **17** (1966), 1192.
W. Albrecht, H. J. Behrend, H. Dorner, W. Flauger and H. Hultshig, Phys. Rev. Letters **18** (1967), 1014.

6) K. Fujimura, Ts. Kobayashi, Te. Kobayashi and M. Namiki, Prog. Theor. Phys. **38** (1967), 210.

7) For example, see W. K. H. Panofsky's report presented at the XIVth International Conference on High Energy Physics in Vienna in 1968.

The Behavior of Hadron Collisions at Extreme Energies

RICHARD P. FEYNMAN

California Institute of Technology, Pasadena, California

Talk given at

Third International Conference
on High Energy Collisions

State University of New York, Stony Brook

September 5-6, 1969

C.N. Yang *et al.* Eds.
Gordon and Breach
(New York, 1969)

Reprinted from *High Energy Collisions* (1969).

There are several reasons to be interested in this problem of very high energy hadron scattering. Firstly, most theoretical inventions are based on analysis of simple collisions, in which only a small number of particles come out. But it is at once realized that questions of unitarity, the asymptotic behavior for high energy in dispersion integrals, etc, require some *ansatz* be made for the higher energy collisions, in order to close the infinite hierarchy of equations which result. Secondly, experiments at high energies usually yield many particles, and only by selecting the rare collision can we find those about which the theorist has been speaking. For the highly multiple inelastic collisions (to which the major part of the inelastic cross section is due) so many variables are involved that it is not known how to organize or present this data. Any theoretical suggestion (even if it proves to be not quite right) suggests a way that this vast amount of data may be analyzed. For this reason I shall present here some preliminary speculations on how these collisions might behave even though I have not yet analyzed them as fully as I would like.

Hadron phenomena possess a number of remarkably simple properties. Besides the well-known agreements with relativity, analyticity, unitarity, etc., there are of course the conservation rules of isospin and strangeness and an approximate agreement with SU_3 symmetry. In addition to these, however, we have some special regularities which appear to be true empirically which apply to very high-energy behavior and which may point the way to an ultimate dynamical theory. A partial list of these regularities at very high energy are:

1. Total cross sections are constant. The elastic part appears to be a constant fraction of this.

2. Exchange reaction cross sections fall with a power of the energy $s^{2\alpha-2}$.

3. This power, α, seems to be t-dependent, and for those $t = m^2$ where $\alpha(t)$ is an appropriate integer (or half-integer) there are resonant particles of mass m (t is the square of the four-vector momentum transfer in the collision).

4. $\alpha(t)$ varies with t as a straight line $\alpha(t) = \alpha_0 + \gamma t$. Although α_0 varies with the quantum numbers which must exchanged, the value of γ is the same (0.95 per $(GeV)^2$).

5. Cross sections fall very rapidly with transverse momentum transfer. The average transverse momenta of the particles in inelastic collisions are limited (to about 0.35 GeV).

There are, of course, very many rough approximations in this brief summary. For example, we do not know if the total x-section might not vary very slowly (for example, like ln s); all the slopes of Regge trajectories are not exactly equal; whether the pion nonet is on the trajectory of such a slope is unknown; there are corrections to the simple "regge expectations" (2), (3), presumably for absorption; some of the inelastic cross sections can be associated with diffraction dissociation of the elastic part; etc. Nevertheless, the list above contains a number of main phenomena whose general behavior must ultimately be understood. It will be noticed that in discussing the power laws associated with Regge behavior, I have explicitly separated the behavior of the total x-section which is often described as the Pomeranchuk trajectory, for it is not certain that this is a typical trajectory. In trying to understand the meaning of these regularities of high energy behavior, I have been led to suggest certain further regularities accompanying them. I should like to present these guesses here to see if they are possibly true, or, if some of them are obviously in disagreement with experiment, to learn where I may have already gone off the track in my thinking.

I shall, for completeness, first explain how I am trying to go about analyzing these things; second, describe some special considerations dealing with Regge exchange; and finally, present my suggestions for the limiting behavior of cross sections at high energies.

I. FIELD THEORY AT HIGH ENERGY

In order to think about these questions I wish to use concepts which will immediately insure that the most fundamental properties of relativistic invariance, quantum superposition, unitarity, etc., will automatically be satisfied. the only theoretical structure I know which has a chance of doing this is a quantum field theory (I say, a chance, because we are not sure if unrenormalized field theories can give finite answers, or if renormalized theories are still unitary). It is not that I believe that the observed high-energy phenomena are necessarily a consequence of field theory. Even less do I know what specific field theory could yield them. But, rather, I believe that they share some of the properties of field theory, so they might share others. Therefore, I wish to study the behavior expected from field theory for collisions of very high energy. What field theory shall I choose? What shall be the fundamental bare particles that the theory begins with? I do not know and perhaps I do not care. I shall try at first to get results which are more general and characteristic of a wide class of theories and which can be stated in a way independent of the field theory which served as a logical crutch for their discovery. Only later, possibly, might it be worth trying to see if certain special experimental details imply something about a special theory which underlies all these phenomena.

In the meantime I call the fundamental bare particles of my underlying field theory "partons" (which may be of several kinds, of course). For example, in quantum electrodynamics the partons are bare electrons and bare photons. Imagine this theory to have a Hamiltonian H which may be separated into two pieces $H = H_{free} + H_{int}$, one to represent free partons and the other the interaction between them. This H is, at first, expressed in terms of creation and annihilation operators (a^*,a) of these partons (or, if you prefer, local field operators of the parton fields in space). Next, for special application to collisions of extremely high energy, W, incoming in the z-direction in the center-of-mass system, only the operators of finite transverse momenta (i.e., x,y components) are kept as $W \to \infty$. The ones with positive z-component of momentum of order W (i.e., $P_z = xW$, where x is a finite quantity as we take a limit as $W \to \infty$) are separated from those of negative z-component. The first are called right movers (a_R) and the second left movers (a_L). If this is done, and the Hamiltonian reexpressed, we get $H = H_R + H_L + H_I$ where H_R is the Hamiltonian involving right movers (a_R^*, a_R) only (containing, of course, interaction terms among these right movers coming from H_{int}, H_L involves left movers (a_L^*, a_L) and H_I contains both a_R and a_L, and represents an interaction between the objects moving [to] the left and those moving to the right. But H_I, as $W \to \infty$, becomes a very simple expression (depending on the theory of course). For example, a collision of a right moving proton with a left moving neutron yielding right and left moving particles can then be analyzed in a simple way. The proton is an eigenfunction of H_R, the neutron of H_L. Neglecting H_I, the system of states before and after collision are complete eigenfunctions (not simple partons) of $H_R + H_L$. The operator H_I makes the transition between them (not necessarily in first order, of course).

I leave to a later publication a more detailed description of how this might be carried out, but here I need only make some remarks about the variables on which things appear to depend in the limit as $W \to \infty$.

To describe a proton of momentum $\vec{P}_0$, energy E_0, ordinary field theory gives a state function or wave function giving the amplitude that a number of partons of 3-momentum $\vec{P}_1, \vec{P}_2, \ldots$ etc., are to be found in it. The total momentum of these partons $\sum_i \vec{P}_i$, equals $\vec{P}_0$, the momentum of the proton, but their total energy $\sum_i E_i$ (where each energy E_i is calculated from the mass μ_i of the parton via $E_i = \sqrt{\mu_i^2 + \vec{P}_i \cdot \vec{P}_i}$ is not equal to that of the final proton E_o. In fact, the amplitude to find this state contains, among many other factors, one which is inversely proportional to this energy difference

$$A \sim (E_o - \sum_i E_i)^{-1} \tag{1}$$

Knowing this wave function completely for some P_o, say at rest, how can we find it at some other momentum? It is very difficult to do, and in fact requires knowledge of the entire Hamiltonian operator H, for the wave function is not a relativistic invariant. This is emphasized by the point that the momentum is the sum of the momenta of the parts but the energy is not. The wave functions that should be useful for us are those in which $\vec{P}_o$ is very large in the z-direction and finite in the directions perpendicular to that. If we take $P_{oz} = x_o W$ and measure the parton's momentum in the z-direction in the same scale $P_{iz} = x_i W$, then the wave function has a definite limiting form as $W \to \infty$ for x_o, x_i finite. (x_o, of course, is arbitrary; it may be taken to be unity, for example.) We have

$$x_o = \sum x_i \tag{2}$$

Let the components of momentum perpendicular to z be called $\vec{Q}$ a two-dimensional vector; then

$$\vec{Q}_o = \sum \vec{Q}_i \tag{3}$$

Finally we can see how A varies (insofar as its denominator behaves) by writing it as

$$A \sim ((E_o - P_{oz}) - \sum_i (E_i - P_{iz}))^{-1}$$

which equals (1) since the total z-momentum is the sum of its parts. However, for large z-momentum,

$$\begin{aligned} E - P_z &= (m^2 + P^2 + P_z^2)^{1/2} - P_z \\ &= (m^2 + Q^2 + x^2 W^2)^{1/2} - xW \\ &\approx \frac{1}{2W}\frac{m^2 + Q^2}{x} , \end{aligned}$$

where m is the mass of the particle. Therefore, the amplitude becomes

$$A \sim 2W\left[\frac{m_o^2 + Q_o^2}{x_o} - \sum_i \frac{m_i^2 + Q_i^2}{x_i}\right]^{-1} . \tag{4}$$

When the numerator as well as the denominator are expressed in these variables x, Q only a simple power of W appears in a which can be removed by a renormalization of the scale of P_z in phase space integrals. The conclusion that I wish to remark for our present purposes is: when expressed in terms of Q, the transverse momentum in absolute scale, and x, the longitudinal momentum in relative scale, the wave functions have definite limits as W, the energy scale of the longitudinal momentum of the state goes to ∞.[1] In this limit the values of x are positive only. They represent the fractions of the momentum x_o which each parton has (thus, if $x_o = 1$, all x_i run from 0 to 1). The reason x_i must be positive is that if a parton has a negative momentum $P_z = Wx$ with x negative or $P_z = -|x|\,W$ the energy E_i is approximately $+|x|\,W$ so $E - P_z$ is $2|x|\,W$ which is of order W^2 larger than the $E - P_z$ of positively moving partons. Appearing in the denominator of amplitudes, such amplitudes are of order (for $W \to \infty$) when $|x|$ is as small as order c/W. The masses and transverse momenta seem to be of the order of 1 GeV so we shall call an $|x|$ of order (1 GeV)/W a "wee" x. For example, for $P_o = W = 100$ GeV in the center-of-mass system (taking $x_o = 1$), $x = 1/10$ is a small x, but it is not wee; a wee x would be $x = 0.005$ say. The behavior of the amplitude for wee x is not without interest, as we shall see, but for the present we shall discuss only x which is not wee. Then the amplitude as $W \to \infty$ appears to be a function of the Q's and x's of the parts.

In a high-energy collision, the initial state consists of one of these groups of partons moving to the right interacting with another similar group moving to the left. What is the interaction? It will not be enough to just naively say that one parton has a cross section for colliding with another, for, in field theory, interaction is represented by mediation of a field; in fact, by the exchange of just another parton coupled by the piece H_I of the Hamiltonian. But this H_I in its form, is not entirely independent of the form of H_R for they both come from operations on the original Hamiltonian H. Thus the amplitude to interact via the exchange [of] a parton is closely related to the amplitude that there is some parton in the right moving system in the initial state that can be "mistakenly" considered as really being a parton belonging to the left moving system. (Just as two electrons interact in first order by exchange of a photon, so it is also true that a right moving physical electron of $x_o = 1$ has a first order amplitude

[1]The statement is not precisely correct. What is meant is the density matrix has definite limits.

to be an ideal electron of longitudinal momentum 1-x (times W) and a photon of momentum x. Interaction results if we consider that the left moving electron (insofar as it is bare) has some amplitude to be a left moving outgoing physical electron if we make it up of a left bare electron and the aforementioned photon.) However, a parton of momentum xW to the right would be moving backwards in the left system and would have practically no amplitude (as $W \to \infty$) to be "mistakenly" considered as belonging to this left system. This is true, of course, only if x is not wee (of order 1 GeV/W). If x is wee, a right moving parton and a left moving parton are very similar in appearance. Thus interaction occurs only through exchange of partons or systems of partons of wee longitudinal momentum.

The energy dependence of reactions thus depends upon the probability of finding wee partons of a certain nature. A great deal of information on wee partons can be gotten from a knowledge of the partons where x is not wee, but only small. For the small x and wee x behavior must join in a continuous fashion. For example, suppose, for small x the amplitude to find a parton system with x small varies as $x^{\alpha}dx$ where x[α?] is some constant ($\alpha < 1$). Then the amplitude to find a wee parton of momentum ~ 1/W in dx would, to fit on, have to vary as $(1/W)^{-\alpha}$, but the range of x that such wee partons occupy is of order 1/W so that the amplitude to find a wee parton must vary as $W^{\alpha-1}$, if W is the momentum of the right moving object. If E_1 is the energy of the right mover, this amplitude is $(E_1)^{\alpha-1}$. If this is to exchange with a similar system moving to the left with energy E_2 the amplitude that this parton system is acceptable to the other system is $(E_2)^{\alpha-1}$. The amplitude for exchange therefore is $(E_1)^{\alpha-1}(E_2)^{\alpha-1}$ or varies as $S^{\alpha-1}$ since $S = E_1 E_2$. The cross section (there is always a problem of convention of the normalization of amplitudes) varies as the square of this, or $S^{2\alpha-2}$. A constant cross section means $\alpha = 1$ or the amplitude to find partons of small x must vary as dx/x. The amplitude to find wee parton is not $\int_{0?}^{1/W} dx/x$ because this dx/x law fails below 1/W. The amplitude to find a wee parton is just constant, independent of energy since the curve 1/x cuts off for x below 1/W at a value of order W and the integral below x = 1/W is finite in this event. Since cross sections (such as the total x section) are constant, we see that this must actually happen. It is, therefore, not strictly true that as $W \to \infty$ if we keep all x and Q constant there is a definite limiting wave function (as we said earlier) for there is always a finite amplitude for wee partons. However, the finite x part of the probability distribution of partons has a definite limit, in this limit there are probabilities of finding partons varying as dx/x. The apparently diverging character of this distribution for small x is cut off at wee x (that is at x of order 1/W). (A complete theory would have to describe this cutoff region in detail. We shall say more about it later.)

The equations for x not wee simplify if one concentrates on the small x part. It is then seen that there is an approximate scaling law for small x (the approximation improving as x decreases) so that solutions with special distributions of partons with a power law scale dependence ($x^{-\alpha}$) are eigenfunctions natural to field theory.

It may help to give a few, nearly trivial examples. First, according to first order perturbation theory in the expression (4) for the amplitude, the numerator does not depend on x, Q for scalar partons (couplings involve no momenta). If one of the partons has an especially low x, the the term $(\mu^2 + Q^2)/x$ belonging to it dominates and we get an amplitude proportional to x (times the scale dP/E, or dx/x, of relativistic phase space). This corresponds to $\alpha = 0$ for the scalar meson. Likewise, it can be shown that the amplitude for (longitudinally polarized) vector partons varies as constant (times dx/x). In this case a factor 1/x comes from the numerator couplings. For spin 1/2 particles coupled in the simplest ways, the amplitude varies as $x^{1/2}$ (times dx/x). In general, α equals the spin of the particle. In perturbation theory, these agree with well-known results for the energy dependence of x sections, in particular that vector meson exchange as in electrodynamics lead to constant cross sections in perturbation theory.

The deep inelastic behavior of electron-proton scattering has been looked at from the point of view described here. It can be argued that the curve of $\frac{\mu^2}{Q^2}W_2$ vs. $Q^2/2M\nu$ (in the variables of Bjorken[2]) is the distribution in x of charged partons (each weighed by the square of its charge). A behavior like dx/x for the low momentum partons is indicated by experiment.

II. REGGE BEHAVIOR

By "Regge behavior" is meant the second item of our list of regularities, that the cross section for exchange reactions vary as an (inverse) power of the energy, which power depends on the momentum transfer.

---

[2]SLAC publication No. 571

The original expectation of Regge behavior were the results of a brilliant induction from Regge's non-relativistic studies by Gell-Mann and Frautschi. Now, however, I should like to consider it as an established empirical fact and to try (in this section) to understand physically why it should be so.

Let us consider a typical exchange reaction, for example, a charge exchange reaction like $p + \pi^- \rightarrow n + \pi^o$, in which the π^-, π^o are right moving, and the p, n left moving. The easiest view of this is that a negative charge has been exchanged from the π system to the nucleon system. The cross section falls about as s^{-1} or according to the best estimates as $s^{2\alpha_o-2}$ with $\alpha_o = 0.43$ or $s^{-1.14}$. It might at first be thought that an exchange via a vector meson such as a ρ^- would lead (as it does in perturbation theory) to a constant cross section. However, it is to be noted that an important current density (the 3 component of isotopic spin) has been suddenly reversed. Initially (π^-,p) this current density had fast moving components of -1 to the right, +1/2 to the left. Afterwards (π^o, n) it has 0 to the right, -1/2 to the left. Although the total 3 component of isospin is not changed, a motion of a part of it (-1 unit) is suddenly changed from right to left motion. In electrodynamics we are aware that a sudden reversal of electric current density induces a copious Bremsstrahlung -- a sudden reversal of the direction of an electron carrying a photon field to the right, leaves the field coasting on to the right in the form of photons (and, of course, the new motion of the charge to the left generates left moving photon Bremsstrahlung).

The hadrons may act similarly. These currents are of considerable importance in our present theories and in fact we believe there are particles (ρ mesons in fact) strongly coupled to them. Thus the strong current reversal in a charge exchange will tend to shed ρ mesons. Perhaps we can guess some of the behavior of the high-energy inelastic collisions by working by analogy to Bremsstrahlung.

In studying the pure reaction $p + \pi^- \rightarrow n + \pi^o$ at high energy, we insist, first that a current be suddenly reversed and, second, that no Bremsstrahlung actually occur. This latter is because we insist that the reaction have only the two particles n, π^o in the fixed state. We can interpret the fact that in such an exchange the cross section falls (relative to the main behavior of the majority of inelastic scatterings -- for which a constant cross section is empirically more appropriate) as the energy rises, by the observation that as the energy rises it becomes increasingly less likely that the current reversal can be accomplished without

Bremsstrahlung.

The theory of Bremsstrahlung with strong coupling and with the "photons" of the field carrying the very type of currents which are sources of further Bremsstrahlung has not been worked out in detail. Nevertheless, we may boldly try to guess that certain analogies to electromagnetic weak interaction Bremsstrahlung exist. Some hope for sense here comes from noting that many features can be be seen from a classical view which takes $\hbar \rightarrow 0$ so $e^2/\hbar c$ large. Therefore some properties are understandable both for $e^2/\hbar c$ large, and for $e^2/\hbar c$ small, may have more general validity. This is especially likely if we understand the reasons for them clearly.

First, the spectrum of the particles in longitudinal momentum is dP_z/E (or dx/x for x's which are not wee). This is because the Lorentz contracted field is so sharp in z that the energy in it is distributed uniformly in P_z (the Fourier transform of a pulse being a constant). The energy distribution of the radiated particle is therefore dP_z, or if the individual particles have energies E their longitudinal momenta are distributed as dP_z/E. Also, such a distribution is stable under further disintegration of the particles, or of interaction between the particles. If we write tanh w + P_z/E for z component of the velocity of a particle, the relativistic rule for the addition of velocities becomes, as is well known, simply the addition of rapidities. Suppose a particle in its rest system can disintegrate or yield a new particle with rapidity u. Then if the old particle has rapidity w, instead, the new particle appears with rapidity $w'=w+u$. Therefore, if the old particles are distributed uniformly in w (as dw) the new particles appear with a distribution also uniform in rapidity because $dw'=dw$. Thus this feature (a spectrum dw = dP_z/E) is to be expected independently of whether we are seeing what was originally radiated (say, ρ mesons) or are observing some other secondary particles that these may have changed into (e.g., if ρ's go into kions). Hence I believe we should expect it for our inelastic distribution of particles of small (and wee) x in our strongly interacting systems also. (We cannot expect this to be valid for large x, say x = 1/2 also, because if that much energy is taken from the primary particle by radiation of one emitted particle, subsequent emissions are severely affected. The dP_z/E spectrum for photons in electrodynamics is precisely valid only for smaller values of x.)

Next, the energy in the field this radiated is some fraction of the energy of the particle which radiates. Thus the particle may be found after the radiation to have lost on the average some fixed fraction of its energy.

This is found experimentally, in some cases. For example, in pp collisions which yield a forward proton, its average momentum is about 0.60 of the incident momentum[3] (in individual collisions, its value fluctuates widely).

For weak coupling electrodynamics, the vector field particles are emitted independently into a Poisson distribution with mean number $\bar{n}$ emitted. The probability that none are emitted is $e^{-\bar{n}}$. The sum of the chance of emitting none, one, two, etc., (that is, the total cross section) is much like it would be without coupling to the photons. Here we know the total x section is constant, and so can try to interpret the energy fall- off $s^{2\alpha_o-2}$ of the pure two body charge exchange reaction as the factor $e^{-\bar{n}}$ for the probability of no emission, where $\bar{n}$ is the expected mean number of primary particles emitted. This multiplicity $\bar{n}$ must rise logarithmically with energy then as $\bar{n} = (2-2\alpha_o) \ln s$. The particles we observe are not, of course, the primary field particles emitted, but rather the observed particles are secondary disintegration products of these unknown primaries. But if each primary produces on the average a fixed number of secondaries, we see that the expectation is that the multiplicity grows logarithmically with E.

This is necessary if our various ideas are to fit together. Because we have already suggested that the mean number of any kind of particle emitted is to vary with x as c dx/x for small x and, for a given x, not to vary otherwise with the energy W of the collision (so that c is a constant). The total mean number emitted, then, is $c\int dx/x$. The upper limit of x is of finite order (for the formula fails as $x \to 1$ and x cannot exceed 1) but the lower limit is of order of wee x, (i.e., order 1/W) where the dx/x fails. Thus the mean number emitted to the right must vary as c(ln W + const). (Actually we can do the integral all the way to zero, for we expect the integrand to be $c\, dP_z/\sqrt{\mu^2 + Q^2 + P_z^2}$ where μ is the mass and Q is the transverse momentum of a typical particle. Putting $P_z = xW$, this is

$$\int_0^{x_1} c \frac{dx}{\sqrt{x^2 + (\mu^2 + Q^2)/W^2}} = c \, ln \frac{2Wx_1}{(\mu^2 + Q^2)^{1/2}}$$

for finite x_1. To go further, we should have to know the transverse

---

[3] Report on the Topical Conference on High Energy Collisions, CERN 68-7, February 1968, Turkot, p.316.

momentum distribution.)

The one respect in which the electrodynamic analogy leads us astray is in the transverse momentum distribution of the photons. These large transverse momenta fall off slowly (like $\frac{dP_z d^2Q}{E\ Q^2}$ for sudden current reversals) but in the strong collisions this is empirically not true. Some characteristics of the distribution of charge across the face of the of the interacting particle is involved here.

I have not yet studied the regularities involving the transverse momenta (items (4) and (5) in our Introduction) from the viewpoint being developed here. In the meantime, we can take these as empirical facts to be included in any expectations. In the same way we leave for further research strangeness and isospin character of these effects. We should notice, however, that, although we discussed a charge exchange arising from an exchange of a particle of the quantum numbers of the ρ^-, the exchange of any current of the usual octet would have analogous effects on the possible Bremsstrahlung of particles coupled to other (non-commuting) currents.

In a pure two-body exchange reaction, since the probability of not Bremsstrahlung depends on exactly what currents are reversed, the value of α_0 will, from this point of view, depend on the quantum numbers of the particle exchanged (which, of course, it does). The α_0 here referred to is evidently only the largest for a given set of quantum numbers exchanged.

The quantum numbers exchanged may involve not only currents of unitary symmetry, for baryon number (and possibly spin) may be exchanged, and we do not know if there are special couplings to baryon currents (or spin currents) which are also involved in determining α_0. However, I should like to hazard the guess that baryon number cannot be exchanged without the transfer of a fundamental part of spin 1/2. Such an ideal part already has $\alpha = 1/2$ which would imply a $1/\sqrt{s}$ behavior of amplitudes before corrections to Bremsstrahlung. Therefore, if we do an experiment which freely allows the emission of wee mesons, except that the quantum numbers are controlled so that a baryon must be exchanged between right and left systems this cross section probably approaches a 1/s behavior, instead of the constant expected for similar experiments in which no baryons need be exchanged.

A final question is that of the distribution of correlations among the various emitted particles in the average inclusive collision. In the perturbation theory these are emitted independently and at random in a Poisson distribution so the probability that there will be k of them of momenta $x_1, x_2, \ldots x_k$ is just $(cdx_1/x_1), (cdx_2/x_2), \ldots(cdx_k/x_k)e^{-\bar{n}}/k!$ where $\bar{n}$ is the mean number emitted. I have not yet found a good reason whether this would be true in our non-perturbative case or not, but if one insists on comparing the experimental distributions and correlations to some theory, perhaps this is the first thing to try: that the pion distribution results from an original Poisson distribution of ρ's, each distributed for small and wee x as c dP_z/E_z with c near 1.1 or 1.2 (c is $2\text{-}2\alpha_0$ for the ρ trajectory). In fact, if we suppose two pions for every ρ, the multiplicity of ρ's would be c ln s, and the multiplicity of the pions twice this, or about 2.3 ln E_{LAB}(GeV). Surprisingly, this fits observations very well (see Table 1). As an additional coincidence, this value c is what one gets from perturbation theory if one uses the coupling constant determined for ρ nucleon coupling. It must be admitted that in this paragraph we have gone much further than we should -- for our precise numerical result depends upon a choice of which particles are fundamental and which they disintegrate into. All our other predictions were of those features which were independent of such specific choices.

The probability that the total momentum of all the emitted right moving ρ's is less than y is proportional to y^c so that (aside from diffraction dissociation) the momentum distribution of the ongoing particle, when it takes a fraction of momentum x close to 1 should vary as $(1\text{-}x)^c$ where 1-x is small.

In Section I we discussed the longitudinal momentum distribution of partons expected in a hadron wave function, but we have not seen how this might lead us to expectations for the distributions of momenta of real hadrons in a collision, for partons are not real hadrons. Nevertheless, we shall suppose that when a hadron is disturbed via interaction, so that its distribution of partons is no longer exactly that of a single real hadron state, it must be compounded of a series of real hadron states, but the distribution of longitudinal momentum of these real hadrons is qualitatively like those of the partons which we described before (in Section I). I have no way to see why this must be true, but the features of the distributions discussed in that section seem, firstly, to rely mainly only on qualities of relativistic transformation; secondly, the principle is right in perturbation theory; and thirdly, the results of assuming this fit very well with the qualitative predictions of the Bremsstrahlung analogy. Finally, for one reason or another -- empirical or theoretical, good or bad -- I suspect that the high energy collisions to have a number of features which we summarize in the next section.

III. EXPECTED BEHAVIOR AT HIGH ENERGY COLLISIONS

In a collision of very high energy between two hadrons each of momentum W, in the center-of-mass system, the outgoing particles of the collision should be described by two variables Q, the transverse momenta in absolute units and x the longitudinal momentum as a proportion of W (so that $P_z = xW$). We intend to describe cross sections for various processes as W increases without limit, keeping the x, Q's of the particles constant. If for large W an x is of the order 1 GeV/W (so that its momentum in the C.M. system is in the BeV range or less) we say the particle has a wee momentum. Small x simply means a value of x much below one, but higher than order 1 GeV/W for extremely large W. Finally we should like to characterize experiments as being of two classes -- exclusive and inclusive.

An exclusive experiment is one in which it is asked that only certain particular particles of fixed Q, x, and character be found in the final state, and no others. In particular, it excludes the emission of particles with wee momenta in the limit. Examples are: two-body reactions $AB \to CD$; an experiment which uses missing mass to try to select events which are virtually two-body reaction, etc. For such collisions, the cross sections should fall off inversely as a power of $s = 2W^2$, the power being $2\alpha(t) - 2$ where $\alpha(t)$ is the α appropriate to the highest Regge trajectory capable of carrying the necessary exchange of quantum numbers between the right moving and left moving system, and where t is the negative square of the transverse momentum which must be exchanged. (With these x, Q {these} variables fixed, the longitudinal momentum transfer and the energy transfer go to zero inversely {as W} as $W \to \infty$.) In the special case that no quantum numbers need to be exchanged, the cross section is constant (empirically, at least, if $t = 0$); it does not fall as $W \to \infty$. This phenomena is called diffraction dissociation, and is sometimes expressed as the exchange of the "Pomeranchuk trajectory".

An inclusive experiment is one in which certain particles are looked for at given Q, x, but one also permits any number of additional particles. More precisely, it does not in any way exclude the emission of arbitrary numbers and kinds of particles with wee momenta. Examples are the total inelastic cross section, the mean number of Ω mesons emitted with momentum x in range dx, the probability that no single particle is moving right nor left in the (center-of-mass system) with more than 1/2 the original momentum W, etc. <u>Such cross sections should approach constant finite values as $W \to \infty$.</u>

The mean number of mesons of a given kind formed with small x in a high-energy <u>inclusive</u> experiment should vary as dx/x. This should extrapolate right through the wee x region in the dP_z/E where E is the energy $\sqrt{\mu^2 + Q^2 + P_z^2}$ of the meson momentum P_z, at fixed Q. (This suggests that more appropriate variables for the small x region, would be w, Q where w is the z-component rapidity $w = \tanh^{-1}(P_z/E)$. The distribution should be uniform in dw (for each Q) and ultimately independent of W.) As a consequence, the multiplicity of a given kind of hadron should rise logarithmically with W.

It is this dx/x distribution with its logarithmically divergent character for small x which makes it possible that the probability of finding any specific set of particles with given x, Q values (except the elastic or diffraction dissociation ones) falls with energy as a power, and yet the total cross section can be constant.

In an inclusive experiment of A colliding from the right, with B from the left, the probability that some particle C comes out moving to the right with an x close to unity should vary as $(1-x)^{2-2\alpha(t)}$, as long as $(1-x)^{1/2}$ is not wee. Here $\alpha(t)$ is the value appropriate to the trajectory of highest α (excluding the Pomeranchon) which could, upon emission, carry away the quantum numbers and transverse momentum needed to turn A to C.

In a special kind of partially exclusive process in which a baryon must be exchanged to get the reactants of finite x, but no wee baryon{s} appears among the particles of wee momentum, then I believe the cross section will vary as 1/W but this is not on as firm a basis as the other suggestions.

TABLE 1

Multiplicity of Pions in High Energy Collisions

E_{LAB} (BeV)	Mult.	2.3 ln E_{LAB}
30	7	7.8
470	13 ± 1	14.1
1500	18 ± 2	16.8
12300	24 ± 4	21.6

PHYSICAL REVIEW VOLUME 185, NUMBER 5 25 SEPTEMBER 1969

Inelastic Electron-Proton and γ-Proton Scattering and the Structure of the Nucleon*

J. D. BJORKEN AND E. A. PASCHOS
Stanford Linear Accelerator Center, Stanford University, Stanford, California 94305
(Received 10 April 1969)

A model for highly inelastic electron-nucleon scattering at high energies is studied and compared with existing data. This model envisages the proton to be composed of pointlike constituents ("partons") from which the electron scatters incoherently. We propose that the model be tested by observing γ rays scattered inelastically in a similar way from the nucleon. The magnitude of this inelastic Compton-scattering cross section can be predicted from existing electron-scattering data, indicating that the experiment is feasible, but difficult, at presently available energies.

I. INTRODUCTION

ONE of the most interesting results emerging from the study of inelastic lepton-hadron scattering at high energies and large momentum transfers is the possibility of obtaining detailed information about the structure, and about any fundamental constituents, of hadrons. We discuss here an intuitive but powerful model, in which the nucleon is built of fundamental pointlike constituents. The important feature of this model, as developed by Feynman, is its emphasis on the infinite-momentum frame of reference.

It is argued that when the inelastic scattering process is viewed from this frame, the proper motion of the constituents of the proton is slowed down by the relativistic time dilatation, and the proton charge distribution is Lorentz-contracted as well. Then, under appropriate experimental conditions, the incident lepton scatters instantaneously and incoherently from the individual constituents of the proton, assuming such a concept makes sense.

We were greatly motivated in this investigation by Feynman, who put the above ideas into a highly workable form. In Sec. II, we discuss the basic ideas and equations for the model as they apply to electron-proton scattering. Two models are then discussed in detail, with interesting consequences for the ratio of electron-proton and electron-neutron scattering. For a broad class of such models, we find a sum rule which indicates that, although it is not difficult to fit the data within $\sim 50\%$, it is more difficult to do better; the observed cross section is uncomfortably small.

In Sec. III, we look for stringent tests of Feynman's picture. We propose that, under similar experimental conditions, inelastic Compton scattering can also be calculated within the model. It is shown that the ratio of inelastic electron-proton to inelastic γ-proton scattering, under identical kinematical conditions, is model-independent and of order unity, provided the proton constituents (which Feynman calls "partons") possess unit charge and spin 0 or $\frac{1}{2}$. We propose experiments which can measure inelastic Compton scattering. To this end we have estimated the yield and background curves, and we conclude that such experiments may be feasible at energies available to SLAC.

* Work supported by the U. S. Atomic Energy Commission.

II. INELASTIC e-p SCATTERING

The basic idea in the model is to represent the inelastic scattering as quasifree scattering from pointlike constituents within the proton, when viewed from a frame in which the proton has infinite momentum. The electron-proton center-of-mass frame is, at high energies, a good approximation of such a frame. In the infinite-momentum frame, the proton is Lorentz-contracted into a thin pancake, and the lepton scatters instantaneously. Furthermore, the proper motion of the constituents, of partons, within the proton is slowed down by time dilatation. We can estimate the interaction time and the lifetime of the virtual states within the proton. By using the notation of Fig. 1, we find the following.

Time of interaction:

$$\tau \approx 1/q^0 = 4P/(2M\nu - Q^2), \tag{2.1}$$

where q^0 was calculated in the lepton-proton center-of-mass frame.

Lifetime of virtual states:

$$\begin{aligned} T &= \{[(xP)^2+\mu_1^2]^{1/2} \\ &\quad + [(1-x)^2P^2+\mu_2^2]^{1/2} - [P^2+M_p^2]^{1/2}\}^{-1} \\ &= \frac{2P}{(\mu_1^2+p_{1\perp}^2)/x + (\mu_2^2+p_{2\perp}^2)/(1-x) - M_p^2}. \end{aligned} \tag{2.2}$$

If we now require that

$$\tau \ll T, \tag{2.3}$$

then we can consider the partons, contained in the proton, as free during the interaction. Furthermore, if we consider large momentum transfers $-q^2 \gg M^2$, then we expect the scattering from the individual partons to be incoherent. The above conditions appear to be satisfied in the high-energy, large-momentum-transfer experiments at SLAC.

The kinematics for e-p inelastic scattering have been discussed in many places, in as many different nota-

Reprinted from *Phys. Rev.* **185**, 1975 (1969).

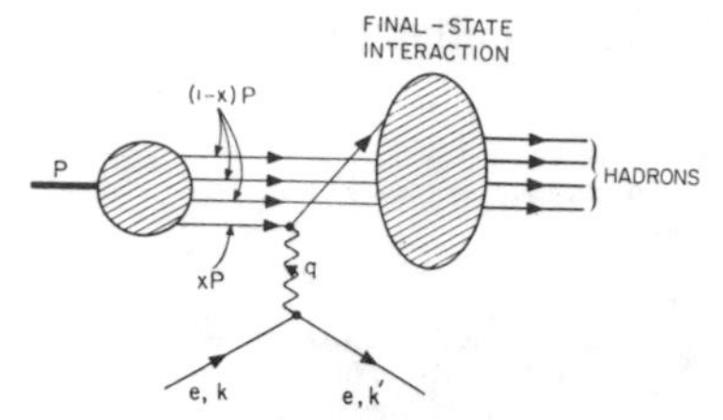

FIG. 1. Kinematics of lepton-nucleon scattering in the parton model.

tions.[1] We recall here that

$$\frac{EE'}{\pi}\frac{d\sigma}{dq^2d\nu}=\frac{d\sigma}{d\Omega dE'}=\frac{\alpha^2}{4E^2\sin^4(\frac{1}{2}\theta)}$$
$$\times[W_2\cos^2(\tfrac{1}{2}\theta)+2W_1\sin^2(\tfrac{1}{2}\theta)], \quad (2.4)$$

where W_1 and W_2 are functions of the two invariants

$$\nu=(E-E')=q\cdot P/M,$$
$$Q^2=-q^2=4EE'\sin^2(\tfrac{1}{2}\theta), \quad (2.5)$$

evaluated in the laboratory frame. The ratio W_1/W_2 is bounded. Using

$$\frac{W_1}{W_2}=\left(1+\frac{\nu^2}{Q^2}\right)\frac{\sigma_t}{\sigma_t+\sigma_l}, \quad (2.6)$$

and the approximation $\frac{1}{2}\theta\ll1$, we can write

$$\frac{d\sigma}{d\Omega dE'}\cong\frac{\alpha^2}{4E^2\sin^4(\frac{1}{2}\theta)}W_2(q^2,\nu)\left[1+\left(\frac{\sigma_t}{\sigma_t+\sigma_l}\right)\frac{\nu^2}{2EE'}\right], \quad (2.7)$$

where σ_t and σ_l (≥0) are the absorption cross sections for transverse and longitudinal photons.[2]

It remains to calculate the invariant functions W_1 and W_2, especially W_2. Within the model, the virtual photon interacts with one of the partons, while the rest remain undisturbed during the interaction. The interaction with the parton is as if the parton were a free, structureless particle. The cross section $d\sigma/d\Omega dE'$ is then a sum over individual electron-parton interactions appropriately weighted by the parton charge and momentum. For a free particle of any spin and unit charge, elementary calculation yields

$$W_2(\nu,q^2)=\delta(\nu-Q^2/2M)=M\delta(q\cdot P-\tfrac{1}{2}Q^2), \quad (2.8)$$

while for W_1, we have

$$\sigma_t=0 \text{ for spin } 0,$$
$$\sigma_l=0 \text{ for spin } \tfrac{1}{2}, \quad (2.9)$$

with an indeterminate result for higher spins.

At infinite momentum, we visualize the intermediate state from which the electron scatters as follows:

(a) It consists of a certain number N of free partons (with probability P_N).

(b) The longitudinal momentum of the ith parton is a fraction x_i of the total momentum of the proton:

$$\mathbf{p}_i=x_i\mathbf{P}. \quad (2.10)$$

(c) The mass of the parton, before and after the collision, is small (or does not significantly change).

(d) The transverse momentum of the parton before the collision can be neglected, in comparison with $\sqrt{(Q^2)}$, the transverse momentum imparted as $p\to\infty$.

With these assumptions, it should be a good approximation to write, at infinite momentum,

$$p_i^\mu\cong x_iP^\mu. \quad (2.11)$$

The question of corrections to this approximation has been studied by Drell, Levy, and Yan.[3] We shall not consider them further here.

The contribution to W_2 from a single parton of momentum $x\mathbf{P}^\mu$ and charge Q_i is then

$$W_2^{(i)}=x_iQ_i^2M\delta(q\cdot x_iP-\tfrac{1}{2}Q^2)$$
$$=Q_i^2M\delta\left(q\cdot P-\frac{Q^2}{2x_i}\right)=Q_i^2\delta\left(\nu-\frac{Q^2}{2Mx}\right). \quad (2.12)$$

The factor x_i in front is necessary to ensure that

$$\lim_{E\to\infty}\frac{d\sigma^{(i)}}{dq^2}=Q_i^2\left(\frac{4\pi\alpha^2}{q^4}\right), \quad (2.13)$$

consistent with the Rutherford formula. For a general distribution of partons in the proton, we have

$$W_2(\nu,q^2)=\sum_N P(N)\langle\sum_i Q_i^2\rangle_N$$
$$\times\int_0^1 dx\, f_N(x)\delta\left(\nu-\frac{Q^2}{2xM}\right). \quad (2.14)$$

Here $P(N)$ is the probability of finding a configuration of N partons in the proton, $\langle\sum_i Q_i^2\rangle_N$ equals the average value of $\sum_i Q_i^2$ in such configurations, and $f_N(x)$ is the probability of finding in such configurations a parton with longitudinal fraction x of the proton's momentum, that is, with four-momentum xP^μ.

Upon integrating over x, we find

$$\nu W_2(\nu,q^2)=\sum_N P(N)\langle\sum_i Q_i^2\rangle_N xf_N(x)\equiv F(x), \quad (2.15)$$

with

$$x=Q^2/2M\nu. \quad (2.16)$$

Therefore νW_2 is predicted to be a function of a *single*

[1] S. Drell and J. Walecka, Ann. Phys. (N. Y.) **28**, 18 (1964); J. D. Bjorken, Phys. Rev. **179**, 1547 (1969).

[2] L. Hand, in *Proceedings of the Third International Symposium on Electron and Photon Interaction at High Energies, Stanford Linear Accelerator Center, 1967* (Clearing House of Federal Scientific and Technical Information, Washington, D. C., 1968).

[3] S. D. Drell, D. J. Levy, and T.-M. Yan (private communication).

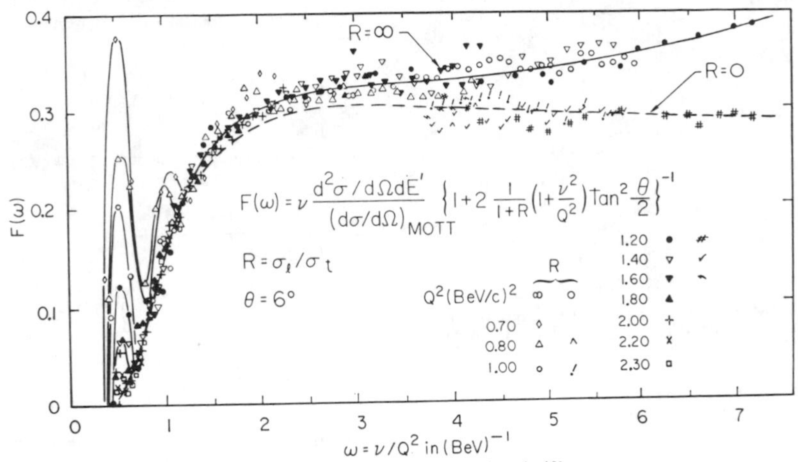

FIG. 2. Plot of the data as a function of ν/Q^2.

variable ν/Q^2, a feature apparently satisfied by the data.[4] Furthermore, the model provides an interpretation of the nature of the function $F(x)/x$: It is the mean square of the charge of partons with four-momentum xP^μ. The experimental[4] determination of $F(x)$ is shown in Fig. 2.

Before going into detailed models for $f_N(x)$, we notice that

$$f_N(x_1)=\int dx_2\cdots dx_N f_N(x_1,\cdots,x_N)\delta(1-\sum_i x_i),\tag{2.17}$$

$$\int_0^1 dx_1 f_N(x_1)=1,$$

where $f_N(x_1,\cdots,x_N)$ is the joint probability of finding partons (irrespective of charge) with longitudinal fractions $x_1,\cdots,x_N$. It follows that f_N is a symmetric function of its arguments. Therefore,

$$\int_0^1 x_1dx_1f_N(x_1)=\frac{1}{N}\int dx_1\cdots dx_N(\sum_i x_i)\times f_N(x_1,\cdots,x_N)\delta(1-\sum_i x_i)=1/N.\tag{2.18}$$

Putting together (2.18) and (2.15), we obtain a sum rule

$$\int dx\,F(x)=\sum_N P(N)\langle\sum_i Q_i^2\rangle_N/N$$
$$=\text{mean-square charge per parton.}\tag{2.19}$$

[4] See the rapporteur talk of W. H. K. Panofsky, in *Proceedings of the Fourteenth International Conference on High-Energy Physics, Vienna, 1968* (CERN, Geneva, 1968), pp. 36–37, based on the work of E. Bloom, D. Coward, H. DeStaebler, J. Drees, J. Litt, G. Miller, L. Mo, R. Taylor, M. Breidenbach, J. Friedman, G. Hartmann, H. Kendall, and S. Loken.

Numerically,

$$\frac{Q^2}{2M}\int\frac{d\nu}{\nu}W_2=\int dx\,F(x)\approx 0.16,\tag{2.20}$$

yielding a rather small mean-square charge per parton.

In the following models, out of ignorance, we shall choose one-dimensional phase space for the distribution function $f_N(x_1\cdots x_N)$; that is,

$$f_N(x_1\cdots x_N)=\text{const}.\tag{2.21}$$

An elementary calculation yields

$$f_N(x)=(N-1)(1-x)^{N-2}.\tag{2.22}$$

A. Three-Quark Model

Assuming that the proton is made up of three quarks with the usual charges,[5] we obtain

$$\nu W_2=f_3(x)=2x(1-x),\quad x=Q^2/2M\nu.\tag{2.23}$$

While the data support $\nu W_2\to$ const as $\nu\to\infty$ (or $x\to 0$), the model predicts that νW_2 should vanish, a result not dependent on the specific choice of $f_3(x)$, but only on the fact that f_3 is normalizable. In fact, within our one-dimensional model, if the number of partons is held finite, then the cross section vanishes as $x\to 0$. If, and only if,

$$\lim_{N2\to\infty}N^2P(N)=\text{const}\neq 0\tag{2.24}$$

will νW_2 approach a constant as $x\to 0$ ($\nu/Q^2\to\infty$). This is shown in the Appendix.

[5] M. Gell-Mann, Phys. Letters 8, 214 (1964); C. Zweig, CERN Report Nos. TH 401, 402, 1964 (unpublished).

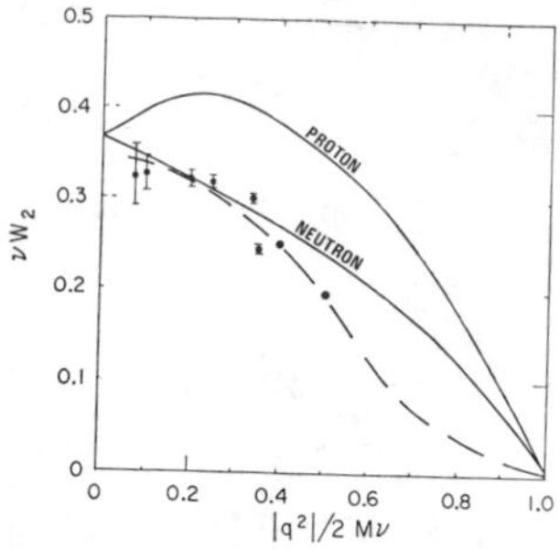

FIG. 3. Plot of the results for a model of three quarks in a sea of quark-antiquark pairs. The dashed line is visual fit through the experimental points of Ref. 4.

B. Three Quarks in a Background of Quark-Antiquark Pairs

In order to try to improve the model, we assume that in addition to the three quarks there is a distribution of quark-antiquark pairs (the "pion cloud"?). The mean-square charge of the cloud we take to be statistical

$$\langle \sum Q_i^2\rangle_{\text{cloud}}/N=\tfrac{1}{3}[(\tfrac{2}{3})^2+(\tfrac{1}{3})^2+(\tfrac{1}{3})^2]=2/9. \quad (2.25)$$

Therefore,

$$\langle \sum_i Q_i^2\rangle_N=1+2/9(N-3)=2/9N+\tfrac{1}{3} \quad \text{for the proton}$$
$$=\tfrac{2}{3}+2/9(N-3)=2/9N \quad \text{for the neutron.} \quad (2.26)$$

For $P(N)$, we choose

$$P(N)=C/N(N-1), \quad (2.27)$$

on the grounds that it is simple and has the asymptotic behavior which makes $\nu W_2 \to$ const as $x\to 0$. Before we begin, we note, from (2.26) and (2.19), that

$$\int_0^1 dx\,F(x)=\int_0^1 dx\,\nu W_2=2/9+\tfrac{1}{3}\langle 1/N\rangle>0.22 \text{ (proton)}$$
$$=2/9=0.22 \text{ (neutron)}$$
$$\approx 0.16 \text{ (expt.)}. \quad (2.28)$$

Thus we cannot expect a fit better than $\sim 50\%$ to the data. Inserting (2.27), (2.26), and (2.22) into the expression (2.15) for $F(x)$, we can perform the sum. For the proton,

$$F(x)=x\sum_{N=3,5\cdots} P(N)(N-1)(1-x)^{N-2}(2/9N+\tfrac{1}{3})$$
$$=C\sum_{N=3,5\cdots}\left(\frac{2}{9}+\frac{1}{3N}\right)x(1-x)^{N-2}$$
$$=C\left\{\frac{2}{9}\frac{(1-x)}{(2-x)}+\frac{1}{6}\frac{x}{(1-x)^2}\left[\ln\left(\frac{2-x}{x}\right)-2(1-x)\right]\right\}. \quad (2.29)$$

The normalization constant C is determined from the condition

$$1=\sum_{N=3,5\cdots} P_N(x)=C\sum_{N=3,5\cdots}\frac{1}{N(N-1)}=C(1-\ln 2). \quad (2.30)$$

For the neutron, the term in square brackets is omitted; thus the final expressions are

$$F_p(x)=\frac{1}{1-\ln 2}\left\{\frac{2}{9}\frac{(1-x)}{(2-x)}+\frac{1}{6}\frac{x}{(1-x)^2}\times\left[\ln\left(\frac{2-x}{x}\right)-2(1-x)\right]\right\}, \quad (2.31)$$

$$F_n(x)=\frac{2}{9}\left(\frac{1}{1-\ln 2}\right)\left(\frac{1-x}{2-x}\right).$$

It is clear that the results are model-dependent and not to be taken too seriously. There is a need for a model-independent check of the basic assumptions. In Sec. III we discuss the corresponding process with electrons replaced by γ rays as a test of the basic idea of the calculation.

In Fig. 3, we compare Eqs. (2.31) with the experiment. The shape of the curve is in fair agreement, and could be improved by suppressing the contribution of three-parton configurations. On the other hand, the over-all normalization is off, as discussed below [(2.28)]. Readjustment of the coefficients P_N or the distributions $f_N(x)$ of longitudinal fraction cannot improve this feature. The ratio of neutron cross section to proton remains nearly constant and about 0.8 over a large range of x, although the ratio approaches 1 as $x\to 0$.

According to (2.9), if the partons all have spin $\frac{1}{2}$, we expect $\sigma_l/\sigma_t\to 0$; if they are spinless, we find instead $\sigma_t/\sigma_l\to 0$. A finite ratio would indicate that both kinds are present.

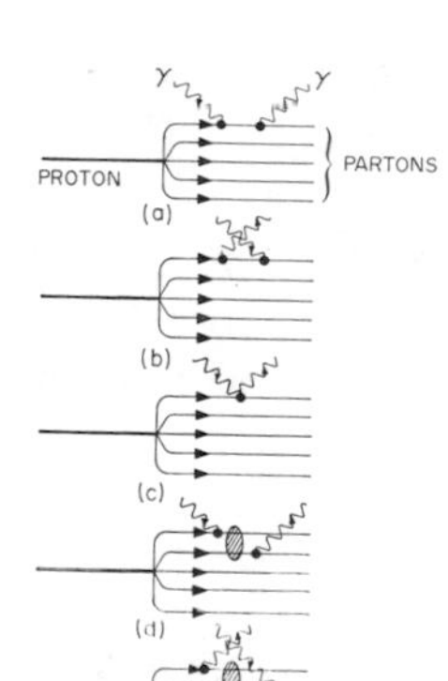

FIG. 4. Diagrams contributing to inelastic Compton scattering.

III. INELASTIC COMPTON SCATTERING

In this section, we suggest that a good way to find out about the internal structure of the proton is to look at it. That is, one should measure the inelastic scattering of photons from protons, yielding a photon plus anything else. The yield can be predicted in terms of this model. We visualize the inelastic scattering again as incoherent scattering from the partons in the proton according to the point cross section. The elementary photon-parton scattering goes by the diagrams shown in Fig. 4. Diagram (c) in Fig. 4, does not occur for spin-$\frac{1}{2}$ partons. Its contribution is evidently very similar to the case of electron scattering. We argue that exchange-terms such as in diagrams (d) and (e) can be ignored on the basis that the lifetime of the intermediate states, between absorption and emission of the photon, is of order $E_{c.m.}^{-1}$ in the γ-proton center-of-mass system and much less than the lifetime of the virtual-parton states of the proton estimated in (2.2). Furthermore, because we require the momentum transfer between the photons to be large (>1 BeV/c), the parton will necessarily have transverse momentum much larger than the average momentum within the proton and the probability that it interacts with another parton within the proton at such high $P_\perp$ is very small.

The kinematics for the process is illustrated in Fig. 5, with the value

$$\begin{aligned} s &= (k+p)^2 \approx 2k\cdot p = (2Mk)_{\text{lab}} \gg M^2\,, \\ t &= (k-k')^2 = -2k\cdot k' = -[4kk'\sin^2(\tfrac{1}{2}\theta)]_{\text{lab}}\,, \\ M\nu &= (k-k')\cdot p = M(k-k')_{\text{lab}}\,, \\ u &\approx -s-t\,. \end{aligned} \tag{3.1}$$

We require $-t$ to be large, as we already mentioned, so that the process is incoherent, and only Figs. 4(a)–4(c) (for integer spin) need be taken into account. The method for calculation is the same as in Sec. II: We take the Compton cross section from point partons and average over the parton momentum distributions and proton configurations. This elementary photon-parton interaction is given (for spin-$\frac{1}{2}$ partons) by the Klein-Nishina formula written in terms of the Mandelstam variables s, t, u. If the parton carries the full momentum $P(x=1)$, we have

$$\begin{aligned} \frac{d\sigma}{dtd\nu} &= \frac{4\pi\alpha^2}{s^2}\delta\left(\nu+\frac{t}{2M}\right)Q_i^4 \quad \text{for spin } 0 \\ &= \frac{-2\pi\alpha^2}{s^2}\left(\frac{s}{u}+\frac{u}{s}\right)\delta\left(\nu+\frac{t}{2M}\right)Q_i^4 \quad \text{for spin } \tfrac{1}{2}\,, \end{aligned} \tag{3.2}$$

which can be combined into the form

$$\frac{d\sigma}{dtd\nu} = \frac{4\pi\alpha^2}{s^2}\left(1-R\frac{(s+u)^2}{2su}\right)\delta\left(\nu+\frac{t}{2M}\right)Q_i^4\,, \tag{3.3}$$

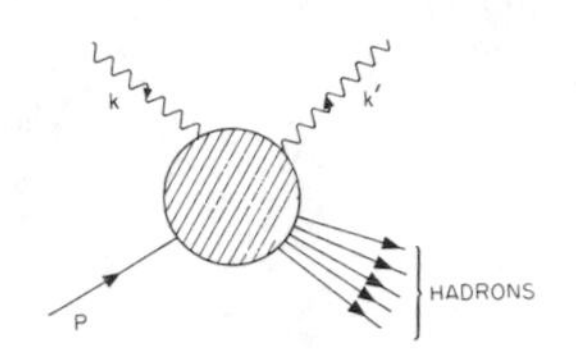

FIG. 5. Kinematics of Compton scattering.

with

$$\begin{aligned} R &= 1 \quad \text{for spin } \tfrac{1}{2} \\ &= 0 \quad \text{for spin } 0\,. \end{aligned} \tag{3.4}$$

If the parton has longitudinal fraction x, we make the replacements (we are here in the infinite-momentum frame)

$$s \to xs\,,\quad u \to xu\,,\quad t \to t\,,\quad \delta(\nu+t/2M) \to \delta(\nu+t/2Mx)\,. \tag{3.5}$$

We then multiply by the distribution function $f_N(x)$ and by $P(N)$, integrate over x, and sum over N [cf. Eqs. (2.12)–(2.15)] to obtain

$$\frac{d\sigma}{dtd\nu} = \sum_N \int_0^1 dx\, P(N) f_N(x) \frac{4\pi\alpha^2}{x^2s^2} \times \left[1-R\frac{(s+u)^2}{2su}\right]\delta\left(\nu+\frac{t}{2Mx}\right)\langle\sum_i Q_i^4\rangle_N\,. \tag{3.6}$$

Integrating over x and expressing the result in laboratory variables, Eq. (3.1) gives the final result:

$$\frac{d\sigma}{d\Omega dk'} = \frac{\alpha^2}{4k'\sin^4(\frac{1}{2}\theta)}\frac{\nu}{kk'}\left[1+R\frac{\nu^2}{2kk'}\right] \times \sum_N P(N) x f_N(x) \langle\sum_i Q_i^4\rangle_N\,, \tag{3.7}$$

with

$$x = -t/2M\nu\,. \tag{3.8}$$

Making the identification

$$k \leftrightarrow E\,,\quad k' \leftrightarrow E'\,,\quad -t \leftrightarrow Q^2\,,\quad R \leftrightarrow \sigma_t/(\sigma_t+\sigma_l)\,, \tag{3.9}$$

we find a remarkable correspondence between (3.7) and (2.14), (2.9), and (2.7), the corresponding cross section for electrons. The only changes are the additional factor ν^2/kk' and the replacement $Q_i^2 \to Q_i^4$. Even the factor in square brackets dependent on parton spin is the same. We conclude that within the validity of this parton model, for partons of unit charge ($Q^2=Q^4$) and spin 0 or $\frac{1}{2}$, the ratio of electron scattering to γ scattering is a model-independent number of order

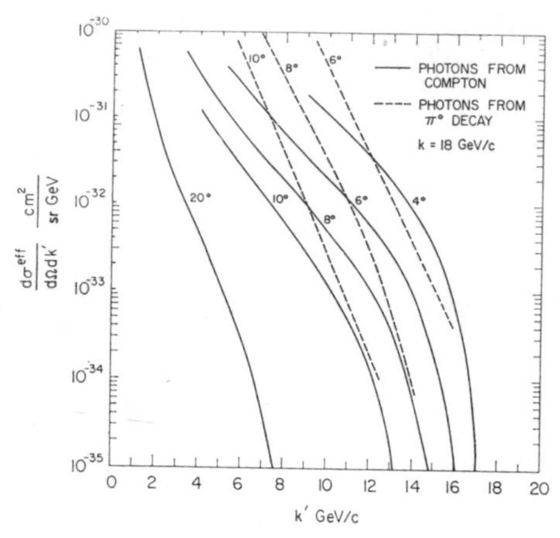

FIG. 6. Double-differential laboratory cross section for inelastic Compton scattering for an 18-GeV incident bremsstrahlung spectrum. The solid curve corresponds to the signal. The dotted curve is the background of γ's from π^0's using the data of Ref. 8 as discussed in the text.

unity. In general,[6]

$$\left(\frac{d\sigma}{d\Omega dE'}\right)_{\gamma p}=\frac{\nu^2}{kk'}\left(\frac{d\sigma}{d\Omega dE'}\right)_{ep}\langle\sum_i Q_i^4\rangle/\langle\sum_i Q_i^2\rangle, \quad (3.10)$$

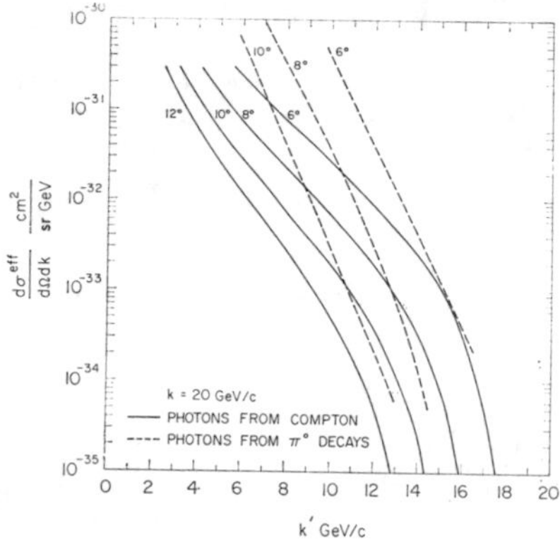

FIG. 7. Inelastic Compton scattering for a 20-GeV incident bremsstrahlung spectrum. The background curves are the same as in Fig. 6.

[6] We point out that the argument used in Sec. II in deriving the sum rule can be repeated here, giving

$$Q^2\int\left(\frac{d\sigma}{d\Omega dE}\right)_{\gamma p}\frac{kk'}{\nu^2\sigma_R}d\nu=\sum_N P(N)\langle\sum_i Q_i^4\rangle_N/N,$$

where $\sigma_R=[\alpha^2/4k^2\sin^4(\frac{1}{2}\theta)](1+R\nu^2/2kk')$. We have also assumed the same momentum distribution $f_N(x)$ for the spin-0 and spin-$\frac{1}{2}$ partons.

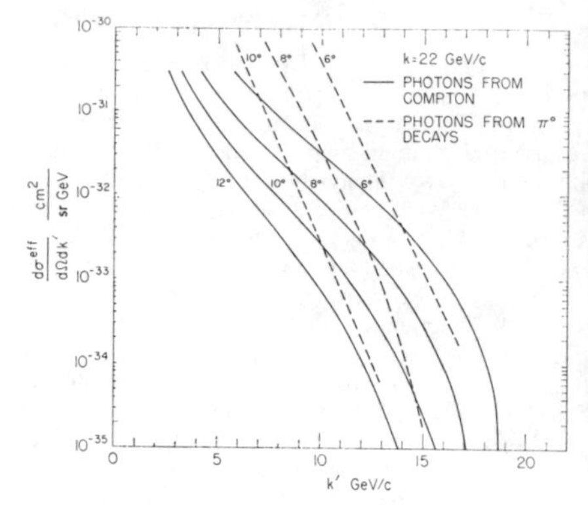

FIG. 8. Inelastic Compton scattering for a 22-GeV incident bremsstrahlung spectrum. The background curves are the same as in Fig. 6.

where for any operator $O(N)$ we have

$$\langle O\rangle=\sum_N P(N)O(N)f_N(x). \quad (3.11)$$

For our model of three quarks in a cloud of quark-antiquark pairs, there exist upper and lower limits for $\langle\sum Q^4\rangle/\langle\sum Q^2\rangle$. We note from (2.25) and (2.26), for the proton, that

$$\langle\sum_i Q_i^4\rangle_N=\frac{11}{27}+\frac{2}{27}(N-3)=\frac{1}{3}\langle\sum_i Q_i^2\rangle_N+\frac{2}{27}$$

$$=\frac{5}{9}\langle\sum_i Q_i^2\rangle_N-\frac{4}{81}N. \quad (3.12)$$

Therefore, for identical kinematical regions[7] we have

$$\frac{1}{3}\frac{\nu^2}{EE'}\left(\frac{d\sigma}{d\Omega dE'}\right)_{ep}<\left(\frac{d\sigma}{d\Omega dE'}\right)_{\gamma p}<\frac{5}{9}\frac{\nu^2}{EE'}\left(\frac{d\sigma}{d\Omega dE'}\right)_{ep}. \quad (3.13)$$

In principle it should be possible, if the model is correct, to distinguish between fractional and integer-charged partons.

IV. BACKGROUND AND RATE ESTIMATES FOR INELASTIC γ-RAY EXPERIMENT

We consider an experiment in which a bremsstrahlung beam is sent through hydrogen and the inelastically scattered γ ray is detected with momentum k' at angle θ. The main problem, in principle, is to differentiate the Compton γ rays from the γ rays coming from the decays of photoproduced π^0's. Here, we calculate the effective

[7] These bounds depend on the implicit assumptions of Eq. (2.25). A smaller lower bound of $\frac{1}{9}$ is obtained if we assume that all the pairs are made up of charges $\frac{1}{3}$ and $-\frac{1}{3}$.

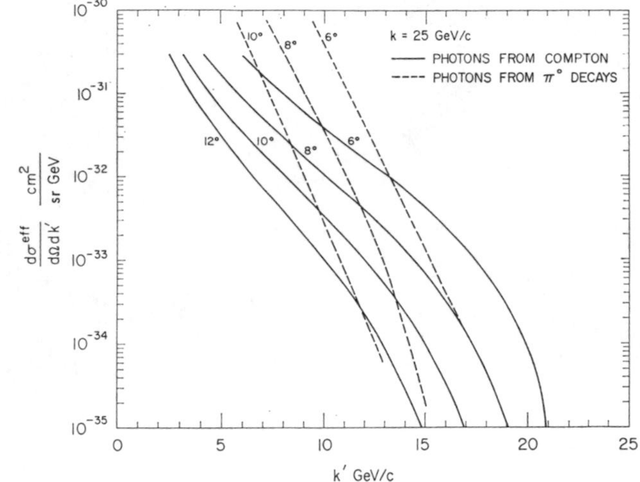

FIG. 9. Inelastic Compton scattering for a 25-GeV incident bremsstrahlung spectrum. The background curves are the same as in Fig. 6.

cross section for the production of Compton-scattered γ rays by folding the Compton cross section over the bremsstrahlung spectrum. We then calculate for comparison the corresponding background γ rays, estimated at SLAC from the beam survey experiments[8] for charged π's. We define

$$\frac{d\sigma^{\text{eff}}}{d\Omega dk'}=\int_{k'+|t|/2M}^{E\text{ electron}}\frac{dk}{k}\frac{d\sigma^{\gamma p}}{d\Omega dk'}. \tag{4.1}$$

We assume, optimistically, that the partons have unit charge and spin $\frac{1}{2}$; from (3.3) and (4.1) we obtain

$$\frac{d\sigma^{\text{eff}}}{d\Omega dk'}=\frac{4\alpha^2}{M^2k'}\int_{1/2}^{M(E-k')/4Ek'\sin^2(\frac{1}{2}\theta)}\lambda d\lambda F(\lambda)$$

$$\times\left[1-\frac{4k'\lambda\sin^2(\frac{1}{2}\theta)}{M}+\frac{8k'^2\lambda^2\sin^4(\frac{1}{2}\theta)}{M^2}\right], \tag{4.2}$$

with

$$\lambda=M\nu/Q^2. \tag{4.3}$$

We have calculated this expression for several incident electron energies as a function of k' and θ, using for $F(\lambda)$ the values given in Ref. 2. The results are shown as Figs. 6–9. In the same figures are shown our estimates of the corresponding background from the decay of photoproduced π^0's into γ rays. The estimate was made by assuming that the yield of π^+ measured in the SLAC beam survey experiment[8] equals the π^0 yield. We thereby obtain, for the effective cross section per nucleon,

$$\left(\frac{d\sigma_\pi}{d\Omega dk'}\right)_{\text{eff}}=(\text{yield})_{\pi^+}\Big/0.7\int_{t=0}^{0.3}tdt\ (\text{g/rad length of Be})$$

$$\times(\text{Avogadro No.})$$

$$=8.2\times10^{-25}(\text{yield})^{\pi+}\ \text{cm}^2/\text{sr BeV}. \tag{4.4}$$

The terms in the denominator have the following origin: The thin-target bremsstrahlung spectrum is tdk/k, where t is the thickness in radiation lengths (r.l.) (the target was 0.3 r.l. Be). The factor 0.7 is a thick-target correction calculated by Tsai and Van Whitis.[9] $(\text{Yield})^{\pi+}$ is taken from the SLAC User's Handbook[8] and the γ-ray flux is obtained by folding the π^0-decay spectrum into (4.4):

$$\left(\frac{d\sigma}{d\Omega dk_\gamma'}\right)^\gamma=2\int_{k_\gamma}^{E}\left(\frac{d\sigma_\pi}{d\Omega dk'}\right)_{\text{eff}}\frac{dk'}{k'}\approx\frac{2}{k_\gamma}\int_{k_\gamma}^{\infty}\left(\frac{d\sigma_\pi}{d\Omega dk'}\right)_{\text{eff}}dk'$$

$$\approx(2E_0/k_\gamma\theta)(8.2\times10^{-25})(\text{yield})^{\pi+}$$

$$(\text{as function of }k_\gamma), \tag{4.5}$$

where in the last step we have used the empirical observation that

$$d\sigma_\pi/d\Omega dk'\sim e^{-k'\theta/E_0}, \tag{4.6}$$

with

$$E_0\approx0.154\text{ BeV}.$$

In Figs. 6–9, the background is that from an 18-BeV bremsstrahlung beam. It is expected that this background increases slowly with beam energy, and keeps

[8] SLAC User's Handbook, Sec. D.1, Figs. 1 and 2 (unpublished).

[9] Y. S. Tsai and Van Whitis, Phys. Rev. 149, 1948 (1966).

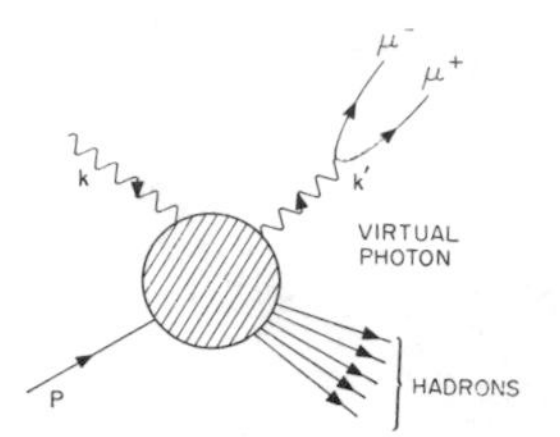

FIG. 10. Muon-pair production by inelastic Compton scattering.

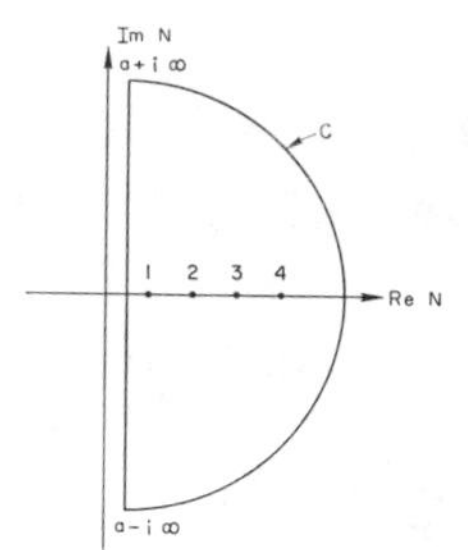

FIG. 11. Contour of integration for the Sommerfeld-Watson transform.

the same shape, in particular, the exponential dependence on transverse momentum. However, as the primary electron energy increases, the γ-ray spectrum from the inelastic Compton process, if it exists, is displaced upward in energy, so that for 20-BeV electrons and above, there is a region where the Compton signal dominates the π^0 noise. In any case, it will certainly be necessary to compare the γ-ray spectrum with that of the π^+ under the same conditions as reassurance that Compton γ rays are indeed being seen.

A variation of this experiment is to consider the inelastic Compton terms in μ-pair photoproduction (see Fig. 10). We have not analyzed this process in detail. The rate is diminished by a factor roughly[10] $\sim(2\alpha/3\pi)[\ln(E_{\max}/m_\mu)-3.5]\sim 1/350$ but if the charged pions are absorbed immediately downstream from the target, the background muon flux from $\pi^\pm$ decay can be reduced by a factor $\sim 1/700$ as well. Furthermore, the two muons are strongly correlated in angle, providing a quite unique signature. All this is encouragement that perhaps the background is manageable. The "singles" background from the Bethe-Heitler diagrams, for which the undetected muon predominantly goes in the forward direction, is interesting, as well, and very likely exceeds the singles rate from Compton μ pairs. But this is also of interest in testing μ-e universality at very high q^2. The "Bethe-Heitler" muons probably dominate the background muons from π^+ decay, but the necessary estimates have not yet been made.

APPENDIX

We attain a condition between $P(N)$ and $\langle\sum Q_i^2\rangle_N$ which guarantees that $F(x)$ is analytic at $x=0$ and equal to a nonzero constant. Using the Sommerfeld-Watson transformation, we rewrite (2.15) as an integral over the contour shown in Fig. 11:

$$F(x)=2x\int_c NP(2N+1)\langle\sum_i Q_i^2\rangle_{(2N+1)}\frac{(1-x)^{2N-1}e^{i\pi N}}{\sin\pi N}dN. \tag{A1}$$

The contribution of the semicircle at infinity is negligible. For what remains in the integral, we can use (for $x\to 0$)

$$x(1-x)^{2N-1}\approx xe^{-(2N-1)x}=-\tfrac{1}{2}\partial e^{-(2N-1)x}/\partial N, \tag{A2}$$

and integrate by parts:

$$F(x)=\int_{a-i\infty}^{a+i\infty}e^{-(2N-1)x}\frac{\partial}{\partial N}\times\left[\frac{NP(2N+1)\langle\sum_i Q_i^2\rangle_{(2N+1)}e^{i\pi N}}{\sin\pi N}\right]dN. \tag{A3}$$

In the limit of $x\to 0$, we have

$$F(x)\longrightarrow\left.\frac{NP(2N+1)\langle\sum_i Q_i^2\rangle_{(2N+1)}e^{i\pi N}}{\sin\pi N}\right|_{a-i\infty}^{a+i\infty}\xrightarrow[N\to a-i\infty]{}NP(2N+1)\langle\sum_i Q_i^2\rangle_{(2N+1)}. \tag{A4}$$

Therefore $F(x)$ approaches a constant if and only if $P(N)\to c/N\langle\sum Q_i^2\rangle_N$. For $\langle\sum_i Q_i^2\rangle_N$ linear in N, this reduces to the condition mentioned in the text.

ACKNOWLEDGMENTS

We thank R. P. Feynman, J. Weyers, and our colleagues at SLAC for many helpful conversations.

[10] R. H. Dalitz, Proc. Phys. Soc. (London) A64, 667 (1951).

PHYSICAL REVIEW D VOLUME 15, NUMBER 1 1 JANUARY 1977

Covariant harmonic oscillators and the parton picture

Y. S. Kim
Center for Theoretical Physics, Department of Physics and Astronomy, University of Maryland, College Park, Maryland 20742

Marilyn E. Noz
Department of Radiology, New York University Medical Center, New York, New York 10016
(Received 23 February 1976)

It is shown that the covariant-harmonic-oscillator wave function exhibits the peculiarities of the Feynman parton picture in the infinite-momentum frame.

In our previous publications,[1,2] we discussed both the conceptual and phenomenological aspects of the covariant-harmonic-oscillator formalism. Based on the Lorentz-invariant differential equation proposed by Yukawa in connection with Born's reciprocity hypothesis,[3] our starting point was a technical innovation over the work of Feynman *et al.*[4] Our solutions to the same oscillator equation satisfy all the requirements of nonrelativistic quantum mechanics in a given Lorentz frame, and satisfy the requirement of Lorentz-contracted probability interpretation for different Lorentz frames. We contend that our oscillator model is the first formalism since the invention of quantum mechanics in which the wave functions carry a covariant probability interpretation.[5]

The real strength of our oscillator model lies in the fact that one and the same wave function can provide the languages for both slow and fast hadrons. Our formalism can be applied to the quark-model calculations in the low-, intermediate-, and high-energy regions.[1,2,6] However, one of the most challenging questions in high-energy physics has been how to explain Feynman's parton[7,8] picture in terms of a formalism which can also describe the static properties of the hadron.

Another approach to this problem has been to explain Bjorken scaling in terms of the light-cone commutators and the initial hadron in its rest frame.[9] Here, one promising line of reasoning has been that the hadron is a composite particle and that its distribution function eliminates all the singularities which cause deviations from Bjorken scaling.[10] Our oscillator model does not contradict this physical picture.

Perhaps the most puzzling and irritating questions in Feynman's parton picture[7] have been the following problems:

(a) The picture is valid only in the infinite-momentum frame.

(b) Partons behave as free independent particles.

(c) While the hadron moves fast, there are wee partons.

(d) The longitudinal parton momenta are light-like.

(e) The number of partons seems much larger than the number of quarks inside the hadron.

The purpose of this paper is to provide qualitative answers to all of the above questions. Our starting point is the system of two bound quarks in the rest frame which can be described by a covariant-harmonic-oscillator wave function. We shall then boost this covariant bound system to an infinite-momentum frame and show that the peculiarities of the covariant oscillator coincide exactly with the parton properties mentioned above.

Following Feynman *et al.*[4] we call these two quarks a and b. In the harmonic-oscillator formalism,[1,2,3] the quark momenta p_a and p_b are not on the mass shell, but the total hadronic momentum

$$P = p_a + p_b \tag{1}$$

is on the mass shell. It is convenient to use the

Reprinted from *Phys. Rev. D* **15**, 335 (1977).

four-momentum difference

$$q = \sqrt{2}\,(p_b - p_a)\ . \tag{2}$$

We can also assign space-time coordinates to these quarks. Let us denote their coordinates by x_a and x_b, and introduce the relative coordinate

$$x = \frac{1}{2\sqrt{2}}\,(x_b - x_a)\ . \tag{3}$$

The transverse variables play only trivial roles in the harmonic-oscillator formalism and also in the parton picture. For this reason, we shall omit the transverse part of the wave function in the following discussion.

If the hadron moves along the z axis with velocity β, the ground-state wave function for this two-quark system can be written as[1,11]

$$\psi(x,\beta) = \frac{\omega}{2\pi}\exp\left\{-\frac{\omega}{2}\left[\left(\frac{1-\beta}{1+\beta}\right)\xi_+^{\,2} + \left(\frac{1+\beta}{1-\beta}\right)\xi_-^{\,2}\right]\right\}, \tag{4}$$

where

$$\xi_+ = \frac{1}{\sqrt{2}}(t+z), \quad \xi_- = \frac{1}{\sqrt{2}}(t-z)\ .$$

ω is the "spring constant" of the oscillator system. We can construct the momentum wave function by taking the Fourier transform of the above expression,

$$\phi(q,\beta) = \left(\frac{1}{2\pi}\right)^2 \int d^4x e^{-iq\cdot x}\,\psi(x,\beta)\ . \tag{5}$$

This momentum wave function takes the form

$$\phi(q,\beta) = \frac{1}{2\pi\omega}\exp\left\{-\frac{1}{2\omega}\left[\left(\frac{1-\beta}{1+\beta}\right)q_+^{\,2} + \left(\frac{1+\beta}{1-\beta}\right)q_-^{\,2}\right]\right\}, \tag{6}$$

where

$$q_+ = \frac{1}{\sqrt{2}}\,(q_0 + q_z), \quad q_- = \frac{1}{\sqrt{2}}\,(q_0 - q_z)\ .$$

According to Eq. (5), we have

$$q_z = -i\frac{\partial}{\partial z}, \quad q_0 = i\frac{\partial}{\partial t}\ . \tag{7}$$

Because of the above asymmetry in sign,

$$q_- = i\frac{\partial}{\partial \xi_+}, \quad q_+ = i\frac{\partial}{\partial \xi_-}\ . \tag{8}$$

This means that q_+ is conjugate to ξ_-, and q_- is conjugate to ξ_+. In terms of these variables, the above β-dependent wave functions generate the following β-independent (Lorentz-invariant) minimum-uncertainty product.[11]

$$\langle \xi_+^{\,2}\rangle\langle q_-^{\,2}\rangle = \tfrac{1}{4}\ ,$$
$$\langle \xi_-^{\,2}\rangle\langle q_+^{\,2}\rangle = \tfrac{1}{4}\ . \tag{9}$$

Let us go back to the wave functions of Eqs. (4) and (5). If $\beta = 0$, the wave functions correspond to those in the rest frame. As the momentum of the hadron becomes large,

$$\left(\frac{1+\beta}{1-\beta}\right) \to \left(\frac{2P_0}{M}\right)^2, \tag{10}$$

where M is the mass of the hadron. As $P_0 \to \infty$, the width of the ξ_- (and q_-) distribution becomes vanishingly small. Consequently,

$$\xi_- = 0 \ \text{ and } \ q_- = 0\ . \tag{11}$$

This means that both ξ and q are lightlike vectors, and

$$\xi_+ = \sqrt{2}\,z = \sqrt{2}\,t\ ,$$
$$q_+ = \sqrt{2}\,q_z = \sqrt{2}\,q_0\ . \tag{12}$$

In the infinite-momentum limit, the effective spring constant associated with the ξ_+ motion becomes vanishingly small. The motion along the ξ_+ axis therefore becomes like that of a free lightlike particle.

The behavior of the q_- distribution and that of the q_+ distribution are illustrated in Fig. 1. The width of the q_+ distribution becomes large when P_0 becomes large. This may appear as a violation of the uncertainty relation, but it is not. q_+ and ξ_+ are not conjugate variables. The precise uncertainty relation was derived in Ref. 11 and is stated in Eq. (9) of the present paper.

We can now associate the above-mentioned pecularity with the puzzling features of Feynman's parton picture. Let us first observe that the hadronic four-momentum P becomes lightlike in the infinite-momentum limit, and consider the four-

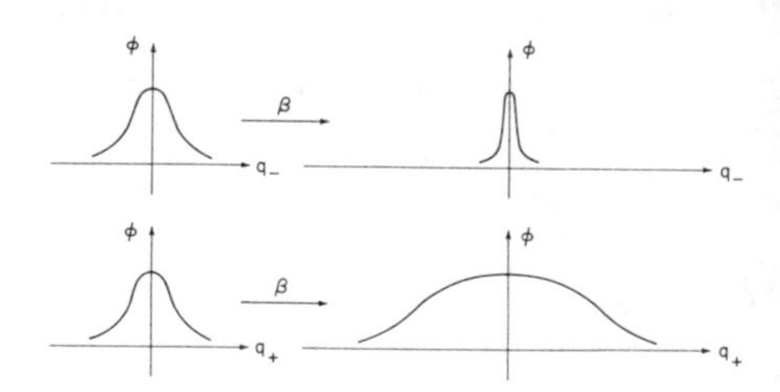

FIG. 1. Momentum wave functions in the rest frame and in a large-momentum frame. In the rest frame where $\beta = 0$, the q_+ and q_- distributions take the same form. When the hadron moves fast and $\beta \to 1$, the q_- distribution becomes narrower while the q_+ distribution becomes wide spread. This wide-spread q_+ distribution corresponds to the parton distribution.

momentum of the constituent quark

$$p_a = \frac{1}{2}P - \frac{1}{2\sqrt{2}}q \ . \tag{13}$$

Since the four-vector q is lightlike, and we are considering here only longitudinal momenta, p_a is also lightlike.

Considering the width of the Gaussian function for the q_+ distribution, which is also the $\sqrt{2}\,q_0$ distribution, we can say that the momentum of the constituent quark mostly lies in the interval defined by the following limits:

$$p_{\max} = P_0\left(\frac{1}{2} + \frac{\sqrt{\omega}}{2M}\right),$$
$$p_{\min} = P_0\left(\frac{1}{2} - \frac{\sqrt{\omega}}{2M}\right). \tag{14}$$

The quantity $(\sqrt{\omega}/2M)$ is of the same order of magnitude as $\frac{1}{2}$. For this reason, the lightlike four-momentum p_a can be written as

$$p_a = \alpha P \ , \tag{15}$$

with α ranging approximately from zero to one. This wide-spread distribution and division of the four-momentum are exactly like those of the parton model.

Let us go back to the ξ_+ distribution, which is also the $\sqrt{2}z$ distribution. We noted above that the motion along this axis should be almost free. Then the momentum has to be sharply defined, and the momentum cannot have a wide-spread distribution. Therefore the momentum distribution we noted in Eqs. (14) and (15) should be regarded as a distribution of free particles which are lightlike. This is exactly what we have in the original form of Feynman's parton model, as well as being characteristic of the quantum-mechanical picture of blackbody radiation. In both cases, the number of lightlike particles is not conserved.

Finally, let us consider the time interval during which the above-mentioned partons behave as free particles. According to Eq. (12), the ξ_+ axis is also the $\sqrt{2}t$ axis. Therefore, the time duration is of the order of $(P_0/M\sqrt{\omega})$. This interval increases as P_0 becomes large. If this interval is much larger than the characteristic time of electromagnetic interaction, then the partons of the present paper will indeed behave as Feynman's partons.

We have shown above qualitatively how the covariant oscillator produces Feynman's parton picture in the infinite-momentum limit. The next question then is how we can use this formalism to carry out the parton-model calculations.

In order to answer this question, we note first of all that the above two-body formalism can be easily generalized to the three-quark nucleon system.[4] In performing the parton-model calculations, we have to square the wave function to get the probability-density function. The Gaussian form remains Gaussian during the squaring process. The Lorentz-contraction property of the Gaussian probability distribution is identical to that of the wave function except for the factor of 2 in the exponent. In fact, the width quoted in Eq. (14) is derived from the width of the probability function.

As was noted earlier in this paper, the probability function exhibits a δ function in the $(q_z - q_0)$ variable in the infinite-momentum limit. We can now eliminate the q_0 dependence by integrating over this variable. The resulting function becomes the parton distribution function in the three-dimensional space.

The immediate calculations we can do using the above-mentioned procedure have already been carried out by Le Yaouanc *et al.*[12] Starting from the three-dimensional parton distribution function which we could obtain by following the procedure outlined above, Le Yaouanc *et al.* indeed carried out a comprehensive phenomenological analysis of all interesting physical quantities in the inelastic electron-nucleon scattering.

---

[1]Y. S. Kim and M. E. Noz, Phys. Rev. D 8, 3521 (1973).

[2]Y. S. Kim and M. E. Noz, Phys. Rev. D 12, 129 (1975). For orthogonality and Lorentz-contraction properties of the harmonic-oscillator wave functions, see M. J. Ruiz, Phys. Rev. D 10, 4306 (1974).

[3]H. Yukawa, Phys. Rev. 91, 416 (1953).

[4]R. P. Feynman, M. Kislinger, and F. Ravndal, Phys. Rev. D 3, 2706 (1971).

[5]P. A. M. Dirac, *The Development of Quantum Theory* (Gordon and Breach, New York, 1971).

[6]Y. S. Kim and M. E. Noz, Phys. Rev. D 12, 122 (1975); M. J. Ruiz, *ibid.* 12, 2922 (1975); Y. S. Kim, *ibid.* 14, 273 (1976); Y. S. Kim and M. E. Noz, *ibid.* (to be published).

[7]R. P. Feynman, in *High Energy Collisions*, proceedings of the Third International Conference, Stony Brook, New York, edited by C. N. Yang *et al.* (Gordon and Breach, New York, 1969).

[8]J. D. Bjorken and E. A. Paschos, Phys. Rev. 185, 1975 (1969).

[9]R. A. Brandt, Phys. Rev. Lett. 22, 1149 (1969); 23, 1260 (1969). For a review article, see Y. Frishman, in *Proceedings of the XVI International Conference on High Energy Physics, Chicago-Batavia, Ill., 1972,* edited by J. D. Jackson and A. Roberts (NAL, Batavia, Ill., 1973), Vol. 4, p. 119.

[10]S. D. Drell and T. D. Lee, Phys. Rev. D 5, 1738 (1972); C. H. Woo, Phys. Rev. D 6, 1128 (1972). We would

like to thank C. H. Woo for explaining the content of his paper.

[11]Y. S. Kim, Univ. of Maryland CTP Tech. Report No. 76-008, 1975 (unpublished). This paper contains also a more precise explanation of the Lorentz-contraction property of excited-state harmonic-oscillator wave functions. The contraction property is explained in terms of the step-up operator which transforms like the longitudinal coordinate.

[12]A. Le Yaouanc, L. Oliver, O. Pène, and J.-C. Raynal, Phys. Rev. D 12, 2137 (1975).

PHYSICAL REVIEW D VOLUME 23, NUMBER 11 1 JUNE 1981

Valons and harmonic oscillators

Paul E. Hussar
Center for Theoretical Physics, Department of Physics and Astronomy, University of Maryland, College Park, Maryland 20742
(Received 19 January 1981)

The valon distribution derived by Hwa is compared with the valence-quark distribution from the covariant-harmonic-oscillator wave function which correctly describes the proton-form-factor behavior, and which provides a covariant representation of the hadron mass spectra. It is shown that the harmonic-oscillator curve closely approximates the valon distribution for $x > 0.25$. For $0 < x < 0.25$, the agreement is reasonable.

Since the introduction of the parton model,[1] one of the central issues in high-energy physics has been the relationship between the partons and the valence quarks which seem responsible for other high-energy properties such as mass spectra and form factors. It was once naively believed that the proton structure function could be calculated from the valence-quark distributions inside the hadron.[2] However, it is by now firmly established that quantum-chromodynamics (QCD) processes stand between the valence quarks and the observed structure functions.

With this point in mind, Hwa recently developed an appealing method for dealing with nucleon structure functions. Hwa's approach separates out a component of the structure functions which is completely determined by QCD renormalization, and uses the data to calculate a momentum-fraction distribution for three constituent quark clusters or "valons".[3] The purpose of this paper is to point out that Hwa's valon distribution is close to the valence-quark distribution derivable from the covariant harmonic-oscillator model which has been effective in explaining the nucleon mass spectra and form factors.

The valon picture[3,4] is basically an attempt to establish a connection between the quark model in which hadrons are bound states of their constituent quarks and the parton model which seems necessary if we are to explain the observed structure functions. When probed at high Q^2, each of the valence quarks will itself be resolved into infinitely many constituents due to the fact that each of the valence quarks will be accompanied by a cloud of quarks, antiquarks, and gluons produced in ongoing QCD processes. We can calculate the evolution of the valence quarks involved here to leading order in QCD using the renormalization-group methods.[5] The nucleon structure functions in the valon model are then given by the corresponding functions for each of the valons (the valence quark plus its cloud) smeared by the momentum-fraction distribution of the valons. The assumption is that the interaction which confines the valons, that is, the gluons which are exchanged among them, will not play so large a role in scattering processes as to make the above analysis insufficient. The consistency with which the valon picture appears to model the actual behavior of nucleon structure functions[4] is a persuasive argument that the assumption made here is a good one.

The nucleon structure functions in the valon model take the form[4]

$$F^N(x, Q^2) = \sum_v \int_x^1 dy\, G_{v/N}(y) F^v(x/y, Q^2)\,, \tag{1}$$

where $F^N(x, Q^2)$ is a nucleon structure function (either F_2 or xF_3), $F^v(x, Q^2)$ is the corresponding function for a valon v, and $G_{v/N}(y)dy$ is the probability of the valon having momentum fraction between y and $y+dy$. The sum is taken over the three valons which constitute the nucleon. From the definition, we must have

$$\int_0^1 G_{v/N}(y)dy = 1 \tag{2}$$

and

$$\sum_v \int_0^1 y G_{v/N}(y)dy = 1\,. \tag{3}$$

In the expression of Eq. (1) the nucleon structure functions $F^N(x, Q^2)$ are well known from experiment. For high-Q^2 processes, the renormalization-group methods in QCD allow a description of $F^v(x, Q^2)$ in terms of its moments. Making use of data from neutrino and muon scattering, and known results from QCD, Hwa has obtained

$$G_{u/p}(y) = 8y^{0.65}(1-y)^2\,, \tag{4}$$

$$G_{d/p}(y) = 6y^{0.35}(1-y)^{2.3}\,, \tag{5}$$

$$G_{0/p}(y) = \tfrac{105}{16} y^{1/2}(1-y)^2\,, \tag{6}$$

Reprinted from *Phys. Rev. D* **23**, 2781 (1981).

where $G_{u/p}(y)$ and $G_{d/p}(y)$ are the momentum-fraction distributions for the u valon and the d valon in a proton, respectively, and $G_{0/p}(y)$ is the momentum-fraction distribution obtained assuming that these distributions are flavor independent. The $G_{u/p}$ and $G_{0/p}$ functions are plotted in Fig. 1.

In spite of QCD's effectiveness in dealing with the Q^2 evolution of the hadronic structure functions, it has not yet been helpful in determining what should really be the starting point of this evolution, that is, the distribution of the valence quarks inside the hadron. In the meantime, we are allowed to consider other models which are consistent with existing rules of quantum mechanics and special relativity. The covariant harmonic oscillator is such a model.

The relativistic oscillator model existed long before the quark model was invented.[6] The early applications of the oscillator model in the quark picture of hadrons include the study of hadronic mass spectra. Its effectiveness in the relativistic domain was demonstrated first by Fujimura *et al.* in their successful calculation of the proton form factor.[7] Fujimura *et al.* used normalizable relativistic wave functions for three valence quarks. We propose to use the same relativistic wave function to determine the valence-quark distribution in the parton regime.

The covariant harmonic oscillator has been discussed extensively in the literature. In particular, it was shown by Kim and Noz[8] that a rapidly moving hadron in this model has a broad longitudinal-momentum distribution while the spring constant of the oscillator becomes weak to the laboratory-frame observer. It was pointed out in Ref. 8 that this behavior of the oscillator

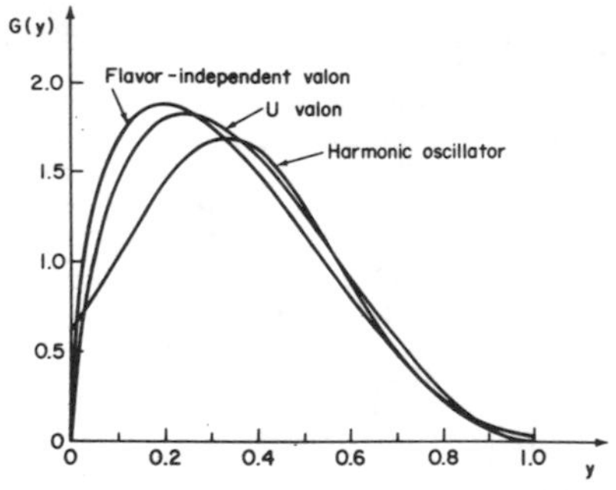

FIG. 1. The experimental and calculated $G(y)$ functions. The G function calculated in the covariant oscillator model is compared with the valon distribution functions obtained by Hwa from experimental data.

function can explain the peculiarities in the original version of Feynman's parton model.[1]

The three-particle kinematics associated with the proton wave function has been studied extensively.[9] It is, then, a straightforward matter to derive the momentum-fraction distribution function in this scheme.[10] The result is

$$G_{\rm osc}(y)=[3m/(2\pi\omega)^{1/2}]\exp[-(m^2/2\omega)(1-3y)^2]\,. \tag{7}$$

In the above expression, m represents the nucleon mass, while ω is the oscillator spring constant. $G_{\rm osc}(y)$ is also plotted in Fig. 1. The value of ω is taken to be $m^2/2$, which is the most acceptable value in the calculations of the mass spectrum[11] and the g_A/g_v ratio.[12]

Noteworthy, perhaps, is the fact that $G_{\rm osc}(y)$ is normalized over the whole real line, while Hwa's distribution extended only from $y=0$ to $y=1$. The oscillator wave function is not restricted to the region of physically observable constituent energies. However, this is nothing strange, inasmuch as the nonrelativistic oscillator exhibits the same property. It should also be pointed out that the integral given in Eq. (1) which determines the structure functions is taken only over values from 0 to 1 so that a nonzero value of G outside of this region will play no role. At the present time, we do not have enough experimental accuracy to decide whether the curvature is of the polynomial type given by Hwa or of the Gaussian form.

The agreement between $G(y)$ from the data and the momentum-fraction distribution function for the constituent quarks in the covariant oscillator model is surprisingly good. Clearly for $y>0.25$, the oscillator distribution is very close to the phenomenological curve given by Hwa. For $y<0.25$, the numerical agreement is not as good as in the larger-y region. However, we have to accept the fact that there are still large experimental uncertainties in this small-y region, and it would be difficult to trust at this time any closer agreement than that given in Fig. 1.

As for the flavor dependence indicated by Hwa's valon curves, it is not yet clear to us to what extent the difference between $G_{u/p}$ and $G_{0/p}$ is actually required by experimental evidence since this feature is observed mostly at smaller values of y where, again, the experimental uncertainties are the greatest. In any case, the covariant oscillator is not designed to account for such a difference. If the difference really exists, there must be additional dynamical effects, beyond what we can account for with the oscillator model and QCD, to explain the flavor dependence. This is beyond the scope of this paper.

This paper is based on a part of the author's dissertation to be submitted to the faculty of the University of Maryland in partial fulfillment of the requirements for a degree of Doctor of Philosophy. The author would like to thank Professor Y. S. Kim for suggesting this research. The services of the University of Maryland Computer Science Center are also appreciated.

---

[1]R. P. Feynman, in *Third Topical Conference on High Energy Collisions*, edited by C. N. Yang (Gordon and Breach, New York, 1969).
[2]J. D. Bjorken and E. A. Paschos, Phys. Rev. 185, 1975 (1969).
[3]R. C. Hwa, Phys. Rev. D 22, 759 (1980).
[4]R. C. Hwa and M. S. Zahir, Phys. Rev. D 23, 2539 (1981).
[5]T. A. DeGrand, Nucl. Phys. B151, 485 (1970).
[6]H. Yukawa, Phys. Rev. 91, 461 (1953); M. Markov, Nuovo Cimento Suppl. 3, 760 (1956). For the latest reinterpretation of these early papers, see D. Han and Y. S. Kim, Prog. Theor. Phys. 64, 1852 (1980).
[7]K. Fujimura, T. Kobayashi, and M. Namiki, Prog. Theor. Phys. 43, 73 (1970). See also R. G. Lipes, Phys. Rev. D 5, 2849 (1972).
[8]Y. S. Kim and M. E. Noz, Phys. Rev. D 15, 335 (1977). See also Y. S. Kim and M. E. Noz, Found. Phys. 9, 375 (1979).
[9]R. P. Feynman, M. Kislinger, and F. Ravndal, Phys. Rev. D 3, 2706 (1971).
[10]Y. S. Kim and M. E. Noz, Prog. Theor. Phys. 60, 801 (1978).
[11]P. E. Hussar, Y. S. Kim, and M. E. Noz, Am. J. Phys. 48, 1043 (1980).
[12]M. Ruiz, Phys. Rev. D 12, 2922 (1975).

Chapter VI

Massless Particles and Gauge Transformations

It was shown in Wigner's 1939 paper that the little group for massless particles is isomorphic to the two-dimensional Euclidean group consisting of rotations around the origin and translations along the two perpendicular directions. It is not difficult to associate the rotation with the helicity. However, the physical interpretation of the translation-like degrees of freedom had been an unsolved problem. Fortunately, two of Weinberg's 1964 papers started breaking ground on this problem, and led to the clue that the translation-like transformations are gauge transformations.

In 1982, Han, Kim, and Son used the Lorentz condition to reduce the complicated transformation matrix of the little group into the transpose of the coordinate transformation matrix on the two-dimensional plane. They confirmed therefore, that the translation-like transformations are gauge transformations. These authors extended their study to the symmetry of massless particles with spin 1/2 and concluded that the polarization of neutrinos is due the requirement of gauge invariance.

PHYSICAL REVIEW VOLUME 134, NUMBER 4B 25 MAY 1964

Feynman Rules for Any Spin. II. Massless Particles*

STEVEN WEINBERG†
Department of Physics, University of California, Berkeley, California
(Received 13 January 1964)

The Feynman rules are derived for massless particles of arbitrary spin j. The rules are the same as those presented in an earlier article for $m>0$, provided that we let $m\to 0$ in propagators and wave functions, and provided that we keep to the $(2j+1)$-component formalism [with fields of the $(j,0)$ or $(0,j)$ type] or the $2(2j+1)$-component formalism [with $(j,0)\oplus(0,j)$ fields]. But there are other field types which cannot be constructed for $m=0$; these include the $(j/2,j/2)$ tensor fields, and in particular the vector potential for $j=1$. This restriction arises from the non-semi-simple structure of the little group for $m=0$. Some other subjects discussed include: **T**, **C**, and **P** for massless particles and fields; the extent to which chirality conservation implies zero physical mass; and the Feynman rules for massive particles in the helicity formalism. Our approach is based on the assumption that the S matrix is Lorentz invariant, and makes no use of Lagrangians or the canonical formalism.

I. INTRODUCTION

THIS article will develop the relativistic field theory of massless particles with general spin, along the lines followed in an earlier work[1] on massive particles. Our chief aim is, again, to derive the Feynman rules.

We assume that the S matrix can be calculated from Dyson's formula

$$S=\sum_{n=0}^{\infty}\frac{(-i)^n}{n!}\int d^4x_1\cdots d^4x_n T\{\mathcal{H}(x_1)\cdots\mathcal{H}(x_n)\}. \quad (1.1)$$

Here, $\mathcal{H}(x)$ is the interaction energy density in the interaction representation. In general, it would be the 00 component $\mathcal{T}^{00}(x)$ of a tensor $\mathcal{T}^{\mu\nu}(x)$, but in order that S be Lorentz-invariant it is necessary that $\mathcal{T}^{\mu\nu}(x)$ be of the form

$$\mathcal{T}^{\mu\nu}(x)=-g^{\mu\nu}\mathcal{H}(x), \quad (1.2)$$

with $\mathcal{H}(x)$ a scalar. Lorentz invariance also dictates that $\mathcal{H}(x)$ commute with $\mathcal{H}(y)$ for $x-y$ space-like, in order that the θ functions implicit in the time-ordered product in (1.1) not destroy the Lorentz invariance of S.

We also assume that $\mathcal{H}(x)$ is built out of the creation and annihilation operators of the free particles appearing in the unperturbed Hamiltonian. In order that $\mathcal{H}(x)$ transform properly we construct it as an invariant polynomial in various free fields $\psi_n(x)$, which behave as usual under translations, and which transform according to various representations of the homogeneous Lorentz group

$$U[\Lambda]\psi_n(x)U[\Lambda]^{-1}=\sum_m D_{nm}[\Lambda^{-1}]\psi_m(\Lambda x). \quad (1.3)$$

In order that $\mathcal{H}(x)$ commute with itself outside the light cone, we require that the $\psi_n(x)$ have causal commutation or anticommutation rules: for $x-y$ space-like,

$$[\psi_n(x),\psi_m(y)]_\pm=0. \quad (1.4)$$

* Research supported in part by the U. S. Air Force Office of Scientific Research, Grant No. AF-AFOSR-232-63.
† Alfred P. Sloan Foundation Fellow.
[1] S. Weinberg, Phys. Rev. **133**, B1318 (1964).

These assumptions will be sufficient for all our purposes. In particular, we will have no need of Lagrangians and the canonical formalism, nor will we need to start with any preconceptions about the form or even the existence of the field equations.

We begin in Sec. II with a review of the transformation properties of massless particle states and creation and annihilation operators. This information is used in Secs. III and IV to construct $(2j+1)$-component fields transforming according to the $(j,0)$ and $(0,j)$ representations. Condition (1.4) is used in Sec. V to complete the construction of the fields, and to prove the spin-statistics theorem and crossing symmetry. The Feynman rules are presented in Secs. VI, VII, and VIII. The inversions **P**, **C**, and **T** are discussed in Sec. IX.

In Sec. X we attack a separate problem: To what extent does chirality conservation guarantee the existence of a particle of zero physical mass? Our conclusion [for general $j\geq\frac{1}{2}$] is that this theorem can probably only be proved in the context of perturbation theory. But if parity as well as chirality is conserved, then it is possible to prove the nonexistence of a nondegenerate particle of finite mass.

The chief conclusion of this work is that the Feynman rules for massless particles in the $(2j+1)$-component or $2(2j+1)$-component formalisms are precisely the same as for $m>0$, except, of course, that we must pass to the limit $m\to 0$ in wave functions and propagators.[2] In this limit it becomes impossible to produce or destroy particles with helicity other than $\pm j$.

But there is still one important qualitative distinction between $m=0$ and $m>0$. We prove in Sec. III that not all of the field types which can be constructed out of the creation and annihilation operators for $m>0$ can be so constructed for $m=0$. Specifically, the annihilation operator for a massless particle of helicity λ and the

[2] This conclusion is in agreement with the theorem that the decomposition of the S matrix into invariant amplitudes takes the same form for $m=0$ and $m>0$, proven by D. Zwanziger, Phys. Rev. **133**, B1036 (1964). Neither Zwanziger's work nor the present article offer any understanding of the fact that photons and gravitons interact with conserved quantities at zero-momentum transfer. This point will be the subject of further articles, to be published in Phys. Letters and in Phys. Rev.

Reprinted from *Phys. Rev.* **134**, B882 (1964).

creation operator for the antiparticle with helicity $-\lambda$ can only be used to form a field transforming as in (1.3) under those representations (A,B) of the homogeneous Lorentz group such that $\lambda=B-A$. This limitation arises purely because of the non-semi-simple structure of the little group for $m=0$. The difficulties (indefinite metric, negative energies, etc.) encountered in previous attempts to represent the photon by a quantized vector potential $A^\mu(x)$ can therefore now be understood as due to the fact that such a field transforms according to the $(\frac{1}{2},\frac{1}{2})$ representation, which is not one of the representations allowed by the theorem of Sec. III for helicity $\lambda=\pm1$. On the other hand, the $(j,0)$ and $(0,j)$ representations used in this article (corresponding for $j=1$ to the field strengths) are allowed by our theorem, and they cause no trouble.[3] In a future article we shall show that it is in fact possible to evade our theorem, and that the Lorentz invariance of the S matrix then forces us to the principle of extended gauge invariance.

In Ref. 1 we gave the Feynman rules for initial and final states specified by the z components of the massive particle spins. In order to facilitate the comparison with the case of zero mass, and for the sake of completeness, we present in Sec. VIII the corresponding Feynman rules in the helicity formalism of Jacob and Wick.[4] The external-line wave functions are much simpler, though of course the propagators are the same.

II. TRANSFORMATION OF STATES

The starting point in our approach is a statement of the Lorentz transformation properties of massless particle *states*. The transformation rules have been completely worked out by Wigner,[5] but it will be convenient to review them here, particularly as there are some little known but extremely important peculiarities that are special to the case of zero mass.

Consider a massless particle moving in the z direction with energy κ. It may have several possible spin states, which we denote $|\lambda\rangle$, the significance of the label λ to be determined by examining the transformation properties of these states. Wigner defines the "little group" as the subgroup of the Lorentz group consisting of all homogeneous proper Lorentz transformations $\mathcal{R}^\mu{}_\nu$ which do not alter the four-momentum k^μ of our particle.

$$\mathcal{R}^\mu{}_\nu k^\nu=k^\mu\,, \tag{2.1}$$

$$k^1=k^2=0;\quad k^3=k^0=\kappa. \tag{2.2}$$

[3] As a case in point, there does not seem to be any obstacle to the construction of field theories for massless charged particles of arbitrary spin j, provided that we use only proper field types, like $(j,0)$ or $(0,j)$. The trouble encountered for $j\geq1$ by K. M. Case and S. G. Gasiorowicz [Phys. Rev. **125**, 1055 (1962)], can be ascribed to their use of improper field types, such as $(\frac{1}{2},\frac{1}{2})$. We plan to discuss this in more detail in a later article on the electromagnetic interactions of particles of any spin.

[4] M. Jacob and G. C. Wick, Ann. Phys. (N. Y.) **7**, 404 (1959).

[5] E. P. Wigner, in *Theoretical Physics* (International Atomic Energy Agency, Vienna, 1963), p. 59.

The states $|\lambda\rangle$ must furnish a representation of the little group. That is, the unitary operator $U[\mathcal{R}]$ corresponding to $\mathcal{R}^\mu{}_\nu$ does not change the momentum of the states $|\lambda\rangle$, and thus must just induce a linear transformation:

$$U[\mathcal{R}]|\lambda\rangle=\sum_{\lambda'} d_{\lambda'\lambda}[\mathcal{R}]|\lambda'\rangle\,, \tag{2.3}$$

with

$$\sum_{\lambda''} d_{\lambda\lambda''}[\mathcal{R}_1]d_{\lambda''\lambda'}[\mathcal{R}_2]=d_{\lambda\lambda'}[\mathcal{R}_1\mathcal{R}_2]. \tag{2.4}$$

Therefore, we can catalog the various possible spin states $|\lambda\rangle$ by studying the representations $d[\mathcal{R}]$ of the little group.

This is most easily accomplished by examining the infinitesimal transformations of the little group. They take the form

$$\mathcal{R}^\mu{}_\nu=\delta^\mu{}_\nu+\Omega^\mu{}_\nu\,, \tag{2.5}$$

where $\Omega^\mu{}_\nu$ is infinitesimal and annihilates k:

$$\Omega^\mu{}_\nu k^\nu=0. \tag{2.6}$$

In order that (2.5) be a Lorentz transformation we must also require that

$$\Omega^{\mu\nu}=-\Omega^{\nu\mu}\,, \tag{2.7}$$

the index ν being raised in the usual way with the metric tensor $g^{\mu\nu}$, defined here to have nonzero components:

$$g^{11}=g^{22}=g^{33}=1\,,\quad g^{00}=-1. \tag{2.8}$$

Inspection of (2.6) and (2.7) shows that the general $\Omega^{\mu\nu}$ is a function of three parameters θ, χ_1, χ_2, with nonzero components given by

$$\Omega^{12}=-\Omega^{21}=\theta\,, \tag{2.9}$$

$$\Omega^{10}=-\Omega^{01}=\Omega^{13}=-\Omega^{31}=\chi_1\,, \tag{2.10}$$

$$\Omega^{20}=-\Omega^{02}=\Omega^{23}=-\Omega^{32}=\chi_2. \tag{2.11}$$

The Lie algebra generated by these transformations can be determined by recalling the algebra generated by the full homogeneous Lorentz group, of which the little group is a subgroup. An infinitesimal Lorentz transformation $\Lambda^\mu{}_\nu$ can be written as in (2.5), with $\Omega^\mu{}_\nu$ subject only to (2.7). The corresponding unitary operator takes the form

$$U[1+\Omega]=1+(i/2)\Omega^{\mu\nu}J_{\mu\nu}\,, \tag{2.12}$$

$$J_{\mu\nu}=-J_{\nu\mu}=J_{\mu\nu}{}^\dagger. \tag{2.13}$$

It is conventional to group the six components of $J_{\mu\nu}$ into two three-vectors:

$$J_i=\tfrac{1}{2}\epsilon_{ijk}J_{jk}\,, \tag{2.14}$$

$$K_i=J_{i0}=-J_{0i}\,, \tag{2.15}$$

with commutation rules

$$[J_i,J_j]=i\epsilon_{ijk}J_k, \tag{2.16}$$

$$[J_i,K_j]=i\epsilon_{ijk}K_k, \tag{2.17}$$

$$[K_i,K_j]=-i\epsilon_{ijk}J_k. \tag{2.18}$$

We see that the unitary operator corresponding to the general infinitesimal transformation (2.9)–(2.11) of the little group is

$$U[\mathcal{R}(\theta,\chi_1,\chi_2)]=1+i\theta J_3+i\chi_1L_1+i\chi_2L_2, \tag{2.19}$$

where

$$L_1\equiv K_1-J_2, \tag{2.20}$$

$$L_2\equiv K_2+J_1. \tag{2.21}$$

The commutation rules for the three generators of the little group are given by (2.16)–(2.18) as

$$[J_3,L_1]=iL_2, \tag{2.22}$$

$$[J_3,L_2]=-iL_1, \tag{2.23}$$

$$[L_1,L_2]=0. \tag{2.24}$$

We can now find all the representations of the little group by finding the representations of this Lie algebra. But it strikes one immediately that this algebra is not semi-simple because the elements L_1 and L_2 form an invariant Abelian subalgebra. [In fact, Wigner[5] points out that (2.22)–(2.24) identify this algebra as that of all rotations *and translations* in two-dimensions, a fact of no known physical significance.] In order that the states $|\lambda\rangle$ form a finite set, it is necessary to represent the "translations" by zero, i.e.,

$$L_1|\lambda\rangle=L_2|\lambda\rangle=0\,. \tag{2.25}$$

Therefore, a general $\mathcal{R}^\mu{}_\nu$ in the little group transforms $|\lambda\rangle$ into

$$U[\mathcal{R}]|\lambda\rangle=\exp\{i\Theta[\mathcal{R}]J_3\}|\lambda\rangle, \tag{2.26}$$

the angle $\Theta[\mathcal{R}]$ being some more or less complicated real function of the $\mathcal{R}^\mu{}_\nu$, which is given for infinitesimal $\mathcal{R}$ by (2.19) as

$$\Theta[\mathcal{R}(\theta,\chi_1,\chi_2)]\to\theta. \tag{2.27}$$

If we now identify the states $|\lambda\rangle$ as eigenstates with definite helicity λ,

$$J_3|\lambda\rangle=\lambda|\lambda\rangle, \tag{2.28}$$

we see that the physically permissible irreducible representations of the little group are all one dimensional:

$$U[\mathcal{R}]|\lambda\rangle=\exp\{i\lambda\Theta[\mathcal{R}]\}|\lambda\rangle. \tag{2.29}$$

Comparing with (2.3) and (2.4) shows that Θ must satisfy the group property

$$\Theta[\mathcal{R}_1]+\Theta[\mathcal{R}_2]=\Theta[\mathcal{R}_1\mathcal{R}_2]. \tag{2.30}$$

For global reasons it is necessary to restrict the helicity λ to be a positive or negative integer or half-integer $\pm j$. We *define* a right- or left-handed particle of spin $j\geqq 0$ as one with helicity λ equal to $+j$ or $-j$, respectively.

It is, of course, very well known that a spinning massless particle need not occur in more than one spin state (or two, if parity is conserved). The restriction (2.25) is much less familiar, but we shall see that it is responsible for the *dynamical* peculiarities of massless particle field theories.

A particle of general momentum $\mathbf{p}$ and helicity λ may now be defined by a Lorentz transformation

$$|\mathbf{p},\lambda\rangle\equiv[\kappa/|\mathbf{p}|]^{1/2}U[\mathcal{L}(\mathbf{p})]|\lambda\rangle, \tag{2.31}$$

where $U[\mathcal{L}(\mathbf{p})]$ is the unitary operator corresponding to the Lorentz transformation $\mathcal{L}_\nu{}^\mu(\mathbf{p})$ which takes our "standard" four-momentum k^μ into p^μ:

$$p^\mu=\mathcal{L}^\mu{}_\nu(\mathbf{p})k^\nu,\qquad p^\mu=\{\mathbf{p},|\mathbf{p}|\};\quad k^\mu=\{0,0,\kappa,\kappa,\}. \tag{2.32}$$

There are various ways of making the definition of $\mathcal{L}(\mathbf{p})$ unambiguous, but we will find it convenient to define $\mathcal{L}$ as

$$\mathcal{L}^\mu{}_\nu(\mathbf{p})=R^\mu{}_\lambda(\hat{p})B^\lambda{}_\nu(|\mathbf{p}|). \tag{2.33}$$

Here, $B(|\mathbf{p}|)$ is a "boost" along the z axis with nonzero components

$$B^1{}_1(|\mathbf{p}|)=B^2{}_2(|\mathbf{p}|)=1,$$

$$B^3{}_3(|\mathbf{p}|)=B^0{}_0(|\mathbf{p}|)=\cosh\phi(|\mathbf{p}|), \tag{2.34}$$

$$B^3{}_0(|\mathbf{p}|)=B^0{}_3(|\mathbf{p}|)=\sinh\phi(|\mathbf{p}|),$$

$$\phi(|\mathbf{p}|)\equiv\ln(|\mathbf{p}|/\kappa). \tag{2.35}$$

Since $B^\mu{}_\nu$ takes k^μ into $\{0,0,|\mathbf{p}|,|\mathbf{p}|\}$, we choose $R(\hat{p})$ as the rotation (say, in the plane containing $\hat{p}$ and the z axis) which takes the z axis into the unit vector $\hat{p}=\mathbf{p}/|\mathbf{p}|$. The factor $[\kappa/|\mathbf{p}|]^{1/2}$ is inserted in (2.31) to keep the normalization conventional,

$$\langle\mathbf{p}',\lambda'|\mathbf{p},\lambda\rangle=\delta^3(\mathbf{p}-\mathbf{p}')\delta_{\lambda\lambda'}. \tag{2.36}$$

Having defined helicity states of arbitrary momentum in terms of states $|\lambda\rangle$ of a fixed standard four-momentum k^μ, it is now quite easy to find their transformation properties. A general Lorentz transformation $\Lambda^\mu{}_\nu$, represented on Hilbert space by a unitary operator $U[\Lambda]$, will transform $|\mathbf{p},\lambda\rangle$ into

$$\begin{aligned}U[\Lambda]|\mathbf{p},\lambda\rangle&=[\kappa/|\mathbf{p}|]^{1/2}U[\Lambda]U[\mathcal{L}(\mathbf{p})]|\lambda\rangle\\&=[\kappa/|\mathbf{p}|]^{1/2}U[\mathcal{L}(\Lambda\mathbf{p})]U[\mathcal{L}^{-1}(\Lambda\mathbf{p})\Lambda\mathcal{L}(\mathbf{p})]|\lambda\rangle.\end{aligned} \tag{2.37}$$

But the transformation $\mathcal{L}^{-1}(\Lambda\mathbf{p})\Lambda\mathcal{L}(\mathbf{p})$ leaves k^μ unchanged, and hence belongs to the little group. Equation (2.29) then lets us write (2.37) as

$$U[\Lambda]|\mathbf{p},\lambda\rangle=[\kappa/|\mathbf{p}|]^{1/2}\times\exp\{i\lambda\Theta[\mathcal{L}^{-1}(\Lambda\mathbf{p})\Lambda\mathcal{L}(\mathbf{p})]\}U[\mathcal{L}(\Lambda\mathbf{p})]|\lambda\rangle,$$

and finally

$$U[\Lambda]|\mathbf{p},\lambda\rangle=[|\Lambda\mathbf{p}|/|\mathbf{p}|]^{1/2}\times\exp\{i\lambda\Theta[\mathcal{L}^{-1}(\Lambda\mathbf{p})\Lambda\mathcal{L}(\mathbf{p})]\}|\Lambda\mathbf{p},\lambda\rangle. \tag{2.38}$$

A general state containing several free particles will transform like (2.38), with a factor $[|\mathbf{p}'|/|\mathbf{p}|]^{1/2}e^{i\Theta\lambda}$ for each particle. These states can be built up by acting on the bare vacuum with creation operators $a^*(\mathbf{p},\lambda)$ which satisfy either the usual Bose or Fermi rules:

$$[a(\mathbf{p},\lambda),a^*(\mathbf{p}',\lambda')]_\pm=\delta_{\lambda\lambda'}\delta^3(\mathbf{p}-\mathbf{p}'), \qquad (2.39)$$

so the general transformation law can be summarized in the statement

$$U[\Lambda]a^*(\mathbf{p},\lambda)U^{-1}[\Lambda]=[|\Lambda\mathbf{p}|/|\mathbf{p}|]^{1/2} \times\exp\{i\lambda\Theta[\mathfrak{L}^{-1}(\Lambda\mathbf{p})\Lambda\mathfrak{L}(\mathbf{p})]\}a^*(\Lambda\mathbf{p},\lambda). \qquad (2.40)$$

Taking the adjoint and using the property [see (2.30)]

$$\Theta[R]=-\Theta[R^{-1}] \qquad (2.41)$$

gives the transformation rule of the annihilation operator

$$U[\Lambda]a(\mathbf{p},\lambda)U^{-1}[\Lambda]=[|\Lambda\mathbf{p}|/|\mathbf{p}|]^{1/2} \times\exp\{i\lambda\Theta[\mathfrak{L}^{-1}(\mathbf{p})\Lambda^{-1}\mathfrak{L}(\Lambda\mathbf{p})]\}a(\Lambda\mathbf{p},\lambda). \qquad (2.42)$$

We speak of one massless particle as being the antiparticle of another if their spins j are the same, while all their charges, baryon numbers, etc., are equal and opposite. Whether or not every massless particle has such an antiparticle is an open question, to be answered affirmatively in Sec. V. But if an antiparticle exists, then its creation operator $b^*(\mathbf{p},\lambda)$ will transform just like $a^*(\mathbf{p},\lambda)$, and $b^*(\mathbf{p},-\lambda)$ will transform just like $a(\mathbf{p},\lambda)$:

$$U[\Lambda]b^*(\mathbf{p},-\lambda)U^{-1}[\Lambda]=[|\Lambda\mathbf{p}|/|\mathbf{p}|]^{1/2} \times\exp\{i\lambda\Theta[\mathfrak{L}^{-1}(\mathbf{p})\Lambda^{-1}\mathfrak{L}(\Lambda\mathbf{p})]\}b^*(\Lambda\mathbf{p},-\lambda). \qquad (2.43)$$

If a particle is its own antiparticle,[6] then we just set $b(\mathbf{p},\lambda)=a(\mathbf{p},\lambda)$.

III. A THEOREM ON GENERAL FIELDS

As a first step, let us try to construct the "annihilation fields" $\psi_n^{(+)}(x;\lambda)$, as linear combinations of the annihilation operators $a(p,\lambda)$, with fixed helicity λ. We require that the $\psi_n^{(+)}$ transform as usual under translations

$$i[P_\mu,\psi_n^{(+)}(x;\lambda)]=\partial_\mu\psi_n^{(+)}(x;\lambda) \qquad (3.1)$$

and transform according to some irreducible representation $D[\Lambda]$ of the *homogeneous* proper orthochronous Lorentz group:

$$U[\Lambda]\psi_n^{(+)}(x;\lambda)U[\Lambda]^{-1} =\sum_m D_{nm}[\Lambda^{-1}]\psi_m^{(+)}(\Lambda x;\lambda). \qquad (3.2)$$

It is well known that the various representations $D[\Lambda]$ can be cataloged by writing the matrices J and K, which represent the rotation generator $\mathbf{J}$ and the boost generator $\mathbf{K}$ as

$$J=A+B;\quad K=-i(A-B). \qquad (3.3)$$

Since J and K satisfy the same commutation rules (2.16)–(2.18) as $\mathbf{J}$ and $\mathbf{K}$, the A and B satisfy decoupled commutation rules

$$A\times A=iA;\quad B\times B=iB,\quad [\mathcal{A}_i,\mathcal{B}_j]=0. \qquad (3.4)$$

The general $(2A+1)(2B+1)$-dimensional irreducible representation (A,B) is conventionally defined for integer values of $2A$ and $2B$ by

$$\mathbf{A}_{ab,a'b'}=\delta_{bb'}\mathbf{J}_{aa'}^{(A)},\quad \mathbf{B}_{ab,a'b'}=\delta_{aa'}\mathbf{J}_{bb'}^{(B)}, \qquad (3.5)$$

where a and b run by unit steps from $-A$ to $+A$ and from $-B$ to $+B$, respectively, and $J^{(j)}$ is the usual $2j+1$-dimensional representation of the angular momentum

$$[J_1^{(j)}\pm iJ_2^{(j)}]_{\sigma'\sigma}=\delta_{\sigma',\sigma\pm1}[(j\mp\sigma)(j\pm\sigma+1)]^{1/2},\quad [J_3^{(j)}]_{\sigma'\sigma}=\sigma\delta_{\sigma'\sigma}. \qquad (3.6)$$

For massive particles of spin j, we have already seen in Sec. VIII of Ref. 1 that a field $\psi^{(+)}(x)$ can be constructed out of the $2j+1$ annihilation operators $a(\mathbf{p},\sigma)$, which will satisfy the transformation requirements (3.1) and (3.2), for any representation (A,B) that "contains" j, i.e., such that

$$j=A+B \text{ or } A+B-1 \text{ or } \cdots \text{ or } |A-B|. \qquad (3.7)$$

[A spin-one field could be a four-vector $(\frac{1}{2},\frac{1}{2})$, a tensor (1,0) or (0,1), etc.] We might expect the same to be true for mass zero, *but this is not the case.* We will prove in this section that a massless particle operator $a(\mathbf{p},\lambda)$ of helicity λ can only be used to construct fields which transform according to representations (A,B) such that

$$B-A=\lambda. \qquad (3.8)$$

For instance, a left-circularly polarized photon with $\lambda=-1$ can be associated with (1,0), $(\frac{3}{2},\frac{1}{2})$, (2,1), $\cdots$ fields but *not* with the vector potential $(\frac{1}{2},\frac{1}{2})$, at least until we broaden our notion of what we mean by a Lorentz transformation. It will be seen that the restriction (3.8) arises because of the non-semi-simple structure of the little group.

[6] It is not so obvious what is meant by a massless particle being its own antiparticle. If charge conjugation were conserved, then we would call a particle purely neutral if it were invariant (up to a phase) under **C**. But if we take weak interactions into account then only **CP** and **CPT** are available, and they convert a particle into the antiparticle with opposite helicity. For massless particles there is no way of deciding whether a particle is the "same" as another of opposite helicity, since one cannot be converted into the other by a rotation. This point has been thoroughly explored with regard to the neutrino by J. A. McLennan, Phys. Rev. **106**, 821 (1957) and K. M. Case, *ibid.* **107**, 307 (1957). See also C. Ryan and S. Okubo, Rochester Preprint URPA-3 (to be published). Even if a massless particle carries some quantum number (like lepton number), we can still call it purely neutral if we let its quantum number depend on the helicity; however, in this case it seems more natural to adopt the convention that the particle is different from its antiparticle, with $b(\mathbf{p},\lambda)\neq a(\mathbf{p},\lambda)$.

The condition (3.1) requires that $\psi_n^{(+)}$ be constructed as a Fourier transform

$$\psi_n^{(+)}(x;\lambda)=\frac{1}{(2\pi)^{3/2}}\times\int\frac{d^3p}{[2|p|]^{1/2}}e^{ip\cdot x}a(\mathbf{p},\lambda)u_n(\mathbf{p},\lambda)\,,\quad(3.9)$$

the factor $(2\pi)^{-3/2}[2|\mathbf{p}|]^{-1/2}$ being extracted from the "wave function" $u_n(\mathbf{p},\lambda)$ for later convenience. The condition (3.2) together with the transformation rule (2.42) then requires that $u_n(\mathbf{p},\lambda)$ satisfy

$$\exp\{i\lambda\Theta[\mathcal{L}^{-1}(\mathbf{p})\Lambda^{-1}\mathcal{L}(\Lambda\mathbf{p})]\}u_n(\mathbf{p},\lambda)=\sum_m D_{nm}[\Lambda^{-1}]u_m(\Lambda\mathbf{p},\lambda)\,.\quad(3.10)$$

We will now show that this determines $u_m(\mathbf{p},\lambda)$ uniquely.

In particular (3.10) must be satisfied if we choose

$$\mathbf{p}=\mathbf{k}\equiv\{0,0,\kappa\}\,;\ \Lambda=\mathcal{L}(\mathbf{q})\,,$$

where $\mathbf{q}$ is some arbitrary momentum. In this case (3.10) reads

$$u_n(\mathbf{q},\lambda)=\sum_m D_{nm}[\mathcal{L}(\mathbf{q})]u_m(\lambda)\,,\quad(3.11)$$

where $u_m(\lambda)$ is the wave function for our "standard" momentum $\mathbf{k}$

$$u_m(\lambda)\equiv u_m(\mathbf{k},\lambda)\,.\quad(3.12)$$

Insertion of (3.11) into both sides of (3.10) shows that (3.10) is satisfied by (3.11) if and only if the $u_m(\lambda)$ satisfy

$$\exp\{i\lambda\Theta[\mathcal{L}^{-1}(\mathbf{p})\Lambda^{-1}\mathcal{L}(\Lambda\mathbf{p})]\}\sum_m D_{nm}[\mathcal{L}(\mathbf{p})]u_m(\lambda)=\sum_m D_{nm}[\Lambda^{-1}\mathcal{L}(\Lambda\mathbf{p})]u_m(\lambda)\,,$$

or in other words, if and only if

$$\sum_m D_{nm}[\mathcal{R}]u_m(\lambda)=\exp\{i\lambda\Theta[\mathcal{R}]\}u_n(\lambda)\quad(3.13)$$

for any Lorentz transformation $\mathcal{R}$ of the form

$$\mathcal{R}=\mathcal{L}^{-1}(\mathbf{p})\Lambda^{-1}\mathcal{L}(\Lambda\mathbf{p})\,.\quad(3.14)$$

But these $\mathcal{R}$'s, for general $\mathbf{p}$ and Λ, just constitute the little group discussed in Sec. II. In order that (3.13) be satisfied for all such $\mathcal{R}$ it is necessary and sufficient that it be satisfied for all infinitesimal transformations

$$\mathcal{R}^\mu{}_\nu=\delta^\mu{}_\nu+\Omega^\mu{}_\nu(\theta,\chi_1,\chi_2)\,,\quad(3.15)$$

the nonvanishing components of Ω being given by (2.9)–(2.11). The matrix $D[\mathcal{R}]$ corresponding to (3.15) is obtained by replacing $\mathbf{J}$ and $\mathbf{K}$ in (2.19) by their matrix representatives $\boldsymbol{J}$ and $\boldsymbol{K}$:

$$D[\mathcal{R}(\theta,\chi_1,\chi_2)]=1+i\theta\mathcal{J}_3+i\chi_1(\mathcal{K}_1-\mathcal{J}_2)+i\chi_2(\mathcal{K}_2+\mathcal{J}_1)\,,\quad(3.16)$$

or, using (3.3),

$$D[\mathcal{R}(\theta,\chi_1,\chi_2)]=1+i\theta(\mathcal{A}_3+\mathcal{B}_3)+(\chi_1+i\chi_2)(\mathcal{A}_1-i\mathcal{A}_2)+(\chi_1-i\chi_2)(\mathcal{B}_1+i\mathcal{B}_2)\,.\quad(3.17)$$

Recalling from (2.27) that $\Theta\to\theta$, our condition (3.13) is now split into three independent conditions:

$$[\mathcal{A}_3+\mathcal{B}_3]u(\lambda)=\lambda u(\lambda)\,,\quad(3.18)$$

$$[\mathcal{A}_1-i\mathcal{A}_2]u(\lambda)=0\,,\quad(3.19)$$

$$[\mathcal{B}_1+i\mathcal{B}_2]u(\lambda)=0\,.\quad(3.20)$$

Of these three conditions, (3.18) could certainly have been anticipated as necessary to a field of helicity λ. The other two arise from the detailed structure of the little group, but are equally important, for they force $u(\lambda)$ to be an eigenvector of $\mathcal{A}_3$ and $\mathcal{B}_3$, with

$$\mathcal{A}_3u(\lambda)=-Au(\lambda)\,,\quad(3.21)$$

$$\mathcal{B}_3u(\lambda)=+Bu(\lambda)\,,\quad(3.22)$$

or more explicitly

$$u_{ab}(\lambda)=\delta_{a,-A}\delta_{b,B}\,.\quad(3.23)$$

Using (3.18) now gives the promised restriction on A and B:

$$-A+B=\lambda\,.\quad(3.8)$$

For a left-handed particle with $\lambda=-j$, the various possible fields are

$$[\text{left}]\quad(j,0),\ (j+\tfrac{1}{2},\tfrac{1}{2}),\ (j+1,1),\ \cdots,\quad(3.24)$$

while a right-handed particle with $\lambda=+j$ can be associated with a field transforming like

$$[\text{right}]\quad(0,j),\ (\tfrac{1}{2},j+\tfrac{1}{2}),\ (1,j+1),\ \cdots,\quad(3.25)$$

If parity is conserved, then the particle must exist in both states $\lambda=\pm j$, and the field *must* then transform reducibly, for example, like $(j,0)\oplus(0,j)$.

Our theorem certainly applies to the *in* and *out* fields, since they are constructed just like free fields. It must then also apply to the Heisenberg representation field that interpolates between *in* and *out* fields if we insist that they all behave in the same way under Lorentz transformations. Furthermore, the only "M functions"[7] that can generally be formed from the S matrix are those corresponding to the representations (3.24) and (3.25).

In a forthcoming article we shall see what goes wrong when we try to construct a field with A and B violating (3.8).

[7] H. Stapp, Phys. Rev. **125**, 2139 (1962); A. O. Barut, I. Muzinich, and D. N. Williams, *ibid.* **130**, 442 (1963).

IV. $(2j+1)$-COMPONENT FIELDS

For a left- or right-handed particle with $\lambda=-j$ or $\lambda=+j$, the simplest field type listed in (3.24) or (3.25) is, respectively, $(j,0)$ or $(0,j)$. The corresponding $(2j+1)$-component annihilation fields will be called $\varphi_\sigma^{(+)}(x)$ and $\chi_\sigma^{(+)}(x)$. They are given by (3.9), (3.11), and (3.23), as

$$\varphi_\sigma^{(+)}(x)=\frac{1}{(2\pi)^{3/2}}\int\frac{d^3p}{[2|\mathbf{p}|]^{1/2}}\times e^{ip\cdot x}D_{\sigma,-j}{}^{(j)}[\mathcal{L}(\mathbf{p})]a(\mathbf{p},-j)\,, \quad (4.1)$$

$$\chi_\sigma^{(+)}(x)=\frac{1}{(2\pi)^{3/2}}\int\frac{d^3p}{[2|\mathbf{p}|]^{1/2}}\times e^{ip\cdot x}\bar{D}_{\sigma j}{}^{(j)}[\mathcal{L}(\mathbf{p})]a(\mathbf{p},j)\,, \quad (4.2)$$

and they transform according to

$$U[\Lambda]\varphi_\sigma^{(+)}(x)U^{-1}[\Lambda]=\sum_{\sigma'}D_{\sigma\sigma'}{}^{(j)}[\Lambda^{-1}]\varphi_{\sigma'}^{(+)}(\Lambda x)\,, \quad (4.3)$$

$$U[\Lambda]\chi_\sigma^{(+)}(x)U^{-1}[\Lambda]=\sum_{\sigma'}\bar{D}_{\sigma\sigma'}{}^{(j)}[\Lambda^{-1}]\chi_{\sigma'}^{(+)}(\Lambda x)\,. \quad (4.4)$$

Here $D^{(j)}[\Lambda]$ and $\bar{D}^{(j)}[\Lambda]$ are the nonunitary $(2j+1)\times(2j+1)$-dimensional matrices corresponding to Λ in the $(j,0)$ and $(0,j)$ representations, respectively. They are the same as used in Ref. 1, and can be defined by taking $\mathcal{B}=0$ or $\mathcal{A}=0$, or, equivalently, by representing the generators $\mathbf{J}$, $\mathbf{K}$ with

$$D^{(j)}:\ \mathbf{J}=\mathbf{J}^{(j)},\quad \mathbf{K}=-i\mathbf{J}^{(j)}\,, \quad (4.5)$$

$$\bar{D}^{(j)}:\ \mathbf{J}=\mathbf{J}^{(j)},\quad \mathbf{K}=+i\mathbf{J}^{(j)}\,, \quad (4.6)$$

where $\mathbf{J}^{(j)}$ is the usual spin-j representation of the angular momentum, defined by (3.6). In particular, the transformation $\mathcal{L}(\mathbf{p})$ defined by (2.33) is represented on Hilbert space by

$$U[\mathcal{L}(\mathbf{p})]=U[R(\hat{p})]\exp\{-i\phi(|\mathbf{p}|)K_3\}\,, \quad (4.7)$$

$$\phi(|\mathbf{p}|)=\ln[|\mathbf{p}|/\kappa]\,, \quad (2.35)$$

and therefore the wave functions appearing in (4.1) and (4.2) are

$$D_{\sigma,-j}{}^{(j)}[\mathcal{L}(\mathbf{p})]=\sum_{\sigma'}D_{\sigma\sigma'}{}^{(j)}[R(\hat{p})][\exp\{-\phi(|\mathbf{p}|)J_3{}^{(j)}\}]_{\sigma',-j}=D_{\sigma,-j}{}^{(j)}[R(\hat{p})](|\mathbf{p}|/\kappa)^j\,, \quad (4.8)$$

$$D_{\sigma,j}{}^{(j)}[\mathcal{L}(\mathbf{p})]=\sum_{\sigma'}\bar{D}_{\sigma\sigma'}{}^{(j)}[R(\hat{p})][\exp\{\phi(|\mathbf{p}|)J_3{}^{(j)}\}]_{\sigma',j}=D_{\sigma,j}{}^{(j)}[R(\hat{p})](|\mathbf{p}|/\kappa)^j\,. \quad (4.9)$$

Note that the matrices $D^{(j)}[R]$ and $\bar{D}^{(j)}[R]$ for a pure rotation R are both equal, being given by the familiar $2j+1$-dimensional unitary representation[8] of the ordinary rotation group. [Note, also, that if we tried to construct a $(j,0)$ field for a right-handed particle, or a $(0,j)$ field for a left-handed particle, we would not only fail to get the desired Lorentz transformation property, but we would also find a catastrophic factor $|\mathbf{p}|^{-j}$ in the wave function.]

Using the wave functions (4.8) and (4.9) in (4.1) and (4.2), the annihilation fields now take the form

$$\varphi_\sigma^{(+)}(x)=\frac{1}{(2\pi)^{3/2}}\int d^3p[2|\mathbf{p}|]^{j-1/2}\times e^{ip\cdot x}D_{\sigma,-j}{}^{(j)}[R(\hat{p})]a(\mathbf{p},-j)\,, \quad (4.10)$$

$$\chi_\sigma^{(+)}(x)=\frac{1}{(2\pi)^{3/2}}\int d^3p[2|\mathbf{p}|]^{j-1/2}\times e^{ip\cdot x}D_{\sigma,j}{}^{(j)}[R(\hat{p})]a(\mathbf{p},j)\,. \quad (4.11)$$

We have redefined their normalization by replacing the factor κ^{-j} by 2^j. We see that only the ordinary unitary rotation matrices[8] are needed; $R(\hat{p})$ is the rotation that carries the z axis into the direction of $\mathbf{p}$.

If our particle has an antiparticle (perhaps itself), then there is available another operator $b^*(\mathbf{p},-\lambda)$ which transforms just like $a(\mathbf{p},\lambda)$ [see (2.43)], and which carries the same charge, baryon number, etc. It is then possible to define creation fields

$$\varphi_\sigma^{(-)}(x)=\frac{1}{(2\pi)^{3/2}}\int d^3p[2|\mathbf{p}|]^{j-1/2}\times e^{-ip\cdot x}D_{\sigma,-j}{}^{(j)}[R(\hat{p})]b^*(\mathbf{p},j)\,, \quad (4.12)$$

$$\chi_\sigma^{(-)}(x)=\frac{1}{(2\pi)^{3/2}}\int d^3p[2|\mathbf{p}|]^{j-1/2}\times e^{-ip\cdot x}D_{\sigma,j}{}^{(j)}[R(\hat{p})]b^*(\mathbf{p},-j)\,, \quad (4.13)$$

which satisfy (3.1), which transform according to (4.3) and (4.4), respectively, and which also transform like $\varphi^{(+)}$ and $\chi^{(+)}$ under gauge transformations of the first kind. [For a "purely neutral" particle,[6] $b^*(\mathbf{p},\lambda)$ is to be replaced by $a^*(\mathbf{p},\lambda)$.]

The most general fields satisfying all these conditions are linear combinations of creation and annihilation fields.

$$\varphi_\sigma(x)=\xi_L\varphi_\sigma^{(+)}(x)+\eta_R\varphi_\sigma^{(-)}(x)\,, \quad (4.14)$$

$$\chi_\sigma(x)=\xi_R\chi_\sigma^{(+)}(x)+\eta_L\chi_\sigma^{(-)}(x)\,. \quad (4.15)$$

They again transform as in (4.3) and (4.4):

$$U[\Lambda]\varphi_\sigma(x)U^{-1}[\Lambda]=\sum_{\sigma'}D_{\sigma\sigma'}{}^{(j)}[\Lambda^{-1}]\varphi_{\sigma'}(\Lambda x)\,, \quad (4.16)$$

$$U[\Lambda]\chi_\sigma(x)U^{-1}[\Lambda]=\sum_{\sigma'}\bar{D}_{\sigma\sigma'}{}^{(j)}[\Lambda]\chi_{\sigma'}(\Lambda x)\,. \quad (4.17)$$

[8] See, for example, M. E. Rose, *Elementary Theory of Angular Momentum* (J. Wiley & Sons, Inc., New York, 1957), p. 48 ff.

If these particles have no antiparticles (including themselves), then we have to take $\eta_L=\eta_R=0$. We will see in the next section that, instead, requirement (1.4) (and hence the Lorentz invariance of the S matrix) dictates full crossing symmetry, with $|\eta_R|=|\xi_L|$, $|\eta_L|=|\xi_R|$.

The fields obviously obey the Klein-Gordon equation

$$\square^2\varphi_\sigma(x)=0;\quad \square^2\chi_\sigma(x)=0. \tag{4.18}$$

However, they are $(2j+1)$-component objects constructed out of just two independent operators $a(\mathbf{p},\lambda)$, $b^*(\mathbf{p},-\lambda)$, and so they have a chance of obeying other field equations as well. It is not hard to see from (4.10)–(4.13) that they do indeed satisfy the additional field equations

$$[\mathbf{J}^{(j)}\cdot\nabla-j(\partial/\partial t)]\varphi(x)=0, \tag{4.19}$$

$$[\mathbf{J}^{(j)}\cdot\nabla+j(\partial/\partial t)]\chi(x)=0. \tag{4.20}$$

For $j=\frac{1}{2}$ these are the Weyl equations for the left- and right-handed neutrino fields, while for $j=1$ they are just Maxwell's free-space equations for left- and right-circularly polarized radiation:

$$\nabla\times[\mathbf{E}-i\mathbf{B}]+i(\partial/\partial t)[\mathbf{E}-i\mathbf{B}]=0, \tag{4.21}$$

$$\nabla\times[\mathbf{E}+i\mathbf{B}]-i(\partial/\partial t)[\mathbf{E}+i\mathbf{B}]=0. \tag{4.22}$$

The fact that these field equations are of first order for any spin seems to me to be of no great significance, since in the case of massive particles we can get along perfectly well with $(2j+1)$-component fields which satisfy only the Klein-Gordon equation.

V. CROSSING AND STATISTICS

We are assuming that the a's and b's satisfy the usual commutation (or anticommutation) rules (2.39), so it is easy to work out the commutators or anticommutators of the fields φ_σ and χ_σ defined by (4.10)–(4.15):

$$[\varphi_\sigma(x),\varphi_{\sigma'}{}^\dagger(y)]_\pm=\frac{1}{(2\pi)^3}\int\frac{d^3p}{2|\mathbf{p}|}\pi_{\sigma\sigma'}(p)\times[|\xi_L|^2e^{ip\cdot(x-y)}\pm|\eta_R|^2e^{-ip\cdot(x-y)}], \tag{5.1}$$

$$[\chi_\sigma(x),\chi_{\sigma'}{}^\dagger(y)]_\pm=\frac{1}{(2\pi)^3}\int\frac{d^3p}{2|\mathbf{p}|}\bar{\pi}_{\sigma\sigma'}(p)\times[|\xi_R|^2e^{ip\cdot(x-y)}\pm|\eta_L|^2e^{-ip\cdot(x-y)}], \tag{5.2}$$

where

$$\pi_{\sigma\sigma'}(p)=|2\mathbf{p}|^{2j}D_{\sigma,-j}{}^{(j)}[R(\hat{p})]D_{\sigma',-j}{}^{(j)*}[R(\hat{p})], \tag{5.3}$$

$$\bar{\pi}_{\sigma\sigma'}(p)=|2\mathbf{p}|^{2j}D_{\sigma,j}{}^{(j)}[R(\hat{p})]D_{\sigma',j}{}^{(j)*}[R(\hat{p})]. \tag{5.4}$$

These are the only nonvanishing commutators (or anticommutators) among the φ, $\varphi^\dagger$, χ, and $\chi^\dagger$ (except for a "purely neutral" particle, in which case χ is proportional to $\varphi^\dagger$; see Sec. IX).

The matrices π and $\bar{\pi}$ can be easily calculated by use of the obvious formulas

$$\delta_{\sigma,-j}\delta_{\sigma',-j}=\frac{1}{(2j)!}[\prod_{\lambda=-j+1}^{j}(\lambda-J_3)]_{\sigma\sigma'}, \tag{5.5}$$

$$\delta_{\sigma,j}\delta_{\sigma',j}=\frac{1}{(2j)!}[\prod_{\lambda=-j}^{j-1}(J_3-\lambda)]_{\sigma\sigma'}. \tag{5.6}$$

Applying the rotation matrix $D^{(j)}[R(\mathbf{p})]$ and multiplying by $|2\mathbf{p}|^{2j}$ gives

$$\pi(p)=\frac{2^{2j}}{(2j)!}\prod_{\lambda=-j+1}^{j}(\lambda p^0-\mathbf{p}\cdot\mathbf{J}), \tag{5.7}$$

$$\bar{\pi}(p)=\frac{2^{2j}}{(2j)!}\prod_{\lambda=-j}^{j-1}(\mathbf{p}\cdot\mathbf{J}-\lambda p^0). \tag{5.8}$$

These are monomials of order $2j$ in the light-like four-vector p^μ, so (5.1) and (5.2) now become

$$[\varphi_\sigma(x),\varphi_{\sigma'}{}^\dagger(y)]_\pm=\frac{1}{(2\pi)^3}\pi_{\sigma\sigma'}(-i\partial)\int\frac{d^3p}{2|\mathbf{p}|}\times[|\xi_L|^2e^{ip\cdot(x-y)}\pm(-)^{2j}|\eta_R|^2e^{-ip\cdot(x-y)}], \tag{5.9}$$

$$[\chi_\sigma(x),\chi_{\sigma'}{}^\dagger(y)]_\pm=\frac{1}{(2\pi)^3}\bar{\pi}_{\sigma\sigma'}(-i\partial)\int\frac{d^3p}{2|\mathbf{p}|}\times[|\xi_R|^2e^{ip\cdot(x-y)}\pm(-)^{2j}|\eta_L|^2e^{-ip\cdot(x-y)}]. \tag{5.10}$$

In order that (5.9) and (5.10) vanish for $x-y$ space-like, it is necessary and sufficient that $\exp[ip\cdot(x-y)]$ and $\exp[-ip\cdot(x-y)]$ have equal and opposite coefficients

$$|\xi_L|^2=\mp(-)^{2j}|\eta_R|^2, \tag{5.11}$$

$$|\xi_R|^2=\mp(-)^{2j}|\eta_L|^2. \tag{5.12}$$

So we must have the usual connection between spin and statistics

$$(\pm)=-(-)^{2j}, \tag{5.13}$$

and furthermore, every left- or right-handed particle must be associated, respectively, with a right- or left-handed antiparticle (perhaps itself) which enters into interactions with equal strength:

$$|\xi_L|=|\eta_R|,\quad |\xi_R|=|\eta_L|. \tag{5.14}$$

By redefining the phases of the a's and b's, and the normalization of φ and χ, we can therefore set

$$\xi_L=\eta_L=\xi_R=\eta_R=1 \tag{5.15}$$

with no loss of generality. The fields are now in their

final form:

$$\varphi_\sigma(x)=\frac{1}{(2\pi)^{3/2}}\int d^3p[2|\mathbf{p}|]^{j-1/2}D_{\sigma,-j}{}^{(j)}[R(\hat{p})]$$
$$\times[a(\mathbf{p},-j)e^{ip\cdot x}+b^*(\mathbf{p},j)e^{-ip\cdot x}], \quad (5.16)$$

$$\chi_\sigma(x)=\frac{1}{(2\pi)^{3/2}}\int d^3p[2|\mathbf{p}|]^{j-1/2}D_{\sigma,j}{}^{(j)}[R(\hat{p})]$$
$$\times[a(\mathbf{p},j)e^{ip\cdot x}+b^*(\mathbf{p},-j)e^{-ip\cdot x}]. \quad (5.17)$$

The commutator or anticommutators are

$$[\varphi_\sigma(x),\varphi_{\sigma'}{}^\dagger(y)]_\pm=i\pi_{\sigma\sigma'}(-i\partial)\Delta(x-y), \quad (5.18)$$

$$[\chi_\sigma(x),\chi_{\sigma'}{}^\dagger(y)]_\pm=i\bar{\pi}_{\sigma\sigma'}(-i\partial)\Delta(x-y), \quad (5.19)$$

where $i\Delta(x-y)$ is the commutator for zero mass and $j=0$.

$$\Delta(x)=\frac{-i}{(2\pi)^3}\int\frac{d^3p}{2|\mathbf{p}|}[e^{ip\cdot x}-e^{-ip\cdot x}]$$
$$=-(1/2\pi)\delta(x_\mu x^\mu)\epsilon(x).$$

If a particle has no additive quantum numbers like the photon, we must[6] set $b(\mathbf{p},\lambda)$ equal to $a(\mathbf{p},\lambda)$, and "causality" then tells us through (5.14) that the particle must exist in both left- and right-handed helicity states. Both fields $\varphi_\sigma(x)$ and $\chi_\sigma(x)$ can be constructed, and in fact we shall see in Sec. IX that φ is just proportional to $\chi^\dagger$.

On the other hand, a particle which carries some additive quantum number that distinguishes it from its antiparticle can possibly exist in only the left- *or* the right-handed helicity state, and "causality" only requires that it has an antiparticle of opposite helicity. (A familiar example is the neutrino.) In this case only one of the fields φ_σ and χ_σ can be constructed. Of course, if parity of charge conjugation are conserved, then both particle and antiparticle must exist in both left- and right-handed states, and both φ_σ and χ_σ exist.

VI. LORENTZ INVARIANCE

Our formulas (5.18) and (5.19) for the commutators or anticommutators were derived in a Lorentz invariant manner, but they do not look like invariant equations. It will be necessary to see how their invariance comes about before we are able to derive the Feynman rules.

It was shown in Appendix A of Ref. 1 that the familiar angular momentum matrices $J^{(j)}$ can be used to construct a pair of scalar $(2j+1)\times(2j+1)$ matrices Π and $\bar{\Pi}$, as monomials in a general four-vector q^μ:

$$\Pi_{\sigma\sigma'}(q)=(-)^{2j}t_{\sigma\sigma'}{}^{\mu_1\mu_2\cdots\mu_{2j}}q_{\mu_1}q_{\mu_2}\cdots q_{\mu_{2j}}, \quad (6.1)$$

$$\bar{\Pi}_{\sigma\sigma'}(q)=(-)^{2j}\bar{t}_{\sigma\sigma'}{}^{\mu_1\mu_2\cdots\mu_{2j}}q_{\mu_1}q_{\mu_2}\cdots q_{\mu_{2j}}, \quad (6.2)$$

with properties:

(a) Π and $\bar{\Pi}$ are scalars, in the sense that

$$D^{(j)}[\Lambda]\Pi(q)D^{(j)}[\Lambda]^\dagger=\Pi(\Lambda q), \quad (6.3)$$

$$\bar{D}^{(j)}[\Lambda]\bar{\Pi}(q)\bar{D}^{(j)}[\Lambda]^\dagger=\bar{\Pi}(\Lambda q). \quad (6.4)$$

(b) t and $\bar{t}$ are symmetric and traceless in $\mu_1\mu_2\cdots\mu_{2j}$.

(c) Π and $\bar{\Pi}$ are related by an inversion

$$\bar{\Pi}(-\mathbf{q},q^0)=\Pi(q). \quad (6.5)$$

(d) Π and $\bar{\Pi}^*$ are related by a similarity transformation

$$\bar{\Pi}^*(q)=C\Pi(q)C^{-1}, \quad (6.6)$$

where

$$-\mathbf{J}^{(j)*}=C\mathbf{J}^{(j)}C^{-1}. \quad (6.7)$$

(e) Π and $\bar{\Pi}$ are further related by

$$\Pi(q)\bar{\Pi}(q)=\bar{\Pi}(q)\Pi(q)=(-q^2)^{2j}. \quad (6.8)$$

(f) If q is in the forward light cone then

$$\Pi(q)=(-q^2)^j\exp[-2\theta(q)\hat{q}\cdot\mathbf{J}^{(j)}], \quad (6.9)$$

$$\bar{\Pi}(q)=(-q^2)^j\exp[2\theta(q)\hat{q}\cdot\mathbf{J}^{(j)}], \quad (6.10)$$

$$\sinh\theta(q)\equiv[|\mathbf{q}|^2/-q^2]^{1/2}. \quad (6.11)$$

(g) For integer j and arbitrary q

$$\Pi^{(j)}(q)=(-q^2)^j+[(-q^2)^{j-1}/2!](2\mathbf{q}\cdot\mathbf{J}^{(j)})(2\mathbf{q}\cdot\mathbf{J}^{(j)}-2q^0)$$
$$+[(-q^2)^{j-2}/4!](2\mathbf{q}\cdot\mathbf{J}^{(j)})[(2\mathbf{q}\cdot\mathbf{J}^{(j)})^2-(2\mathbf{q})^2]$$
$$\times[2\mathbf{q}\cdot\mathbf{J}^{(j)}-4q^0]+[(-q^2)^{j-3}/6!](2\mathbf{q}\cdot\mathbf{J}^{(j)})$$
$$\times[(2\mathbf{q}\cdot\mathbf{J}^{(j)})^2-(2\mathbf{q})^2][(2\mathbf{q}\cdot\mathbf{J}^{(j)})^2-(4\mathbf{q})^2]$$
$$\times[2\mathbf{q}\cdot\mathbf{J}^{(j)}-6q^0]+\cdots, \quad (6.12)$$

the series cutting itself off automatically after $j+1$ terms.

(h) For half-integer j and arbitrary q

$$\Pi^{(j)}(q)=(-q^2)^{j-1/2}[q^0-2\mathbf{q}\cdot\mathbf{J}^{(j)}]+(1/3!)(-q^2)^{j-3/2}$$
$$\times[(2\mathbf{q}\cdot\mathbf{J}^{(j)})^2-\mathbf{q}^2][3q^0-2\mathbf{q}\cdot\mathbf{J}^{(j)}]$$
$$+(1/5!)(-q^2)^{j-5/2}[(2\mathbf{q}\cdot\mathbf{J}^{(j)})^2-\mathbf{q}^2]$$
$$\times[(2\mathbf{q}\cdot\mathbf{J}^{(j)})^2-(3\mathbf{q})^2]$$
$$\times[5q^0-2\mathbf{q}\cdot\mathbf{J}^{(j)}]+\cdots, \quad (6.13)$$

the series cutting itself off automatically after $j+\frac{1}{2}$ terms.

It follows from (6.12), (6.13), and (6.5) [or, more directly, from (6.9) and (6.10)] that for a light-like vector p^μ the monomials Π and $\bar{\Pi}$ simplify to

$$\Pi(p)=\frac{2^{2j}}{(2j)!}\prod_{\lambda=-j+1}^{j}(\lambda p^0-\mathbf{p}\cdot\mathbf{J}^{(j)}), \quad (6.14)$$

$$\bar{\Pi}(p)=\frac{2^{2j}}{(2j)!}\prod_{\lambda=-j+1}^{j}(\lambda p^0+\mathbf{p}\cdot\mathbf{J}^{(j)}), \quad (6.15)$$

or in terms of the matrices (5.7), (5.8)

$$\Pi_{\sigma\sigma'}(p)=\pi_{\sigma\sigma'}(p)\quad[p\text{ light-like}], \quad (6.16)$$

$$\bar{\Pi}_{\sigma\sigma'}(p)=\bar{\pi}_{\sigma\sigma'}(p)\quad[p\text{ light-like}]. \quad (6.17)$$

The Lorentz invariance of formulas (5.18) and (5.19) for the commutators or anticommutators now follows immediately from (6.3) and (6.4).

VII. THE FEYNMAN RULES

The Hamiltonian density $\mathcal{H}(x)$ is to be constructed as an invariant polynomial in the $(2j+1)$-component fields $\varphi_\sigma(x)$ and $\chi_\sigma(x)$, without any distinction made between zero and nonzero mass. In each term of $\mathcal{H}(x)$ all σ indices on the $\varphi_\sigma(x)$ are to be coupled together to form a scalar, using Clebsch-Gordan coefficients in the familiar way. The same is to be done *independently* with the indices on the $\chi_\sigma(x)$. If adjoint fields enter in $\mathcal{H}(x)$ then $C_{\sigma\sigma'}^{-1}\chi_{\sigma'}^\dagger(x)$ is to be treated like $\varphi_\sigma(x)$ and $C_{\sigma\sigma'}^{-1}\varphi_{\sigma'}(x)^\dagger$ is to be treated like $\chi_\sigma(x)$; the matrix C is defined by

$$\bar{D}^{(j)}[\Lambda]^* = CD^{(j)}[\Lambda]C^{-1}, \tag{7.1}$$

or more specifically,

$$-\mathbf{J}^{(j)*} = C\mathbf{J}^{(j)}C^{-1}. \tag{7.2}$$

[We use an asterisk for the ordinary complex conjugate of a matrix.] If derivatives appear they will enter as a 2×2 matrix:

$$\partial_{\sigma\sigma'} = \sigma_{\sigma\sigma'}{}^i(\partial/\partial x^i) - \delta_{\sigma\sigma'}(\partial/\partial t), \tag{7.3}$$

where σ^i are the usual Pauli spin matrices; the indices σ and σ' are to be treated as if they appeared respectively on $j=\frac{1}{2}$ fields φ_σ and $\chi_{\sigma'}$.

We list below some typical examples of possible invariant terms in $\mathcal{H}(x)$:

$$\sum_{\sigma_1\sigma_2\sigma_3} \begin{pmatrix} j_1 & j_2 & j_3 \\ \sigma_1 & \sigma_2 & \sigma_3 \end{pmatrix} \varphi_{\sigma_1}{}^{(j_1)}(x)\varphi_{\sigma_2}{}^{(j_2)}(x)\varphi_{\sigma_3}{}^{(j_3)}(x), \tag{7.4}$$

$$\sum_{\sigma_1\sigma_2\sigma_3\sigma_3'} \begin{pmatrix} j_1 & j_2 & j_3 \\ \sigma_1 & \sigma_2 & \sigma_3 \end{pmatrix} \times \varphi_{\sigma_1}{}^{(j_1)}(x)\varphi_{\sigma_2}{}^{(j_2)}(x)C_{\sigma_3\sigma_3'}{}^{-1}\chi_{\sigma_3'}{}^{(j_3)}(x)^\dagger, \tag{7.5}$$

$$\sum_{\sigma_1\sigma_2\sigma_3\sigma\sigma'} \begin{pmatrix} j_1 & j_2 & \frac{1}{2} \\ \sigma_1 & \sigma_2 & \sigma \end{pmatrix}\begin{pmatrix} \frac{1}{2} & \frac{1}{2} & 0 \\ \sigma' & \sigma_3 & 0 \end{pmatrix} \times \varphi_{\sigma_1}{}^{(j_1)}(x)\varphi_{\sigma_2}{}^{(j_2)}(x)\partial_{\sigma\sigma'}\chi_{\sigma_3}{}^{(\frac{1}{2})}(x), \tag{7.6}$$

etc. The fields φ_σ and χ_σ appearing here may be either of zero or of nonzero mass.

The S matrix can be calculated from $\mathcal{H}(x)$ by using Wick's theorem to derive the Feynman rules, as we did in Sec. V of Ref. 1. The only additional information needed here is a statement of the wave functions corresponding to external mass zero lines, and a formula for the propagators corresponding to internal mass zero lines.

The factor arising from the destruction or creation at x of a massless particle or antiparticle of helicity $\lambda=\pm j$ can be determined from (5.16) and (5.17) as the coefficient of the appropriate creation or annihilation operator in φ_σ, χ_σ, $\varphi_\sigma^\dagger$, or $\chi_\sigma^\dagger$:

$$(2\pi)^{-3/2}(2|\mathbf{p}|)^{j-1/2}D_{\sigma,\lambda}{}^{(j)}[R(\hat{p})]e^{ip\cdot x} \quad \text{[particle destroyed]}, \tag{7.7}$$

$$(2\pi)^{-3/2}(2|\mathbf{p}|)^{j-1/2}D_{\sigma,\lambda}{}^{(j)}[R(\hat{p})]^*e^{-ip\cdot x} \quad \text{[particle created]}, \tag{7.8}$$

$$(2\pi)^{-3/2}(2|\mathbf{p}|)^{j-1/2}D_{\sigma,-\lambda}{}^{(j)}[R(\hat{p})]e^{-ip\cdot x} \quad \text{[antiparticle created]}, \tag{7.9}$$

$$(2\pi)^{-3/2}(2|\mathbf{p}|)^{j-1/2}D_{\sigma,-\lambda}{}^{(j)}[R(\hat{p})]^*e^{ip\cdot x} \quad \text{[antiparticle destroyed]}. \tag{7.10}$$

We remind the reader that $D^{(j)}[R]$ is the usual $(2j+1)\times(2j+1)$ unitary matrix[8] corresponding to an ordinary rotation R, and that $R(\hat{p})$ is the rotation that carries the z axis into the direction of $\mathbf{p}$.

The "raw" propagator corresponding to an internal massless particle line running from x to y is

$$\langle T\{\varphi_\sigma(x),\varphi_{\sigma'}{}^\dagger(y)\}\rangle_0 = \theta(x-y)\langle\varphi_\sigma(x)\varphi_{\sigma'}{}^\dagger(y)\rangle_0 + (-)^{2j}\theta(y-x)\langle\varphi_{\sigma'}{}^\dagger(y)\varphi_\sigma(x)\rangle_0, \tag{7.11}$$

or

$$\langle T\{\chi_\sigma(x),\chi_{\sigma'}{}^\dagger(y)\}\rangle_0 = \theta(x-y)\langle\chi_\sigma(x)\chi_{\sigma'}{}^\dagger(y)\rangle_0 + (-)^{2j}\theta(y-x)\langle\chi_{\sigma'}{}^\dagger(y)\chi_\sigma(x)\rangle_0. \tag{7.12}$$

An elementary calculation using (5.16), (5.17), (5.3), (5.4), (6.16), and (6.17) gives the vacuum expectation values as

$$\langle\varphi_\sigma(x)\varphi_{\sigma'}{}^\dagger(y)\rangle_0 = i\Pi_{\sigma\sigma'}(-i\partial)\Delta_+(x-y), \tag{7.13}$$

$$(-)^{2j}\langle\varphi_{\sigma'}{}^\dagger(y)\varphi_\sigma(x)\rangle_0 = i\Pi_{\sigma\sigma'}(-i\partial)\Delta_+(y-x), \tag{7.14}$$

and

$$\langle\chi_\sigma(x)\chi_{\sigma'}{}^\dagger(y)\rangle_0 = i\bar{\Pi}_{\sigma\sigma'}(-i\partial)\Delta_+(x-y), \tag{7.15}$$

$$(-)^{2j}\langle\chi_{\sigma'}{}^\dagger(y)\chi_\sigma(x)\rangle_0 = i\bar{\Pi}_{\sigma\sigma'}(-i\partial)\Delta_+(y-x), \tag{7.16}$$

where

$$i\Delta_+(x) \equiv \frac{1}{(2\pi)^3}\int \frac{d^3p}{2|\mathbf{p}|}e^{ip\cdot x} = \frac{1}{4\pi^2}\left[\frac{1}{x^2} - i\pi\delta(x^2)\epsilon(x)\right]. \tag{7.17}$$

As discussed in Ref. 1, the presence of the θ functions in (7.11) and (7.12) makes these propagators noncovariant at the point $x=y$, for spins $j\geq 1$. In order that the S matrix be Lorentz invariant, it is necessary to assume that noncovariant contact interactions appear in $\mathcal{H}(x)$ which cancel the noncovariant terms in (7.11) and (7.12). (The Coulomb interaction in Coulomb gauge is such a contact interaction, made necessary by the unit spin rather than by the zero mass of the photon.) With this understanding, we can move the derivative operators $\Pi(-i\partial)$ and $\bar{\Pi}(-i\partial)$ in (7.13)–

(7.16) to the left of the θ functions in (7.11) and (7.12), obtaining the propagators

$$S_{\sigma\sigma'}(x-y)=-i\Pi_{\sigma\sigma'}(-i\partial)\Delta^c(x-y)$$
$$=-i^{2j+1}t_{\sigma\sigma'}{}^{\mu_1\mu_2\cdots\mu_{2j}}\partial_{\mu_1}\partial_{\mu_2}\cdots$$
$$\times\partial_{\mu_{2j}}\Delta^c(x-y), \quad (7.18)$$

$$\bar{S}_{\sigma\sigma'}(x-y)=-i\bar{\Pi}_{\sigma\sigma'}(-i\partial)\Delta^c(x-y)$$
$$=-i^{2j+1}\bar{t}_{\sigma\sigma'}{}^{\mu_1\mu_2\cdots\mu_{2j}}\partial_{\mu_1}\partial_{\mu_2}\cdots$$
$$\times\partial_{\mu_{2j}}\Delta^c(x-y), \quad (7.19)$$

where $-i\Delta^c(x-y)$ is the usual propagator for spin zero and mass zero

$$-i\Delta^c(x)=i\theta(x)\Delta_+(x)+i\theta(-x)\Delta_+(-x)$$
$$=+[1/4\pi^2(x^2+i\epsilon)]. \quad (7.20)$$

Equations (6.3) and (6.4) show that these propagators are covariant in the sense that

$$D^{(j)}[\Lambda]S(x)D^{(j)}[\Lambda]^\dagger=S(\Lambda x), \quad (7.21)$$
$$\bar{D}^{(j)}[\Lambda]\bar{S}(x)\bar{D}^{(j)}[\Lambda]^\dagger=\bar{S}(\Lambda x). \quad (7.22)$$

The propagators in momentum space are given by the Fourier transforms of (7.18) and (7.19)

$$S(q)=\int d^4x e^{iq\cdot x}S(x)=-i\Pi(q)/q^2-i\epsilon, \quad (7.23)$$

$$\bar{S}(q)=\int d^4x e^{-iq\cdot x}\bar{S}(x)=-i\bar{\Pi}(q)/q^2-i\epsilon. \quad (7.24)$$

Explicit formulas for the monomials $\Pi(q)$ and $\bar{\Pi}(q)$ are given in Eqs. (6.12), (6.13), and (6.5), or for $j\leqq 3$ in Table I of Ref. 1.

VIII. GENERAL HELICITY AMPLITUDES AND THE LIMIT $m\rightarrow 0$

The Feynman rules were given in Ref. 1 for incoming and outgoing massive particles having prescribed values for the z components of their spins. It turns out, however, that the external-line wave functions are much simpler in the Jacob-Wick formalism,[4] where initial and final states are labeled instead by the particle helicities. For $m=0$, of course, we have had no choice, since only the helicity amplitudes are physically meaningful. We will first derive the helicity wave functions for $m>0$, and then use them to show how the Feynman rules given here for $m=0$ can be obtained by taking the limit $m\rightarrow 0$ of the Feynman rules for positive m.

According to the Feynman rules of Ref. 1, the wave function for a particle of spin j, $J_z=\mu$, momentum $\mathbf{p}$, and mass m, destroyed by $\varphi_\sigma(x)$, is

$$u_\sigma(x;\mathbf{p},\mu)=(2\omega)^{-1/2}(2\pi)^{-3/2}$$
$$\times[\exp(-\mathbf{p}\cdot\mathbf{J}^{(j)}\theta)]_{\sigma\mu}e^{ip\cdot x}, \quad (8.1)$$

where

$$\omega=[\mathbf{p}^2+m^2]^{1/2},$$
$$\sinh\theta=|\mathbf{p}|/m. \quad (8.2)$$

(It should be kept in mind that the index σ, which is of no direct physical significance, will appear on some other wave function or propagator, and eventually be summed over.) The corresponding wave function for a particle of definite helicity λ is

$$U_\sigma(x;\mathbf{p},\lambda)=\sum_\mu D_{\mu\lambda}{}^{(j)}[R(\hat{p})]u_\sigma(x;\mathbf{p},\mu), \quad (8.3)$$

where $R(\hat{p})$, as always, is the rotation that carries the z axis into the direction of $\mathbf{p}$. Using (8.1) in (8.3) gives

$$U_\sigma(x;\mathbf{p},\lambda)=(2\omega)^{-1/2}(2\pi)^{-3/2}$$
$$\times\{\exp(-\hat{p}\cdot\mathbf{J}^{(j)}\theta)D^{(j)}[R(\hat{p})]\}_{\sigma\lambda}e^{ip\cdot x}$$
$$=(2\omega)^{-1/2}(2\pi)^{-3/2}$$
$$\times\{D^{(j)}[R(\hat{p})]\exp(-J_3{}^{(j)}\theta)\}_{\sigma\lambda}e^{ip\cdot x}$$
$$=(2\omega)^{-1/2}(2\pi)^{-3/2}D_{\sigma\lambda}{}^{(j)}[R(\hat{p})]e^{-\lambda\theta}e^{ip\cdot x}. \quad (8.4)$$

Furthermore we see from (8.2) that

$$e^{-\lambda\theta}=[\omega(\mathbf{p})+|\mathbf{p}|/m]^{-\lambda}. \quad (8.5)$$

In order to avoid m's appearing in the denominator of U_σ for negative helicity, it will be convenient to renormalize all fields of mass m by multiplying them with a factor m^j. With this understanding, the wave function for a particle of spin j, helicity λ, momentum $\mathbf{p}$, and mass m, destroyed by $\varphi_\sigma(x)$, is

$$U_\sigma(x;\mathbf{p},\lambda)=(2\omega)^{-1/2}(2\pi)^{-3/2}D_{\sigma\lambda}{}^{(j)}[R(\hat{p})]$$
$$\times m^{j+\lambda}(\omega+|\mathbf{p}|)^{-\lambda}e^{ip\cdot x}. \quad (8.6)$$

The wave function for the creation of the same particle by $\varphi_\sigma{}^\dagger(x)$ is just the complex conjugate

$$U_\sigma{}^*(x;\mathbf{p},\lambda)=(2\omega)^{-1/2}(2\pi)^{-3/2}D_{\sigma\lambda}{}^{(j)*}[R(\hat{p})]$$
$$\times m^{j+\lambda}(\omega+|\mathbf{p}|)^{-\lambda}e^{-ip\cdot x}. \quad (8.7)$$

The wave function for the creation by $\varphi_\sigma(x)$ of the antiparticle with helicity λ and spin $\mathbf{p}$ can be easily obtained in the same way from Eq. (5.4) of Ref. 1, by using the relations

$$D^{(j)*}[R(\hat{p})]=CD^{(j)}[R(\hat{p})]C^{-1},$$
$$C_{\mu\lambda}{}^{-1}=(-)^{-j+\lambda}\delta_{\mu,-\lambda}.$$

We find that the antiparticle creation wave function is

$$V_\sigma(x;\mathbf{p},\lambda)=(2\omega)^{-1/2}(2\pi)^{-3/2}(-)^{j+\lambda}D_{\sigma,-\lambda}{}^{(j)}[R(\hat{p})]$$
$$\times m^{j-\lambda}(\omega+|\mathbf{p}|)^{\lambda}e^{-ip\cdot x}, \quad (8.8)$$

and the wave function for destruction of the same antiparticle by $\varphi_\sigma{}^\dagger(x)$ is the complex conjugate

$$V_\sigma{}^*(x;\mathbf{p},\lambda)=(2\omega)^{-1/2}(2\pi)^{-3/2}(-)^{j+\lambda}D_{\sigma,-\lambda}{}^{(j)*}[R(\hat{p})]$$
$$\times m^{j-\lambda}(\omega+|\mathbf{p}|)^{\lambda}e^{+ip\cdot x}. \quad (8.9)$$

A massive particle can be created or destroyed in any helicity state by either the $(j,0)$ field $\varphi_\sigma(x)$ or the $(0,j)$ field $\chi_\sigma(x)$. Inspection of the field $\chi_\sigma(x)$ given in Eq. (6.9) of Ref. 1 shows that the wave functions corresponding to (8.6)–(8.9) are given by replacing θ by $-\theta$, and supplying a sign $(-)^{2j}$ for antiparticles:

$$\bar{U}_\sigma(x;\mathbf{p},\lambda)=(2\omega)^{-1/2}(2\pi)^{-3/2}D_{\sigma\lambda}{}^{(j)}[R(\hat{p})]\times m^{j-\lambda}(\omega+|\mathbf{p}|)^{\lambda}e^{ip\cdot x}$$
[particle destroyed], (8.10)

$$\bar{U}_\sigma{}^*(x;\mathbf{p},\lambda)=(2\omega)^{-1/2}(2\pi)^{-3/2}D_{\sigma\lambda}{}^{(j)*}[R(\hat{p})]\times m^{j-\lambda}(\omega+|\mathbf{p}|)^{\lambda}e^{-ip\cdot x}$$
[particle created], (8.11)

$$\bar{V}_\sigma(x;p\lambda)=(2\omega)^{-1/2}(2\pi)^{-3/2}(-)^{j+\lambda}D_{\sigma,-\lambda}{}^{(j)}[R(\hat{p})]\times m^{j+\lambda}(\omega+|\mathbf{p}|)^{-\lambda}e^{-ip\cdot x}$$
[antiparticle created], (8.12)

$$\bar{V}_\sigma{}^*(x;\mathbf{p},\lambda)=(2\omega)^{-1/2}(2\pi)^{-3/2}(-)^{j+\lambda}D_{\sigma,-\lambda}{}^{(j)*}[R(\hat{p})]\times m^{j+\lambda}(\omega+|\mathbf{p}|)^{-\lambda}e^{+ip\cdot x}$$
[antiparticle destroyed]. (8.13)

Now suppose that $m\to 0$, or, more precisely, that $|\mathbf{p}|/m\to\infty$. The only wave functions among (8.6)–(8.13) that survive in this limit are (8.6), (8.7), (8.12), and (8.13) for $\lambda=-j$, and (8.8), (8.9), (8.10), and (8.11) for $\lambda=+j$. This agrees with the situation for $m=0$, in which case we know that φ_σ and $\varphi_\sigma{}^\dagger$ can only create and destroy particles with $\lambda=-j$ and antiparticles with $\lambda=+j$, while χ_σ and $\chi_\sigma{}^\dagger$ only create and destroy particles with $\lambda=+j$ and antiparticles with $\lambda=-j$. Furthermore, if we set $\lambda=-j$ in (8.6) or $\lambda=+j$ in (8.10) we see that these wave functions reduce for $|\mathbf{p}|/m\to\infty$ to the particle destruction wave function given for $m=0$ by (7.7). The same agreement is obtained on comparison of (8.7) and (8.11) with (7.8), (8.8), and (8.12) with (7.9), and (8.9), and (8.13) with (7.10). [The observation that particles described only by $\varphi_\sigma(x)$ are difficult to create or destroy for $|\mathbf{p}|\gg m$ in any helicity state other than $\lambda=-j$ is very familiar for electrons in beta decay.]

The propagators for an internal φ or χ line are given in Ref. 1 as

$$S_{\sigma\sigma'}(x-y)=-i\Pi_{\sigma\sigma'}(-i\partial)\Delta^c(x-y;m), \quad (8.14)$$

$$\bar{S}_{\sigma\sigma'}(x-y)=-i\bar{\Pi}_{\sigma\sigma'}(-i\partial)\Delta^c(x-y;m). \quad (8.15)$$

[Recall that we are now using fields renormalized by a factor m^j, so the factor m^{-2j} in Eq. (5.7) of Ref. 1 is absent here.] We see that the propagators given for $m=0$ by (7.18) and (7.19) are the limits respectively of (8.14) and (8.15) as $m\to 0$. For $m\neq 0$ there is also a "transition propagator" between φ_σ and $\chi_{\sigma'}{}^\dagger$, but it is proportional to m^{2j} and disappears as $m\to 0$.

In contrast, the Feynman rules for $m=0$ could *not* be obtained as the limit as $m\to 0$ of the corresponding rules for $m>0$, if we used one of the field types like $(j/2,j/2)$ which are forbidden by the theorem of Sec. III. For example, it is well known that the propagator for a vector field has a longitudinal part which blows up as m^{-2} for $m\to 0$; this is just our punishment for attempting to use the forbidden $(\frac{1}{2},\frac{1}{2})$ field type for $j=1$ particles of zero mass.[3]

IX. T, C, AND P

Time-reversal (**T**) and space inversion (**P**) are classically defined as transforming a particle of momentum **p** and helicity λ into

$$\mathbf{T}|\mathbf{p},\lambda\rangle\propto|-\mathbf{p},\lambda\rangle, \quad (9.1)$$

$$\mathbf{P}|\mathbf{p},\lambda\rangle\propto|-\mathbf{p},-\lambda\rangle, \quad (9.2)$$

while charge conjugation (**C**) just changes all particles into antiparticles, with no change in **p** and λ. However, in quantum mechanics there appear phases in (9.1) and (9.2), which we shall see are necessarily *momentum-dependent* for massless particles. In order to get these phases right it is necessary first to define the action of **T** and **P** on our standard states $|\lambda\rangle$ of momentum $k=\{0,0,\kappa\}$, and then use the definition (2.31) of $|\mathbf{p},\lambda\rangle$.

We will define "standard phases" $\eta_\lambda(\mathbf{T})$ and $\eta_\lambda(\mathbf{P})$ by

$$\mathbf{T}|\lambda\rangle=\eta_\lambda{}^*(\mathbf{T})U[R_c]|\lambda\rangle, \quad (9.3)$$

$$\mathbf{P}|\lambda\rangle=(-)^{j+\lambda}\eta_\lambda{}^*(\mathbf{P})U[R_c]|-\lambda\rangle, \quad (9.4)$$

where R_c is some fixed but arbitrary rotation such that

$$R_c\{0,0,1\}=\{0,0,-1\}, \quad (9.5)$$

so that $U[R_c]|\lambda\rangle$ is a state of momentum $\{0,0,-\kappa\}$. [The factor $(-)^{j+\lambda}$ is extracted from $\eta_\lambda{}^*(\mathbf{P})$ for convenience later.] In order to calculate the effect of **T** and **P** on $|\mathbf{p},\lambda\rangle$ we need the well-known formulas

$$\mathbf{T}J_i\mathbf{T}^{-1}=-J_i, \quad (9.6)$$

$$\mathbf{T}K_i\mathbf{T}^{-1}=K_i, \quad (9.7)$$

$$\mathbf{P}J_i\mathbf{P}^{-1}=J_i, \quad (9.8)$$

$$\mathbf{P}K_i\mathbf{P}^{-1}=-K_i. \quad (9.9)$$

[It is easy to check that (9.6)–(9.9) are consistent with the commutation relations (2.16)–(2.18), if we recall that **T** is antiunitary.] According to (2.31) and (4.7), the state $|\mathbf{p},\lambda\rangle$ is

$$|\mathbf{p},\lambda\rangle\equiv[\kappa/|\mathbf{p}|]^{1/2}U[R(\hat{p})]\exp[-i\phi(|\mathbf{p}|)K_3]|\lambda\rangle, \quad (9.10)$$

so therefore

$$\mathbf{T}|\mathbf{p},\lambda\rangle=\eta_\lambda{}^*(\mathbf{T})[\kappa/|\mathbf{p}|]^{1/2}U[R(\hat{p})]\times\exp[i\phi(|\mathbf{p}|)K_3]U[R_c]|\lambda\rangle,$$

$$\mathbf{P}|\mathbf{p},\lambda\rangle=(-)^{j+\lambda}\eta_\lambda{}^*(\mathbf{P})[\kappa/|\mathbf{p}|]^{1/2}U[R(\hat{p})]\times\exp[i\phi(|\mathbf{p}|)K_3]U[R_c]|-\lambda\rangle.$$

But

$$U^{-1}[R_c]K_3U[R_c]=-K_3,$$

and thus

$$\mathbf{T}|\mathbf{p},\lambda\rangle=\eta_\lambda{}^*(\mathbf{T})[\kappa/|\mathbf{p}|]^{1/2}U[R(\hat{p})R_c]\times\exp[-i\phi(|\mathbf{p}|)K_3]|\lambda\rangle, \quad (9.11)$$

$$\mathbf{P}|\mathbf{p},\lambda\rangle=(-)^{j+\lambda}\eta_\lambda{}^*(\mathbf{P})[\kappa/|\mathbf{p}|]^{1/2}U[R(\hat{p})R_c]\times\exp[-i\phi(|\mathbf{p}|)K_3]|-\lambda\rangle. \quad (9.12)$$

The rotation $R(\hat{p})R_c$ carries the z axis into the direction of $-\mathbf{p}$, and must therefore be the product of $R(-\hat{p})$

times a rotation of $\Phi(\hat{p})$ degrees about the z axis

$$U[R(\hat{p})R_c]=U[R(-\hat{p})]\exp[i\Phi(\hat{p})J_3]. \quad (9.13)$$

The angle $\Phi(\hat{p})$ depends on how we standardize R_c and $R(\hat{p})$, but we will fortunately not need to calculate it, as it will cancel in the field transformation laws. Using (9.13) in (9.11) and (9.12), and recalling that J_3 commutes with K_3, we have at last

$$\mathbf{T}|\mathbf{p},\lambda\rangle=\eta_\lambda^*(\mathbf{T})\exp[i\lambda\Phi(\hat{p})]|-\mathbf{p},\lambda\rangle, \quad (9.14)$$

$$\mathbf{P}|\mathbf{p},\lambda\rangle=(-)^{j+\lambda}\eta_\lambda^*(\mathbf{P})\exp[-i\lambda\Phi(\hat{p})]|-\mathbf{p},-\lambda\rangle. \quad (9.15)$$

These one-particle transformation equations can be translated immediately into transformation rules for the annihilation operator:

$$\mathbf{T}a(\mathbf{p},\lambda)\mathbf{T}^{-1}=\eta_\lambda(\mathbf{T})\exp[-i\lambda\Phi(\hat{p})]a(-\mathbf{p},\lambda), \quad (9.16)$$

$$\mathbf{P}a(\mathbf{p},\lambda)\mathbf{P}^{-1}=(-)^{j+\lambda}\eta_\lambda(\mathbf{P}) \times\exp[i\lambda\Phi(\hat{p})]a(-\mathbf{p},-\lambda). \quad (9.17)$$

The antiparticle operators will transform similarly, but perhaps with different "standard" phases $\bar{\eta}_\lambda(\mathbf{T})$ and $\bar{\eta}_\lambda(\mathbf{p})$:

$$\mathbf{T}b(\mathbf{p},\lambda)\mathbf{T}^{-1}=\bar{\eta}_\lambda(\mathbf{T})\exp[-i\lambda\Phi(\hat{p})]b(-\mathbf{p},\lambda), \quad (9.18)$$

$$\mathbf{P}b(\mathbf{p},\lambda)\mathbf{P}^{-1}=(-)^{j+\lambda}\bar{\eta}_\lambda(\mathbf{P}) \times\exp[i\lambda\Phi(\hat{p})]b(-\mathbf{p},-\lambda). \quad (9.19)$$

And, of course, $\mathbf{C}$ just changes a's into b's and vice versa.

$$\mathbf{C}a(\mathbf{p},\lambda)\mathbf{C}^{-1}=\eta_\lambda(\mathbf{C})b(\mathbf{p},\lambda), \quad (9.20)$$

$$\mathbf{C}b(\mathbf{p},\lambda)\mathbf{C}^{-1}=\bar{\eta}_\lambda(\mathbf{C})a(\mathbf{p},\lambda). \quad (9.21)$$

The phases $\eta_\lambda(\mathbf{T,C,P})$, $\bar{\eta}_\lambda(\mathbf{T,C,P})$ are partly arbitrary,[9] partly determined by the structure of the Hamiltonian, and partly fixed by the specifically field-theoretic considerations below.

In order to calculate the effect of $\mathbf{T}$, $\mathbf{C}$, and $\mathbf{P}$ on the fields $\varphi_\sigma(x)$ and $\chi_\sigma(x)$, it will be necessary to use the well-known reality property of the rotation matrices

$$D^{(j)}[R]^*=CD^{(j)}[R]C^{-1}, \quad (9.22)$$

where, with the usual phase conventions,

$$C_{\sigma'\sigma}=(-)^{j+\sigma}\delta_{\sigma',-\sigma}=[\exp(i\pi J_2^{(j)})]_{\sigma'\sigma}. \quad (9.23)$$

We shall fix the rotation R_c introduced in Eq. (9.5) as a rotation of 180° about the y axis, such that

$$D^{(j)}[R_c]=C^{-1}=(-)^{2j}C. \quad (9.24)$$

Another needed relation then follows from (9.13).

$$D_{\sigma\lambda}^{(j)}[R(\hat{p})] =(-)^{j+\lambda}\exp[-i\lambda\Phi(\hat{p})]D_{\sigma,-\lambda}^{(j)}[R(-\hat{p})]. \quad (9.25)$$

The effect of $\mathbf{T}$, $\mathbf{C}$, and $\mathbf{P}$ on the fields (5.16) and (5.17) can now be easily determined by using (9.16)–(9.25):

$$\mathbf{T}\varphi_\sigma(x)\mathbf{T}^{-1}=\eta_{-j}(\mathbf{T})\sum_{\sigma'}C_{\sigma\sigma'}\varphi_{\sigma'}(\mathbf{x},-x^0), \quad (9.26)$$

$$\mathbf{T}\chi_\sigma(x)\mathbf{T}^{-1}=\eta_j(\mathbf{T})\sum_{\sigma'}C_{\sigma\sigma'}\chi_{\sigma'}(\mathbf{x},-x^0), \quad (9.27)$$

$$\mathbf{C}\varphi_\sigma(x)C^{-1}=\eta_{-j}(\mathbf{C})\sum_{\sigma'}C_{\sigma\sigma'}^{-1}\chi_{\sigma'}^\dagger(x), \quad (9.28)$$

$$\mathbf{C}\chi_\sigma(x)\mathbf{C}^{-1}=\eta_j(\mathbf{C})(-)^{2j}\sum_{\sigma'}C_{\sigma\sigma'}^{-1}\varphi_{\sigma'}^\dagger(x), \quad (9.29)$$

$$\mathbf{P}\varphi_\sigma(x)\mathbf{P}^{-1}=\eta_{-j}(\mathbf{P})\chi_\sigma(-\mathbf{x},x^0), \quad (9.30)$$

$$\mathbf{P}\chi_\sigma(x)\mathbf{P}^{-1}=\eta_j(\mathbf{P})\varphi_\sigma(-\mathbf{x},x^0). \quad (9.31)$$

In deriving (9.26)–(9.31) it is necessary to fix the antiparticle inversion phases as

$$\bar{\eta}_\lambda(\mathbf{T})=\eta_{-\lambda}^*(\mathbf{T}), \quad (9.32)$$

$$\bar{\eta}_\lambda(\mathbf{C})=\eta_{-\lambda}^*(\mathbf{C}), \quad (9.33)$$

$$\bar{\eta}_\lambda(\mathbf{P})=(-)^{2j}\eta_{-\lambda}^*(\mathbf{P}), \quad (9.34)$$

because any other choice of the $\bar{\eta}_\lambda$ would result in the creation and annihilation parts of the field transforming with different phases, and would therefore destroy the possibility of simple transformation laws.

It is interesting that the transformation rules (9.26)–(9.31) turn out to be identical with those derived in Sec. 6 of Ref. 1 for the case of massive particles, though the derivation has been different in many respects. The same is true of the phase relations (9.32)–(9.34), except that the only correlated particle and antiparticle inversion phases are those of opposite helicity. In particular, (9.34) tells us that a left- or right-handed particle plus a right- or left-handed antiparticle together have intrinsic parity

$$\eta_{-\lambda}(\mathbf{P})\bar{\eta}_\lambda(\mathbf{P})=(-)^{2j}, \quad (9.35)$$

while the intrinsic parity of a massless particle antiparticle pair of the same helicity is not fixed by these general field-theoretic arguments.

If a particle is its own antiparticle[6] then we must set

$$b(\mathbf{p},\lambda)=a(\mathbf{p},\lambda). \quad (9.36)$$

In this special case, the $(j,0)$ and $(0,j)$ fields are related by

$$\chi_\sigma^\dagger(x)=\sum_{\sigma'}C_{\sigma\sigma'}\varphi_{\sigma'}(x), \quad (9.37)$$

$$\varphi_\sigma^\dagger(x)=(-)^{2j}\sum_{\sigma'}C_{\sigma\sigma'}\chi_{\sigma'}(x). \quad (9.38)$$

Also (9.36) requires that the antiparticle inversion phases $\bar{\eta}_\lambda$ be equal to the corresponding η_λ, and therefore (9.32)–(9.34) provide relations between η_λ and $\eta_{-\lambda}$:

$$\eta_\lambda(\mathbf{T})=\eta_{-\lambda}^*(\mathbf{T}), \quad (9.39)$$

$$\eta_\lambda(\mathbf{C})=\eta_{-\lambda}^*(\mathbf{C}), \quad (9.40)$$

$$\eta_\lambda(\mathbf{P})=(-)^{2j}\eta_{-\lambda}^*(\mathbf{P}). \quad (9.41)$$

[9] For a general discussion, see G. Feinberg and S. Weinberg, Nuovo Cimento **14**, 571 (1959).

However, there is still no necessity for any of these phases to be real.

Observe that (9.17) and (9.19)–(9.21) make sense only if both the particle and its antiparticle each exist in both helicity states $\lambda=\pm j$. For a particle not identical with its antiparticle, this is now a part of the assumption of **C** or **P** invariance, whereas in the case of massive particles it followed directly from the Lorentz invariance of the S matrix.

In contrast, **T** conservation leaves open the possibility that the particle exists in only one of the two helicity states, with an antiparticle of the opposite helicity. This is consistent with (9.26) and (9.27), which show that **T** does not mix φ_σ and χ_σ. The same is true of the combined inversion **CP**.

$$\mathbf{CP}\varphi_\sigma(x)\mathbf{P}^{-1}\mathbf{C}^{-1}=\eta_j(\mathbf{C})\eta_{-j}(\mathbf{P})\sum_{\sigma'} C_{\sigma\sigma'}{}^{-1}\varphi_{\sigma'}{}^\dagger(-\mathbf{x},x^0), \quad (9.42)$$

$$\mathbf{CP}\chi_\sigma(x)\mathbf{P}^{-1}\mathbf{C}^{-1}=\eta_{-j}(\mathbf{C})\eta_j(\mathbf{P})\sum_{\sigma'} C_{\sigma\sigma'}{}^{-1}\chi_{\sigma'}{}^\dagger(-\mathbf{x},x^0), \quad (9.43)$$

and of course it is also true of **CPT**.

X. CHIRALITY AND RENORMALIZED MASS

We have not made any distinction, either here or in Ref. 1, between the mass characterizing the free field and the mass of the physical particles. This was purposeful, because it is always possible and preferable to arrange that the unperturbed and the full Hamiltonians have the same spectrum. But there still remains the question: Under what circumstances will the physical particle mass in fact be zero? The classic conditions are gauge invariance or chirality [i.e., "γ_5"] conservation. Gauge invariance is without content for the $(j,0)$ and $(0,j)$ fields discussed in this article, so we are led to consider the implications of chirality conservation. Our work in this section is entirely academic except for $j=\frac{1}{2}$, but even in this familiar case our conclusions are not quite in accord with public opinion.

For definiteness we will understand chirality conservation as invariance under a continuous transformation

$$\varphi_\sigma(x)\rightarrow e^{i\epsilon}\varphi_\sigma(x); \quad \chi_\sigma(x)\rightarrow e^{-i\epsilon}\chi_\sigma(x). \quad (10.1)$$

In the $2(2j+1)$-component formalism[10] we unite the $(j,0)$ and $(0,j)$ fields $\varphi_\sigma(x)$ and $\chi_\sigma(x)$ into a $(j,0)\oplus(0,j)$ field $\psi(x)$:

$$\psi(x)=\begin{bmatrix}\varphi(x)\\ \chi(x)\end{bmatrix} \quad (10.2)$$

[10] See Ref. 1. Many features of this formalism have been worked out independently in unpublished work by D. N. Williams.

and we write the transformation (10.1) as

$$\psi(x)\rightarrow \exp(i\epsilon\gamma_5)\psi(x). \quad (10.3)$$

$$\gamma_5\equiv\begin{bmatrix}1 & 0\\ 0 & -1\end{bmatrix}.$$

There are other possible discrete or continuous chirality transformations, but our discussion will apply equally to all of them.

The question, of whether chirality conservation implies zero physical mass, can be asked on two different levels:

(1) Suppose that H_0 is chosen so the interaction representation fields $\varphi_\sigma(x)$ and/or $\chi_\sigma(x)$ describe free particles of zero mass, and suppose that the interaction density $\mathcal{H}(x)$ is invariant under the transformation (10.1). Is the renormalized mass then zero in each order of perturbation theory?

(2) Suppose that there exists a unitary operator which induces the transformation (10.1) on the Heisenberg representation fields, and which leaves the physical vacuum invariant. Can we then prove anything about the physical mass spectrum?

Our answers to these two questions are (1) yes, and (2) not necessarily. Let us consider perturbation theory first. The bare momentum-space propagator of the φ_σ field is given by (7.23) as

$$S(q)=-i\Pi(q)/(q^2-i\epsilon). \quad (10.4)$$

The exact propagator is

$$S'(q)=S(q)+S(q)\Sigma^{(*)}(q)S'(q)=[S^{-1}(q)-\Sigma^{(*)}(q)]^{-1}. \quad (10.5)$$

The $(2j+1)\times(2j+1)$ matrix $\Sigma^{(*)}(q)$ is the sum of all proper diagrams with one φ_σ line coming in and one going out, with no propagators on these lines. Stripping away its external propagators changes the Lorentz transformation behavior of $\Sigma_{\sigma\sigma'}{}^{(*)}$ from that of $\varphi_\sigma\varphi_{\sigma'}{}^*$ to that of $\chi_\sigma\chi_{\sigma'}{}^*$, so Lorentz invariance dictates its form as

$$\Sigma_{\sigma\sigma'}{}^{(*)}(q)=i\bar{\Pi}_{\sigma\sigma'}(q)F(-q^2). \quad (10.6)$$

Using (6.8) now gives the exact propagator (10.5) as

$$S'(q)=\frac{-i\Pi(q)}{[1-(-q^2)^jF(-q^2)][q^2-i\epsilon]}. \quad (10.7)$$

We have not used chirality yet. In general the self-energy part $\Sigma^{(*)}(q)$, and hence the function $F(-q^2)$, may have a pole at $q^2=0$, due to graphs with one intermediate χ_σ line. But under any form of chirality conservation such graphs are forbidden. (For example, there is no neutrino χ_σ field.) Hence $F(-q^2)$ has no pole at $q^2=0$, and therefore $S'(q)$ does have such a pole, corresponding to a particle of zero renormalized mass.

Of course there may also be another particle with nonzero mass m given by

$$1=m^{2j}F(m^2).$$

But such a particle would have to be unstable so m would lie off the physical sheet.

Now let us turn to the second question. We assume that there exists a unitary chirality operator $X(\epsilon)$ which transforms the Heisenberg representation fields into

$$X(\epsilon)\varphi_\sigma{}^H(x)X^{-1}(\epsilon)=e^{i\epsilon}\varphi_\sigma{}^H(x)\,, \qquad (10.8)$$

$$X(\epsilon)\chi_\sigma{}^H(x)X^{-1}(\epsilon)=e^{-i\epsilon}\chi_\sigma{}^H(x)\,, \qquad (10.9)$$

and which leaves the physical vacuum invariant. It is certain that this assumption alone is not sufficient, in itself, to allow us to prove anything about physical particle masses, because we have not yet said anything to connect the fields $\varphi_\sigma(x)$ and $\chi_\sigma(x)$ with each other. For instance, we might choose $\varphi_\sigma(x)$ as $(1+\gamma_5)/2$ times the electron field, and $\chi_\sigma(x)$ as $(1-\gamma_5)/2$ times the muon field. Then (10.8) and (10.9) are obviously satisfied if we choose the chirality operator as

$$X(\epsilon)=\exp\{i\epsilon\,[\text{electron number} - \text{muon number}]\}\,. \qquad (10.10)$$

But we can hardly conclude from this that the electron or muon is massless.

Clearly, the only information that can be gleaned solely from the existence of $X(\epsilon)$ is just what would follow from any ordinary additive conservation law. Namely, the propagator of $\varphi_\sigma(x)$ or $\chi_\sigma(x)$ can receive no contribution from any *massive purely neutral* one-particle state that has no degeneracy beyond the $(2j+1)$-fold degeneracy associated with its spin.[11] For any such state $|\mathbf{p},\mu\rangle$ would have to be a chirality eigenstate

$$X(\epsilon)|\mathbf{p},\mu\rangle=e^{i\xi\epsilon}|\mathbf{p},\mu\rangle \quad (\mu=-j,\cdots,j)\,, \qquad (10.11)$$

and thus

$$\langle 0|\varphi_\sigma{}^H(x)|\mathbf{p},\mu\rangle=0 \quad \text{unless} \quad \xi=1\,, \qquad (10.12)$$

$$\langle 0|\varphi_\sigma{}^{H\dagger}(x)|\mathbf{p},\mu\rangle=0 \quad \text{unless} \quad \xi=-1\,. \qquad (10.13)$$

But **CP** or **CPT** conservation tells us that these two matrix elements are proportional to each other, and hence must both vanish. [Observe that we cannot forbid a massless purely neutral particle from contributing to the propagator of $\varphi_\sigma(x)$ or $\chi_\sigma(x)$, since **CP** and **CPT** reverse its helicity, and its two helicity states might have opposite chirality. This is consistent with the remark[6] that it is only a matter of convention whether we call a massless particle purely neutral or not.]

[11] This is an abbreviated version of a proof given by B. Touschek, in *Lectures on Field Theory and the Many-Body Problem*, edited by E. R. Caianiello (Academic Press Inc., New York, 1961), p. 173. It is not clear from Touschek's article whether he feels that this theorem implies that the neutrino cannot have finite mass. As indicated herein, I do not.

Unfortunately this theorem offers no proof that the accepted chirality-conserving weak interactions do not give a massive neutrino, with a distinct massive antineutrino. It should be kept in mind that we cannot decide just by looking at a Lagrangian whether the physical one-particle states will be purely neutral or not. Of course, any massless particle can be called purely neutral, but this is not relevant if what we want is to prove the absence of massive particles.

We can say somewhat more about the mass spectrum if we are willing to assume parity conservation [which links $\varphi_\sigma(x)$ with $\chi_\sigma(x)$ by (9.30) and (9.31)] as well as chirality conservation. In this case the propagator of $\varphi_\sigma(x)$ or $\chi_\sigma(x)$ can receive no contribution from any massive one-particle state that has no degeneracy, beyond the $(2j+1)$-fold degeneracy associated with its spin, and an additional 2-fold degeneracy if it happens to have a distinct antiparticle. For it would then be possible to form a one-particle chirality eigenstate $|\mathbf{p},\mu\rangle$:

$$X(\epsilon)|\mathbf{p},\mu\rangle=\exp(i\epsilon\xi)|\mathbf{p},\mu\rangle \qquad (10.14)$$

by taking $|\mathbf{p},\mu\rangle$ as either the one-particle state itself or some linear combination of it and its charge conjugate. Lorentz invariance requires that

$$\langle 0|\varphi_\sigma{}^H(x)|\mathbf{p},\mu\rangle=N_\varphi(2\omega)^{-1/2}D_{\sigma\mu}{}^{(j)}[L(\mathbf{p})]e^{ip\cdot x}\,, \qquad (10.15)$$

$$\langle 0|\chi_\sigma{}^H(x)|\mathbf{p},\mu\rangle=N_\chi(2\omega)^{-1/2}\bar{D}_{\sigma\mu}{}^{(j)}[L(\mathbf{p})]e^{ip\cdot x}\,. \qquad (10.16)$$

Parity conservation tells us further that

$$|N_\varphi|=|N_\chi|\equiv N\,. \qquad (10.17)$$

This is just to say that the matrix element of the $2(2j+1)$-component field $\psi(x)$ satisfies the generalized Dirac equation [Eq. (7.19) of Ref. 1], which is to be expected under the assumption of parity conservation. But (10.8) and (10.14) give $N=0$ unless $\xi=+1$, while (10.9) and (10.14) give $N=0$ unless $\xi=-1$, so we may conclude that $N=0$. Again, this proof does not apply for zero mass, because the two helicity states are unconnected by space rotations and hence may have different ξ's.

[It might at first sight appear that the free fields constructed in Ref. 1 provide a counter-example to this proof. In the absence of interactions they certainly describe nondegenerate particles with nonvanishing bare and physical masses, and yet there is no coupling that violates either parity or chirality. The trouble with this argument is that no operator $X(\epsilon)$ can be constructed; in fact Eqs. (7.23) and (7.25) of Ref. 1 show that

$$\langle T\{\varphi_\sigma(x),\chi_{\sigma'}{}^\dagger(y)\}\rangle_0\neq 0\,. \qquad (10.18)$$

This point is more transparent in the conventional language in which we would just say that the free-field Lagrangian does not conserve chirality. As $m\to 0$, (10.18) vanishes as m^{2j}, and for $m=0$ it is easy to construct $X(\epsilon)$ explicitly.]

The last proof is of some interest, because it shows that unless the vacuum or electron is degenerate, the mass of the electron cannot arise entirely from electromagnetic interactions, which conserve both parity and chirality. But it is useless for the neutrino, and we are forced to conclude that only perturbation theory can account for its zero mass.

XI. CONCLUSIONS

The Feynman rules for massless particles in the $(2j+1)$-component formalism are identical with those derived in Ref. 1 for particles of mass $m>0$. It is only necessary to pass to the limit $m\rightarrow 0$ to obtain the correct propagators for internal lines, and wave functions for external lines. Also, the various possible invariant Hamiltonians $\mathcal{H}(x)$ can be constructed out of the fields $\varphi_\sigma(x)$ and $\chi_\sigma(x)$, with no distinction between massive and massless particle fields.

Furthermore, the transformation properties of $\varphi_\sigma(x)$ and $\chi_\sigma(x)$ under **T**, **C**, and **P** are the same for $m>0$ and $m=0$. If **P** and/or **C** are conserved it is very convenient to unite $\varphi_\sigma(x)$ and $\chi_\sigma(x)$ into a $2(2j+1)$-component field $\psi(x)$, which transforms according to the reducible $(j,0)\oplus(0,j)$ representation; for $j=\frac{1}{2}$ this yields the Dirac formalism, while for $j=1$ it corresponds to the union of the irreducible fields $\mathbf{E}\pm i\mathbf{B}$ into a six-vector $\{\mathbf{E},\mathbf{B}\}$. Here again there is no distinction to be made between zero and nonzero mass, so we need not repeat here the details of the $2(2j+1)$-component formalism[10] constructed in Ref. 1.

We have seen no hint of anything like gauge invariance in our work so far. In fact, the really significant distinctions between field theories for zero and nonzero mass arise when we try to go beyond the $(2j+1)$- or $2(2j+1)$-component formalisms. In particular, for $m>0$ there is no difficulty in constructing tensor fields transforming according to the $(j/2,j/2)$ representations, while for $m=0$ this is strictly forbidden by the theorem proven in Sec. III. We will see in a forthcoming article that the attempt to evade this prohibition and yet keep the S matrix Lorentz-invariant yields all the results usually associated with gauge invariance.

PHYSICAL REVIEW VOLUME 134, NUMBER 4B 25 MAY 1964

Possible Effects of Strong Interactions in Feinberg-Pais Theory of Weak Interactions. II

N. P. CHANG*
Institute for Advanced Study, Princeton, New Jersey

AND

H. S. MANI†
Physics Department, Columbia University, New York, New York
(Received 23 December 1963)

In a previous paper, a simplified model was used to study the effects of strong interactions on the weak interaction theory of Feinberg and Pais. In this paper, we use a more general argument, a power count based upon the Ward-Takahashi-Nishijima multimeson vertex function identity, to show that the same conclusion remains valid even when crossed ladder graphs are included. Our conclusion may not apply, however, to the modified program of peratization where $W-W$ scattering plays an essential role.

I. INTRODUCTION

IN a previous paper,[1] the possible effects of strong interactions on the peratization theory of Feinberg and Pais[2] were studied in a simplified model where the strong interactions acted through modifications only of the baryon vertices and propagators. It was shown there that the final "peratized" nuclear vector β-decay coupling strength $G_\beta{}^v$ is no longer equal to the "peratized" μ-decay coupling strength, G_μ if the vector current is conserved. In this paper, we wish to present an argument which shows that the same power counting conclusion holds when all possible effects of strong interactions, within the framework of peratization theory, are taken into account. Furthermore, the very nature of our argument shows that the same conclusion holds even when one includes, in peratization theory, the sum over the crossed ladder graphs so long as power counting is valid. That is to say, if we define the peratized (crossed+uncrossed ladder graphs) μ-decay constant by $G_\mu=(g^2/m^2)(1-\eta)$, then the corresponding peratized nuclear vector β-decay constant is $G_\beta{}^v=(g^2/m^2)(1-Z\eta)$, where Z is the strong interaction nucleon renormalization factor. Thus, unless peratization vanishes ($\eta=0$) when all graphs are included, the situation remains that $G_\mu\neq G_\beta{}^v$ when the vector current is conserved. This makes it hard to understand the

* Supported in part by a grant from the National Science Foundation.
† Boese predoctoral fellow.
[1] N. P. Chang, Phys. Rev. **133**, B454 (1964).
[2] G. Feinberg and A. Pais, Phys. Rev. **131**, 2724 (1963); **133**, 477 (1964).

PHYSICAL REVIEW VOLUME 135, NUMBER 4B 24 AUGUST 1964

Photons and Gravitons in S-Matrix Theory: Derivation of Charge Conservation and Equality of Gravitational and Inertial Mass*

STEVEN WEINBERG†
Physics Department, University of California, Berkeley, California
(Received 13 April 1964)

We give a purely S-matrix-theoretic proof of the conservation of charge (defined by the strength of soft photon interactions) and the equality of gravitational and inertial mass. Our only assumptions are the Lorentz invariance and pole structure of the S matrix, and the zero mass and spins 1 and 2 of the photon and graviton. We also prove that Lorentz invariance alone requires the S matrix for emission of a massless particle of arbitrary integer spin to satisfy a "mass-shell gauge invariance" condition, and we explain why there are no macroscopic fields corresponding to particles of spin 3 or higher.

I. INTRODUCTION

It is not yet clear whether field theory will continue to play a role in particle physics, or whether it will ultimately be supplanted by a pure S-matrix theory. However, most physicists would probably agree that the place of local fields is nowhere so secure as in the theory of photons and gravitons, whose properties seem indissolubly linked with the space-time concepts of gauge invariance (of the second kind) and/or Einstein's equivalence principle.

The purpose of this article is to bring into question the need for field theory in understanding electromagnetism and gravitation. We shall show that there are no general properties of photons and gravitons, which *have* been explained by field theory, which cannot also be understood as consequences of the Lorentz invariance and pole structure of the S matrix for massless particles of spin 1 or 2.[1] We will also show why there can be no macroscopic fields whose quanta carry spin 3 or higher.

What are the special properties of the photon or graviton S matrix, which might be supposed to reflect specifically field-theoretic assumptions? Of course, the usual version of gauge invariance and the equivalence principle cannot even be stated, much less proved, in terms of the S matrix alone. (We decline to turn on external fields.) But there are two striking properties of the S matrix which *seem* to require the assumption of gauge invariance and the equivalence principle:

(1) The S matrix for emission of a photon or graviton can be written as the product of a polarization "vector" ϵ^μ or "tensor" $\epsilon^\mu\epsilon^\nu$ with a covariant vector or tensor amplitude, and it vanishes if any ϵ^μ is replaced by the photon or graviton momentum q^μ.

(2) Charge, defined dynamically by the strength of soft-photon interactions, is additively conserved in all reactions. Gravitational mass, defined by the strength of soft graviton interactions, is equal to inertial mass for all nonrelativistic particles (and is twice the total energy for relativistic or massless particles).

Property (1) is actually a straightforward consequence of the well-known[2,3] Lorentz transformation properties of massless particle states, and is proven in Sec. II for massless particles of arbitrary integer spin. (It has already been proven for photons by D. Zwanziger.[4])

Property (2) does not at first sight appear to be derivable from property (1). Even in field theory (1) does not prove that the photon and graviton "currents" $J_\mu(x)$ and $\theta_{\mu\nu}(x)$ are conserved, but only that their matrix elements are conserved for light-like momentum transfer, so we cannot use the usual argument that $\int d^3x J^0(x)$ and $\int d^3x\theta^{0\mu}(x)$ are time-independent. And in pure S-matrix theory it is not even possible to define what we mean by the operators $J^\mu(x)$ and $\theta^{\mu\nu}(x)$.

We overcome these obstacles by a trick, which replaces the operator calculus of field theory with a little simple polology. After defining charge and gravitational mass as soft photon and graviton coupling constants in Sec. III, we prove in Sec. IV that if a reaction violates charge conservation, then the same process with inner bremsstrahlung of a soft extra photon would have an S matrix which does not satisfy property (1), and hence would not be Lorentz invariant; similarly, the inner bremstrahlung of a soft graviton would violate Lorentz invariance if any particle taking part in the reaction has an anomalous ratio of gravitational to inertial mass.

Appendices A, B, and C are devoted to some technical problems: (A) the transformation properties of polarization vectors, (B) the construction of tensor amplitudes for massless particles of general integer spin, and (C) the presence of kinematic singularities in the conventional $(2j+1)$-component "M functions."

A word may be needed about our use of S-matrix theory for particles of zero mass. We do not know whether it will ever be possible to formulate S-matrix

* Research supported by the U. S. Air Force Office of Scientific Research, Grant No. AF-AFOSR-232-63.
† Alfred P. Sloan Foundation Fellow.
[1] Some of the material of this article was discussed briefly in a recent letter [S. Weinberg Phys. Letters 9, 357 (1964)]. We will repeat a few points here, in order that the present article be completely self-contained.
[2] E. P. Wigner, in *Theoretical Physics* (International Atomic Energy Agency, Vienna, 1963), p. 59. We have repeated Wigner's work in Ref. 3.
[3] S. Weinberg, Phys. Rev. 134, B882 (1964).
[4] D. Zwanziger, Phys. Rev. 113, B1036 (1964). Zwanziger omits some straightforward details, which are presented here in Appendix B.

Reprinted from *Phys. Rev.* **135**, B1049 (1964).

theory as a complete dynamical theory even for strong interactions alone, and the presence of massless particles will certainly add a formidable technical difficulty, since every pole sits at the beginning of an infinite number of branch cuts. All such "infrared" problems are outside the scope of the present work. We shall simply make believe that there does exist an S-matrix theory, and that one of its consequences is that the S matrix has the same poles that it has in perturbation theory, with residues that factor in the same way as in perturbation theory. (We will lapse into the language of Feynman diagrams when we do our 2π bookkeeping in Sec. IV, but the reader will recognize in this the effects of our childhood training, rather than any essential dependence on field theory.)

When we refer to the "photon" or the "graviton" in this article, we assume no properties beyond their zero mass and spin 1 or 2. We will not attempt to explain why there should exist such massless particles, but may guess from perturbation theory that zero mass has a special kind of dynamical self-consistency for spins 1 and 2, which it would not have for spin 0.

Most of our work in the present article has a counterpart in Feynman-Dyson perturbation theory. In a future paper we will show how the Lorentz invariance of the S matrix forces the coupling of the photon and graviton "potentials" to take the same form as required by gauge invariance and the equivalence principle.

II. TENSOR AMPLITUDES FOR MASSLESS PARTICLES OF INTEGER SPIN

Let us consider a process in which a massless particle is emitted with momentum $\mathbf{q}$ and helicity $\pm j$. We shall call the S-matrix element simply $S_{\pm j}(\mathbf{q},p)$, letting p stand for the momenta and helicities of all other particles participating in the reaction. The Lorentz transformation property of S can be inferred from the well-known transformation law for one-particle states[2]; we find that

$$S_{\pm j}(\mathbf{q},p)=(|\Lambda\mathbf{q}|/|\mathbf{q}|)^{1/2}\times\exp[\pm ij\Theta(\mathbf{q},\Lambda)]S_{\pm j}(\Lambda\mathbf{q},\Lambda p). \quad (2.1)$$

The angle Θ is given in Appendix A as a function of the momentum $\mathbf{q}$ and the Lorentz transformation $\Lambda^\mu{}_\nu$.

We prove in Appendix B that, in consequence of (2.1), it is always possible for integer j to write $S_{\pm j}$ as the scalar product of a "polarization tensor" and what Stapp[5] would call an "M function":

$$S_{\pm j}(\mathbf{q},p)=(2|\mathbf{q}|)^{-1/2}\epsilon_\pm^{\mu_1*}(\mathbf{q})\cdots\times\epsilon_\pm^{\mu_j*}(\mathbf{q})M_{\pm\mu_1\cdots\mu_j}(\mathbf{q},p) \quad (2.2)$$

with M a symmetric tensor,[6] in the sense that

$$M_\pm^{\mu_1\cdots\mu_j}(\mathbf{q},p)=\Lambda_{\nu_1}{}^{\mu_1}\cdots\Lambda_{\nu_j}{}^{\mu_j}M_\pm^{\nu_1\cdots\nu_j}(\Lambda\mathbf{q},\Lambda p). \quad (2.3)$$

[5] M functions for massive particles were introduced by H. Stapp, Phys. Rev. **125**, 2139 (1962). See also A. O. Barut, I. Muzinich, and D. N. Williams, Phys. Rev. **130**, 442 (1963).

[6] We use a real metric, with signature $\{+++-\}$. Indices are raised and lowered in the usual way. The inverse of the Lorentz transformation $\Lambda^\mu{}_\nu$ is $[\Lambda^{-1}]^\mu{}_\nu=\Lambda_\nu{}^\mu$.

The polarization $\epsilon_\pm^\mu(\hat{q})$ is defined by

$$\epsilon_\pm^\mu(\hat{q})\equiv R(\hat{q})^\mu{}_\nu\epsilon_\pm^\nu, \quad (2.4)$$

where $R(\hat{q})$ is a standard rotation that carries the z axis into the direction of $\mathbf{q}$, and $\epsilon_\pm^\mu$ is the polarization for momentum in the z direction:

$$\epsilon_\pm^\mu\equiv\{1,\pm i,0,0\}/\sqrt{2}. \quad (2.5)$$

Some properties of $\epsilon_\pm^\mu(\hat{q})$ are obvious:

$$\epsilon_{\pm\mu}^*(\hat{q})\epsilon_\pm^\mu(\hat{q})=1, \quad (2.6)$$

$$\epsilon_{\pm\mu}(\hat{q})\epsilon_\pm^\mu(\hat{q})=0, \quad (2.7)$$

$$\epsilon_\pm^{\mu*}(\hat{q})=\epsilon_\mp^\mu(\hat{q}), \quad (2.8)$$

$$\epsilon_\pm^0(\hat{q})=0, \quad (2.9)$$

$$q_\mu\epsilon_\pm^\mu(\hat{q})=0, \quad (2.10)$$

$$\sum_\pm\epsilon_\pm^\mu(\hat{q})\epsilon_\pm^{\nu*}(\hat{q})=\Pi^{\mu\nu}(\hat{q})\equiv g^{\mu\nu}+(\bar{q}^\mu q^\nu+\bar{q}^\nu q^\mu)/|\mathbf{q}|^2, \quad [\bar{q}^\mu\equiv\{-\mathbf{q},|\mathbf{q}|\}], \quad (2.11)$$

$$\sum_\pm\epsilon_\pm^{\mu_1}(\hat{q})\epsilon_\pm^{\mu_2}(\hat{q})\epsilon_\pm^{\nu_1*}(\hat{q})\epsilon_\pm^{\nu_2*}(\hat{q})=\tfrac{1}{2}\{\Pi^{\mu_1\nu_1}(\hat{q})\Pi^{\mu_2\nu_2}(\hat{q})+\Pi^{\mu_1\nu_2}(\hat{q})\Pi^{\mu_2\nu_1}(\hat{q})-\Pi^{\mu_1\mu_2}(\hat{q})\Pi^{\nu_1\nu_2}(\hat{q})\}. \quad (2.12)$$

We also note the very important transformation rule, proved in Appendix A,

$$(\Lambda_\nu{}^\mu-q^\mu\Lambda_\nu{}^0/|\mathbf{q}|)\epsilon_\pm^\nu(\Lambda\hat{q})=\exp\{\pm i\Theta[\mathbf{q},\Lambda]\}\epsilon_\pm^\mu(\hat{q}), \quad (2.13)$$

with Θ the same angle as in (2.1).

If it were not for the q^μ term in (2.13), the polarization "tensor" $\epsilon_\pm^{\mu_1}\cdots\epsilon_\pm^{\mu_j}$ would be a true tensor, and the tensor transformation law (2.3) for $M_\pm^{\mu_1\cdots\mu_j}$ would be sufficient to ensure the correct behavior (2.1) of the S matrix. But $\epsilon_\pm^\mu$ is not a vector,[7] and (2.3) and (2.13) give the S-matrix transformation rule

$$S_{\pm j}(\mathbf{q},p)=(2|\mathbf{q}|)^{-1/2}\exp\{\pm ij\Theta(\mathbf{q},\Lambda)\}\times[\epsilon_\pm^{\mu_1}(\Lambda\hat{q})-(\Lambda q)^{\mu_1}\Lambda_\nu{}^0\epsilon_\pm^\nu(\Lambda\hat{q})/|\mathbf{q}|]^*\cdots\times[\epsilon_\pm^{\mu_j}(\Lambda\hat{q})-(\Lambda q)^{\mu_j}\Lambda_\nu{}^0\epsilon_\pm^\nu(\Lambda\hat{q})/|\mathbf{q}|]^*\times M_{\pm,\mu_1\cdots\mu_j}(\Lambda\mathbf{q},\Lambda p). \quad (2.14)$$

For an infinitesimal Lorentz transformation $\Lambda^\mu{}_\nu=\delta^\mu{}_\nu+\omega^\mu{}_\nu$, we can use (2.2) and the symmetry of M to put (2.14) in the form

$$S_{\pm j}(\mathbf{q},p)=(|\Lambda\mathbf{q}|/|\mathbf{q}|)^{1/2}\exp\{\pm ij\Theta(\mathbf{q},\Lambda)\}S_{\pm j}(\Lambda\mathbf{q},\Lambda p)-j(2|q|^3)^{-1/2}(\omega_\nu{}^0\epsilon_\pm^{\nu*}(\hat{q}))q^{\mu_1}\epsilon_\pm^{\mu_2*}(\hat{q})\cdots\times\epsilon_\pm^{\mu_j*}(\hat{q})M_{\pm,\mu_1\cdots\mu_j}(\mathbf{q},p). \quad (2.15)$$

Hence the necessary and sufficient condition that (2.14) agree with the correct Lorentz transformation property (2.1), is that $S_\pm$ vanish when one of the $\epsilon_\pm^\mu$ is replaced with q^μ:

$$q^{\mu_1}\epsilon_\pm^{\mu_2*}(\hat{q})\cdots\epsilon_\pm^{\mu_j*}(\hat{q})M_{\pm,\mu_1\cdots\mu_j}(\mathbf{q},p)=0. \quad (2.16)$$

For $j=1$ this may be expressed as the conservation

[7] The transformation rule (2.13) shows that $\epsilon_\pm^\mu(\hat{q})$ transforms according to one of the infinite-dimensional representations of the Lorentz group discussed by V. Bargmann and E. P. Wigner, Proc. Natl. Acad. Sci. 34, 211 (1948).

condition

$$q_\mu M_\pm^{\mu}(\mathbf{q},p)=0. \tag{2.17}$$

For $j=2$ we conclude that

$$q_\mu M_\pm^{\mu\nu}(\mathbf{q},p)\propto q^\nu. \tag{2.18}$$

However, (2.7) shows that the subtraction of a term proportional to $g^{\mu\nu}$ from $M_\pm^{\mu\nu}$ will not alter the S matrix (2.2), so $M_\pm^{\mu\nu}$ can always be defined in such a way that (2.18) becomes

$$q_\mu M_\pm^{\mu\nu}(\mathbf{q},p)=0. \tag{2.19}$$

The condition (2.16) may look empty, since it can always be satisfied by a suitable adjustment of $M_\pm^{0\mu_2\cdots\mu_j}$, which in light of (2.9) will have no effect on the S matrix. But we cannot play with the time-like components of $M_\pm^{\mu_1\cdots\mu_j}$ and still keep it a tensor in the sense of (2.3). Neither (2.3) nor (2.16) is alone sufficient for Lorentz invariance, and together they constitute a nontrivial condition on $M_\pm^{\mu_1\cdots\mu_j}$.

Condition (2.16) may, if we wish, be described as "mass-shell gauge invariance," because it implies that the S matrix is invariant under a regauging of the polarization vector

$$\epsilon_\pm^{\mu}(\hat{q})\rightarrow\epsilon_\pm^{\mu}(\hat{q})+\lambda_\pm(\mathbf{q})q^\mu, \tag{2.20}$$

with $\lambda_\pm(\mathbf{q})$ arbitrary. It was purely for convenience that we started with the "Coulomb gauge" in (2.4), (2.5). [However, the theorem in Sec. III of Ref. 3 shows that it is *not* possible to construct an $\epsilon_\pm^{\mu}(\hat{q})$ which would satisfy (2.13) without any q^μ term.]

The S matrix for emission and absorption of several massless particles can be treated in the same way, except that $\epsilon^{\mu *}$ is replaced by ϵ^μ when a massless particle is absorbed.

III. DYNAMIC DEFINITION OF CHARGE AND GRAVITATIONAL MASS

We are going to define the charge and gravitational mass of a particle as its coupling constants to very-low-energy photons and gravitons, with "coupling constant" understood in the same sense as the Watson-Lepore pion-nucleon coupling constant. In general, such definitions are based on the fact that the S matrix has poles, corresponding to Feynman diagrams in which a virtual particle is exchanged between two sets of A and B of incoming and outgoing particles, with four-momentum nearly on its mass shell. The residue at the pole factors into Γ_A and Γ_B, the two "vertex amplitudes" Γ_A and Γ_B depending respectively only upon the quantum numbers of the particles in sets A and B, and of the exchanged particle. Hence it is possible to give a purely S-matrix-theoretic definition of the vertex amplitude Γ for any set of physical particles, as a function of their momenta and helicities; the coupling constant or constants define the magnitude of Γ. (As discussed in the introduction, we will not be concerned in this article with whether the above remarks can be proven rigorously in S-matrix theories involving massless particles, or with the related question of whether $m=0$ poles can really be separated from the branch cuts on which they lie. Our purpose is to explore the implications of the generally accepted ideas about the pole structure.)

Let us first consider the vertex amplitude for a very-low-energy massless particle of integer helicity $\pm j$, emitted by a particle of spin $J=0$, mass m (perhaps zero), and momentum $p^\mu=\{\mathbf{p},E\}$, with $E=(\mathbf{p}^2+m^2)^{1/2}$. (We are restricting ourselves here to very soft photons and gravitons, because we only want to define the charge and gravitational mass, and not the other electromagnetic and gravitational multipole moments.) The only tensor which can be used to form $M_\pm^{\mu_1\cdots\mu_j}$ is $p^{\mu_1}\cdots p^{\mu_j}$ [note that terms involving $g^{\mu\mu'}$ do not contribute to the S matrix, because of (2.7)] so the tensor character of $M_\pm^{\mu_1\cdots\mu_j}$ dictates the form of the vertex amplitude as

$$p_{\mu_1}\cdots p_{\mu_j}\epsilon_\pm^{\mu_1 *}(\hat{q})\cdots\epsilon_\pm^{\mu_j *}(\hat{q})/2E(\mathbf{p})(2|\mathbf{q}|)^{1/2}. \tag{3.1}$$

If the emitting particle has spin $J>0$, with initial and final helicities σ and σ' then (3.1) still gives a tensor M function if we multiply it by $\delta_{\sigma\sigma'}$; this is because the unit matrix has the Lorentz transformation property

$$\delta_{\sigma\sigma'}\rightarrow D_{\sigma\sigma''}^{(J)}(\mathbf{p},\Lambda)D_{\sigma'\sigma'''}^{(J)*}(\mathbf{p},\Lambda)\delta_{\sigma''\sigma'''}=\delta_{\sigma\sigma'}, \tag{3.2}$$

where $D^{(J)}(\mathbf{p},\Lambda)$ is the unitary spin-J representation of the Wigner rotation[8] (or its analog,[2] if $m=0$) associated with momentum $\mathbf{p}$ and Lorentz transformation Λ. However, the vertex amplitude so obtained is not unique. For instance if $J=\frac{1}{2}$ and $m>0$ then we get (3.1) times $\delta_{\sigma\sigma'}$ if we use a "current"[9]

$$\bar{\psi}\{\gamma_{\mu_1}p_{\mu_2}\cdots p_{\mu_j}+\text{permutations}\}\psi, \tag{3.3}$$

while using $\gamma_5\gamma_\mu$ in place of γ_μ would give a helicity-flip vertex amplitude.

At the end of the next section we will see that these other possibilities are prohibited by the Lorentz invariance of the total S matrix. Indeed, the only allowed vertex functions for soft massless particles of spin j are of the form (3.1) times $\delta_{\sigma\sigma'}$ for $j=1$ and $j=2$ (and none at all for $j\geqq 3$). We may therefore define the soft photon coupling constant e, by the statement that the $j=1$ vertex amplitude is[10]

$$\frac{2ie(2\pi)^4\delta_{\sigma\sigma'}p_\mu\epsilon_\pm^{\mu *}(\hat{q})}{(2\pi)^{9/2}[2E(\mathbf{p})](2|\mathbf{q}|)^{1/2}}, \tag{3.4}$$

[8] E. P. Wigner, Ann. Math. **40**, 149 (1939). For a review, see S. Weinberg, Phys. Rev. **133**, B1318 (1964).

[9] For $j=2$, see I. Y. Kobsarev and L. B. Okun, Dubna (unpublished).

[10] Proper Lorentz invariance alone would allow different charges $e_\pm$ for photon helicities ± 1. Parity conservation would normally require that $e_+=e_-$ (with an appropriate convention for the photon parity). However if space inversion takes some particle into its antiparticle then its "right charge" e_+ will be equal to the "left charge" $\bar{e}_-$ of its antiparticle, and we will see in the next section that this gives $e_+=\bar{e}_-=-e_-$. In this case we speak of a magnetic monopole rather than a charge. The same conclusions can be drawn from CP conservation. We will not consider magnetic monopoles in this paper, though in fact none of our work in Sec. IV will depend on any relation between e_+ and e_-. Time-reversal invariance allows us to take e as real.

the factors 2, i, and π being separated from e in obedience to convention. And in the same way we may define a "gravitational charge" f, by the statement that the $j=2$ vertex amplitude is[11]

$$\frac{2if(8\pi G)^{1/2}(2\pi)^4\delta_{\sigma\sigma'}(p_\mu\epsilon_\pm{}^{\mu*}(\hat{q}))^2}{(2\pi)^{9/2}[2E(\mathbf{p})](2|\mathbf{q}|)^{1/2}}, \quad (3.5)$$

the extra factor $(8\pi G)^{1/2}$ (where G is Newton's constant) being inserted to make f dimensionless.

In order to see how e and f are related to the usual charge and gravitational mss, let us consider the near forward scattering of two particles with masses m_a and m_b, spins J_a and J_b, photon coupling constants e_a and e_b, and graviton coupling constants f_a and f_b. As the invariant momentum transfer $t=-(p_a-p_a')^2$ goes to zero, the S matrix becomes dominated by its one-photon-exchange and one-graviton-exchange poles. An elementary calculation[12] using (2.11) and (2.12) shows that for $t\to 0$, the S matrix becomes

$$\frac{\delta_{\sigma_a\sigma_a'}\delta_{\sigma_b\sigma_b'}}{4\pi^2E_aE_bt}[e_ae_b(p_a\cdot p_b) +8\pi Gf_af_b\{(p_a\cdot p_b)^2-m_a^2m_b^2/2\}]. \quad (3.6)$$

If particle b is at rest, this gives

$$\frac{\delta_{\sigma_a\sigma_a'}\delta_{\sigma_b\sigma_b'}}{\pi t}\left[-\frac{e_ae_b}{4\pi}+Gf_a\left\{2E_a-\frac{m_a^2}{E_a}\right\}f_bm_b\right]. \quad (3.7)$$

Hence we may identify e_a as the *charge* of particle a, while its effective *gravitational mass* is

$$\tilde{m}_a=f_a\{2E_a-(m_a^2/E_a)\}. \quad (3.8)$$

If particle a is nonrelativistic, then $E_a\cong m_a$, and (3.8) gives its gravitational rest mass as

$$\tilde{m}_a=f_am_a. \quad (3.9)$$

On the other hand, if a is massless or extremely relativistic, then $E_a\gg m_a$ and (3.8) gives

$$\tilde{m}_a=2f_aE_a. \quad (3.10)$$

[Formulas (3.8) or (3.10) should not of course be understood to mean anything more than already stated in (3.7). However, they serve to remind us that the response of a massless particle to a static gravitational field is finite, and proportional to f.]

The presence of massless particles in the initial or final state will also generate poles in the S matrix, which, like that in (3.7), lie on the edge of the physical region. It is therefore possible to measure the coupling constants e and f in a variety of process, such as Thomson scattering or soft bremsstrahlung, or their analogs for gravitons. All these different experiments will give the same value for any given particle's e or f, for purely S-matrix-theoretic reasons. The task before us is to show how the e's and f's are related for different particles.

IV. CONSERVATION OF e AND UNIVERSALITY OF f

Let $S_{\beta\alpha}$ be the S matrix for some reaction $\alpha\to\beta$, the states α and β consisting of various charged and uncharged particles, perhaps including gravitons and photons. The same reaction can also occur with emission of a very soft extra photon or graviton of momentum $\mathbf{q}$ and helicity ±1, or ±2, and we will denote the corresponding S-matrix element as $S_{\beta\alpha}{}^{\pm1}(\mathbf{q})$ or $S_{\beta\alpha}{}^{\pm2}(\mathbf{q})$.

These emission matrix elements will have poles at $\mathbf{q}=0$, corresponding to the Feynman diagrams in which the extra photon or graviton is emitted by one of the incoming or outgoing particles in states α or β. The poles arise because the virtual particle line connecting the photon or graviton vertex with the rest of the diagram gives a vanishing denominator

$$1/[(p_n+q)^2+m_n^2]=1/2p_n\cdot q \quad \text{(particle } n \text{ outgoing)},$$
$$1/[(p_n-q)^2+m_n^2]=-1/2p_n\cdot q \quad \text{(particle } n \text{ incoming)}. \quad (4.1)$$

For $|\mathbf{q}|$ sufficiently small, these poles will completely dominate the emission-matrix element. The singular factor (4.1) will be multiplied by a factor $-i(2\pi)^{-4}$ associated with the extra internal line, a factor

$$\frac{2ie[p_n\cdot\epsilon_\pm{}^*(\hat{q})](2\pi)^4}{(2\pi)^{3/2}(2|\mathbf{q}|)^{1/2}} \quad (4.2)$$

or

$$\frac{2if(8\pi G)^{1/2}[p_n\cdot\epsilon_\pm{}^*(\hat{q})]^2(2\pi)^4}{(2\pi)^{3/2}(2|\mathbf{q}|)^{1/2}} \quad (4.3)$$

arising from the vertices (3.4) or (3.5), and a factor $S_{\beta\alpha}$ for the rest of the diagram. Hence the S matrix for soft photon or graviton emission is given in the limit

[11] Proper Lorentz invariance alone would not rule out different values for the gravitational charges $f_\pm$ for gravitons of helicity ±2. Parity conservation (with an appropriate convention for the graviton parity) requires that $f_+=f_-$. This conclusion holds even for the magnetic monopole case discussed in footnote 10, since then $f_+=\bar{f}_-$, and we will see in Sec. IV that the antiparticle has "left gravitational charge" $\bar{f}_-$ equal to f_-. The same conclusions can be drawn from CP conservation. Time-reversal invariance allows us to take f as real.

[12] The residue of the pole at $t=0$ can be most easily calculated by adopting a coordinate system in which $q\equiv p_{a'}-p_a=p_b-p_{b'}$ is a finite real light-like four-vector, while p_a, p_b, $p_{a'}$, $p_{b'}$ are on their mass shells, and hence necessarily complex. Then the gradient terms in (2.11) and (2.12) do not contribute, because $q\cdot p_a=q\cdot p_b=0$, so that $\Pi_{\mu\nu}$ may be replaced by $g_{\mu\nu}$, yielding (3.6). We are justified in using (3.6) in the physical region (where p_a, p_b, $p_{a'}$, $p_{b'}$ are real and q is small, though *not* in the direction of the light cone) because Lorentz invariance tells us that the matrix element depends only upon s and t. Lorentz invariance is actually far from trivial in a perturbation theory based on physical photons and gravitons, since then the Coulomb force and Newtonian attraction must be explicitly introduced into the interaction in order to get the invariant S matrix (3.6). (Such a perturbation theory will be discussed in an article now in preparation.) The Lorentz-invariant extrapolation of (3.6) into the physical region of small t is the analog, in S-matrix theory, of the introduction of the Coulomb and Newton forces in perturbation theory.

$\mathbf{q}\to 0$ by[13–15]

$$S_{\beta\alpha}{}^{\pm 1}(\mathbf{q}) \to (2\pi)^{-3/2}(2|\mathbf{q}|)^{-1/2} \times\left[\sum_n \eta_n e_n \frac{[p_n\cdot\epsilon_\pm{}^*(\hat{q})]}{(p_n\cdot q)}\right] S_{\beta\alpha} \quad (4.4)$$

or

$$S_{\beta\alpha}{}^{\pm 2}(\mathbf{q}) \to (2\pi)^{-3/2}(2|\mathbf{q}|)^{-1/2}(8\pi G)^{1/2} \times\left[\sum_n \eta_n f_n \frac{[p_n\cdot\epsilon_\pm{}^*(q)]^2}{(p_n\cdot q)}\right] S_{\beta\alpha}, \quad (4.5)$$

the sign η_n being $+1$ or -1 according to whether particle n is outgoing or incoming.

These emission matrices are of the general form (2.2), i.e.,

$$S_{\beta\alpha}{}^{\pm 1}(\mathbf{q}) \to (2|\mathbf{q}|)^{-1/2}\epsilon_\pm{}^{\mu *}(\hat{q})M_\mu(\mathbf{q},\alpha\to\beta), \quad (4.6)$$

$$S_{\beta\alpha}{}^{\pm 2}(\mathbf{q}) \to (2|\mathbf{q}|)^{-1/2}\epsilon_\pm{}^{\mu *}(\hat{q})\epsilon_\pm{}^{\nu *}(\hat{q})M_{\mu\nu}(\mathbf{q},\alpha\to\beta), \quad (4.7)$$

where M_μ and $M_{\mu\nu}$ are tensor M functions

$$M^\mu(\mathbf{q},\alpha\to\beta) = (2\pi)^{-3/2}[\sum_n \eta_n e_n p_n{}^\mu/(p_n\cdot q)]S_{\beta\alpha}, \quad (4.8)$$

$$M^{\mu\nu}(\mathbf{q},\alpha\to\beta) = (2\pi)^{-3/2}(8\pi G)^{1/2} \times[\sum_n \eta_n f_n p_n{}^\mu p_n{}^\nu/(p_n\cdot q)]S_{\beta\alpha}. \quad (4.9)$$

However, we have learned in Sec. II that the covariance of M_μ and $M_{\mu\nu}$ is not sufficient by itself to guarantee the Lorentz invariance of the S matrix; Lorentz invariance also requires the vanishing of (2.2) when any one $\epsilon_\pm{}^\mu(\hat{q})$ is replaced with q^μ. For photons this implies (2.17), i.e.,

$$0 = q^\mu M_\mu(\mathbf{q},\alpha\to\beta) = (2\pi)^{-3/2}[\sum_n \eta_n e_n]S_{\beta\alpha}, \quad (4.10)$$

so if $S_{\beta\alpha}$ is not to vanish, *the transition* $\alpha\to\beta$ *must conserve charge,* with

$$\sum_n \eta_n e_n = 0. \quad (4.11)$$

For gravitons Lorentz invariance requires (2.18), which here takes the simpler form (2.19)

$$0 = q_\mu M^{\mu\nu}(\mathbf{q},\alpha\to\beta) = (2\pi)^{-3/2}(8\pi G)^{1/2}[\sum_n \eta_n f_n p_n{}^\nu]S_{\beta\alpha}. \quad (4.12)$$

But the $p_n{}^\mu$ are arbitrary four-momenta, subject only to the condition of energy momentum conservation:

$$\sum_n \eta_n p_n{}^\mu = 0. \quad (4.13)$$

The requirement that (4.12) vanish for all such $p_n{}^\mu$, can be met if and only if *all particles have the same gravitational charge.* The conventional definition of Newton's constant G is such as to make the common value of the f_n unity, so

$$f_n = 1 \quad (\text{all } n) \quad (4.14)$$

and (3.8) then tells us that any particle with inertial mass m and energy E has effective gravitational mass

$$\tilde{m} = 2E - m^2/E. \quad (4.15)$$

In particular, a particle at rest has gravitational mass $\tilde{m}$ equal to its inertial mass m.

It seems worth emphasizing that our proof also applies when some particle n in the initial or final state is itself a graviton. Hence the graviton must emit and absorb single soft gravitons (and therefore respond to a uniform gravitational field) with gravitational mass $2E$. It would be conceivable to have a universe in which all f_n vanish, but since we know that soft gravitons interact with matter, they must also interact with gravitons.

Having reached our goal, we may look back, and see that no other vertex amplitudes could have been used for $\mathbf{q}\to 0$ except (3.4) and (3.5). A helicity-flip or helicity-dependent vertex amplitude could never give rise to the cancellations between different poles [as in (4.10) and (4.12)] needed to satisfy the Lorentz invariance conditions (2.17) and (2.19). It is also interesting that such cancellations cannot occur for massless particles of integer spin higher than 2. For suppose we take the vertex amplitude for emission of a soft massless particle of helicity $\pm j$ ($j=3, 4, \cdots$) as

$$\frac{2ig^{(j)}(2\pi)^4(\epsilon_\pm{}^*(\hat{q})\cdot p)^j\delta_{\sigma\sigma'}}{(2\pi)^{9/2}[2E(\mathbf{p})](2|\mathbf{q}|)^{1/2}} \quad (4.16)$$

in analogy with (3.4) and (3.5), the S matrix $S_{\beta\alpha}{}^{\pm j}(\mathbf{q})$ for emission of this particle in a reaction $\alpha\to\beta$ will be given in the limit $\mathbf{q}\to 0$ by

$$S_{\beta\alpha}{}^{\pm j}(\mathbf{q}) \to (2\pi)^{-3/2}(2|\mathbf{q}|)^{-1/2} \times[\sum_n \eta_n g_n{}^{(j)}[p_n\cdot\epsilon_\pm{}^*(\hat{q})]^j/(p_n\cdot q)]S_{\beta\alpha}. \quad (4.17)$$

This is only Lorentz invariant if it vanishes when any one $\epsilon_\pm{}^\mu$ is replaced with q^μ, so we must have

$$\sum_n \eta_n g_n{}^{(j)}[p_n\cdot\epsilon_\pm{}^*(\hat{q})]^{j-1} = 0. \quad (4.18)$$

But there is no way that this can be satisfied for all momenta p_n obeying (4.13), unless $j=1$ or $j=2$. This

[13] Formula (4.4) is well known to hold to all orders in quantum electrodynamic perturbation theory. See, for example, J. M. Jauch and F. Rohrlich, *Theory of Photons and Electrons* (Addison-Wesley Publishing Company, Inc., Reading, Massachusetts, 1955), p. 392, and F. E. Low, Ref. 14.

[14] It has been shown by F. E. Low, Phys. Rev. **110**, 974 (1958), that the next term in an expansion of the S matrix in powers of $|\mathbf{q}|$ is uniquely determined by the electromagnetic multipole moments of the participating particles and by $S_{\beta\alpha}$. However, this next (zeroth-order) term is Lorentz-invariant for any values of the multipole moments.

[15] Relations like (4.4) and (4.5) are also valid if $S_{\beta\alpha}{}^{\pm 1}(\mathbf{q})$, $S_{\beta\alpha}{}^{\pm 2}(\mathbf{q})$, and $S_{\beta\alpha}$ are interpreted as the effective matrix elements for the transition $\alpha\to\beta$, respectively, with or without one extra soft photon or graviton of momentum $\mathbf{q}$, *plus* any number of unobserved soft photons or gravitons with total energy less than some small resolution ΔE. [For a proof in quantum-electrodynamic perturbation theory, see, for example, D. R. Yennie and H. Suura, Phys. Rev. **105**, 1378 (1957). The same is undoubtedly true also for gravitons, and in pure S-matrix theory.]

is not to say that massless particles of spin 3 or higher cannot exist, but only that they cannot interact at zero frequency, and hence cannot generate macroscopic fields. And similarly, the uniqueness of the vertex amplitudes (3.4) and (3.5) does not show that electromagnetism and gravitation conserve parity, but only that parity must be conserved by zero-frequency photons and gravitons.

The crucial point in our proof is that the emission of soft photons or gravitons generates poles which individually make non-Lorentz-invariant contributions to the S matrix. Only the sum of the poles is Lorentz-invariant, and then only if e is conserved and f is universal. Just as the universality of f can be expressed as the equality of gravitational and inertial mass, the conservation of e can be stated as the equality of charge defined dynamically, with a quantum number defined by an additive conservation law. But, however, we state them, these two facts are the outstanding dynamical peculiarities of photons and gravitons, which until now have been proven only under the *a priori* assumption of a gauge-invariant or generally covariant Lagrangian density.

ACKNOWLEDGMENTS

I am very grateful for helpful conversations with N. Cabibbo, E. Leader, S. Mandelstam, H. Stapp, E. Wichmann, and C. Zemach.

APPENDIX A: POLARIZATION VECTORS AND THE LITTLE GROUP

In this Appendix we shall discuss the "little group"[2] for massless particles, with the aim of defining the angle $\Theta(\mathbf{q},\Lambda)$, and of determining the transformation properties of the polarization vectors $\epsilon_\pm(\hat{q})$.

The little group is defined as consisting of all Lorentz transformations $\mathcal{R}^\mu{}_\nu$ which leave invariant a standard light-like four-vector K^μ:

$$\mathcal{R}^\mu{}_\nu K^\nu = K^\mu\,, \tag{A1}$$

$$K^1=K^2=0\,,\quad K^3=K^0=\kappa>0\,. \tag{A2}$$

It is a matter of simple algebra to show that the most general such $\mathcal{R}^\mu{}_\nu$ can be written as a function of three parameters Θ, X^1, X^2:

$$\mathcal{R}^\mu{}_\nu=\begin{bmatrix} \cos\Theta & \sin\Theta & -X_1\cos\Theta-X_2\sin\Theta & X_1\cos\Theta+X_2\sin\Theta \\ -\sin\Theta & \cos\Theta & X_1\sin\Theta-X_2\cos\Theta & -X_1\sin\Theta+X_2\cos\Theta \\ X_1 & X_2 & 1-X^2/2 & X^2/2 \\ X_1 & X_2 & -X^2/2 & 1+X^2/2 \end{bmatrix}. \tag{A3}$$

$$X^2\equiv X_1{}^2+X_2{}^2\,.$$

(The rows and columns are in order 1, 2, 3, 0.) Wigner[2] has noted that this group is isomorphic to the group of rotations (by angle Θ) and translations (by vector $\{X_1,X_2\}$) in the Euclidean plane. In particular the "translations" form an invariant Abelian subgroup, defined by the condition $\Theta=0$, and are represented on the physical Hilbert space by unity. It is possible to factor any $\mathcal{R}^\mu{}_\nu$ into

$$\mathcal{R}^\mu{}_\nu=\begin{bmatrix} \cos\Theta & \sin\Theta & 0 & 0 \\ -\sin\Theta & \cos\Theta & 0 & 0 \\ 0 & 0 & 1 & 0 \\ 0 & 0 & 0 & 1 \end{bmatrix}\begin{bmatrix} 1 & 0 & -X_1 & X_1 \\ 0 & 1 & -X_2 & X_2 \\ X_1 & X_2 & 1-X^2/2 & X^2/2 \\ X_1 & X_2 & -X^2/2 & 1+X^2/2 \end{bmatrix}. \tag{A4}$$

The representation of $\mathcal{R}^\mu{}_\nu$ on physical Hilbert space is determined solely by the first factor, so

$$U[\mathcal{R}]=\exp(i\Theta[\mathcal{R}]J_3)\,. \tag{A5}$$

In discussing the transformation rules for massless particles it is necessary to consider members of the little group defined by

$$\mathcal{R}(\mathbf{q},\Lambda)=\mathcal{L}^{-1}(\mathbf{q})\Lambda^{-1}\mathcal{L}(\Lambda\mathbf{q})\,. \tag{A6}$$

Here Λ is an arbitrary Lorentz transformation, and $\mathcal{L}(\mathbf{q})$ is the Lorentz transformation:

$$\mathcal{L}^\mu{}_\nu(\mathbf{q})=R^\mu{}_\rho(\hat{q})B^\rho{}_\nu(|\mathbf{q}|)\,, \tag{A7}$$

where $B(|\mathbf{q}|)$ is a "boost" along the z axis, with nonzero components

$$\begin{aligned} B^1{}_1&=B^2{}_2=1\,,\\ B^3{}_3&=B^0{}_0=\cosh\varphi\,,\\ B^3{}_0&=B^0{}_3=\sinh\varphi\,,\\ \varphi&\equiv\log(|q|/\kappa)\,, \end{aligned} \tag{A8}$$

and $R(\hat{q})$ is the rotation introduced in (2.4), which takes the z axis into the direction of $\mathbf{q}$. The transformation $\mathcal{L}(\mathbf{q})$ takes the standard four-momentum K^μ [see (A2)] into $q^\mu\equiv\{\mathbf{q},|\mathbf{q}|\}$:

$$\mathcal{L}^\mu{}_\nu(\mathbf{q})K^\nu=q^\mu \tag{A9}$$

so therefore,

$$\begin{aligned}\mathcal{R}^\mu{}_\nu(\mathbf{q},\Lambda)K^\nu&=[\mathcal{L}^{-1}(\mathbf{q})\Lambda^{-1}]^\mu{}_\nu(\Lambda q)^\nu\\ &=[\mathcal{L}^{-1}(\mathbf{q})]^\mu{}_\nu q^\nu=K^\mu\,.\end{aligned} \tag{A10}$$

Hence $\mathcal{R}(\mathbf{q},\Lambda)$ does belong to the little group.

It was shown in Ref. 3 that, as a consequence of (A5), the S matrix obeys the transformation rule (2.1), with $\Theta(\mathbf{q},\Lambda)$ given as the Θ angle of $\mathcal{R}(\mathbf{q},\Lambda)$:

$$\Theta(\mathbf{q},\Lambda)=\Theta[\mathcal{L}^{-1}(\mathbf{q})\Lambda^{-1}\mathcal{L}(\Lambda\mathbf{q})]. \quad \text{(A11)}$$

We now turn to the polarization "vectors" $\epsilon_\pm{}^\mu(\hat{q})$, defined in Sec. II by

$$\epsilon_\pm{}^\mu(\mathbf{q})=R^\mu{}_\nu(\hat{q})\epsilon_\pm{}^\nu, \quad \text{(A12)}$$

$$\epsilon_\pm{}^\mu\equiv\{1,\pm i,0,0\}/\sqrt{2}. \quad \text{(A13)}$$

Observe that we could just as well write (A12) as

$$\epsilon_\pm{}^\mu(q)=\mathcal{L}^\mu{}_\nu(\mathbf{q})\epsilon_\pm{}^\nu \quad \text{(A14)}$$

since $B(|\mathbf{q}|)$ has no effect on $\epsilon_\pm$.

An arbitrary $\mathcal{R}^\mu{}_\nu$ of the form (A3) will transform $\epsilon_\pm{}^\nu$ into

$$\mathcal{R}^\mu{}_\nu\epsilon_\pm{}^\nu=\exp(\pm i\Theta[\mathcal{R}])\epsilon_\pm{}^\mu+X_\pm[\mathcal{R}]K^\mu, \quad \text{(A15)}$$

where

$$X_\pm[\mathcal{R}]=\frac{X_1[\mathcal{R}]\pm iX_2[\mathcal{R}]}{\kappa\sqrt{2}}. \quad \text{(A16)}$$

If we let $\mathcal{R}$ be the transformation (A6), and use (A14), then (A15) gives

$$[\mathcal{L}^{-1}(\mathbf{q})\Lambda^{-1}]^\mu{}_\nu\epsilon_\pm{}^\nu(\Lambda\mathbf{q})=\exp[\pm i\Theta(\mathbf{q},\Lambda)]\epsilon_\pm{}^\mu+X_\pm(\mathbf{q},\Lambda)K^\mu, \quad \text{(A17)}$$

where

$$X_\pm(\mathbf{q},\Lambda)\equiv\frac{X_1[\mathcal{L}^{-1}(\mathbf{q})\Lambda^{-1}\mathcal{L}(\Lambda\mathbf{q})]\pm iX_2[\mathcal{L}^{-1}(\mathbf{q})\Lambda^{-1}\mathcal{L}(\Lambda\mathbf{q})]}{\kappa\sqrt{2}}. \quad \text{(A18)}$$

Multiplying (A17) by $\mathcal{L}(\mathbf{q})$, we have the desired result

$$\Lambda_\nu{}^\mu\epsilon_\pm{}^\nu(\Lambda\mathbf{q})=\exp[\pm i\Theta(\mathbf{q},\Lambda)]\epsilon_\pm{}^\mu(\mathbf{q})+X_\pm(\mathbf{q},\Lambda)q^\mu. \quad \text{(A19)}$$

Note that it is the "translations" which at the same time make the little group non-semi-simple, and which yield the gradient term in (A19).

The quantity $X_\pm(\mathbf{p},\Lambda)$ may be found in terms of $\epsilon_\pm(\mathbf{q})$ by setting $\mu=0$ in (A19):

$$X_\pm(\mathbf{q},\Lambda)|\mathbf{q}|=\Lambda_\nu{}^0e_\pm{}^\nu(\Lambda\mathbf{q}). \quad \text{(A20)}$$

Hence we may rewrite (A19) as a homogeneous transformation rule:

$$(\Lambda_\nu{}^\mu-\Lambda_\nu{}^0q^\mu/|\mathbf{q}|)\epsilon_\pm{}^\nu(\Lambda\mathbf{q})=\exp[\pm i\Theta(\mathbf{q},\Lambda)]\epsilon_\pm{}^\mu(\hat{q}) \quad \text{(A21)}$$

or, recalling that $\epsilon_\pm{}^0\equiv0$,

$$(\Lambda_i{}^j-\Lambda_i{}^0\hat{q}^j)e_\pm{}^i(\Lambda\mathbf{q})=\exp[\pm i\Theta(\mathbf{q},\Lambda)]\epsilon_\pm{}^j(\hat{q}).$$

This also incidentally shows that $\Theta(\mathbf{q},\Lambda)$ does not depend on $|\mathbf{q}|$.

We have not had to define the rotation $R(\hat{q})$ any further than by just specifying that it carries the z axis into the direction of q. However, the reader may wish to see explicit expressions for the polarization vectors, so we will consider one particular standardization of $R(\hat{q})$. Write $\hat{q}$ in the form

$$\hat{q}=\{-\sin\beta\cos\gamma,\ \sin\beta\sin\gamma,\ \cos\beta\} \quad \text{(A22)}$$

and let $R(\hat{q})$ be the rotation with Euler angles $0, \beta, \gamma$:

$$R^\mu{}_\nu(\hat{q})=\begin{bmatrix}\cos\beta\cos\gamma & \sin\gamma & -\sin\beta\cos\gamma & 0\\ -\cos\beta\sin\gamma & \cos\gamma & \sin\beta\sin\gamma & 0\\ \sin\beta & 0 & \cos\beta & 0\\ 0 & 0 & 0 & 1\end{bmatrix}. \quad \text{(A23)}$$

Then (2.4) and (2.5) give

$$\epsilon_\pm{}^\mu(\hat{q})=\{\cos\beta\cos\gamma\pm i\sin\gamma,\ -\cos\beta\sin\gamma\pm i\cos\gamma,\ \sin\beta,\ 0\}/\sqrt{2} \quad (\mu=1,2,3,0). \quad \text{(A24)}$$

We can easily check (2.6)–(2.12) explicitly for (A24).

APPENDIX B: CONSTRUCTION OF TENSOR AMPLITUDES

We consider a reaction in which is emitted a massless particle of momentum $\mathbf{q}$ and integer helicity $\pm j$, all other particle variables being collected in the single symbol p. Let us first divide the set of all possible $\{\mathbf{q},p\}$ into disjoint equivalence classes, $\{\mathbf{q},p\}$ being equivalent to $\{\mathbf{q}',p'\}$ if one can be transformed into the other by a Lorentz transformation. (This is an equivalence relation, because the Lorentz group is a group.) The axiom of choice allows us to make an arbitrary selection of one set of standard values $\{\mathbf{q}_c,p_c\}$ from each equivalence class, so any $\{\mathbf{q},p\}$ determines a unique standard $\{\mathbf{q}_c,p_c\}$, such that for some Lorentz transformation $L^\mu{}_\nu$ we have

$$\mathbf{q}=L\mathbf{q}_c,\quad p=Lp_c. \quad \text{(B1)}$$

It will invariably be the case in physical processes that the only $\Lambda^\mu{}_\nu$ leaving both $\mathbf{q}$ and p invariant is the identity $\delta^\mu{}_\nu$, so the $L^\mu{}_\nu$ in (B1) is uniquely determined by $\mathbf{q}$ and p. (This is true, for instance, if p stands for two or more general four-momenta.) Hence the arguments $\{\mathbf{q},p\}$ stand in one-to-one relation to the variables $\{\mathbf{q}_c,p_c,L\}$.

Now let us construct an $M_\pm{}^{\mu_1\cdots\mu_j}(\mathbf{q}_c,p_c)$ satisfying (2.2) for each standard $\{\mathbf{q}_c,p_c\}$. A suitable choice is

$$M_\pm{}^{\mu_1\cdots\mu_j}(\mathbf{q}_c,p_c)\equiv(2|\mathbf{q}_c|)^{1/2}\epsilon_\pm{}^{\mu_1}(\hat{q}_c)\cdots\epsilon_\pm{}^{\mu_j}(\hat{q}_c)S_{\pm j}(\mathbf{q}_c,p_c), \quad \text{(B2)}$$

which satisfies (2.2) because of (2.6). The tensor amplitude for a general $\mathbf{q}$, p is then defined by

$$M_\pm{}^{\mu_1\cdots\mu_j}(\mathbf{q},p)\equiv L^{\mu_1}{}_{\nu_1}(\mathbf{q},p)\cdots L^{\mu_j}{}_{\nu_j}(\mathbf{q},p)M_\pm{}^{\nu_1\cdots\nu_j}(\mathbf{q}_c,p_c), \quad \text{(B3)}$$

where $\mathbf{q}_c$, p_c, and $L(\mathbf{q},p)$ are the standard variables and Lorentz transformation defined by (B1). With this definition we can easily show that $M_{\pm}{}^{\mu_1\cdots\mu_j}$ is a tensor, because

$$M_{\pm}{}^{\mu_1\cdots\mu_j}(\mathbf{q},p)=L^{\mu_1}{}_{\nu_1}(\mathbf{q},p)\cdots L^{\mu_j}{}_{\nu_j}(\mathbf{q},p)L_{\rho_1}{}^{\nu_1}(\Lambda\mathbf{q},\Lambda p)\cdots L_{\rho_j}{}^{\nu_j}(\Lambda\mathbf{q},\Lambda p)M_{\pm}{}^{\rho_1\cdots\rho_j}(\Lambda\mathbf{q},\Lambda p)$$
$$=\Lambda_{\rho_1}{}^{\mu_1}\cdots\Lambda_{\rho_j}{}^{\mu_j}M_{\pm}{}^{\rho_1\cdots\rho_j}(\Lambda\mathbf{q},\Lambda p), \quad \text{(B4)}$$

the latter equality holding because $L(\mathbf{q},p)L^{-1}(\Lambda\mathbf{q},\Lambda p)$ induces the transformation $\{\Lambda\mathbf{q},\Lambda p\}\rightarrow\{\mathbf{q}_c,p_c\}\rightarrow\{\mathbf{q},p\}$ and hence must be just Λ^{-1}.

We must now show that (B.3) satisfies (2.2) for all $\{\mathbf{q},p\}$. The Lorentz transformation property (2.13) of $\epsilon_{\pm}{}^{\mu}$ can be written as

$$\epsilon_{\pm}{}^{\mu}(\hat{q})=\exp\{\mp i\Theta(\hat{q},L^{-1}(\mathbf{q},p))\}[L^{\mu}{}_{\nu}(\mathbf{q},p)-q^{\mu}L^{0}{}_{\nu}(\mathbf{q},p)/|\mathbf{q}|]\epsilon_{\pm}{}^{\nu}(\hat{q}_c).$$

Hence, (B.3) gives

$$\epsilon_{\pm}{}^{\mu_1 *}(\hat{q})\cdots\epsilon_{\pm}{}^{\mu_j *}(\hat{q})M_{\mu_1\cdots\mu_j}(\hat{q},p)=\exp\{\pm ij\Theta(\hat{q},L^{-1}(\hat{q},p))\}[\epsilon_{\pm}{}^{\mu_1}(\hat{q}_c)-q_c{}^{\mu_1}\epsilon_{\pm}{}^{\nu_1}(\hat{q}_c)L^{0}{}_{\nu_1}(\mathbf{q},p)/|\mathbf{q}|]^*\cdots$$
$$\times[\epsilon_{\pm}{}^{\mu_j}(\hat{q}_c)-q_c{}^{\mu_j}\epsilon_{\pm}{}^{\nu_j}(\hat{q}_c)L^{0}{}_{\nu_j}(\mathbf{q},p)/|\mathbf{q}|]^*M_{\mu_1\cdots\mu_j}(\mathbf{q}_c,p_c). \quad \text{(B5)}$$

But (B2) and (2.10) show that all $q_c{}^{\mu}$ terms may be dropped, because

$$q_c{}^{\mu_r}M_{\mu_1\cdots\mu_j}(\mathbf{q}_c,p_c)=0, \quad \text{(B6)}$$

so (B5) simplifies to

$$\epsilon_{\pm}{}^{\mu_1 *}(\hat{q})\cdots\epsilon_{\pm}{}^{\mu_j *}(\hat{q})M_{\mu_1\cdots\mu_j}(\mathbf{q},p)=\exp\{\pm ij\Theta(\hat{q},L^{-1}(\mathbf{q},p))\}\epsilon_{\pm}{}^{\mu_1 *}(\hat{q}_c)\cdots\epsilon_{\pm}{}^{\mu_j *}(\hat{q}_c)M_{\mu_1\cdots\mu_j}(\mathbf{q}_c,p_c) \quad \text{(B7)}$$

or, using (B2) and (2.6),

$$(2|\mathbf{q}|)^{-1/2}\epsilon_{\pm}{}^{\mu_1 *}(\hat{q})\cdots\epsilon_{\pm}{}^{\mu_j *}(\hat{q})M_{\mu_1\cdots\mu_j}(\mathbf{q},p)=(|\mathbf{q}_c|/|\mathbf{q}|)^{1/2}\exp\{\pm ij\Theta(\hat{q},L^{-1}(\mathbf{q},p))\}S_{\pm j}(q_c,p_c). \quad \text{(B8)}$$

The right-hand side is just the formula for $S_{\pm j}(q,p)$ obtained by setting $\Lambda=L^{-1}(\mathbf{q},p)$ in (2.1), so (B8) gives finally

$$S_{\pm j}(\mathbf{q},p)=(2|\mathbf{q}|)^{-1/2}\epsilon_{\pm}{}^{\mu_1 *}(\hat{q})\cdots$$
$$\times\epsilon_{\pm}{}^{\mu_j *}(\hat{q})M_{\mu_1\cdots\mu_j}(\mathbf{q},p). \quad \text{(B9)}$$

It should be noted that (B2) is *not* valid for all $\mathbf{q}$, p, since then $M_{\pm}{}^{0\mu_2\cdots\mu_j}(\mathbf{q},p)$ would vanish in all Lorentz frames, and $M_{\pm}$ could hardly then be a tensor.

APPENDIX C: $(2j+1)$-COMPONENT M FUNCTIONS

It has become customary[5] to write the S matrix for massive particles of spin j in terms of $2j+1$-component M functions, which transform under the $(j,0)$ or $(0,j)$ representation of the homogeneous Lorentz group. In contrast, the symmetric-tensor M functions used here transform according to the $(j/2, j/2)$ representation. The massless-particle S matrix could also have been written in terms of a conventional $(2j+1)$-component M function, but only at the price of giving the M function a very peculiar pole structure.

To see what sort of peculiarities can occur for zero mass, let us consider the emission of a very soft photon in a reaction like Compton scattering, in which there is only one charged particle in the initial state α and in the final state β. The S-matrix element is then given by (4.4) as

$$S_{\beta\alpha}{}^{\pm1}(\mathbf{q})\rightarrow(2\pi)^{-3/2}(2|\mathbf{q}|)^{-1/2}$$
$$\times e\left[\frac{p_\mu}{(p\cdot q)}-\frac{p'_\mu}{(p'\cdot q)}\right]\epsilon_{\pm}{}^{\mu *}(\hat{q})S_{\beta\alpha}, \quad \text{(C1)}$$

where p and p' are the initial and final charged-particle momenta. This may be rewritten as

$$S_{\beta\alpha}{}^{\pm}(\mathbf{q})\rightarrow(2|\mathbf{q}|)^{-1/2}M_{[\mu,\nu]}(\mathbf{q},\alpha\rightarrow\beta)$$
$$\times\{q^{\nu *}\epsilon_{\pm}{}^{\mu}(\hat{q})-q^{\mu}\epsilon_{\pm}{}^{\nu *}(\hat{q})\}, \quad \text{(C2)}$$

where $M_{[\mu,\nu]}$ is a $(1,0)\oplus(0,1)$ M function

$$M_{[\mu,\nu]}(\mathbf{q},\alpha\rightarrow\beta)=\frac{e[p_\mu p'_\nu-p_\nu p'_\mu]S_{\beta\alpha}}{(2\pi)^{3/2}(p\cdot q)(p'\cdot q)}. \quad \text{(C3)}$$

It can be shown that $S_{\beta\alpha}{}^{+}$ and $S_{\beta\alpha}{}^{-}$ receive contributions, respectively, only from the self-dual and anti-self dual parts of $M_{[\mu,\nu]}$, which transform according to the three-component $(0,1)$ and $(1,0)$ representations. But (C3) shows that *these conventional M functions have a double pole*, arising simultaneously from the incoming and outgoing charged particle propagators. This singularity is partly kinematic, since the S matrix (C1) involves a sum of single poles, but certainly no double pole. The presence of kinematic singularities in $M_{[\mu,\nu]}$ makes it an inappropriate covariant photon amplitude. Similar remarks apply to gravitons, but not to any other massless particles like the neutrino, for which there is no analog to charge.

PHYSICAL REVIEW D VOLUME 26, NUMBER 12 15 DECEMBER 1982

E(2)-like little group for massless particles and neutrino polarization as a consequence of gauge invariance

D. Han
Systems and Applied Sciences Corporation, Riverdale, Maryland 20737

Y. S. Kim and D. Son
Center for Theoretical Physics, Department of Physics and Astronomy, University of Maryland, College Park, Maryland 20742
(Received 18 January 1982)

The content of the isomorphism between the two-dimensional Euclidean group and the E(2)-like little group for massless particles is studied in detail. Representations of the E(2) group are explicitly constructed. The finite-dimensional representations which correspond to physical massless particles are discussed in detail, particularly for the cases of spin 1 and $\frac{1}{2}$. For photons, the little-group transformation matrix is reduced to the transpose of the coordinate transformation matrix in the E(2) plane. In the case of spin-$\frac{1}{2}$ particles, it is shown that the polarization of neutrinos is a consequence of the requirement of gauge invariance.

I. INTRODUCTION

The basic spacetime symmetry for relativistic particles is that of the inhomogeneous Lorentz group.[1] In constructing representations of this group, we usually write down its ten generators, and then construct the Casimir operators which commute with all of the ten generators. When studying symmetry properties of free particles, we specify first the four-momentum of the particle, and then ask which operators commute with the four-momentum operator.

The above-mentioned procedure is known as the method based on "little groups."[1] The little group is defined to be a group whose transformations do not change the four-momentum. The little groups for massive and massless particles are known to be isomorphic to O(3) and E(2), respectively. The study of the isomorphism between the O(3)-like little group for massive particles and the rotation group is straightforward, and has thus been thoroughly examined in the literature. However, for the E(2)-like little group for massless particles, there are still problems which can and should be examined.

The study of massless particles starts from Secs. 6-D and 7-B of Wigner's paper.[1] In these sections, Wigner shows that the generators of the little group for massless particles satisfy the commutation relations for the E(2) group. Wigner notes also that there is an E(2)-like subgroup of SL(2,C) which is isomorphic to the group of Lorentz transformations. He then attempted to construct E(2) eigenfunctions in a two-dimensional space spanned by the parameters of the translational degrees of freedom in the E(2) group. However, for the E(2) representatives diagonal in the helicity operator, Wigner gives only the O(2)-symmetric part in his Eq. (93).[1]

In our previous paper,[2] we gave an E(2) representative which contains Wigner's O(2) function.[3] It is a solution of Laplace's equation in the two-dimensional space spanned by the parameters of the translational degrees of freedom, which, in the case of photons, correspond to gauge transformation parameters.[2,4] However, in Ref. 2, we gave only a general form of the E(2) eigenfunction, and made no attempt to exploit physical implications of the result.

The purpose of the present paper is to study the representations of the E(2) group and those for the little group for massless particles more carefully. We shall show first that the polar-coordinate form given in Ref. 2 is applicable only to massless particles with integer spin, and that the E(2) generators applicable to the coordinate variables should be supplemented by the elements of the E(2)-like subgroup of the SL(2,C) group, as in the case of

Reprinted from *Phys. Rev. D* **26**, 3717 (1982).

the rotation group where the rotation operator consists of the orbital and spin parts.

We shall then show that the SL(2,*C*) part corresponds to spin-$\frac{1}{2}$ massless particles. From our analysis, we conclude that neutrino polarization is a consequence of the requirement of the invariance under the translationlike transformation of the E(2)-like little group, which in the case of photons is a gauge transformation.

In Sec. II, we reorganize Wigner's work[1] on massless particles into a form suitable for studying the content of the isomorphism between the E(2) and E(2)-like little groups. It is emphasized in Sec. III that the case of the O(3)-like little group for massive particles will be helpful in understanding the relation between the little group and E(2). Section IV contains a detailed discussion of the E(2) group. It is pointed out that the E(2) representative given in our previous paper[2] is adequate only for integer-spin particles, and that the E(2)-like subgroup of SL(2,*C*) should be used for spin-$\frac{1}{2}$ massless particles.

In Sec. V, the representation of the E(2)-like little group is studied in detail. The four-by-four little-group transformation matrix is reduced to a form similar to the three-by-three regular representation of the E(2) group. The relationship between these two matrices is worked out in detail. In Sec. VI, the algorithm developed for photons is applied to the case of neutrinos. It is shown that the requirement of gauge invariance leads to the polarization of neutrinos.

In Appendix A, it is shown that the gauge transformation on photon or neutrino wave functions is a transformation within an equivalence class defined by a given rotation angle in the E(2) plane. In Appendix B, the connection between the photon wave function and the E(2) coordinate is discussed in detail. It is pointed out that, although there is a one-to-one correspondence between these two quantities, one cannot be transformed to the other through a linear transformation.

II. E(2)-LIKE LITTLE GROUP

Let us go back to 1939[1] and write down the ten generators of the inhomogeneous Lorentz group

$$P_\mu \text{ and } M_{\mu\nu}\,, \tag{1}$$

which generate translations and Lorentz transformations, respectively. Their commutation relations are well known. The Casimir operators are

$$P^2=P^\mu P_\mu,\quad W^2=W^\mu W_\mu\,, \tag{2}$$

where

$$W^\mu=\tfrac{1}{2}\epsilon_{\mu\nu\alpha\beta}P^\nu M^{\alpha\beta}\,.$$

We are considering in this paper massless particles, and assume without loss of generality that the momentum of a given particle is along the *z* direction. Then

$$p_1=p_2,\quad p_3=-p_0\,, \tag{3}$$

where p_μ are the eigenvalues of the operators P_μ. The generators of the little group which commute with P_μ in this case are

$$N_1=K_1-J_2,\quad N_2=K_2+J_1,\quad J_3\,, \tag{4}$$

where

$$J_i=\tfrac{1}{2}\epsilon_{ijk}M^{jk},\quad K_i=M_{i0}\,.$$

These operators satisfy the commutation relations

$$\begin{aligned}[][J_3,N_1]&=iN_2\,,\\ [J_3,N_2]&=-iN_1\,,\\ [N_1,N_2]&=0\,,\end{aligned} \tag{5}$$

which are like those for the generators of the two-dimensional Euclidean group which is often called the E(2) group.[1] The little group for massless particles is therefore locally isomorphic to the E(2) group.

The study of isomorphism does not stop at the commutation relations. As in the case of the O(3)-like little group for massive particles, the study should include explicit construction of representations, and this construction starts with the choice of commuting operators.

The above generators of the little group commute with N^2, where

$$N^2=N_1{}^2+N_2{}^2\,, \tag{6}$$

and the Casimir operator W^2 of Eq. (2) takes the form

$$W^2=p_0{}^2N^2\,. \tag{7}$$

The set of commuting operators will therefore consist of either

$$P_0,P_3,W^2,N_1,N_2$$

or

$$P_0,P_3,W^2,J_3\,. \tag{8}$$

It is important to keep in mind that the above

operators are still the generators of Lorentz transformations. However, since the commutation relations of Eq. (5) are exactly like those for the generators of the E(2) group, we can learn lessons from this simpler group.[1]

If we use the four-vector convention[1,4]

$$x^{\mu}=(x,y,z,t)\ , \tag{9}$$

then the generators of the coordinate transformation take the form

$$N_1=\begin{bmatrix} 0 & 0 & -i & i \\ 0 & 0 & 0 & 0 \\ i & 0 & 0 & 0 \\ i & 0 & 0 & 0 \end{bmatrix},$$

$$N_2=\begin{bmatrix} 0 & 0 & 0 & 0 \\ 0 & 0 & -i & i \\ 0 & i & 0 & 0 \\ 0 & i & 0 & 0 \end{bmatrix}, \tag{10}$$

$$J_3=\begin{bmatrix} 0 & -i & 0 & 0 \\ i & 0 & 0 & 0 \\ 0 & 0 & 0 & 0 \\ 0 & 0 & 0 & 0 \end{bmatrix}.$$

The above generators lead to the transformation matrices

$$D(u,v,\theta)=D_1(u)D_2(v)D_3(\theta)\ , \tag{11}$$

where

$$D_1(u)=\exp(-iuN_1)\ ,$$

$$D_2(v)=\exp(-ivN_2)\ ,$$

$$D_3(\theta)=\exp(-i\theta J_3)\ .$$

After a straightforward algebra, we can write the D matrix as

$$D(u,v,\theta)=\begin{bmatrix} \cos\theta & -\sin\theta & -u & u \\ \sin\theta & \cos\theta & -v & v \\ u^* & v^* & 1-r^2/2 & r^2/2 \\ u^* & v^* & -r^2/2 & 1+r^2/2 \end{bmatrix}, \tag{12}$$

where $r^2=u^2+v^2$;

$$\begin{bmatrix} u^* \\ v^* \end{bmatrix}=\begin{bmatrix} \cos\theta & \sin\theta \\ -\sin\theta & \cos\theta \end{bmatrix}\begin{bmatrix} u \\ v \end{bmatrix} \tag{13}$$

Similar explicit forms for the big matrix of Eq. (12) are given in the literature.[4,7–10] Compared with the O(3) case, this matrix appears to be complicated, and this probably was the reason why not many authors were encouraged in the past to study the E(2) problem for massless particles. In the following sections, we shall examine whether this matrix can be reduced to a transformation matrix in a two-dimensional Euclidean space.

III. LESSONS FROM THE O(3)-LIKE LITTLE GROUP FOR MASSIVE PARTICLES

There are enough books and papers on the three-dimensional rotation group, and we are quite familiar with the language developed for studying this group. Therefore, the most effective way to study the E(2) group and its isomorphism with the little group for massless particles is to organize the material in a way parallel to the case of the O(3)-like little group for massive particles.

Also for the case of massive particles, the little group consists of four-by-four Lorentz transformation matrices. However, in the Lorentz frame where the particle is at rest, the four-by-four matrix reduces to a three-by-three rotation matrix and a one-by-one unit matrix. From this reduced expression, we can immediately see the content of the isomorphism between the O(3) group and the little group. In the E(2) case, with the form given in Eq. (12), it is not easy to see the correspondence.

The physical quantities associated with the O(3) degrees of freedom are well known. The O(3) group has three parameters. One of them is used for the amount of rotation around a given axis, and two of them are for the orientation of the axis. The direction of the rotation axis is the direction of the spin. All rotations with the same amount of rotation, but with different axis orientations, belong to the same equivalence class. Therefore, the reorientation of the axis, without changing the amount of the rotation, is a transformation within an equivalence class.[5] Is there this kind of reasoning for the E(2)-like little group?

Again for the O(3)-like little group, both O(3) and SU(2) groups are needed. The SU(2) group is needed for specifying particles with half-integer spin, particularly the electron. The O(3) group is needed for the description of orbital motion of quarks inside an extended hadron.[6] In studying the E(2)-like little group, it is reasonable to expect both single- and double-valued representations. Indeed, in view of the recent progress made on understanding the little groups of the Poincaré

TABLE I. Table of the little groups for massive and massless particles. Both the O(3) and E(2)-like little groups are subgroups of the O(3,1) or SL(2,C) groups depending on the spin. The little groups for electrons and hadrons have been studied in Refs. 1 and 6, respectively. The little group for photons has also been studied in Refs. 1–4 and 7–10, but there is enough room for further investigation. The little group for neutrinos is expected to be a subgroup of SL(2,C)

	O(3,1)	SL(2,C)
Massive: $p^2>0$	O(3)-like subgroup of O(3,1): hadrons	SU(2)-like subgroup of SL(2,C): electrons
Massless: $p^2=0$	E(2)-like subgroup of O(3,1): photons	E(2)-like subgroup of SL(2,C): neutrinos

group, we can summarize what has been done and what to expect in Table I. In the following sections, we shall use the above-mentioned parallelism with the familiar O(3) group to exploit the contents of the E(2)-like little group for massless particles.

IV. REPRESENTATIONS OF THE E(2) GROUP

As was pointed out in Sec. II, the four-by-four matrix of Eq. (12) is too complicated to manage, although its algebra is known to be isomorphic to that of the E(2) group. We are therefore led to look for a simpler way to approach the internal degrees of freedom for massless particles, and to study the E(2) group carefully.

For this purpose, let us consider a two-dimensional Cartesian plane with coordinate variable u and v. In Ref. 2, we used the following forms as the generators of the E(2) group:

$$N_1=-i\partial/\partial u\ ,\quad N_2=-i\partial/\partial v\ ,\quad J_3=-i(u\partial/\partial v-v\partial/\partial u)\ . \tag{14}$$

We noted in Ref. 2 that the operator N^2 commutes with all three of the above generators. Thus we have to solve the eigenvalue equation

$$N^2\psi(u,v)=b^2\psi(u,v)\ , \tag{15}$$

where

$$N^2=-[(\partial/\partial u)^2+(\partial/\partial v)^2]\ .$$

Reference 2 contains a detailed discussion of solutions of the above differential equation for both zero and nonzero values of b^2, and the result is summarized in Table II. As is well known, the $b^2=0$ solutions correspond to physical states. If $b^2=0$, the E(2) representative becomes

$$\psi(r,\theta)=[Ar^m+B(1/r)^m]\exp(\pm im\theta)\ . \tag{16}$$

The expression given in Eq. (16) leads us to the temptation to say that m should take either integer or half-integer values,[2] in view of the form given in Eq. (93) of Wigner's paper.[1,11] However, the continuity of the transformation requires that group representatives be analytic. For this reason, we have to write the $b^2=0$ solution as

$$\psi(r,\theta)=r^m\exp(\pm im\theta)\ , \tag{17}$$

only for integer values of m. Then, where are the

TABLE II. Solutions of Laplace's equation in two-dimensional space for E(2) representatives. Physical particles correspond to the finite-dimensional representation diagonal in J_3 with $b^2=0$.

Commuting set of operators	Infinite-dimensional representation	Finite-dimensional representation
N^2,N_1,N_2	$\exp[i(b_1u+b_2v)]$	1, with $b_1=b_2=0$
N^2,J_3	$J_m(br)\exp(\pm im\theta)$	$(Ar^m+Br^{-m})\exp(\pm im\theta)$ with $b^2=0$

E(2) representatives for spin-$\frac{1}{2}$ or half-integer-spin particles?

In order to answer this question, we should look at the case of the O(3) little group for massive particles, in which the SU(2) group is needed in addition to the orbital group, and this SU(2) group is a subgroup of $SL(2,C)$ which is isomorphic to the group of Lorentz transformations. Therefore we are led to ask the question of whether there is an E(2)-like little subgroup in the $SL(2,C)$ group. The answer to this question is definitely "Yes." The E(2)-like subgroup, which was discussed in Wigner's paper,[1] is generated by

$$T_1=\tfrac{1}{2}(i\sigma_1-\sigma_2)\ ,$$
$$T_2=\tfrac{1}{2}(i\sigma_2+\sigma_1)\ , \tag{18}$$
$$S_3=\tfrac{1}{2}\sigma_3\ ,$$

which satisfy the commutation relations for the E(2) group given in Eq. (5). Thus the generators of the E(2)-like group should in general take the form

$$N_1=-i\partial/\partial u+T_1\ ,$$
$$N_2=-i\partial/\partial v+T_2\ , \tag{19}$$
$$J_3=-i(u\partial/\partial v-v\partial/\partial u)+S_3\ .$$

We should note here that J_3 in the above expression is identical to that for the rotation group.

The above generators lead to the following transformation matrices:

$$E_1(u)=\exp(-iuN_1)\ ,$$
$$E_2(v)=\exp(-ivN_2)\ , \tag{20}$$
$$E_3(\theta)=\exp(-i\theta J_3)\ .$$

The most general form of the transformation matrix then is

$$E(u,v,\theta)=E_1(u)E_2(v)E_3(\theta)\ , \tag{21}$$

whose algebraic properties have been discussed in the literature.[1,7,8,10]

For photons, only the "orbital" parts are needed in Eq. (19), and we can still use the function given in Eq. (17) as the state vector. Since $m=1$, $\psi(r,\theta)$ of Eq. (17) is the coordinate variable in the two-dimensional Euclidean space. The E(2) transformation in this case can be achieved through the three-parameter three-by-three matrix for the regular representation of the E(2) group.

The explicit form of this regular representation is given in Sec. V. The similarity between the E(2) geometry and that of the familiar O(3) group is discussed in Appendix A. In Appendix B, we study the vector spaces in which the E(2) transformations are performed.

For neutrinos, only the two-by-two matrix parts are needed for the E(2) generators given in Eq. (19). The representations for neutrinos are discussed in detail in Sec. VI.

V. PHOTONS

If we use the four-vector form for the photon wave function, the little-group transformation matrix is the four-by-four matrix given in Eqs. (11) and (12). The matrix performs both gauge and Lorentz transformations while leaving the four-momentum invariant.[10] Its algebraic properties are isomorphic to those of the E(2) group.

However, the explicit form of Eq. (12) is rather unattractive and does not appear like any of the standard form of E(2) transformation matrices, in contrast to the case of massive particles where the identification of the little group with the O(3) group is straightforward.[1,6] This is the issue we would like to address in this section.

The E(2) representative for photons is given in Eq. (17). The generators of the E(2) group in this case are the differential operators given in Eq. (14) or those in Eq. (19) without the $SL(2,C)$ part. Since m in Eq. (17) is 1 for the photon case, the transformation matrix is a linear coordinate transformation matrix. In this case, the E(2) transformation matrix acting on the column vector $(u_0,v_0,1)$ takes the following matrix form:

$$E(u,v,\theta)=\begin{pmatrix}\cos\theta & -\sin\theta & u\\ \sin\theta & \cos\theta & v\\ 0 & 0 & 1\end{pmatrix}\ . \tag{22}$$

It is easy to show, if not well known, that this matrix is the three-parameter regular representation of the E(2) group. The geometrical properties of this E(2) matrix are discussed in Appendix A. Appendix B contains a discussion of the vector spaces to which this matrix is to be applied.

It is also easy to calculate the inverse of the above form:

$$E^{-1}(u,v,\theta)=\begin{pmatrix}\cos\theta & \sin\theta & -u^*\\ -\sin\theta & \cos\theta & -v^*\\ 0 & 0 & 1\end{pmatrix}\ , \tag{23}$$

where u^* and v^* are given in Eq. (13). This matrix also has vanishing elements in the lower left corner. Unlike the case of the O(3) group, the inverse of the E(2) matrix is not its transpose, and the third row of E^{-1} is identical to that of E.

The problem is whether the D matrix of Eq. (12) can be reduced to a form similar to the E matrix. The first obstacle we have to face is that the D matrix given in Eq. (12) is quadratic while the E matrix of Eq. (22) is linear in the u and v parameters. In order to overcome this obstacle, let us go back to the Lorentz condition on the photon four-vector, which imposes the restriction that the third and fourth components be identical. We note further that, in the expression for the D matrix of Eq. (12), the nontrivial parts of the third and fourth columns have opposite signs, and they vanish when the matrix is applied to the photon four-vector. We can therefore write the D matrix as

$$D(u,v,\theta)=\begin{bmatrix}\cos\theta & -\sin\theta & 0 & 0\\ \sin\theta & \cos\theta & 0 & 0\\ u^* & v^* & 1 & 0\\ u^* & v^* & 0 & 1\end{bmatrix}. \qquad (24)$$

This form is linear in u and v which are gauge transformation parameters,[2,4,10] and is similar to the E matrix of Eq. (20).[12] We can reduce this matrix further by noting that the fourth column and fourth row are redundant. Thus the reduced D matrix can be written as

$$D(u,v,\theta)=\begin{bmatrix}\cos\theta & -\sin\theta & 0\\ \sin\theta & \cos\theta & 0\\ u^* & v^* & 1\end{bmatrix}. \qquad (25)$$

We are using the same notation for the four-by-four matrix of Eq. (24) and for the three-by-three matrix of Eq. (25), but this should not cause any confusion. We can now identify the above expression with the E^{-1} matrix given in Eq. (22):

$$D(u,v,\theta)=[E^{-1}(-u,-v,\theta)]^{\dagger}. \qquad (26)$$

Let us next consider antiparticles. Wigner's original work[1] includes discussions of the little groups for particles with negative energies. If the energy is negative, N_1 and N_2 of Eq. (10) should be replaced by their respective Hermitian conjugates. J_3 is Hermitian. This replacement does not change the E(2) commutation relations given in Eq. (5).

Thus the little group for antiphotons is also isomorphic to the E(2) group. We can then construct the four-by-four D matrix,[8] and then reduce it to a three-by-three form. The result is

$$\bar{D}(u,v,\theta)=\begin{bmatrix}\cos\theta & -\sin\theta & 0\\ \sin\theta & \cos\theta & 0\\ u^* & v^* & 1\end{bmatrix}. \qquad (27)$$

We have used $\bar{D}$, instead of D, to specify that the matrix is for antiphotons. The above expression is identical to the form given in Eq. (25) for photons. Therefore, photons and antiphotons have the same gauge transformation property.

The D matrix is to be applied to photon polarization vectors while the E matrix is for the two-dimensional Cartesian plane. Then, how are these two vector spaces related? As is well known, the translation subgroup of the E(2) group is Abelian and invariant. Therefore E(2) is neither simple nor semisimple. As was pointed out by Racah,[13] groups containing Abelian invariant subgroups are "most troublesome," and require a careful examination. In Appendix B, we shall see how this "trouble" affects the relation between the photon wave function and the E(2) tranformation in the Cartesian plane.

VI. NEUTRINOS

In the case of neutrinos, we should discard the differential operators in Eq. (19), and retain only the SL(2,C) part. If we choose the representation given in Ref. 1 for neutrinos, the E(2) transformation matrix takes the form

$$E(u,v,\theta)=\begin{bmatrix}\exp(-i\theta/2) & (u-iv)\exp(i\theta/2)\\ 0 & \exp(i\theta/2)\end{bmatrix}. \qquad (28)$$

According to the isomorphism established for photons in Eq. (26), we have to obtain the D matrix by

$$D(u,v,\theta)=[E^{-1}(-u,-v,\theta)]^{\dagger}. \qquad (29)$$

Thus

$$D(u,v,\theta)=\begin{bmatrix}\exp(-i\theta/2) & 0\\ (u+iv)\exp(-i\theta/2) & \exp(i\theta/2)\end{bmatrix}. \qquad (30)$$

If this matrix is applied to the spin-up and spin-down states we get

$$D(u,v,\theta)\begin{bmatrix}1\\0\end{bmatrix}=\begin{bmatrix}\exp(-i\theta/2)\\(u+iv)\exp(i\theta/2)\end{bmatrix}, \qquad (31)$$

$$D(u,v,\theta)\begin{bmatrix}0\\1\end{bmatrix}=\begin{bmatrix}0\\ \exp(i\theta/2)\end{bmatrix}. \qquad (32)$$

The spin-down state remains invariant under the u and v transformations, while the spin-up state undergoes spin flips.

In the case of photons, the parameters u and v generate gauge transformations.[2,4,10] Then, according to Eq. (31), the gauge transformation changes the spin orientation if the spin is parallel to the direction of momentum. However, according to Eq. (32), the gauge transformation does not change the spin state if the spin is antiparallel to the momentum.

The above analysis therefore leads us to the conclusion that the spin of the spin-$\frac{1}{2}$ massless particle should be antiparallel to the momentum in order that the spin state be gauge invariant. For the case of spin-$\frac{1}{2}$ antiparticles, we construct the E matrix using the Hermitian conjugates of T_1, T_2, and S_3 given in Eq. (18), and calculate the D matrix using again Eq. (29). Consequently, the spin of the antiparticle has to be parallel to the momentum.

However, it is important to realize that the above conclusions on the directions of neutrino and antineutrino spins depend on the choice of the E(2) representation. The E(2) matrix which was given in Ref. 1 and which we used in the above analysis has a vanishing element in the lower left corner, similar to the regular representation of the E(2) group given in Eq. (22). However, this is not the only E(2) for which can be constructed as a subgroup of SL$(2,C)$.[14] We can also consider

$$E'(u,v,\theta)=[E^{-1}(u,v,\theta)]^{\dagger} . \tag{33}$$

If we use the matrix in Eq. (28) and follow the same reasoning as before, the spins of neutrinos and antineutrinos would be parallel and antiparallel to the momentum, respectively.

It is clear in either case that the polarization of neutrinos is a consequence of the requirement of gauge invariance. Let us translate this conclusion into the familiar language of the Dirac equation. It is easy to construct the N_1 and N_2 operators applicable to the Dirac spinors for massless particles. It then turns out that the Dirac spinors are invariant under the N_1 and N_2 transformations for both polarizations. However, γ_5 commutes with the Hamiltonian, and this allows us to choose a definite eigenvalue of γ_5. If the eigenvalue is -1, neutrinos and antineutrinos are left- and right-handed, respectively.[15,16] If the eigenvalue of γ_5 is $+1$, then the polarizations are opposite to those for the $\gamma_5=-1$ case. Experimentally, the eigenvalue of γ_5 is known to be -1.[17] As is well known, neutrino theory based on a definite eigenvalue of γ_5 is called the "two-component theory of neutrinos." It is interesting to note that this two-component theory is a gauge-invariant theory.

VII. CONCLUDING REMARKS

In this paper, it was noted first that the little-group transformation matrix applicable to the photon four-vector is somewhat complicated. We have reduced this unattractive form into a three-by-three matrix which can be compared with the regular representation of the E(2) group. This reduced form allows us to compare the little-group parameters for massless particles with those in the well-known O(3)-like little group for massive particles.

The explicit construction of the isomorphism between the little group and E(2) allows us to study internal space-time parameters for neutrinos. We have shown in this paper that the polarization of neutrinos is a consequence of the requirement of invariance under the translationlike transformation of the little group of the Poincaré group which, in the case of photons, is a gauge transformation.

As is well known, the subject of neutrino polarization has a stormy history. It was Weyl who first proposed the two-component theory of neutrinos, but this suggestion was rejected by Pauli on the grounds that the theory does not preserve parity invariance.[18] Since 1956,[19] we have understood neutrino polarization as a manifestation of parity violation. The time has come for us to ask what space-time invariance principle is responsible for the polarization of neutrinos. In this paper, we have provided an answer to this question.

ACKNOWLEDGMENT

We would like to thank Professor George A. Snow for very helpful criticisms and stimulating discussions.

APPENDIX A

In this appendix, we discuss the similarity between the E(2) and O(3) groups using the concept of equivalence class.[5]

Both the O(3) and E(2) groups are three-parameter groups. We can obtain the E(2) group from O(3) through a process of group contraction.[20] The group contraction process goes as follows. Every rotation can be regarded as a rotation around a given axis.[5] The orientation of this rotation axis can be specified by two angular variables. This can also be achieved through the coordinate specification on a spherical surface. The reorientation of the rotation axis in this case can be specified by a movement of a point on the spherical surface. If the radius of this sphere becomes sufficiently large, and if the reorientation is sufficiently localized, the axis reorientation would appear like a motion of a point on a flat surface. As is described in the literature,[20] the translation on the E(2) plane is the limiting case of the axis reorientation in the O(3) group.

The amount of rotation around a given axis is an independent quantity. All rotations with the same amount of rotation, but not necessarily around the same axis, belong to the same "equivalence class."[5] The traces of transformation matrices belonging to the same equivalence class are known to be the same. Likewise, for photons and neutrinos, it is easy to see from the expression given in this paper that the trace of the transformation matrix is independent of the u and v variable which only change the location of the rotation axis on the E(2) plane. Since u and v are the gauge transformation parameters, the gauge transformation is a transformation within the same equivalence class.

Let us translate what we said above into formulas. As is well known, every rotation matrix R can be brought to the form

$$R=A\exp(-i\alpha J_z)A^{-1}\,, \tag{A1}$$

where α is the rotation angle, and A is the two-parameter matrix which brings the rotation axis to the desired direction from the z axis. The trace R is independent of the parameters of the A matrix.

In the E(2) case, it is always possible to write

$$E(u,v,\theta)=E(\xi,\eta,0)E(0,0,\theta)E^{-1}(\xi,\eta,0)\,. \tag{A2}$$

$E(0,0,\theta)$ is a rotation around the origin. The $E(\xi,\eta,0)$ in the above expression moves the rotation axis from the origin to (ξ,η). The trace of the above matrix depends only on the rotation parameter. The translation of the axis to (ξ,η) is a transformation within the same equivalence class.

APPENDIX B

We study in this appendix the vector spaces to which the D and E matrices are to be applied. The effect of the E matrix given in Eq. (22) on the column vector $(u_0,v_0,1)$ is well known. However, an interesting case here is the action of E^{-1} in view of Eq. (26):

$$E^{-1}(-u,-v,\theta)=E(u^*,v^*,-\theta)\,. \tag{B1}$$

In order to see the effect of the above matrix on the column vector $(u_0,v_0,1)$ let us carry out explicitly the following matrix multiplication:

$$\begin{pmatrix} u' \\ v' \\ 1 \end{pmatrix}=\begin{pmatrix} \cos\theta & \sin\theta & u^* \\ -\sin\theta & \cos\theta & v^* \\ 0 & 0 & 1 \end{pmatrix}\begin{pmatrix} 1 & 0 & u_0 \\ 0 & 1 & v_0 \\ 0 & 0 & 1 \end{pmatrix}\begin{pmatrix} 0 \\ 0 \\ 1 \end{pmatrix}. \tag{B2}$$

From this matrix algebra, u' and v' can be written out as

$$\begin{aligned} u'&=(u_0+u)\cos\theta+(v_0+v)\sin\theta\,, \\ v'&=-(u_0+u)\sin\theta+(v_0+v)\cos\theta\,. \end{aligned} \tag{B3}$$

The above E(2) geometry is easy to understand, and does not require any further explanation.

If we insist on doing the same matrix algebra using the D matrix, applicable to the photon polarization vector,

$$\begin{pmatrix} e^{\mp i\theta} \\ \pm ie^{\mp i\theta} \\ u'\pm iv' \end{pmatrix}=\begin{pmatrix} 1 & 0 & 0 \\ 0 & 1 & 0 \\ u_0 & v_0 & 1 \end{pmatrix}\begin{pmatrix} \cos\theta & -\sin\theta & 0 \\ \sin\theta & \cos\theta & 0 \\ u^* & v^* & 1 \end{pmatrix}\begin{pmatrix} 1 \\ \pm i \\ 0 \end{pmatrix}. \tag{B4}$$

For convenience, we shall hereafter call the column vectors, to which the E and D matrices are applicable, the E and D vectors, respectively. The geometry of the E vector is well known. Its third component is trivial. Its first two components specify the coordinate position in the two-dimensional space spanned by the u and v variables.

The physics of the D vector is well known. Its first two components specify the photon polarization state, and its third component is parallel to the direction of the momentum and is an unmeasurable gauge parameter. The question then is how the D vector is related to the geometry of the E vector.

The matrix algebras of Eqs. (B2) and (B4) allow us to see the correspondence between the D and E

vectors. The matrix algebra for D given in Eq. (B3) is simply the Hermitian conjugate of the algebra of Eq. (B4) for the E matrix. However, it is not possible to construct a similarity transformation matrix which will bring the E matrix to its Hermitian conjugate, and this is one of the complications in groups containing Abelian invariant subgroups.[13] Therefore, the D vector is not a linear transformation of the E vector, although there is a one-to-one correspondence between them.

---

[1]E. P. Wigner, Ann. Math. 40, 149 (1939).

[2]D. Han, Y. S. Kim, and D. Son, Phys. Rev. D 25, 461 (1982).

[3]The fact that the compact O(2) group is only a subgroup of the full noncompact E(2) group was pointed out by Gottlieb. See H. P. W. Gottlieb, Proc. R. Soc. London A368, 429 (1979).

[4]S. Weinberg, Phys. Rev. 134, B882 (1964); 135, B1049 (1964).

[5]E. P. Wigner, *Group Theory and Atomic Spectra*, translated by J. J. Griffin (Academic, New York, 1959).

[6]Y. S. Kim, M. E. Noz, and S. H. Oh, J. Math Phys. 20, 1341 (1979); D. Han and Y. S. Kim, Am. J. Phys. 49, 1157 (1981); D. Han, M. E. Noz, Y. S. Kim, and D. Son, Phys. Rev. D 25, 1740 (1982).

[7]E. P. Wigner, Z. Phys. 124, 665 (1948).

[8]F. R. Halpern, *Special Relativity and Quantum Mechanics* (Prentice-Hall, Englewood Cliffs, New Jersey, 1968).

[9]J. Kupersztych, Nuovo Cimento 31B, 1 (1976).

[10]D. Han and Y. S. Kim, Am. J. Phys. 49, 348 (1981).

[11]What we did in Ref. 2 was to replace Wigner's Eq. (93) (Ref. 1), which takes the form $\exp(\pm im\theta)$ for both integer and half-integer values of m, by Eq. (16) of the present paper. Our statement in Ref. 2 about the integer values of m is correct. However, what we said there about the half-integer values is incorrect

[12]The D matrix of Eq. (12) performs both Lorentz and gauge transformations. Once reduced to the form of Eq. (24), D is no longer a Lorentz transformation matrix. It only performs gauge transformations.

[13]G. Racah, CERN Report No. 1961-8 (unpublished).

[14]M. Flato and P. Hillion, Phys. Rev. D 1, 1667 (1970).

[15]T. D. Lee and C. N. Yang, Phys. Rev. 105, 1671 (1957); A. Salam, Nuovo Cimento 5, 299 (1957); L. D. Landau, Nucl. Phys. 3, 127 (1957).

[16]For a density matrix formulation of this problem , see A. S. Wightman, in *Dispersion Relations and Elementary Particles*, edited by C. DeWitt and R. Omnes (Hermann, Paris, 1960).

[17]M. Goldhaber, L. Grodzins, and A. W. Sunyar, Phys. Rev. 109, 1015 (1958).

[18]W. Pauli, Ann. Inst. Henri Poincaré 6, 137 (1936).

[19]The violation of parity invariance was discovered in 1956. See C. S. Wu, E. Ambler, R. W. Hayward, D. D. Hoppes, and R. P. Hudson, Phys. Rev. 105, 1413 (1957).

[20]R. Gilmore, *Lie Groups and Lie Algebras, and Some of Their Applications* (Wiley, New York, 1974).

Chapter VII

Group Contractions

While the earth is like a sphere, its surface is like a flat plane in a reasonably confined area. This raises the question of whether the symmetry governing the sphere can be made that of a flat plane in certain limits. Indeed, in 1953, Inonu and Wigner formulated this problem as the contraction of the three-dimensional rotation group into the two- dimensional Euclidean group. In this case, the limiting parameter is the radius of the sphere. When the radius becomes very large, the area element on the surface becomes flat.

Since the little groups of massive and massless particles are locally isomorphic to the three-dimensional rotation group and the two-dimensional Euclidean group respectively, it is natural to suspect that the little group for massless particles is a limiting case of that of massive particles. Then, what is the limiting parameter. This has been shown to be the momentum/mass. The 1984 paper of Han, Kim, Noz, and Son explains why the momentum/mass acts as the radius of the sphere in the limiting process.

While the above-mentioned procedures constitute applications of the group contraction, it is possible to achieve the same purpose by obtaining the little group for massless particles from kinematical considerations. In 1986, Han, Kim, and Son observed that the little group transformation is the transformation which does not change the momentum, and constructed a set of non-colinear transformations whose net effect is to leave the momentum invariant. This matrix is analytic in the $(\text{mass})^2$ variable in the neighborhood of $(\text{mass})^2 = 0$. Therefore, it is possible to obtain the zero-mass limit using the explicit expression for the little group transformation matrix. The role of Wigner's little groups is summarized in Figure 3.

	Massive Slow	between	Massless Fast
Energy Momentum	$E = \frac{p^2}{2m}$	Einstein's $E = \sqrt{m^2 + p^2}$	$E = p$
Spin, Gauge Helicity	S_3 S_1 S_2	Wigner's Little Group	S_3 Gauge Trans.

FIG. 3. Physical implications of Wigner's little groups. Einstein's $E = mc^2$ unifies the energy-momentum relation for massive and slow particles and that for massless particles. Wigner's little group unifies the internal space-time symmetries of massive and massless particles.

ON THE CONTRACTION OF GROUPS AND THEIR REPRESENTATIONS

By E. Inonu and E. P. Wigner

Palmer Physical Laboratory, Princeton University

Communicated April 21, 1953

Introduction. Classical mechanics is a limiting case of relativistic mechanics. Hence the group of the former, the Galilei group, must be in some sense a limiting case of the relativistic mechanics' group, the representations of the former must be limiting cases of the latter's representations. There are other examples for similar relations between groups. Thus, the inhomogeneous Lorentz group must be, in the same sense, a limiting case of the de Sitter groups. The purpose of the present note is to investigate, in some generality, in which sense groups can be limiting cases of other groups (Section I), and how their representations can be obtained from the representations of the groups of which they appear as limits (Section II). Section III deals briefly with the transition from inhomogeneous Lorentz group to Galilei group. It shows in which way the representation up to a factor of the Galilei group, embodied in

Reprinted from *Proc. Nat. Acad. Sci. (U.S.A.)* 39, 510 (1953).

the Schrödinger equation, appears as a limit of a representation of the inhomogeneous Lorentz group and also gives the reason why no physical interpretation is possible for the real representations of that group.

I. Contraction of Groups

Let us consider an arbitrary Lie group with n parameters a^i and infinitesimal operators I_i. These shall be given, as usual by

$$I_i = \lim_{h \to 0} \frac{g(he_i) - g(0)}{h} \tag{1}$$

where 0 are the parameters of the unit element and e_i differs from 0 by a unit increase of a_i. The I_j are skew hermitean matrices if the group consists of unitary matrices; instead of them one often uses the hermitean quantities iI_j. However, all our equations remain somewhat simpler if expressed in terms of the I_j. The structure constants C are defined by

$$[I_i, I_j] = \sum_k C_{ij}^k I_k . \tag{2}$$

If we subject the I_i to a linear homogeneous non-singular transformation, the C will be replaced by other constants. These are obtained from the C by contragradient transformations of its upper and lower indices. However, such a transformation has, naturally, no effect on the structure of the group. Let us denote the transformation in question by

$$J_\nu = \sum_i I_i U^i{}_\nu. \tag{3}$$

It corresponds to the transformation

$$a^i = \sum_\kappa U^i{}_\kappa b^\kappa \tag{3a}$$

according to which the J are obtained by the same equation (1) as the I, except that the e_i have to be replaced in it by a similar quantity, defined with respect to the b.

The above transformation may lead to a new group only if the matrix U of (3) is singular. We shall call the operation of obtaining a new group by a singular transformation of the infinitesimal elements of the old group a *contraction* of the latter. The reason for this term will become clear below. The singular matrix will be a limiting case of a non-singular matrix. The latter will depend linearly on a parameter ϵ which will tend to zero:

$$U^i{}_\nu = u^i{}_\nu + \epsilon w^i{}_\nu. \tag{4}$$

For $0 < \epsilon < \epsilon_0$ the determinant of (4) is different from zero, it vanishes for $\epsilon = 0$.

We shall transform (4) into a normal form by a non-singular and ϵ

independent transformation of both the J_ν and I_t. If the matrices of the corresponding transformations are denoted by α and β, respectively, u and w will be replaced by $\beta u \alpha^{-1}$ and $\beta w \alpha^{-1}$. It is possible, by such a transformation, to give u and w the form

$$u = \begin{Vmatrix} 1 & 0 \\ 0 & 0 \end{Vmatrix} \qquad w = \begin{Vmatrix} v & 0 \\ 0 & 1 \end{Vmatrix}. \tag{5}$$

The number of rows and columns in the unit matrix in u, and in v, is equal to the rank r of u. It is advantageous to label the transformed J and I with a pair of indices, the first referring to the subdivision of u given in (5), the second specifying the various J and I within that subdivision. Hence, (3) assumes the form

$$\begin{aligned} J_{1\nu} &= I_{1\nu} + \epsilon \sum_{\mu=1}^{r} v_{\mu\nu} I_{1\mu} & (\nu = 1, 2, \ldots, r) \\ J_{2\nu} &= \epsilon I_{2\nu}. & (\nu = 1, 2, \ldots, n - r) \end{aligned} \tag{6}$$

The corresponding transformation of the group parameters is

$$\begin{aligned} a_{1\nu} &= b_{1\nu} + \epsilon \sum_{\mu=1}^{r} v_{\nu\mu} b_{1\mu} & (\nu = 1, 2, \ldots, r) \\ a_{2\nu} &= \epsilon b_{2\nu}. & (\nu = 1, 2, \ldots, n - r) \end{aligned} \tag{6a}$$

It is well to remember that the parameters a lead to the infinitesimal elements I, the parameters b to the J. The last equation shows that a given set of parameters b correspond, with decreasing ϵ, to smaller and smaller values of the parameters a_2. In the limit $\epsilon = 0$ (if such a limit exists), one will have contracted the whole group to an infinitesimally small neighborhood of the group defined by the $a_{1\nu}$ alone. This justifies the name given to the process considered.

The transformation of the infinitesimal elements which we carried out also changes the structure constants and we shall write for (2)

$$[I_{\alpha\nu}, I_{\beta\mu}] = \sum_{\kappa=1}^{r} C_{\alpha\nu.\,\beta\mu}{}^{1\kappa} I_{1\kappa} + \sum_{\kappa=1}^{n-r} C_{\alpha\nu.\,\beta\mu}{}^{2\kappa} I_{2\kappa} \tag{7}$$

wherein α and β can assume the values 1 and 2. This gives for

$$\begin{aligned} [J_{1\nu}, J_{1\mu}] &= [I_{1\nu}, I_{1\mu}] + \epsilon \sum (v_{\nu\nu'} \delta_{\mu\mu'} + \delta_{\nu\nu'} v_{\mu\mu'} + \epsilon v_{\nu\nu'} v_{\mu\mu'}) [I_{1\nu'}, I_{1\mu'}] \\ &= \sum_{\kappa} C_{1\nu.\,1\mu}{}^{1\kappa} J_{1\kappa} + \frac{1}{\epsilon} \sum_{\kappa} C_{1\nu.\,1\mu}{}^{2\kappa} J_{2\kappa} + 0(1). \end{aligned}$$

Hence, if the commutator of $J_{1\nu}$ and $J_{1\mu}$ is to converge, as $\epsilon \to 0$, to a linear combination of the J, the structure constants

VOL. 39, 1953 *MATHEMATICS: INONU AND WIGNER* 513

$$C_{1\nu,\,1\mu}{}^{2\kappa} = 0 \tag{8}$$

i.e., the $I_{1\nu}$ must span a subgroup. On the other hand, if this happens to be the case, the structure constants will converge to definite values $c_{\alpha\nu,\,\beta\mu}{}^{\gamma\kappa}$ as $\epsilon \rightarrow 0$

$$\begin{aligned} c_{1\nu,\,1\mu}{}^{1\kappa} &= C_{1\nu,\,1\mu}{}^{1\kappa} & c_{1\nu,\,1\mu}{}^{2\kappa} &= C_{1\nu,\,1\mu}{}^{2\kappa} = 0 \\ c_{1\nu,\,2\mu}{}^{1\kappa} &= 0 & c_{1\nu,\,2\mu}{}^{2\kappa} &= C_{1\nu,\,2\mu}{}^{2\kappa} \\ c_{2\nu,\,2\mu}{}^{1\kappa} &= c_{2\nu,\,2\mu}{}^{2\kappa} = 0. \end{aligned} \tag{9}$$

These structure constants satisfy Jacobi's identities since the structure constants for the J do this for non-vanishing ϵ. We shall say that the above operation is a contraction of the group with respect to the infinitesimal elements $I_{1\nu}$ or that the infinitesimal elements $I_{2\mu}$ are contracted. We then have, from (9).

THEOREM 1. *Every Lie group can be contracted with respect to any of its continuous subgroups and only with respect to these.* The subgroup with respect to which the contraction is undertaken will be called S. *The contracted infinitesimal elements form an abelian invariant subgroup of the contracted group. The subgroup S with respect to which the contraction was undertaken is isomorphic with the factor group of this invariant subgroup. Conversely, the existence of an abelian invariant subgroup and the possibility to choose from each of its cosets an element so that these form a subgroup S, is a necessary condition for the possibility to obtain the group from another group by contraction.*

It is easy to visualize now the effect of the contraction on the whole group. The subgroup S with respect to which the contraction is undertaken remains unchanged and it is advantageous to choose the group parameters in such a way that $a_{2\nu} = 0$ throughout S. Then (6a) can be replaced by

$$a_{1\nu} = b_{1\nu} \qquad a_{2\nu} = \epsilon b_{2\nu} \tag{6b}$$

and this can be assumed to be valid throughout the whole group, not only in the neighborhood of the unit element. As ϵ decreases, a fixed range of the parameter b will describe an increasingly small surrounding of S. As ϵ tends to 0, the range of the $b_{2\nu}$ will become infinite and describe only those group elements which differ infinitesimally from the elements of S. The elements which are in the neighborhood of the unit element of the original group but have finite parameters $b_{2\nu}$ will commute and form the aforementioned commutative invariant subgroup. Naturally, the elements of this invariant subgroup will not commute, in general, with the elements of the subgroup S: the change of the parameters $a_{2\nu} = \epsilon b_{2\nu}$ will be, upon transformation by finite elements of S, of the same order of magnitude as these parameters themselves. Naturally, the convergence of the original group toward the contracted group is a typically non-uniform convergence.

Every Lie group can be contracted with respect to any of its one parametric subgroups. If the three dimensional rotation group is contracted in this way, one obtains the Euclidean group for two dimensions. Contraction of the homogeneous Lorentz group with respect to the subgroup which leaves the time coordinate invariant yields the homogeneous Galilei group, contraction of the inhomogeneous Lorentz group with respect to the group generated by spatial rotations and time displacements yields the full Galilei group. Contraction of the de Sitter groups yields the inhomogeneous Lorentz group. It should be remarked, finally, that if a group, obtained by contraction of another group with respect to the subgroup S, is contracted again with respect to S, the second contraction remains without effect.

The above considerations show a certain similarity with those of I. E. Segal.[1] However, Segal's considerations are more general than ours as he considers a sequence of Lie groups the structure constants of which converge toward the structure constants of a non-isomorphic group. In the above, we have considered only one Lie group but have introduced a sequence of coordinate systems therein and investigated the limiting case of these coordinate systems becoming singular. As a result of our problem being more restricted, we could arrive at more specific results.

II. Contraction of Representations

If one applies the transformation (6) to the infinitesimal elements of a representation of the group to be contracted, and lets ϵ tend to zero, the $J_{2\nu}$ will also tend to zero. The representation will become isomorphic to the representation of the subgroup S, i.e., will be a representation of the factor group of the invariant subgroup. In order to obtain a faithful representation, one must *either subject the $J_{2\nu}$ to an ϵ dependent transformation, or consider the $J_{2\nu}$ which correspond to different representations*, e.g., go to higher and higher dimensional representations as ϵ decreases. We shall give examples for both procedures.

(*a*) *Representations of the Contracted Group by Means of ϵ Dependent Transformations*.—The first procedure is applicable only if the infinitesimal elements are not bounded operators, i.e., as far as irreducible representations are concerned, only if the group is not compact. The simplest non-compact non-commutative group is that of the linear transformations $x' = e^{\alpha}x + \beta$. The general group element is $O_{\alpha,\beta} = T(\beta)R_{\alpha}$ with the group relations $T(\beta)T(\beta') = T(\beta + \beta')$; $R_{\alpha}R_{\alpha'} = R_{\alpha+\alpha'}$ and $R_{\alpha}T(\beta) = T(e^{\alpha}\beta)R_{\alpha}$. The only faithful irreducible unitary representation of this group can be given in the Hilbert space of square integrable functions of $0 < x < \infty$

$$R_\alpha \psi(x) = e^{(1/2)\alpha}\psi(e^\alpha x) \qquad T_\beta \psi(x) = e^{i\beta x}\psi(x). \tag{10}$$

The infinitesimal operators are

$$I_\alpha = I_1 = {}^1/_2 + x\frac{d}{dx} \qquad I_\beta = I_2 = ix \tag{10a}$$

with the structure relation

$$[I_1, I_2] = I_2. \tag{10b}$$

Contraction with respect to the group of transformations $x' = e^{\alpha x}$ leaves the group unchanged: the only non-vanishing structure constant is $C_{12}{}^2$ and (9) shows that this does not change. We can try, therefore, to transform I_1 and ϵI_2 with an ϵ dependent unitary matrix so that I_1 remain unchanged, the transformed ϵI_2 converge to I_2

$$S_\epsilon^{-1} I_1 S_\epsilon = I_1 \qquad \epsilon S_\epsilon^{-1} I_2 S_\epsilon \to I_2. \tag{11}$$

This is indeed possible: one has to choose $S_\epsilon = R_{\ln \epsilon}$. This commutes with I_1. It follows from the group relation that $S_\epsilon^{-1}T(\beta)S_\epsilon = R_{-\ln\epsilon}T(\beta) R_{\ln \epsilon} = T(\beta/\epsilon)$ and one has, hence, as $h \to 0$

$$\begin{aligned}\epsilon S_\epsilon^{-1} I_2 S_\epsilon &= \epsilon S_\epsilon^{-1} \lim h^{-1}(T(h) - 1)S_\epsilon \\ &= \lim \epsilon h^{-1}(T(h/\epsilon) - 1) = I_2\end{aligned} \tag{11a}$$

Hence $\epsilon S_\epsilon^{-1} I_2 S_\epsilon$ not only converges to $J_2 = I_2$ but remains equal to it for all ϵ.

It is more surprising, perhaps, to see that the same device is possible also if one contracts the group with respect to the subgroup of the $T(\beta)$. The contracted group is, in this case, the two parametric abelian group. We demand, in this case, that S_ϵ commute with I_β and hence, by (10a), that it be multiplication with a function of x. Because of S' unitary nature, we can give it the form $\exp(if(x, \epsilon))$. Transformation of the ϵI_α of (10a) with this gives

$$\begin{aligned}\epsilon S_\epsilon^{-1} I_\alpha S_\epsilon &= \epsilon e^{-if(x,\epsilon)}({}^1/_2 + x\,d/dx)e^{if(x,\epsilon)} \\ &= \epsilon({}^1/_2 + x\,d/dx) + \epsilon ix\,d/dx\,f(x, \epsilon).\end{aligned} \tag{12}$$

The first part of this converges to 0 as it should since (12) should converge to an operator which commutes with ix. The second part converges to $J_2 = ixf'(x) = ig(x)$ if one sets

$$f(x, \epsilon) = \epsilon^{-1}f(x). \tag{12a}$$

Hence, the transformations of the contracted group corresponding to the parameters α, β is multiplication with

$$e^{i\alpha g(x) + i\beta x} \tag{13}$$

which is again a faithful (reducible) representation of the contracted group.

The operators corresponding to finite group elements could have been obtained directly by transforming $O_{\alpha,\epsilon\beta} = T(\epsilon\beta)R_\alpha$ with $R_{\ln\epsilon}$ in the case of contraction with respect to R_α. Similarly, (13) could have been obtained directly by transforming $O_{\epsilon\alpha,\beta} = T(\beta)R_{\epsilon\alpha}$ with $\exp(\epsilon^{-1}if(x))$.

It is not clear how generally one can obtain a faithful representation of the contracted group as a limit of an ϵ dependent transform of a faithful representation of the original group and the substitution (6b) of its parameters. Certainly, the procedure is not applicable to irreducible representations of compact groups or, more generally, if the infinitesimal operators are bounded.

(*b*) *Representations of the Contracted Group from a Sequence of Representations.*—We shall now give a few examples for the second procedure, i.e., obtaining a representation of the contracted group by choosing a sequence of unitary representations $D^{(1)}, D^{(2)}, \ldots, D^{(l)}, \ldots$ so that each of the operators

$$I_{1\nu}^{(l)}, \epsilon I_{2\mu}^{(l)} \qquad (\nu = 1, 2, \ldots, r; \ \mu = 1, 2, \ldots, n - r) \qquad (15)$$

converge to a finite operator as $\epsilon \rightarrow 0$ and $l \rightarrow \infty$. Alternately, we can ask that the transformation corresponding to finite group elements

$$D^{(l)}(b_{1\nu}, \epsilon b_{2\mu}) \qquad (15a)$$

converge to a unitary representation of the contracted group as $\epsilon \rightarrow 0$, $l \rightarrow \infty$. The b are the parameters of the contracted group; the operator (15) corresponds, in the representation l, to the group element of the original group the parameters a of which are given by (6b). The formulation making use of the finite group elements (15a) is unambiguous because it deals with the convergence of unitary operators; the first one is usually easier to attack directly.

The convergence of the sequences of (15) and (15a) will depend not only on the values which ϵ assumes and on the corresponding representations $D^{(l)}$, it will also depend in which form that representation is assumed. Hence, method (*a*) can be considered as a special case of the present method in which all $D^{(l)}$ are unitary equivalent.

The contracted group is always an open group because the variability domain of the b is infinite. Hence its representations are, as a rule, infinite dimensional. If the $D^{(l)}$ are finite dimensional, they should be considered to affect only a finite number of the coordinates of Hilbert space.

The simplest non-commutative compact group is the three-dimensional rotation group. All its subgroups are one parametric, we shall contract it with respect to the rotations about the z axis of a rectangular coordinate system of ordinary space. The contracted group is the Euclidean group of the plane, i.e., the inhomogeneous two-dimensional rotation group. We shall choose for $D^{(l)}$ the representation which is usually denoted[2] by $D^{(l)}$: it is $2l + 1$ dimensional (l being any integer) and is usually described in a space the coordinate axes of which are labeled with $m = -l, -l + 1, \ldots, l - 1, l$. Hence, we label the coordinate axes of Hilbert space with all integers m from $-\infty$ to ∞. In keeping with our previous notation, we call the infinitesimal element which corresponds to rotations about the z axis $M_z = I_1$. One can then write for $|m| \leq l$, $|m'| \leq l$

$$(I_1^{(l)})_{mm'} = (M_z^{(l)})_{mm'} = im\delta_{mm'} \tag{16}$$

$$(I_{2x}^{(l)})_{mm'} = (M_x^{(l)})_{mm'} = -\frac{1}{2}\sqrt{(l-m)(l+m')}\,\delta_{m'm+1} + \frac{1}{2}\sqrt{(l-m')(l+m)}\delta_{m'm-1}$$

$$(I_{2y}^{(l)})_{mm'} = (M_y^{(l)})_{mm'} = \frac{i}{2}\sqrt{(l-m)(l+m')}\,\delta_{m'm+1} + \frac{i}{2}\sqrt{(l-m')(l+m)}\,\delta_{m'm-1}.$$

All matrix elements vanish if either $|m|$ or $|m'|$ is larger than l. As $l \rightarrow \infty$, the I_1 converge to a definite operator

$$(J_1)_{mm'} = im\,\delta_{mm'}. \qquad (-\infty < m, m' < \infty) \tag{17a}$$

In fact, the convergence is strong in the sense that $I_1\varphi$ converges strongly to $J_1\varphi$ if φ is in the definition domain of J_1. If ϵ and l converge to zero and infinity, respectively, in such a way that $l\epsilon \rightarrow \Xi$, the other two infinitesimal elements will also converge to J_{2x}, J_{2y}, respectively, where

$$(J_{2x})_{mm'} = \frac{1}{2}\,\Xi(\delta_{m'm-1} - \delta_{m'm+1}) \tag{17b}$$

$$(J_{2y})_{mm'} = \frac{i}{2}\,\Xi\,(\delta_{m'm-1} + \delta_{m'm+1}).$$

It follows from the fact that there are only a finite number (two) non-vanishing matrix elements in both I_{2x} and I_{2y} that the operators satisfy the commutation relations of the contracted group

$$[J_{2x}, J_{2y}] = 0 \qquad [J_1, J_{2x}] = J_{2y} \qquad [J_1, J_{2y}] = -J_{2x}. \tag{18}$$

It further follows from

$$(I_1^{(l)})^2 + (I_{2x}^{(l)})^2 + (I_{2y}^{(l)})^2 = -l(l+1)$$

by multiplication with ϵ^2 and going to the limit in the above way that

$$J_{2x}^2 + J_{2y}^2 = -\Xi^2. \tag{18a}$$

It was important, for obtaining convergent sequences of $I_1^{(l)}$ and $\epsilon I_2^{(l)}$, to have assumed the $D^{(l)}$ in the form given by (16). This form was reduced out with respect to the subgroup S and this caused, in this case, the convergence of the sequence $I_1^{(l)}$. The convergence of the $\epsilon I_2^{(l)}$ does not actually follow from the convergence of the $I_1^{(l)}$, and hence from the reduced out form of the representations $D^{(l)}$, but was made at least possible by this circumstance.

Before going over to the investigation of the Lorentz groups, it may be worth while to make a final remark about the above contraction, even though it has little to do with our subject. We shall determine, first, the matrices which correspond to finite group elements. For this purpose, it is useful to consider that form of the above representation in which the Hilbert space consists of functions of x and y and the infinitesimal operators have the natural form for infinitesimal operators of the Euclidean group (α and r are polar coordinates $x = r\cos\alpha$, $y = r\sin\alpha$)

$$J_1' = -\partial/\partial\alpha = y\partial/\partial x - x\partial/\partial y \tag{19}$$

$$J_{2x}' = \partial/\partial y \qquad J_{2y}' = -\partial/\partial x. \tag{19a}$$

One should keep in mind that J_{2x}' arose from I_{2x} which is the infinitesimal rotation about the x axis and corresponds to a displacement in the $-y$ direction. Similarly J_{2y}' corresponds to a displacement in the x direction.

The function $\varphi(\alpha, r)$ corresponds in the new Hilbert space to the vector which has the components φ_m in the Hilbert space of (17a), (17b). It further follows from (18a) that

$$\left(\frac{\partial^2}{\partial x^2} + \frac{\partial^2}{\partial y^2}\right)\varphi = -\Xi^2\varphi \tag{20}$$

whence one can write

$$\begin{aligned}\varphi(x, y) &= \int e^{-i\Xi(x\cos\alpha' + y\sin\alpha')}g(\alpha')d\alpha' \\ &= \int e^{-i\Xi r\cos(\alpha-\alpha')}g(\alpha')d\alpha' \\ &= \int e^{-i\Xi r\cos\alpha'}g(\alpha-\alpha')d\alpha'.\end{aligned} \tag{20a}$$

All integrations are from 0 to 2π. Expanding $g(\alpha - \alpha')$ into a Fourier series of $\alpha - \alpha'$, one finds

$$\varphi(x, y) = \int e^{-i\Xi r\cos\alpha'}\sum g_m e^{-im(\alpha-\alpha')}d\alpha' \tag{20b}$$

m runs from $-\infty$ to ∞. The last form makes it easy to calculate J_1' by (19), the first form permits one to calculate J_{2x}' and J_{2y}' easily. Comparison of the expressions obtained in this way with (17a) and (17b) shows that $g_m = \varphi_m$. In order to remain in keeping with the usual notation, we define[3]

$$2\pi J_m(z) = i^m \int e^{-iz\cos\alpha'} e^{im\alpha'} d\alpha'.$$

J_m is then the ordinary Bessel function of order m. This permits one to write for (20b)

$$\varphi(\alpha, r) = 2\pi \sum \varphi_m i^{-m} e^{-im\alpha} J_m(\Xi r). \quad (21)$$

Because of (19a), one can write down at once the finite group operations. In particular, for the displacement $T(\xi, \eta)$ by ξ and η in the x and y directions one has

$$T(\xi, \eta)\varphi(x, y) = \varphi(x - \xi, y - \eta). \quad (22)$$

Hence, denoting the matrix for the same operation in the original Hilbert space by $T(\xi, \eta)_{mm'}$, one has

$$\varphi(x - \xi, y - \eta) = 2\pi \sum_m \sum_m T(\xi, \eta)_{mm'} \varphi_{m'} i^{-m} e^{-im\alpha} J_m(\Xi r). \quad (22a)$$

This permits an explicit determination of the $T(\xi, \eta)_{mm'}$. We shall not carry this out completely but set only $r = 0$ in (22a). Since all $J_m(0) = 0$ except $J_0(0) = 1$, the summation over m disappears on the right side. The left side becomes, at the same time by (21)

$$\varphi(-\xi, -\eta) = 2\pi \sum \varphi_m i^{-m} e^{-im(\beta + \pi)} J_m(\Xi\rho) \quad (22b)$$

where β, ρ are the polar coordinates for ξ, η. Comparing (22a) and (22b) one finds, with $\xi = 0$, $\beta = {}^1/_2\pi$, $\eta = \rho$

$$J_m(\Xi\rho) = T(0, \rho)_{0m}. \quad (23)$$

The group relations and the form of the infinitesimal operators (19a) gives at once the most important relations for Bessel functions, such as the addition theorem, differential equation (cf. (20)), etc. Up to this point the argument is not new but merely a repetition, for the two dimensional Euclidean group, of a similar reasoning given before[4] for the rotation group. This led to the equation[4]

$$D^{(l)}(0, \beta 0)_{0m} = \bar{P}^m(\cos\beta) = \left(\frac{(l - m)!}{(l + m)!}\right)^{1/2} P_l^m(\cos\beta) \quad (24)$$

in which P_l^m is Legendre's associated function of the first kind, $\bar{P}_l^m$ is normalized to the same value as $P_l^0 = P_l$. Furthermore, 0, β, 0 is the rotation about the x axis so that, by the definition of $T(0, \rho)$

$$T(0, \rho)_{m'm} = (-)^{m'} \lim D^{(l)} \quad (0, \Xi\rho/l, 0)_{m'm}. \quad (25)$$

This, together with (23), gives the asymptotic expression for the associated Legendre functions[5]

$$\lim_{l \to \infty} \bar{P}_l^m(\cos(\rho/l)) = J_m(\rho). \quad (25a)$$

III. Contraction of Lorentz Groups

Let us consider, first, the inhomogeneous Lorentz group with one space-like, one time-like dimension. It is given by the transformations

$$\begin{aligned} x' &= x \cosh \lambda + t \sinh \lambda + a_x \\ t' &= x \sinh \lambda + t \cosh \lambda + a_t. \end{aligned} \tag{26}$$

We wish to contract it with respect to the subgroup of time displacements $t' = t + a_t$. The infinitesimal elements of (26) are: time displacement I_1, space displacement I_{2x} and "rotation" in space-time $I_{2\lambda}$. Their commutation relations read

$$[I_1, I_{2x}] = 0 \qquad [I_1, I_{2\lambda}] = -I_{2x} \qquad [I_{2x}, I_{2\lambda}] = -I_1. \tag{27}$$

Hence, by (9), the commutation relations of the contracted group are

$$[J_1, J_{2x}] = 0 \qquad [J_1, J_{2\lambda}] = -J_{2x} \qquad [J_{2x}, J_{2\lambda}] = 0. \tag{27a}$$

The "rotations" in space-time, together with the displacements in space, form a commutative invariant subgroup.

The matrices

$$\begin{Vmatrix} \cosh \lambda & \sinh \lambda & a_x \\ \sinh \lambda & \cosh \lambda & a_t \\ 0 & 0 & 1 \end{Vmatrix} \tag{26a}$$

form a natural, though not unitary, representation of the group of transformations (26). We can carry out the contraction by setting $a_t = b_t$, $\lambda = \epsilon v$, $a_x = \epsilon b_x$ or $\lambda = v/c$, $a_x = b_x/c$ and letting ϵ converge to 0, or c converge to infinity. If we do this directly in (26a), the representation will not remain faithful for the contracted group. We shall transform therefore (26a) with a suitable ϵ (or c) dependent matrix: multiply the first row with c, the first column with $1/c$. If c goes to infinity in the matrix obtained in this way, one obtains the transformations of the contracted group

$$\begin{aligned} x' &= x + vt + b_x \\ t' &= \qquad t + b_t. \end{aligned} \tag{27a}$$

It is the inhomogeneous Galilei group with one spatial dimension. The transformations $x' = x + vt + b_x$, $t' = t$ form the commutative invariant subgroup.

The same contraction can be carried out for an inhomogeneous Lorentz group with an arbitrary number of spatial dimensions. The only difference is that the subgroup S, with respect to which the contraction is carried out, contains not only the displacements in time as in the above example, but also all purely spatial rotations, i.e., all homogeneous transformations which leave t invariant. The invariant subgroup of the contracted group consists of all spatial displacements and Galilei transformations: $x_i' = x_i + v_i t + b_i$, $t' = t$.

Contraction of the Unitary Representations of the Lorentz Groups.—We shall be principally concerned here with the group of the special theory of relativity, i.e., the inhomogeneous Lorentz group with three space-like and one time-like dimension. The subgroup S with respect to which we shall contract it contains the displacements in time, the spatial rotations, and the products of these operations. The contracted group is the ordinary Galilei group, i.e., the group of classical mechanics.

We shall denote the displacement operators in the direction of the three space-like axes by I_k ($k = 1, 2, 3$), the displacement in the direction of the time axis by I_0. The rotations in the kl plane will be denoted by I_{kl}, the acceleration in the direction of the k axis by I_{k0}. I_0 and I_{kl} span the subgroup S.

The quantity

$$I_1^2 + I_2^2 + I_3^2 - I_0^2 = P \tag{28}$$

is a constant in every irreducible representation and the irreducible representations can be divided into three classes according to the value of this constant.[6] In the first class, $P < 0$ and the momenta (which are $-i$ times the infinitesimal operators) are space-like. In the second class $P = 0$ and the momenta form a null vector, $P > 0$ in the third class.

It is generally admitted that the representations of the first class have no physical significance because the momenta of all observed particles are time-like or null vectors. Hence we shall investigate only the simplest one of these representations. Its operators are most easily given in the Hilbert space of functions of three variables p_1, p_2, p_3 which are restricted to the outside of a sphere of radius $\sqrt{-P}$. The expressions for the infinitesimal operators are

$$I_k = ip_k \qquad I_0 = -i(p_1^2 + p_2^2 + p_3^2 + P)^{1/2}. \tag{28a}$$

This last equation also shows the reason for the variables p_k to be restricted by $p_1^2 + p_2^2 + p_3^2 > -P$: if this inequality is not fulfilled, I_0 ceases to be skew hermitean and, hence, the representation is not unitary. Further

$$I_{kl} = p_l \partial/\partial p_k - p_k \partial/\partial p_l \tag{28b}$$

and

$$I_{k0} = -(p_1^2 + p_2^2 + p_3^2 + P)^{1/2} \partial/\partial p_k. \tag{28c}$$

It is useful to introduce new variables instead of the p_l both in order to simplify the definition domain of the variables and also to bring the operations of the subgroup S into a form which is independent of P. This can be done, most simply, by introducing as new variables

$$p_0 = (p_1^2 + p_2^2 + p_3^2 + P)^{1/2} \qquad \Omega_k = p_k/(p_0^2 - P)^{1/2}. \quad (29)$$

The Ω_k are restricted to a unit sphere, p_0 changes from 0 to ∞. In these variables, the scalar product between φ and ψ reduces to

$$(\varphi, \psi) = \int d\Omega \int dp_0 (p_0^2 - P)^{1/2} \bar{\varphi}\psi \quad (29a)$$

$d\Omega$ being the surface element of the unit sphere over which the integration with respect to the Ω is to be extended. In terms of the new variables, the infinitesimal operators assume the form

$$I_{kl} = \Omega_l \partial/\partial\Omega_k - \Omega_k \partial/\partial\Omega_l \qquad I_0 = -ip_0 \quad (30)$$

and

$$I_k = i\Omega_k(p_0^2 - P)^{1/2} \quad (30a)$$

$$I_{k0} = -(p_0^2 - P)^{1/2}\Omega_k \partial/\partial p_0 - \frac{p_0}{(p_0^2 - P)^{1/2}} \sum_l \Omega_l(\Omega_l \partial/\partial\Omega_k - \Omega_k \partial/\partial\Omega_l).$$

If we now set $J_k = \epsilon I_k$, the J_k will converge to zero unless $-P$ becomes inversely proportional to ϵ^2, i.e., unless $-\epsilon^2 P$ converges to a definite limit P. If this is assumed, the second term of ϵI_{k0} will converge to zero and the infinitesimal elements of the representation of the contracted group become

$$J_{kl} = \Omega_l \partial/\partial\Omega_k - \Omega_k \partial/\partial\Omega_l \qquad J_0 = -ip_0 \quad (31)$$

$$J_{k0} = -\mathrm{P}\Omega_k \partial/\partial p_0 \qquad J_k = i\,\mathrm{P}\Omega_k. \quad (31a)$$

These operators indeed span a unitary representation of the Galilei group: they correspond to case II with $m = 0$ of a recent determination of these representations.[7] The p_k of this article correspond to our $\mathrm{P}\Omega_k$, the variable s is given by $i\,\mathrm{P}\,\partial/\partial p_0$. One also understands now why it was impossible to find a physical interpretation to this representation: it is the limiting case of a representation of the relativistic group with imaginary mass $P^{1/2}$. The same is probably true of the other true representations[7] of the Galilei group.

We go over now to the investigation of the simplest representation with positive P, i.e., the representation of the Klein-Gordon equation. The infinitesimal operators are again given by (28a), (28b), (28c). However, since P is positive, the variability domain of the p extends over the whole three dimensional space. The scalar product of two functions φ and ψ is now given by

$$(\varphi, \psi) = \int\int\int dp_1\, dp_2\, dp_3\, p_0^{-1}\, \bar{\varphi}\psi \quad (32)$$

in which p_0 is still given by (29).

If we set, in the sense of (6),

$$J_k = \epsilon I_k = \epsilon i p_k \tag{33}$$

the operation of time displacement becomes

$$iJ_0 = iI_0 = (-I_1^2 - I_2^2 - I_3^2 + P)^{1/2} = \epsilon^{-1}(-J_1^2 - J_2^2 - J_3^2 + \epsilon^2 P)^{1/2} \tag{33a}$$

Since the J_k are skew hermitean, their squares are positive definite hermitean operators. Since $\epsilon^2 P$ is also positive the expectation value $(\varphi, iJ_0\varphi)$ of iJ_0 is, for any state φ, greater than ϵ^{-1} times the expectation value of iJ_1. It follows that if $J_1\varphi$ converges to a vector in Hilbert space as $\epsilon \to 0$, the vector $J_0\varphi$ must grow beyond all limits. The same is true, of course, for the other J_k. It follows that the representations considered cannot be contracted in the sense discussed in the previous sections and the same is true of all representations of the classes $P \geq 0$.

It is possible, however, to contract these representations to representations up to a factor of the Galilei group. The commutation relations of the infinitesimal elements of representations up to a factor differ from the commutation relations of real representations by the appearance of a constant in the structural relations. Hence

$$[J_\alpha, J_\beta] = \sum c_{\alpha\beta}^\gamma J_\gamma + b_{\alpha\beta} 1 \tag{34}$$

where $c_{\alpha\beta}^\gamma$ are the structure constants of the group to be represented (in our case, the inhomogeneous Galilei group) and the $b_{\alpha\beta}1$ are multiples of the unit operator. One will, therefore, obtain infinitesimal elements of representations up to a factor if one sets, instead of (6)

$$J_{1\nu} = I_{1\nu} - a_{1\nu}1 \qquad J_{2\nu} = \epsilon I_{2\nu} - a_{2\nu}1 \tag{34a}$$

in which all a may depend on ϵ. Since the additional terms in (34a) commute with all other operators, these additional terms will not affect the left side of (34). Hence, they must be compensated also on the right side and this is done by the additional terms $b_{\alpha\beta}1$. The point of introducing the terms $a1$ in (34a), which then necessitates the introduction of the b in (34), is that the right sides of (34a) may converge to finite non-vanishing operators even if the $I_{1\nu}$, $\epsilon I_{2\nu}$ cannot be made to converge.

The above generalization of the concept of contraction indeed allows a contraction of the representations given by (28a), (28b), (28c) also for $P > 0$. As we let P go to infinity, I_0 will also tend to infinity ($(I_0\varphi, I_0\varphi)$ converges to infinity for all φ). However, subtracting $-iP^{1/2}1$ from I_0, it will converge to

$$\begin{aligned} J_0 &= \lim -i(-I_1^2 - I_2^2 - I_3^2 + P)^{1/2} + iP^{1/2} \\ &= (i/2P^{1/2})(I_1^2 + I_2^2 + I_3^2) = (i/2P^{1/2}\epsilon^2)(J_1^2 + J_2^2 + J_3^2). \end{aligned} \tag{35}$$

This shows that J_0 will converge to a finite operator if the $J_k = \epsilon I_k$ do and if $P^{1/2}\epsilon^2$ converges to a finite constant m as $P \to \infty$, $\epsilon \to 0$. Both can be accomplished by assuming the representation in such a form that, instead of (28)

$$I_k = ip_k/\epsilon \qquad I_0 = -i(P + (p_1^2 + p_2^2 + p_3^2)/\epsilon^2)^{1/2}. \tag{36}$$

This is indeed possible because the variability domain of the p_k is unrestricted and the above form of the infinitesimal elements can be obtained by unitary transformation of the operators given in (28a). Such a transformation leaves the I_{kl} of (28b) unchanged but transforms the I_{k0} of (28c) into

$$I_{k0} = -\epsilon(P + (p_1^2 + p_2^2 + p_3^2)/\epsilon^2)^{1/2}\partial/\partial p_k. \tag{36b}$$

Hence we shall have

$$J_0 = \lim -i(P + (p_1^2 + p_2^2 + p_3^2)/\epsilon^2)^{1/2} + iP^{1/2} = -(i/2m)(p_1^2 + p_2^2 + p_3^2)$$

$$J_{kl} = p_l\partial/\partial p_k - p_k\partial/\partial p_l \tag{37.1}$$

$$J_k = \lim \epsilon(ip_k/\epsilon) = ip_k$$

$$J_{k0} = \lim -\epsilon(P + (p_1^2 + p_2^2 + p_3^2)/\epsilon^2)^{1/2}\epsilon\partial/\partial p_k$$

$$= \lim -(P\epsilon^4 + \epsilon^2(p_1^2 + p_2^2 + p_3^2))^{1/2}\partial/\partial p_k = -m\partial/\partial p_k. \tag{37.2}$$

The reader familiar with the transition from the Klein-Gordon to the Schrödinger equation will recognize the increase of the rest mass with increasing c and the elimination of this rest mass by the subtraction of $-iP^{1/2}1$ from the infinitesimal operator of the time-displacement operator. The infinitesimal operators (37.1), (37.2) for the contracted group are in fact those of Schrödinger's theory. It is likely that a similar contraction is possible also for the other representations with positive rest mass (i.e., $P > 0$) but this and the behavior of the representations with $P = 0$ will not be further discussed here.

[1] Segal, I. E., *Duke Math. J.*, **18**, 221 (1951).

[2] Cf. e.g. Wigner, E., *Gruppentheorie und ihre Auwendiengen* etc., Friedr. Vieweg, Braunschweig (1931) and Edwards Brothers, Ann Arbor (1944). Chapter XV.

[3] Cf. Jahnke, E., and Emde, F., *Tables of Functions*, Dover Publications, 1943, p. 149; or Watson, G. N., *Treatise on Bessel Functions*, Cambridge Univ. Press, 1922, p. 19ff.

[4] Reference 2, Chapter XIX, particularly p. 230, 232. Cf. also Wigner, E. P., *J. Franklin Inst.*, **250**, 477 (1950), and Godement, R., *Trans. Amer. Math. Soc.* **73**, 496 (1952).

[5] For $m = 0$ this is given on p. 65, of Watson's *Bessel Functions* (ref. 3).

[6] Cf. e.g. Bargmann, V., and Wigner, E. P., these PROCEEDINGS, **34**, 211 (1948) and further literature quoted there.

[7] Inonu, E., and Wigner, E. P., *Nuovo cimento*, **9**, 705 (1952).

Internal space-time symmetries of massive and massless particles

D. Han
Systems and Applied Sciences Corporation, Hyattsville, Maryland 20784
Y. S. Kim
Center for Theoretical Physics, Department of Physics and Astronomy, University of Maryland, College Park, Maryland 20742
Marilyn E. Noz
Department of Radiology, New York University, New York, New York 10016
D. Son
Department of Physics, Columbia University, New York, New York 10027

(Received 20 October 1983; accepted for publication 20 February 1984)

A unified description is given for internal space-time symmetries for massive and massless particles. It is noted that the little groups governing the internal symmetries of massive and massless particles are locally isomorphic to the three-dimensional rotation group or $O(3)$ and the two-dimensional rotation group or $E(2)$, respectively. It is noted also that the $E(2)$ group can be obtained from a large-radius/flat-surface approximation of $O(3)$. This procedure is then shown to be directly applicable to that of obtaining the $E(2)$-like little group for massless particles from the $O(3)$-like little group for massive particles in the infinite-momentum/zero-mass limit.

I. INTRODUCTION

One of the beauties of Einstein's special relativity is the unified description of the energy-momentum relation for massive and massless particles through

$$E = [(cP)^2 + (Mc^2)^2]^{1/2}. \quad (1)$$

In addition to mass, energy, and momentum, relativistic

Reprinted from *Am. J. Phys.* **52**, 1037 (1984).

Table I. Little groups for massive and massless particles.

P: four-momentum	Subgroup of $O(3,1)$	Subgroup of $SL(2,c)$
Massive: $P^2>0$	$O(3)$-like subgroup of $O(3,1)$: *hadrons*	$SU(2)$-like subgroup of $SL(2,c)$: *electrons*
Massless: $P^2=0$	$E(2)$-like subgroup of $O(3,1)$: *photons*	$E(2)$-like subgroup of $SL(2,c)$: *neutrinos*

particles have internal space-time degrees of freedom. For instance, a massive particle has rotational degrees of freedom in the Lorentz frame in which the particle is at rest. On the other hand, free massless particles have the helicity and gauge degrees of freedom. We are therefore led to the question of whether the internal symmetry for massless particles can be obtained as an infinite-momentum/zero-mass limit of the space-time symmetry for massive particles, as in the case of the energy-momentum relation.

In order to study internal space-time symmetries of relativistic particles, Wigner in 1939 formulated a method based on the little groups of the Poincaré group.[1] The little group is a subgroup of the Lorentz group which leaves the four-momentum of a given particle invariant. The little groups for massive and massless particles are locally isomorphic to the three-dimensional rotation group and the two-dimensional Euclidean group, respectively. For convenience, we shall use the word "like" in order to indicate that two groups have the same algebraic properties. The internal symmetries of massive and massless particles are dictated by the $O(3)$-like and $E(2)$-like little groups, respectively.

The first step in obtaining a unified picture of both massive and massless particles is to gain thorough understanding of each of the four cases listed in Table I. The representation suitable for electrons and positrons was discussed in Wigner's original paper.[1] The representations for relativistic extended hadrons in the quark model have been discussed extensively in the literature.[2,3] The representations suitable for photons and neutrinos have also been worked out.[4,5]

The purpose of the present paper is to discuss in detail the problem of showing that the $O(3)$-like little group for massive particles becomes the $E(2)$-like little group for massless particles in the infinite-momentum/zero-mass limit. In order to deal with this problem, we have to show first that the $E(2)$ group can be regarded as a limiting case of $O(3)$. We are quite familiar with the three-dimensional rotation group. However, the $E(2)$ group is largely unknown to us, in spite of the fact that this group can serve as a good illustrative example for many important aspects of geometry and group theory.

Indeed, from a pedagogical point of view, the $E(2)$ group has its own merit. When we study the three-dimensional rotation group, it is quite natural for us to ask the following questions.

(a) We discuss $O(3)$ repeatedly in the established curriculum, because it describes an important aspect of physics, *and* because it generates a beautiful mathematics. Then, is the rotation group the only interesting example in the existing curriculum?

(b) When we commute from home to school, we are making translations and rotations on a two-dimensional plane. Why can we not formulate a group theory based on this daily experience?

(c) Strictly speaking, when we travel on the surface of the Earth, we are performing rotations around the center of the Earth. Can the $E(2)$ group be regarded as a limiting case of the rotation group?

We shall therefore start this paper with a discussion of the $E(2)$ group.

In Sec. II, transformations on the two-dimensional Euclidean plane are discussed. It is shown that solutions of the two-dimensional Laplace equation form the basis for finite-dimensional representations of the $E(2)$ group. The 3×3 matrices representing coordinate transformations on the $E(2)$ plane are discussed in detail. In Sec. III, the $E(2)$ group is discussed as a contracted form of $O(3)$. It is noted that we can achieve this purpose by looking into a small and almost flat portion of a spherical surface whose radius becomes large.

We discuss in Sec. IV the internal space-time symmetries of massive and massless particles. It is shown that the $E(2)$-like symmetry of massless particles may be regarded as a limiting case of the $O(3)$-like symmetry for massive particles. In Sec. V, we discuss further applications of group contractions. It is shown in the Appendix that the $E(2)$ group can serve as a useful example for illustrating the difference between active and passive transformations.

II. WHAT IS THE $E(2)$ GROUP?

The two-dimensional Euclidean group, often called $E(2)$, consists of rotations and translations on a two-dimensional Euclidean plane. The coordinate transformation takes the form

$$x' = x\cos\theta - y\sin\theta + u,$$
$$y' = x\sin\theta + y\cos\theta + v. \tag{2}$$

This transformation can be written in matrix form as

$$\begin{pmatrix} x' \\ y' \\ 1 \end{pmatrix} = \begin{pmatrix} \cos\theta & -\sin\theta & u \\ \sin\theta & \cos\theta & v \\ 0 & 0 & 1 \end{pmatrix} \begin{pmatrix} x \\ y \\ 1 \end{pmatrix}. \tag{3}$$

The algebraic properties of the above transformation matrix have been discussed in Ref. 4.

The 3×3 matrix in Eq. (3) can be exponentiated as

$$D(\theta,u,v) = \exp[-i(uP_1 + vP_2)]\exp(-i\theta L_3). \tag{4}$$

The generators in this case are

$$L_3 = \begin{pmatrix} 0 & -i & 0 \\ i & 0 & 0 \\ 0 & 0 & 0 \end{pmatrix},$$
$$P_1 = \begin{pmatrix} 0 & 0 & i \\ 0 & 0 & 0 \\ 0 & 0 & 0 \end{pmatrix}, \quad P_2 = \begin{pmatrix} 0 & 0 & 0 \\ 0 & 0 & i \\ 0 & 0 & 0 \end{pmatrix}. \tag{5}$$

These generators satisfy the following commutation relations:

$$[P_1,P_2] = 0,$$
$$[L_3,P_1] = iP_2, \tag{6}$$
$$[L_3,P_2] = -iP_1.$$

The transformation described in Eqs. (2) and (3) and generated by the matrices of Eq. (5) is "active" in the sense that it transforms the object, as is described in Eq. (2).

Let us next consider transformations of functions of x and y, and continue to use P_1 and P_2 as the generators of translations and L_3 as the generator of rotations. These generators take the form

$$P_1 = -i\frac{\partial}{\partial x},\quad P_2 = -i\frac{\partial}{\partial y},\quad L_3 = -i\left(x\frac{\partial}{\partial y} - y\frac{\partial}{\partial x}\right). \tag{7}$$

The above operators satisfy the commutation relations of Eq. (6). The transformation using these differential operators is "passive" in the sense that it is achieved through a coordinate transformation which is the inverse of that given in Eq. (2). Although they achieve the same purpose, the active and passive transformations result in coordinate transformations in the opposite directions. In the Appendix, the effects of active and passive transformations are discussed in detail.

As in the case of the rotation group, the standard method of studying this group is to find an operator which commutes with all three of the above generators. It is easy to check that P^2, defined as

$$P^2 = P_1^2 + P_2^2, \tag{8}$$

commutes with all three generators. Thus one way to construct representations of the $E(2)$ group is to solve the equation

$$[P_1^2 + P_2^2]\psi(x,y) = k^2\psi(x,y), \tag{9}$$

using the differential forms of P_1 and P_2 given in Eq. (7). This partial differential equation can be separated in the polar, Cartesian, parabolic, or elliptic coordinate system.[6]

If we are interested in constructing representations diagonal in P^2 and P_1 and P_2, we use the Cartesian coordinate system. On the other hand, if we are interested in representations diagonal in P^2 and L_3, the polar coordinate system is appropriate. These possibilities have been considered in the past as summarized in Table II.

Inonu and Wigner constructed infinite-dimensional unitary representations by solving the differential equation of Eq. (9) with nonvanishing k^2.[7] On the other hand, the finite-dimensional nonunitary representations are based on the solutions with $k^2 = 0$. We are interested here in the representations diagonal in L_3 with $k^2 = 0$, because they describe the wave functions for massless particles observed in the real world.[4,5] In addition, the mathematics required for studying this case is much easier than that for the general case discussed in the original paper of Inonu and Wigner,[7]

Table II. Representations of the $E(2)$ group.

Diagonal in	Unitary infinite dimensional	Nonunitary finite dimensional
P_1 and P_2	Wigner in 1939 (Ref. 1)	Trivial
L_3	Inonu and Wigner in 1953 (Ref. 7)	Han *et al.* in 1982 (Ref. 5)

and is suitable for exercise problems even in the undergraduate curriculum.

If $k^2 = 0$, the differential equation of Eq. (9) takes the form

$$\left[\left(\frac{\partial}{\partial x}\right)^2 + \left(\frac{\partial}{\partial y}\right)^2\right]\psi(x,y) = 0. \tag{10}$$

This is a two-dimensional Laplace equation, and its solutions are quite familiar to us. The analytic solution of this equation takes the form

$$\psi = r^m \exp[\pm im\phi] = (x \pm iy)^m, \tag{11}$$

where

$$\phi = \tan^{-1}(y/x).$$

This is an eigenstate of L_3 or a rotation around the origin. The effect of the rotation operator

$$R(\theta) = \exp[-i\theta L_3] \tag{12}$$

on ψ of Eq. (10) is well known. If we translate the expression of Eq. (11) by applying the operator

$$T(u,v) = \exp[-i(uP_1 + vP_2)], \tag{13}$$

then the translated form becomes

$$T(u,v)\psi(x,y) = [(x-u) \pm i(y-v)]^m. \tag{14}$$

This is an eigenstate of a rotation around the point $x = u$ and $y = v$.

If $m = 1$, the basis vector for the representation diagonal in L_3 is

$$W_1 = \begin{pmatrix} x+iy \\ x-iy \\ 1 \end{pmatrix}. \tag{15}$$

The generators of the $E(2)$ transformation matrices take the form

$$L_3 = \begin{pmatrix} 1 & 0 & 0 \\ 0 & -1 & 0 \\ 0 & 0 & 0 \end{pmatrix},$$
$$P_1 = \begin{pmatrix} 0 & 0 & -i \\ 0 & 0 & -i \\ 0 & 0 & 0 \end{pmatrix},\quad P_2 = \begin{pmatrix} 0 & 0 & 1 \\ 0 & 0 & -1 \\ 0 & 0 & 0 \end{pmatrix}. \tag{16}$$

There is a nonsingular matrix which transforms the column vector in Eq. (2) to W_1 of Eq. (15). The above 3×3 matrices are applicable to functions and not to the coordinates. For this reason, each of them is related to the negative of its counterpart in Eq. (5) through a similarity transformation.

The matrices of Eq. (16) satisfy the commutation relations for the $E(2)$ group given in Eq. (6), and

$$P_1^2 + P_2^2 = 0, \tag{17}$$

which is a reflection of Eq. (5). In addition, P_1 and P_2 satisfy

$$P_1^2 = P_2^2 = P_1P_2 = 0, \tag{18}$$

because

$$\left(\frac{\partial}{\partial x}\right)^2 (x \pm iy) = \left(\frac{\partial}{\partial y}\right)^2 (x \pm iy) = \frac{\partial^2}{\partial x \partial y}(x \pm iy) = 0. \tag{19}$$

We expect that the procedure of constructing representations for larger values of m will be similar to the $m = 1$

case. It would be an interesting exercise to construct explicit matrices for an arbitrary integer value of m.[8] The basic vector W_1 given in Eq. (15) is not unlike the spherical vector whose components are $(x \pm iy)$ and z in which z is replaced by 1.

III. $E(2)$ GROUP AS A LIMITING CASE OF $O(3)$

The discussion given in Sec. II on $E(2)$ is quite similar to the case of $O(3)$. Like $O(3)$, the $E(2)$ group has three generators, and its coordinate transformation matrices are 3×3. Then how are these two groups related? This fundamental question was addressed by Inonu and Wigner in their theory of group contraction.[7,9]

One way to define this problem is to consider a sphere with a large radius. Imagine a football field on the surface of the Earth covering the north pole. A player can run from east to west, and from north to south. He can also turn around at any point in the field. Indeed, the player can perform $E(2)$ transformations on himself. Strictly speaking, however, these translations and rotations are all rotations on the spherical surface of the Earth.

Let us start with the familiar $O(3)$ rotation operator which can be written as

$$\exp[-i(\alpha L_1 + \beta L_2 + \theta L_3)]. \tag{20}$$

We are interested in the effect of this rotation on the flat football field. L_3 generates rotations around the north pole. L_2, which generates rotation around the y axis, takes the form

$$L_2 = -i\left(z\frac{\partial}{\partial x} - x\frac{\partial}{\partial z}\right). \tag{21}$$

Therefore, for large values of the radius R, $z = R$ and

$$L_2 = -iR\frac{\partial}{\partial x} = RP_1 \text{ or } P_1 = (1/R)L_2. \tag{22}$$

If we rotate the system by angle β around the y axis, the resulting translation on the $E(2)$ plane is

$$\beta L_2 = -i\beta R\frac{\partial}{\partial x} = uP_1, \tag{23}$$

with

$$u = \beta R \text{ or } \beta = u/R.$$

Likewise

$$L_1 = -RP_2, \quad \text{and } \alpha = v/R. \tag{24}$$

The parameters u and v are discussed in Sec. II.

If we write the commutation relations for the $O(3)$ group as

$$[L_3,(1/R)L_1] = i(1/R)L_2,$$
$$[L_3,(1/R)L_2] = -i(1/R)L_1, \tag{25}$$
$$[(1/R)L_1,(1/R)L_2] = i(1/R)^2L_3,$$

it is easy to see that these expressions, in the large R limit, become the commutation relations for the $E(2)$ group given in Eq. (2).

Let us translate the above limiting procedure into the language of matrices. 3×3 rotation matrices applicable to coordinate variables (x,y,z) are well known. They are generated by L_3 of Eq. (5) and

$$L_1 = \begin{pmatrix} 0 & 0 & 0 \\ 0 & 0 & -i \\ 0 & i & 0 \end{pmatrix}, \quad L_2 = \begin{pmatrix} 0 & 0 & i \\ 0 & 0 & 0 \\ -i & 0 & 0 \end{pmatrix}. \tag{26}$$

For the present purpose, we can consider the case where z is large and approximately equal to the radius of the sphere, and write

$$\begin{pmatrix} x \\ y \\ z \end{pmatrix} = \begin{pmatrix} 1 & 0 & 0 \\ 0 & 1 & 0 \\ 0 & 0 & R \end{pmatrix}\begin{pmatrix} x \\ y \\ 1 \end{pmatrix}. \tag{27}$$

The column vectors on the left- and right-hand sides are, respectively, the coordinate vectors on which the $O(3)$ and $E(2)$ transformations are applicable. We shall use A for the 3×3 matrix on the right-hand side. Then, in the limit of large R,

$$L_3 = A^{-1}L_3A,$$
$$P_1 = (1/R)A^{-1}L_2A, \tag{28}$$
$$P_2 = -(1/R)A^{-1}L_1A,$$

where L_3, P_1, and P_2 are given in Eq. (5). This limiting procedure is called the contraction of $O(3)$ to $E(2)$.

During the contraction process, L_3 remains invariant. For L_1 and L_2, the upper-right parts of the above matrices remain unchanged, except for a sign change in L_2 due to Eq. (24). However, the lower-left parts become zero. In terms of the spherical harmonics $Y_l^m(\theta,\phi)$, the above limiting procedure is the same as replacing $\cos\theta$ by 1, and $(e^{\pm i\phi}\sin\theta)$ by $(x \pm iy)$.

Another interesting property of $E(2)$ which is inherited from $O(3)$ is the concept of equivalence class.[10,11] In $O(3)$, rotations by the same angle around different axes belong to the same equivalence class. This notion is translated to rotations by the same angle around different points on the xy plane forming an equivalence class. It is not difficult to form a geometrical visualization of the concept of equivalence classes applicable to $O(3)$ and $E(2)$.

IV. INTERNAL SPACE-TIME SYMMETRIES OF RELATIVISTIC PARTICLES

In describing a free relativistic particle, we specify first its mass, momentum, and energy. After determining its four-momentum, we should ask what other space-time degrees of freedom the particle has. This question was systematically formulated by Wigner in his 1939 paper on representations of the inhomogeneous Lorentz group or the Poincaré group.[1] The subgroups of the Poincaré group governing internal space-time symmetries are called the little groups.[1]

The little group is generated by a maximal subset of J_i and K_i which leaves the four-momentum invariant, where J_i is the generator of rotations around the ith axis, and K_i is the boost generator along the ith axis. These generators satisfy the commutation relations

$$[J_i,J_j] = i\epsilon_{ijk}J_k,$$
$$[J_i,K_j] = i\epsilon_{ijk}K_k, \tag{29}$$
$$[K_i,K_j] = -i\epsilon_{ijk}J_k.$$

The little groups for massive and massless particles are locally isomorphic to $O(3)$ and $E(2)$, respectively, and they have been discussed in separate papers in this Journal.[2,4] After studying the procedure of obtaining the $E(2)$ group as a limiting case of $O(3)$, we are naturally led to consider the internal symmetry group of massless particles as a limiting case of the $O(3)$-like little group for massless particles.

If a massive particle is at rest, the symmetry group is

generated by the angular momentum operators J_1,[1] J_2, and J_3.[1] These operators do not change the four-momentum of the particle at rest. If this particle moves along the z direction, J_3 remains invariant, and its eigenvalue is the helicity. However, we have been avoiding in the past the question of what happens to J_1 and J_2, particularly in the infinite-momentum limit.

There are no Lorentz frames in which massless particles are at rest. The little group for a massless particle moving along the z direction is generated by J_3, N_1, and N_2,[1] where

$$N_1 = K_1 - J_2,$$
$$N_2 = K_2 + J_1. \tag{30}$$

The four-momentum of the massless particle remains invariant under transformations generated by these operators. These generators satisfy the commutation relations

$$[N_1,N_2] = 0,$$
$$[J_3,N_1] = iN_2, \tag{31}$$
$$[J_3,N_2] = -iN_1,$$

which are identical to those for the $E(2)$ group given in Eq. (6). J_3 is like the generator of rotation while N_1 and N_2 are like the generators of translations in the two-dimensional plane. These translationlike operators are known to generate gauge transformations.[4]

Einstein's energy-momentum relation of Eq. (1) clearly indicates that a massive particle becomes like a massless particle as the momentum/mass ratio becomes infinite. We are thus led to the suspicion that the $O(3)$-like internal symmetry for massive particles will become the $E(2)$-like symmetry for massless particles, and that this limiting procedure will be like the group contraction procedure discussed in Sec. III.

If we boost the massive particle along the z direction, J_3 will remain invariant. The analysis of Sec. III leads us to expect that J_2 and J_1 for massive particle at rest will become N_1 and $-N_2$, respectively, in the infinite-momentum/zero-mass limit, just like the contraction of $O(3)$ resulting in $E(2)$.[12]

Let us carry out an explicit calculation to justify the above expectation starting with a massive particle at rest. If we use the four-vector convention

$$x^\mu = (x,y,z,t), \quad x_\mu = (x,y,z,-t), \tag{32}$$

with $c = 1$, the generators of Lorentz transformations applicable to this coordinate space are

$$J_1 = \begin{pmatrix} 0 & 0 & 0 & 0 \\ 0 & 0 & -i & 0 \\ 0 & i & 0 & 0 \\ 0 & 0 & 0 & 0 \end{pmatrix}, \quad K_1 = \begin{pmatrix} 0 & 0 & 0 & i \\ 0 & 0 & 0 & 0 \\ 0 & 0 & 0 & 0 \\ i & 0 & 0 & 0 \end{pmatrix},$$
$$J_2 = \begin{pmatrix} 0 & 0 & i & 0 \\ 0 & 0 & 0 & 0 \\ -i & 0 & 0 & 0 \\ 0 & 0 & 0 & 0 \end{pmatrix}, \quad K_2 = \begin{pmatrix} 0 & 0 & 0 & 0 \\ 0 & 0 & 0 & i \\ 0 & 0 & 0 & 0 \\ 0 & i & 0 & 0 \end{pmatrix}, \tag{33}$$
$$J_3 = \begin{pmatrix} 0 & -i & 0 & 0 \\ 0 & 0 & 0 & 0 \\ i & 0 & 0 & 0 \\ 0 & 0 & 0 & 0 \end{pmatrix}, \quad K_3 = \begin{pmatrix} 0 & 0 & 0 & 0 \\ 0 & 0 & 0 & 0 \\ 0 & 0 & 0 & i \\ 0 & 0 & i & 0 \end{pmatrix},$$

where J_i and K_i generate rotations and boosts, respectively.

Let us start with a massive particle at rest with its mass M. Then the little group is isomorphic to the $O(3)$ group generated by J_1, J_2, and J_3. If we boost this massive particle along the z direction, its momentum and energy will become P and $E = [P^2 + M^2]^{1/2}$, respectively. The boost matrix is

$$B(P) = \begin{pmatrix} 1 & 0 & 0 & 0 \\ 0 & 1 & 0 & 0 \\ 0 & 0 & E/M & P/M \\ 0 & 0 & P/M & E/M \end{pmatrix}. \tag{34}$$

Under this boost operation, J_3 given in Eq. (33) remains invariant:

$$J'_3 = BJ_3B^{-1} = J_3. \tag{35}$$

However, the boosted J_2 and J_1 become

$$J'_2 = (E/M)J_2 - (P/M)K_1,$$
$$J'_1 = (E/M)J_1 + (P/M)K_2. \tag{36}$$

Because the Lorentz boosts in Eqs. (35) and (36) are similarity transformations, the J' operators still satisfy the $O(3)$ commutation relations:

$$[J'_i,J'_j] = i\epsilon_{ijk}J'_k. \tag{37}$$

Since the quantities in Eq. (36) become very large as the momentum increases, we introduce new operators:

$$G_1 = -(M/E)J'_2,$$
$$G_2 = (M/E)J'_1. \tag{38}$$

In terms of these new operators, we can write the $O(3)$ commutation relations of Eq. (37) as

$$[J_3,G_1] = -iG_2,$$
$$[J_3,G_2] = iG_1, \tag{39}$$
$$[G_1,G_2] = -(M/E)^2J_3.$$

The quantity $(M/E)^2$ becomes vanishingly small if the mass becomes small or the momentum becomes very large. In this limit,

$$G_1 \to N_1 \text{ and } G_2 \to N_2, \tag{40}$$

where

$$N_1 = \begin{pmatrix} 0 & 0 & -i & i \\ 0 & 0 & 0 & 0 \\ i & 0 & 0 & 0 \\ i & 0 & 0 & 0 \end{pmatrix},$$
$$N_2 = \begin{pmatrix} 0 & 0 & 0 & 0 \\ 0 & 0 & -i & i \\ 0 & i & 0 & 0 \\ 0 & i & 0 & 0 \end{pmatrix}: \tag{41}$$

The fact that the above N matrices generate gauge transformations has been extensively discussed in the literature.[4,5,13]

Indeed, rotations around the axes perpendicular to the momentum become gauge transformations in the infinite-momentum/zero-mass limit.

V. FURTHER PHYSICAL APPLICATIONS OF GROUP CONTRACTIONS

The purpose of this section is to indicate that there are many other interesting applications of group contractions.

In Sec. III, we studied the contraction of $O(3)$ to $E(2)$ using the notion of a plane tangent to a spherical surface. The concept of this tangent plane plays a very important role in many branches of physics and engineering dealing with curved surfaces.

For example, let us consider the surface of the hyperbola in a three-dimensional space spanned by x, y, and t:

$$(ct)^2 - x^2 - y^2 = \text{const}, \tag{42}$$

where c is a constant and may become very large. This is a description of the $O(2,1)$ group consisting of Lorentz boosts along the x and y directions and rotations on the xy plane. The pedagogical value of this group has been amply discussed in Ref. 14.

We can now consider a plane tangent to the surface at $x = y = 0$. As the constant c becomes very large, the portion of surface in which $(x^2 + y^2)$ is finite becomes flat and coincides with the tangent plane. It is then not difficult to imagine that transformations on this tangent plane are Galilean transformations.

In order to see this point, let us start with coordinate transformations of the group $O(2,1)$ generated by L_3 of Eq. (5), and

$$K_1 = \begin{pmatrix} 0 & 0 & 0 \\ 0 & 0 & -i \\ 0 & -i & 0 \end{pmatrix}, \quad K_2 = \begin{pmatrix} 0 & 0 & i \\ 0 & 0 & 0 \\ i & 0 & 0 \end{pmatrix}, \tag{43}$$

applicable to the column vector (x,y,ct).[14] K_1 and K_2 are the generators of Lorentz boosts along the x and y directions, respectively.

The column vector (x,y,ct) can be written as

$$\begin{pmatrix} x \\ y \\ ct \end{pmatrix} = \begin{pmatrix} 1 & 0 & 0 \\ 0 & 1 & 0 \\ 0 & 0 & c \end{pmatrix} \begin{pmatrix} x \\ y \\ t \end{pmatrix}. \tag{44}$$

Then, as c becomes very large, the circumstance is identical to the case of Eqs. (26) and (28). The resulting transformation matrix becomes

$$\begin{pmatrix} x' \\ y' \\ t' \end{pmatrix} = \begin{pmatrix} \cos\theta & -\sin\theta & u \\ \sin\theta & \cos\theta & v \\ 0 & 0 & 1 \end{pmatrix} \begin{pmatrix} x \\ y \\ t \end{pmatrix}. \tag{45}$$

This form is a rotation on the xy plane followed by Galilean boosts along the x and y directions. Indeed, special relativity becomes Galilean relativity in the limit of large c.[7,9]

Let us consider another example. In scattering processes in which two incoming particles collide with each other resulting in two particles moving in different directions, we commonly use the Legendre polynomials $P_l(\cos\theta)$ to describe the dependence on the scattering angle. The quantum number l is the angular momentum around the scattering center, and can be regarded as a measure of the incoming momentum multiplied by the impact parameter.

When particles move slowly, it is sufficient to consider only two or three lowest values l. On the other hand, when the particles move with speed very close to that of light, the scattering becomes predominantly forward and becomes like the Fraunhofer diffraction. In this case, we have to deal with large values of l. One way to approach this problem is to start from the operator

$$(\mathbf{L})^2 = L_x^2 + L_y^2 + L_z^2, \tag{46}$$

with the eigenvalue $l(l+1)$. For large values of l, we can ignore L_3 whose eigenvalue is usually not larger than 1, and let

$$l(l+1) \simeq (l+\tfrac{1}{2})^2. \tag{47}$$

In their 1976 paper,[15] Misra and Maharana made an interesting observation that, when the scattering angle is very small, we can replace $(L_x^2 + L_y^2)$ by

$$R^2(P_x^2 + P_y^2) = (l+\tfrac{1}{2})^2, \tag{48}$$

in the spirit of Eqs. (22) and (24) with suitable redefinitions for R and P_i. In view of the discussion given in Sec. IV, we can readily let R be (P_0/M).[15] Thus, for a given value of P_0, the eigenvalue of

$$(P_1^2 + P_2^2) \tag{49}$$

will give a measure of l. In view of the discussion given in Sec. II, this new parameter will be that of the Bessel function. Thus, for large values of l, the Legendre polynomial becomes the Bessel function[7,9]:

$$P_l(\cos\theta) \to J_0(\alpha q\theta), \tag{50}$$

where

$$\alpha = P_0/M.$$

The parameter q now measures l, and becomes continuous for large values of P_0.

The above Bessel-function form is commonly used for studying high-energy data.[15,16] It is interesting to note that the transition from the use of the Legendre polynomials for low-energy processes to that of the Bessel functions in high-energy scattering is a group contraction of $O(3)$ to $E(2)$.

ACKNOWLEDGMENTS

We are grateful to O. W. Greenberg for providing the following information. In 1962, Wigner gave a series of lectures on the representations of the Poincaré group at Trieste and Istanbul. At one of his lectures, the problem of obtaining the $E(2)$-like little group for massless particles as a limiting case of the $O(3)$-like little group for massive particles was informally discussed as an unsolved problem, although this discussion was not included in Wigner's lecture notes published in Ref. 17. We would like to thank M. Parida for bringing Ref. 15 to our attention and for explaining its content to us.

APPENDIX

We study in this Appendix active and passive transformations on the $E(2)$ plane. Let us define the transformations given in Eqs. (2) and (3) to be active. This transformation first rotates the coordinate point (x,y) by angle θ around the origin. It then translates the rotated point by u and v along the x and y directions, respectively.

On the other hand, if we perform the same rotation on the function

$$g(x,y) = (x+iy)^m = r^m e^{im\phi}, \tag{A1}$$

using L_3 given in Eq. (7),

$$(e^{-i\theta L_3})g(x,y) = r^m e^{im(\phi-\theta)}. \tag{A2}$$

If we apply the translation operators on the above expression,

$$(e^{-i(uP_1+vP_2)}e^{-i\theta L_3})g(x,y) = (x''+iy'')^m = g(x'',y''), \tag{A3}$$

where

$$x'' = (x-u)\cos\theta + (y-v)\sin\theta,$$
$$y'' = -(x-u)\sin\theta + (y-v)\cos\theta.$$

The above linear transformation can also be written as

$$\begin{pmatrix} x'' \\ y'' \\ 1 \end{pmatrix} = \begin{pmatrix} \cos\theta & \sin\theta & -(u\cos\theta + v\sin\theta) \\ -\sin\theta & \cos\theta & (u\sin\theta - v\cos\theta) \\ 0 & 0 & 1 \end{pmatrix} \begin{pmatrix} x \\ y \\ 1 \end{pmatrix}. \quad \text{(A4)}$$

The matrix in this expression is precisely the inverse of that of the active transformation matrix of Eq. (3).

[1]E. P. Wigner, Ann. Math. **149**, 40 (1939). See also V. Bargmann and E. P. Wigner, Proc. Natl. Acad. Sci. U. S. A. **34**, 211 (1946).

[2]Y. S. Kim, M. E. Noz, and S. H. Oh, Am. J. Phys. **47**, 892 (1979); D. Han and Y. S. Kim, *ibid*. **49**, 1157 (1981).

[3]Y. S. Kim, M. E. Noz, and S. H. Oh, J. Math. Phys. **20**, 1341 (1979); D. Han, M. E. Noz, Y. S. Kim, and D. Son, Phys. Rev. D **25**, 1740 (1982).

[4]D. Han and Y. S. Kim, Am. J. Phys. **49**, 348 (1981).

[5]D. Han, Y. S. Kim, and D. Son, Phys. Rev. D **25**, 461 (1982); **26**, 3717 (1982).

[6]P. Winternitz and I. Fris, Yad. Fiz. **1**, 889 (1965) [Sov. J. Nucl. Phys. **1**, 636 (1965)].

[7]E. Inonu and E. P. Wigner, Proc. Natl. Acad. Sci. U. S. A. **39**, 510 (1953).

[8]D. Han, Y. S. Kim, and D. Son, University of Maryland Physics Publication No. 83-141 (1983).

[9]For a pedagogical reformulation of the theory of group contraction, including a discussion of the $E(2)$ group as a contraction of the $O(3)$ group, see R. Gilmore, *Lie Groups and Lie Algebras, and Some of Their Applications* (Wiley, New York, 1974).

[10]A. S. Wightman, in *Dispersion Relations and Elementary Particles*, edited by C. De Witt and R. Omnes (Hermann, Paris, 1960).

[11]The concept of equivalence class was discussed in detail for the $O(3)$ case by Wigner. This concept survives in $E(2)$ after contraction. See E. P. Wigner, *Group Theory, and Its Applications to the Quantum Theory of Atomic Spectra* (Academic, New York, 1959).

[12]D. Han, Y. S. Kim, and D. Son, Phys. Lett. **131B**, 327 (1983).

[13]S. Weinberg, Phys. Rev. **134**, B882 (1964); **135**, B1049 (1964).

[14]Y. S. Kim and M. E. Noz, Am. J. Phys. **51**, 368 (1983).

[15]S. P. Misra and J. Maharana, Phys. Rev. D **14**, 133 (1976).

[16]R. Blankenbecler and M. L. Goldberger, Phys. Rev. **126**, 766 (1962). See also S. J. Wallace, Phys. Rev. D **8**, 1846 (1973) and D **9**, 406 (1974).

[17]E. P. Wigner, in *Group Theoretical Concepts and Methods in Elementary Particle Physics*, edited by F. Gürsey (Gordon and Breach, New York, 1962); and in *Theoretical Physics*, edited by A. Salam (International Atomic Energy Agency, Vienna, 1962).

Eulerian parametrization of Wigner's little groups and gauge transformations in terms of rotations in two-component spinors

D. Han
SASC Technologies, Inc., 5809 Annapolis Road, Hyattsville, Maryland 20784
Y. S. Kim
Department of Physics and Astronomy, University of Maryland, College Park, Maryland 20742
D. Son
Department of Physics, Kyungpook National University, Daegu 635, Korea

(Received 20 February 1986; accepted for publication 30 April 1986)

A set of rotations and Lorentz boosts is presented for studying the three-parameter little groups of the Poincaré group. This set constitutes a Lorentz generalization of the Euler angles for the description of classical rigid bodies. The concept of Lorentz-generalized Euler rotations is then extended to the parametrization of the E(2)-like little group and the O(2,1)-like little group for massless and imaginary-mass particles, respectively. It is shown that the E(2)-like little group for massless particles is a limiting case of the O(3)-like or O(2,1)-like little group. A detailed analysis is carried out for the two-component SL(2,c) spinors. It is shown that the gauge degrees of freedom associated with the translationlike transformation of the E(2)-like little group can be traced to the SL(2,c) spins that fail to align themselves to their respective momenta in the limit of large momentum and/or vanishing mass.

I. INTRODUCTION

The Euler angles constitute a convenient parametrization of the three-dimensional rotation group. The Euler kinematics consists of two rotations around the z axis with one rotation around the y axis between them. The first question we would like to address in this paper is what happens if we add a Lorentz boost along the z direction to this traditional procedure. Since the rotation around the z axis is not affected by the boost along the same axis, we are asking what is the Lorentz-generalized form of the rotation around the y axis.

Since the publication of Wigner's fundamental paper on the Poincaré group in 1939,[1] a number of mathematical techniques have been developed to deal with the three-parameter little groups that leave a given four-momentum invariant. Our second question is why we do not yet have a standard set of transformations for Wigner's little groups.

In this paper, we combine the first and second questions. One of Wigner's little groups is locally isomorphic to O(3). Furthermore, the Euler angles constitute the natural language for spinning tops in classical mechanics, while Wigner's little groups describe the internal space-time symmetries of relativistic particles, including spins. It is thus quite natural for us to look for a possible Eulerian parametrization of the three-parameter little groups.

As far as massive particles are concerned, the traditional approach to this problem is to go to the Lorentz frame in which the particle is at rest, and then perform rotations there.[1] Then, its four-momentum is not affected, but the direction of its spin becomes changed. This operation, however, is not possible for massless or imaginary-mass particles.

In order to construct a Lorentz kinematics that includes both massive and massless particles, we observe that the transformation that changes a given four-momentum can be carried out in many different ways. However, as Wigner observed in 1957, the resulting spin orientation depends on the way in which the transformation is performed and on the mass of the particle.[2] For instance, when a particle with positive helicity is rotated, the helicity remains unchanged. As far as the momentum is concerned, we can achieve the same purpose by performing a simple boost. However, this boost does not leave the helicity invariant. Furthermore, the change in the direction of spin depends on the mass.

Indeed, the difference between the rotation and boost was studied for massless photons by Kupersztych,[3] who observed that this difference amounts to a gauge transformation. In this paper, we extend the kinematics of Kupersztych to include massive and imaginary-mass particles. We shall show that this extended kinematics constitutes the abovementioned Lorentz generalization of the Euler rotations.

We then study the extended Kupersztych kinematics using the SL(2,c) spinors. Among the four two-component SL(2,c) spinors, two of them preserve the helicity under boosts in the zero-mass limit, as was noted by Wigner in 1957. However, the remaining two do not preserve the helicity in the same limit. We show that these helicity nonpreserving spinors are responsible for gauge degrees of freedom contained in the E(2)-like little group for photons.

In Sec. II, we work out the Kupersztych kinematics for massive particles. It is pointed out that this new kinematics is equivalent to the traditional O(3)-like kinematics in which the particle is rotated in its rest frame. We show in Sec. III that the E(2)-like little group for massless particles is the infinite-momentum/zero-mass limit of the O(3)-like little group discussed in Sec. II. In Sec. IV, we discuss the continuation of the transformation matrices for the O(3)-like little

Reprinted from *J. Math. Phys.* **27**, 2228 (1986).

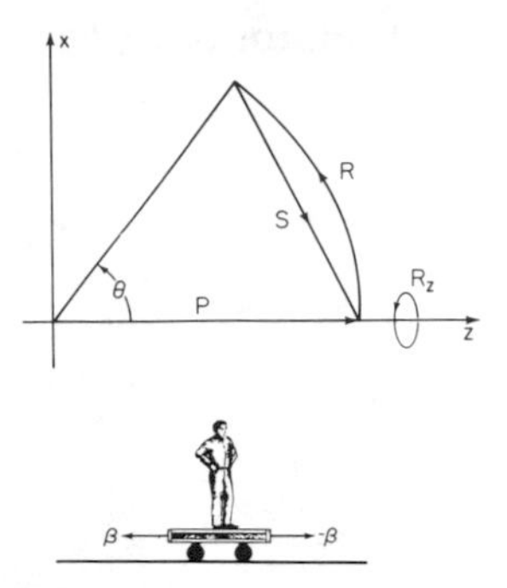

FIG. 1. Lorentz-generalized Euler rotations. The traditional Euler parametrization consists of two rotations around the z axis with one rotation around the y axis between them. If we add a Lorentz boost along the z axis, the two rotations around the z axis are not affected. The rotation around the y axis can be Lorentz-generalized in the following manner. If we boost the system along the z direction, we are dealing with the system with a nonzero four-momentum along the same direction. The four-momentum p can be rotated around the y axis by angle θ. The same result can be achieved by boost S^{-1}. However, these two transformations do not produce the same effect on the spin. The most effective way of studying this difference is to study the transformation SR, which leaves the initial four-momentum invariant.

group to the case of imaginary-mass particles.

In Sec. V, we study the transformation properties of the four two-component spinors in the SL(2,c) regime. It is shown that in the limit of infinite momentum and/or zero mass, two of the SL(2,c) spinors preserve their respective helicities, while the remaining two do not. We note, in Sec. VI, that four-vectors can be constructed from the four two-component SL(2,c) spinors. It is shown that the origin of the gauge degrees of freedom for photons can be traced to the spinors that refuse to align themselves to the momentum in the infinite-momentum/zero-mass limit.

II. KINEMATICS OF THE O(3)-LIKE LITTLE GROUP

The Euler rotation consists of a rotation around the y axis preceded *and* followed by rotations around the z axis. If the boost is made along the z axis, the rotations around the z axis are not affected. In this section, we discuss a Lorentz generalization of the rotation around the y axis and its relation to the O(3)-like little group for massive particles.

Let us start with a massive particle at rest whose four-momentum is

$$(0,0,0,m) . \tag{1}$$

We use the four-vector convention: $x^\mu = (x, y, z, t)$. We can boost the above four-momentum along the z direction with velocity parameter α:

$$P = m(0,0,\alpha/(1-\alpha^2)^{1/2},1/(1-\alpha^2)^{1/2}) . \tag{2}$$

The four-by-four matrix which transforms the four-vector of Eq. (1) to that of Eq. (2) is

$$A(\alpha) = \begin{pmatrix} 1 & 0 & 0 & 0 \\ 0 & 1 & 0 & 0 \\ 0 & 0 & 1/(1-\alpha^2)^{1/2} & \alpha/(1-\alpha^2)^{1/2} \\ 0 & 0 & \alpha/(1-\alpha^2)^{1/2} & 1/(1-\alpha^2)^{1/2} \end{pmatrix} . \tag{3}$$

Let us next rotate the four-vector of Eq. (2) using the rotation matrix:

$$R(\theta) = \begin{pmatrix} \cos\theta & 0 & \sin\theta & 0 \\ 0 & 1 & 0 & 0 \\ -\sin\theta & 0 & \cos\theta & 0 \\ 0 & 0 & 0 & 1 \end{pmatrix} . \tag{4}$$

This rotation does not alter the helicity of the particle.[2]

As is specified in Fig. 1, we can achieve the same result on the four-momentum by applying a boost matrix. However, unlike the rotation of Eq. (4), this boost is not a helicity-preserving transformation.[2] We can study the difference between these two transformations by taking the product of the rotation and the inverse of the boost. This inverse boost is illustrated in Fig. 1, and is represented by

$$S = \begin{pmatrix} 1 + 2(\sinh(\lambda/2)\cos(\theta/2))^2 & 0 & -(\sinh(\lambda/2))^2 \sin\theta & -(\sinh\lambda)\cos(\theta/2) \\ 0 & 1 & 0 & 0 \\ -(\sinh(\lambda/2))^2\sin\theta & 0 & 1 + 2(\sinh(\lambda/2)\sin(\theta/2))^2 & (\sinh\lambda)\sin(\theta/2) \\ -(\sinh\lambda)\cos(\theta/2) & 0 & (\sinh\lambda)\sin(\theta/2) & \cosh\lambda \end{pmatrix}, \tag{5}$$

where

$$\lambda = 2[\tanh^{-1}(\alpha\sin(\theta/2))] . \tag{6}$$

This matrix depends on the rotation angle θ and the velocity parameter α, and becomes an identity matrix when the particle is at rest with $\alpha = 0$.

Indeed, the rotation $R(\theta)$ followed by the boost $S(\alpha,\theta)$ leaves the four-momentum p of Eq. (2) invariant:

$$P = D(\alpha,\theta)P , \tag{7}$$

where

$$D(\alpha,\theta) = S(\alpha,\theta)R(\theta) .$$

The multiplication of the two matrices is straightforward, and the result is

$$D(\alpha,\theta)=\begin{pmatrix}1-(1-\alpha^2)u^2/2T & 0 & -u/T & \alpha u/T\\ 0 & 1 & 0 & 0\\ u/T & 0 & 1+u^2/2T & \alpha u^2/2T\\ \alpha u/T & 0 & -\alpha u^2/2T & 1+\alpha u^2/2T\end{pmatrix}. \quad (8)$$

where

$u=-2(\tan(\theta/2))$ and $T=1+(1-\alpha^2)(\tan(\theta/2))^2$.

This complicated expression leaves the four-momentum P of Eq. (2) invariant. Indeed, if the particle is at rest with vanishing velocity parameter α, the above expression becomes a rotation matrix. As the velocity parameter α increases, this D matrix performs a combination of rotation and boost, but leaves the four-momentum invariant.

Let us approach this problem in the traditional framework.[1] The above transformation is clearly an element of the O(3)-like little group that leaves the four-momentum P invariant. Then we can boost the particle with its four-momentum P by A^{-1} until the four-momentum becomes that of Eq. (1), rotate it around the y axis, and then boost it by A until the four-momentum becomes P of Eq. (2). It is appropriate to call this rotation in the rest frame the *Wigner rotation*.[4] The transformation of the O(3)-like little group constructed in this manner should take the form

$$D(\alpha,\theta)=A(\alpha)W(\theta^*)[A(\alpha)]^{-1}, \quad (9)$$

where W is the Wigner rotation matrix

$$W(\theta^*)=\begin{pmatrix}\cos\theta^* & 0 & \sin\theta^* & 0\\ 0 & 1 & 0 & 0\\ -\sin\theta^* & 0 & \cos\theta^* & 0\\ 0 & 0 & 0 & 1\end{pmatrix}. \quad (10)$$

We may call θ^* the *Wigner angle*. The question then is whether D of Eq. (9) is the same as D of Eq. (8). In order to answer this question, we first take the trace of the expression given in Eq. (9). The similarity transformation of Eq. (9) assures us that the trace of W be equal to that of D. This leads to

$$\theta^*=\cos^{-1}\left(\frac{1-(1-\alpha^2)(\tan(\theta/2))^2}{1+(1-\alpha^2)(\tan(\theta/2))^2}\right). \quad (11)$$

It is then a matter of matrix algebra to confirm that D of Eq. (9) and that of Eq. (8) are identical.

We have plotted in Fig. 2 the Wigner rotation angle θ^* as a function of the velocity parameter α. Here θ^* becomes θ when $\alpha=0$, and remains approximately equal to θ when α is smaller than 0.4. Then θ^* vanishes when $\alpha\to 1$. Indeed, for a given value of θ, it is possible to determine the value of θ^* that is the rotation angle in the Lorentz frame in which the particle is at rest.

The D matrix in the traditional form of Eq. (9) is well known.[1] However, the fact that it can also be derived from the closed-loop $R(\theta)$ and $S(\alpha,\theta)$ suggests that it has a richer content. For instance, the closed-loop kinematics does not have to be unique. There is at least one other closed-loop kinematics that leaves the four-momentum invariant.[5] The Kupersztych kinematics, which we are using in this paper, is convenient for studying the relation between the Euler angles and the parameters of the O(3)-like little group.

We have so far discussed the transformations in the x-z plane. It is quite clear that the same analysis can be carried out in the y-z plane or any other plane containing the z axis. This means that we can perform rotations $R_z(\phi)$ and $R_z(\psi)$, respectively, before and after carrying out the transformations in the x-z plane. Indeed, together with the velocity parameter α, the three parameters θ, ϕ, and ψ constitute the Eulerian parametrization of the O(3)-like little group.

III. E(2)-LIKE LITTLE GROUP FOR MASSLESS PARTICLES

Let us study in this section the D matrix of Eq. (8) as the particle mass becomes vanishingly small, by taking the limit of $\alpha\to 1$. In this limit, the D matrix of Eq. (8) becomes

$$D(u)=\begin{pmatrix}1 & 0 & -u & u\\ 0 & 1 & 0 & 0\\ u & 0 & 1-u^2/2 & u^2/2\\ u & 0 & -u^2/2 & 1+u^2/2\end{pmatrix}. \quad (12)$$

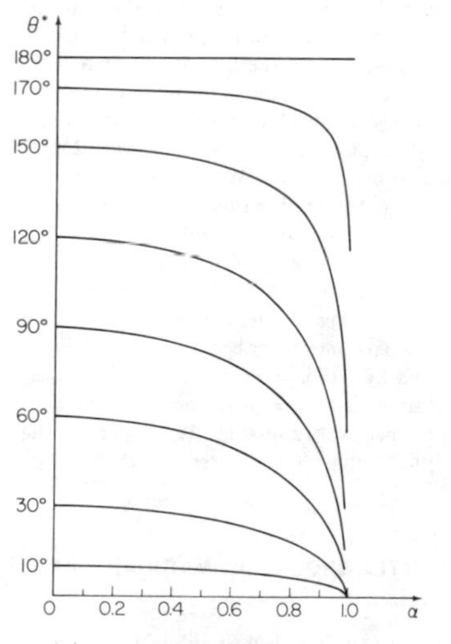

FIG. 2. Wigner rotation angle versus lab-frame rotation angle. We have plotted θ^* as a function of α for various values of θ using Eq. (11). $\theta=\theta^*$ at $\alpha=0$. θ^* is nearly equal to θ for moderate values of α, but it rapidly approaches 0 as α becomes 1.

After losing the memory of how the zero-mass limit was taken, it is impossible to transform this matrix into a rotation matrix. There is no Lorentz frame in which the particle is at rest. If we boost this expression along the z direction using the boost matrix

$$B(\beta)=\begin{pmatrix}1 & 0 & 0 & 0\\ 0 & 1 & 0 & 0\\ 0 & 0 & 1/(1-\beta^2)^{1/2} & \beta/(1-\beta^2)^{1/2}\\ 0 & 0 & \beta/(1-\beta^2)^{1/2} & 1/(1-\beta^2)^{1/2}\end{pmatrix}, \tag{13}$$

D remains form-invariant:

$$D'(u)=B(\beta)D(u)[B(\beta)]^{-1}=D(u'), \tag{14}$$

where

$$u'=[(1+\beta)/(1-\beta)]^{1/2}u.$$

The matrix of Eq. (12) is the case where the Kupersztych kinematics is performed in the x-z plane. This kinematics also can be performed in the y-z plane. Thus the most general form for the D matrix is

$$D(u,v)=\begin{pmatrix}1 & 0 & -u & u\\ 0 & 1 & -v & v\\ u & v & 1-(u^2+v^2)/2 & (u^2+v^2)/2\\ u & v & -(u^2+v^2)/2 & 1+(u^2+v^2)/2\end{pmatrix}. \tag{15}$$

The algebraic property of this expression has been discussed extensively in the literature.[1,5-8] If applied to the photon four-potential, this matrix performs a gauge transformation.[5,7] The reduction of the above matrix into the three-by-three matrix representing a finite-dimensional representation of the two-dimensional Euclidean group has also been discussed in the literature.[8]

Let us go back to Eq. (9). We have obtained the above gauge transformation by boosting the rotation matrix W given in Eq. (10). This means that the Lorentz-boosted rotation becomes a gauge transformation in the infinite-momentum and/or zero-mass limit. This observation was made earlier in terms of the *group contraction* of O(3) to E(2),[9,10] which is a singular transformation. We are then led to the question of how the method used in this section can be analytic, while the traditional method is singular.

The answer to this question is very simple. The group contraction is a language of Lie groups.[9,10] The parameter α we use in this paper is not a parameter of the Lie group. If we use η as the Lie-group parameter for boost along the z direction, it is related to α by $\sinh\eta=\alpha/(1-\alpha^2)^{1/2}$. However, this expression is singular at $\alpha=\pm 1$. Therefore, the continuation in α is not necessarily singular. We shall continue the discussion of this limiting process in terms of the SL(2,c) spinors in Sec. VI.

IV. O(2,1)-LIKE LITTLE GROUP FOR IMAGINARY-MASS PARTICLES

We are now interested in transformations that leave the four-vector of the form

$$P=im(0,0,\alpha/(\alpha^2-1)^{1/2},1/(\alpha^2-1)^{1/2}) \tag{16}$$

invariant, with α greater than 1. Although particles with imaginary mass are not observed in the real world, the transformation group that leaves the above four-momentum invariant is locally isomorphic to O(2,1) and plays a pivotal role in studying noncompact groups and their applications in physics. This group has been discussed extensively in the literature.[11]

We are interested here in the question of whether the D matrix constructed in Secs. II and III can be analytically continued to $\alpha>1$. Indeed, we can perform the rotation and boost of Fig. 1 to obtain the D matrix of the form given in Eq. (8), if α is smaller than α_0 where

$$\alpha_0^2=[1+(\tan(\theta/2))^2]/(\tan(\theta/2))^2. \tag{17}$$

As α increases, some elements of the D matrix become singular when T vanishes or $\alpha=\alpha_0$. Mathematically, this is a simple pole that can be avoided either clockwise or counterclockwise. However, the physics of this continuation process requires a more careful investigation.

One way to study the D transformation more effectively is to boost the spacelike four-vector of Eq. (16) along the z direction to a simpler vector

$$(0,0,im,0), \tag{18}$$

using the boost matrix of Eq. (13) with the boost parameter $\beta=1/\alpha$. Consequently, the D matrix is a Lorentz-boosted form of a simpler matrix F:

$$D=B(1/\alpha)F(\lambda)[B(1/\alpha)]^{-1}. \tag{19}$$

Here F is a boost matrix along the x direction:

$$F(\lambda)=\begin{pmatrix}\cosh\lambda & 0 & 0 & \sinh\lambda\\ 0 & 1 & 0 & 0\\ 0 & 0 & 1 & 0\\ \sinh\lambda & 0 & 0 & \cosh\lambda\end{pmatrix}, \tag{20}$$

where

$$\tanh\lambda=\frac{-2(\alpha^2-1)^{1/2}\tan(\theta/2)}{1+(\alpha^2-1)(\tan(\theta/2))^2},\qquad \cosh\lambda=\frac{1+(\alpha^2-1)(\tan(\theta/2))^2}{1-(\alpha^2-1)(\tan(\theta/2))^2}. \tag{21}$$

If we add the rotational degree of freedom around the z axis, the above result is perfectly consistent with Wigner's original observation that the little group for imaginary-mass particles is locally isomorphic to O(2,1).[1]

We have observed earlier that the D matrix of Eq. (8) can be analytically continued from $\alpha=1$ to $1<\alpha<\alpha_0$. At $\alpha=\alpha_0$, some of its elements are singular. If $\alpha>\alpha_0$, $\cosh\lambda$ in Eqs. (20) and (21) become negative, and this is not acceptable.

One way to deal with this problem is to take advantage of the fact that the expression for $\tanh\lambda$ in Eq. (21) is never singular for real α greater than 1. This is possible if we change the signs of both $\sinh\lambda$ and $\cosh\lambda$ when we jump from $\alpha<\alpha_0$ to $\alpha>\alpha_0$. Indeed, the continuation is possible if it is accompanied by the reflection of x and t coordinates. After taking into account the reflection of the x and t coordinates, we can construct the D matrix by boosting F of Eq. (20). The expression for the D matrix for $\alpha>\alpha_0$ becomes

$$D=\begin{pmatrix} 1-2/T & 0 & u/T & -\alpha u/T \\ 0 & 1 & 0 & 0 \\ -u/T & 0 & 1+2/[(\alpha^2-1)T] & 2\alpha/[(\alpha^2-1)T] \\ -\alpha u/T & 0 & -2\alpha/[(\alpha^2-1)T] & 1-2\alpha^2[(\alpha^2-1)T] \end{pmatrix}. \qquad (22)$$

This expression cannot be used for the $\alpha\to 1$ limit, but can be used for the $\alpha\to\infty$ limit. In the limit $\alpha\to\infty$, P of Eq. (16) becomes identical to Eq. (18), and the above expression becomes an identity matrix. As for the question of whether D of Eq. (22) is an analytic continuation of Eq. (8), the answer is "no," because the transition from Eq. (22) to Eq. (8) requires the reflection of the x and t axes.

V. PARTICLES WITH SPIN-$\frac{1}{2}$

The purpose of this section is to study the D kinematics of spin-$\frac{1}{2}$ particles within the framework of SL(2,c). Let us study the Lie algebra of SL(2,c) (see Refs. 12 and 13):

$$[S_i,S_j]=i\epsilon_{ijk}S_k\,,\quad [S_i,K_j]=i\epsilon_{ijk}\,K_k\,,$$
$$[K_i,K_j]=-i\epsilon_{ijk}S_k\,, \qquad (23)$$

where S_i and K_i are the generators of rotations and boosts, respectively. The above commutation relations are not invariant under the sign change in S_i, but they remain invariant under the sign change in K_i. For this reason, while the generators of rotations are $S_i=\frac{1}{2}\sigma_i$, the boost generators can take two different signs $K_j=(\pm)(i/2)\sigma_j$.

Let us start with a massive particle at rest, and the usual normalized Pauli spinors χ_+ and χ_- for the spin in the positive and negative z directions, respectively. If we take into account Lorentz boosts, there are four spinors. We shall use the notation $\chi_\pm$ to which the boost generators $K_i=(i/2)\sigma_i$ are applicable, and $\dot\chi_\pm$ to which $K_i=-(i/2)\sigma_i$ are applicable. There are therefore four independent SL(2,c) spinors.[12,13] In the conventional four-component Dirac equation, only two of them are independent, because the Dirac equation relates the dotted spinors to the undotted counterparts. However, the recent development in supersymmetric theories,[14] as well as some of more traditional approaches,[15] indicates that both physics and mathematics become richer in the world where all four of SL(2,c) spinors are independent. In the Appendix, we examine the nature of the restriction the Dirac equation imposes on the four SL(2,c) spinors.

As Wigner did in 1957,[2] we start with a massive particle whose spin is initially along the direction of the momentum. The boost matrix, which brings the SL(2,c) spinors from the zero-momentum state to that of p, is

$$A^{(\pm)}(\alpha)=\begin{pmatrix}((1\pm\alpha)/(1\mp\alpha))^{1/4} & 0\\ 0 & ((1\mp\alpha)/(1\pm\alpha))^{1/4}\end{pmatrix}, \qquad (24)$$

where the superscripts (+) and (−) are applicable to the undotted and dotted spinors, respectively. In the Lorentz frame in which the particle is at rest, there is only one rotation applicable to both sets of spinors. The rotation matrix corresponding to W of Eq. (10) is

$$W(\theta^*)=\begin{pmatrix}\cos(\theta^*/2) & -\sin(\theta^*/2)\\ \sin(\theta^*/2) & \cos(\theta^*/2)\end{pmatrix}, \qquad (25)$$

where the rotation angle θ^* is given in Eq. (11).

Using the formula of Eq. (9), we can calculate the D matrix for the SL(2,c) spinors. The D matrix applicable to the undotted spinors is

$$D^{(+)}(\alpha,\theta)=\begin{pmatrix}1/\sqrt{T} & (1+\alpha)u/2\sqrt{T}\\ -(1-\alpha)u/2\sqrt{T} & 1/\sqrt{T}\end{pmatrix}, \qquad (26)$$

where T and u are given in Eq. (8). The D matrix applicable to the dotted spinors is

$$D^{(-)}(\alpha,\theta)=\begin{pmatrix}1/\sqrt{T} & (1-\alpha)u/2\sqrt{T}\\ -(1+\alpha)u/2\sqrt{T} & 1/\sqrt{T}\end{pmatrix}. \qquad (27)$$

We can obtain $D^{(-)}$ from $D^{(+)}$ by changing the sign of α. Both $D^{(+)}$ and $D^{(-)}$ become W of Eq. (25) when $\alpha=0$.

If the D transformation is applied to the $\chi_\pm$ and $\dot\chi_\pm$ spinors,

$$\chi'_\pm=D^{(+)}\chi_\pm\,,\quad \dot\chi'_\pm=D^{(-)}\dot\chi_\pm\,, \qquad (28)$$

the angle between the momentum and the directions of the spins represented by χ_+ and $\dot\chi_-$ is

$$\theta'=\tan^{-1}((1-\alpha)\tan(\theta/2))\,, \qquad (29)$$

which becomes zero as $\alpha\to 1$. On the other hand, in the case of χ_- and $\dot\chi_+$, the angle becomes

$$\theta''=\tan^{-1}((1+\alpha)\tan(\theta/2))\,. \qquad (30)$$

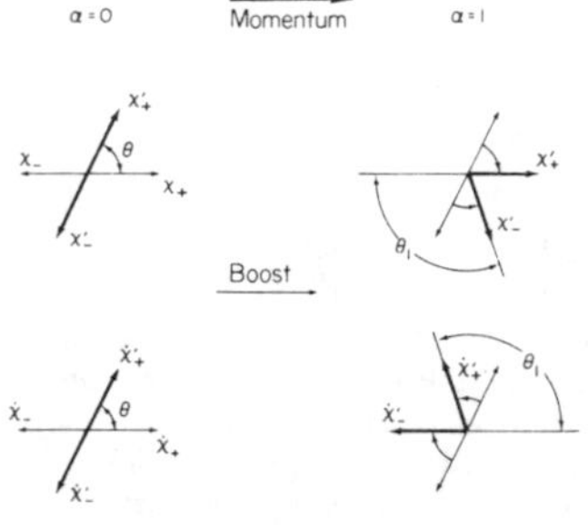

FIG. 3. Lorentz-boosted rotations of the four SL(2,c) spinors. If the particle velocity is zero, all the spinors rotate like the Pauli spinors. As the particle speed approaches that of light, two of the spins line up with the momentum, while the remaining two refuse to do so. Those spinors that line up are gauge-invariant spinors. Those that do not are not gauge invariant, and they form the origin of the gauge degrees of freedom for photon four-potentials.

In the limit of $\alpha \to 1$, this angle becomes θ_1, where

$$\theta_1 = \tan^{-1}(2(\tan(\theta/2))) . \quad (31)$$

Indeed, the spins represented by χ_- and $\dot{\chi}_+$ refuse to align themselves with the momentum. This result is illustrated in Fig. 3.

There are D transformations for the $\alpha > 1$ case. In the special Lorentz frame in which the four-momentum takes the form of Eq. (18), the D transformation becomes that of a pure boost along the x axis:

$$F^{(\pm)}(\lambda) = \begin{pmatrix} \cosh(\lambda/2) & \pm\sinh(\lambda/2) \\ \pm\sinh(\lambda/2) & \cosh(\lambda/2) \end{pmatrix}, \quad (32)$$

where λ is given in Eq. (21).

For $\alpha < \alpha_0$, we can continue to use $D^{(+)}$ and $D^{(-)}$ given in Eq. (26) and Eq. (27), respectively. However, for $\alpha > \alpha_0$, the D matrix is

$$D^{(\pm)}(\alpha,\theta) = \begin{pmatrix} (\alpha^2-1)^{1/2}(\tan(\theta/2))/\sqrt{-T} & \pm((\alpha\pm 1)/(\alpha\mp 1))^{1/2}/\sqrt{-T} \\ \pm((\alpha\mp 1)/(\alpha\pm 1))^{1/2}/\sqrt{-T} & (\alpha^2-1)^{1/2}(\tan(\theta/2))/\sqrt{-T} \end{pmatrix}. \quad (33)$$

The above expression becomes an identity matrix when $\alpha \to \infty$, as is expected from the result of Sec. IV. The D matrices of Eq. (33) are not analytic continuations of their counterparts given in Eqs. (26) and (27), because the continuation procedure, which we adopted in Sec. IV and used in this section, involves reflections in the x and t coordinates.

VI. GAUGE TRANSFORMATIONS IN TERMS OF ROTATIONS OF SPINORS

It is clear from the discussions of Secs. III–V that the limit $\alpha \to 1$ can be defined from both directions, namely from $\alpha < 1$ and from $\alpha > 1$. In the limit $\alpha \to 1$, $D^{(+)}$ and $D^{(-)}$ of Eq. (26) and Eq. (27) become

$$D^{(+)} = \begin{pmatrix} 1 & u \\ 0 & 1 \end{pmatrix}, \quad D^{(-)} = \begin{pmatrix} 1 & 0 \\ -u & 1 \end{pmatrix}. \quad (34)$$

After going through the same procedure as that from Eq. (12) to Eq. (15), we arrive at the gauge transformation matrices[8]

$$D^{(+)}(u,v) = \begin{pmatrix} 1 & u-iv \\ 0 & 1 \end{pmatrix}, \quad D^{(-)}(u,v) = \begin{pmatrix} 1 & 0 \\ -u-iv & 1 \end{pmatrix}, \quad (35)$$

applicable to the SL(2,c) spinors, where the $D^{(\pm)}$ are applicable to undotted and dotted spinors, respectively.

The SL(2,c) spinors are gauge invariant in the sense that

$$D^{(+)}(u,v)\chi_+ = \chi_+, \quad D^{(-)}(u,v)\dot{\chi}_- = \dot{\chi}_- . \quad (36)$$

On the other hand, the SL(2,c) spinors are gauge dependent in the sense that

$$D^{(+)}(u,v)\chi_- = \chi_- + (u-iv)\chi_+, \quad D^{(-)}(u,v)\dot{\chi}_+ = \dot{\chi}_+ - (u+iv)\dot{\chi}_- . \quad (37)$$

The gauge-invariant spinors of Eq. (36) appear as polarized neutrinos in the real world. However, where do the above gauge-dependent spinors stand in the physics of spin-$\frac{1}{2}$ particles? Are they really responsible for the gauge dependence of electromagnetic four-potentials when we construct a four-vector by taking a bilinear combination of spinors?

The relation between the SL(2,c) spinors and the four-vectors has been discussed for massive particles. However, it is not yet known whether the same holds true for the massless case. The central issue is again the gauge transformation. The four-potentials are gauge dependent, while the spinors allowed in the Dirac equation are gauge invariant. Therefore, it is not possible to construct four-potentials from the Dirac spinors.

On the other hand, there are gauge-dependent SL(2,c) spinors, which are given in Eq. (37). They disappear from the Dirac spinors because N_- vanishes in the $\alpha \to 1$ limit. However, these spinors can still play an important role if they are multiplied by N_+, which neutralizes N_-. Indeed, we can construct unit vectors in the Minkowskian space by taking the direct products of two SL(2,c) spinors

$$-\chi_+\dot{\chi}_+ = (1,i,0,0), \quad \chi_-\dot{\chi}_- = (1,-i,0,0), \quad \chi_+\dot{\chi}_- = (0,0,1,1), \quad \chi_-\dot{\chi}_+ = (0,0,1,-1). \quad (38)$$

These unit vectors in one Lorentz frame are not the unit vectors in other frames. For instance, if we boost a massive particle initially at rest along the z direction, $|\chi_+\dot{\chi}_+\rangle$ and $|\chi_-\dot{\chi}_-\rangle$ remain invariant. However, $|\chi_+\dot{\chi}_-\rangle$ and $|\chi_-\dot{\chi}_+\rangle$ acquire the constant factors $[(1+\alpha)/(1-\alpha)]^{1/2}$ and $[(1-\alpha)/(1+\alpha)]^{1/2}$, respectively. We can therefore drop $|\chi_-\dot{\chi}_+\rangle$ when we go through the renormalization process of replacing the coefficient $[(1+\alpha)/(1-\alpha)]^{1/2}$ by 1 for particles moving with the speed of light.

The $D(u,v)$ matrix for the above spinor combinations should take the form

$$D(u,v) = D^{(+)}(u,v)D^{(-)}(u,v), \quad (39)$$

where $D^{(+)}$ and $D^{(-)}$ are applicable to the first and second spinors of Eq. (38), respectively. Then

$$D(u,v)(-|\chi_+\dot{\chi}_+\rangle) = |\chi_+\dot{\chi}_+\rangle + (u+iv)|\chi_+\dot{\chi}_-\rangle, \quad D(u,v)|\chi_-\dot{\chi}_-\rangle = |\chi_-\dot{\chi}_-\rangle + (u-iv)|\chi_+\dot{\chi}_-\rangle, \quad D(u,v)|\chi_+\dot{\chi}_-\rangle = |\chi_+\dot{\chi}_-\rangle . \quad (40)$$

The first two equations of the above expression correspond to the gauge transformations on the photon polarization vectors. The third equation describes the effect of the D transformation on the four-momentum, confirming the fact that $D(u,v)$ is an element of the little group. The above operation is identical to that of the four-by-four D matrix of Eq. (15) on photon polarization vectors.

VII. CONCLUDING REMARKS

We studied in this paper Wigner's little groups by constructing a Lorentz kinematics that leaves the four-momen-

	Massive Slow	between	Massless Fast
Energy Momentum	$E=\frac{p^2}{2m}$	Einstein's $E=\sqrt{m^2+p^2}$	E = p
Spin, Gauge Helicity	S_3 S_1 S_2	Wigner's Little Group	S_3 Gauge Trans.

FIG. 4. Significance of the concept of Wigner's little groups. The beauty of Einstein's special relativity is that the energy-momentum relation for massive and slow particles and that for massless particles can be unified. Wigner's concept of the little groups unifies the internal space-time symmetries of massive and massless particles.

tum of a particle invariant. This kinematics consists of one rotation followed by one boost. Although the net transformation leaves the four-momentum invariant, the particle spin does not remain unchanged. The departure from the original spin orientation is studied in detail.

For a massive particle, this departure can be interpreted as a rotation in the Lorentz frame in which the particle is at rest. For massless particles with spin-1, the net result is a gauge transformation. For a spin-$\frac{1}{2}$ particle, there are four independent spinors as the Dirac equation indicates. As the particle mass approaches zero, the spin orientations of two of the spinors remain invariant. However, the remaining two spinors do not. It is shown that this noninvariance is the cause of the gauge degrees of freedom massless particles with spin-1.

In 1957,[2] Wigner considered the possibility of unifying the internal space-time symmetries of massive and massless particles by noting the difference between rotations and boosts. Wigner considered the scheme of obtaining the internal symmetry by taking the massless limit of the internal space-time symmetry groups for massive particles. In the present paper, we have added the gauge degrees of freedom and spinors that refuse to align themselves to the momentum in the massless limit. The result of the present paper can be summarized in Fig. 4. While Einstein's special relativity unifies the energy-momentum relations for massive and massless particles, Wigner's little group unifies the internal space-time symmetries of massive and massless particles.

ACKNOWLEDGMENTS

We are grateful to Professor Eugene P. Wigner for a very illuminating discussion on his 1957 paper[2] on transformations that preserve helicity and those that do not. We would like to thank Dr. Avi I. Hauser for explaining to us the content of his paper on possible imaginary-mass neutrinos.[14]

APPENDIX: SL(2,*c*) SPINORS IN THE DIRAC SPINORS

We pointed out in Sec. V that the four-component Dirac equation puts a restriction on the SL(2,c) spinors. Let us see how this restriction manifests itself in the limit procedure of $\alpha \to 1$. In the Weyl representation of the Dirac equation, the rotation and boost generators take the form

$$S_i = \begin{pmatrix} \frac{1}{2}\sigma_i & 0 \\ 0 & \frac{1}{2}\sigma_i \end{pmatrix}, \quad K_i = \begin{pmatrix} (i/2)\sigma_i & 0 \\ 0 & -(i/2)\sigma_i \end{pmatrix}. \tag{A1}$$

These generators accommodate both signs of the boost generators for the SL(2,c) spinors. In this representation, γ_5 is diagonal, and its eigenvalue determines the sign of the boost generators.

In the Weyl representation, the D matrix should take the form

$$D(u,v) = \begin{pmatrix} D^{(+)}(u,v) & 0 \\ 0 & D^{(-)}(u,v) \end{pmatrix}, \tag{A2}$$

applicable to the Dirac spinors, which, for the particle moving along the z direction with four-momentum p, are

$$U(\mathbf{p}) = \begin{pmatrix} N_+\chi_+ \\ \vdots \\ \pm N_-\chi_+ \end{pmatrix}, \quad V(\mathbf{p}) = \begin{pmatrix} \pm N_-\chi_- \\ \vdots \\ N_+\chi_- \end{pmatrix}, \tag{A3}$$

where the + and − signs in the above expression specify positive and negative energy states, respectively. Here N_+ and N_- are the normalization constants, and

$$N_\pm = ((1 \pm \alpha)/(1 \mp \alpha))^{1/4}. \tag{A4}$$

As the momentum/mass becomes very large, N_-/N_+ becomes very small. From Eqs. (36) and (37), we can see that the large components are gauge invariant while the small components are gauge dependent. The gauge-dependent component of the Dirac spinor disappears in the $\alpha \to 1$ limit; the Dirac equation becomes a pair of the Weyl equations. If we renormalize the Dirac spinors of Eq. (A3) by dividing them by N_+, they become

$$U(\mathbf{p}) = \begin{pmatrix} \chi_+ \\ \vdots \\ 0 \end{pmatrix}, \quad V(\mathbf{p}) = \begin{pmatrix} 0 \\ \vdots \\ \chi_- \end{pmatrix}, \tag{A5}$$

For $\gamma_5 = \pm 1$, respectively. The gauge-dependent spinors disappear in the large-momentum/zero-mass limit. This is precisely why we do not talk about gauge transformations on neutrinos in the two-component neutrino theory.

The important point is that we can obtain the above decoupled form of spinors immediately from the most general form of spinors by imposing the gauge invariance. This means that the requirement of gauge invariance is equivalent to $\gamma_5 = 1$, as was suspected in Ref. 8.

[1]E. P. Wigner, Ann. Math. **40**, 149 (1939); V. Bargmann and E. P. Wigner, Proc. Natl. Acad. Sci. (U.S.A.) **34**, 211 (1946); E. P. Wigner, in *Theoretical Physics*, edited by A. Salam (I.A.E.A., Vienna, 1963).

[2]E. P. Wigner, Rev. Mod. Phys. **29**, 255 (1957). See also C. Kuang-Chao and L. G. Zastavenco, Zh. Exp. Teor. Fiz. **35**, 1417 (1958) [Sov. Phys. JETP **8**, 990 (1959)]; M. Jacob and G. C. Wick, Ann. Phys. (NY) **7**, 404 (1959).

[3]J. Kupersztych, Nuovo Cimento B **31**, 1 (1976); Phys. Rev. D **17**, 629 (1978).

[4]The Wigner rotation is frequently mentioned in the literature because two successive boosts result in a boost preceded or followed by a rotation. We believe, however, that the Wigner rotation should be defined in the Lorentz frame in which the particle is at rest, in view of the fact that the O(3)-like little group is the rotation group in the rest frame. See R. Gilmore, *Lie Groups, Lie Algebras, and Some of Their Applications in Physics* (Wiley, New York, 1974); A. Le Yaouanc, L. Oliver, O. Pene, and J. C. Raynal, Phys. Rev. D **12**, 2137 (1975); A. Ben-Menahem, Am. J. Phys. **53**, 62 (1985). The concept of rotations in the rest frame played an important

role in the development of quantum mechanics and atomic spectra. See L. H.Thomas, Nature **117**, 514 (1926); Philos. Mag. **3**, 1 (1927).
[5]D. Han and Y. S. Kim, Am. J. Phys. **49**, 348 (1981); D. Han, Y. S. Kim, and D. Son, Phys. Rev. D **31**, 328 (1985).
[6]E. P. Wigner, Z. Phys. **124**, 665 (1948); A. S. Wightman, in *Dispersion Relations and Elementary Particles*, edited by C. De Witt and R. Omnes (Hermann, Paris, 1960); M. Hamermesh, *Group Theory* (Addison–Wesley, Reading, MA, 1962); E. P. Wigner, in *Theoretical Physics*, edited by A. Salam (I.A.E.A., Vienna, 1962); A. Janner and T. Jenssen, Physica **53**, 1 (1971); **60**, 292 (1972); J. L. Richard, Nuovo Cimento A **8**, 485 (1972); H. P. W. Gottlieb, Proc. R. Soc. London Ser. A **368**, 429 (1979).
[7]S. Weinberg, Phys. Rev. **134**, B 882 (1964); **135**, B1049 (1964).
[8]D. Han, Y. S. Kim, and D. Son, Phys. Rev. D **26**, 3717 (1982).
[9]E. Inonu and E. P. Wigner, Proc. Natl. Acad. Sci. (U.S.A.) **39**, 510 (1953); D. W. Robinson, Helv. Phys. Acta **35**, 98 (1962); D. Korff, J. Math. Phys. **5**, 869 (1964); S. Weinberg, in *Lectures on Particles and Field Theory, Brandeis 1964*, Vol. 2, edited by S. Deser and K. W. Ford (Prentice–Hall, Englewood Cliffs, NJ, 1965); J. D. Talman, *Special Functions, A Group Theoretical Approach Based on Lectures by E. P. Wigner* (Benjamin, New York, 1968); S. P. Misra and J. Maharana, Phys. Rev. D **14**, 133 (1976).
[10]D. Han, Y. S. Kim, and D. Son, Phys. Lett. B **131**, 327 (1983); D. Han, Y. S. Kim, M. E. Noz, and D. Son, Am. J. Phys. **52**, 1037 (1984).
[11]V. Bargmann, Ann. Math. **48**, 568 (1947); L. Pukanszky, Trans. Am. Math. Soc. **100**, 116 (1961); L. Serterio and M. Toller, Nuovo Cimento **33**, 413 (1964); A. O. Barut and C. Fronsdal, Proc. R. Soc. London Ser. A **287**, 532 (1965); M. Toller, Nuovo Cimento **37**, 631 (1968); W. J. Holman and L. C. Biedenharn, Ann. Phys. (NY) **39**, 1 (1966); **47**, 205 (1968); N. Makunda, J. Math. Phys. **9**, 50, 417 (1968); **10**, 2068, 2092 (1973); K. B. Wolf, J. Math. Phys. **15**, 1295, 2102 (1974); S. Lang, SL(2,*r*) (Addison–Wesley, Reading, MA, 1975).
[12]M. A. Naimark, Am. Math. Soc. Transl. **6**, 379 (1957); I. M. Gel'fand, R. A. Minlos, and Z. Ya. Shapiro, *Representations of the Rotation and Lorentz Groups and their Applications* (MacMillan, New York, 1963).
[13]Yu. V. Novozhilov, *Introduction to Elementary Particle Theory* (Pergamon, Oxford, 1975).
[14]S. J. Gates, M. T. Grisaru, M. Rocek, and W. Siegel, *Superspaces* (Benjamin/Cummings, Reading, MA, 1983). See also A. Chodos, A. I. Hauser, and V. A. Kostelecky, Phys. Lett. B **150**, 431 (1985); H. van Dam, Y. J. Ng, and L. C. Biedenharn, *ibid.* **158**, 227 (1985).
[15]L. C. Biedenharn, M. Y. Han, and H. van Dam, Phys. Rev. D **6**, 500 (1972).

Cylindrical group and massless particles

Y. S. Kim
Department of Physics and Astronomy, University of Maryland, College Park, Maryland 20742
E. P. Wigner
Joseph Henry Laboratories, Princeton University, Princeton, New Jersey 08544

(Received 30 September 1986; accepted for publication 31 December 1986)

It is shown that the representation of the E(2)-like little group for photons can be reduced to the coordinate transformation matrix of the cylindrical group, which describes movement of a point on a cylindrical surface. The cylindrical group is isomorphic to the two-dimensional Euclidean group. As in the case of E(2), the cylindrical group can be regarded as a contraction of the three-dimensional rotation group. It is pointed out that the E(2)-like little group is the Lorentz-boosted O(3)-like little group for massive particles in the infinite-momentum/zero-mass limit. This limiting process is shown to be identical to that of the contraction of O(3) to the cylindrical group. Gauge transformations for free massless particles can thus be regarded as Lorentz-boosted rotations.

I. INTRODUCTION

In their 1953 paper,[1] Inonu and Wigner discussed the contraction of the three-dimensional rotation group [or O(3)] to the two-dimensional Euclidean group [or E(2)]. Since the little groups governing the internal space-time symmetries of massive and massless particles are locally isomorphic to O(3) and E(2) respectively,[2] it is quite natural for us to expect that the E(2)-like little group is a limiting case of the O(3)-like little group.[3]

The kinematics of the O(3)-like little group for a massive particle is well understood. The identification of this little group with O(3) can best be achieved in the Lorentz frame in which the particle is at rest.[2] In this frame, we can rotate the direction of the spin without changing the momentum. Indeed, for a massive particle, the little group is for the description of the spin orientation in the rest frame.

The kinematics of the E(2)-like little group has been somewhat less transparent, because there is no Lorentz frame in which the particle is at rest. While the geometry of E(2) can best be understood in terms of rotations and translations in two-dimensional space, there is no physical reason to expect that the translationlike degress of freedom in the E(2)-like little group represent translations in an observable space. In fact, the translationlike degrees of freedom in the little group are the gauge degrees of freedom.[4] Therefore, in the past, the correspondence between the E(2)-like little group and the two-dimensional Euclidean group has been strictly algebraic.

In this paper, we formulate a group theory of a point moving on the surface of a circular cylinder. This group is locally isomorphic to the two-dimensional Euclidean group. We show that the transformation matrix of the little group for photons reduces to that of the coordinate transformation matrix of the cylindrical group. The cylindrical group therefore bridges the gap between E(2) and the E(2)-like little group.

As in the case of E(2), we can obtain the cylindrical group by contracting the three-dimensional rotation group. While the contraction of O(3) to E(2) is a tangent-plane approximation of a spherical surface with large radius,[1] the contraction to the cylindrical group is a tangent-cylinder approximation. Using this result, together with the fact that the representation of the E(2)-like little group reduces to that of the cylindrical group, we show that the gauge degree of freedom for massless particles comes from Lorentz-boosted rotations.

In Sec. II, we discuss the cylindrical group and its isomorphism to the two-dimensional Euclidean group. Section III deals with the E(2)-like little group for photons and its isomorphism to the cylindrical group. It is shown in Sec. IV that the cylindrical group can be regarded as an equatorial-belt approximation of the three-dimensional rotation group, while E(2) can be regarded as a north-pole approximation. In Sec. V, we combine the conclusions of Sec. III and Sec. IV to show that the gauge degrees of freedom for free massless particles are Lorentz-boosted rotational degrees of freedom.

II. TWO-DIMENSIONAL EUCLIDEAN GROUP AND CYLINDRICAL GROUP

The two-dimensional Euclidean group, often called E(2), consists of rotations and translations on a two-dimensional Euclidian plane. The coordinate transformation takes the form

$$x' = x\cos\alpha - y\sin\alpha + u,$$
$$y' = x\sin\alpha + y\cos\alpha + v. \tag{2.1}$$

This transformation can be written in the matrix form as

$$\begin{bmatrix} x' \\ y' \\ 1 \end{bmatrix} = \begin{bmatrix} \cos\alpha & -\sin\alpha & u \\ \sin\alpha & \cos\alpha & v \\ 0 & 0 & 1 \end{bmatrix} \begin{bmatrix} x \\ y \\ 1 \end{bmatrix}. \tag{2.2}$$

The three-by-three matrix in the above expression can be exponentiated as

$$E(u, v, \alpha) = \exp[-i(uP_1 + vP_2)]\exp(-i\alpha L_3), \tag{2.3}$$

where L_3 is the generator of rotations, and P_1 and P_2 generate translations. These generators take the form

Reprinted from *J. Math. Phys.* **28**, 1175 (1986).

$$L_3 = \begin{bmatrix} 0 & -i & 0 \\ i & 0 & 0 \\ 0 & 0 & 0 \end{bmatrix},$$

$$P_1 = \begin{bmatrix} 0 & 0 & i \\ 0 & 0 & 0 \\ 0 & 0 & 0 \end{bmatrix}, \quad P_2 \begin{bmatrix} 0 & 0 & 0 \\ 0 & 0 & i \\ 0 & 0 & 0 \end{bmatrix}, \tag{2.4}$$

and satisfy the commutation relations

$$[P_1,P_2] = 0, \quad [L_3,P_1] = iP_2, \quad [L_3,P_2] = -iP_1, \tag{2.5}$$

which form the Lie algebra for E(2).

The above commutation relations are invariant under the sign change in P_1 and P_2. They are also invariant under Hermitian conjugation. Since L_3 is Hermitian, we can replace P_1 and P_2 by

$$Q_1 = -(P_1)^\dagger, \quad Q_2 = -(P_2)^\dagger, \tag{2.6}$$

respectively, to obtain

$$[Q_1,Q_2] = 0, \quad [L_3,Q_1] = iQ_2, \quad [L_3,Q_2] = -iQ_1. \tag{2.7}$$

These commutation relations are identical to those for E(2) given in Eq. (2.5). However, Q_1 and Q_2 are not the generators of Euclidean translations in the two-dimensional space. Let us write their matrix forms:

$$Q_1 = \begin{bmatrix} 0 & 0 & 0 \\ 0 & 0 & 0 \\ i & 0 & 0 \end{bmatrix}, \quad Q_2 = \begin{bmatrix} 0 & 0 & 0 \\ 0 & 0 & 0 \\ 0 & i & 0 \end{bmatrix}. \tag{2.8}$$

Here L_3 is given in Eq. (2.4). As in the case of E(2), we can consider the transformation matrix

$$C(u,v,\alpha) = C(0,0,\alpha)C(u,v,0), \tag{2.9}$$

where $C(0,0,\alpha)$ is the rotation matrix and takes the form

$$C(0,0,\alpha) = \exp(-i\alpha L_3) = \begin{bmatrix} \cos\alpha & -\sin\alpha & 0 \\ \sin\alpha & \cos\alpha & 0 \\ 0 & 0 & 1 \end{bmatrix}, \tag{2.10}$$

$$C(u,v,0) = \exp[-i(uQ_1 + vQ_2)] = \begin{bmatrix} 1 & 0 & 0 \\ 0 & 1 & 0 \\ u & v & 1 \end{bmatrix}. \tag{2.11}$$

The multiplication of the above two matrices results in the most general form of $C(u,v,\alpha)$. If this matrix is applied to the column vector (x,y,z), the result is

$$\begin{bmatrix} \cos\alpha & -\sin\alpha & 0 \\ \sin\alpha & \cos\alpha & 0 \\ u & v & 1 \end{bmatrix} \begin{bmatrix} x \\ y \\ z \end{bmatrix} = \begin{bmatrix} x\cos\alpha - y\sin\alpha \\ x\sin\alpha + y\sin\alpha \\ z + ux + vy \end{bmatrix}. \tag{2.12}$$

This transformation leaves $(x^2 + y^2)$ invariant, while z can vary from $-\infty$ to $+\infty$. For this reason, it is quite appropriate to call the group of the above linear transformation the *cylindrical group*. This group is locally isomorphic to E(2).

If, for convenience, we set the radius of the cylinder to be unity,

$$(x^2 + y^2) = 1, \tag{2.13}$$

then x and y can be written as

$$x = \cos\phi, \quad y = \sin\phi, \tag{2.14}$$

and the transformation of Eq. (2.12) takes the form

$$\begin{bmatrix} \cos\alpha & -\sin\alpha & 0 \\ \sin\alpha & \cos\alpha & 0 \\ u & v & 1 \end{bmatrix} \begin{bmatrix} \cos\phi \\ \sin\phi \\ z \end{bmatrix} = \begin{bmatrix} \cos(\phi+\alpha) \\ \sin(\phi+\alpha) \\ z + u\cos\phi + v\sin\phi \end{bmatrix}. \tag{2.15}$$

We shall see in the following sections how this cylindrical group describes gauge transformations for massless particles.

III. E(2)-LIKE LITTLE GROUP FOR PHOTONS

Let us consider a single free photon moving along the z direction. Then we can write the four-potential as

$$A^\mu(x) = A^\mu e^{i\omega(z-t)}, \tag{3.1}$$

where

$$A^\mu = (A_1,A_2,A_3,A_0).$$

The momentum four-vector is clearly

$$p^\mu = (0,0,\omega,\omega). \tag{3.2}$$

Then, the little group applicable to the photon four-potential is generated by

$$J_3 = \begin{bmatrix} 0 & -i & 0 & 0 \\ i & 0 & 0 & 0 \\ 0 & 0 & 0 & 0 \\ 0 & 0 & 0 & 0 \end{bmatrix},$$

$$N_1 = \begin{bmatrix} 0 & 0 & -i & i \\ 0 & 0 & 0 & 0 \\ i & 0 & 0 & 0 \\ i & 0 & 0 & 0 \end{bmatrix}, \quad N_2 = \begin{bmatrix} 0 & 0 & 0 & 0 \\ 0 & 0 & -i & i \\ 0 & i & 0 & 0 \\ 0 & i & 0 & 0 \end{bmatrix}. \tag{3.3}$$

These matrices satisfy the commutation relations:

$$[J_3,N_1] = iN_2, \quad [J_3,N_2] = -iN_1, \quad [N_1,N_2] = 0, \tag{3.4}$$

which are identical to those for E(2). From these generators, we can construct the transformation matrix:

$$D(u,v,\alpha) = D(0,0,\alpha)D(u,v,0), \tag{3.5}$$

where

$$D(u,v,0) = \exp[-i(uN_1 + vN_2)],$$

$$D(0,0,\alpha) = R(\alpha) = \exp[-i\alpha J_3].$$

We can now expand the above formulas in power series, and the results are

$$R(\alpha) = \begin{bmatrix} \cos\alpha & -\sin\alpha & 0 & 0 \\ \sin\alpha & \cos\alpha & 0 & 0 \\ 0 & 0 & 1 & 0 \\ 0 & 0 & 0 & 1 \end{bmatrix}, \tag{3.6}$$

and

$$D(u,v,0) = \begin{bmatrix} 1 & 0 & -u & u \\ 0 & 1 & -v & v \\ u & v & 1-(u^2+v^2)/2 & (u^2+v^2)/2 \\ u & v & -(u^2+v^2)/2 & 1+(u^2+v^2)/2 \end{bmatrix}. \quad (3.7)$$

When applied to the four-potential, the above D matrix performs a gauge transformation,[4] while $R(\alpha)$ is the rotation matrix around the momentum.

The D matrices of Eq. (3.5) have the same algebraic property as that for the E matrices discussed in Sec. II. Why, then, do they look so different? In the case of the O(3)-like little group, the four-by-four matrices of the little group can be reduced to a block diagonal form consisting of the three-by-three rotation matrix and one-by-one unit matrix.[2] Is it then possible to reduce the D matrices to the form which can be directly compared with the three-by-three E or C matrices discussed in Sec. II?

One major problem in bringing the D matrix to the form of the E matrix is that the D matrix is quadratic in the u and v variables. In order to attack this problem, let us impose the Lorentz condition on the four-potential:

$$\frac{\partial}{\partial x^{\mu}}(A^{\mu}(x)) = p^{\mu}A_{\mu}(x) = 0, \quad (3.8)$$

resulting in $A_3 = A_0$. Since the third and fourth components are identical, the N_1 and N_2 matrices of Eq. (3.3) can be replaced, respectively, by

$$N_1 = \begin{bmatrix} 0 & 0 & 0 & 0 \\ 0 & 0 & 0 & 0 \\ i & 0 & 0 & 0 \\ i & 0 & 0 & 0 \end{bmatrix}, \quad N_2 = \begin{bmatrix} 0 & 0 & 0 & 0 \\ 0 & 0 & 0 & 0 \\ 0 & i & 0 & 0 \\ 0 & i & 0 & 0 \end{bmatrix}. \quad (3.9)$$

At the same time, the $D(u,v,0)$ of Eq. (3.7) becomes

$$D(u,v,0) = \begin{bmatrix} 1 & 0 & 0 & 0 \\ 0 & 1 & 0 & 0 \\ u & v & 1 & 0 \\ u & v & 0 & 1 \end{bmatrix}. \quad (3.10)$$

This matrix has some resemblance to the representation of the cylindrical group given in Eq. (2.11).[5]

In order to make the above form identical to Eq. (2.11), we use the light cone coordinate system in which the combinations x, y, $(z+t)/\sqrt{2}$, and $(z-t)/\sqrt{2}$ are used as the coordinate variables.[6] In this system the four-potential of Eq. (3.1) is written as

$$A^{\mu} = (A_1, A_2, (A_3 + A_0)/\sqrt{2}, (A_3 - A_0)/\sqrt{2}). \quad (3.11)$$

The linear transformation from the four-vector of Eq. (3.1) to the above expression is straightforward. According to the Lorentz condition, the fourth component of the above expression vanishes. We are thus left with the first three components.

During the transformation into the light-cone coordinate system, J_3 remains the same. If we take into account the fact that the fourth component of A^{μ} vanishes, N_1 and N_2 become

$$N_1 = \frac{1}{\sqrt{2}}\begin{bmatrix} 0 & 0 & 0 & 0 \\ 0 & 0 & 0 & 0 \\ i & 0 & 0 & 0 \\ 0 & 0 & 0 & 0 \end{bmatrix}, \quad N_2 = \frac{1}{\sqrt{2}}\begin{bmatrix} 0 & 0 & 0 & 0 \\ 0 & 0 & 0 & 0 \\ 0 & i & 0 & 0 \\ 0 & 0 & 0 & 0 \end{bmatrix}. \quad (3.12)$$

As a consequence, $D(u,v)$ takes the form

$$D(u,v,0) = \begin{bmatrix} 1 & 0 & 0 & 0 \\ 0 & 1 & 0 & 0 \\ u/\sqrt{2} & v/\sqrt{2} & 1 & 0 \\ 0 & 0 & 0 & 1 \end{bmatrix}, \quad (3.13)$$

and $R(\alpha)$ remains the same as before. It is now clear that the four-by-four representation of the little group is reduced to one three-by-three matrix and one trivial one-by-one matrix. If we use $\tilde{J}_3$, $\tilde{N}_1$, and $\tilde{N}_2$ for the three-by-three portion of the four-by-four J_3, N_1, and N_2 matrices, respectively, then

$$\tilde{J}_3 = L_3, \quad \tilde{N}_1 = (1/\sqrt{2})Q_1, \quad \tilde{N}_2 = (1/\sqrt{2})Q_2. \quad (3.14)$$

Now the identification of E(2)-like little group with the cylindrical group is complete.

IV. THE CYLINDRICAL GROUP AS A CONTRACTION OF O(3)

The contraction of O(3) to E(2) is well known and discussed widely in the literature.[1] The easiest way to understand this procedure is to consider a sphere with large radius, and a small area around the north pole. This area would appear like a flat surface. We can then make Euclidean transformations on this surface, consisting of translations along the x and y directions and rotations around any point within this area. Strictly speaking, however, these Euclidean transformations are SO(3) rotations around the x axis, y axis, and around the axis which makes a very small angle with the z axis.

Let us start with the generators of O(3), which satisfy the commutation relations:

$$[L_i, L_j] = i\epsilon_{ijk}L_k. \quad (4.1)$$

Here L_3 generates rotations around the north pole, and its matrix form is given in Eq. (2.4). Also, L_1 and L_2 take the form

$$L_1 = \begin{bmatrix} 0 & 0 & 0 \\ 0 & 0 & -i \\ 0 & i & 0 \end{bmatrix}, \quad L_2 = \begin{bmatrix} 0 & 0 & i \\ 0 & 0 & 0 \\ -i & 0 & 0 \end{bmatrix}. \quad (4.2)$$

For the present purpose, we can restrict ourselves to a small region near the north pole, where z is large and is equal to the radius of the sphere R, and x and y are much smaller than the radius. We can then write

$$\begin{bmatrix} x \\ y \\ 1 \end{bmatrix} = \begin{bmatrix} 1 & 0 & 0 \\ 0 & 1 & 0 \\ 0 & 0 & 1/R \end{bmatrix}\begin{bmatrix} x \\ y \\ z \end{bmatrix}. \quad (4.3)$$

The column vectors on the left- and right-hand sides are, respectively, the coordinate vectors on which the E(2) and O(3) transformations are applicable. We shall use the notation A for the three-by-three matrix on the right-hand side. In the limit of large R,

$$L_3 = AL_3A^{-1},$$
$$P_1 = (1/R)AL_2A^{-1}, \quad (4.4)$$
$$P_2 = -(1/R)AL_1A^{-1}.$$

This procedure leaves L_3 invariant. However, L_1 and L_2 become the P_1 and P_2 matrices discussed in Sec. II. Furthermore, in terms of P_1, P_2 and L_3, the commutation relations for O(3) given in Eq. (4.1) become

$$[L_3,P_1] = iP_2, \quad [L_3,P_2] = -iP_1,$$
$$[P_1,P_2] = -i(1/R)^2L_3. \quad (4.5)$$

In the large-R limit, the commutator $[P_1, P_2]$ vanishes, and the above set of commutators becomes the Lie algebra for E(2).

We have so far considered the area near the north pole where z is much larger than $(x^2+y^2)^{1/2}$. Let us next consider the opposite case, in which $(x^2+y^2)^{1/2}$ is much larger than z. This is the equatorial belt of the sphere. Around this belt, x and y can be written as

$$x = R\cos\phi, \quad y = R\sin\phi. \quad (4.6)$$

We can now write

$$\begin{bmatrix}\cos\phi\\ \sin\phi\\ z\end{bmatrix} = \begin{bmatrix}1/R & 0 & 0\\ 0 & 1/R & 0\\ 0 & 0 & 1\end{bmatrix}\begin{bmatrix}x\\ y\\ z\end{bmatrix}, \quad (4.7)$$

to obtain the vector space for the cylindrical group discussed in Sec. II. The three-by-three matrix on the right-hand side of the above expression is proportional to the inverse of the matrix A given in Eq. (4.3). Thus in the limit of large R,

$$L_3 = A^{-1}L_3A,$$
$$Q_1 = -(1/R)A^{-1}L_2A, \quad (4.8)$$
$$Q_2 = (1/R)A^{-1}L_1A.$$

In terms of L_3, Q_1, and Q_2, the commutation relations for O(3) given in Eq. (4.1) become

$$[L_3,Q_1] = iQ_2, \quad [L_3,Q_2] = -iQ_1,$$
$$[Q_1,Q_2] = -i(1/R)^2L_3, \quad (4.9)$$

which become the Lie algebra for E(2) in the large-R limit. The contraction of O(3) to E(2) and to the cylindrical group is illustrated in Fig. 1.

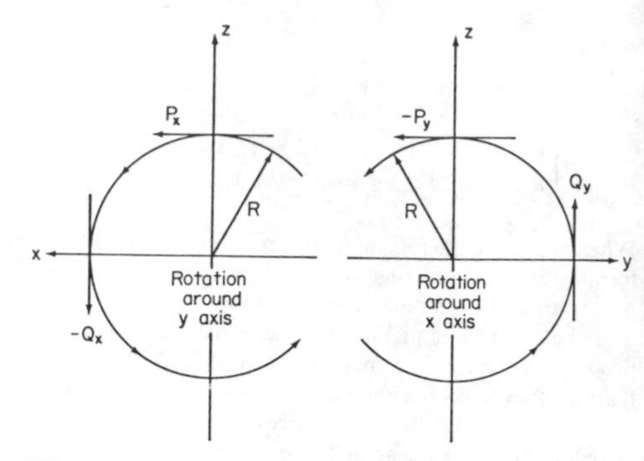

FIG. 1. Contraction of the three-dimensional rotation group to the two-dimensional Euclidean group and to the cylindrical group. The rotation around the z axis remains unchanged as the radius becomes large. In the case of E(2), rotations around the y and x axes become translations in the x and $-y$ directions, respectively, within a flat area near the north pole. In the case of the cylindrical group, the rotations around the y and x axes result in translations in the negative and positive z directions, respectively, within a cylindrical belt around the equator.

V. E(2)-LIKE LITTLE GROUP AS AN INFINITE-MOMENTUM/ZERO-MASS LIMIT OF THE O(3)-LIKE LITTLE GROUP FOR MASSIVE PARTICLES

If a massive particle is at rest, the symmetry group is generated by the angular momentum operators J_1, J_2, and J_3. If this particle moves along the z direction, J_3 remains invariant, and its eigenvalue is the helicity. However, what happens to J_1 and J_2, particularly in the infinite-momentum limit?

In order to tackle this problem, let us summarize the results of the preceding sections. The generators of the E(2)-like little group can be reduced to those of the cylindrical group. The cylindrical group can be obtained from the three-dimensional rotation group through a large-radius approximation. Therefore if the boost matrix takes a diagonal form as in the case of Eq. (4.3) or Eq. (4.7), we should be able to obtain N_1 and N_2 by boosting J_2 and J_1, respectively, along the z direction.[7]

Indeed, in the light-cone coordinate system, the boost matrix takes the form

$$B(P) = \begin{bmatrix}1 & 0 & 0 & 0\\ 0 & 1 & 0 & 0\\ 0 & 0 & R & 0\\ 0 & 0 & 0 & 1/R\end{bmatrix}, \quad (5.1)$$

with

$$R = \left(\frac{1+\beta}{1-\beta}\right)^{1/2},$$

where β is the velocity parameter of the particle. Under this boost, J_3 will remain invariant:

$$J_3' = BJ_3B^{-1} = J_3. \quad (5.2)$$

Here J_1 and J_2 in the light-cone coordinate system take the form

$$J_1 = \frac{1}{\sqrt{2}}\begin{bmatrix}0 & 0 & 0 & 0\\ 0 & 0 & -i & i\\ 0 & i & 0 & 0\\ 0 & -i & 0 & 0\end{bmatrix},$$
$$J_2 = \frac{1}{\sqrt{2}}\begin{bmatrix}0 & 0 & i & -i\\ 0 & 0 & 0 & 0\\ -i & 0 & 0 & 0\\ i & 0 & 0 & 0\end{bmatrix}. \quad (5.3)$$

If we boost this massive particle along the z direction, the boosted J_1 and J_2 become

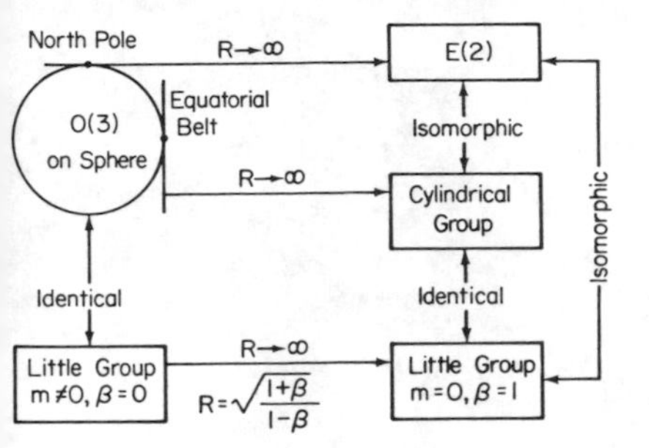

FIG. 2. Here are E(2), the E(2)-like little group for massless particles, and the cylindrical group. The correspondence between E(2) and the E(2)-like little group is isomorphic but not identical. The cylindrical group is identical to the E(2)-like little group. Both E(2) and the cylindrical group can be regarded as contractions of O(3) in the large-radius limit. The Lorentz boost of the O(3)-like little group for a massive particle at rest to the E(2)-like little group for a massless particle is exactly the same as the contraction of O(3) to the cylindrical group. The radius of the sphere in this case can be identified as $((1+\beta)/(1-\beta))^{1/2}$.

$$J_1' = BJ_1B^{-1} = \frac{1}{\sqrt{2}}\begin{bmatrix} 0 & 0 & 0 & 0 \\ 0 & 0 & -i/R & iR \\ 0 & iR & 0 & 0 \\ 0 & -i/R & 0 & 0 \end{bmatrix},$$

$$J_2' = BJ_2B^{-1} = \frac{1}{\sqrt{2}}\begin{bmatrix} 0 & 0 & i/R & -iR \\ 0 & 0 & 0 & 0 \\ -iR & 0 & 0 & 0 \\ i/R & 0 & 0 & 0 \end{bmatrix}. \quad (5.4)$$

Because of the Lorentz condition, the iR terms in the fourth column of the above matrices can be dropped. Therefore, in the large-R limit which is the limit of large momentum,

$$N_1 = -(1/R)J_2', \quad N_2 = (1/R)J_1', \quad (5.5)$$

where N_1 and N_2 are given in Eq. (3.12). This completes the proof that the gauge degrees of freedom in the E(2)-like little group for photons are Lorentz-boosted rotational degrees of freedom. The limiting process is the same as the contraction of the three-dimensional rotation group to the cylindrical group.

VI. CONCLUDING REMARKS

The isomorphism between the two-dimensional Euclidean group and the little group for massless particles is well known and well understood. However, the isomorphism in this case does not mean that they are identical. We have shown in this paper that the E(2)-like little group can be reduced to the identity group and the cylindrical group which is isomorphic to E(2). As in the case of E(2), we can obtain the cylindrical group by contracting the three-dimensional rotation group. This contraction procedure is identical to the Lorentz boost of the O(3)-like little group for a massive particle at rest to the E(2)-like little group for a massless particle. The result of the present paper is summarized in Fig. 2.

[1] E. Inonu and E. P. Wigner, Proc. Natl. Acad. Sci. USA **39**, 510 (1953); J. D. Talman, *Special Functions, A Group Theoretical Approach Based on Lectures by E. P. Wigner* (Benjamin, New York, 1968). See also R. Gilmore, *Lie Groups, Lie Algebras, and Some of Their Applications in Physics* (Wiley, New York, 1974).

[2] E. P. Wigner, Ann. Math. **40**, 149 (1939); V. Bargmann and E. P. Wigner, Proc. Natl. Acad. Sci. USA **34**, 211 (1946); E. P. Wigner, Z. Phys. **124**, 665 (1948); A. S. Wightman, in *Dispersion Relations and Elementary Particles*, edited by C. De Witt and R. Omnes (Hermann, Paris, 1960); M. Hamermesh, *Group Theory* (Addison-Wesley, Reading, MA, 1962); E. P. Wigner, in *Theoretical Physics*, edited by A. Salam (IAEA, Vienna, 1962); A. Janner and T. Jenssen, Physica **53**, 1 (1971); **60**, 292 (1972); J. L. Richard, Nuovo Cimento A **8**, 485 (1972); H. P. W. Gottlieb, Proc. R. Soc. London Ser. A **368**, 429 (1979); H. van Dam, Y. J. Ng, and L. C. Biedenharn, Phys. Lett. B **158**, 227 (1985). For a recent textbook on this subject, see Y. S. Kim and M. E. Noz, *Theory and Applications of the Poincaré Group* (Reidel, Dordrecht, Holland, 1986).

[3] E. P. Wigner, Rev. Mod. Phys. **29**, 255 (1957). See also D. W. Robinson, Helv. Phys. Acta **35**, 98 (1962); D. Korff, J. Math. Phys. **5**, 869 (1964); S. Weinberg, in *Lectures on Particles and Field Theory, Brandeis 1964*, edited by S. Deser and K. W. Ford (Prentice-Hall, Englewood Cliffs, NJ, 1965) Vol. 2; S. P. Misra and J. Maharana, Phys. Rev. D **14**, 133 (1976); D. Han, Y. S. Kim, and D. Son, J. Math. Phys. **27**, 2228 (1986).

[4] S. Weinberg, Phys. Rev. B **134**, 882 (1964); B **135**, 1049 (1964); J. Kuperzstych, Nuovo Cimento B **31**, 1 (1976); D. Han and Y. S. Kim, Am. J. Phys. **49**, 348 (1981); J. J. van der Bij, H. van Dam, and Y. J. Ng, Physica A **116**, 307 (1982); D. Han, Y. S. Kim, and D. Son, Phys. Rev. D **31**, 328 (1985).

[5] D. Han, Y. S. Kim, and D. Son, Phys. Rev. D **26**, 3717 (1982). For an earlier effort to study the E(2)-like little group in terms of the cylindrical group, see L. J. Boya and J. A. de Azcarraga, An. R. Soc. Esp. Fis. Quim. A **63**, 143 (1967). We are grateful to Professor Azcarraga for bringing this paper to our attention.

[6] P. A. M. Dirac, Rev. Mod. Phys. **21**, 392 (1949); L. P. Parker and G. M. Schmieg, Am. J. Phys. **38**, 218, 1298 (1970); Y. S. Kim and M. E. Noz, J. Math. Phys. **22**, 2289 (1981).

[7] D. Han, Y. S. Kim, and D. Son, Phys. Lett. B **131**, 327 (1983); D. Han, Y. S. Kim, M. E. Noz, and D. Son, Am. J. Phys. **52**, 1037 (1984). These authors studied the correspondence between the contraction of O(3) to E(2) and the Lorentz boost of the O(3)-like little group.

Chapter VIII

Localization Problems

In quantum mechanics, solutions of the Schrödinger equation carry a localizable probability interpretation. On the other hand, solutions of Maxwell's equations do not necessarily have a probability interpretation. It was in fact pointed out by Newton and Wigner in 1949 that there is no covariant Hermitian position operator for photons. In 1962, Wightman established this as a mathematical theorem. This is known as the photon localization problem.

On the other hand, from oscilloscope screens, we observe what can be described as localized photons. Are they really photons? The answer to this question is "No." They are localized light waves. It was shown by Han, Kim, and Noz in 1987 that it is possible to localize light waves in a covariant manner with a probability interpretation. However, this light wave cannot be given a particle interpretation in terms of the creation and annihilation operators. Two different mathematical algorithms are employed for photons and light waves, and the difference can be seen very easily.

Reprinted from REVIEWS OF MODERN PHYSICS, Vol. 21, No. 3, pp. 400–406, July, 1949
Printed in U. S. A.

Localized States for Elementary Systems

T. D. NEWTON AND E. P. WIGNER
Palmer Physical Laboratory, Princeton University, Princeton, New Jersey

It is attempted to formulate the properties of localized states on the basis of natural invariance requirements. Chief of these is that a state, localized at a certain point, becomes, after a translation, orthogonal to all the undisplaced states localized at that point. It is found that the required properties uniquely define the set of localized states for elementary systems of non-zero mass and arbitrary spin. The localized functions belong to a continuous spectrum of an operator which it is natural to call the position operator. This operator has automatically the property of preserving the positive energy character of the wave function to which it is applied (and it should be applied only to such wave functions). It is believed that the development here presented may have applications in the theory of elementary particles and of the collision matrix.

INTRODUCTION

IT is well known that invariance arguments suffice to obtain an enumeration of the relativistic equations for elementary systems.[1] The concept of an "elementary system" is, however, not quite identical with the intuitive concept of an elementary particle. Intuitively, we consider a particle "elementary" if it does not appear to be useful to attribute a structure to it. The definition under which the aforementioned enumeration can be made is a more explicit one: it requires that all states of the system be obtainable from the relativistic transforms of *any* state by superpositions. In other words, there must be no relativistically invariant distinction between the various states of the system which would allow for the principle of superposition. This condition is often referred to as irreducibility condition. Relativistic transform is meant to include in the above connection not only the customary Lorentz transformations but also rotations and displacements in space and time.

The role of elementary systems as initial and final states of collision phenomena, and hence their connection with the theory of the collision matrix, will be discussed at the end of this article. We wish to turn now to the connection of elementary systems with elementary particles.

Two conditions seem to play the most important role in the concept of an elementary particle. The first one is that its states shall form an elementary system in the sense given above. This condition is quite unambiguous. The second condition is less clear cut: it is that it should not be useful to consider the particle as a union of other particles. In the case of an electron or a proton both conditions are fulfilled and there is no question as to the elementary nature of these particles. Only the first condition is fulfilled for a hydrogen atom in its normal state and we do not consider it to be an elementary particle.

The situation is more ambiguous, for instance, in the case of a π-meson. Qualitatively, a π-meson differs in no way from a very sharp resonance state, formed by the collision of a μ-meson and a neutrino. Strictly speaking, the states of a π-meson do not form an elementary system because, after a sufficiently long time, it can be either in the dissociated or in the undissociated state and the distinction between these is surely invariant relativistically. Nevertheless, the life time of the π-meson is very long as compared with any relevant unit of time (such as h/mc^2) and within this time interval its states do form an elementary system. On the other hand, the properties of a π-meson are very different from what one would expect from a compound consisting of a μ-meson and a neutrino. Thus the second condition for an elementary particle is fulfilled. It is this condition which has no counterpart in the definition of an elementary system. As a result of this circumstance the concept of an elementary system is much broader than that of an elementary particle; as was mentioned above, a hydrogen atom in its normal state forms an elementary system.

Every system, even one consisting of an arbitrary number of particles, can be decomposed into elementary systems. These elementary systems can be specified in a relativistically invariant manner, as containing only certain states. Thus, the restriction to the normal state of the hydrogen atom selected an elementary system from all the states of the hydrogen atom, which, together, do not form an elementary system. The usefulness of the decomposition into elementary systems depends on how often one has to deal with linear combinations containing several elementary systems.

The great drawback of using the elementary systems as a basis of the theory is that their existence follows by a rather abstract argument from the principles of quantum mechanics. As a result, the expressions for some of the most important operators "get lost" in the process. The only physical quantities for which the theory directly provides expressions are the basic quantities of the components of the momentum-energy vector and the six components of the relativistic angular momentum tensor. The subject of the present article is an attempt to find general, invariant theoretic principles

[1] E. P. Wigner, Ann. of Math. **40**, 149 (1939). The concept of an elementary system, which will be explained below, is a description of a set of states which forms, in mathematical language, an irreducible representation space for the inhomogeneous Lorentz group.

Reprinted from *Rev. Mod. Phys.* **21**, 400 (1949).

on the basis of which operators for the position coordinates can be found.

If we restrict ourselves to an elementary system, the physical interpretation of the operators to be found is unique: they will correspond to the position of the particle if we deal with an elementary particle. Otherwise they may correspond to the center of mass of the system. If the system is not elementary, the interpretation will not be unique and neither will our postulates lead us to a uniquely determined set of operators.

Before proceeding with our argument, we wish to refer to other investigations with somewhat similar objectives. The problem of the center of mass in relativity theory has been treated particularly by Eddington[2] and by Fokker[3] on the basis of non-quantum mechanics. Their work was evaluated and a quantum mechanical generalization thereto given by Pryce.[4] We shall have frequent occasion to refer to his results. Ideas related to Pryce's work have been first put forward by Schrödinger[5] and, more recently, by Finkelstein[6] and also by Møller.[7]

The present paper arose from a reinvestigation of the irreducible representations[8] of de Sitter space which was undertaken by one of us.[9] These representations are in a one to one correspondence with relativistically invariant wave equations for elementary systems in de Sitter space. At the conclusion of the investigation it appeared that the physical content of the equations which were obtained could be understood much more readily if position operators could be defined on an invariant theoretic basis. As an introduction to this, a similar investigation was undertaken in flat space with the results given in the following sections.

POSTULATES FOR LOCALIZED STATES AND POSITION OPERATORS

The position operator could easily be written down if the wave function of the state (or the states) were known for which the three space coordinates are zero at $t=0$. If ψ is such a function and $T(a)$ the operator of displacement by a_x, a_y, a_z, a_t, the wave function $T(a)^{-1}\psi$ represents a state for which the space coordinates are a_x, a_y, a_z at time a_t. Thus the knowledge of the wave functions corresponding to the state $x=y=z=0$ at $t=0$ (and the knowledge of the displacement operators) entails the knowledge of all localized states, i.e., of all characteristic functions of the position operators. From these, the position operators are easily obtained. For this reason we concentrated on obtaining the wave functions of those states which are, at time $t=0$, localized at the origin of the coordinate system.

We postulate that the states which represent a system localized at time $t=0$ at $x=y=z=0$: (a) form a linear set S_0, i.e., that the superposition of two such localized states be again localized in the same manner; (b) that the set S_0 be invariant under rotations about the origin and reflections both of the spatial and of the time coordinate; (c) that if a state ψ is localized as above, a spatial displacement of ψ shall make it orthogonal to all states of S_0; (d) certain regularity conditions, amounting essentially to the requirement that all the infinitesimal operators of the Lorentz group be applicable to the localized states, will be introduced later.

It is to be expected that the states localized at a certain point have the same properties as characteristic functions of a continuous spectrum, i.e., they will not be square integrable but the limits of square integrable functions. It seems to us that the above postulates are a reasonable expression for the localization of the system to the extent that one would naturally call a system unlocalizable if it should prove to be impossible to satisfy these requirements.

We shall carry out our calculations in the realization of the elementary systems which was described by Bargmann and Wigner[10] and will proceed with the calculation.

Particle with no spin (Klein-Gordon particle)

The determination of the localized state is particularly simple in this case. It will be carried out in some detail in spite of this, because the same steps occur in the consideration of systems with spin.

The wave functions are defined, in this case, on the positive shell of a hyperboloid $p_0^2=p_1^2+p_2^2+p_3^2+\mu^2$ and we shall use p^1, p^2, p^3 as independent variables. In any formula, p_0 is an abbreviation for $(p_1^2+p_2^2+p_3^2+\mu^2)^{\frac{1}{2}}$. The invariant scalar product is

$$(\psi,\phi)=\int\int\int\psi(p_1p_2p_3)^*\phi(p_1p_2p_3)dp_1dp_2dp_3/p_0. \quad (1)$$

The wave function Φ in coordinate space becomes

$$\Phi(x^1,x^2,x^3,x^0)=(2\pi)^{-\frac{3}{2}}\int\phi(p_1,p_2,p_3)\times\exp(-i\{x,p\})dp_1dp_2dp_3/p_0, \quad (2)$$

where

$$\{x,p\}=x^0p^0-x^1p^1-x^2p^2-x^3p^3=x_0p_0-x_1p_1-x_2p_2-x_3p_3, \quad (3)$$

is the Lorentz invariant scalar product. Throughout this paper, the covariant and contravariant components are equal for the time (0) coordinate, oppositely equal for the space (1, 2, 3) coordinates. This governs the

[2] A. S. Eddington, *Fundamental Theory* (Cambridge University Press, London, 1946).
[3] A. D. Fokker, *Relativitatstheorie* (Groningen, Noordhoff, 1929).
[4] M. H. L. Pryce, Proc. Roy. Soc. **195A**, 62 (1948).
[5] E. Schrödinger, Berl. Ber. 418 (1930); 63 (1931).
[6] R. J. Finkelstein, Phys. Rev. **74**, 1563A (1948).
[7] Chr. Møller, Comm. Dublin Inst. for Adv. Studies A, No. 5 (1949); also A. Papapetrou, Acad. Athens **14**, 540 (1939).
[8] L. H. Thomas, Ann. of Math. **42**, 113 (1941).
[9] T. D. Newton, Princeton Dissertation (1949).
[10] V. Bargmann and E. P. Wigner, Proc. Nat. Acad. Sci. **34**, 211 (1948).

raising and lowering of all indices. Occasionally, we shall use for the scalar product of two space-like vectors the notation (x, p) so that, e.g., $\{x, p\} = x^0p^0 - (x \cdot p)$.

The linear manifolds which are invariant with respect to rotations about $p_1 = p_2 = p_3 = 0$ are, for any integer j, the $2j+1$ functions

$$P_m^j(\vartheta, \phi) f(p) \quad (m = -j, -j+1, \cdots j-1, j), \quad (4)$$

where p, ϑ and ϕ are polar coordinates for p_1, p_2, p_3 and f is an arbitrary function. The P_m^j are the well-known spherical harmonics. The sets (4) are also invariant with respect to inversion, i.e., replacement of p_1, p_2, p_3 by $-p_1$, $-p_2$, $-p_3$. Naturally, not only a single set (4) has these properties of invariance but the sum of an arbitrary number of such sets as long as one includes with one function (4) all $2j+1$ functions and their linear combinations. The $f(p)$ could be different for different j.

Under time reversal $\psi(p_1p_2p_3)$ goes over into[11]

$$\Theta\psi(p_1, p_2, p_3) = \psi(-p_1, -p_2, -p_3)^*. \quad (5)$$

We understand by time reversal the operation which makes out of a wave function ψ the wave function $\theta\psi$ on which every experiment, if carried out at $-t$, yields same results as the same experiment carried out on ψ at time t. Because of (5) and our postulate (b), if the $P_m^j(\theta, \phi)f(p)$ are localized at the origin, the set $P_{-m}^j(\theta, \phi)f(p)^*$, i.e., the $P_{+m}^j(\theta, \phi)f(p)^*$ are also localized. The same is then true for the sum and difference of the corresponding pairs of functions which shows that the $f(p)$ can be assumed to be real without loss of generality.

The displacement operator in momentum space is simply multiplication with $\exp(-i\{a, p\})$;

$$T(a)\psi = \exp(-i\{a, p\})\psi. \quad (6)$$

We shall have to consider purely space-like displacements, i.e., assume that $a^0 = 0$. It then follows from our postulate (c) that, in particular, $\exp(i(a, p))\psi$ is orthogonal to ψ if ψ is localized, or that

$$\iiint |\psi(p_1p_2p_3)|^2 \exp i(a_1p_1 + a_2p_2 + a_3p_3) dp_1dp_2dp_3/p_0 = 0, \quad (6a)$$

for any non-vanishing vector a. This shows that only the zero wave number part occurs in the expansion of $|\psi|^2/p_0$ into a Fourier integral. Hence $|\psi|^2/p_0$ is a constant, $|\psi|$ proportional to $p_0^{\frac{1}{2}}$. Comparing this with (4), we see that only the set $j=0$ can be chosen. Since, furthermore, we saw that $f(p)$ can be assumed to be real, we have

$$\psi^2 = (2\pi)^{-3} p_0. \quad (7)$$

As was anticipated, (ψ, ψ) is infinite, the localized function is part of a continuous spectrum.

[11] E. P. Wigner, Gottinger Nachrichten 546 (1932).

As far as postulates (a), (b), (c) are concerned, ψ could be a discontinuous function, being $+p_0^{\frac{1}{2}} = (p^2+\mu^2)^{\frac{1}{4}}$ for some p, and $-(p^2+\mu^2)^{\frac{1}{4}}$ for the remaining p. However, no matter how ψ is chosen, consistently with (7), there is, in this case, only one state localized at the origin because if there were two, say ψ_1, and ψ_2, the ψ_1 would have to be orthogonal not only to $\psi_1 \exp(-i(a, p))$ but also to $\psi_2 \exp(-i(a, p))$ from which not only $|\psi|^2 \sim p_0$ but also $\psi_1^*\psi_2 \sim p_0$ and hence the proportionality of ψ_1 to ψ_2 follows.

In order to eliminate the discontinuous ψ as localized state, we introduce the further regularity condition that

$$(M^{0k}\psi_n, M^{0k}\psi_n)/(\psi_n, \psi_n) \quad (8)$$

shall remain finite as the normalizable wave functions ψ_n approach ψ. The M_{0k} is the infinitesimal operator of a proper Lorentz transformation in the x^0x^k plane, its operator is[10]

$$M^{0k} = ip^0\partial/\partial p_k. \quad (8a)$$

This further postulate eliminates all discontinuous ψ and we obtain for the wave function of the only state which is localized at the origin

$$\psi = (2\pi)^{-\frac{3}{2}} p_0^{\frac{1}{2}}. \quad (9)$$

The regularity requirement (d) actually asks for the finiteness of (8) for all M^{kl}. However, if one substitutes M^{23}, M^{31} or M^{12} for M^{0k} in (8), the resulting expression is automatically bounded—in fact their sum is $j(j+1)$. Hence requiring the applicability of the M^{23}, M^{31}, M^{12} to ψ does not introduce a new condition.

The localized wave function in coordinate space is obtained by (2). It is, apart from a constant[12]

$$\Psi(r) = (\mu/r)^{5/4} H_{5/4}^{(1)}(i\mu r). \quad (9a)$$

It goes to zero at $r=\infty$ as $e^{-\mu r}$, at $r=0$ it becomes infinite as $r^{-5/2}$. It is, of course, not square integrable since it is part of a continuous spectrum.

Applying the operator of displacement to (9) we obtain for the wave function of the state which is localized at x^1, x^2, x^3 at time $t=0$

$$T(-x)\psi = (2\pi)^{\frac{3}{2}} p_0^{\frac{1}{2}} \exp -i(p^1x^1 + p^2x^2 + p^3x^3) = (2\pi)^{-\frac{3}{2}} p_0^{\frac{1}{2}} e^{-i(p\cdot x)}.$$

This must be a characteristic function of the operator q^k for the k-coordinate with the characteristic value x^k. The operator q^k is therefore defined by

$$q^k\phi(p) = (2\pi)^{-3} \int p_0^{\frac{1}{2}} e^{-i(p\cdot x)} x^k (p_0')^{\frac{1}{2}} \times e^{i(p'\cdot x)} \phi(p') dx dp'/p_0'; \quad (10)$$

dx and dp' stand for $dx^1dx^2dx^3$ and $dp_1'dp_2'dp_3'$. One

[12] G. A. Campbell and R. M. Foster, "Fourier Integrals for Practical Applications," American Telephone and Telegraph Company (1931).

can transform (10) in a well-known fashion

$$q^k\phi(p)=\left(i\frac{\partial}{\partial p^k}-\frac{i}{2}\frac{p^k}{p_0{}^2}\right)(2\pi)^{-3}$$
$$\times\int p_0{}^{\frac{1}{2}}e^{i(p'-p)\cdot x}(p_0')^{-\frac{1}{2}}\phi(p')dxdp'$$
$$=-i\left(\frac{\partial}{\partial p_k}+\frac{p^k}{2p_0{}^2}\right)\phi(p). \tag{11}$$

These expressions are valid for finite as well as for vanishing rest mass. It is remarkable that the operator q^k can be transformed into coordinate space and retains a relatively simple form

$$q^k\Phi(x)=x^k\Phi(x)+\frac{1}{8\pi}\int\frac{\exp(-\mu|x-y|)}{|x-y|}\frac{\partial\Phi(y)}{\partial y_k}dy. \tag{12}$$

x and y stand for the spatial part of the four vectors x^μ and y^μ and dy indicates integration over y^1, y^2, y^3. The customary q^k operator contains only the first term of (11).

It may be well to remember at this point that the position operators to which our postulates lead necessarily commute with each other so that only Pryce's case (*e*) can be used for comparison. In fact, our q^k is identical with his $\bar{q}^k$. It may be pointed out, second, that a state which is localized at the origin in one coordinate system, is not localized in a moving coordinate system, even if the origins coincide at $t=0$. Hence our operators q^k have no simple covariant meaning under relativistic transformations. This is not the case for the customary operators q^k either. Furthermore, even though it appears that $\Phi(x)=\delta(x)$ is invariant under relativistic transformations which leave the origin unchanged, this is not much more than a mathematical quirk. One sees this best by transforming the δ-function to momentum space through the inversion of (2). The result, p_0, seems to have a simple covariant meaning. However, it does not represent a square integrable function and if one approximates it by one, say by $\psi_\alpha=p_0\exp(-\alpha^2p_0{}^2)$, the Lorentz transform of ψ_α will not approach ψ_α with decreasing α. In fact, as soon as $\alpha\mu\ll 1$, the scalar product of ψ_α and its transform will be independent of α and smaller than the norm of ψ_α.

Particles with spin and finite mass

We again use the description given in reference 10, i.e., define wave functions on the positive hyperboloid $p_0{}^2=p_1{}^2+p_2{}^2+p_3{}^2+\mu^2$ and use in addition to p_1, p_2, p_3 the $2s$ spin variables $\xi_1, \xi_2, \cdots\xi_{2s}$ all of them four-valued. The wave functions which describe the possible states of the system will be symmetric functions of the ξ and satisfy the $2s$ equations

$$\textstyle\sum_\kappa\gamma_\alpha{}^\kappa p_\kappa\psi=\mu\psi\quad \alpha=1, 2, \cdots 2s. \tag{13}$$

The consistency of these was shown before.[10] The γ_α apply to ξ_α, two γ with different first indices commute, with the same first index they satisfy the well-known relations

$$\gamma_\alpha{}^\kappa\gamma_\alpha{}^\lambda+\gamma_\alpha{}^\lambda\gamma_\alpha{}^\kappa=2g^{\kappa\lambda}. \tag{13a}$$

The great difference between the present case and that of zero spin consists in the limitation (13) of the permissible wave functions, in addition to the limitation to the positive hyperboloid. This latter limitation can be taken care of by using only the p_1, p_2, p_3 as independent variables, the former limitation cannot be taken care of in an equally simple fashion. We shall make extensive use, however, of a device, most successfully employed by Schrödinger[5] and define operators

$$E_\alpha=\tfrac{1}{2}(p_0)^{-1}(\textstyle\sum_\kappa\gamma_\alpha{}^\kappa p_\kappa+\mu)\gamma_\alpha{}^0. \tag{14}$$

This is a projection operator: $E_\alpha{}^2=E_\alpha$ and $E_\alpha\psi$ automatically satisfies the corresponding Eq. (13). Denoting the product of all the E_α by E

$$E=E_1E_2\cdots E_{2s} \tag{14a}$$

any $E\psi$ is a permissible wave function, satisfying all Eqs. (13).

For the scalar product, we shall use the expression

$$(\psi,\phi)=\int p_0{}^{-2s-1}\textstyle\sum_\xi\psi^*\phi dp. \tag{15}$$

It follows from this at once, because of our postulate (*c*), and since (6) is valid in this case also, that every wave function which is localized at the origin satisfies the analogon of (7):

$$\textstyle\sum_\xi|\psi|^2=(2\pi)^{-3}p_0{}^{2s+1}. \tag{16}$$

The operator for time reversal is

$$\Theta\psi(p_1,p_2,p_3)=C\psi(-p_1,-p_2,-p_3)^*, \tag{17}$$

where C is a matrix which operates on the ξ coordinates and satisfies the equations

$$\begin{aligned}C\gamma_\alpha{}^{0*}&=\gamma_\alpha{}^0C; &&(\alpha=1,2,\cdots,2s)\\ C\gamma_\alpha{}^{k*}&=-\gamma_\alpha{}^kC. &&(\alpha=1,2,\cdots,2s;\,k=1,2,3).\end{aligned} \tag{17a}$$

If γ^0, γ^2, γ^3 are real, γ^1 imaginary

$$C=\prod_{\alpha=1}^{2s}\gamma_\alpha{}^2\gamma_\alpha{}^3;\quad C^2=(-)^{2s}. \tag{17b}$$

Since C, as defined above, is a real matrix we also have $\Theta^2=(-)^{2s}$ which is true independently of the choice of the γ-matrices. The operator for the inversion of the space coordinates is

$$I\psi(p_1,p_2,p_3)=\gamma_1{}^0\gamma_2{}^0\cdots\gamma_{2s}{}^0\psi(-p_1,-p_2,-p_3); \tag{18}$$

it commutes with the E_α of (14).

In order to determine the sets of wave functions which are invariant under rotations, we first define the analogue of the pure spin function for the relativistic

Eqs. (13). For this purpose we define auxiliary functions v_m which are independent of p_1, p_2, p_3 and functions of the ξ only. They satisfy the equations

$$\gamma_\alpha^0 v_m = v_m \quad (\alpha=1, 2, \cdots, 2s) \tag{19}$$

and

$$\tfrac{1}{2}i\sum_\alpha \gamma_\alpha^1\gamma_\alpha^2 v_m = m v_m \quad (m=-s, -s+1, \cdots, s-1, s). \tag{19a}$$

Since the γ^0 and the $i\gamma^1\gamma^2$ commute, it is possible to assume temporarily that they are all diagonal. Equation (19) then demands that they all belong to the characteristic value $+1$ of each γ^0; there are 2^{2s} such functions. However, we are interested only in symmetric functions of the ξ and there are only $2s+1$ of these. They are distinguished by the index m: the v_m has non-zero components only for those ξ for which $s+m$ of the $i\gamma^1\gamma^2$ are $+1$, the remaining $s-m$ are -1. For these ξ the value of v_m is $((s+m)!(s-m)!/(2s)!)^{\frac{1}{2}}2^{-s}$ so that v_m is normalized in the sense

$$\sum_\xi |v_m|^2 = \sum_\xi v_m^2 = 1. \tag{19b}$$

Physically, m corresponds to the spin angular momentum about the x^3 axis, the parity of v_m is even because of (19) and (18).

The v_m are not permissible wave functions because they do not satisfy the wave Eqs. (13). We therefore define as spin functions

$$V_m(p_1, p_2, p_3, \xi_1, \cdots, \xi_{2s}) = E v_m \quad (m=-s, \cdots, s). \tag{20}$$

They are permissible wave functions of even parity and V_m represents a state of angular momentum $m\hbar$ about the third coordinate axis. Their normalization is, instead of (19b)

$$\sum_\xi |V_m|^2 = ((p_0+\mu)/2p_0)^{2s}. \tag{20a}$$

The most general solution of (13) is a linear combination of the V_m multiplied with arbitrary functions of the p_1, p_2, p_3. A set of wave functions which is invariant under rotations and reflections contains wave functions of the form

$$\psi_{jm} = \sum_{l, m'} S(l, s)_{j, m-m', m'} P^l_{m-m'}(\vartheta, \phi) f_l(p) V_{m'}. \tag{21}$$

The p, ϑ, ϕ are again polar coordinates for p^1, p^2, p^3; the f_l are arbitrary unknown functions of the length of p. However, if one function of the form (21) occurs in the set, all others with different m but the same f_l also occur. The summation over l is to be extended over all even values between $|j-s|$ and $j+s$ if the parity of the ψ_j is to be even, over all odd values of l if the parity of ψ_j is odd. The $S(l, s)$ are the customary coefficients[13] giving a total j from wave functions with given "orbital momentum" l and "spin momentum" s.

Since the polar angles ϑ, ϕ are indeterminate for $p=0$, the $f_l(p)$ must vanish for $p=0$ unless $l=0$. Otherwise, the ψ_{jm} would become singular at $p=0$ and the M^{0k} could not be applied to them in the sense of the boundedness of (8). (Actually it is necessary to postulate this equation for the square of M^{0k} instead only for M^{0k}.) It follows that ψ_{jm} vanishes at $p=0$ unless the series (21) contains a term with $l=0$. However, (16) shows that ψ cannot vanish at this point if the rest mass is finite and that, hence, every localizable wave function must have an $l=0$ term in its expansion. This happens only if $j=s$ and the wave function has even parity. If the parity of ψ_{jm} is odd, only f_1, f_3, etc., enter (21) and these still vanish at $p=0$. It follows that the wave functions which are localized at the origin all have angular momentum $j=s$ and the form

$$\psi_m = \sum_{l=0}^{2s}\sum_{m'} S(l, s)_{s, m-m', m'} P^l_{m-m'}(\vartheta, \phi) f_l V_{m'}. \tag{21a}$$

We now skip the part of the calculation which deals with the determination of the f_l and give only the result: The wave functions localized at the origin are the $2s+1$ functions

$$\psi_m = (2\pi)^{-3/2} 2^s p_0^{2s+\frac{1}{2}} (p_0+\mu)^{-s} \times V_m(p_1 p_2 p_3; \xi_1, \cdots \xi_{2s}) \tag{21b}$$

(i.e., the $l=0$ term alone remains from (21a)). Actually, this result is far from being surprising.[14]

The operator for the position coordinate can be calculated in exactly the same way as this was done in the case of zero spin, and gives

$$q^k = E\prod_{\alpha=1}^{2s}(1+\gamma_\alpha^0)\frac{p_0^{2s+\frac{1}{2}}}{(p_0+\mu)^s}\left(-i\frac{\partial}{\partial p_k}\right)\frac{p_0^{-\frac{1}{2}}}{(p_0+\mu)^s}E. \tag{22}$$

For $s=\frac{1}{2}$ this again agrees with Pryce's result[4] for his case (e), i.e., for his operator $\bar{q}$.

The significance of the projection operators E in (22) is only to annihilate any negative energy part of the wave function to which it is applied and to produce a purely positive energy wave function. Since q^k is the position operator only for wave functions which are

[13] See e.g., E. P. Wigner, *Gruppentheorie, etc.* (V. Vieweg & Sohn, Braunschweig, 1931). The composition of the V and the spherical harmonics P^l to the ψ_j is the same operation as the composition of the spin functions with a definite S and the space coordinate functions with a definite L, to functions of both, with definite J. This composition is explained in Chapter XXII. The coefficients of the composition, i.e., our $S(l, s)$ are calculated p. 202 ff. (they are denoted by $s^{(LS)}$).

[14] The proof runs as follows. One first shows, by considering $\psi_m \pm \Theta\psi_{-m}$ that the f_l of (21a) can all be assumed to be real. One then subdivides ψ_m into two parts: the $l=0$ part of the sum (21a) will be denoted by ψ^0, the rest ψ^r. As we have seen, ψ^0 is finite at $p=0$, while ψ^r vanishes at that point. The proof then consists in showing that there can be no region in which ψ^r is finite but very much smaller than ψ^0. It then follows from the continuity of both ψ^0 and ψ^r that the latter vanishes everywhere. Inserting $\psi^0+\psi^r$ for ψ in (16), one can neglect in the aforementioned region the square of ψ^r as compared with the other terms. The right side, as well as the term from the square of ψ^0, are independent of ϑ and φ. This must be true, therefore, also for the term arising from the cross product of ψ^0 and ψ^r. This term is, however, a sum of expressions $P_m^l(\vartheta,\varphi)f_0 f_l$ which cannot be independent of ϑ,φ except if all f_l with $l>0$ vanish (f_0 is finite by assumption). It then follows that the f_l vanish everywhere and (21a) reduces to a single term. This can be obtained from (16) by taking the square root on both sides.

defined on the positive hyperboloid alone, the E on the right could logically be omitted. Both E can be omitted[5] if one calculates a matrix element between two purely positive energy wave functions. The factors involving p_0 are necessary in order to make $i\partial/\partial p_j$ hermitian: because of the factor p_0^{-2s-1} in the volume element (15), an operator is hermitian if it looks hermitian after multiplication with $p_0^{s+\frac{1}{2}}$ on the right and division with the same factor on the left. The operator $\prod_{\alpha=1}^{2s} \frac{1}{2}(1+\gamma_\alpha{}^0)$ is a projection operator, i.e., it is identical with its square and could therefore be inserted into (22) once more before the second E, thus making (22) somewhat more symmetric. The position operator (11) for the Klein-Gordon particle is a special case of (22) and can be obtained from (22) by setting $s=0$.

If one displaces a state by a and measures its x^k coordinate afterwards, the result will be greater by a^k than the x^k coordinate measured on the undisplaced state. This leads to the relation

$$T(-a)q^kT(a)=q^k+a^k \tag{23}$$

for $a^0=0$. Inserting the expression (6) for $T(a)$ and going to the limit of very small a^k, one obtains

$$(q^kp^l-p^lq^k)\phi=-i\delta_{kl}\phi \tag{23a}$$

where ϕ is any permissible wave function. Actually one obtains by direct calculation, using in particular the identity

$$E_\alpha(1+\gamma_\alpha{}^0)E_\alpha=p_0^{-1}(p_0+\mu)E_\alpha \tag{24}$$

the commutation relation

$$q^kp^l-p^lq^k=-i\delta_{kl}E. \tag{25}$$

The commutation relations of the q^k with p_0 are naturally also the usual ones as p_0 is a function of the p^k alone. Since the q^k are the components of a vector operator in three-dimensional space, their commutation relations with the spatial components of M^{kl} are also the usual ones.

We wish to remark, finally, that a consideration, similar to the above, has been carried out also for the equations with zero mass. In the case of spin 0 and $\frac{1}{2}$, we were led back to the expressions for localized systems which were given in (9) and (21b). However, for higher but finite s, beginning with $s=1$ (i.e., Maxwell's equations), we found that no localized states in the above sense exist. This is an unsatisfactory, if not unexpected, feature of our work. The situation is not entirely satisfactory for infinite spin either.

DISCUSSION

One might wonder, first, what the reason is that our localized states are not the δ-functions in coordinate space which are usually considered to represent localized states. The reason is, naturally, that all our wave functions represent pure positive energy states. This is not true of the δ-function. Similarly, our operator (22) transforms positive energy functions into positive energy functions.

It is often stated that a measurement of the position of a particle, such as an electron, if carried out with a greater precision than the Compton wave-length, would lead to pair production and that it is, therefore, natural that the position operators do not preserve the positive energy nature of a wave function. Since a position measurement on a particle should result in a particle at a definite position, and not in a particle and some pairs, this consideration really denies the possibility of the measurement of the position of the particle. If this is accepted it still remains strange that pair creation renders the position measurement impossible to the same degree in such widely different systems as an electron, a neutron and even a neutriono. The calculations given above prove, at any rate, that there is nothing absurd in assuming the measurability of the position, and the existence of localized states, of elementary systems of non-zero mass. Moreover, the postulates (*a*), (*b*), (*c*) and (*d*), which are based on considerations of invariance, define the localized states and position operators uniquely for all non-zero mass elementary systems.

No similarly unique definition of localized states is possible for composite systems. Although it remains easy to show that definite total angular momenta j can be attributed to localized states, one soon runs into difficulties with the rest of the argument. In particular, the summation in (16) must be extended not only over the spin coordinates ξ but also over all states with different total rest mass and different intrinsic spin. As a result, one can, e.g., find states which can coexist as localized states in the sense of our axioms even though their j values are different. This is also what one would expect on ordinary reasoning since, if the system contains several particles, the states in which any one of them is localized at the origin satisfy our postulates. This holds also for the states in which another one of the particles is so localized or for states in which an arbitrary linear combination of the coordinates is zero. As a result, not only is the number of localized states greatly augmented but, further, one must expect to find many such large sets for which our postulates hold, although no two sets can be considered to be localized simultaneously. In other words, each set of localized states is not only much larger for composite systems but one also has to make a choice between many sets all of which satisfy our postulates by themselves. It does not appear that one can proceed much further in the definition of localized states for composite systems without making much more specific assumptions. Naturally, one can define as localized states those, which, in any of the elementary parts of the composite system, appear localizable. It appears reasonable to assume that this definition corresponds to the center of mass of the whole system.

One may wonder, even in the case of elementary particles, whether the determination of the localized states and position operators has much significance. Such doubts might arise particularly strongly if one is inclined to consider the collision matrix as the future form of the theory. One must not forget, however, that the customary exposition of this theory refers only to questions about cross sections. There is another interesting set of questions referring to the position of the scattered particles: how much further back (i.e., closer to the scattering center) are they than if they had gone straight to the scattering center and then continued in the new direction without any delay.[15] In order to answer such questions in the relativistic region, one will need some definition of localized states for elementary systems. From this point of view it is satisfactory that the localized states could be defined without ambiguity just for these systems.

[15] L. Eisenbud, Princeton Dissertation (1948).

REVIEWS OF MODERN PHYSICS VOLUME 34, NUMBER 4 OCTOBER 196

On the Localizability of Quantum Mechanical Systems*

A. S. WIGHTMAN
Princeton University, Princeton, New Jersey

1. INTRODUCTION

FROM the very beginning of quantum mechanics, the notion of the position of a particle has been much discussed. In the nonrelativistic case, the proof of the equivalence of matrix and wave mechanics, the discovery of the uncertainty relations, and the development of the statistical interpretation of the theory led to an understanding which, within the inevitable limitations of the nonrelativistic theor may be regarded as completely satisfactory.

Historically, confusion reigned in the relativist case, because situations requiring a description i terms of many particles were squeezed into a fo malism built to describe a single particle. I have i mind the difficulties with wave functions for a sing particle which seem to yield nonzero probability f finding it in a state of negative energy. Soon attentic shifted to the problems of the quantum theory fields and the question of the status of positic

* Dedicated to Eugene Wigner on his sixtieth birthday.

Reprinted from *Rev. Mod. Phys.* **34**, 845 (1962).

operators for relativistic particles was left without a clear resolution. That does not mean that papers were not written on the subject, but that those papers had completely different objectives in mind: They permitted the particles in question to be in nonphysical (negative energy) states or they studied operators which could not serve as position *observables* since their three components did not commute.

In the opinion of the present author, the decisive clarification of the relativistic case occurs in a paper of Newton and Wigner.[1] These authors show that, if the notion of localized state satisfies certain nearly inevitable requirements, for a single free particle it is uniquely determined by the transformation law of the wave function under inhomogeneous Lorentz transformations. The resulting position observables turn out, in the case of spin-$\frac{1}{2}$, to be identical with the Foldy-Wouthuysen "mean position" operators.[2] An analogous investigation for the case of Galilean relativity was carried out by Inönü and Wigner.[3]

The essential result of Newton and Wigner is that for single particles a notion of localizability and a corresponding position observable are uniquely determined by relativistic kinematics when they exist at all. Whether, in fact, the position of such a particle is observable in the sense of the quantum theory of measurement is, of course, a much deeper problem; *that* probably can only be decided within the context of a specific consequent dynamical theory of particles. All investigations of localizability for relativistic particles up to now, including the present one, must be regarded as preliminary from this point of view: They construct position observables consistent with a given transformation law. It remains to construct complete dynamical theories consistent with a given transformation law and then to investigate whether the position observables are indeed observable with the apparatus that the dynamical theories themselves predict.

In Newton and Wigner's formulation, the set $S_{\mathbf{a}}$ of states localized at a point $\mathbf{a}$ of the three-dimensional space at a given time, must satisfy the following axioms:

(a) $S_{\mathbf{a}}$ is a linear manifold;

(b) $S_{\mathbf{a}}$ is invariant under rotations about $\mathbf{a}$, reflections in $\mathbf{a}$, and time inversions;

(c) $S_{\mathbf{a}}$ is orthogonal to all its space translates;

(d) certain regularity conditions.

The solutions of (a) . . . (d) for elementary systems, i.e., for systems whose states transform according to an irreducible representation of the inhomogeneous Lorentz group, turn out to be continuum wave functions when they exist at all, i.e., according to the usual definitions of Hilbert space, there is no manifold $S_{\mathbf{a}}$. However, it is physically and mathematically clear that Newton and Wigner's formulation ought to be regarded as the limiting case of a notion of localizability in a region.

In the present paper, I propose a reformulation of the physical ideas of (a) . . . (d) in terms of a notion of localizability in a region. When the ideas are so formulated, one sees that the existence and uniqueness of a notion of localizability for a physical system are properties which depend only on the transformation law of the system under the Euclidean group, i.e., the group of all space translations and rotations. The analysis of localizability in the Lorentz and Galilei invariant cases is then just a matter of discussing what representations of the Euclidean group can arise there. To obtain uniqueness, one must add invariance under time inversion and an analogy of Newton and Wigner's regularity assumption. As would be expected, all the results obtained earlier in the old formulation come out. One can ask what is the point of the present extended footnote to Newton and Wigner's paper. First, it seems worthwhile to me to have a mathematically rigorous proof of the fundamental result of Newton and Wigner that a single photon is *not* localizable. Second, the work of Newton and Wigner can be regarded as a contribution to the general problem of determining what physical characteristics of a quantum mechanical system are consistent with a given relativistic transformation law. In this connection, it is interesting to regard the axioms I . . . V below for localizability in a region as a very special case of the notion of particle observables for a quantum theory. Elsewhere[4] I gave a set of axioms for the notion of a particle interpretation which yield I . . . V when specialized to the case of a single particle. One of the main reasons for giving full mathematical detail in

[1] T. D. Newton and E. P. Wigner, Revs. Modern Phys. **21**, 400 (1949).

[2] L. Foldy and S. Wouthuysen, Phys. Rev. **78**, 29 (1950). This paper was widely read because of its exceptional clarity. The mean position operators themselves were discussed before by A. Papapetrou, Acad. Athens **14**, 540 (1939); R. Becker, Gött. Nach. p. 39 (1945); and M. H. L. Pryce, Proc. Roy. Soc. (London) **A150**, 166 (1935); **A195**, 62 (1948). For further references and discussion see A. S. Wightman and S. Schweber, Phys. Rev. **98**, 812 (1955).

[3] E. Inönü and E. Wigner, Nuovo cimento **9**, 705 (1952). The main point of this paper is that laws of transformation of the states of a particle under the inhomogeneous Galilei group other than those in the ordinary Schrödinger mechanics are inconsistent with localizability.

[4] See, *Les problèmes mathématiques de la théorie quantique des champs*, (CNRS, Paris, 1959), especially pages 36–38.

the present simple case is in preparation for the problem of determining particle interpretations.

It turns out that the natural mathematical tool for the analysis of localizability as understood here is the theory of imprimitive representations of the Euclidean group. The notion of imprimitivity was introduced for finite groups early in the history of group theory. It was generalized to the case of a large class of topological groups by Mackey.[5] From a mathematical point of view, the present paper merely writes out Mackey's theory in detail for the case of the Euclidean group. However, I decided to make the exposition as self-contained as possible, and to incorporate certain elegant ideas of Loomis in the proofs.[6] The purpose of this expository account is to make it possible for the reader to understand how the mathematical arguments go for the Euclidean group without having to work through the general case, however character building that experience might be.

2. MATHEMATICAL FORMULATION OF THE AXIOMS AND PRELIMINARY HEURISTIC DISCUSSION

The axioms are formulated in terms of projection operators $E(S)$, where S is some subset of Euclidean space at a given time. The $E(S)$ are supposed to be observables. They must be projection operators because they are supposed to describe a *property* of the system, the property of being localized in S. That is, if Φ is a vector in a separable Hilbert space, $\mathcal{H}$, describing a state in which the system lies in S, then $E(S)\Phi = \Phi$. If the system does not lie in S then $E(S)\Phi = 0$. $E(S)$ can therefore only have proper values one or zero and, as an observable, must be self-adjoint. Thus, it is a projection operator.[7]

The axioms are:

I. For every Borel set, S, of three-dimensional Euclidean space, $\mathbf{R}^3$, there is a projection operator $E(S)$ whose expectation value is the probability of finding the system in S.

II. $E(S_1 \cap S_2) = E(S_1)E(S_2)$.

III. $E(S_1 \cup S_2) = E(S_1) + E(S_2) - E(S_1 \cap S_2)$.

If $S_i, i = 1, 2, \cdots$ are disjoint Borel sets then

$E(\bigcup S_i) = \sum_{i=1} E(S_i)$.

IV. $E(\mathbf{R}^3) = 1$.

V′. $E(RS + \mathsf{a}) = U(\mathsf{a},R)E(S)U(\mathsf{a},R)^{-1}$,

where $RS + \mathsf{a}$ is the set obtained from S by carrying out the rotation R followed by the translation a, and $U(\mathsf{a},R)$ is the unitary operator whose application yields the wave function rotated by R and translated by a.

The notation $S_1 \cap S_2$ and $S_1 \cup S_2$ is used to indicate the common part and union, respectively, of the sets S_1 and S_2. $\bigcup S_i$ is the union of the sets S_i.

The physical significance of these axioms is as follows.

The Borel sets form the smallest family of sets which includes cubes and is closed under the operations of forming complements and denumerable unions. One might try to replace the Borel sets by all sets obtained by forming complements and *finite* unions starting from cubes and require III only for *finite* sums. However, it can be shown that any such $E(S)$ could be extended to one defined on the Borel sets and satisfying III as it stands. (See Appendix I for further discussion of this point.) In fact, $E(S)$ can be extended even further to all Lebesgue measurable sets, but this extension will not be needed here.[8]

II states that a system which is in *both* S_1 and S_2 is in $S_1 \cap S_2$. It is immediately clear from II that $E(S_1)E(S_2) = E(S_2)E(S_1)$.

III states that the set of states of the system for which it is localized in $S_1 \cup S_2$ is the closed linear manifold spanned by the states localized in S_1 and those localized in S_2.

IV says that the system has probability one of being somewhere.

V′ says that if Φ is a state in which the system is localized in S, then $U(\mathsf{a},R)\Phi$ is a state in which the system is localized in $RS + \mathsf{a}$.

I venture to say that any notion of localizability in three-dimensional space which does *not* satisfy I . . . V′ will represent a radical departure from present physical ideas.

The $E(S)$ define a set of commuting coordinate

[5] G. W. Mackey, Proc. Natl. Acad. Sci. U.S. **35**, 537 (1949); Ann. Math. **55**, 101 (1952); **58**, 193 (1953); Acta Math. **99**, 265 (1958). That Mackey's theory applies to localizability in quantum mechanics was independently realized by Mackey himself. I thank Professor Mackey for correspondence on the subject. Mackey's treatment is summarized in his Colloquium Lectures to the American Mathematical Society, Stillwater, Oklahoma Aug. 29–Sept. 1, 1961. It is a part of a coherent axiomatic treatment of quantum mechanics given in his unpublished Harvard lectures 1960–61.

[6] L. H. Loomis, Duke Math. J. **27**, 569 (1960).

[7] For a general discussion of observables describing a property see J. von Neumann, *Mathematical Foundations of Quantum Mechanics* (Princeton University Press, Princeton, New Jersey, 1955), pp. 247–254.

[8] An argument that the Lebesgue measurable sets form a physically natural class is contained in J. von Neumann, Ann. Math. **33**, 595 (1932).

operators q_1,q_2,q_3 which form a vector in 3-space. In fact,

$$q_i = \int_{-\infty}^{\infty} \lambda dE(\{x_i \leqslant \lambda\}) , \tag{2.1}$$

where $E(\{x_i \leqslant \lambda\})$ is the projection operator for the set $\{x_i \leqslant \lambda\}$ of all points of three-space whose ith coordinates satisfy $x_i \leqslant \lambda$. Of course, (2.1) has to be interpreted as meaning that the Stieltjes integral

$$(\Phi,q_i\Psi) = \int_{-\infty}^{\infty} \lambda d(\Phi,E(\{x_i \leqslant \lambda\})\Psi)$$

holds for all Φ and all Ψ on which q_i can be defined. Thus, any set of $E(S)$ uniquely determines a position operator $\mathbf{q}$. Conversely, one can regard the requirement that the $E(S)$ exist as a precise way of stating that $\mathbf{q}$ exists and its components are simultaneously observable. A notion of localizability for which $[q_i,q_j] \neq 0$ does not fall under the above scheme if, indeed, such a notion makes sense at all.

Axiom V′ has been stated in terms of the unitary operators $U(\mathbf{a},R)$. It is well known that without loss of physical generality these can be assumed to form a representation up to a $\pm$ sign,[9] i.e.,

$$U(\mathbf{a}_1,R_1)U(\mathbf{a}_2,R_2) = \omega(\mathbf{a}_1,R_1;\mathbf{a}_2R_2)U(\mathbf{a}_1 + R_1\mathbf{a}_2,R_1R_2) ,$$

where $\omega = \pm 1$. It is more convenient, from a mathematical point of view, to deal with a true representation for which $\omega = +1$. It is also well known that this can be arranged by passing to the two-sheeted covering group of the Euclidean group $\mathcal{E}_3$.[10] It may be defined as the set of pairs $\mathbf{a}$, A, where $\mathbf{a}$ is again a three-dimensional translation vector and A is a 2×2 unitary matrix of determinant one. The matrices $\pm A$ determine the same rotation given by

$$A\mathbf{x}\cdot\boldsymbol{\tau}A^* = (R(\pm A)\mathbf{x})\cdot\boldsymbol{\tau}$$

Here $\boldsymbol{\tau}$ stands for the Pauli matrices

$$\tau^1 = \begin{bmatrix} 0 & 1 \\ 1 & 0 \end{bmatrix}, \quad \tau^2 = \begin{bmatrix} 0 & -i \\ i & 0 \end{bmatrix}, \quad \tau^3 = \begin{bmatrix} 1 & 0 \\ 0 & -1 \end{bmatrix}.$$

The multiplication law of $\mathcal{E}_3$ is

$$\{\mathbf{a},A_1\}\{\mathbf{a}_2,A_2\} = \{\mathbf{a}_1 + A_1\mathbf{a}_2,A_1A_2\} .$$

Here, for brevity, instead of writing $R(A_1)\mathbf{a}_2$ we write $A_1\mathbf{a}_2$. This will be done throughout the following. Thus, V′ is replaced by

V. $U(\mathbf{a},A)E(S)U(\mathbf{a},A)^{-1} = E(AS + \mathbf{a})$ for all Borel sets S of $\mathbf{R}^3$, and all $\{\mathbf{a},A\} \in \mathcal{E}_3$. Here $AS + \mathbf{a}$ is the set obtained from S by the transformation $\{\mathbf{a},A\}$ and $\{\mathbf{a},A\} \to U(\mathbf{a},A)$ is the representation of $\mathcal{E}_3$ belonging to the physical system in question.

In the terminology of Mackey, I . . . V state that the set of operators $\{E(S)\}$ are a *system of imprimitivity* for the representation $U(\mathbf{a},A)$ of $\mathcal{E}_3$ with *base* $\mathbf{R}^3$. In order to see the present problem in the context of Mackey's general theory, recall that he considers a topological group G and two continuous realizations of G, one a representation by homeomorphisms of a topological space $M : x \to h(g)x$ (homeomorphism means one-to-one mapping continuous both ways) and the other a unitary representation of G in a Hilbert space $\mathcal{H} : g \to U(g)$. Then a system of imprimitivity with base M is a family of projection operators in $\mathcal{H}$ which satisfy I, II, III, IV with the sets S interpreted as Borel sets of M, and, in addition, the appropriate modification of V:

$$U(g)E(S)U(g)^{-1} = E(h(g)S) .$$

A representation $U(g)$ which has at least one system of imprimitivity (with respect to M) is said to be imprimitive (with respect to M). A system of imprimitivity is *transitive* if the group of homeomorphisms $g \to h(g)$ is, i.e., if each point x is carried into every other by a suitable h.

In the case of a transitive system of imprimitivity, the space M can be replaced by a coset space as follows. Let G_x be the subgroup of all $g \in G$ such that $h(g)x = x$. Notice that if $h(g_1)x = y = h(g_2)x$ then $h(g_2^{-1}g_1)x = x$, so $g_2g = g_1$ where $g \in G_x$. The set of all elements of the form g_2g, $g \in G_x$ is denoted g_2G_x and called the left coset of G_x belonging to g_2. Thus, each left coset corresponds to a point of M, distinct cosets corresponding to distinct points, and by a mere change of names M can be replaced by the space of left cosets, usually denoted G/G_x. In the more general case of a nontransitive system the space M will split into orbits and the points of an orbit can be labeled by the points of G/G_x where x is any point of the orbit.

In the problem of localizability considered here, the system of imprimitivity is transitive but for momentum observables and particle observables, in general, the system of imprimitivity is not transitive.

Mackey's theory shows that the transitive system of imprimitivity and its associated representation

[9] The argument (originally due to E. P. Wigner) is outlined in *Dispersion Relations and Elementary Particles* (John Wiley & Sons, Inc., 1961), pp. 176–181.

[10] The argument (originally due to E. P. Wigner for the rotation group and Poincaré group) is given for the Euclidean group in V. Bargmann, Ann. Math. **59**, 1 (1954).

can be brought into a standard form by a suitably chosen unitary transformation, V:

$$\{E(S),U(g)\} \to \{VE(S)V^{-1},VU(g)V^{-1}\} .$$

In this standard form the $VU(g)V^{-1}$ becomes a so-called induced representation associated with a unitary representation of G_x where x is some fixed point of M. Two pairs $\{E_1(S),U_1(g)\}$ and $\{E_2(S), U_2(g)\}$ are unitary equivalent:

$$E_1(S) = VE_2(S)V^{-1},U_1(g) = VU_2(g)V^{-1} ,$$

if and only if the unitary representations of G_x are equivalent.

Detailed proofs of these assertions of Mackey's theory for the special case of $\mathcal{E}_3$ will be offered in the following sections. For the moment, the results will be taken for granted and used to discuss the uniqueness of $E(S)$ for given $U(g)$. Clearly, for $U(g)$ given the only unitary transformations, V, which can give new $VE(S)V^{-1} \neq E(S)$ are ones which commute with the given $U(g)$ but not with the $E(S)$.

That this possibility is actually realized in simple physical examples can be seen by considering a compound system of two free spinless Schrödinger particles with wave function $\psi(\mathbf{x}_1,\mathbf{x}_2)$. Let the corresponding representation of the Euclidean group be $U(\mathbf{a},R)$:

$$\begin{aligned}\psi(\mathbf{x}_1,\mathbf{x}_2) &\to (U(\mathbf{a},R)\psi)(\mathbf{x}_1,\mathbf{x}_2)\\ &= \psi(R^{-1}(\mathbf{x}_1 - \mathbf{a}),R^{-1}(\mathbf{x}_2 - \mathbf{a})) .\end{aligned}$$

Define the operators $\mathbf{X}^{(\alpha)}$ by

$$\mathbf{X}^{(\alpha)} = \alpha\mathbf{x}_1^{\text{op}} + (1 - \alpha)\mathbf{x}_2^{\text{op}}$$

where α is any real number, and by definition

$$\begin{aligned}(\mathbf{x}_1^{\text{op}}\Phi)(\mathbf{y}_1,\mathbf{y}_2) &= \mathbf{y}_1\Phi(\mathbf{y}_1,\mathbf{y}_2)\\ (\mathbf{x}_2^{\text{op}}\Phi)(\mathbf{y}_1,\mathbf{y}_2) &= \mathbf{y}_2\Phi(\mathbf{y}_1,\mathbf{y}_2) .\end{aligned}$$

Then, for each α, $\mathbf{X}^{(\alpha)}$ defines a possible position operator (the spectral representation of $\mathbf{X}_j^{(\alpha)}$, $j = 1$, 2, 3 yields the projections appearing in (2.1), and the general $E(S)$ can be found from these). In particular, $\mathbf{X}^{(0)} = \mathbf{x}_2^{\text{op}}$ and $\mathbf{X}^{(1)} = \mathbf{x}_1^{\text{op}}$, are possible position operators.

Now there exists a unitary operator, V, which commutes with the representation of the Euclidean group

$$[V,U(\mathbf{a},R)]_- = 0$$

and carries $\mathbf{X}^{(\alpha)}$ into $\mathbf{X}^{(\beta)}$

$$V\mathbf{X}^{(\alpha)}V^{-1} = \mathbf{X}^{(\beta)} .$$

To obtain V, one may note first that the operator $T^{(\alpha)}$ defined by

$$\begin{aligned}(T^{(\alpha)}\Phi)(\mathbf{x}_1,\mathbf{x}_2) = \Phi(\alpha\mathbf{x}_1 &+ (1 - \alpha)\mathbf{x}_2,(\alpha - 1)\mathbf{x}_1\\ &+ (2 - \alpha)\mathbf{x}_2)\end{aligned}$$

is unitary and satisfies

$$T^{(\alpha)}\mathbf{x}_1^{\text{op}}T^{(\alpha)-1} = \mathbf{X}^{(\alpha)} .$$

Then V is given by

$$V = T^{(\beta)}T^{(\alpha)-1} .$$

Clearly, the kind of nonuniqueness appearing in this example may be expected to be absent only when one is dealing with a single particle. Theorem 4 obtained below gives a precise criterion for uniqueness and a parametrization of the possible answers when more than one exists.

The uniqueness of the notion of localizability for given representation of the Euclidean group has been discussed assuming Mackey's theory. Now I attempt to give an intuitive idea of the circumstances in which a notion of localizability exists.

Since all the $E(S)$ commute, diagonalize them. Then the state vectors are represented by quantities $\Phi(\mathbf{x})$ defined on space and with a number of components which may vary with $\mathbf{x}$. [In fact, these $\Phi(\mathbf{x})$ for $\mathbf{x} = \mathbf{a}$ are just Newton and Wigner's linear manifold $S_{\mathbf{a}}$.] In this realization the scalar product of two vectors Φ and Ψ is

$$(\Phi,\Psi) = \int d\mathbf{x}(\Phi(\mathbf{x}),\Psi(\mathbf{x})) ,$$

where the scalar product appearing under the integral sign is in the components of $\Phi(\mathbf{x})$ and $\Psi(\mathbf{x})$ for fixed $\mathbf{x}$. The operators $E(S)$ take the form

$$(E(S)\Phi)(\mathbf{x}) = \chi_S(\mathbf{x})\Phi(\mathbf{x}) ,$$

where $\chi_S(\mathbf{x}) = 1$ if $\mathbf{x} \in S$, 0 if $\mathbf{x} \notin S$. From the transformation law of $E(S)$ it is plausible that by a suitable choice of basis it can be arranged that

$$(U(\mathbf{a},1)\Phi)(\mathbf{x}) = \Phi(\mathbf{x} - \mathbf{a}) .$$

From this equation, it follows that the number of components of $\Phi(\mathbf{x})$ is the same for all $\mathbf{x}$. It is also plausible that by a suitable choice of basis the transformation law under rotation can be made to look the same for each $\mathbf{x}$:

$$(U(0,A)\Phi)(\mathbf{x}) = \mathfrak{D}(A)\Phi(A^{-1}\mathbf{x}) ,$$

where $\mathfrak{D}(A)$ acts on the components of $\Phi(A^{-1}\mathbf{x})$ at each point. Once these results are accepted, one can pass by Fourier transform to momentum space amplitudes. There one has

$$(U(\mathbf{a},A)\Phi)(\mathbf{p}) = e^{-i\mathbf{p}\cdot\mathbf{a}}\mathfrak{D}(A)\Phi(A^{-1}\mathbf{p}) \quad (2.2)$$

with the scalar product

$$(\Phi,\Psi) = \int d\mathbf{p}(\Phi(\mathbf{p}),\Psi(\mathbf{p}))\ . \quad (2.3)$$

The canonical form (2.2) is to be compared with

$$(U(\mathbf{a},A)\Phi)(\mathbf{p}) = e^{-i\mathbf{p}\cdot\mathbf{a}}Q(\mathbf{p},A)\Phi(A^{-1}\mathbf{p})\ , \quad (2.4)$$

where

$$Q(\mathbf{p},A)Q(A^{-1}\mathbf{p},B) = Q(\mathbf{p},AB)\ ,$$

and the scalar product is

$$(\Phi,\Psi) = \int d\mu(\mathbf{p})(\Phi(\mathbf{p}),\Psi(\mathbf{p}))\ ,$$

a form which will be derived in Sec. 3.

The comparison shows:

(i) When the representation is in the canonical form (2.4) the measure $d\mu(\mathbf{p})$ on momentum space is just Lebesgue measure $d\mathbf{p}$.

(ii) The dimension of the vectors $\Phi(\mathbf{p})$ is the same for all $\mathbf{p}$.

(iii) The operators $Q(\mathbf{p},A)$ are of the form $\mathfrak{D}(A)$, where $A \rightarrow \mathfrak{D}(A)$ is a representation of the unitary unimodular group.

Intuitively (i) and (ii) are accounted for because, if one makes any state whose $\mathbf{x}$ dependence is a δ function one gets all momenta. Thus, one would expect to have the same number of linearly independent states for each $\mathbf{p}$. (iii) is essentially a consequence of the rotational invariance of the states localized at a point.

All three restrictions are nontrivial if applied to an arbitrary representation of $\mathcal{E}_3$. However, as will be seen in Secs. 5 and 6, (i) and (ii) are always satisfied in any relativistic theory (provided one leaves out the vacuum state). (iii) excludes a very important physical system, the single photon. One can see this immediately by looking at the $Q(\mathbf{p},A)$ for those A which leave $\mathbf{p}$ invariant. Such Q's have two eigenvectors corresponding to right-circularly and left-circularly polarized photons having angular momentum along $\mathbf{p}$, $\pm\hbar$, respectively. On the other hand, in $\mathfrak{D}(A)$ one cannot have states with angular momentum $\pm\hbar$ along $\mathbf{p}$ without also having states with zero component of angular momentum along $\mathbf{p}$. The nonlocalizability of the photon (and all other particles of spin $\geqslant \frac{1}{2}$ and mass zero) is a consequence of this simple kinematical fact.[11] For spin-0, (iii) is satisfied and so the phonon is localizable.[12] It is an oddity that the same is not true for Wigner's particles of infinite spin,[13] as will be seen in Sec. 5, even though in that case each angular momentum along $\mathbf{p}$ appears just once.

There is one paradox to which the preceding discussion might appear to give rise. Suppose one describes a photon by a real-valued three-component field $\mathbf{B}(\mathbf{x})$ satisfying

$$\operatorname{div}\mathbf{B} = 0\ , \quad (2.5)$$

defines a scalar product (this is a *real* Hilbert space) by

$$(\mathbf{B}_1,\mathbf{B}_2) = \int \mathbf{B}_1(\mathbf{x})\cdot\mathbf{B}_2(\mathbf{x})d\mathbf{x}\ ,$$

and a representation of the Euclidean group

$$(U(\mathbf{a},R)\mathbf{B})(\mathbf{x}) = R\mathbf{B}(R^{-1}(\mathbf{x}-\mathbf{a}))\ .$$

Attempt to define projection operators by the equation

$$(E(S)\mathbf{B})(\mathbf{x}) = \chi_S(\mathbf{x})\mathbf{B}(\mathbf{x})\ .$$

Why does not this describe the photon as a localizable system? The answer is that the $E(S)$ carry vectors satisfying the condition (2.5) into vectors which do not satisfy it, so $E(S)$ is not a well-defined operator in the manifold of states and the $\mathbf{x}$ in $\mathbf{B}(\mathbf{x})$ has nothing to do with localizability.

The notion of localizability discussed here is concerned with states localized in space at a given time. It is natural to inquire whether there exists a corresponding property in space-time. Then the $E(S)$ would satisfy

$$U(a,\Lambda)E(S)U(a,\Lambda)^{-1} = E(\Lambda S + a)\ ,$$

where S is a Borel set of space-time and $\{a,\Lambda\}$ is an inhomogeneous Lorentz transformation of space-time translation, a, and homogeneous Lorentz transformation, Λ. However, a requirement analogous to (i) follows from Mackey's theory: All four-momenta must occur in the theory. This is in flat violation of the physical requirement that there be a lowest

[11] That the photon was nonlocalizable was stated and believed long before reference 1 was written. See, for example, L. Landau and R. Peierls Z. Physik **62**, 188 (1930); **69**, 56 (1931); especially p. 67 of the latter. While the arguments given could possibly be regarded as plausible, they do not make clear what is the heart of the problem.

[12] If the neutrino had turned out to possess states of both helicities, i.e., states with components $\pm\frac{1}{2}\hbar$ of the component of angular momentum along $\mathbf{p}$, then it too would be localizable. A neutrino of definite helicity is not localizable.

[13] E. P. Wigner, Ann. Math. **40**, 149 (1939); Z. Physik **124**, 665 (1947–8).

energy state. Thus, a sensible notion of localizability in space-time does not exist.

3. RECAPITULATION OF THE UNITARY REPRESENTATIONS OF $\mathcal{E}_3$ THE UNIVERSAL COVERING GROUP OF THE EUCLIDEAN GROUP

In this section a canonical form of the representations of $\mathcal{E}_3$ will be derived in which the translation subgroup is diagonalized.

Any continuous unitary representation of $\mathcal{E}_3$: $\{\mathbf{a},A\} \to U(\mathbf{a},A)$ gives rise to a continuous unitary representation of the translation group $\mathfrak{T}_3$: $\mathbf{a} \to U(\mathbf{a},1)$. The first step in the analysis is to describe all such representations. By a unitary transformation the $U(\mathbf{a},1)$ are to be diagonalized, i.e., brought into the form

$$(U(\mathbf{a},1)\Phi)(\mathbf{p}) = e^{-i\mathbf{p}\cdot\mathbf{a}}\Phi(\mathbf{p}) . \quad (3.1)$$

(The minus sign in the exponent is a matter of convention; it is adopted to conform with custom in quantum mechanics.)

For this purpose, the notion of direct integral of Hilbert spaces and representations is needed. It will be described briefly in the present special context.[14]

Let μ be a positive measure on three-dimensional (momentum) space $\mathfrak{T}_3^*$. For each point $\mathbf{p}$ of $\mathfrak{T}_3^*$, let there be given a Hilbert space $\mathcal{H}_\mathbf{p}$ whose dimension $\nu(\mathbf{p})$ is a μ-measurable function of $\mathbf{p}$. Then the direct integral of the $\mathcal{H}_\mathbf{p}$ with respect to μ is a Hilbert space denoted $\int_{\mathfrak{T}_3^*}^{\oplus} d\mu(\mathbf{p})\mathcal{H}_\mathbf{p}$ whose elements are functions defined on $\mathfrak{T}_3^*$, with values satisfying $\Phi(\mathbf{p}) \in \mathcal{H}_\mathbf{p}$. Furthermore, the elements must satisfy $(\Phi_1(\mathbf{p}),\Phi_2(\mathbf{p}))$ is a μ-measurable function of $\mathbf{p}$ for any two

$$\Phi_1,\Phi_2 \in \int_{\mathfrak{T}_3^*}^{\oplus} d\mu(\mathbf{p})\mathcal{H}_\mathbf{p} \quad (3.2)$$

[here $(\Phi_1(\mathbf{p}),\Phi_2(\mathbf{p}))$ is the scalar product in $\mathcal{H}_\mathbf{p}$], and

$$\int (\Phi(\mathbf{p}),\Phi(\mathbf{p}))d\mu(\mathbf{p}) < \infty . \quad (3.3)$$

The scalar product in $\int_{\mathfrak{T}_3^*}^{\oplus} d\mu(\mathbf{p})\mathcal{H}_\mathbf{p}$ is defined by

$$(\Phi_1,\Phi_2) = \int d\mu(\mathbf{p})(\Phi_1(\mathbf{p}),\Phi_2(\mathbf{p})) . \quad (3.4)$$

With this notation, the following theorem holds.

Theorem 1. Every continuous unitary representation of the three-dimensional translation group $\mathfrak{T}_3$ is unitary equivalent to one of the following form:

$$(U(\mathbf{a})\Phi)(\mathbf{p}) = e^{-i\mathbf{p}\cdot\mathbf{a}}\Phi(\mathbf{p}) ,$$

where Φ is an element of a direct integral, $\int_{\mathfrak{T}_3^*}^{\oplus} d\mu(\mathbf{p})\mathcal{H}_\mathbf{p}$, over $\mathfrak{T}_3^*$ with measure μ and multiplicity function $\nu(\mathbf{p}) = \dim \mathcal{H}_\mathbf{p}$.

A bounded operator, B, which commutes with the operators of the representation can be written in the form

$$(B\Phi)(\mathbf{p}) = B(\mathbf{p})\Phi(\mathbf{p}) ,$$

where $B(\mathbf{p})$ is a bounded operator in $\mathcal{H}_\mathbf{p}$ and such that for all Φ_1 and $\Phi_2 \in \int_{\mathfrak{T}_3^*}^{\oplus} d\mu(\mathbf{p})\mathcal{H}_\mathbf{p}$,

$(\Phi_1(\mathbf{p}), B(\mathbf{p})\Phi_2(\mathbf{p}))$ is measurable in $\mathbf{p}$.

Two such representations $\mathbf{a} \to U_1(\mathbf{a})$ and $\mathbf{a} \to U_2(\mathbf{a})$, with measures μ_1 and μ_2 and multiplicity functions $\nu_1(\mathbf{p})$ and $\nu_2(\mathbf{p})$, respectively, are unitary equivalent if and only if

(1) $\mu_1 \equiv \mu_2$, i.e., μ_1 and μ_2 give zero measure for the same sets of $\mathfrak{T}_3^*$.

(2) $\nu_1(\mathbf{p}) = \nu_2(\mathbf{p})$ except, perhaps, in a set of μ_1 measure zero.

For a sketch of a proof of Theorem 1, the reader is referred to Appendix II, and the references quoted there.

The measures μ and multiplicity functions ν appearing in a general representation of the translation group are completely arbitrary. Those which can appear in a representation of $\mathfrak{T}_3$ obtained by restriction from a representation of $\mathcal{E}_3$ are quite special. This comes about because $\mathbf{a} \to U(A\mathbf{a},1)$ defines a representation of $\mathfrak{T}_3$ which is unitary equivalent to $\mathbf{a} \to U(\mathbf{a},1)$ as a consequence of

$$U(0,A)U(\mathbf{a},1)U(0,A)^{-1} = U(A\mathbf{a},1) .$$

Now when $U(\mathbf{a},1)$ is brought into the diagonal form (3.1) by an appropriate unitary transformation, the representation $\mathbf{a} \to U(A\mathbf{a},1)$ takes the form

$$(U(A\mathbf{a},1)\Phi)(\mathbf{p}) = e^{-i(A^{-1}\mathbf{p})\cdot\mathbf{a}}\Phi(\mathbf{p})$$

and this in turn can be brought into the standard form by the unitary transformation

$$(W\Phi)(\mathbf{p}) = \Phi(A\mathbf{p})$$

which yields

$$(WU(A\mathbf{a},1)W^{-1})(W\Phi))(\mathbf{p}) = e^{-i\mathbf{p}\cdot\mathbf{a}}(W\Phi)(\mathbf{p})$$

and carries the direct integral $\int_{\mathfrak{T}_3^*}^{\oplus} d\mu(\mathbf{p})\mathcal{H}_\mathbf{p}$ into

[14] For a full account of the notion of direct integral see J. Dixmier, *Les algèbres d'opérateurs dans l'espace Hilbertien* (Gauthier-Villars, Paris, 1957). The theory gives a precise mathematical meaning to the Dirac formalism of "representations" in quantum mechanics. See P. A. M. Dirac, *The Principles of Quantum Mechanics* (Oxford University Press, New York, 1947), 3rd ed., Chap. III.

$\int_{\mathfrak{T}_3^*}^{\oplus} d\mu(\mathbf{p})\mathcal{H}_{A\mathbf{p}}$ where $d\mu_A(\mathbf{p}) = d\mu(A\mathbf{p})$. The unitary equivalence criterion given in Theorem 1 then implies

$$\mu \equiv \mu_A \quad (3.5)$$

$\nu(\mathbf{p}) = \nu(A\mathbf{p})$ for all $\mathbf{p}$ except possibly on a set of μ measure zero. (3.6)

Now in Appendix 2, it is shown that the only measures on $\mathfrak{T}_3^*$ satisfying (3.5) are equivalent to ones of the form

$$\mu_0\delta(\mathbf{p}) + d\rho(|\mathbf{p}|)d\omega(\mathbf{p}) , \quad (3.7)$$

where $\mu_0 \geqslant 0$, $d\omega(\mathbf{p})$ is the area on the sphere of radius $|\mathbf{p}|$ and $d\rho$ is a measure on the positive real axis. Since, if $\mu \equiv \mu_1$, the unitary mapping

$$(W\Phi)(\mathbf{p}) = \Phi(\mathbf{p})\left[\frac{d\mu(\mathbf{p})}{d\mu_1(\mathbf{p})}\right]^{1/2}$$

carries the direct integral $\int_{\mathfrak{T}_3^*}^{\oplus} d\mu(\mathbf{p})\mathcal{H}_{\mathbf{p}}$ into $\int_{\mathfrak{T}_3^*}^{\oplus} d\mu_1(\mathbf{p})\mathcal{H}_{\mathbf{p}}$, one may for convenience choose μ in the form (3.7).[15] Later on μ will be taken in this form but for the moment a general μ satisfying (3.5) will be carried along. Furthermore since any two Hilbert space of the same dimension can be mapped on one another by unitary transformation, there is no loss in generality in taking $\mathcal{H}_{\mathbf{p}} = \mathcal{H}_{A\mathbf{p}}$ for all A.

The next task is to put the operators $U(0,A)$ in standard form. They will be written as a product $U(0,A) = Q(A)T(A)$ where $T(A)$ is defined by

$$(T(A)\Phi)(\mathbf{p}) = \Phi(A^{-1}\mathbf{p})\left[\frac{d\mu(A^{-1}\mathbf{p})}{d\mu(\mathbf{p})}\right]^{1/2} .$$

(Here the convention $\mathcal{H}_{\mathbf{p}} = \mathcal{H}_{A\mathbf{p}}$ has made it possible to equate vectors from two different Hilbert spaces.) It is easy to verify that $T(A)$ is unitary, with an adjoint given by

$$(T(A)^*\Phi)(\mathbf{p}) = \Phi(A\mathbf{p})\left[\frac{d\mu(A\mathbf{p})}{d\mu(\mathbf{p})}\right]^{1/2} . \quad (3.8)$$

An elementary computation shows that

$$T(A)U(\mathbf{a},1)T(A)^{-1} = U(A\mathbf{a},1) ,$$

so $U(0,A)$ and $T(A)$ satisfy the same commutation relation with $U(\mathbf{a},1)$. Therefore $Q(A)$ commutes with $U(\mathbf{a},1)$. Thus, by Theorem 1, $Q(A)$ can be written in the form

$$(Q(A)\Phi)(\mathbf{p}) = Q(\mathbf{p},A)\Phi(\mathbf{p}) \quad (3.9)$$

Since $Q(A)$ is unitary $Q(\mathbf{p},A)$ must be unitary for almost all $\mathbf{p}$. Furthermore, the group multiplication law implies

$$Q(A)T(A)Q(B)T(B) = Q(AB)T(AB) ,$$

which yields

$$Q(\mathbf{p},A)Q(A^{-1}\mathbf{p},B) = Q(p,AB) \quad (3.10)$$

for each A and B and almost all $\mathbf{p}$.

At this point a measure-theoretic technicality arises. It is possible *a priori*, that the set of measure zero on which (3.10) does not hold could depend on A and B in such a way that when one took the union over all such sets one would get a set of measure greater than zero. Actually, one can show that one can alter $Q(\mathbf{p},A)$ on a set of measure zero in $\mathbf{p}$ so that $Q(A)$ is unaffected, but (3.10) holds for all $\mathbf{p},A,B$ and $Q(\mathbf{p},A)$ is measurable in both variables. This argument is deferred to Appendix IV, because of its technical character. The result will be assumed in what follows.

The representation has now been reduced to the standard form

$$(U(a,A)\Phi)(\mathbf{p}) = e^{-i\mathbf{p}\cdot\mathbf{a}}Q(\mathbf{p},A) \times \Phi(A^{-1}\mathbf{p})\left[\frac{d\mu(A^{-1}\mathbf{p})}{d\mu(\mathbf{p})}\right]^{1/2} . \quad (3.11)$$

To understand the physical meaning of the $Q(\mathbf{p},A)$ it is helpful to consider some elementary examples. For a single free particle in Schrödinger theory, the wave function may be taken as a complex-valued function of $\mathbf{p}$, the scalar product is

$$(\Phi,\Psi) = \int d\mathbf{p}\ \Phi(\mathbf{p})^*\Psi(\mathbf{p}) \quad (3.12)$$

and the representation of the Euclidean group is

$$(U(\mathbf{a},A)\Phi)(\mathbf{p}) = e^{-i\mathbf{p}\cdot\mathbf{a}}\Phi(A^{-1}\mathbf{p}) . \quad (3.13)$$

Thus, for a single free particle of spin zero $Q(\mathbf{p},A) = 1$. On the other hand, for a single free particle of spin-$\frac{1}{2}$ (Pauli theory), Φ has two components, the integrand in the formula (3.4) for the scalar product is

$$\sum_{i=1}^{2} \Phi_i(\mathbf{p})^*\Psi_i(\mathbf{p}) ,$$

and the transformation law (3.11) becomes

$$(U(\mathbf{a},A)\Phi)(\mathbf{p}) = e^{-i\mathbf{p}\cdot\mathbf{a}}A\Phi(A^{-1}\mathbf{p}) .$$

Here, evidently $Q(\mathbf{p},A) = A$ and describes the transformation properties of the spin degree of freedom. In this case $Q(\mathbf{p},A)$ is independent of $\mathbf{p}$. To get an example in which $Q(\mathbf{p},A)$ cannot be brought

[15] We here use the Radon-Nikodym theorem which asserts that if two measures μ_1 and μ_2 are equivalent, i.e., take the value zero for the same sets, then there exists a positive measurable function $\rho(\mathbf{p})$ such that $d\mu_1(\mathbf{p}) = \rho(\mathbf{p})d\mu_2(\mathbf{p})$. $\rho(\mathbf{p})$ is customarily denoted $(d\mu_1/d\mu_2)(\mathbf{p})$. See, for example, P. R. Halmos, *Measure Theory* (D. Van Nostrand Company, Inc., Princeton, New Jersey, 1950), p. 128.

by unitary transformation of $U(\mathbf{a},A)$ to a form independent of $\mathbf{p}$ one can consider the case of a single photon described in Sec. 5. [One of the results of section 4 is that for a localizable system $Q(\mathbf{p},A)$ can always be chosen independent of $\mathbf{p}$.] Clearly, in all these examples the $Q(\mathbf{p},A)$ gives the transformation law of the internal degrees of freedom of the system under rotations.

A detailed analysis of the consequences of the multiplication law of the Q's, Eq. (3.10), will be undertaken shortly. For the moment, only the fact that for those A which satisfy $A\mathbf{p} = \mathbf{p}$, (3.10) implies

$$Q(\mathbf{p},A)Q(\mathbf{p},B) = Q(\mathbf{p},AB) \qquad (3.14)$$

is needed. Such A form a group called the *little group of* $\mathbf{p}$, and (3.14) means that $A \rightarrow Q(\mathbf{p},A)$ defines a continuous unitary representation of the little group of $\mathbf{p}$. (Again see Appendix IV for a proof that every measurable unitary representation is continuous.) Evidently, when $\mathbf{p} = 0$ the little group of $\mathbf{p}$ is the group of all A, i.e., the unitary unimodular group itself. On the other hand, when $\mathbf{p} \neq 0$, the little group is the two sheeted covering group of the group of rotations around a fixed axis. It is therefore isomorphic to the multiplicative group of the complex numbers $e^{i\theta/2}$, $0 \leqslant \theta < 4\pi$.

The problem of determining when two representations of $\mathcal{E}_3$ are unitary equivalent can now be reduced to a related problem for their $Q(\mathbf{p},A)$. For, if $\{\mathbf{a},A\} \rightarrow U_1(\mathbf{a},A)$ and $\{\mathbf{a},A\} \rightarrow U_2(\mathbf{a},A)$ are equivalent representations, Theorem 1 implies $\mu_1 \equiv \mu_2$ and $\nu_1 = \nu_2$ almost everywhere. Thus, by a unitary transformation one can bring $U_1(\mathbf{a},A)$ into a form where $U_1(\mathbf{a},1) = U_2(\mathbf{a},1)$. Then U_1 and U_2 differ only in their $Q(\mathbf{p},A)$. If

$$U_1(\mathbf{a},A) = VU_2(\mathbf{a},A)V^{-1}, \qquad (3.15)$$

where V is a unitary operator, then, applying Theorem 1, one finds that V is of the form

$$(V\Phi)(\mathbf{p}) = V(\mathbf{p})\Phi(\mathbf{p}) \qquad (3.16)$$

and (3.15) reduces to

$$Q_1(\mathbf{p},A) = V(\mathbf{p})Q_2(\mathbf{p},A)V(A^{-1}\mathbf{p})^{-1}. \qquad (3.17)$$

If $\mathbf{p}$ rather than $A^{-1}\mathbf{p}$ occurred in the last factor, this would describe unitary equivalence of $Q_1(\mathbf{p},A)$ and $Q_2(\mathbf{p},A)$. When A belongs to the little group of $\mathbf{p}$, $A^{-1}\mathbf{p} = \mathbf{p}$ and that is indeed the case.

Again at this point a measure-theoretic technicality arises. Equation (3.17) holds for almost all $\mathbf{p}$, for each A. Again the reader is referred to Appendix IV for a proof that there is a fixed set of measure zero in $|\mathbf{p}|$ such that for all other $\mathbf{p}$ and all A, (3.17) holds.

Next, it will be shown that, if there exists a $V(\mathbf{p})$ for a single $\mathbf{p}$ which satisfies

$$Q_1(\mathbf{p},A) = V(\mathbf{p})Q_2(\mathbf{p},A)V(\mathbf{p})^* \qquad (3.18)$$

for all A in the little group of $\mathbf{p}$, then $V(\mathbf{q})$ can be extended to all $\mathbf{q}$ with $|\mathbf{q}| = |\mathbf{p}|$ so that (3.17) holds. (The statement holds trivially for $\mathbf{p} = 0$ so $\mathbf{p} \neq 0$ is assumed.) Solved for $V(A^{-1}\mathbf{p})$, (3.17) reads

$$V(A^{-1}\mathbf{p}) = Q_1(\mathbf{p},A)^{-1}V(\mathbf{p})Q_2(\mathbf{p},A). \qquad (3.19)$$

This will be consistent as a definition of V at $A^{-1}\mathbf{p}$ only if the right-hand side is constant on right cosets of the little group of $\mathbf{p}$, i.e., only if $A_1^{-1} = A_2^{-1}A_3^{-1}$ with A_3^{-1} in the little group of $\mathbf{p}$ implies that the right-hand side of (3.19) takes the same value for $A = A_1$ and A_2:

$$\begin{aligned} Q_1(\mathbf{p},A_1)^{-1}V(\mathbf{p})Q_2(\mathbf{p},A_1) &\\ = [Q_1(\mathbf{p},A_3)Q_1(A_3^{-1}\mathbf{p},A_2)]^{-1}\mathrm{V}(\mathbf{p}) &\\ \times [Q_2(\mathbf{p},A_3)Q_2(A_3^{-1}\mathbf{p},A_2)] &\\ = Q_1(\mathbf{p},A_2)^{-1}[Q_1(\mathbf{p},A_3)^{-1}V(\mathbf{p})Q_2(\mathbf{p},A_3)]Q_2(\mathbf{p},A_2) &\\ = Q_1(\mathbf{p},A_2)^{-1}V(\mathbf{p})Q_2(\mathbf{p},A_2). & \end{aligned}$$

This defines $V(\mathbf{q})$ for all $\mathbf{q}$ with $|\mathbf{q}| = |\mathbf{p}|$. Next, it has to be verified that V so defined satisfies

$$Q_1(\mathbf{q}, A) = V(\mathbf{q})Q_2(\mathbf{q},A)V(A^{-1}\mathbf{q})^{-1}. \qquad (3.20)$$

Suppose that $\mathbf{q} = B^{-1}\mathbf{p}$. Then, the right-hand side of (3.20) is

$$\begin{aligned} &[Q_1(\mathbf{p},B)^{-1}V(\mathbf{p})Q_2(\mathbf{p},B)]Q_2(B^{-1}\mathbf{p},A)\\ &\quad\times [Q_1(\mathbf{p},BA)^{-1}V(\mathbf{p})Q_2(\mathbf{p},BA)]^{-1}\\ &= Q_1(\mathbf{p},B)^{-1}Q_1(\mathbf{p},BA) = Q_1(\mathbf{q},A), \end{aligned}$$

where, in the last step, the identity $Q_1(\mathbf{p},B)^{-1} = Q_1(B^{-1}\mathbf{p},B^{-1})$ which follows from (3.10), has been used.

Therefore, a necessary and sufficient condition that U_1 be unitary equivalent to U_2 is $\mu_1 = \mu_2$, $\nu_1 = \nu_2$ almost everywhere and the representations of the little groups $A \rightarrow Q_1(\mathbf{p},A)$, $A \rightarrow Q_2(\mathbf{p},A)$ be unitary equivalent for almost all $|\mathbf{p}|$ and at least one $\mathbf{p}$ for each $|\mathbf{p}|$.

Incidentally, in the course of the argument, it has been established that the little groups for $\mathbf{p}$ and $\mathbf{q}$ have unitary equivalent representations if $|\mathbf{p}| = |\mathbf{q}|$. Explicitly, if $\mathbf{q} = B\mathbf{p}$ and $A\mathbf{q} = \mathbf{q}$, then $B^{-1}AB\mathbf{p} = \mathbf{p}$ and

$$Q(\mathbf{q},A) = Q(\mathbf{p},B^{-1})^{-1}Q(\mathbf{p},B^{-1}AB)Q(\mathbf{p},B^{-1}). \qquad (3.21)$$

The mapping $A \rightarrow B^{-1}AB$ is an isomorphism between the little groups of $\mathbf{q}$ and $\mathbf{p}$ and (3.21) displays the unitary equivalence of the corresponding representations.

The classification of the unitary inequivalent representations of the little groups is well known. For $\mathbf{p} = 0$, they are labeled by giving an integer-valued multiplicity function n_{0j} for $j = 0, +\frac{1}{2}, 1, \frac{3}{2}, \cdots$. n_{0j} is the number of times the irreducible representation of angular momentum j appears. For $\mathbf{p} \neq 0$ the unitary inequivalent representations are labeled by an integer or + infinity valued function, $n_{\mathbf{p}m}, m = 0, \pm\frac{1}{2}, \pm 1, \cdots$ where $n_{\mathbf{p}m}$ is the number of times the one-dimensional irreducible representation $\phi \to e^{im\varphi}$ occurs.

All these results are collected in Theorem 2.

Theorem 2. Every continuous unitary representation of $\mathcal{E}_3$, the universal covering group of the Euclidean group, is unitary equivalent to one of the following form.

Let $\mathcal{H}_\mathbf{p}$ be a family of Hilbert spaces, one for each $\mathbf{p} \in \mathfrak{T}_3^*$ identical for all $\mathbf{p}$ with the same $|\mathbf{p}|$. Let

$$\mathcal{H} = \int^{\oplus} d\rho(|\mathbf{p}|) d\omega(\mathbf{p}) \mathcal{H}_\mathbf{p} \oplus \int \mathcal{H}_0$$

where $d\rho$ is a non-negative measure on the positive real axis, such that $\nu(\mathbf{p}) = \dim \mathcal{H}_\mathbf{p}$ is measurable, and $d\omega(\mathbf{p})$ is the measure on the unit sphere of the vectors $\mathbf{p}/|\mathbf{p}|$, invariant under rotations. $\mathcal{H}_0$, the contribution from $\mathbf{p} = 0$, may or may not occur.

The representation is defined by

$$(U(\mathbf{a},A)\Phi)(\mathbf{p}) = e^{-i\mathbf{p}\cdot\mathbf{a}} Q(\mathbf{p},A)\Phi(A^{-1}\mathbf{p}) \quad (3.22)$$

where $Q(\mathbf{p},A)$ is a unitary operator in $\mathcal{H}_\mathbf{p}$ satisfying $Q(\mathbf{p},1) = 1$ and

$$Q(\mathbf{p},A)Q(A^{-1}\mathbf{p},B) = Q(\mathbf{p},AB)$$

for all A, B.

Two representations U_1 and U_2 are unitary equivalent if and only if

(1) $\rho_1 = \rho_2$, i.e., the measures ρ_1 and ρ_2 have the same null sets as measures on the positive real axis.

(2) $\mathcal{H}_0$ either occurs or not in both representations

(3) $\nu_1(\mathbf{p}) = \nu_2(\mathbf{p})$ for almost all $\mathbf{p}$.

(4) the representations of the little groups whose elements are all A such that $A\mathbf{p} = \mathbf{p}$ given by

$$A \to Q_1(\mathbf{p},A) \quad A \to Q_2(\mathbf{p},A)$$

are unitary equivalent for almost all $|\mathbf{p}|$.

The conditions (2), (3), and (4) are satisfied if the multiplicity functions of the representations of the little groups satisfy

$$n_{|\mathbf{p}|m}{}^{(1)} = n_{|\mathbf{p}|m}{}^{(2)} \quad \text{for all } m = 0, \pm\tfrac{1}{2}, \pm 1, \cdots$$

and almost all $|\mathbf{p}|$

$$n_{0j}{}^{(1)} = n_{0j}{}^{(2)} \quad \text{for all} \quad j = 0, \tfrac{1}{2}, 1, \tfrac{3}{2}, \cdots.$$

Any Euclidean invariant theory has a manifold of states whose transformation law is unitary equivalent to one of this form. It is to be expected (and may be seen in detail from the discussion of Secs. 6 and 7) that the imposition of requirements of relativistic invariance will eliminate some of these representations.

Up to this point, the only assumption that has been made about the quantum mechanical system under consideration is its invariance under the Euclidean group. Now the operation of time inversion I_t will be adjoined. It is well known that I_t has to be represented by an antiunitary operator, $U(I_t)$, whose square is $\omega(I_t) = \pm 1$, and that by suitable choice of phase it can be arranged that[16]

$$U(I_t)U(\mathbf{a},A)U(I_t)^{-1} = U(\mathbf{a},A)$$

$$U(\mathbf{a},A)U(I_t) = U(\{\mathbf{a},A\}I_t)$$

$$U(\{\mathbf{a}_1,A_1\}I_t)U(\{\mathbf{a}_2,A_2\}I_t)$$

$$= \omega(I_t)U(\{\mathbf{a}_1,A\}I_t\{\mathbf{a}_2,A\}I_t)$$

$$U(\{\mathbf{a}_1,A_1\}I_t)U(\{\mathbf{a}_2,A_2\}) = U(\{\mathbf{a}_1,A_1\}I_t\{\mathbf{a}_2,A_2\}) .$$

Notice that if $\omega(I_t) = -1$, this is a representation only up to a sign.

To get a standard form for $U(I_t)$ when $U(\mathbf{a},A)$ is in the form (3.12), an extension of Theorem 1 to the case of antiunitary operators is needed. It will be assumed here. The result is

$$(U(I_t)\phi)(\mathbf{p}) = Q(\mathbf{p},I_t)\phi(-\mathbf{p})^* , \quad (3.23)$$

with the Q unitary operators satisfying

$$Q(\mathbf{p},I_t)Q(-\mathbf{p},I_t)^* = \omega(I_t)$$

and

$$Q(\mathbf{p},I_t)Q(-\mathbf{p},A)^* = Q(\mathbf{p},A)Q(A^{-1}\mathbf{p},I_t) .$$

The full analysis of these equations yields a proof that two representations including time inversion are unitary equivalent if and only if their measures μ are equivalent, they have the same multiplicity functions, and their representations of the little group extended by time inversion:

$$A \to Q(\mathbf{p},A) \quad A\mathbf{p} = \mathbf{p}$$

$$I_t \to Q(\mathbf{p},I_t)K$$

are equivalent for at least one $\mathbf{p}$ and almost all $|\mathbf{p}|$.

[16] See E. P. Wigner, *Group Theory and Its Application to the Quantum Mechanics of Atomic Spectra* (Academic Press Inc., 1959), Chap. 26.

Here K stands for complex conjugation. Only a special case will be considered here, namely, that in which $Q(\mathbf{p},A) = \mathfrak{D}(A)$ where $A \rightarrow \mathfrak{D}(A)$ is a continuous unitary representation of the unimodular group. As will be shown in the next section, for localizable systems this can always be arranged. A second specialization will be made. Only time inversion transformation laws for which

$$Q(\mathbf{p},I_t) = \mathfrak{D}(\tau^2) \tag{3.24}$$

will be considered. This amounts to considering the case of ordinary type.[17] Time inversion invariance will be used only to get Theorem 4 on the uniqueness of the position observables.

4. REPRESENTATIONS OF $\mathcal{E}_3$ WHICH POSSESS A TRANSITIVE SYSTEM OF IMPRIMITIVITY[5]

The discussion of this section is in three parts. First, Mackey's standard form of an imprimitive representation is given and shown to be equivalent, in the special case at hand, to a simpler form which will be more convenient for present purposes. Second, for a given imprimitive representation a unitary transformation is found which brings it into Mackey's form. Third, the unitary transformations which commute with an imprimitive representation $U(\mathbf{a},A)$ but not with its system of imprimitivity $E(S)$ are parametrized. This yields a parametrization of the nonuniqueness in the definition of a position operator.

Suppose there is given a continuous unitary representation $A \rightarrow \mathfrak{D}(A)$ of the 2×2 unitary unimodular group in a Hilbert space $\mathcal{H}(\mathfrak{D})$. Then the representation of $\mathcal{E}_3$ *induced by* $\mathfrak{D}(A)$ is denoted $U^{\mathfrak{D}}$ and constructed as follows. Consider functions $\Phi(\mathbf{a},A)$ which are defined on $\mathcal{E}_3$, whose values lie in $\mathcal{H}(\mathfrak{D})$, and which satisfy

(a) $(\Phi(\mathbf{a},A),\chi)$ is a measurable function of $\{\mathbf{a},A\}$, for every $\chi \in \mathcal{H}(\mathfrak{D})$. [The indicated scalar product is in $\mathcal{H}(\mathfrak{D})$.]

$$\text{(b)}\quad \Phi(A\mathbf{b},AB) = \mathfrak{D}(A)\Phi(\mathbf{b},B) \tag{4.1}$$

$$\text{(c)}\quad \int ||\Phi(\mathbf{a},A)||^2 d\mathbf{a} < \infty$$

Notice that (b) implies

$$(\Phi(\mathbf{a},A),\Psi(\mathbf{a},A)) = (\Phi(B\mathbf{a},BA),\Psi(B\mathbf{a},BA))$$

so the integral in (c) is independent of A. Clearly, any linear combination of functions satisfying (a), (b), (c) also satisfies (a), (b), (c) so these functions form a vector space. If a scalar product of Φ and Ψ is defined

$$(\Phi,\Psi) = \int d\mathbf{a}(\Phi(\mathbf{a},A)\ \Psi(\mathbf{a},A))$$

the vector space becomes a Hilbert space $\mathcal{H}$.[18] The representation $U^{\mathfrak{D}}$ is defined in $\mathcal{H}$ by

$$(U(\mathbf{a},A)\Phi)(\mathbf{b},B) = \Phi(\mathbf{b} + B\mathbf{a},BA)\,. \tag{4.2}$$

This representation possesses a transitive system of imprimitivity defined by

$$(E(S)\Phi)(\mathbf{a},A) = \chi_S(\mathbf{a})\Phi(\mathbf{a},A)$$

defined for Borel sets S of $\mathfrak{T}_3$ where, as usual, χ_S is the characteristic function of S: $\chi_S(\mathbf{a}) = 1$ if $\mathbf{a} \in S$, 0 if $\mathbf{a} \notin S$. It is easy to verify using (4.2) that the $E(S)$ transform correctly under $U(\mathbf{a},A)$, i.e., satisfy V.

Because of the smooth fashion in which A acts on $\mathbf{a}$ this representation can be put in a simpler form. If, for the moment, attention is restricted to continuous functions $\Phi(\mathbf{a},A)$, Eq. (4.1) can be used to write

$$\Phi(\mathbf{a},A) = \mathfrak{D}(A)\Phi(A^{-1}\mathbf{a},1) \tag{4.3}$$

which expresses $\Phi(\mathbf{a},A)$ for general values of A in terms of its value for $A = 1$. Conversely, given any continuous function $\Phi(\mathbf{a})$ with values in $\mathcal{H}(\mathfrak{D})$, one can define a continuous $\Phi(\mathbf{a},A)$ by (4.3) and it will then satisfy (4.1). The scalar product of two such $\Phi(\mathbf{a})$ and $\Psi(\mathbf{a})$, $\int d\mathbf{a}(\Phi(\mathbf{a}),\Psi(\mathbf{a}))$ is equal to that of the corresponding $\Phi(\mathbf{a},A),\Psi(\mathbf{a},A)$ so the one to one correspondence can be extended by continuity to a unitary mapping between the Hilbert space $\mathcal{H}$ and the Hilbert space of the measurable square integrable $\Phi(\mathbf{a})$.

The representation (4.2) determines a corresponding representation on the $\Phi(\mathbf{a})$ given by

$$\mathfrak{D}(B)(U(\mathbf{a},A)\Phi)(B^{-1}\mathbf{b}) = \mathfrak{D}(BA)\Phi((BA)^{-1}(\mathbf{b} + B\mathbf{a}))$$

or

$$(U(\mathbf{a},A)\Phi)(\mathbf{b}) = \mathfrak{D}(A)\Phi(A^{-1}(\mathbf{b} + \mathbf{a}))\,.$$

Now this looks just like the standard form of Euclidean transformation appearing in Schrödinger

[17] See reference 16, especially pp. 343–344.

[18] The details of the proof involve identifying functions which differ only on a set of measure zero and establishing that the space is closed. For a proof which is easily adapted to the present circumstances see M. H. Stone, *Linear Transformations in Hilbert Space* (American Mathematical Society, Providence, Rhode Island, 1932), pp. 23–32.

theory except that there one has $-\mathbf{a}$ instead of $\mathbf{a}$ on the right-hand side. That just means that one uses as representative of the function $\Phi(-\mathbf{b})$ instead of $\Phi(\mathbf{b})$. This will be done from this point on. Thus, in the present context, Mackey's form of the imprimitive representation induced by $\mathfrak{D}$ may be taken as

$$(U(\mathbf{a},A)\Phi)(\mathbf{b}) = \mathfrak{D}(A)\Phi(A^{-1}(\mathbf{b}-\mathbf{a})) \quad (4.4)$$

$$(E(S)\Phi)(\mathbf{b}) = \chi_S(\mathbf{b})\Phi(\mathbf{b}) \quad (4.5)$$

with the scalar product

$$(\Phi,\Psi) = \int d\mathbf{b}(\Phi(\mathbf{b}),\Psi(\mathbf{b}))\,. \quad (4.6)$$

Now, the second step of the argument is undertaken; it is to be shown that for each pair consisting of a continuous unitary representation $\{\mathbf{a},A\} \to U(\mathbf{a},A)$ and a system of imprimitivity $E(S)$, there exists a unitary operator V such that $VU(\mathbf{a},A)V^{-1}$ and $VE(S)V^{-1}$ are of the form (4.4) and (4.5), respectively. Available to show this are several lines of argument, not one of them trivial. Here the elegant proof of Loomis[19] will be written out for the present simple case.

The first step in the argument is to express the problem in terms of certain complex-valued functions defined on the group. This is quite analogous to the study of general unitary representations in terms of positive definite functions on the group. To motivate Loomis' method, a brief sketch will first be given of the relation of positive definite functions and representations.

A function φ defined on a group G is *positive definite* if for each $n = 1,2,\cdots$ and all complex numbers $\alpha_1\cdots\alpha_n$ and $g_1\cdots g_n \in G$

$$\sum_{i,j=1}^{n} \alpha_i^*\varphi(g_i^{-1}g_j)\alpha_j \geqslant 0$$

Clearly, taking $n = 1$, one gets

$$\varphi(e) \geqslant 0\,. \quad (4.7)$$

For $n = 2$,

$$(|\alpha_1|^2 + |\alpha_2|^2)\varphi(e) + \alpha_1^*\alpha_2\varphi(g_1^{-1}g_2) + \alpha_2^*\alpha_1\varphi(g_2^{-1}g_1) \geqslant 0\,.$$

From the reality of the left-hand side one concludes $\varphi(g_1^{-1}g_2) = \varphi(g_2^{-1}g_1)^*$ which is equivalent to

$$\varphi(g^{-1}) = \varphi(g)^* \quad \text{all} \quad g \in G\,, \quad (4.8)$$

since G is a group. The positivity of the quadratic form then implies that the determinant of its matrix is positive, i.e.,

$$|\varphi(g)| \leqslant \varphi(e)\,.$$

Any unitary representation of G, $g \to U(g)$, yields examples of positive definite functions[20]

$$\varphi(g) = (\Phi,U(g)\Phi)$$

because, in this case,

$$\sum \alpha_i^*\alpha_k\varphi(g_i^{-1}g_k) = ||\sum \alpha_i U(g_i)\Phi||^2 \geqslant 0\,.$$

If the representation is continuous then $\varphi(g)$ is continuous.

Conversely, given a continuous positive definite function one can construct a continuous representation of G. Let r, s be complex-valued functions on G which are different from zero only at a finite number of points. (Such functions form a vector space.) Introduce the form

$$(r,s) = \sum_{g,h\in G} \overline{r(g)}\,\varphi(g^{-1}h)s(h) \quad (4.9)$$

(r,s) is sesqui-linear, i.e.,

$$(r,s_1+s_2) = (r,s_1) + (r,s_2)\,, \quad (r,\alpha s) = \alpha(r,s) \quad (4.10)$$

$$(r,s) = \overline{(s,r)} \quad (4.11)$$

by virtue of (4.8), and

$$(r,r) \geqslant 0\,. \quad (4.12)$$

Now it may happen that there are some r for which

$$(r,r) = 0\,.$$

If so, it is easy to see that they form a linear subspace and the components orthogonal to this linear subspace form a vector space on which (r,s) again satisfies (4.10), (4.11), and (4.12) but, in addition, $(r,r) = 0$ implies $r = 0$. This space may or may not be complete. If not, complete it and get a Hilbert space H_φ. To get a continuous representation of G in H_φ, define, first on functions with only a finite number of values different from zero,

$$(U(g)r)(h) = r(g^{-1}h) \quad (4.13)$$

with the inverse $U(g^{-1})$.

So defined $U(g)$ leaves the scalar product invariant

$$(U(g)r,U(g)s) = \sum_{h,k} r(g^{-1}h)^*\varphi(h^{-1}k)s(g^{-1}k) = \sum r(h')^*\varphi((gh')^{-1}(gk'))s(k') = (r,s)$$

[19] See reference 6. One of the main virtues of Loomis' treatment is that it applies to nonseparable Hilbert spaces. Since separability is assumed here this advantage will not be apparent.

[20] Positive definite functions were used in the proofs of the celebrated theorems of Bochner and Gelfand-Raikov for Abelian and locally-compact groups, respectively. A systematic account of their properties is found in R. Godement, Trans. Am. Math. Soc. **63**, 1 (1948).

so the subspace of those r for which $(r,r) = 0$ is left invariant by $U(g)$. Therefore, so is its orthogonal complement. Because $U(g)$ is therefore defined and continuous on a dense subset of H_φ it can be extended by continuity to be a unitary operator in H_φ. Clearly, on the original functions

$$U(g_1)U(g_2) = U(g_1g_2)\,,$$

so by continuity, $g \to U(g)$ defines a representation. To prove $U(g)$ is continuous in g, consider

$$\begin{aligned}\|(Ug) - U(g'))r\|^2 &= \|(U(g^{-1}g') - 1)r\|^2 \\ &= 2[(r,r) - \operatorname{Re}\,(U(g^{-1}g')r,r)]\,.\end{aligned}$$

Clearly, this equation implies that it suffices to verify $(U(g)r,r)$ is continuous in g at $g = e$ for all $r \in \mathsf{H}_\varphi$. For r of the special kind appearing in (4.9), which only take values different from zero at a finite number of points the continuity is easy to verify:

$$\begin{aligned}(U(g)r,r) &= \textstyle\sum_{h,k} r(g^{-1}h)^*\varphi(h^{-1}k)r(k) \\ &= \textstyle\sum_{h,k} r(h)^*\varphi(gh)^{-1}k)r(k)\,,\end{aligned}$$

which clearly converges to (r,r) as $g \to e$ because φ is continuous and there is only a finite number of terms in the sum. For a general r, there always exists an s of the above form so that $||r - s|| < \epsilon/3$. By the above argument a neighborhood of e can be found so that $||U(g)s - s|| < \epsilon/3$. Then

$$\begin{aligned}||U(g)r - r|| < ||U(g)r - U(g)s|| + ||U(g)s - s|| \\ + ||s - r||\,,\end{aligned}$$

which completes the proof that $g\ U(g)$ is a continuous unitary representation of G.

Actually, if the continuous positive definite function from which one starts is of the form $(\Phi,V(g)\Phi)$, the representation constructed by the above process will be closely related to V itself. For, if the subspace (of the Hilbert space $\mathcal{H}$ in which φ lies) spanned by vectors of the form $V(g)\Phi$ is denoted $\tilde{\mathcal{H}}$, the constructed representation as unitary equivalent to the restriction of V to $\tilde{\mathcal{H}}$. The required unitary equivalence is obtained by making $\sum r(g)V(g)$ correspond to r, for r differing from zero only at a finite number of points. Equation (4.9) is just arranged to make scalar products correspond. Clearly $(U(g)r)$ corresponds to $V(g) \sum r(h)V(h)\Phi$. The correspondence can be extended by continuity to yield the required unitary equivalence.

A representation V for which there is a vector Φ such that the $V(g)\Phi$ span the representation space is called *cyclic* and Φ is then a *cyclic vector*. Note that the function which is one at $g = e$ and zero elsewhere is a cyclic vector for the representation defined above. Thus, what has been established in the preceding paragraphs is that all cyclic representations are unitary equivalent to those of the form (4.9) and (4.13). Since any representation can be written as a direct sum of cyclic representations, it suffices for many purposes to study cyclic representations.

In the present case, there is a system of imprimitivity $E(S)$ in addition to the group representation $U(g)$ so one has to consider cyclic vectors and representations of $E(S)$ and $U(g)$ together. This suggests studying the function $(E(S)\Phi,U(g)\Phi) = \varphi_S(g)$ and using it to construct a pair unitary equivalent to $\{E(S),U(g)\}$ and in Mackey's form.

Now return to the special case of $\mathcal{E}_3$. When the representation and system of imprimitivity is in Mackey's form (4.4) and (4.5), the function $\varphi_S(\mathbf{a},A)$ is

$$\begin{aligned}\varphi_S(\mathbf{a},A) &= \int d\mathbf{b}((E(S)\Phi)(\mathbf{b}),(U(\mathbf{a},A)\Phi)(\mathbf{b})) \\ &= \int_S d\mathbf{b}(\Phi(\mathbf{b}),\mathfrak{D}(A)\Phi(A^{-1}(\mathbf{b} - \mathbf{a}))) \quad (4.14)\end{aligned}$$

The next task is to show that $\varphi_S(\mathbf{a},A)$ has a form closely related to this for any representation and transitive system of imprimitivity.

Before the discussion can begin a preliminary remark is necessary. Extensive use is going to be made of the part of the Radon-Nikodym theorem which says that if, for two measures μ_1 and μ_2, $\mu_1(S) = 0$ implies $\mu_2(S) = 0$, then there exists a measurable function $\rho(x)$ such that $d\mu_2(x) = \rho(x)d\mu_1(x)$. To make these applications it is essential to know that $E(S) = 0$ for all Borel sets S of Lebesgue measure zero. To obtain this result, it is convenient to use the fact that the $E(S)$ possess a *separating vector*, i.e., a vector Φ such that $E(S)\Phi = 0$ implies $E(S) = 0$. Although this is a standard result[21] a proof will be outlined. Choose an arbitrary unit vector Φ_1, and let $\mathcal{H}_1$ be the subspace spanned by the $E(S)\Phi_1$. Choose a unit vector Φ_2 orthogonal to $\mathcal{H}_1$ and let $\mathcal{H}_2$ be the subspace spanned by the $E(S)\Phi_2$. Continuing in this way one gets a family of orthogonal subspaces such that $\mathcal{H}$ is the direct sum of the $\mathcal{H}_i$ and Φ_i is a cyclic unit vector for $\mathcal{H}_i$. Take as separating vector $\Phi = \sum_n 2^{-n}\Phi_n$. Clearly, if $E(T)\Phi = 0$ then $E(T)\Phi_i = 0$ for all i. Consequently, $E(T)$ yields zero when applied to a dense set of vectors, the linear combinations of the $E(S)\Phi_i$. It is therefore zero and Φ is a separating vector. Note first that if $E(S) = 0$, then $E(AS + \mathbf{a}) = U(\mathbf{a},A)E(S)U(\mathbf{a},A)^{-1} = 0$. Thus if Φ

[21] See reference 14, p. 20.

is a separating vector, $(\Phi,E(S)\Phi) = ||E(S)\Phi||^2$ is quasi-invariant under Euclidean transformation, i.e., for all $\{\mathbf{a},A\}$, $(\Phi,E(S)\Phi) = 0$ if and only if $(\Phi,E(AS + \mathbf{a})\Phi) = 0$. Furthermore, $(\Phi,E(S)\Phi)$ defines a σ-additive positive measure on the Borel sets S of $\mathbf{R}^3$. Now in Appendix II it is shown that any measure defined on the Borel sets of $\mathbf{R}^3$ and quasi-invariant under translations is equivalent to Lebesgue measure. That implies in particular that $(\Phi,E(S)\Phi)$ and therefore $E(S) = 0$ whenever S is a Borel set of measure zero. Thus, the Radon-Nikodymn theorem implies that if Φ is any vector there exists a non-negative measurable function ρ such that

$$(\Phi,E(S)\Phi) = \int_S \rho(\mathbf{b})d\mathbf{b} \qquad (4.15)$$

$\rho(\mathbf{b})$ is clearly integrable over all space.

This equation can be used to get an expression for $\varphi_S(\mathbf{a},A)$ which is the first step in proving that it can always be arranged to have the form (4.14). Note that

$$|\varphi_S(\mathbf{a},A)| = |(E(S)\Phi,U(\mathbf{a},A)\Phi)| \leqslant ||E(S)\Phi||$$

where for convenience it has been assumed that $||\Phi|| = 1$. It therefore follows that if T is any Borel set of $\mathcal{E}_3$

$$\int_T \varphi_S(\mathbf{a},A)d\mathbf{a}dA \leqslant \int_T d\mathbf{a}dA \int_S \rho(\mathbf{b})d\mathbf{b}\,. \qquad (4.16)$$

Now the left-hand side as a function of S and T is initially defined for products of rectangles $S \times T$, and is bounded by the product $\mu(S)\nu(T)$ where $\mu(S) = \int_S\rho(\mathbf{b})d\mathbf{b}$ and $\nu(T)$ is the Lebesgue measure of T. It is finitely additive on such products in the sense that if $S \times T = \bigcup S_i \times T_i$ where $(S_i \times T_i) \cap (S_j \times T_j) = 0$ for $i \neq j$, then its value on $S \times T$ is the sum of its values on the $S_i \times T_i$. Consequently, it has a unique extension to a Borel measure on $\mathbf{R}^3 \times \mathcal{E}_3$. [The necessary argument first shows that it can be extended to be finitely additive on sets which are arbitrary finite unions of disjoint products $S_i \times T_i$. Second, it shows that the boundedness described by (4.16) implies that this extension is actually σ additive. Finally, it uses a standard extension theorem[22] to assert that the resulting set function has a unique extension to be a Borel measure.] Clearly, this measure is bounded by the product measure $\mu \times \nu$. Thus again by the Radon Nikodym theorem this time for complex valued measures there exists a measurable function q of absolute value less than or equal to 1 such that

$$\int d\mathbf{a}dA\varphi_S(\mathbf{a},A) = \int_T\int_S d\mathbf{a}dAd\mathbf{b}q(\mathbf{a},A;b)\rho(\mathbf{b})\,. \qquad (4.17)$$

From (4.17), it follows that

$$\varphi_S(\mathbf{a},A) = \int_S d\mathbf{b}\mathbf{q}(\mathbf{a},A;\mathbf{b})\rho(\mathbf{b})$$

for almost all $\{\mathbf{a},A\}$ which begins to look like (4.14). This completes the first stage of the proof.

The next stage is the construction of the Hilbert space of the $\Phi(\mathbf{a})$ which appears in (4.4) . . . (4.6). This is done in close analogy with the construction carried out in connection with (4.9) but for technical reasons which will appear in the proof it is convenient to consider continuous functions of compact support on $\mathcal{E}_3$ rather than the functions differing from zero only at a finite number of points, which were used there. Therefore let f and g be continuous complex-valued functions of compact support on $\mathcal{E}_3$ and define

$$U(f) = \int d\mathbf{b}dBf(\mathbf{b},B)U(\mathbf{b},B)$$

$$U(g) = \int d\mathbf{c}dCg(\mathbf{c},C)U(\mathbf{c},C)\,. \qquad (4.18)$$

Then

$$\begin{aligned}(E(S)U(f)\Phi,U(\mathbf{a},A)U(g)\Phi) &= \int d\mathbf{b}dB\int d\mathbf{c}dCf(\mathbf{b},B)^*\\ &\quad\times g(\mathbf{c},C)(E(S)U(\mathbf{b},B)\Phi,U(\mathbf{a},A)U(\mathbf{c},C)\Phi)\\ &= \int d\mathbf{b}dB\int d\mathbf{c}dCf(\mathbf{b},B)^*g(\mathbf{c},C)(E(B^{-1}S - B^{-1}\mathbf{b})\Phi\\ &\quad\times U(\{\mathbf{b},B\}^{-1}\{\mathbf{a},A\}\{\mathbf{c},C\})\Phi) = \int d\mathbf{r}\int d\mathbf{b}dB\\ &\quad\times\int_{B^{-1}S-B^{-1}\mathbf{b}} d\mathbf{c}dCf(\mathbf{b},B)^*g(\mathbf{c},C)\\ &\quad\times q(\{\mathbf{b},B\}^{-1}\{\mathbf{a},A\}\{\mathbf{c},C\};\mathbf{r})\rho(\mathbf{r})\,,\\ &= \int_S d\mathbf{r}\int d\mathbf{b}dB\int d\mathbf{c}dCf(\mathbf{b},B)^*g(\mathbf{c},C)\\ &\quad\times q(\{\mathbf{b},B\}^{-1}\{\mathbf{a},A\}\{\mathbf{c},C\};B^{-1}(\mathbf{r}-\mathbf{b}))\\ &\quad\times\rho(B^{-1}(\mathbf{r}-\mathbf{b}))\,.\end{aligned} \qquad (4.19)$$

For $\{\mathbf{a},A\} = \{0,1\}$ this reduces to

$$\begin{aligned}(E(S)U(f)\Phi,U(g)\Phi) &= \int_S d\mathbf{r}\int d\mathbf{b}dB\int d\mathbf{c}dC\\ &\quad\times f(\mathbf{b}+\mathbf{r},B)^*g(\mathbf{c}+\mathbf{r},C)q(\{\mathbf{b},B\}^{-1}\{\mathbf{c},C\};-B^{-1}\mathbf{b})\\ &\quad\times\rho(-B^{-1}\mathbf{b})\,,\end{aligned} \qquad (4.20)$$

which suggests introducing $(U(f)\Phi)(\mathbf{r})$ as the function $f(\mathbf{b}+\mathbf{r},B)$ of $\mathbf{b}$ and B.

[22] See reference 15, p. 54, Theorem A.

Then the form

$$((U(f)\Phi)(\mathbf{r}),(U(g)\Phi)(\mathbf{r})) = \int d\mathbf{b}dB \int d\mathbf{c}dC$$
$$\times f(\mathbf{b} + \mathbf{r},B)^* g(\mathbf{c} + \mathbf{r},C) q(\{\mathbf{b},B\}^{-1}\{\mathbf{c},C\}, -B^{-1}\mathbf{b})$$
$$\times \rho(-B^{-1}\mathbf{b}) \tag{4.21}$$

is suggested as the scalar product appearing in the integrand of (4.14).

With these definitions, one has

$$(E(S)U(f)\Phi,U(\mathbf{a},A)U(g)\Phi) = \int_S d\mathbf{r}$$
$$\times ((U(f)\Phi)(\mathbf{r}),W(\mathbf{a},A)(U(f)\Phi))(\mathbf{r})), \tag{4.22}$$

where $W(\mathbf{a},A)$ is the operator defined by

$$f(\mathbf{b} + \mathbf{r},B) \rightarrow f(A^{-1}\mathbf{b} + A^{-1}(\mathbf{r} - \mathbf{a}),A^{-1}B) .$$

Notice that if linear transformation $\mathfrak{D}(A)$ is defined by the correspondence

$$f(\mathbf{b} + \mathbf{r},B) \rightarrow f(A^{-1}\mathbf{b} + \mathbf{r},A^{-1}B) . \tag{4.23}$$

Then W may be written

$$(W(\mathbf{a},A)U(f)\Phi)(\mathbf{r}) = \mathfrak{D}(A)(U(f)\Phi)(A^{-1}(\mathbf{r} - \mathbf{a}))$$

so that W is precisely of the form (4.4). It is obvious (by a simple change of variable) that $\mathfrak{D}(A)$ leaves the scalar product (4.21) invariant.

Now it has to be verified that (4.21) does indeed define a scalar product. First note that it is linear in g and conjugate linear in f. Furthermore, because (4.20) holds for every Borel set S,

$$(E(S)U(f)\Phi,U(g)\Phi) = [(E(S)U(g)\Phi,U(f)\Phi)]$$

and

$$(E(S)U(f)\Phi,U(f)\Phi) \geqslant 0$$

imply

$$((U(f)\Phi)(\mathbf{r}),(U(g)\Phi)(\mathbf{r}))$$
$$= [((U(g)\Phi)(\mathbf{r}),(U(f)\Phi)(\mathbf{r}))]^* \tag{4.24}$$

and

$$((U(f)\Phi)(\mathbf{r}),(U(f)\Phi)(\mathbf{r})) \geqslant 0 \tag{4.25}$$

for almost all $\mathbf{r}$. However, since f and g are continuous and of compact support the integral appearing in $\mathbf{r}$ is continuous in $\mathbf{r}$. Therefore (4.24) and (4.25) hold for all $\mathbf{r}$. Now, just as in the case of (4.12), one can introduce components of vectors orthogonal to the subspace for which (4.25) is an equality, and complete the resulting space to get a Hilbert space $\mathcal{K}$ the same for each $\mathbf{r}$. $A \rightarrow \mathfrak{D}(A)$ is then a continuous unitary representation in $\mathcal{K}$. Since the correspondence $U(f)\Phi \rightarrow (U(f)\Phi)(\mathbf{r})$ carries a dense set of vectors in the subspace spanned by the $U(f)\Phi$ into a dense set of vectors in the Hilbert space spanned by the functions of $\mathbf{r}$: $(U(f)\Phi)(\mathbf{r})$ and preserves scalar products it can be extended by continuity to become a unitary transformation V.

All this discussion is collected in Theorem 3.

Theorem 3. Let $\{\mathbf{a},A\} \rightarrow U(\mathbf{a},A)$ be a continuous unitary representation of $\mathcal{E}_3$ with a transitive system of imprimitivity, $E(S)$, based on $\mathbf{R}^3$. Then there exists a unitary transformation V, such that $VU(\mathbf{a},A)V^{-1} = W(\mathbf{a},A)$ and $VE(S)V^{-1} = F(S)$, respectively, given by

$$(W(\mathbf{a},A)\Phi)(\mathbf{b}) = \mathfrak{D}(A)\Phi(A^{-1}(\mathbf{b} - \mathbf{a})) \tag{4.26}$$

$$(F(S)\Phi)(\mathbf{b}) = \chi_S(\mathbf{b})\Phi(\mathbf{b}) . \tag{4.27}$$

Here $A \rightarrow \mathfrak{D}(A)$ is a continuous unitary representation of the 2×2 unitary unimodular group in a separable Hilbert space $\mathcal{K}$ and the $\Phi(\mathbf{b})$ are functions on $\mathbf{R}^3$ with values in $\mathcal{K}$ which are measurable in the sense that for all pairs of such functions $(\Phi(\mathbf{b}),\Psi(\mathbf{b}))$ is a measurable function of $\mathbf{b}$. In symbols

$$\mathcal{K} = \int_{\mathbf{R}^3}^{\oplus} d\mathbf{b}\mathcal{K}_{\mathbf{p}} \quad \text{with} \quad \mathcal{K}_{\mathbf{p}} = \mathcal{K} .$$

The remaining task of this section is to examine the arbitrariness in the definition of the position observable. For this purpose, one can bring the pair $\{E(S),U(\mathbf{a},A)\}$ into the form (4.26) and (4.27), and then determine all unitary operators which commute with $U(\mathbf{a},A)$ but not with $E(S)$. It is convenient for this purpose to rewrite (4.26) in momentum space

$$(U(\mathbf{a},A)\Phi)(\mathbf{p}) = e^{-i\mathbf{p}\cdot\mathbf{a}}\mathfrak{D}(A)\Phi(A^{-1}\mathbf{p}) .$$

If B is a unitary operator such that $[B,U(\mathbf{a},1)] = 0$, Theorem 1 shows that B can be written in the form

$$(B\Phi)(\mathbf{p}) = B(\mathbf{p})\Phi(\mathbf{p})$$

where $B(\mathbf{p})$ is a unitary operator in $\mathcal{K}_{\mathbf{p}} = \mathcal{K}$. The commutativity with $U(0,A)$ then implies

$$B(\mathbf{p})\mathfrak{D}(A) = \mathfrak{D}(A)B(A^{-1}\mathbf{p}) \tag{4.28}$$

for almost all $\mathbf{p}$.

This equation can be discussed along lines familiar from Sec. 3 and Appendix IV. For those A which satisfy $A\mathbf{p} = \mathbf{p}$, i.e., for A in the little group of $\mathbf{p}$, (4.28) reduces to

$$B(\mathbf{p})\mathfrak{D}(A) = \mathfrak{D}(A)B(\mathbf{p}) . \tag{4.29}$$

The set of all $B(\mathbf{p})$ satisfying this equation is easy to compute. Supposing them known one gets the general solution of (4.28) by using it as a definition:

$$B(\mathbf{p}) = \mathfrak{D}(A_{\mathbf{q}\leftarrow\mathbf{p}})^{-1}B(\mathbf{q})\mathfrak{D}(A_{\mathbf{q}\leftarrow\mathbf{p}}) . \qquad (4.30)$$

Here the $A_{\mathbf{q}\leftarrow\mathbf{p}}$ satisfy $A_{\mathbf{q}\leftarrow\mathbf{p}}\mathbf{p} = \mathbf{q}$ and parametrize the cosets of the little group of $\mathbf{q}$. By virtue of (4.29) at $\mathbf{q}$ any parametrization yields the same $B(\mathbf{p})$. An argument just like that following equation (3.20) shows that the $B(\mathbf{p})$ defined for a fixed $\mathbf{q}/|\mathbf{q}|$ and all $\mathbf{p}$ by (4.30) satisfies (4.28).

To obtain all solutions of (4.29), one can decompose $A \to \mathfrak{D}(A)$ into irreducible representations of the 2×2 unitary unimodular group, and these in turn into irreducible representations of the little group

$$\mathfrak{D} = \sum_{j=0,1/2,1,\cdots} n_j\mathfrak{D}^{(j)} ,$$

$$\mathfrak{D}^{(j)} = \sum_{m=-j,-j-1,\cdots j} e^{im\phi} .$$

Thus $A \to \mathfrak{D}(A)$ restricted to the little group of $\mathbf{p}$ is unitary equivalent to

$$\sum n_m e^{im\phi} \qquad (4.31)$$

with

$$n_m = \sum_{j\geqq|m|} n_j .$$

Here the summation over j is over integers if m is integral and half-odd integers if m is half an odd integer.

The $B(\mathbf{p})$ corresponding to a given set $\{n_m\}$, $m = 0, \pm\frac{1}{2}, \pm 1, \cdots$ is a direct sum of unitary operators acting in the subspaces of vectors with a definite value of m, and any such defines a possible $B(\mathbf{p})$. The number of real parameters free in an arbitrary $n_m \times n_m$ unitary matrix is n_m^2 so that $B(\mathbf{p})$ contains $\sum_m n_m^2$ arbitrary real parameters, each of which could be a function of $|\mathbf{p}|$.

Collecting the information acquired in the preceding discussion one has Theorem 4.

Theorem 4. If $E(S)$ is a system of imprimitivity for the unitary representation $\{\mathbf{a},A\} \to U(\mathbf{a},A)$ of $\mathcal{E}_3$ in the standard form (4.26), (4.27), then all other systems of imprimitivity consistent with U are given by

$$F(S) = BE(S)B^{-1} ,$$

where B is a unitary operator given by

$$(B\Phi)(\mathbf{p}) = \mathfrak{D}(A_{\mathbf{q}\leftarrow\mathbf{p}})^{-1}B(\mathbf{q})\mathfrak{D}(A_{\mathbf{q}\leftarrow\mathbf{p}})\Phi(\mathbf{p}) \quad (4.32)$$

so that

$$(F(S)\Phi)(\mathbf{p}) = \mathfrak{D}(A_{\mathbf{q}\leftarrow\mathbf{p}})^{-1}B(\mathbf{q})^{-1}\mathfrak{D}(A_{\mathbf{q}\leftarrow\mathbf{p}})$$
$$\times (2\pi^{-3/2} \int \tilde{\chi}_S(\mathbf{p}-\mathbf{r})d\mathbf{r}\mathfrak{D}(A_{\mathbf{q}\leftarrow\mathbf{r}})^{-1}B(\mathbf{q})^{-1}$$
$$\times \mathfrak{D}(A_{\mathbf{q}\leftarrow\mathbf{r}})\Phi(\mathbf{r}) . \qquad (4.33)$$

Here

$$\tilde{\chi}(\mathbf{p}) = [2\pi]^{-3/2} \int e^{i\mathbf{p}\cdot\mathbf{b}}\chi_S(\mathbf{b})d\mathbf{b}$$

and $B(\mathbf{q})$ is a solution of

$$[B(\mathbf{q}),\mathfrak{D}(A)] = 0$$

for all A satisfying $A\mathbf{q} = \mathbf{q}$.

In the discussion up to this point, symmetry under time inversion and any analog of the regularity assumption of Newton and Wigner have been ignored. This is natural in the case of Theorem 3 because the canonical form of a transitive system of imprimitivity can be obtained without the use of these additional assumptions. However, for Theorem 4, they are of decisive importance. Even in this case $\mathfrak{D}(A)$ one dimensional, (4.33) would give a wide variety of distinct position observable ($B(\mathbf{q})$) is then a complex-valued function of the form $B(\mathbf{q}) = e^{i\eta(|\mathbf{q}|)}$, η real. In this case, the effect of the assumption of time inversion invariance is to force B to be real, and therefore to be equal to $+1$. However, it could be $+1$ for some $|\mathbf{p}|$ and -1 for others without violating either Euclidean or time inversion invariance. It is here that Newton and Wigner's assumption of regularity has the effect of making B a constant and $F(S) = E(S)$. They require (in a Lorentz invariant theory) that the infinitesimal Lorentz transformation operators be applicable to localized states in the sense that if Φ_n is a sequence of vectors which converge to a state localized at a point $\mathbf{a}$, as $n \to \infty$, then $\lim_{\infty\to n} ||M_{\mu\nu}\Phi_n||/||\Phi_n|| < \infty$. Since $M_{0i}, i = 1,2,3$ are essentially differentiation operators this forces continuity on the momentum space representation of Newton and Wigner's localized (continuum) state. An analogous requirement in the present formulation has an analogous consequence. The details are as follows.

According to (3.14), the transformation law of states under time inversion is of the form

$$(U(I_t)\Phi)(\mathbf{p}) = \mathfrak{D}(\tau^2)\Phi(-\mathbf{p})^*$$

The requirement that B commute with $U(I_t)$ then forces

$$\mathfrak{D}(\tau^2)\overline{B(-\mathbf{p})} = B(\mathbf{p})\mathfrak{D}(\tau^2) \qquad (4.34)$$

which is

$$\mathfrak{D}(\tau^2)\overline{\mathfrak{D}(A_{\mathbf{q}\leftarrow-\mathbf{p}})^{-1}B(\mathbf{q})\mathfrak{D}(A_{\mathbf{q}\leftarrow-\mathbf{p}})}$$
$$= \mathfrak{D}(A_{\mathbf{q}\leftarrow\mathbf{p}})^{-1}B(\mathbf{q})\mathfrak{D}(A_{\mathbf{q}\leftarrow\mathbf{p}})\mathfrak{D}(\tau^2)$$

or using

$$\overline{\mathfrak{D}(A)} = \mathfrak{D}(\tau^2)^{-1}\mathfrak{D}(A)\mathfrak{D}(\tau^2),$$
$$\mathfrak{D}(A_{\mathbf{q}\leftarrow-\mathbf{p}}A^{-1}_{\mathbf{q}\leftarrow-\mathbf{p}})\mathfrak{D}(\tau^2)\overline{B(\mathbf{q})}$$
$$= B(\mathbf{q})\mathfrak{D}(A_{\mathbf{q}\leftarrow\mathbf{p}}A^{-1}_{\mathbf{q}\leftarrow\mathbf{p}})\mathfrak{D}(\tau^2)\,. \quad (4.35)$$

For suitably chosen $\mathbf{p}$ the factors in $\mathfrak{D}$ cancel and one gets

$$\overline{B(\mathbf{q})} = B(\mathbf{q})\,, \quad (4.36)$$

provided that $\mathbf{q}$ is not along the 2 axis as will be assumed. The remaining condition on $B(\mathbf{q})$ says that it commutes with all $\mathfrak{D}(A_{\mathbf{q}\leftarrow\mathbf{p}}A^{-1}_{\mathbf{q}\leftarrow-\mathbf{p}}\, i\tau^2)$. This will be no further restriction since we may for convenience choose $\mathbf{q}$ along the 3-axis and then every such transformation is an element of the little group of $\mathbf{q}$, and $B(\mathbf{q})$ already commutes with them. [To see this it is convenient to choose a particular form for the $A_{\mathbf{p}\leftarrow\mathbf{q}}$:

$$A_{\mathbf{q}\leftarrow\mathbf{p}} = 2\left(1 + \frac{\mathbf{q}\cdot\mathbf{p}}{|\mathbf{p}||\mathbf{q}|}\right)^{-1/2} \times\left[1 + \frac{\mathbf{p}\cdot\mathbf{q}}{|\mathbf{p}||\mathbf{q}|} - i\frac{\mathbf{p}\times\mathbf{q}}{|\mathbf{p}||\mathbf{q}|}\cdot\boldsymbol{\tau}\right].$$

This is well defined for all $\mathbf{p} \neq -\mathbf{q}$.
Then

$$A_{\mathbf{q}\leftarrow\mathbf{p}}A^{-1}_{\mathbf{q}\leftarrow-\mathbf{p}} = -i\frac{(\mathbf{p}\times\mathbf{q})}{|\mathbf{p}||\mathbf{q}|}\cdot\boldsymbol{\tau}\,.$$

It is easy to choose $\mathbf{p}$ so that $(\mathbf{p}\times\mathbf{q}/|\mathbf{p}||\mathbf{q}|\cdot\boldsymbol{\tau})\tau^2 = 1$; then (4.36) follows. However $\mathbf{p}$ is chosen provided $\mathbf{q}$ is along the 3 axis $(\mathbf{q}\times\mathbf{p}/|\mathbf{q}||\mathbf{p}|)\cdot\boldsymbol{\tau}\tau^2$ leaves $\mathbf{q}$ invariant. This proves the second statement.]

A comparison of these statements with the discussion just before Theorem 4 shows that the effect of time inversion invariance on the arbitrariness of $B(\mathbf{q})$ is to reduce the number of arbitrary real parameters from $\sum n_m^2$ to $\sum n_m(n_m - 1)$ each of which could be on a function of $|\mathbf{q}|$. It is clear that the position observable will be nonunique as long as $\mathfrak{D}(A)$ is not irreducible. If $\mathfrak{D}(A)$ is irreducible and the elements of the little group have $\mathfrak{D}(A)$ reduced to diagonal form $B(\mathbf{q})$ is diagonal with diagonal elements which are real functions of $|\mathbf{q}|$ of square 1; the position observable is still not unique. However, unless $B(\mathbf{q})$ is the constant matrix ± 1, the formula (4.33) will yield discontinuous functions of $\mathbf{p}$. [Take a compact set S, then the integral in (4.33) is differentiable, so discontinuities in the function outside the integral are discontinuities of $(F(S)\Phi)(\mathbf{p})$.] Such discontinuities will appear at any value of $\mathbf{q}$ where $B(|\mathbf{q}|)$ jumps so $B(|\mathbf{q}|)$ must be constant in $|\mathbf{q}|$. It must be a constant multiple of the identity if it is to be differentiable at $\mathbf{p} = -\mathbf{q}$. In summary, we have Theorem 5.

Theorem 5. In a Euclidean invariant system with time inversion symmetry the possible observables $F(S)$ which describe localization are given by (4.33) with $B(\mathbf{q})$ a real unitary operator.

If localized states are differentiable in $\mathbf{p}$ and $A \to \mathfrak{D}(A)$ is an irreducible representation of the 2×2 unitary unimodular group then $B(\mathbf{p})$ is a constant multiple of the identity and $F(S) = E(S)$. Conversely, if the $F(S)$ are unique $A \to \mathfrak{D}(A)$ must be irreducible and localized states differentiable.

In the following in Theorems 6 and 7, the regularity and time inversion invariance requirements assumed in Theorem 5 will always be taken for granted.

5. REPRESENTATIONS OF $\mathcal{E}_3$ WHICH ARE RESTRICTIONS OF REPRESENTATIONS OF THE COVERING GROUP OF THE INHOMOGENEOUS LORENTZ GROUP

It is well known that every continuous unitary representation of the covering group of the Poincaré group is unitary equivalent to one of the form

$$(U(\mathbf{a},A)\Phi)(\mathbf{p}) = e^{i\mathbf{p}\cdot\mathbf{a}}Q(\mathbf{p},A)\Phi(A^{-1}\mathbf{p})$$

in $\mathfrak{H}$,

$$\mathfrak{H} = \int^{\oplus} d\mu(p)\mathfrak{H}_p$$

with

$$d\mu(p) = \mu_0\delta(p)dp + d\rho_+(m)d\Omega_{m+}(p) + d\rho_-(m)d\Omega_{m-}(p) + d\rho(im)d\Omega_{im}(p)\,,$$

$d\Omega_{m\pm}(p) = d\,\mathbf{p}/[m^2 + \mathbf{p}^2]^{1/2}$ being the invariant measure on the hyperboloids $p^2 = m^2$, $p^0 \gtrless 0$, respectively. $d\Omega_{im}(p)$ is the invariant measure on the hyperboloid $p^2 = -m^2$. $Q(p,A)$ is unitary and satisfies

$$Q(p,A)Q(A^{-1}p,B) = Q(p,AB)\,.$$

For the subrepresentations with $m^2 > 0$, $Q(p,A)$ can be chosen in the form

$$Q(p,A) = \mathfrak{Q}(A^{-1}_{p\leftarrow k}AA_{A^{-1}p\leftarrow k}) \quad (5.1)$$

where $k = (m,0,0,0)$ and $A_{q\leftarrow k}$ is given by

$$A_{q\leftarrow k} = [2(q^0 + m)m]^{-1/2}[m1 + \tilde{q}]\,,\ \tilde{q} = q^0 + \mathbf{q}\cdot\boldsymbol{\tau}$$

and $A \to \mathfrak{Q}(A)$ is a continuous unitary representation of the unitary unimodular group. For $m = 0$, the $Q(p,A)$ are a direct sum of two parts, the first of which contains all the finite spin constituents while the second contains all infinite spin constituents. For both of these (5.1) again holds but k is some standard light-like vector, say (1,0,0,1), and $A_{q\leftarrow k}$ is a parametrization of the cosets of the little group of k. That little group is isomorphic to the two-sheeted covering group of the Euclidean group of the plane and $A \to \mathcal{Q}(A)$ is a continuous unitary representation of it. For the finite spin part this representation is trivial for the "translations" while for the infinite spin part it is not. The subspace of the mass zero representations can be written as a direct integral over two-dimensional Ξ space

$$\mathcal{H}_p = \int^{\oplus} d\sigma(\Xi)\mathcal{H}_{p\Xi}\,, \quad p^2 = 0\,,$$

with the scalar product

$$(\Phi,\Psi)_{m=0} = \int d\Omega_0(p) \int d\sigma(\Xi)\,(\Phi(p,\Xi),\Psi(p,\Xi))$$

where

$$d\sigma(\Xi) = \sigma_0\delta(\Xi)d\Xi + d\sigma_1(|\Xi|)d\varphi$$

with $\Xi = \Xi_1 + i\Xi_2 = |\Xi|e^{i\varphi}$ and

$$(Q(k,A)\Phi)(k,\Xi) = \exp\,(i\Xi\cdot\mathbf{t})Q_1(k,\Xi,A)\Phi(k,e^{-i\theta}\Xi)$$

for

$$A = [1 + \tfrac{1}{2}\,t(\mathbf{e}_1 + i\mathbf{e}_2)\cdot\boldsymbol{\tau}] \times [\cos\theta/2 - i\sin\,(\theta/2)(\mathbf{k}/k^0)\cdot\boldsymbol{\tau}]$$

with $\mathbf{e}_1^2 = 1 = \mathbf{e}_2^2$, $\mathbf{e}_1\cdot\mathbf{e}_2 = 0 = \mathbf{e}_1\cdot\mathbf{k} = \mathbf{e}_2\cdot\mathbf{k}$, $t = t_1 + it_2$. Here $Q_1(k,0,A)$ may be expressed in terms of a representation $\mathcal{Q}_1$ of the above A leaving k fixed.

$$Q_1(k,0,A) = \mathcal{Q}_1[\cos\theta/2 - i\sin\,(\theta/2)(\mathbf{k}/k^0)\cdot\boldsymbol{\tau}]$$

$Q_1(k,\Xi,A)$, $\Xi \neq 0$ may be expressed in terms of a representation $\mathcal{Q}$ of the two element groups $A = \pm 1$, which is the subgroup of those unitary unimodular A which leave k and some Ξ, say Ξ_1 fixed:

$$Q_1(k,\Xi,A) = \mathcal{Q}(A_{\Xi\leftarrow\Xi_1}AA_{A^{-1}\Xi\leftarrow\Xi_1})$$

where A is a transformation of the form $\cos\theta/2 - i\sin\,(\theta/2)(\mathbf{k}/k^0)\cdot\boldsymbol{\tau}$ carrying Ξ_1 into Ξ.

The representations of imaginary mass and null four-momentum (apart from the identity representation) will be ignored here as being irrelevant to the transformation properties of physical systems.

Clearly, when $\{a,A\}$ is restricted to lie in $\mathcal{E}_3$, the subrepresentation which comes from mass 0 is in precisely the form (2.2) and Theorems 4 and 5 apply directly. For the case of mass zero and $\Xi = 0$, the system is localizable if the representation of the little group $A \to Q_1(k,0,A)$ is the restriction of a representation of the unitary unimodular group. This happens for the spin-zero case but for no other irreducible representation. For the case of mass zero and $\Xi \neq 0$, the representation of the little group is a direct integral over irreducible representations which are determined by the value of $|\Xi|$ and the representation of the little group of the little group, $A = \pm 1$. The representatives of the state vectors, $\Phi(k,\Xi)$ can be expanded in Fourier series on the circle $|\Xi| = \text{const}$. This corresponds to a decomposition into irreducible representations of the subgroup of the unitary unimodular group that leaves k fixed. In case the little group of the little group is trivially represented, each integer angular momentum along $\mathbf{k}$ appears exactly once. In case it is nontrivially represented, each half odd integer angular momentum along $\mathbf{k}$ appears twice. Such representations can never be the restriction of a representation of the full unimodular group. Thus elementary systems with $\Xi \neq 0$ are never localizable. Reducible systems are localizable only if each representation $|\Xi|$ appears with infinite multiplicity or not at all.

Theorem 6. Lorentz invariant systems of $m^2 > 0$ are always localizable. Their position observables are unique if the systems are elementary, i.e., their representations are irreducible.

For $m = 0$, the only localizable elementary system has spin zero. For a reducible system to be localizable it is necessary and sufficient that each irreducible representation of infinite spin appear with zero or infinite multiplicity, and the finite spin parts contribute states of angular momentum along a fixed direction whose multiplicities coincide with those of the restriction of a representation of the unimodular group.

The identity representation for which $p = 0$ can not appear in the transformation law of any localizable system.

6. REPRESENTATIONS OF $\mathcal{E}_3$ ARISING IN GALILEI-INVARIANT SYSTEMS

Unlike the case of Lorentz invariance where all representations up to a factor are physically equivalent to representations of the covering group, Galilei invariance leads to factors which cannot be got rid of by passing to the covering group. However, as Bargmann showed,[23] one can regard them as true repre-

[23] V. Bargmann, Ann. Math. 59, 1 (1954), especially pp. 38–43.

sentations of a certain extension of the covering group of the Galilei group. The first task of this section is to express this statement in explicit formulas and summarize the classification of the representations.

The Galilei transformations will be denoted (a,Γ) or in more detail $(\tau,\mathbf{a},\mathbf{v},R)$ where $(0,\Gamma) = (0,0,\mathbf{v},R)$ $(a,1) = (\tau,\mathbf{a},0,1)$ and

$$(\tau,\mathbf{a})\begin{Bmatrix} t \\ \mathbf{x} \end{Bmatrix} = \begin{Bmatrix} t+\tau \\ \mathbf{x}+\mathbf{a} \end{Bmatrix}, \quad (\mathbf{v},R)\begin{Bmatrix} t \\ \mathbf{x} \end{Bmatrix} = \begin{Bmatrix} t \\ R\mathbf{x}+\mathbf{v}t \end{Bmatrix}.$$

R is a rotation of the three space of the $\mathbf{x}$. The group multiplication law is

$$(\tau_1,\mathbf{a}_1,\mathbf{v}_1,R_1)(\tau_2,\mathbf{a}_2,\mathbf{v}_2,R_2) = (\tau_1+\tau_2,\ \mathbf{a}_1+\mathbf{v}_1\tau_2,\ \mathbf{v}_1+R_1\mathbf{v}_2,\ R_1R_2)\ .$$

The covering group is obtained by replacing R by A, a 2×2 unitary matrix of determinant 1, just as in (2.1). For simplicity, $\{a,\Gamma\}$ will be written for the group elements in this case also.

Bargmann showed that by physically inessential changes of phase, one could bring all the factors into the following form:

$$\omega(a_1,\Gamma_1;a_2,\Gamma_2) = \exp i(M/\hbar) \times (\mathbf{v}_1\cdot A_1\cdot\mathbf{a}_2 + \tfrac{1}{2}\mathbf{v}_1^2\tau_2)\ .$$

Here $M/\hbar$ is a constant of the dimensions time/[length]^2, which has arbitrarily been written as a ratio in order that its interpretation shall come out automatically when applied to Schrödinger theory ($\hbar$ is Planck's constant divided by 2π).

Furthermore, Bargmann pointed out that every representation up to a factor with this factor arises from a representation of the group whose elements are $(\exp i\theta,a,\Gamma)$ $0\le\theta<2\pi$ and whose multiplication law is

$$\begin{aligned}(\exp i\theta_1,a_1,\Gamma_1)\cdot(\exp i\theta_2,a_2,\Gamma_2) &= (\exp i[\theta_1+\theta_2+(M/h)(\mathbf{v}_1\cdot R_1\cdot\mathbf{a}_2+\tfrac{1}{2}\mathbf{v}_1^2\tau_2)],\\ &\quad a_1+\Gamma_1a_2,\ \Gamma_1\Gamma_2)\end{aligned}$$

via the formula

$$U(\exp i\theta,a,\Gamma) = e^{i\theta}U(a,\Gamma)\ . \qquad (6.1)$$

For the case $M=0$, this refinement is unnecessary. That case will be discussed later.

Now the elements of the group of the $(\exp i\theta,a,\Gamma)$ which are of the form $(\exp i\theta,a,1)$ form a normal subgroup and the group is the semi-direct product of this Abelian normal subgroup and the subgroup of the $(\phi,0,\Gamma)$. Just as in the case of the Euclidean group one diagonalizes the Abelian subgroup in terms of a direct integral over the character group whose elements are $\exp i[q\theta+\hbar^{-1}(E\tau-\mathbf{p}\cdot\mathbf{a})]$. The states are then functions $\Phi(q,p)$ labeled by integers q and a real four-component $p=(E/\hbar,\mathbf{p}/\hbar)$. The scalar product is

$$(\Phi,\Psi) = \int d\mu(q,p)(\Phi(q,p),\Psi(q,p))$$

and

$$\begin{aligned}(U(\exp i\theta,a,1)\Phi)(q,p) &= \exp i[q\theta+\hbar^{-1}(E\tau-\mathbf{p}\cdot\mathbf{a})]\Phi(q,p)\ .\end{aligned}$$

The action of Γ on the subgroup $(\exp i\theta,\tau,\mathbf{a},1)$ is

$$\begin{aligned}&(1,0,\Gamma)(\exp i\theta,a,1)(1,0,\Gamma)^{-1}\\ &= (\exp i[\theta+(M/\hbar)(\mathbf{v}\cdot A\cdot\mathbf{a}+\tfrac{1}{2}\mathbf{v}^2\tau)],\Gamma a,1)\ .\end{aligned}$$

It induces a corresponding transformation of the characters

$$\begin{aligned}&q\to q \quad E\to E-\mathbf{p}\cdot\mathbf{v}+\tfrac{1}{2}qM\mathbf{v}^2\\ &\mathbf{p}\to A^{-1}(\mathbf{p}-qM\mathbf{v})\ .\end{aligned}$$

From this, it follows that

$$d\mu(q,E,\mathbf{p}) \equiv d\mu(q,E+\mathbf{v}\cdot A\mathbf{p}+\frac{q}{2}M\mathbf{v}^2,A\mathbf{p}+qM\mathbf{v}).$$

To yield a representation of the form (6.1), $d\mu$ must be a product of δ_{q1} with a measure in (E,p) above, $d\nu(E,p)$, satisfying

$$d\nu(E,\mathbf{p}) \equiv d\nu(E+\mathbf{v}\cdot A\mathbf{p}+\frac{M}{2}\mathbf{v}^2,A\mathbf{p}+m\mathbf{v})\ .$$

This in turn implies that $d\nu$ is equivalent to a measure constant on parabolas

$$d\nu(E,p) = d\rho(E_0)dN_{E_0}(\mathbf{p})$$

where $dN_{E_0}(\mathbf{p})$ is the measure $d\mathbf{p}$ on the parabola

$$E = E_0+\frac{\mathbf{p}^2}{2M}$$

and $d\rho(E_0)$ is a measure on the real axis describing the spectrum of rest energy of the system. Again just in the case of the Euclidean group, there is a canonical form

$$\begin{aligned}U(\exp i\theta,a,\Gamma)\Phi)(E_0,p) &= \exp(i\theta)\\ &\times \exp[(i/\hbar)(E\tau-\mathbf{p}\cdot\mathbf{a})]Q(E_0,\mathbf{p},\Gamma)\\ &\times \Phi(E_0,A^{-1}(\mathbf{p}-M\mathbf{v}))\end{aligned} \qquad (6.2)$$

where

$$\begin{aligned}&Q(E_0,\mathbf{p},\mathbf{v}_1,A_1)Q(E_0,A_1^{-1}(\mathbf{p}-M\mathbf{v}_1),\mathbf{v}_2,A_2)\\ &= Q(E_0,\mathbf{p},\mathbf{v}_1+A_1\mathbf{v}_2,A_1A_2)\end{aligned} \qquad (6.3)$$

and the scalar product is

$$(\Phi,\Psi) = \int d\rho(E_0) \int dN_{E_0}(\mathbf{p})(\Phi(E_0,\mathbf{p}),\Psi(E_0,\mathbf{p})) .$$

The little group of a vector $\mathbf{q}$ consists of all $(\mathbf{v},A)$ of the form $(M^{-1}(\mathbf{q} - A\mathbf{q}),A)$ and is isomorphic to the unitary unimodular group.

The Q's can be brought into the canonical form

$$Q(E_0,\mathbf{p},\mathbf{v},A) = Q(E_0,0, -\mathbf{p}/M,1)^{-1}Q(E_0,0,0,A) \times Q(E_0,0,A^{-1}(\mathbf{v} - \mathbf{p}/M,1) \quad (6.4)$$

where

$$A \rightarrow \mathcal{Q}(A) = Q(E_0,0,0,A)$$

is a continuous unitary representation of the group of unitary unimodular matrices. Evidently, (6.4) just describes a superposition of Schrödinger particles of mass M, and various rest energies (described by E_0), and spins (described by the irreducible constituents of $\mathcal{Q}$).

It is clear that the representation of $\mathcal{E}_3$ that is obtained from (6.4)

$$(U(\mathbf{a},A)\Phi(p)) = e^{-i\mathbf{p}\cdot\mathbf{a}}\mathcal{Q}(A)\Phi(A^{-1}\mathbf{p})$$

with the scalar product

$$(\Phi,\Psi) = \int d\mathbf{p}(\Phi(\mathbf{p}),\Psi(\mathbf{p}))$$

and

$$(\Phi(\mathbf{p}),\Psi(\mathbf{p})) = \int d\rho(E_0)(\Phi(E_0,\mathbf{p}),\Psi(E_0,\mathbf{p})) .$$

Thus for $M > 0$, the situation is essentially identical with that in the Lorentz invariant case. There is always a position operator and the arbitrariness in it is that associated with the representation of the unimodular group which describes the transformation properties of the system under rotations in the rest system.

For $M < 0$, localizability still makes sense but such representations are rejected on the physical ground that the kinetic energy of a particle is negative.

For representations with $M = 0$, the preceding argument has to be reexamined. There is no need to introduce θ as in (6.1). The diagonalization of $U(a,1)$ leads to

$$(U(a,1)\Phi)(E,\mathbf{p}) = \exp\,[i(E\tau - \mathbf{p}\cdot\mathbf{a})/\hbar]\Phi(E,\mathbf{p})$$

with a scalar product

$$(\Phi,\Psi) = \int d\mu(E,\mathbf{p})(\Phi(E,\mathbf{p}),\Psi(E,\mathbf{p}))$$

but the measure μ now satisfies

$$d\mu(E,\mathbf{p}) \equiv d\mu(E + \mathbf{v}\cdot A\cdot\mathbf{p},A\mathbf{p})$$

and this implies that $d\mu$ is equivalent to

$$d\omega(\mathbf{p})d\rho(|\mathbf{p}|)dE + \mu_0\delta(\mathbf{p})dEd\mathbf{p} ,$$

where $d\omega(\mathbf{p})$ is the area on the sphere of radius $|\mathbf{p}|$. (The fact that the energy spectrum of the system runs to $-\infty$, makes these representations of dubious physical interest, but does not exclude their being localizable.) The full transformation law is of the form

$$(U(a,\Gamma)\Phi)(E,\mathbf{p}) = \exp\,[i(E\tau - \mathbf{p}\cdot\mathbf{a})/\hbar]Q(E,\mathbf{p},\Gamma) \times \Phi(E - \mathbf{v}\cdot\mathbf{p},A^{-1}\mathbf{p})$$

where $Q(E,p,\Gamma)$ satisfies

$$Q(E,\mathbf{p},\mathbf{v}_1,A_1)Q(E - \mathbf{v}_1\cdot\mathbf{p},A_1^{-1}\mathbf{p},\mathbf{v}_2,A_2) = Q(E,\mathbf{p},\mathbf{v}_1 + A_1\mathbf{v}_2,A_1A_2) .$$

The little group of p, $p \neq 0$ consists of all (ω,A) such that $A\mathbf{p} = \mathbf{p}$ and $\mathbf{v}\cdot\mathbf{p} = 0$. This is a group isomorphic to the two-sheeted covering group of the Euclidean group of the plane. For $\mathbf{p} = 0$, the little group is the full unitary unimodular group. There can be no contribution of this latter kind in any localizable system because the criterion (i) of Sec. 2 is not satisfied so only the former case will be considered. There, the criterion (i) forces $d\rho(|\mathbf{p}|)$ to be equivalent to $|\mathbf{p}|^2d|\mathbf{p}|$. From this, it is clear that no irreducible representation is localizable because an irreducible representation has $d\mu$ concentrated on an orbit $|\mathbf{p}| = \text{const.}$[24]

The general $Q(E,\mathbf{p},\mathbf{v},A)$ is expressed in terms of the representation of the little group of the vector $(0,\mathbf{q})$, where $|\mathbf{p}| = |\mathbf{q}|$, in the following way:

$$Q(E,\mathbf{p},\Gamma) = Q(0,\mathbf{q},\Gamma((E,\mathbf{p}) \leftarrow (0,\mathbf{q}))^{-1}$$
$$\times Q(0,\mathbf{q},\Gamma((E,\mathbf{p}) \leftarrow (0,\mathbf{q}))^{-1}\Gamma\Gamma((\Gamma^{-1}(E,\mathbf{p})) \leftarrow (0,\mathbf{q}))$$
$$\times Q(0,\mathbf{q},\Gamma((\Gamma^{-1}(E,\mathbf{p}) \leftarrow (0,\mathbf{q}))) .$$

Here $\Gamma((E,\mathbf{p}) \leftarrow (0,\mathbf{q}))$ is a Galilei transformation which carries $(0,\mathbf{q})$ into $(E,\mathbf{p})$, so

$$\Gamma((E,\mathbf{p}) \leftarrow (0,\mathbf{q}))^{-1}\Gamma\Gamma((\Gamma^{-1}(E,\mathbf{p})) \leftarrow (0,\mathbf{q}))$$

belongs to the little group of $(0,\mathbf{q})$.

The same procedure can be applied to analyze the representation $(\mathbf{v},A) \rightarrow Q(0,\mathbf{q},\mathbf{v},A)$, $\mathbf{v}\cdot\mathbf{q} = 0$, $A\mathbf{q} = \mathbf{q}$, of the little group of $(0,\mathbf{q})$ as was applied to $\mathcal{E}_3$ itself. One diagonalizes the "translations" $\mathbf{v}$. Then the

[24] This result agrees with that of Inönü and Wigner, reference 3.

representation takes the form

$$Q(0,\mathbf{q},\mathbf{v},A)\Phi)(\mathbf{q},\mathbf{n}) = e^{i\mathbf{v}\cdot\mathbf{n}}Q_1(\mathbf{q},\mathbf{n},A)\Phi(\mathbf{q},A^{-1}\mathbf{n}) .$$

Here $\mathbf{n}$ is a two-component vector in the plane perpendicular to $\mathbf{q}$ which labels the characters of the "translation" subgroup. The scalar product is

$$(\Phi(\mathbf{q}),\Psi(\mathbf{q})) = \int d\sigma(\mathbf{n})(\Phi(\mathbf{q},\mathbf{n}),\Psi(\mathbf{q},\mathbf{n})) ,$$

where the measure σ is equivalent to one of the form

$$d\sigma(\mathbf{n}) = \sigma_0\delta(\mathbf{n})d\mathbf{n} + d\sigma_1(|\mathbf{n}|)d\varphi,\ \mathbf{n}_1 + i\mathbf{n}_2 = |\mathbf{n}|e^{i\varphi} .$$

The little group of the little group is the little group itself if $\mathbf{n} = 0$, while it is the two-element group: $A = \pm 1$ if $\mathbf{n} \neq 0$. In the former case $A \rightarrow Q_1(\mathbf{q},0,A)$ is any continuous unitary representation of the little group of $\mathbf{q}$. In the latter case, $\pm 1 \rightarrow Q_1(\mathbf{q},\mathbf{n},\pm 1)$ is any unitary representation of the 2-element group and the Q_1 of general argument is expressed in terms of the elements of the little group by

$$Q_1(q,\mathbf{n},A) = [Q_1(\mathbf{q},\mathbf{n}_0,A_{\mathbf{n}\leftarrow\mathbf{n}_0}{}^{-1})]^{-1}$$
$$\times\ Q_1(\mathbf{q},\mathbf{n}_0,A_{\mathbf{n}\leftarrow\mathbf{n}_0}AA_{A^{-1}\mathbf{n}\leftarrow\mathbf{n}_0})Q_1(\mathbf{q},\mathbf{n}_0,A_{A^{-1}\mathbf{n}\leftarrow\mathbf{n}_0})$$

The irreducible representations of the little group of the little group have either $\sigma_0 > 0$, $d\sigma_1 = 0$ or $\sigma_0 = 0$, $d\sigma_1(|\mathbf{n}|) = \delta(|\mathbf{n}| - \alpha)d|\mathbf{n}|$, for some $\alpha > 0$. The corresponding Q_1 are one dimensional.

In the case $\sigma_0 > 0$, $Q_1(\mathbf{q},0,A)$ is just $Q(0,\mathbf{q},\mathbf{v},A)$ for $\mathbf{v}\cdot\mathbf{q} = 0$ $A\mathbf{q} = \mathbf{q}$, so the system will be localizable if $A \rightarrow Q(0,\mathbf{q},0,A)$ for $A\mathbf{q} = \mathbf{q}$ defines a representation which is a restriction of a representation of the full unimodular group. In the case $|\mathbf{n}| \neq 0$, $Q_1(q,\mathbf{n},\pm 1) = +1$ yields a $Q(0,\mathbf{q},0,A)$, $A\mathbf{q} = \mathbf{q}$ which contains each integer angular momentum along $\mathbf{q}$ just once, so it is not localizable. A necessary condition for localizability is that the representation $Q_1(q,\mathbf{n},\pm 1) = +1$ have zero or infinite multiplicity. The irreducible representation $Q_1(q,\mathbf{n},\pm 1) = \pm 1$ yields a $Q(0,\mathbf{q},0,A)$, $A\mathbf{q} = \mathbf{q}$ which contains each half-odd integer angular momentum along $\mathbf{q}$ just once so it is not localizable. A necessary condition for localizability is that $Q_1(q,\mathbf{n},\pm 1) = \pm 1$ appear with zero or infinite multiplicity.

All this is summarized in Theorem 7.

Theorem 7. Every Galilei invariant system with $M > 0$ is localizable.

For $M = 0$, no elementary system is localizable because such a system has momentum satisfying $|\mathbf{p}| = \text{const}$. Systems with $M = 0$ and a reducible representation of the Galilei group are localizable if and only if:

(a). The measure on momentum space is equivalent to Lebesgue measure;

(b). The subrepresentation of the little group of $(0,\mathbf{q})$ for which the pure Galilei transformations $\mathbf{\Gamma} = (\mathbf{v},1)$ are trivially represented is, for almost all $|\mathbf{q}|$, the restriction to the group of A such that $A\mathbf{q} = \mathbf{q}$ of a fixed representation of the 2×2 unitary unimodular group;

(c). The subrepresentation of the little group of $(0,q)$ for which the pure Galilei transformations are non-trivially represented contains each irreducible with multiplicity zero or infinity, the same for almost all $|\mathbf{q}|$.

ACKNOWLEDGMENTS

A substantial part of this paper was written in 1952, when the author was a National Research Council Post Doctoral Fellow in Copenhagen. He thanks Professor Niels Bohr for the hospitality of the Institut for Theoretisk Fysik and Professor Lars Gårding for the hospitality of Lunds Matematiska Institution. The paper was completed in 1962 with the support of the National Science Foundation. The author thanks Professor Robert Oppenheimer for the hospitality of the Institute for Advanced Study during the later period.

APPENDIX I. FINITE ADDITIVITY ON FINITE UNIONS OF CUBES

In connection with axioms I, II, III, it was remarked that it might appear more natural from a physical point of view to weaken the axioms so that the existence of the observable $E(S)$ is required only for S a finite union of cubes, and finite additivity is required:

$$E(S_1 \cup S_2) = E(S_1) + E(S_2) \quad \text{if} \quad S_1 \cap S_2 = 0$$

instead of the complete or σ-additivity described in III.

In this Appendix, it is shown that such a weakening of the axioms is only apparent because any $E(S)$ satisfying the weakened axioms can be extended uniquely so as to satisfy I, II, III, as they stand.

Consider the family $\mathfrak{A}$ of all sets of $\mathbf{R}^3$ which are finite unions of half-open intervals. By a half-open interval is meant a set $[\mathbf{a},\mathbf{b})$ of the form

$$\{\mathbf{y};\ a_1 \leqslant y_1 < b_1,\ \ a_2 \leqslant y_2 < b_2,\ \ a_3 \leqslant y_3 < b_3\} ,$$

that is, the set of all y satisfying the listed inequalities. By assumption, the cases $a_j = -\infty$ and $b_j = +\infty$ are also included; in the former case, the equality sign in $a_j \leqslant y_j$ should be ignored. $\mathfrak{A}$ is referred to as an algebra of sets because it is closed under the operations of taking the complement of a set and taking the union of a finite number of sets. A *σ algebra* of sets is one closed under complementation and denumerable unions. A *projection-valued finitely-additive measure* on $\mathfrak{A}$ is a function, E, with values which are projections in a Hilbert space $\mathcal{H}$, defined for all sets of $\mathfrak{A}$ and satisfying II, and

$$\text{III}' \quad E(S_1 \cup S_2) = E(S_1) + E(S_2) - E(S_1 \cap S_2)$$

for any $S_1, S_2 \in \mathfrak{A}$.

A projection-valued finitely additive measure that satisfies in addition

$$E(\bigcup S_i) = \sum_{i=1}^{\infty} E(S_i)$$

for any sequence of $S_i \in \mathfrak{A}, i = 1,2,\cdots$ such that $S_i \cap S_j = 0$, $i \neq j$ and $\bigcup S_i \in \mathfrak{A}$ is called *completely additive or σ additive*. The precise statement of the result of this Appendix is

Theorem A5. Any finitely-additive projection-valued measure on $\mathfrak{A}$ which satisfies

$$E(S + \mathbf{a}) = U(\mathbf{a})E(S)U(\mathbf{a})^{-1}$$

for some continuous unitary representation of the translation group $\mathbf{a} \rightarrow U(\mathbf{a})$ is necessarily completely additive on $\mathfrak{A}$. It then possesses a unique completely additive extension to the σ algebra of all Borel sets on $\mathbf{R}^3$.

Variants of the last statement of the theorem are quite standard in various contexts in measure theory, so it will not be proved here. (In Halmos' book, reference 15, p. 54, the theorem is stated: "If μ is a σ finite measure on a ring $\mathbf{R}$, then there is a unique measure $\bar{\mu}$ on the σ ring, $S(R)$, generated by $\mathbf{R}$ such that for E in $\mathbf{R}$, $\bar{\mu}$ $(E) = \mu(E)$; the measure $\bar{\mu}$ is σ finite." The assumptions of the present Appendix are more general in that one has a projection-valued measure rather than a real-valued measure, but otherwise everything is more special: The ring of sets, $\mathbf{R}$, is an algebra because the whole space is in $\mathbf{R}$, the measure is finite rather than only σ finite.) The first part of the theorem is a consequence of the following chain of four theorems. The argument is a straightforward generalization of one due to Hewitt.[25]

If F is any function on $\mathbf{R}^3$ whose values can be added and subtracted and $[\mathbf{a},\mathbf{b})$ is an interval, define

$$\begin{aligned}\Delta_F[\mathbf{a},\mathbf{b}) = F(b_1,b_2,b_3) - F(a_1,b_2,b_3) - F(b_1,a_2,b_3) \\ - F(b_1,b_2,a_3) + F(a_1,a_2,b_3) + F(a_1,b_2,a_3) \\ + F(b_1,a_2,a_3) - F(a_1,a_2,a_3)\,.\end{aligned}$$

[25] E. Hewitt, Mat. Tidsskrift (1951B), pp. 81–94.

If F is real valued it is said to be *positively monotonic* if $\Delta_F[\mathbf{a},\mathbf{b}) \geqslant 0$ for all $[\mathbf{a},\mathbf{b})$.[26] If the values of F are commuting projections the analogous requirement is that $\Delta_F[\mathbf{a},\mathbf{b})$ be a projection for all intervals $[\mathbf{a},\mathbf{b})$. Notice that if $E(S)$ is any finitely additive projection valued measure defined on $\mathfrak{A}$, it yields such an F from the definition

$$F(x_1,x_2,x_3) = E(\{\mathbf{y}; y_1 < x_1, y_2 < x_2, y_3 < x_3\})\,. \quad \text{(A1)}$$

Conversely, the following theorem holds.

Theorem A1. Let F be a positively monotonic function defined on $\mathbf{R}^3$ with values which are commutative projections. Suppose

$$\begin{aligned}F(-\infty,x_2,x_3) &= F(x_1, -\infty,x_3) \\ &= F(x_1,x_2,-\infty) = 0\,. \quad \text{(A2)}\end{aligned}$$

Then there exists a finitely additive projection valued measure E on $\mathfrak{A}$ satisfying (A1).

The proof is completely elementary and will be omitted.

Now consider the increasing sequence of projections

$$F(x_1 - 1/k, \cdots x_3 - 1/k) \quad k = 1,2,\cdots\,.$$

It converges to a projection $F_-(x_1,\cdots x_3)$ which may or may not be $F(x_1,\cdots x_3)$.

Example. Consider the function E_t defined on $\mathfrak{A}$ which is the projection $E \neq 0$ for a set S if there is an interval of the form $\{\mathbf{y}; t_1 - \epsilon \leqslant y_1 < t_1, t_2 - \epsilon \leqslant y_2 < t_2, t_3 - \epsilon \leqslant y_3 < t_3\}$ which lies in S and zero otherwise. It is easy to see that E_t is a finitely additive projection valued measure on $\mathfrak{A}$. It is not completely additive because the interval $\{\mathbf{y}; t_1 - 1 \leqslant y_1 < t_1, t_2 - 1 \leqslant y_2 < t_2, t_3 - 1 \leqslant y_3 < t_3\}$ can be written as a denumerable union of intervals for which the coordinates y_j lie in intervals where right-hand end points are less than t_j. For each such interval $E_t(S) = 0$ but for the union $E_t(S) = E$. Clearly, the F corresponding to E_t does not satisfy $F(t_1 \cdots t_3) = F_-(t_1 \cdots t_3)$.

If for each $\mathbf{x} \in \mathbf{R}^3$, $F(\mathbf{x}) = F_-(\mathbf{x})$, then the phenomenon occurring in the example cannot happen and the projection valued measure defined by F is σ additive on $\mathfrak{A}$.

Theorem A2. Every projection valued positively monotonic function F on $\mathbf{R}^3$ which satisfies (A2) and

$$\lim_{k\to\infty} F(x_1 - 1/k, \cdots x_3 - 1/k) = F(x_1, \cdots x_3)\,. \quad \text{(A3)}$$

[26] A detailed discussion of positively monotonic functions is given in, E. J. McShane, *Integration* (Princeton University Press, Princeton, New Jersey, 1947), pp. 242–274.

defines a projection valued measure E on $\mathfrak{A}$ which is σ additive.

Proof. Since each element of $\mathfrak{A}$ is a finite union of disjoint intervals and E is finitely additive according to Theorem A2, it suffices to consider the case of a denumerable union of sets in $\mathfrak{A}$ whose union is an interval. But such a union defines a monotonically increasing sequence of projections which converges to the projection belonging to the interval by virtue of (A3). Therefore E is completely additive.

A finitely-additive projection-valued measure E is called *purely finitely additive* if there is no nontrivial σ additive projection-valued measure which is zero on every set S for which $E(S) = 0$. (It is not difficult to see that the example $E_{\mathbf{t}}$ is purely finitely additive.)

Theorem A3. Every finitely additive projection valued measure on $\mathfrak{A}$ is the sum of a purely finitely additive part and a σ additive part. This decomposition is unique.

Proof. The difference $F(\mathbf{x}) - F_{-}(\mathbf{x})$ is a projection, and two such, corresponding to distinct points $\mathbf{x}$ are orthogonal. Because the Hilbert space is separable, there can be at most a denumerable set of points $\mathbf{x}$ where $F(\mathbf{x}) - F_{-}(\mathbf{x}) \neq 0$; call them $\mathbf{t}^{(k)}$. Let $E_{\mathbf{t}^{(k)}}(S)$ be the projection-valued measure given in the example above with $E = F(\mathbf{t}^{(k)}) - F_{-}(\mathbf{t}^{(k)})$. Then

$$E(S) - \textstyle\sum_k E_{\mathbf{t}^{(k)}}(S)$$

defines a finitely-additive projection-valued measure whose F satisfies (A3) for all points $\mathbf{x}$ and so by Theorem A2 is σ additive. Thus

$$E(S) = \textstyle\sum_k E_{\mathbf{t}^{(k)}}(S) + E^{(2)}(S)$$

defines a decomposition into a purely finitely additive part and a σ additive part. For the case in which $E(S)$ is purely finitely additive, $E^{(2)}(S) = 0$ because otherwise $E^{(2)}(S)$ would be a σ additive projection-valued measure vanishing whenever $E(S)$ does in contradiction with the definition of a purely finitely-additive measure. This shows that the purely finitely-additive part of any E is uniquely determined by the discontinuities of the corresponding F.

Now note that if $E(S)$ is quasi-invariant under translations in the sense that $E(S + \mathbf{a}) = 0$ if and only if $E(S) = 0$, then the same applies to the purely finitely-additive part, $E^{(1)}(S)$, of $E(S)$ and the σ additive part of S. [$E(S)$ is surely quasi-invariant if there exists a representation $\mathbf{a} \to U(\mathbf{a})$ of the translation group such that $E(S + \mathbf{a}) = U(\mathbf{a})E(S)U(\mathbf{a})^{-1}$] Furthermore, if $F^{(1)}$ has a nonzero discontinuity $F^{(1)}(\mathbf{x}) - F_{-}^{(1)}(\mathbf{x})$ at $\mathbf{x} = \mathbf{t}$, it must also have a nonzero discontinuity at $\mathbf{x} = \mathbf{t} + \mathbf{a}$. This statement is in conflict with the denumerability of the points of discontinuity unless $F^{(1)} = 0$. Thus, there are no nontrivial purely finitely-additive projection valued measures quasi-invariant under translations.

Theorem A4. Every finitely additive projection valued measure on $\mathfrak{A}$ which is quasi-invariant under translations is σ additive.

From Theorem A4 and the result already cited that σ additive projection-valued measures on $\mathfrak{A}$ have unique extensions to the Borel sets of $\mathbf{R}^3$, Theorem A5 follows.

While the results of this Appendix make it clear that the assumptions of I to V can be weakened without impairing the results of the paper, it should be noted that the particular weakened assumptions used have been chosen primarily for reasons of mathematical elegance. A deeper physical analysis would ask whether the existence of some kind of approximate position measurement implied the existence of precise position measurements in the sense of I to V.

APPENDIX II. SKETCH OF THE DERIVATION OF THE CONTINUOUS UNITARY REPRESENTATIONS OF THE TRANSLATION GROUP

The result of Theorem 1 which describes all unitary representations of the translation groups has been used in physics since the beginning of quantum mechanics, but explicit mathematical statements and proofs of it are relatively recent. The purpose of this Appendix is to outline some of the ideas involved in the proofs.

The translation group of n-dimensional real Euclidean space $\mathbf{R}^n$ will here be denoted $\mathfrak{T}$ with elements $\mathbf{a}$. (The whole machinery works in the same way for any dimension n so the assumption $n = 3$ is dropped.) The derivation of Theorem 1 can be divided into three parts:

(1) Determination of the character group $\mathfrak{T}^*$ of $\mathfrak{T}$,

(2) Derivation of the spectral representation

$$U(\mathbf{a}) = \int_{\mathfrak{T}^*} e^{-i\mathbf{p}\cdot\mathbf{a}} dF(\mathbf{p}) ,$$

(3) Spectral multiplicity theory for the projection valued measure F on $\mathfrak{T}^*$.

These stages actually reflect the historical development of the theorem and I will follow them here at least in part.

A *character* of $\mathfrak{T}$ is a one-dimensional continuous unitary representation of $\mathfrak{T}$, i.e., a complex-valued continuous function χ of modulus one, which satisfies

$$\chi(\mathbf{a} + \mathbf{b}) = \chi(\mathbf{a})\chi(\mathbf{b}) . \qquad \text{(A4)}$$

It is well known that any such χ is of the form $\chi_{\mathbf{p}}$, where

$$\chi_{\mathbf{p}}(\mathbf{a}) = \exp -i(\mathbf{p}\cdot\mathbf{a}) \quad \text{and} \quad \mathbf{p}\cdot\mathbf{a} = \sum_{j=1}^{n} p^j a^j .$$

[The argument goes as follows. From (A4), $\chi(0) = 1$ and $\chi(\mathbf{a})$ can be written

$$\chi(\mathbf{a}) = \chi(a^1,0,\cdots 0)\chi(0,a^2,0,\cdots 0)\cdots\chi(0,0,\cdots a^n) ,$$

where $\chi(0\cdots \mathbf{a}^j\cdots)$ is a character of the one-dimensional translation group of a^j. Thus the problem is reduced to finding all characters for the translation group of the real line. By introducing $i \ln \chi = f$ one reduces the problem to that of finding all real continuous $f(a)$ defined mod 2π such that

$$f(a) + f(b) = f(a+b) \bmod 2\pi \qquad \text{(A.5)}$$

To complete the proof it is convenient to specify $f(a)$ completely instead of mod 2π. Because χ is continuous, a unique specification is obtained in some neighborhood of $a = 0$ by requiring $f(0) = 0$ and $f(a)$ continuous in the neighborhood. From (A5), one then derives $qf(q^{-1}c) = f(c)$ for any c in the neighborhood and any integer q. Thus, again using (A5), $f((p/q)c) = (p/q)f(c)$ for any rational number $p/q < 1$. The continuity of f then implies $f(yc) = yf(c)$ for every real number <1, i.e., $f(y) = yf(c)/c$ for y in the neighborhood. Finally, using (A5) again, one gets $f(y) = yf(c)/c$ mod 2π for all y. Q.E.D.]

The characters clearly form a group under multiplication

$$\chi_{\mathbf{p}_1}(\mathbf{a})\chi_{\mathbf{p}_2}(\mathbf{a}) = \chi_{\mathbf{p}_1+\mathbf{p}_2}(\mathbf{a})$$

and, if the usual topology of Euclidean space is introduced for the **p**'s, the group operations are continuous. The set of all characters (or equivalently the set of all **p**'s) is denoted $\mathfrak{T}^*$ and called the character group of $\mathfrak{T}$[27].

The step (2) alone can be regarded as a decomposition of an arbitrary continuous unitary representation into irreducibles. This operation is familiar in quantum mechanics for the one dimensional translation group as Stone's theorem: Any one-parameter continuous unitary group is of the form

$$U(a) = \exp -iaH , \qquad \text{(A6)}$$

where H is self-adjoint. Then by the spectral theorem for self-adjoint operators $H = \int_{-\infty}^{\infty} p dF(p)$ so (A6) can be written

$$U(a) = \int e^{-ipa} dF(p) . \qquad \text{(A7)}$$

Here F defines a projection valued measure via $F(S) = \int_S dF(p)$. The extension of (A7) to arbitrary Abelian groups was carried out by a number of authors.[28] Since the step from the one-dimensional to n-dimensional translation group is easy, and excellent textbook accounts of Stone's theorem are available,[29] no more details of (2) will be given here.

The problem of determining when two representations are unitary equivalent is reduced by the SNAG theorem to the corresponding problem for their F's. A solution of this problem is provided by (3), the theory of spectral multiplicity. It shows that the unitary equivalence class of an F can be characterized by two objects, a measure class on $\mathfrak{T}^*$ and a multiplicity function on $\mathfrak{T}^*$, which described, respectively (and roughly), tell which irreducible representations of $\mathfrak{T}$ occur in $\mathbf{a} \to U(\mathbf{a})$ and how often. This theory is to the theory of (2) what the Hellinger-Hahn theory of a self-adjoint operator[30] is to the spectral resolution of a self-adjoint operator.

There are available nearly as many approaches to the theory of spectral multiplicity as there are authors who have written on the subject. One may make a direct analysis of the commutative algebra of projections.[31] This leads to a decomposition of the Hilbert space into orthogonal subspaces $\mathcal{H}_j$ on which the projections are uniformly j-dimensional. That means that $\mathcal{H}_j$ is a direct sum of j subspaces $\mathcal{H}_1^j\cdots\mathcal{H}_j^j$ such that the projections E take of the form

$$E(\Phi_1,\cdots\Phi_j) = (E_1\Phi_1,\cdots E_j\Phi_j)$$

and on $\mathcal{H}_k^j$ the E_k are uniformly one dimensional. Finally, a uniformly one-dimensional algebra of projections is one which is maximal Abelian, i.e., any projection which commutes with all the given projections is one of them. It is shown that a uni-

[27] This construction of the character group can be carried out for an arbitrary locally compact Abelian group. See, for example, L. Pontrjagin, *Topological Groups* (Princeton University Press, Princeton, New Jersey, 1939), Chap. V.

[28] Stone's original paper is Ann. Math. **33**, 643 (1932). The extension to any locally compact Abelian group is contained in M. Naĭmark, Izvest. Akad. Nauk U.S.S.R. **7**, 237 (1943); W. Ambrose, Duke Math. J. **11**, 589 (1944); R. Godement, Compt. rend. **218**, 901 (1944). It is sometimes referred to as the SNAG theorem.

[29] See for example F. Riesz and B. Sz.-Nagy, *Leçons d'analyse fonctionelle* (Budapest, 1953), p. 377.

[30] See M. H. Stone, *Linear Transformations in Hilbert Space* (American Mathematical Society, Providence, Rhode Island, 1932), Chap. VII.

[31] See, for example, H. Nakano, Ann. Math. **42**, 657 (1941); I. E. Segal, Memoirs Am. Math. Soc. **9** (1951), Secs. I and II; P. R. Halmos, *Introduction to Hilbert Space and the Theory of Spectral Multiplicity* (Chelsea Publishing Company, New York, 1951).

formly one-dimensional algebra of projections is unitary equivalent to one in which the projections have the form

$$(E(S)\Phi)(x) = \chi_S(x)\Phi(x) ,$$

where $\Phi(x)$ are complex-valued functions square integrable with respect to a measure μ. Reassembling the Hilbert space, one just gets a form of the projection operators just like that indicated for the projection operators on momentum space given in Theorem 1.

Alternatively, one can imbed the projection operators in an appropriately chosen commutative algebra of bounded operators and then use the spectral theory of such commutative algebras to obtain the required canonical form.[32]

APPENDIX III. QUASI-INVARIANT MEASURES

In this Appendix, the structure of quasi-invariant measures defined on the Borel sets of $\mathbf{R}^3$ is determined for two different situations. In the first, the group acting on $\mathbf{R}^3$ is $\mathbf{R}^3$ itself. Then, every finite quasi-invariant measure is equivalent to Lebesgue measure. In the second, the group acting on $\mathbf{R}^3$ is the rotation group. Then the most general finite quasi-invariant measure is equivalent to a measure of the form

$$\mu(S) = \mu_0\chi_S(0) + \int_0^\infty d\rho(a) \int_{S\cap\{\mathbf{p};|\mathbf{p}|=a\}} d\omega_a(\mathbf{p}) , \quad \text{(A8)}$$

where $\mu_0 \geqslant 0, \chi_S$ is the characteristic function of the set S, $d\omega_a(\mathbf{p})$ is the invariant surface element on the sphere $|\mathbf{p}| = a$, and $d\rho(a)$ is a measure on the positive real axis.

The result for the first situation is a special case of the general result that any Borel measure on a locally compact group quasi-invariant with respect to the action of the group on itself is equivalent to Haar measure.[33] The proof of Loomis given in[33] is so simple that it will be repeated here in the special context of $\mathbf{R}^3$.

Let S be any Borel set in $\mathbf{R}^3$. Denote the finite measure quasi-invariant with respect to Lebesgue measure by μ. Let S^* be the set in $\mathbf{R}^6$ defined by $\mathbf{x} - \mathbf{y} \in S$. (It is a Borel set because $\mathbf{x} - \mathbf{y}$ is a continuous function of $\mathbf{x}$ and $\mathbf{y}$. Then the characteristic functions of S^* and S are related by

$$\chi_{S^*}(\mathbf{x},\mathbf{y}) = \chi_S(\mathbf{x} - \mathbf{y})$$

and so because χ_S is positive

$$\iint \chi_{S^*}(\mathbf{x},\mathbf{y})d\mathbf{x}d\mu(\mathbf{y}) = \int\left(\int d\mathbf{x}\chi_S(\mathbf{x} - \mathbf{y})\right) d\mu(\mathbf{y})$$

$$= \int\left(\int d\mu(\mathbf{y})\chi_S(\mathbf{x} - \mathbf{y})\right) d\mathbf{x} .$$

Because $d\mathbf{x}$ is invariant under translation the second expression is $(\int_S d\mathbf{x}) \int d\mu(\mathbf{y})$. On the other hand, $\chi_S(\mathbf{x} - \mathbf{y}) = \chi_{-S+\mathbf{x}}(\mathbf{y})$ so the last equality becomes

$$\left(\int_S d\mathbf{x}\right) \int d\mu(\mathbf{y}) = \int \mu(-S + \mathbf{x})dx . \quad \text{(A9)}$$

From this equality, the required equivalence can be deduced as follows. Note that $\int_S d\mathbf{x} = 0$ if and only if $\int_{-S} d\mathbf{x} = 0$. Thus, from (A9), $\int_S d\mathbf{x} = 0$ implies $\mu(S + \mathbf{x}) = 0$ for almost all $\mathbf{x}$. By the quasi-invariance of μ, this, in turn, implies $\mu(S) = 0$. Conversely, if $\mu(S) = 0$ and therefore $\mu(S + \mathbf{x}) = 0$, (A9) implies $\int_S d\mathbf{x} = 0$.

This completes the proof of the equivalence of μ with Lebesgue measure. The Radon-Nikodym theorem guarantees that $d\mu(\mathbf{x}) = \rho(\mathbf{x})d\mathbf{x}$ where $\rho(\mathbf{x})$ is positive and measurable.

In the second situation, one has a finite measure μ on the Borel sets of $\mathbf{R}^3$ such that for every Borel set S and every rotation R, $\mu(S) = 0$ if and only if $\mu(RS) = 0$.

It is easy to see that any such quasi-invariant measure is equivalent to an invariant measure. In fact, consider the non-negative set function

$$\tilde{\mu}(S) = \int dR\mu(RS) ,$$

where the integration is over all the rotation group and dR is the invariant measure on the rotation group for which $\int dR = 1$. It is not difficult to verify that $\tilde{\mu}$ is σ additive. Furthermore, it is equivalent to μ, because $\tilde{\mu}(S) = 0$ implies $\mu(RS) = 0$ for almost all R, which, because of the quasi-invariance of μ, yields $\mu(S) = 0$. Conversely, $\mu(S) = 0$ implies $\mu(RS) = 0$, which implies $\tilde{\mu}(S) = 0$. Thus, it suffices to consider invariant μ.

It is convenient in completing the proof to use an alternative characterization of a finite measure on $\mathbf{R}^3$ as a non-negative bounded linear functional on the continuous functions of compact support, $\mathcal{C}(\mathbf{R}^3)$. That the functional, μ, is non-negative means $\mu(f)$

[32] R. Godement, Ann. Math. **53**, 68 (1951); J. Dixmier, "Les Algebres d'Operateurs dans l'Espace Hilbertien," *Algebres de von Neumann* (Gauthier-Villars, Paris, 1957), Chap. II; see especially pp. 216–224.

[33] See, for example, G. W. Mackey, Duke Math. J. **16**, 313 (1949), Lemma **33**; J. von Neumann, Bull. Amer. Math. Soc. **42**, 343 (1936).

$\geqslant 0$ for $f \geqslant 0$, $f \in \mathcal{C}$. That μ is bounded means

$$\sup_{|f| \leq 1} |\mu(f)| < \infty, \quad \text{where} \quad |f| = \sup_{\mathbf{x} \in \mathbf{R}^3} |f(\mathbf{x})|.$$

The relation between the functional μ and the corresponding measure μ is simply

$$\mu(f) = \int f(\mathbf{x}) d\mu(\mathbf{x}).$$

Since the measure is uniquely determined by the functional, to verify the equality of two measures it suffices to verify the equality of the corresponding functionals.[34]

Now, for an invariant measure

$$\mu(f) = \int dR\, \mu(Rf) = \mu\left(\int (Rf) dR\right),$$

because the approximating sums to the integral $\int (Rf)(\mathbf{x}) dR$ converge uniformly in $\mathbf{x}$, and $\mu(f)$ is continuous for uniform convergence of its argument. But $f(\mathbf{x}) \to \int (Rf)(\mathbf{x}) dR = \int f(R\mathbf{x}) dR$ maps the continuous functions of compact support on $\mathbf{R}^3$ onto the continuous functions of compact support on $0 \leqslant |\mathbf{x}| < \infty$ and convergence in $\mathcal{C}(\mathbf{R}^3)$ implies convergence in $\mathcal{C}([0, \infty))$. Thus, the functional μ regarded as defined on $\mathcal{C}([0, \infty))$ defines a finite measure on the non-negative real axis. Splitting it into a contribution with support at 0, and the rest, one has just the μ_0 and $d\rho$ of (A8). In fact, (A8) is just an explicit form in terms of measure of

$$\mu(f) = \mu\left(\int Rf dR\right).$$

APPENDIX IV. SOME MEASURE-THEORETIC NICETIES CONNECTED WITH EQS. (3.10) AND (3.17)

This Appendix is devoted to some fine points which arise in the otherwise elementary derivation of Sec. 3.

Recall that Theorem 1 states that if

$$[Q(A), U(\mathbf{a}, 1)] = 0,$$

then $Q(A)$ is of the form

$$(Q(A)\Phi)(\mathbf{p}) = Q(\mathbf{p}, A)\Phi(\mathbf{p}),$$

where for each unitary unimodular A, $Q(\mathbf{p}, A)$ is measurable in $\mathbf{p}$ in the sense that for each $\Psi_1, \Psi_2 \in \mathcal{H}$, $(\Psi_1(\mathbf{p}), Q(\mathbf{p}, A)\Psi_2(\mathbf{p}))$ is μ measurable. The first step in the argument is to prove that $Q(\mathbf{p}, A)$ can, if necessary, be altered on a set of μ-measure zero so that it becomes measurable in both variables relative to the measure $\mu \times \alpha$, where α is the invariant measure on the 2×2 unitary unimodular group.

Let $\Phi_j, j = 1, 2, \cdots$ be a complete orthonormal set in $\mathcal{H}$. Then it suffices to treat the functions $(\Phi_j(\mathbf{p}), Q(\mathbf{p}, A)\Phi_k(\mathbf{p}))$ separately because the general case then follows by the expansions

$$\Psi_1 = \sum a_j \Phi_j, \quad \Psi_2 = \sum b_j \Phi_j, \quad \text{and}$$

$$(\Psi_1(\mathbf{p}), Q(\mathbf{p}, A)\Psi_2(\mathbf{p})) = \sum_{j,k=1} a_j^* b_k \times (\Phi_j(\mathbf{p}), Q(\mathbf{p}, A)\Phi_k(\mathbf{p})).$$

An ugly little lemma is necessary.

Lemma. Let $f(\mathbf{p}, A)$ be a complex-valued function on $\mathbf{R}^3 \times G$ which is μ measurable and μ essentially bounded on $\mathbf{R}^3$ for each $A \in G$, the 2×2 unitary unimodular group. Suppose $\int f(\mathbf{p}, A)\chi_E(\mathbf{p}) d\mu(\mathbf{p})$ is α measurable on G for each μ measurable subset E of $\mathbf{R}^3$ of finite measure. Here, α-measurability on G is with respect to the invariant measure dA.

Then there exists a function, g, $\mu \times \alpha$ measurable on $\mathbf{R}^3 \times G$, and such that for a certain μ-measurable subset N of $\mathbf{R}^3$ of zero measure

$$f(\mathbf{p}, A) = g(\mathbf{p}, A) \quad \text{for all} \quad A \in G \quad \text{and} \quad \mathbf{p} \notin N.$$

This lemma is a special case of Lemma 3.1 of reference 33, and will not be proved here.

The lemma shows that by a suitable redefinition of $Q(\mathbf{p}, A)$ which does not affect the corresponding operator $Q(A)$, one can have $Q(\mathbf{p}, A)$, $\mu \times A$ measurable.

The next step in the argument is to show that in the equation

$$\sum_l (\Phi_j(\mathbf{p}), Q(\mathbf{p}, A)\Phi_l(\mathbf{p}))(\Phi_l(\mathbf{p}), Q(A^{-1}\mathbf{p}, B)\Phi_k(\mathbf{p})) = (\Phi_j(\mathbf{p}), Q(\mathbf{p}, AB)\Phi_k(\mathbf{p})) \quad \text{(A10)}$$

which holds for each $A, B \in G$ and $\mathbf{p} \in \mathbf{R}^3$ such that $\mathbf{p} \notin N_1(A, B)$, $A^{-1}\mathbf{p} \notin N_2(A, B)$ where N_1 and N_2 are μ-measurable sets of μ-measure zero, the right- and left-hand sides are $\mu \times \alpha \times \alpha$ measurable on $\mathbf{R}^3 \times G \times G$. Because a Borel-measurable function of a Borel-measurable function is Borel measurable, it suffices to prove that the mappings $T_1: \{\mathbf{p}, A, B\} \to \{\mathbf{p}, AB\}$ and $T_2: \{\mathbf{p}, A, B\} \to \{A^{-1}\mathbf{p}, B\}$ are Borel-measurable functions.

Now T_1 and T_2 are continuous, and a set F which is $\mu \times \alpha$ measurable in $\mathbf{R}^3 \times G$ differs from a Borel set by a subset of a Borel set of zero $\mu \times \alpha$ measure.[35]

Furthermore, a continuous function has the property that the antecedent of any Borel set of its range

[34] See, for example, P. R. Halmos, *Measure Theory* (D. van Nostrand Company, Inc., Princeton, New Jersey, 1950), pp. 243–9.

[35] See reference 15, pp. 55–56.

is a Borel set of its domain. Thus to prove the $\mu \times \alpha \times \alpha$ measurability of T_1 and T_2, it suffices to show that for any Borel set F of $\mathbf{R}^3 \times G$ of zero $\mu \times \alpha$ measure $T_1^{-1}(F)$ and $T_2^{-1}(F)$ have zero $\mu \times \alpha \times \alpha$ measure. Consider T_2, the proof for T_1 being similar.

Let $F_\mathbf{p}$ denote $\{A;\{\mathbf{p},A\} \in F\}$. Clearly, $\{\mathbf{p},AB\} \in F$ if and only if $AB \in F_\mathbf{p}$ (or $A \in F_\mathbf{p}B^{-1}$) for some $\mathbf{p}$. Now

$$\mu \times \alpha(F) = 0 = \int d\mu(\mathbf{p})dA\chi_{F_\mathbf{p}}(A) = \int d\mu(\mathbf{p})\alpha(F_\mathbf{p})$$

This implies $\alpha(F_\mathbf{p}) = 0$ for μ almost all $\mathbf{p}$. By the invariance of α, $\alpha(F_\mathbf{p}B^{-1}) = 0$ for each B and μ almost all $\mathbf{p}$. Because $\{\mathbf{p},A,B\} \in T_2^{-1}(F)$ if and only if $A \in F_\mathbf{p}B^{-1}$ for some $\mathbf{p}$ and B,

$$(\mu \times \alpha \times \alpha)(T_2^{-1}F) = \int d(\mu \times \alpha \times \alpha)(\mathbf{p},A,B) \times \chi_{T_2^{-1}(F)}(\mathbf{p},A,B) = \int d(\mu \times \alpha)(\mathbf{p},B)\alpha(F_\mathbf{p}B^{-1}) = 0\,.$$

The preceding argument shows that (A10) is a relation between $\mu \times \alpha \times \alpha$ measurable functions, which, for fixed A,B can fail to hold only on a set of $\mathbf{p}$'s of μ-measure zero. However, the union of these null sets as A and B run over G could, *a priori*, be a set of measure greater than zero. That this is, in fact, not the case is seen as follows. Since the set of $\{\mathbf{p},A,B\}$ where (A10) fails is of $(\mu \times \alpha \times \alpha)$-measure zero its section for $\mathbf{p}$ and B fixed is of α measure zero. But, as A runs over a set of α measure zero, $A^{-1}\mathbf{p}$ runs over a set of μ measure zero. [Here the equivalence of μ to a measure of the form (A8) is being used.] Thus, the set of $A^{-1}\mathbf{p}$ where

$$Q(A^{-1}\mathbf{p},B) = Q(\mathbf{p},A)^{-1}Q(\mathbf{p},AB) \qquad \text{(A11)}$$

fails, $\mathbf{p}$ and B being fixed, is a set of μ measure zero. By redefining $Q(A^{-1}\mathbf{p},B)$ at those $A^{-1}\mathbf{p}$ by (A11), one obtains a new family of $Q(\mathbf{q},B)$ measurable in $\{\mathbf{q},B\}$, which still yield the old $Q(B)$ but for which (A11) [or equivalently (A10)] is always valid. This completes the justification of the statement just after Eq. (3.10).

A second measure theoretic point arises in connection with Eq. (3.14). Using the $Q(\mathbf{p},A)$ whose existence has just been established, one gets a measurable but, *a priori*, not necessarily continuous unitary representation of the little group. In fact, every measurable unitary representation of any locally compact group G is continuous, as will now be shown by a well-known argument which has not yet crept into the text books.

Because

$$\|(U(x) - U(y))\Phi\|^2 = \|(U(y^{-1}x) - 1)\Phi\|^2 = 2(\Phi,\Phi) - 2\,\mathrm{Re}\,(U(y^{-1}x)\Phi,\Phi)$$

the strong continuity of $U(x)$, i.e., the requirement that for each $\Phi \in \mathcal{H}$ and $y \in G$ the first of these expressions be small when x is close to y, is implied by the weak continuity of $U(x)$ at the identity, i.e., the requirement that for each $\Phi,\Psi \in \mathcal{H}$, $(\Phi,U(x)\Psi)$ is close to (Φ,Ψ) for x close to the identity. The continuity of $(\Phi,U(x)\Psi)$ at the identity for all Φ,Ψ is implied by the continuity of $(\chi,U(x)\chi)$ for all χ as one sees by considering $\chi = \Phi + \Psi, \Phi + i\Psi, \Phi, \Psi$ in turn, and taking appropriate linear combinations. Because $U(x)$ is unitary, it suffices to verify the weak continuity for the elements of any dense set of vectors in $\mathcal{H}$, say Φ_i. To see this one can look at the identity

$$(\Phi,(U(x) - 1)\Phi) = (\Phi - \Phi_i,(U(x) - 1)\Phi) + (\Phi_i,(U(x) - 1)(\Phi - \Phi_i)) + (\Phi_i,(U(x) - 1)\Phi_i)\,,$$

which yields the estimate

$$(\Phi,(U(x) - 1)\Phi) \leq 2\|\Phi\|\,\|\Phi - \Phi_i\| + 2\|\Phi_i\|\,\|\Phi - \Phi_i\| + |(\Phi_i,(U(x) - 1)\Phi_i)|\,.$$

The first two terms on the right-hand side can be made small by appropriate choice of $\Phi_i \cdot \Phi_i$ having been chosen, the last term can be made small by an appropriate choice of x according to the assumed continuity of $(\Phi_i,U(x)\Phi_i)$.

Since $U(x)$ is measurable and unitary $(\Phi,U(x)\Phi)$ is a bounded measurable function for each Φ. Thus, for each continuous function of compact support, φ, it makes sense to talk about

$$(\Phi,U(\varphi)\Phi) = \int \varphi(x)dx(\Phi,U(x)\Phi)$$

and if φ_y is defined by $\varphi_y(x) = \varphi(y^{-1}x)$,

$$|(\Phi,(U(\varphi_y) - U(\varphi))\Phi)| \leqslant \int |\varphi_y(x) - \varphi(x)|dx\|\Phi\|^2. \qquad \text{(A12)}$$

Here dx is the left invariant integral on G. The right-hand side of this inequality is small for y sufficiently close to the identity. Now $\mathcal{H}$ is a direct sum of subspaces in which there is a vector Φ such that vectors of the form $U(\varphi)\Phi$ are dense, φ being continuous and of compact support so to prove $U(x)$ continuous it suffices to verify that $(U(\varphi)\Phi,U(x)U(\varphi)\Phi)$ is continuous for any such φ and Φ. But

$$(U(\varphi)\Phi,U(x)U(\varphi)\Phi) = (U(\varphi)\Phi,U(\varphi_x)\Phi)$$

so that the required continuity follows from (A12) and the proof is complete.

Finally, there is the matter of sets of measure zero in the criterion for unitary equivalence (3.17). Solved for $V(A^{-1}\mathbf{p})$ it reads

$$V(A^{-1}\mathbf{p}) = Q_1(\mathbf{p},A)^{-1}V(\mathbf{p})Q_2(\mathbf{p},A) .$$

By an argument just like that used in the first few paragraphs of this Appendix, one concludes that both sides of this equality are $(\mu \times \alpha)$ measurable functions of $\mathbf{p}$ and A and the set on which the equality fails is of $\mu \times \alpha$ measure zero. It then follows that, for fixed $\mathbf{p}$, the set of A on which it fails is of α measure zero. That in turn implies that the set of $A^{-1}\mathbf{p}$ for which it fails is of μ measure zero. Picking one $\mathbf{p}$ from each orbit and altering $V(\mathbf{p})$ on the corresponding set of measure zero one gets a new family $V(\mathbf{p})$ which is also measurable and yields the same V but for which (3.17) always holds.

PHYSICAL REVIEW A VOLUME 35, NUMBER 4 FEBRUARY 15, 1987

Uncertainty relations for light waves and the concept of photons

D. Han
National Aeronautics and Space Administration, Goddard Space Flight Center (Code 636), Greenbelt, Maryland 20771

Y. S. Kim
Department of Physics and Astronomy, University of Maryland, College Park, Maryland 20742

Marilyn E. Noz
Department of Radiology, New York University, New York, New York 10016
(Received 25 September 1986)

A Lorentz-covariant localization for light waves is presented. The unitary representation for the electromagnetic four-potential is constructed for a monochromatic light wave. A model for covariant superposition is constructed for light waves with different frequencies. It is therefore possible to construct a wave function for light waves carrying a covariant probability interpretation. It is shown that the time-energy uncertainty relation $(\Delta t)(\Delta\omega)\simeq 1$ for light waves is a Lorentz-invariant relation. The connection between photons and localized light waves is examined critically.

I. INTRODUCTION

For light waves, the Fourier relation $(\Delta t)(\Delta\omega)$ was known before the present form of quantum mechanics was formulated.[1,2] However, the question of whether this is a Lorentz-invariant relation has not yet been properly addressed.[3] Let us consider a blinking traffic light. A stationary observer will insist on $(\Delta t)(\Delta\omega)\simeq 1$. An observer in an automobile moving toward the light will see the same blinking light. This observer will also insist on $(\Delta t^*)(\Delta\omega^*)\simeq 1$ on his or her coordinate system. However, these observers may not agree with each other because neither Δt nor $\Delta\omega$ is a Lorentz-invariant variable. The product of two noninvariant quantities does not always lead to an invariant quantity.

Let us assume that the automobile is moving in the negative z direction with velocity parameter β. Since both t and ω are the timelike components of four-vectors $(\mathbf{x},t)$ and $(\mathbf{k},\omega)$, respectively, a Lorentz boost along the z direction will lead to new variables

$$t^*=(t+\beta z)/(1-\beta^2)^{1/2}, \quad \omega^*=(\omega+\beta k)/(1-\beta^2)^{1/2}, \tag{1}$$

where the light wave is assumed to travel along the z axis with $k=\omega$. In the above transformation, the light wave is boosted along the positive z direction. If the light passes through the point $z=0$ at $t=0$, then $t=z$ on the light front, and the transformations of Eq. (1) become

$$t^*=\left[\frac{1+\beta}{1-\beta}\right]^{1/2} t, \quad \omega^*=\left[\frac{1+\beta}{1-\beta}\right]^{1/2}\omega . \tag{2}$$

These equations will formally lead us to

$$(\Delta t^*)(\Delta\omega^*)=\left[\frac{1+\beta}{1-\beta}\right](\Delta t)(\Delta\omega) , \tag{3}$$

which indicates that the time-energy uncertainty relation is not Lorentz invariant, and that Planck's constant depends on the Lorentz frame in which the measurement is taken. This is not correct, and we need a better understanding of the transformation properties of Δt and $\Delta\omega$.

This problem is related to another fundamental problem in physics. We are tempted to say that the above-mentioned Fourier relation is a time-energy uncertainty relation. However, in order that it be an uncertainty relation, the wave function for the light wave should carry a probability interpretation. This problem has a stormy history and is commonly known as the photon localization problem.[4–6] The traditional way of stating this problem is that there is no self-adjoint position operator for massless particles including photons.

In spite of this theoretical difficulty, it is becoming increasingly clear that single photons can be localized by detectors in laboratories. The question then is whether it is possible to construct the language of the photon localization which we observe through oscilloscopes. Throughout the history of this localization problem, the main issue has been and still is how to construct localized photon wave functions consistent with special relativity.

We do not propose to solve this difficult problem in this paper. We shall instead approach this problem by constructing covariant localized light waves and comparing them with photon field operators. First, we construct a unitary representation for Lorentz transformations for a monochromatic light wave. It is shown then that a Lorentz-covariant superposition of light waves is possible for different frequencies. After constructing the covariant light wave, we shall observe that there is a gap between the concept of photons and that of localized waves. From the physical point of view, this gap is not significant. However, there is a definite distinction between the mathematics of photons and that of light waves.

In approaching the problem of the covariant superposition of light waves, we shall start with the uncertainty re-

Reprinted from *Phys. Rev. A* **5** 1682 (1987).

lation applicable to nonrelativistic quantum mechanics. We shall then borrow the techniques from the covariant harmonic oscillator model which provides a quantification of the uncertainty relations observed in the relativistic quark model.[7-10] Since the uncertainty principle is universal, the uncertainty relation applicable to one specific physical example should be consistent with those for other physical phenomena.

In Sec. II, we start with the motion of free-particle wave packets in the Schrödinger picture of nonrelativistic quantum mechanics. For localized light waves, there is no difficulty in giving a probability interpretation if Lorentz boosts are not considered. It is pointed out that the basic problem for light waves is how to make the probability interpretation Lorentz covariant.

In Sec. III, we discuss Lorentz-transformation properties of the four-vector representation for photons. Section IV examines the time-energy uncertainty relation applicable to the relativistic quark model. It is noted that the uncertainty relation applicable to the time separation variable between the quarks confined in a hadron can be combined covariantly with the position-momentum uncertainty relation.

In Sec. V, based on the lessons we learned in Secs. II, III, and IV, we construct a model of Lorentz-covariant localization of light waves. Finally, in Sec. VI, we examine closely how the concept of photons can emerge from localized light waves.

II. LIGHT WAVES AND WAVE PACKETS IN NONRELATIVISTIC QUANTUM MECHANICS

In this paper we are concerned with the possibility of constructing wave functions with quantum probability interpretation for relativistic massless particles. The natural starting point for tackling this problem is a free-particle wave packet in nonrelativistic quantum mechanics which we pretend to understand. Let us write down the time-dependent Schrödinger equation for a free particle moving in the z direction:

$$i\frac{\partial}{\partial t}\psi(z,t)=\frac{-1}{2m}\left[\frac{\partial}{\partial z}\right]^2\psi(z,t) . \tag{4}$$

The Hamiltonian commutes with the momentum operator. If the momentum is sharply defined, the solution of the above differential equation is

$$\psi(z,t)=\exp[i(pz-p^2t/2m)] . \tag{5}$$

If the momentum is not sharply defined, we have to take the linear superposition

$$\psi(z,t)=\int g(k)\exp[i(kz-k^2t/2m)]dk . \tag{6}$$

The width of the wave function becomes wider as the time variable increases, as is illustrated in Fig. 1(a). This is known as the wave-packet spread.

Let us study transformation properties of this wave function. Rotation and translation properties are trivial. In order to study the boost property within the framework of the Galilei kinematics, let us imagine an observer moving in the negative z direction. To this observer, the

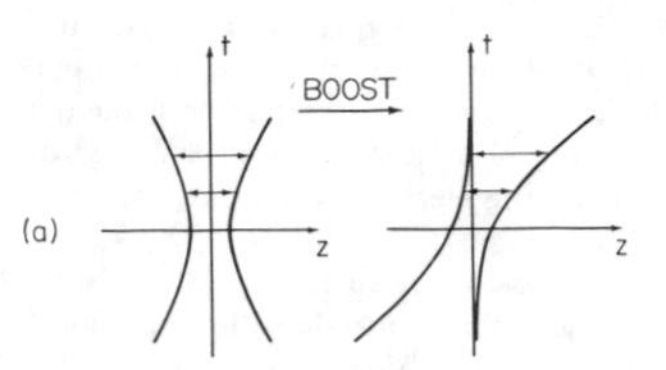

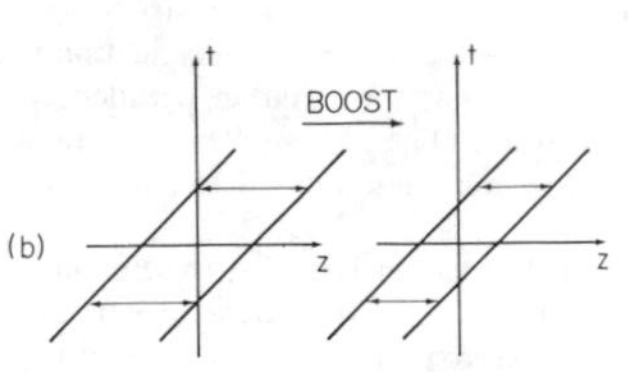

FIG. 1. The time dependence of the wave packets. (a) shows the spread of the Schrödinger wave function. (b) shows the behavior of the light wave which does not spread. However, for an observer moving in the negative z direction, the Schrödinger wave function is boosted according to the Galilei transformation. The quantum probability interpretation is consistent with the Galilean world. On the other hand, the light wave carries the burden of being consistent with the Lorentzian world.

center of the wave function moves along the positive z direction as is specified also in Fig. 1(b). The transformed wave function takes the form

$$\psi_v(z,t)=\exp[-im(vz-\tfrac{1}{2}v^2t)] \times\int g(k-mv)e^{i(kz-k^2t/2m)}dk , \tag{7}$$

where v is the boost velocity. This expression is different from the usual expression in textbooks by an exponential factor in front of the integral sign. The origin of this phase factor is well understood.[11]

In nonrelativistic quantum mechanics, $\psi(z,t)$ has a probability interpretation, and there is no difficulty in giving an interpretation for the transformed wave function in spite of the above-mentioned phase factor. The basic unsolved problem is whether the probabilistic interpretation can be extended into the Lorentzian regime. This has been a fundamental unsolved problem for decades, and we do not propose to solve all the problems in this paper. A reasonable starting point for approaching this problem is to see whether a covariant probability interpretation can be given to light waves.

For light waves, let us start with the usual expression

$$f(z,t)=\left[\frac{1}{2\pi}\right]^{1/2}\int g(k)e^{i(kz-\omega t)}dk . \tag{8}$$

Unlike the case of the Schrödinger wave, ω is equal to k, and there is no spread of wave packet. The velocity of propagation is always that of light, as is illustrated in Fig. 1(b). We might therefore be led to think that the problem for light waves is simpler than that for nonrelativistic Schrödinger waves. This is not the case for the following reasons.

(1) We like to have a quantal wave function for light waves. However, it is not clear which component of the Maxwell wave should be identified with the quantal wave whose absolute square gives a probability distribution. Should this be the electric or magnetic field, or should it be the four-potential?

(2) The expression given in Eq. (8) is valid in a given Lorentz frame. What form does this equation take for an observer in a different frame?

(3) Even if we are able to construct localized light waves, does this solve the photon localization problem?

(4) The photon has spin 1 either parallel or antiparallel to its momentum. The photon also has gauge degrees of freedom. How are these related to the above-mentioned problems?

Indeed, the burden on Eq. (8) is the Lorentz covariance. It is not difficult to carry out a spectral analysis and therefore to give a probability interpretation for the expression of Eq. (8) in a given Lorentz frame. However, this interpretation has to be covariant. This is precisely the problem we are addressing in the present paper.

III. UNITARY REPRESENTATION FOR FOUR-POTENTIALS

One of the difficulties in dealing with the photon problem has been that the electromagnetic four-potential could not be identified with a *unitary* irreducible representation of the Poincaré group.[12–15] The purpose of this section is to resolve this problem. In Ref. 15 we studied unitary transformations associated with Lorentz boosts along the direction perpendicular to the momentum. In this section we shall deal with the most general case of boosting along an arbitrary direction.

Let us consider a monochromatic light wave traveling along the z axis with four-momentum p. The four-potential takes the form

$$A^{\mu}(x)=A^{\mu}e^{i\omega(z-t)}, \tag{9}$$

with

$$A^{\mu}=(A_1,A_2,A_3,A_0) .$$

We use the metric convention $x^{\mu}=(x,y,z,t)$. The momentum four-vector in this convention is

$$p^{\mu}=(0,0,\omega,\omega) . \tag{10}$$

Among many possible forms of the gauge-dependent four-vector A^{μ}, we are interested in the eigenstates of the helicity operator

$$S_3=\begin{bmatrix} 0 & -i & 0 & 0 \\ i & 0 & 0 & 0 \\ 0 & 0 & 0 & 0 \\ 0 & 0 & 0 & 0 \end{bmatrix} . \tag{11}$$

The four-vectors satisfying this condition are

$$A^{\mu}_{\pm}=(1,\pm i,0,0) , \tag{12}$$

where the subscripts + and − specify the positive and negative helicity states, respectively. These are commonly called the photon *polarization vectors.*

In order that the four-vector be a helicity state, it is essential that the time-like and longitudinal components vanish:

$$A_3=A_0=0 . \tag{13}$$

This condition is equivalent to the combined effect of the Lorentz condition

$$\frac{\partial}{\partial x^{\mu}}[A^{\mu}(x)]=0 , \tag{14}$$

and the transversality condition

$$\nabla\cdot\mathbf{A}(x)=0 . \tag{15}$$

As before, we call this combined condition the *helicity gauge.*[15]

While the Lorentz condition of Eq. (14) is Lorentz invariant, the transversality condition of Eq. (15) is not. However, both conditions are invariant under rotations and under boosts along the direction of momentum. We call these helicity preserving transformations. The boost along an arbitrary direction is illustrated in Fig. 2. This is not a helicity preserving transformation. However, according to Ref. 15, we can express this in terms of helicity preserving transformations preceded by a gauge transformation.

Let us consider in detail a boost along the arbitrary direction specified in Fig. 2. This boost will transform the momentum to $\mathbf{p}'$, as is illustrated in Fig. 2;

$$p'^{\mu}=B_{\phi}(\eta)p^{\mu} . \tag{16}$$

However, this is not the only way in which p can be transformed to p'. We can boost p along the z direction and rotate it around the y axis as is shown in Fig. 2. The application of the transformation $[R(\theta)B_z(\xi)]$ on the four-momentum gives the same effect as that of the application of $B_{\phi}(\eta)$. Indeed, the matrix

$$D(\eta)=[B_{\phi}(\eta)]^{-1}R(\theta)B_z(\xi) \tag{17}$$

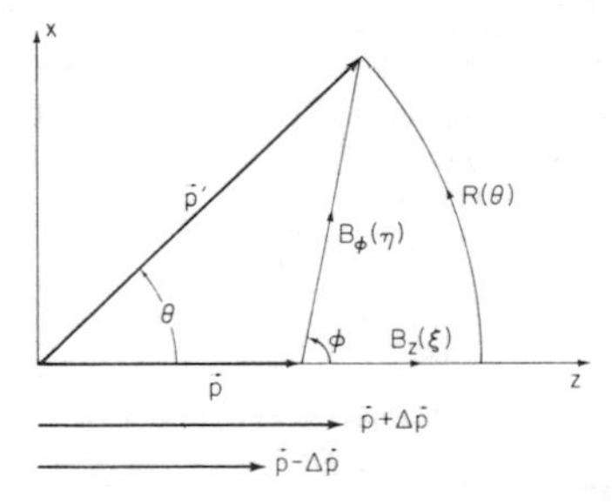

FIG. 2. Lorentz boost along an arbitrary direction of the light wave. The four-momentum can be boosted either directly by B_{ϕ} or through the rotation R_y preceded by B_z along the z direction. These operations produce two different four-vectors when applied to the polarization vector. However, they are connected by a gauge transformation.

leaves the four-momentum invariant, and is therefore an element of the $E(2)$-like little group for photons.

The effect of the above D matrix on the polarization vectors $A^{\mu}_{\pm}$ has been calculated in the Appendix, and the result is

$$D(\eta)A^{\mu}_{\pm}=A^{\mu}_{\pm}+(p^{\mu}/\omega)\mu(\eta,\theta)\,, \tag{18}$$

where

$$u(\eta,\theta)=\frac{-2\left\{\sin\frac{\theta}{2}\cosh\frac{\eta}{2}\right\}}{\left\{\cos\frac{\theta}{2}\right\}\left\{\cos\frac{\eta}{2}\right\}+\left[\left\{\cos\frac{\theta}{2}\cosh\frac{\eta}{2}\right\}^{2}-1\right]^{1/2}}\;.$$

Thus $D(\eta)$ applied to the polarization vector results in the addition of a term which is proportional to the four-momentum. $D(\eta)$ therefore performs a gauge transformation on $A^{\mu}_{\pm}$.[14,15]

With this preparation, let us boost the photon polarization vector

$$\tilde{A}^{\mu}_{\pm}=B_{\phi}(\phi)A^{\mu}_{\pm}\;. \tag{19}$$

The four-vector $\tilde{A}^{\mu}_{\pm}$ satisfies the Lorentz condition $p'^{\mu}\tilde{A}_{\pm\mu}=0$, but its fourth component will not vanish. The four-vector $\tilde{A}^{\mu}_{\pm}$ does not satisfy the helicity condition.

On the other hand, if we boost the four-vector $A^{\mu}_{\pm}$ after performing the gauge transformation $D(\eta)$,

$$\begin{aligned}A'^{\mu}_{\pm}&=B_{\phi}(\eta)A^{\mu}_{\pm}\\&=B_{\phi}(\eta)\{[B_{\phi}(\eta)]^{-1}R(\theta)B_{z}(\xi)\}A^{\mu}_{\pm}\\&=R(\theta)B_{z}(\xi)A^{\mu}_{\pm}\;.\end{aligned} \tag{20}$$

Since $B_z(\xi)$ leaves $A^{\mu}_{\pm}$ invariant, we arrive at the conclusion that

$$A'^{\mu}_{\pm}=R(\theta)A^{\mu}_{\pm}\;. \tag{21}$$

This means

$$A'^{\mu}_{\pm}=B_{x}(\eta)D(\eta)A^{\mu}_{\pm}=(\cos\theta,\pm i,-\sin\theta,0)\,, \tag{22}$$

which satisfies the helicity condition

$$A'^{0}_{\pm}=0$$

and (23)

$$\mathbf{p}'\cdot\mathbf{A}'_{\pm}=0\;.$$

The Lorentz boost $B(\eta)$ on $A^{\mu}_{\pm}$ preceded by the gauge transformation $D(\eta)$ leads to the pure rotation $R(\theta)$. This rotation is a finite-dimensional unitary transformation.

The above result indicates, for a monochromatic wave, that all we have to know is how to rotate. If, however, the photon momentum has a distribution, we have to deal with a linear superposition of waves with different momenta. The photon momentum can have both longitudinal and transverse distributions. In this paper we shall *assume* that there is only longitudinal distribution. This, of course, is a limitation of the model we present. However, our apology is limited in view of the fact that laser beams these days can go to the moon and come back after reflection.[16]

With this point in mind, we note first that the abovementioned unitary transformation preserves the photon polarization. This means that we can drop the polarization index from A^{μ} assuming that the photon has either positive or negative polarization. $A^{\mu}(x)$ can now be replaced by $A(x)$.

Next, the transformation matrices discussed in this section depend only on the direction and the magnitude of the boost but *not* on the photon energy. This is due to the fact that the photon is a massless particle.[17] Indeed, the matrices in Sec. III remain invariant even if ω in Eq. (9) is replaced by a different value. This means that for the superposition of two different frequency states,

$$A(x)=A_{l}e^{i\omega_{l}(z-t)}+A_{2}e^{i\omega_{2}(z-t)}\,, \tag{24}$$

a Lorentz boost along an arbitrary direction results in a rotation preceded by a boost along the z direction. Since neither the rotation nor the boost along the z axis changes the magnitude of $A_i (i=1,2)$, the quantity

$$|A|^{2}=|A_{1}|^{2}+|A_{2}|^{2} \tag{25}$$

remains invariant under the Lorentz transformation. This result can be generalized to the superposition of many different frequencies:

$$A(x)=\sum_{k}A_{k}e^{i\omega_{k}(z-t)}\,, \tag{26}$$

with

$$|A|^{2}=\sum_{k}|A_{k}|^{2}\;. \tag{27}$$

The norm $|A|^2$ remains invariant under the Lorentz transformation in the sense that it is invariant under rotations and is invariant under the boost along the z direction.

Can this sum be transformed into an integral form of Eq. (8)? From the physical point of view, the answer should be yes. Mathematically, the problem is how to construct a Lorentz-invariant integral measure. It is not difficult to see that the norm of Eq. (27) remains invariant under rotations, which perform unitary transformations on the system. The problem is how to construct a measure invariant under the boost along the z direction.

For this purpose, we shall borrow the techniques developed for the covariant harmonic-oscillator formalism which has been very effective in explaining the basic relativistic features in the quark model,[10,18–20] and which enables us to combine covariantly Heisenberg's position-momentum uncertainty relations and the c-number time-energy uncertainty relation.[3,9,21]

IV. LOCALIZATION PROBLEMS IN THE RELATIVISTIC QUARK MODEL

We shall discuss in this section the aspects of the covariant harmonic-oscillator formalism which are useful in converting the sum of Eq. (26) into an integral form while preserving the invariance of the sum of Eq. (27) under the Lorentz boost along the z direction. The covariant oscillator formalism has been extensively discussed in the literature. What we need is here to review its property under the boost along the z direction.

Let us use x_a and x_b for the space-time coordinates of the two quarks bound together in a hadron. Then it is more convenient to use the four-vectors

$$X=(x_a+x_b)/2, \quad x=(x_a-x_b)/2\sqrt{2} . \tag{28}$$

It is not difficult to write down the uncertainty relations for space-time separation variables and to define the region within which the hadronic wave function is localized in the Lorentz frame where the hadron is at rest. However, the crucial question is how these uncertainty relations appear to an observer in the laboratory frame.

The uncertainty principle applicable to the space-time separation of quarks in the hadronic rest frame is the same as the currently accepted form based on the existing theories and observations. The usual Heisenberg uncertainty relation holds for each of the three spatial coordinates. The time-separation variable is a c number and therefore does not cause quantum excitations.[3,21] The question is then whether this peculiar time-energy uncertainty can be combined with Heisenberg's position-momentum uncertainty relation to a covariant form.[9] Such a combination is possible within the framework of the covariant harmonic-oscillator formalism which can explain the basic hadronic features including the mass spectrum,[18] proton form factors,[19] parton picture,[9,22] and jet phenomenon.[20]

We assume throughout this section that the hadron moves along the z axis, and ignore and x and y coordinates which are not affected by the boost along the z direction. If we consider only the ground-state wave function, then the localization dictated by the uncertainty relations associated with both space and time separation variables will lead to a distribution centered around the origin in the hadron-rest frame with the z^* and t^* variables. The ground-state harmonic-oscillator wave function takes the form

$$\psi(z,t)=(1/\sqrt{\pi})\exp\{-[(z^*)^2+(t^*)^2]/2\} . \tag{29}$$

The question then is how this region appears to the laboratory—frame observer, while the coordinates of the two different frames are related by the Lorentz transformation

$$\begin{aligned} z&=(z^*+\beta t^*)/(1-\beta^2)^{1/2} , \\ t&=(t^*+\beta z^*)/(1-\beta^2)^{1/2} , \end{aligned} \tag{30}$$

where β is the velocity parameter of the hadron.

In order to approach this problem, let us employ Dirac's form of Lorentz transformation. In his 1949 paper,[23] Dirac introduced the light-cone coordinate system in which the coordinate variables are

$$z_+=(z+t)/\sqrt{2}, \quad z_-=(z-t)/\sqrt{2} . \tag{31}$$

In terms of these variables, the Lorentz transformation of Eq. (30) takes the form

$$z_+=\left[\frac{1+\beta}{1-\beta}\right]^{1/2} z_+^*, \quad z_-=\left[\frac{1-\beta}{1+\beta}\right]^{1/2} z_-^* . \tag{32}$$

z_+ and z_- are called the light-cone variables, and the product z_+z_- remains invariant under the boost:

$$z_+z_-=z_+^*z_-^* . \tag{33}$$

In the light-cone coordinate system, the ground-state wave function of Eq. (29) takes the form

$$\psi(z,t)=\frac{1}{\sqrt{\pi}}\exp\left[-\left[\frac{1-\beta}{1+\beta}z_+^2+\frac{1+\beta}{1-\beta}z_-^2\right]\right] . \tag{34}$$

This wave function or the probability density is localized in a circular region centered around the origin in the z_+z_- plane when $\beta=0$. As the hadron moves, the region becomes elliptic. This elliptic deformation property is illustrated in Fig. 3.

Let us next consider the momentum-energy wave function. If the quarks have four-momenta p_a and p_b, respec-

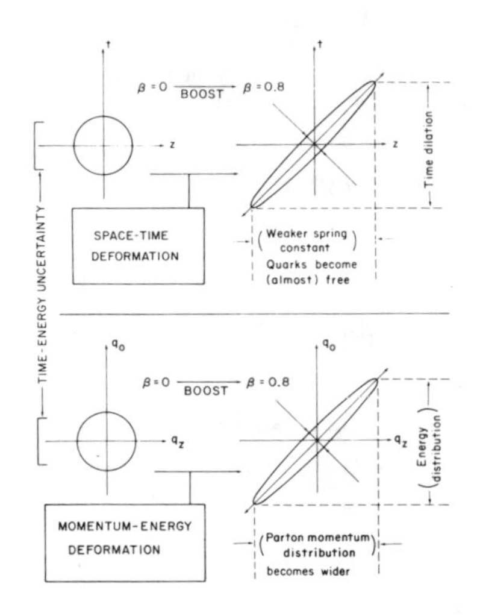

FIG. 3. The Lorentz deformation of a relativistic extended hadron in both the space-time and momentum-energy coordinate systems. Because the Lorentz transformation property of momentum-energy four-vector is the same as that of the space-time four-vector, the Lorentz deformation in the momentum-energy plane is expected to be the same as that in the zt plane. This figure summarizes the content of the earlier paper on the parton picture. (Refs. 9 and 10) in which the hadron, while being a bound state of quarks in its rest frame, appears as a collection of partons to an observer who moves with a speed close to that of light.

tively, then the standard procedure is to introduce P and q

$$P=p_a+p_b,\quad q=\sqrt{2}(p_a-p_b)\ , \tag{35}$$

where P is the four-momentum of the hadron, and q is the momentum-energy separation between the quarks. We are concerned here with the uncertainty relations between the x variables of Eq. (28) and the above q variables. The momentum-energy wave function is

$$\begin{aligned}\phi(q_z,q_0)&=\left[\frac{1}{2\pi}\right]\int \exp[i(q_z z-q_0 t)]\psi(z,t)dt\,dz\\ &=(1/\sqrt{\pi})\exp\{-[(q_z^*)^2+(q_0^*)^2]/2\}\ ,\end{aligned} \tag{36}$$

where q_z and q_0 are the momentum and energy separation variables, respectively. Their Lorentz transformation property is the same as that for z and t. The form of this wave function is identical to that of the space-time wave function. In terms of the light-cone variables:

$$q_+=(q_z+q_0)/\sqrt{2},\quad q_-=(q_z-q_0)/\sqrt{2}\ , \tag{37}$$

the momentum wave function of Eq. (36) takes the form

$$\phi(q_z,q_0)=(1/\sqrt{\pi})\exp\left[\frac{1}{2}\left[\frac{1-\beta}{1+\beta}q_+^2+\frac{1+\beta}{1-\beta}q_-^2\right]\right]. \tag{38}$$

The Lorentz deformation of this wave function is also illustrated in Fig. 3.[24]

The basic advantage of using the light-cone variables is that the coordinate system remains orthogonal, and z_+ and z_- do not become linearly mixed when the system is boosted. The Fourier relations between the space-time and momentum-energy coordinates are

$$q_-=-i(\partial/\partial z_+),\quad q_+=-i(\partial/\partial z_-)\ . \tag{39}$$

This means that the major and minor axes of the momentum-energy coordinates are the Fourier conjugates of the minor and major axes of the space-time coordinates, respectively. Thus we have the following Lorentz-invariant uncertainty relations.[25]

$$\begin{aligned}(\Delta z_+)(\Delta q_-)&=(\Delta z_+^*)(\Delta q_-^*)\simeq 1\ ,\\ (\Delta z_-)(\Delta q_+)&=(\Delta z_-^*)(\Delta q_+^*)\simeq 1\ .\end{aligned} \tag{40}$$

These uncertainty relations are well understood when the hadron is at rest with $\beta=0$. On the other hand, the limit $\beta\to 1$ can teach us many interesting lessons. The connection between this limit and Feynman's original form of the parton model[22] has been discussed repeatedly in the literature.[9,10] As far as the localization of massless particles is concerned, the distribution along one of the light-cone axes becomes so widespread that it loses its localization along the axis.

In Sec. V, we shall "give up" the localization along one of the light-cone axes in order to study photons and light waves. In so doing, we will have the problem of normalizing the wave function by integration. The integration measure dz_+dz_- is a boost-invariant quantity. This means that the normalization integral,

$$\int |\psi(z,t)|^2 dz_+dz_-\ , \tag{41}$$

is independent of β. However, dz_- or dz_+ alone is not. The integration over z_+ gives the factor $[(1+\beta)/(1-\beta)]^{1/2}$, and this factor is compensated by its inverse $[(1-\beta)/(1+\beta)]^{1/2}$ coming from the z_- integration.

We used in this section the Gaussian form of the wave function purely for convenience. The above reasoning is valid for all forms of distributions having the same space-time boundary condition as that of the Gaussian function. Indeed, if we give up the localization along the z_+ axis, then the integration measure along the z_- axis should be compensated by the contraction or elongation along the z_+ direction. If the system is boosted along the z direction, dz_+ and dz_- are transformed as

$$dz_+\to\left[\frac{1+\beta}{1-\beta}\right]^{1/2}dz_+,\quad dz_-\to\left[\frac{1-\beta}{1+\beta}\right]^{1/2}dz_-\ . \tag{42}$$

We can give the same reasoning for the momentum-energy measures dq_+ and dq_-. This transformation property will play an important role in constructing localized light waves.

V. COVARIANT LOCALIZATION OF LIGHT WAVES

We discussed in Sec. IV the relativistic quark model in which the overall hadron four-momentum is well defined, and the internal coordinate system has a momentum-energy distribution. In the case of light waves, the frequency or momentum is not sharply defined, but has a distribution. In this case, we can take the average value of the momentum, and the distribution around this average value. We can treat the average momentum like the hadronic momentum, and the distribution around the average value like the internal momentum distribution.

With this point in mind, let us rewrite Eq. (8) as

$$f(z,t)=\left[\frac{1}{2\pi}\right]^{1/2}\int g(k)e^{i(kz-\omega t)}dk\ . \tag{43}$$

We shall approach this problem using Dirac's light-cone coordinate system discussed in Sec. IV. For convenience, we shall define here the light-cone variables as

$$s=(z+t)/2,\quad u=(z-t)\ . \tag{44}$$

s and u are different from z_+ and z_- of Eq. (31) by a factor of $\sqrt{2}$. but their Lorentz transformation property remains the same. We shall also define the new momentum variables as

$$k_u=(k+\omega)/2,\quad k_s=(k-\omega)\ . \tag{45}$$

In the case of light waves, k_s vanishes and k_u becomes k or ω. In terms of the light-cone variables, the expression of Eq. (43) becomes

$$f(u)=\left[\frac{1}{2\pi}\right]^{1/2}\int g(k)e^{iku}dk\ . \tag{46}$$

For a massive particle, the most convenient Lorentz frame is the frame in which the particle is at rest, as was noted in Sec. IV. For a massless particle, as our study in Sec. III suggests,[26] we can start with a specific Lorentz frame in which the photon momentum has a given magnitude along the z direction. In this Lorentz frame, we assume that the average photon frequency is Ω_0

$$\Omega_0 = \frac{1}{N}\int k\,|g_0(k,\Omega_0)|^2 dk , \quad (47)$$

with

$$N = \int |g_0(k,\Omega_0)|^2 dk . \quad (48)$$

It is important to note that the introduction of this specific Lorentz frame does not cause any loss of generality.[26] We can obtain the most general form by boosting the photon along the z direction. If we use β as the boost parameter, the new photon frequency is

$$\Omega = \left[\frac{1+\beta}{1-\beta}\right]^{1/2} \Omega_0 , \quad (49)$$

and this frequency should be the average value calculated from the new and most general distribution function $g(k,\Omega)$:

$$\Omega = \frac{1}{N}\int k\,|g(k,\Omega)|^2 dk , \quad (50)$$

where N is given in Eq. (48) of $g_0(k,\Omega_0)$. N is a Lorentz-invariant quantity. In order that unitarity be preserved, $g(k,\Omega)$ should satisfy the normalization condition

$$N = \int |g(k,\Omega)|^2 dk . \quad (51)$$

The basic problem here is that the integral measure dk is not Lorentz invariant. One form for $g(k,\Omega)$ which meets the invariance requirement is

$$g(k,\Omega) = (1/\Omega)^{1/2} a(k-\Omega) , \quad (52)$$

where $a(k-\Omega)$ is a scalar function depending only on $(k-\Omega)$. The $(1/\Omega)^{1/2}$ factor is proportional to $[(1-\beta)/(1+\beta)]^{1/4}$, and assures the Lorentz invariance of the normalization integral. This makes $(1/Q)dk$ a Lorentz-invariant measure.

Let us next consider the left-hand side of Eq. (8). If we insist on the Lorentz invariance of the normalization integral

$$\int |f(u)|^2 du , \quad (53)$$

then $(\Omega/\Omega_0)du$ is a Lorentz-invariant measure, and $f(u)$ can take the form

$$f(u) = (\Omega/\Omega_0)^{1/2} A(u) , \quad (54)$$

where $A(u)$ is a scalar quantity. The integral form of Eq. (26) is

$$(\Omega/\Omega_0)^{1/2} A(u) = (\tfrac{1}{2}\pi\Omega)^{1/2} \int a(k-\Omega) e^{iku} dk , \quad (55)$$

with

$$(\Omega/\Omega_0)\int_{-\infty}^{\infty} |A(u)|^2 du = (1/\Omega)\int_{-\infty}^{\infty} |a(k-\Omega)|^2 dk . \quad (56)$$

Indeed, in Eq. (56), we have to multiply and divide the right- and left-hand sides, respectively, by $\sqrt{Q}$. We did this in order to make the system covariant. The net result is that

$$f(u) = \left[\frac{1+\beta}{1-\beta}\right]^{1/4} A(u) \quad (57)$$

and

$$g(k) = \left[\frac{1-\beta}{1+\beta}\right]^{1/4} (1/\Omega_0)^{1/2} a \left[k - \left[\frac{1+\beta}{1-\beta}\right]^{1/2} \Omega_0\right] . \quad (58)$$

The velocity parameter β is zero when the average photon frequency is Ω_0.

Let us examine the problem using a concrete form of $g(k,\Omega)$. The covariant oscillator model discussed in Sec. IV suggests the following normalized Gaussian form for the frequency distribution:

$$g(k) = \left[\frac{1}{\pi b}\right]^{1/4} \left[\frac{1}{\Omega_0}\right]^{1/2} \left[\frac{1-\beta}{1+\beta}\right]^{1/4} \times \exp\left[\frac{-1}{2b}(k-\Omega)^2\right] , \quad (59)$$

where b is a constant and specifies the width of the distribution. The above form describes the distribution in k or k_u around Ω, and there is no localization in the k_s variables. In view of the discussion given in Sec. IV, it is not difficult to understand the origin of the factor $[(1-\beta)/(1+\beta)]^{1/2}$ in Eq. (59). The space-time wave function $f(u)$ takes the form

$$f(u) = \left[\frac{b}{\pi}\right]^{1/4} \left[\frac{1+\beta}{1-\beta}\right]^{1/4} \exp\left[i\Omega(z-t) - \frac{b}{2}(z-t)^2\right] . \quad (60)$$

This function has a distribution along the $u=(z-t)$ axis, but has no localization along the s axis.

Let us go back to the question mentioned in Sec. I. Is the time-frequency uncertainty relation a Lorentz-invariant relation? The wave function of Eq. (59) and Eq. (60) constitute the quantification of the Lorentz-invariant uncertainty relation

$$(\Delta u)(\Delta k) \simeq 1 . \quad (61)$$

From the definition given in Eq. (45), $\Delta k = \Delta\omega$. From Eq. (44), $\Delta u = -\Delta t$ for a fixed value of z. This relation becomes $\Delta u = \Delta t$ when the symbol Δ means the width of distribution. Thus the time-frequency relation $(\Delta\omega)(\Delta t)$ is a Lorentz-invariant relation.

VI. THE CONCEPT OF PHOTONS

We discussed in this paper Lorentz-covariant wave functions for light waves. It is possible to construct a localized wave function for light waves with a Lorentz-invariant normalization. The mathematics of this procedure is not complicated. We are then led to the question of why the photon localization is so difficult, while it is possible to produce photons in states narrowly confined in space and time.[27]

Let us see how the mathematics for the light-wave localization is different from that of quantum electrodynamics (QED) where photons acquire a particle interpretation through second quantization. In QED, we start with the Klein-Gordon equation with its normalization procedure. As a consequence, we use the expression[28]

$$g(k)=(1/\sqrt{k})a(k)\ , \tag{62}$$

instead of the form given in Eq. (52). The Lorentz-transformation property of this quantity is the same as that for $g(k)$ of Eq. (52).

However, the basic difference between the above expression and that of Eq. (52) is that the kinematical factor in front of $a(k)$ is $(1/\sqrt{k})$ in Eq. (62), while that for Eq. (52) is $(1\sqrt{Q})$. There is no concept of the average momentum in quantum field theory, while it was essential for the localized light wave discussed in Sec. V. Numerically, the above expression becomes equal to $g(k)$ of Sec. V when $a(k)$ of Eq. (62) represents a sharp distribution around a fixed value of k. This is why the photon can appear as a light pulse on oscilloscope screens.

On the other hand, the normalization property of Eq. (62) is quite different. In quantum field theory, it is possible to give a particle interpretation in terms of creation and annihilation operators by second-quantizing $a(k)$. In the light-wave normalization, it is very difficult, if not impossible, to give a particle interpretation. This means that, from the mathematical point of view, the gap between photons and localized light waves is real and very serious.

As for the space-time distribution $f(u)$ of Eq. (46), we use the form

$$f(u)=A(u) \tag{63}$$

in QED, without the factor $(\Omega/\Omega_0)^{1/2}$ discussed in Sec. V. As a consequence, the normalization condition is that the integral

$$i\int\left[A^*(u)\frac{\partial}{\partial t}A(u)-A(u)\frac{\partial}{\partial t}A^*(u)\right]dz \tag{64}$$

be Lorentz invariant. If we use this form of normalization, the total probability is not always positive.[28]

We can summarize the discussion of this section in Table I. There definitely is a gap between the concept of localized waves and that of photons. Numerically this gap is not serious. However, we have to cross this gap when we make the transition from localized Maxwell waves to photons through second quantization.

TABLE I. Light waves and photons. Light waves can be localized, but we still do not know how to localize photons. The difference between these two cases is not serious from the physical point of view. The mathematical difference is still a serious problem.

	Probability interpretation	Particle interpretation
Light waves	yes	no
Photons	no	yes

ACKNOWLEDGMENT

We are grateful to Professor Eugene P. Wigner for explaining to us the background of the photon localization problem.

APPENDIX: UNITARY TRANSFORMATIONS OF PHOTON POLARIZATION VECTORS

Let us work out the kinetics of Fig. 2. If we use the four-vector convention $x^\mu=(x,y,z,t)$, the matrix which boosts p to p' is

$$B_\phi(\eta)=\begin{bmatrix} c^2+s^2(\cosh\eta) & 0 & sc(\cosh\eta-1) & s(\sinh\eta) \\ 0 & 1 & 0 & 0 \\ sc(\cosh\eta-1) & 0 & s^2+c^2(\cosh\eta) & c(\sinh\eta) \\ s(\sinh\eta) & 0 & c(\sinh\eta) & \cosh\eta \end{bmatrix}, \tag{A1}$$

where $c=\cos\phi$ and $s=\sin\phi$. The parameters η and ϕ specify the magnitude and direction of the boost, respectively. On the other hand, we can achieve the same purpose on the four-momentum by boosting p along the z direction first and rotating the boosted four-momentum as is specified in Fig. 2. The boost matrix takes the form

$$B_z(\xi)=\begin{bmatrix} 1 & 0 & 0 & 0 \\ 0 & 1 & 0 & 0 \\ 0 & 0 & \cosh\xi & \sinh\xi \\ 0 & 0 & \sinh\xi & \cosh\xi \end{bmatrix}, \tag{A2}$$

with $e^\xi=\cosh\eta+(\cos\phi)\sinh\eta$. The rotation matrix is

$$R(\theta)=\begin{bmatrix} \cos\theta & 0 & \sin\theta & 0 \\ 0 & 1 & 0 & 0 \\ -\sin\theta & 0 & \cos\theta & 0 \\ 0 & 0 & 0 & 1 \end{bmatrix}. \tag{A3}$$

The rotation angle θ is related to the boost parameters η and ϕ by

$$\sin\theta=\frac{(\sin\phi)[(\cosh\eta-1)\cos\phi+\sinh\eta]}{\cosh\eta+(\cos\phi)\sinh\eta}\ ,$$
$$\cos\theta=\frac{1+(\cos\phi)\sinh\eta+(\cos\phi)^2(\cosh\eta-1)}{\cosh\eta+(\cos\phi)\sinh\eta}\ . \tag{A4}$$

The key question then is what is the difference between these two transformations which produce the same result on the four-momentum. The best way to examine this problem is to examine the closed-loop transformation

$$D(u)=[B_\phi(\eta)]^{-1}R(\theta)B_z(\xi)\ . \tag{A5}$$

The matrix algebra is somewhat complicated, but is straightforward. The result is

$$D(u)=\begin{bmatrix} 1 & 0 & -u & u \\ 0 & 1 & 0 & 0 \\ u & 0 & 1-u^2/2 & u^2/2 \\ u & 0 & -u^2/2 & 1+u^2/2 \end{bmatrix}, \quad \text{(A6)}$$

where

$$u=\frac{-(\sin\phi)\sinh\eta}{\cosh\eta+(\cos\phi)\sinh\eta}.$$

We can now write $\sin\phi$ and $\cos\phi$ in terms of θ, and arrive at the expression given in Eq. (18).

Similar calculations exist in the literature, but the previous calculations are only for specified values of ϕ. In Ref. 15 the calculation was made for $\phi=90°$. In a recent paper by Han *et al.*[17] a similar calculation was carried out for the angle which will make $\xi=0$. Here we carried out the calculation for the most general case.

---

[1]W. Heitler, *The Quantum Theory of Radiation,* 3rd ed. (Clarendon, Oxford, 1954).

[2]E. P. Wigner, in *Aspects of Quantum Theory, in Honour of P. A. M. Dirac's 70th Birthday,* edited by A. Salam and E. P. Wigner (Cambridge University Press, London, 1972).

[3]P. A. M. Dirac, Proc. R. Soc. London, Ser. A **114**, 243, (1927); **114**, 710 (1927).

[4]T. D. Newton and E. P. Wigner, Rev. Mod. Phys. **21**, 400 (1949); A. S. Wightman, *ibid.* **34**, Ser. *A* 845 (1962); T. O. Philips, Phys. Rev. **136**, B893 (1964).

[5]J. M. Jauch and C. Piron, Helv. Phys. Acta **40**, 559 (1967); W. O. Amrein, *ibid.* **42**, 149 (1969); K. Kraus, Am. J. Phys. **38**, 1489 (1970); Ann. Phys. (NY), **64**, 311 (1971); H. Neumann, Commun. Math. Phys. **23**, 100 (1971); H. Neumann, Helv. Phys. Acta **45**, 881 (1972); S. T. Ali and G. G. Emch, J. Math. Phys. **15**, 176 (1974); G. C. Hegerfeldt, Phys. Rev. D **10**, 3320 (1974); K. Kraus, in *Foundations of Quantum Mechanics and Ordered Linear Space,* Vol. 29 of *Lecture Notes in Physics,* edited by J. Ehlers *et al.* (Springer-Verlag, Berlin, 1974), p. 206.

[6]For review papers on this subject, see R. Kraus, in *The Uncertainty Principle and Foundations of Quantum Mechanics,* edited by W. C. Price and S. S. Chissick (Wiley, New York, 1977); S. T. Ali, Riv. Nuovo Cimento **8**, 1 (1985).

[7]For early papers on this subject, see P. A. M. Dirac, Proc. R. Soc. London, Ser. A **183**, 284 (1945); H. Yukawa, Phys. Rev. **91**, 416 (1953); M. Markov, Nuovo Cimento Suppl. **3**, 760 (1956); T. Takabayasi, Nuovo Cimento **33**, 668 (1964): S. Ishida, Prog. Theor. Phys. **46**, 1570 (1971); **46**, 1905 (1971).

[8]For some of the recent articles, see R. P. Feynman, M. Kislinger, and F. Ravndal, Phys. Rev. D **3**, 2706 (1971); Y. S. Kim and M. E. Noz, *ibid.* **8**, 3521 (1973); M. J. Ruiz, Phys. Rev. **10**, 4306 (1974); Y. S. Kim, Phys. Rev. D **14**, 273 (1976); Y. S. Kim and M. E. Noz, Found. Phys. **9**, 375 (1979); Y. S. Kim, M. E. Noz, and S. H. Oh, J. Math. Phys. **20**, 1341 (1979); Am. J. Phys. **47**, 892 (1979); J. Math. Phys. **21** 1224 (1980); D. Han, Y. S. Kim, and M. E. Noz, Found. Phys. **11**, 895 (1980); D. Han and Y. S. Kim, Am. J. Phys. **49**, 1157 (1981); **49**, 348 (1981); D. Han, M. E. Noz, Y. S. Kim, and D. Son, Phys. Rev. D **25**, 1740 (1982); D. Han, Y. S. Kim, and D. Son, *ibid.* **26**, 3717 (1982); D. Han, M. E. Noz, Y. S. Kim, and D. Son, *ibid.* **27**, 3032 (1983). For review oriented articles comparing various early approaches to this problem, see T. Takabayasi, Prog. Theor. Phys. Suppl. **67**, 1 (1979); D. Han and Y. S. Kim, Prog. Theor. Phys. **64**, 1854 (1980).

[9]The uncertainty relation applicable to the time separation between the constituent quarks is responsible for the peculiarities in Feynman's parton picture universally observed in high-energy hadronic experiments. See, P. E. Hussar, Phys. Rev. D **23**, 2781 (1981). For papers dealing with qualitative features of the parton picture, see Y. S. Kim and M. E. Noz, Phys. Rev. D **15**, 335 (1977); J. Math. Phys. **22**, 2289 (1981); P. E. Hussar, Y. S. Kim, and M. E. Noz, Am. J. Phys. **53**, 142 (1985). For earlier papers dealing with the dependence of the quark-model wave function on the time-separation variable, see G. Preparata and N. S. Craigie, Nucl. Phys. B **102**, 478 (1976); J. Lukierski and M. Oziewics, Phys. Lett. **69B**, 339 (1977); D. Dominici and G. Longhi, Nuovo Cimento **42A**, 235 (1977); T. Goto, Prog. Theor. Phys. **58**, 1635 (1977); H. Leutwyler and J. Stern, Phys. Lett. **73B**, 75 (1978); Nucl. Phys. B **157**, 327 (1979); I. Fujiwara, K. Wakita, and H. Yoro, Prog. Theor. Phys. **64**, 363 (1980); J. Jersak and D. Rein, Z. Phys. C **3**, 339 (1980); I. Sogami and H. Yabuki, Phys. Lett. **94B**, 157 (1980); M. Pauri, in *Group Theoretical Methods in Physics, Proceedings of the IX International Colloquim, Cocoyoc, Mexico,* edited by K. B. Wolf (Springer-Verlag, Berlin, 1980); G. Marchesini and E. Onofri, Nuovo Cimento A **65**, 298 (1981); E. C. G. Sudarshan, N. Mukunda, and C. C. Chiang, Phys. Rev. D **25**, 3237 (1982). The uncertainty relation applicable to the time-separation between the quarks is not inconsistent with the proposition that the time-energy uncertainty is applicable only to the time separation between two independent events. See L. D. Landau and E. M. Lifschitz, *Quantum Mechanics,* 2nd ed. (Pergamon, New York, 1958).

[10]Y. S. Kim and M. E. Noz, *Theory and Applications of the Poincaré Group* (Reidel, Dordrecht, 1986).

[11]E. Inonu and E. P. Wigner, Nuovo Cimento **9**, 705 (1952); M. Hamermesh, *Group Theory* (Addison-Wesley, Reading, Mass., 1962).

[12]E. P. Wigner, Ann. Math. **40**, 149 (1939); V. Bargmann and E. P. Wigner, Proc. Natl. Acad. Sci. U.S.A. **34**, 211 (1948); E. P. Wigner, Z. Phys. **124**, 665 (1948); E. P. Wigner, in *Theoretical Physics,* edited by A. Salam (International Atomic Energy Agency, Vienna, 1962).

[13]Chou Kuang-Chao and L. G. Zastavenco, Zh. Eksp. Teor. Fiz. **35**, 1417 (1958) [Sov. Phys.—JETP **8**, 990 (1959)]; M. Jacob and G. C. Wick, Ann. Phys. (NY) **7**, 404 (1959); A. S. Wightman, in *Dispersion Relations and Elementary Particles,* edited by C. DeWitt and R. Omnes (Hermann, Paris, 1960); J. Kuperzstych, Nuovo Cimento **31B**, 1 (1976); Phys. Rev. D **17**, 629 (1978); A. Janner and T. Jenssen, Physica **53**, 1 (1971); **60**, 292 (1972); J. L. Richard, Nuovo Cimento **8A**, 485 (1972); H. P. W. Gottlieb, Proc. R. Soc. London, Ser. A **368**, 429 (1979).

[14]S. Weinberg, Phys. Rev. **134**, B 882 (1964); S. Weinberg, Phys. Rev. **135**, B1049 (1964); J. Kupersztych, Phys. Rev. D **17**, 629 (1978); D.Han, Y. S. Kim, and D. Son, *ibid.* **26**, 3717 (1982); D. Han, Y. S. Kim, and D. Son, Phys. Lett. **131B**, 327 (1983); D. Han, Y. S. Kim, M. E. Noz, and D. Son, Am. J. Phys. **52**,

1037 (1984).

[15]D. Han, Y. S. Kim, and D. Son, Phys. Rev. D **31**, 328 (1985).

[16]C. O. Alley, in *Quantum Optics, Experimental Gravity, and Measurement Theory,* edited by P. Meystre and M. O. Scully (Plenum, New York, 1983).

[17]E. P. Wigner, Rev. Mod. Phys. **29**, 255 (1957); D. Han, Y. S. Kim, and D. Son, J. Math. Phys. **27**, 2228 (1986).

[18]For some of the latest papers on hadronic mass spectra, see N. Isgur and G. Karl, Phys. Rev. D **19**, 2653 (1978); D. P. Stanley and D. Robson, Phys. Rev. Lett. **45**, 235 (1980). For review articles written for teaching purposes, see P. E. Hussar, Y. S. Kim, and M. E. Noz. Am. J. Phys. **48**, 1038, (1980); **48**, 1043 (1980). See also O. W. Greenberg, Am. J. Phys. **50**, 1074 (1982).

[19]For papers dealing with form factor behavior, K. Fujimura, T. Kobayashi, and M. Namiki, Prog. Theor. Phys. **43**, 73 (1970); R. G. Lipes, Phys. Rev. D **5**, 2849 (1972); Y. S. Kim and M. E. Noz, *ibid.* **8**, 3521 (1973). See also Ref. 10.

[20]For papers dealing with the jet phenomenon, see T. Kitazoe and S. Hama, Phys. Rev. D **19**, 2006 (1979); Y. S. Kim, M. E. Noz, and S. H. Oh, Found. Phys. **9**, 947 (1979); T. Kitazoe and T. Morii, Phys. Rev. D **21**, 685 (1980); Nucl. Phys. B **164**, 76 (1980).

[21]For continuing debates on the time-energy uncertainty relation, see Y. Aharonov and D. Bohm, Phys. Rev. **122**, 1649 (1961); V. A. Fock, Zh. Eksp. Teor. Fiz. **42**, 1135 (1962) [Sov. Phys.—JETP **15**, 784 (1962)]; J. H. Eberly and L. P. S. Singh, Phys. Rev. D **7**, 359 (1973); M. Bauer and P. A. Mello, Ann. Phys. (N.Y.) **11**, 38 (1978); M. Bauer, *ibid.* **150**, 1 (1975). For some review papers on the time-energy uncertainty relation, see articles by J. Rayski and J. M. Rayski, Jr., E. Recami, and E. W. R. Papp, in *The Uncertainty Principles and Quantum Mechanics,* edited by W. C. Price and S. S. Chissick (Wiley, New York, 1977); C. H. Blanchard, Am. J. Phys. **50**, 642 (1982). For some of the recent articles on the subject, see E. A. Gislason, N. H. Sabelli, and J. W. Wood, Phys. Rev. A **31**, 2078 (1985); M. Hossein Partovi, Phys. Rev. Lett. **23**, 2887 (1986).

[22]R. P. Feynman, in *High Energy Collisions,* Proceedings of The Third International Conference, Stony Brook, New York, 1969, edited by C. N. Yang *et al.* (Gordon and Breach, New York, 1969); J. D. Bjorken and E. A. Paschos, Phys. Rev. **185**, 1975 (1969).

[23]P. A. M. Dirac, Rev. Mod. Phys. **21**, 392 (1949); D. Bohm, *The Special Theory of Relativity* (Benjamin/Cummings, Reading, Mass., 1965); Y. S. Kim and M. E. Noz, Am. J. Phys. **50**, 721 (1982).

[24]For earlier discussions on Lorentz-deformed hadrons, see N. Byers and C. N. Yang, Phys. Rev. **142**, 976 (1966); T. T. Chou and C. N. Yang, *ibid.* **170**, 1591 (1968); J. D. Bjorken and E. A. Paschos, *ibid.* **185**, 1975 (1969); A. L. Licht and A. Pagnamenta, Phys. Rev. D **2**, 1150, (1970); **2**, 1156 (1970). V. N. Gribov, B. L. Ioffe, and I. Ya. Pomeranchuk, J. Nucl. Phys. (USSR) J **2**, 768 (1965) [Sov. J. Nucl. Phys. **2**, 549 (1966)]; B. L. Ioffe, Phys. Lett. B **30**, 123 (1969); Y. S. Kim and R. Zaoui, Phys. Rev. D **4**, 1738 (1971); S. D. Drell and T. M. Yan, Ann. Phys. (N.Y.) **60**, 578 (1971).

[25]Y. S. Kim and M. E. Noz, Found. Phys. **9**, 375 (1979).

[26]This point has been extensively discussed in the literature. See Refs. 10, 12, 13, 14, 15, and 17.

[27]For a pedagogical discussion of the connection between photons and light waves, see E. Gordin, *Waves and Photons* (Wiley, New York, 1982).

[28]For the role of the Klein-Gordon equation in the development of Schrödinger's form of quantum mechanics, see P. A. M. Dirac, *Development of Quantum Theory* (Gordon and Breach, New York, 1972).

Chapter IX

Lorentz Transformations

Let us start with a massive particle at rest with its spin along the x direction. If we boost this particle along the x axis, it will gain a momentum along the same direction. If we boost this moving particle along the y direction, the direction and the magnitude of the momentum will be changed. The resulting transformation will be a Lorentz boost preceded by a rotation. This rotation does not change the momentum of the particle at rest, but will change the direction of the spin. This is called the Wigner rotation and, as was carefully analyzed by Han, Kim and Son in 1987, manifests itself as the Thomas precession in atomic physics.

The Lorentz group is useful also in studying charged particles in electromagnetic fields. In 1959, Bargmann, Michel, and Telegdi studied the precession of the spin of a charged particle in a homogeneous magnetic field. In 1976, Kuperzstych studied a charged electron in a plane-wave electromagnetic field. He discussed the possible origin of the Lorentz force in terms of the little group for photons.

The light-cone coordinate system is very useful in many physical applications. In this system, Lorentz boosts are scale transformations in the light-cone variables. In 1970, Parker and Schmieg studied the fundamental hypotheses of special relativity in terms of the light-cone variables.

The group of Lorentz transformations, while being the basic language for special relativity, is becoming an indispensable theoretical tool many other branches of physics. The (2 + 1)-dimensional Lorentz group is locally isomorphic to the group of homogeneous linear canonical transformations in phase space. As was discussed by Han, Kim, and Noz in 1988, the group of canonical transformations is very useful in studying coherent and squeezed states in terms of the Wigner distribution function. In their 1986 paper Yurke, McCall and Klauder (1986) used the (2 + 1)-dimensional Lorentz group very effectively in their discussion of a new interferometer. Figure 4 illustrates the point that the Lorentz group is useful both in special relativity and modern optics.

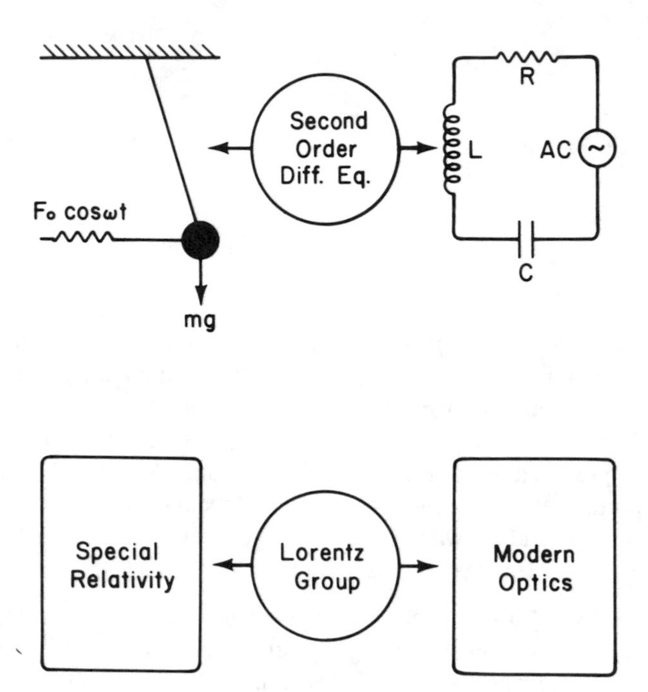

FIG. 4. Analogy of analogies. The analogy between the forced harmonic oscillator and the driven LCR circuit is well known. Since the Lorentz group is rapidly becoming one of the standard languages in optical sciences, there will be many instances in which one formula in the Lorentz group will describe one physics in optics and another physics in special relativity. This figure is from D. Han and Y.S. Kim, Univ. of Maryland P.P. #88-92 (to be published in Phys. Rev. A in 1988).

Reprinted from: *Physical Review Letters, Volume 2, 1959, pp. 435-436*

PRECESSION OF THE POLARIZATION OF PARTICLES MOVING IN A HOMOGENEOUS ELECTROMAGNETIC FIELD*

V. Bargmann
Princeton University, Princeton, New Jersey
Louis Michel
Ecole Polytechnique, Paris, France
and
V.L. Telegdi
University of Chicago, Chicago, Illinois
(Received April 27, 1959)

The problem of the precession of the "spin" of a particle moving in a homogeneous electromagnetic field - a problem which has recently acquired considerable experimental interest - has already been investigated for spin ½ particles in some particular cases.[1] In the literature the results were derived by explicit use of the Dirac equation, with the occasional inclusion of a Pauli term to account for an anomalous magnetic moment. On the other hand, following a remark of Bloch[2] in connection with the nonrelativistic case, the expectation value of the vector operator representing the "spin" will necessarily follow the same time dependence as one would obtain from a classical equation of motion. To solve the problem for arbitrary spin in the relativistic case, it will thus suffice to produce a consistent set of covariant classical equations of motion. Such equations have been indicated a long time ago by Frenkel[3] and are discussed by Kramers.[4] These authors use an antisymmetric tensor M as the relativistic generalization of the intrinsic angular momentum observed in the rest-frame of the particle. A formulation in terms of the (axial) four-vector s which describes the polarization in a covariant fashion[5] - though basically equivalent - is however much more convenient for our problem. We shall therefore derive first the equations of motion directly in terms of this four-vector s.

Let the spin of the particle be represented[6] in the rest-frame (R) by $\vec{s}$. We assume (a) that there exists a four-vector s such that in (R) it coincides with $\vec{s}$:

$$s = (s^0, \vec{s}); \qquad \text{in } (R), \quad s = (0, \vec{s}) . \tag{1}$$

Denoting the four-velocity of the particle by $u = (u^0, \vec{u}) = \gamma(1, \vec{v})$ [where $\vec{v}$ is the ordinary velocity, and $\gamma(v) = (1 - v^2)^{-1/2}$], one has in every frame

Reprinted from *Phys. Rev. Lett.* 2, 435 (1959).

$$s \cdot u = 0, \qquad \text{i.e.,} \qquad s^0 = \vec{s} \cdot \vec{v}\,. \tag{2}$$

We further assume (b) that $\vec{s}$ obeys in (R) the customary equation of motion

$$d\vec{s}/d\tau = (ge/2m)(\vec{s} \times \vec{H}), \qquad (R) \tag{3}$$

where $\vec{H}$, e, and m have their standard meanings, while the gyromagnetic ratio g is defined by this very equation. While s^0 vanishes by hypothesis in any instantaneous rest-frame, $ds^0/d\tau$ need not. In fact, (2) implies

$$ds^0/d\tau = \vec{s} \cdot (d\vec{v}/d\tau), \qquad (R) \tag{4}$$

for such frames. In general, $du/d\tau = f/m$ (where f = four-force), while in a homogeneous external electromagnetic field specified by $F = -(\vec{E}, \vec{H})$

$$du/d\tau = (e/m) F \cdot u\,. \tag{5}$$

The immediate generalization of Eqs. (3) and (4) to arbitrary frames is

$$ds/d\tau = (ge/2m)\ [F \cdot s + (s \cdot F \cdot u)u] - [(du/d\tau) \cdot s]u, \tag{6}$$

as can be checked by reducing to the rest-frame. With (5), one has for homogeneous fields

$$ds/d\tau = (e/m)[(g/2)F \cdot s + (g/2 - 1)\ (s \cdot F \cdot u)u]\,. \tag{7}$$

(5) and (7) constitute, for any value of g and arbitrary spin S, a consistent set of equations of motion; they imply that $s \cdot s$ and $s \cdot u$ are constant, so that condition (2) is maintained.[7] For experiments of current interest, the main use of (7) is in the computation of the rate Ω at which longitudinal polarization is transformed into a transverse one (and vice versa). For this, we express s in the laboratory frame (L) in terms of two unit polarization four-vectors, e_l and e_t:

$$s/S = e_l \cos\phi + e_t \sin\phi,$$

where

$$S = (-s \cdot s)^{1/2},$$

$$e_1 = \gamma\,(v\,, \vec{v}/v) \equiv \gamma\,(v\,, \hat{v}), \quad e_t = (0, \hat{n}),$$

$$\hat{n} \cdot \hat{n} = 1, \quad \hat{n} \cdot \hat{v} = 0\,. \qquad (L) \tag{8}$$

Clearly, $\Omega = d\phi/dt = d\phi/\gamma d\tau$. Introducing (8) into (7), and expressing all quantities as ordinary vectors, we find

$$\Omega = (e/m)\ \{(\vec{E} \cdot \hat{n}/v)[(g/2 - 1) - g/2\gamma^2] + (\hat{v} \cdot \vec{H} \times \hat{n})(g/2 - 1)\}\,. \tag{9}$$

The relevant "anomaly" of spin $-1/2$ particles, $(g/2 - 1)$, is clearly exhibited in (9) although our derivation was classical throughout.

We now specialize (9) to some cases of practical interest (the references are to experiments):

(A) $\vec{E} \times \vec{v} = \vec{H} \times \vec{v} = 0$; $\Omega = 0$. The character of the polarization does not change, but the transverse polarization precesses around $\vec{v}$ in longitudinal fields with an angular frequency $\omega = (ge/2m\gamma)H = (g/2)\omega_L$, as follows readily from (8).

(B)[8] $\vec{H}\cdot\hat{n}\times\hat{v} = H$, $\vec{E} = 0$; $\Omega = \omega_L(g/2-1)\gamma$, where ω_L is the Larmor frequency defined in (A) above.

(C)[9] $\vec{E}\cdot\hat{n} = E$, $\vec{H} = 0$; $\Omega = \omega_p[-g/2\gamma + (g/2-1)\gamma]$, where $\omega_p = eE/m\gamma v$ is the angular frequency of the particle's motion in the laboratory.

(D)[10] $\vec{E}\cdot\vec{H} = 0$, rectilinear motion: $\vec{E} = -\vec{v}\times\vec{H}$; $\vec{H}\cdot\hat{n}\times\hat{v} = H$, $\Omega = \omega_L(g/2\gamma)$.

(E)[11] $\vec{E}\cdot\vec{H} = 0$, $\vec{H}\cdot\hat{n}\times\hat{v} = H$, $\vec{E}\cdot\hat{n} = -Ev_x/v$, trochoidal motion: $E/H \ll 1$; $\Omega = (e/m)[(g/2-1)(H-Ev_x) + Ev_x/(\gamma^2-1)]$,

$$(\Lambda\phi/2\pi) \text{ per loop} = \gamma(E/H)\gamma(v)[1-(E/H)v_x](g/2-1)$$
$$= \gamma(v')(g/2-1),$$

where v' is velocity in a frame where $E' = 0$.

The generalization of (6) to cover particles having an intrinsic electric dipole moment $\vec{\varepsilon} = (g'e/2m)\vec{s}$ may be of interest. In the (R) frame, the effect of $\vec{\varepsilon}$ is taken into account by adding $\vec{\varepsilon}\times\vec{E}$ to the right-hand side of (3), while leaving (4) unchanged. Thus the required change in the right-hand side of (6) is the addition of a term $-(g'e/2m)[(F^*\cdot s) + (s\cdot F^*\cdot u)u]$, denoting by F^* the dual of F, i.e., $F^* = -(\vec{H}, -\vec{E})$. For the experiment (B) above, one obtains then $|\Omega| = \omega_L\gamma\times[g/2-1)^2 + (g'v/2)^2]^{1/2}$.

Discussions leading to this note were initiated when the two last-named authors were visiting Princeton University. They are indebted to J. R. Oppenheimer for making the facilities of the Institute for Advanced Study available to them.

*Research partly assisted by the Office of Scientific Research, Air Research and Development Command, and by the French Service des Poudres.

[1] H. A. Tolhoek and S. R. de Groot, Physica *17*, 17 (1951); H. Mendlowitz and K. M. Case, Phys. Rev. *97*, 33 (1955); L. M. Carrassi, Nuovo cimento 7, 524 (1958).

[2] F. Bloch, Phys. Rev. *70*, 460 (1946).

[3] J. Frenkel, Z. Physik *37*, 243 (1926).

[4] H. A. Kramers, *Quantum Mechanics* (North Holland Publishing Company, Amsterdam, 1957), p. 226 *et seq.* Kramers' Eq. (4) (p. 229) does not correspond to our Eq. (6). His conclusion that already classically $g = 2$ is implied for an electron, based on the relativistic equation he uses, stems from the fact that that equation corresponds in the rest-frame to $ds^0/d\tau = (ge/2m)\vec{s}\cdot\vec{E}$. Comparing with our (4), with $d\vec{v}/d\tau = (e/m)\vec{E}$, one sees that the "derived" result is, in fact, built into the theory from the start. A more general equation is mentioned by Kramers on p. 231, in fine print, and attributed to Frenkel. The inconsistencies arising in Kramers' discussion of what he calls "spin-orbit" forces [i.e., of the form $(\nabla\vec{H})\cdot\vec{u}$ in the rest-frame] are connected with the fact that neither of his equations of motion applies when the field is inhomogeneous. In that case, $du/d\tau$ is not given by (5) alone, but has to include an additional term (the covariant analog of the gradient force just mentioned) before being introduced into (6).

[5] The covariant polarization four-vector s is essentially the expectation value of the operator w used by Bargmann and Wigner [Proc. Natl. Acad. Sci. U. S. *34*, 211 (1948)] to characterize representations of the inhomogeneous Lorentz group: $\langle w\cdot w\rangle = -S(S+1)m^2$, S = spin. s can be

expressed in terms of the skew tensor M of Frenkel (which satisfies $M \cdot u = 0$), and vice versa: $s = M^* \cdot u$, $M^* = s \times u$, i.e., $M^{*ik} = s^i u^k - s^k u^i$. For the quantum-mechanical applications of s see, e.g., C. Bouchiat and L. Michel, Phys. Rev. *106*, 170 (1955).

[6]Our notation is: c = 1, $\hbar$ = 1 throughout; coordinate four-vector of components $x^0 = t, x^1, x^2, x^3$: $x = (x^0, \vec{x})$, $\vec{x} = \{x^\alpha\}$ (α = 1, 2, 3); metric of signature (+ ---); τ = proper time; a dot between symbols, contraction of neighboring indices with the metric tensor, e.g., $x \cdot x = (x^0)^2 - \vec{x}^2$; skew tensor of components T^{ik} indicated as $T = (\vec{T}^t, \vec{T}^s)$, $\vec{T}^t = \{T^{0\alpha}\}$, $\vec{T}^s = \{T^{\beta\gamma}\}$, α, β, γ = 1, 2, 3; its dual by $T^* = (\vec{T}_s, -\vec{T}_t)$.

[7]Equations (5) and (7) can be integrated explicitly by reference to four orthonormal four-vectors $e^{(i)}$ such that each of them obeys (5), and $e^{(0)} = u$.

[8]Crane, Pidd, and Louisell, Bull. Am. Phys. Soc. Ser. II, *3*, 369 (1958).

[9]H. Frauenfelder *et al.*, Phys. Rev. *106*, 386 (1957).

[10]P. E. Cavanagh *et al.*, Phil. Mag. *2*, 1105 (1957).

[11]P. S. Farago, Proc Phys. Soc. (London) *72*, 891 (1958).

Is there a Link between Gauge Invariance, Relativistic Invariance and Electron Spin?

J. KUPERSZTYCH

Commissariat à l'Energie Atomique, Centre d'Etudes de Limeil
B.P. 27, 94190, *Villeneuve-Saint-Georges, France*

(ricevuto il 28 Luglio 1975; manoscritto revisionato ricevuto l'8 Ottobre 1975)

Summary. — Without any assumptions other than the theory of classical electrodynamics applied to the problem of a charged particle interacting with the field of an electromagnetic plane wave, an operator is derived which has no other significance than the motion of the spin of a Dirac particle, given by the solution of the Bargmann-Michel-Telegdi equation. It is shown that the motion of an electron and the motion of its spin in a plane wave are given by an operator of Lorentz type which has the remarkable property of being equivalent to a gauge transformation.

Introduction.

It is well known that the notion of spin is a notion which has no classical equivalent. Historically, the spin appeared as a supplementary degree of freedom (an intrinsic angular moment) necessary to explain the Zeeman effect. One of the most remarkable results of Dirac's electron theory is, perhaps, the theoretical derivation of the correct value of the electron magnetic moment, initially postulated by KRONIG, UHLENBECK and GOUDSMIT from experimental data (1).

The fact that the existence of spin can be revealed by seeking a manifestly Lorentz-invariant theory seems to show that the spin is an essentially relativistic notion. This idea is reinforced by the fact that electron spin occurs when

(1) M. JAMMER: *The Conceptual Development of Quantum Mechanics* (New York, N. Y., 1966).

Reprinted from *Nuovo Cimento* **31B**, 1 (1976).

electromagnetic interactions take place, interactions which do not satisfy Galilean invariance.

In spite of the spin appearing to be a purely quantum mechanical and relativistic notion, it seems reasonable to ask how the spin of a particle which is moving classically according to the Lorentz force law, behaves.

The answer to this question is given, in the case of slowly varying electromagnetic fields, by the Bargmann-Michel-Telegdi (BMT) equation ([2]). In addition, this equation takes into account the influence of the « anomalous part » in the magnetic moment of the particle. The BMT equation can be derived either by calculating the classical limit of the generalized Dirac equation, or more simply, by using relativistic invariance considerations.

Thus, in classical electrodynamics, the spin of a particle seems to be an extraneous concept. Its origin seems only to be justified by the existence of the relativistic quantum theory.

In this paper, using only the laws of classical electrodynamics (Maxwell equations, Lorentz force law) and invariance properties (gauge invariance, relativistic invariance) in the problem of an electron interacting with the field of an electromagnetic plane wave, we shall show that it is possible to derive an operator which cannot be physically understood without introducing the notion of electron spin.

We shall begin by considering only the electromagnetic field and we shall show that, in the case of electromagnetic plane waves, it is possible to find an operator of Lorentz transformation which is equivalent to a gauge transformation.

We shall next consider an electron interacting with that field and we shall show that the above-mentioned operator allows the passage from the frame where the electron was at rest before the field was switched on, to a frame where the particle is at rest at each moment of time (an instantaneous rest frame of the particle).

This operator of Lorentz type will be constructed from the product of two operators $\mathscr{L}$ and $\mathscr{R}$ which will be shown each to have a precise physical significance. We shall show that the operator $\mathscr{L}$ represents the motion of a charged particle in the field and, giving the exact solution of the Bargmann-Michel-Telegdi equation, that the operator $\mathscr{R}$ represents the motion of spin of a particle having a magnetic moment exactly equal to that of a Dirac particle.

1. – Derivation of a Lorentz transformation equivalent to a gauge transformation in the case of electromagnetic plane waves.

In this section it will be shown that it is possible to find a Lorentz transformation which will leave the fields of a plane wave unaltered.

([2]) V. BARGMANN, L. MICHEL and V. L. TELEGDI: *Phys. Rev. Lett.*, **2**, 435 (1959).

Let us consider an electromagnetic plane wave travelling along the x-axis of a frame L which will be called the laboratory frame. The fields of the plane wave are functions of the relativistic invariant $\tau = nr = ct - x$, where n and r are respectively the four-vectors $\begin{pmatrix}1\\ \boldsymbol{n}\end{pmatrix}$ and $\begin{pmatrix}ct\\ \boldsymbol{r}\end{pmatrix}$. $\boldsymbol{n}$ is the unit vector of the x-axis. A unit system where $\hbar = c = 1$ will henceforth be used.

The problem is to find a Lorentz transformation, characterized by a matrix $\mathscr{M}$, such that it will yield the following:

$$\left\{\begin{aligned} \boldsymbol{E}'(\tau') &= \boldsymbol{E}'(\tau) = \boldsymbol{E}(\tau)\,, \\ \boldsymbol{H}'(\tau') &= \boldsymbol{H}'(\tau) = \boldsymbol{H}(\tau)\,, \end{aligned}\right. \tag{1}$$

where $\boldsymbol{E}$, $\boldsymbol{H}$ and $\boldsymbol{E}'$, $\boldsymbol{H}'$ are respectively the electric and magnetic fields of the plane wave, derived from the four-potentials $\begin{pmatrix}\varphi\\ \boldsymbol{A}\end{pmatrix}$ and $\begin{pmatrix}\varphi'\\ \boldsymbol{A}'\end{pmatrix}$ in the frame L and in the transformed frame L'.

If we use the definition of fields from potentials, the system of eqs. (1) involves the following equation:

$$\begin{pmatrix}\varphi'\\ \boldsymbol{A}'\end{pmatrix} \equiv \mathscr{M}\begin{pmatrix}\varphi\\ \boldsymbol{A}\end{pmatrix} = \begin{pmatrix}\varphi - \dfrac{\partial\lambda}{\partial t}\\ \boldsymbol{A} + \dfrac{\partial\lambda}{\partial \boldsymbol{r}}\end{pmatrix}, \tag{2}$$

where λ is an unknown function of time and space.

Equation (2) is obviously significant: while the right member of this equation is a definite gauge transformation of the four-potential $\begin{pmatrix}\varphi\\ \boldsymbol{A}\end{pmatrix}$, the left member is the transformation of this four-potential by the desired Lorentz operator.

In order to derive the operator $\mathscr{M}$, it is necessary to know the form of the most general transformation of the proper Lorentz group.

It is known that a general Lorentz transformation [3], generated by an operator $\mathscr{M}$, can be constructed from a special Lorentz transformation $\mathscr{L}$ (without change in direction of the space co-ordinate axes, but with an arbitrary direction of the velocity $\boldsymbol{\beta}$), followed by an arbitrary rotation $\mathscr{R}$ of the axes in space.

Consequently

$$\mathscr{M}\begin{pmatrix}\varphi\\ \boldsymbol{A}\end{pmatrix} = \mathscr{R}\mathscr{L}\begin{pmatrix}\varphi\\ \boldsymbol{A}\end{pmatrix} = \mathscr{R}\begin{pmatrix}\gamma(\varphi - \boldsymbol{\beta}\cdot\boldsymbol{A})\\ \boldsymbol{A} + (\gamma-1)(\boldsymbol{\beta}\cdot\boldsymbol{A})\beta^{-2}\boldsymbol{\beta} - \gamma\varphi\boldsymbol{\beta}\end{pmatrix}, \tag{3}$$

[3] C. Møller: *The Theory of Relativity*, Second Edition, Subsect. 2.4 (London, 1972).

where

$$\gamma = (1 - \beta^2)^{-\frac{1}{2}}.$$

Because of the axial symmetry around the x-axis, we can choose $\boldsymbol{\beta}$ in the (x, y)-plane without prejudicing the generality of the calculation. Let α be the angle between $\boldsymbol{n}$ and $\boldsymbol{\beta}$.

A priori, the operator $\mathscr{R}$ is a function of three independent parameters (for instance the three Euler angles). However, in order to shorten the calculation, the following must be kept in mind: since the Lorentz transformation $\mathscr{M}$ in question would be equivalent to a gauge transformation, the direction $\boldsymbol{n}$ of propagation of the plane wave must not, of course, be modified. Now, owing to the $\mathscr{L}$-transformation, the α-angle will change in the $(\boldsymbol{n}, \boldsymbol{\beta})$-plane. Therefore to compensate for this change, we can immediately conclude that it is necessary for the desired rotation to be in the $(\boldsymbol{n}, \boldsymbol{\beta})$-plane, that is, around the z-axis. Calling ψ the angle of rotation, we can then write the operator $\mathscr{M}$ in the following form:

$$\mathscr{M} = \mathscr{R}(\psi)\, \mathscr{L}(\beta, \alpha),$$

that is

$$\mathscr{M} = \begin{pmatrix} 1 & 0 & 0 & 0 \\ 0 & \cos\psi & -\sin\psi & 0 \\ 0 & \sin\psi & \cos\psi & 0 \\ 0 & 0 & 0 & 1 \end{pmatrix} \cdot \tag{4}$$

$$\cdot \begin{pmatrix} \gamma & -\gamma\beta\cos\alpha & -\gamma\beta\sin\alpha & 0 \\ -\gamma\beta\cos\alpha & 1 + (\gamma - 1)\cos^2\alpha & (\gamma - 1)\sin\alpha\cos\alpha & 0 \\ -\gamma\beta\sin\alpha & (\gamma - 1)\sin\alpha\cos\alpha & 1 + (\gamma - 1)\sin^2\alpha & 0 \\ 0 & 0 & 0 & 1 \end{pmatrix}.$$

The Lorentz transformation $\mathscr{M}$ which will leave the fields of the plane wave unaltered is obviously gauge invariant. We can therefore require the potentials φ, $\boldsymbol{A}$, φ', $\boldsymbol{A}'$, and the gauge function $\lambda(t, \boldsymbol{r})$ to be functions of τ only. Then, these requirements involves the following relations:

$$\frac{\partial\lambda}{\partial t} + \frac{\partial\lambda}{\partial x} = 0\,, \qquad \frac{\partial\lambda}{\partial y} = \frac{\partial\lambda}{\partial z} = 0\,, \qquad \varphi = A_x\,,$$

which, inserted in the basic equation (2) with $\mathcal{M}$ given by (4), now give the system of two equations

$$[\sin\psi + (\gamma-1)\cos\alpha\sin(\alpha+\psi) - \gamma\beta\sin(\alpha+\psi)]\,\varphi + \\ + [\cos\psi - 1 + (\gamma-1)\sin\alpha\sin(\alpha+\psi)]\,A_y = 0\,,$$

$$[\cos\psi + (\gamma-1)\cos\alpha\cos(\alpha+\psi) + \gamma\beta\cos\alpha - \gamma\beta\cos(\alpha+\psi) - \gamma]\,\varphi + \\ + [-\sin\psi + (\gamma-1)\sin\alpha\cos(\alpha+\psi) + \gamma\beta\sin\alpha]\,A_y = 0\,.$$

Since φ and A_y are linearly independent, all their coefficients must be equal to zero. We finally obtain the solution to our problem in the form of the two following relations:

$$\alpha = \arccos\frac{\gamma-1}{\gamma\beta} = \arccos\frac{\gamma\beta}{\gamma+1} \quad \text{or} \quad 1-\beta\cos\alpha = \frac{1}{\gamma}\,, \tag{5}$$

$$\psi = \pi - 2\alpha\,. \tag{6}$$

The matrix $\mathcal{M}$ of the sought Lorentz transformation is finally

$$\mathcal{M}(\gamma) = \begin{pmatrix} \gamma & -(\gamma-1) & -[2(\gamma-1)]^{\frac{1}{2}} & 0 \\ (\gamma-1) & (2-\gamma) & -[2(\gamma-1)]^{\frac{1}{2}} & 0 \\ -[2(\gamma-1)]^{\frac{1}{2}} & [2(\gamma-1)]^{\frac{1}{2}} & 1 & 0 \\ 0 & 0 & 0 & 1 \end{pmatrix}, \tag{7}$$

and is dependent on the arbitrary parameter $\gamma \geqslant 1$. We can now obtain the gauge function $\lambda(\tau)$ such that eq. (2) has a solution. It is

$$\lambda(\tau,\gamma) = \int_{-\infty}^{\tau} [2(\gamma-1)]^{\frac{1}{2}} A_y(\tau')\,\mathrm{d}\tau'\,. \tag{8}$$

Therefore, when the potentials are functions of the retarded time τ like the fields of the plane wave, the Lorentz transformation $\mathcal{M}(\gamma)$ given by (7) is equivalent to a gauge transformation whose gauge function is given by (8).

Thus, we have shown that it is possible to find a Lorentz transformation which will leave both the electric and magnetic fields of a plane wave unaltered.

2. – On the motion of a charged particle in the field of an electromagnetic plane wave.

Let us consider the equations of motion of a charged particle in the field of a plane wave (supposed to be linearly polarized along the y-axis) in the laboratory frame L where the particle was at rest before the field was switched on.

They are ([4])

$$(9)\qquad \begin{cases} \varepsilon(\tau) = m(1-v^2(\tau))^{-\frac{1}{2}} = m\left(1+\dfrac{\nu^2(\tau)}{2}\right), \\ \boldsymbol{P}(\tau) = m\nu(\tau)\boldsymbol{j} + \dfrac{1}{2}m\nu^2(\tau)\boldsymbol{n}, \end{cases}$$

where $\varepsilon(\tau)$ represents the energy of the particle in the field, $\boldsymbol{P}(\tau)$ is its momentum, $\boldsymbol{v}$ is its velocity, m is its rest mass and $-e$ $(e>0)$ is its charge. $\boldsymbol{j}$ is the unit vector of the y-axis, $\nu = (e/m)(-A^\mu A_\mu)^{\frac{1}{2}}$ is a dimensionless parameter.

If we use the same device as that employed for the definition of proper time of a moving particle ([5]), the motion of a charged particle in the field of a plane wave may be considered as uniform at each moment of time.

Thus, at each moment of time we can use a Lorentz transformation in order to pass from the frame L to the particle instantaneous rest frame R. It will now be shown that the frame R can be deduced from the frame L using precisely a Lorentz transformation of type $\mathscr{M}(\gamma)$ which is equivalent to a gauge transformation.

In order to do this, the parameter γ of the operator $\mathscr{M}(\gamma)$ given by (7), has now to be determined. We have consequently to solve the following equation:

$$(10)\qquad \mathscr{M}(\gamma)\begin{pmatrix}\varepsilon(\tau)\\ \boldsymbol{P}(\tau)\end{pmatrix} = \begin{pmatrix} m \\ \boldsymbol{0}\end{pmatrix},$$

where $\begin{pmatrix}\varepsilon(\tau)\\ \boldsymbol{P}(\tau)\end{pmatrix}$ is the four-momentum of the particle in the frame L given by eqs. (9) and where $\begin{pmatrix} m \\ \boldsymbol{0}\end{pmatrix}$ is this four-momentum in the rest frame R.

From eq. (10) we have immediately

$$(11)\qquad \gamma(\tau) = 1 + \frac{\nu^2(\tau)}{2},$$

([4]) E. S. SARACHIK and G. T. SHAPPERT: *Phys. Rev. D*, **1**, 2738 (1970); L. D. LANDAU and E. M. LIFSHITZ: *The Classical Theory of Fields*, Second Edition, Sect. **47**, Problem 2 (Oxford, 1962).

([5]) L. D. LANDAU and E. M. LIFSHITZ: *The Classical Theory of Fields*, Second Edition, Sect. **3** (Oxford, 1962).

and the matrix $\mathscr{M}(\gamma)$ now takes the following form:

$$\mathscr{M}(\nu(\tau)) = \begin{pmatrix} 1+\frac{\nu^2}{2} & -\frac{\nu^2}{2} & -\nu & 0 \\ \frac{\nu^2}{2} & 1-\frac{\nu^2}{2} & -\nu & 0 \\ -\nu & \nu & 1 & 0 \\ 0 & 0 & 0 & 1 \end{pmatrix}. \tag{12}$$

In other words, the equations of motion (9) of a charged particle in the field of a plane wave can be written in the simple following form:

$$\begin{pmatrix} \varepsilon(\tau) \\ \boldsymbol{P}(\tau) \end{pmatrix} = \mathscr{M}^{-1}(\nu) \begin{pmatrix} m \\ \boldsymbol{0} \end{pmatrix} = \mathscr{M}(-\nu) \begin{pmatrix} m \\ \boldsymbol{0} \end{pmatrix}. \tag{13}$$

Therefore, it may be asserted that *in the particle instantaneous rest frame R, the fields* **E** *and* **H** *of the plane wave are the same as in the laboratory L, where the particle was at rest before the field was switched on.*

Now we may observe that eq. (13) can also be written in the form

$$\begin{pmatrix} \varepsilon(\tau) \\ \boldsymbol{P}(\tau) \end{pmatrix} = \mathscr{M}^{-1}(\nu) \begin{pmatrix} m \\ \boldsymbol{0} \end{pmatrix} = \mathscr{L}^{-1}(\nu)\mathscr{R}^{-1}(\nu) \begin{pmatrix} m \\ \boldsymbol{0} \end{pmatrix} = \mathscr{L}^{-1}(\nu) \begin{pmatrix} m \\ \boldsymbol{0} \end{pmatrix}.$$

It is clear that the operator $\mathscr{L}(\nu)$ is sufficient to bring about the passage from the laboratory L to a frame where the particle is at rest at each moment of time.

We remember that $\mathscr{L}$ is a Lorentz transformation (without change in direction of the space co-ordinate axes) characterized by the following relation:

$$1-\boldsymbol{\beta}\cdot\boldsymbol{n} = \frac{1}{\gamma} = (1-\beta^2)^{\frac{1}{2}}. \tag{5'}$$

It is easy to check that the condition (5′) is satisfied by the velocity $\boldsymbol{v}$ of the particle. As a matter of fact, we have from (9)

$$1-\boldsymbol{v}(\tau)\cdot\boldsymbol{n} = \left(1+\frac{\nu^2}{2}\right)^{-1},$$

which is exactly the relation (5). Moreover, it follows from this above relation that τ is also the proper time of the particle.

The physical significance of the operator $\mathscr{L}$ is therefore obvious: it shows the motion of the particle in the field.

On the other hand, the physical significance of the operator of rotation $\mathscr{R}$ of angle ψ submitted to relation (6) does not seem at all evident. In order to clear up this problem, it is convenient to have the operator $\mathscr{R}(\nu)$ in an explicit form. The matrices $\mathscr{R}$ and $\mathscr{L}$, written as functions of ν, if we start from expression (4) and take into account the relations (5), (6) and (11), becomes

$$\mathscr{R}(\nu(\tau)) = \begin{pmatrix} 1 & 0 & 0 & 0 \\ 0 & \dfrac{1-\nu^2/4}{1+\nu^2/4} & -\dfrac{\nu}{1+\nu^2/4} & 0 \\ 0 & \dfrac{\nu}{1+\nu^2/4} & \dfrac{1-\nu^2/4}{1+\nu^2/4} & 0 \\ 0 & 0 & 0 & 1 \end{pmatrix}, \tag{14}$$

$$\mathscr{L}(\nu(\tau)) = \begin{pmatrix} 1+\dfrac{\nu^2}{2} & -\dfrac{\nu^2}{2} & -\nu & 0 \\ -\dfrac{\nu^2}{2} & 1+\dfrac{\nu^4}{8(1+\nu^2/4)} & \dfrac{\nu^3}{4(1+\nu^2/4)} & 0 \\ -\nu & \dfrac{\nu^3}{4(1+\nu^2/4)} & 1+\dfrac{\nu^2}{2(1+\nu^2/4)} & 0 \\ 0 & 0 & 0 & 1 \end{pmatrix}. \tag{15}$$

So, if the transformation $\mathscr{L}(\nu)$ is applied to the four-vector $\begin{pmatrix} \varepsilon(\tau) \\ \boldsymbol{P}(\tau) \end{pmatrix}$ given by (9), we have, as mentioned above

$$\mathscr{L}(\nu(\tau)) \begin{pmatrix} \varepsilon(\tau) \\ \boldsymbol{P}(\tau) \end{pmatrix} = \begin{pmatrix} m \\ \boldsymbol{0} \end{pmatrix}.$$

Thus, the particle is also at rest in the frame K which is transformed from the frame L by the Lorentz transformation $\mathscr{L}(\nu)$.

We arrive now at the crucial point of the paper. The operator $\mathscr{R}(\nu)$ given by (14), which stems from the calculation, has been derived without any assumption further than the theory of classical electrodynamics.

In order to find its physical significance, we shall now consider a dynamic quantity for the particle other than its four-momentum, namely its intrinsic angular moment, that is to say, its spin (if it exists). In what follows it will be shown that the motion of the spin of a charged particle, the gyromagnetic ratio of which is $g = 2$ as for a Dirac particle, is shown by just the operator $\mathscr{R}(\nu)$ in question.

3. – Motion of the spin of an electron with an anomalous magnetic moment in the field of an electromagnetic plane wave.

Let us consider the particle in its instantaneous rest frame K. We will now look for the solution to the problem of the behaviour of the spin of a charged particle which is executing a given classical motion in the field of a plane wave.

This solution will be given by the solution of the following equation (6), which is derived from the BMT equation:

$$\frac{\mathrm{d}\boldsymbol{\zeta}}{\mathrm{d}t} = \frac{2\mu m + 2\mu'(\varepsilon - m)}{\varepsilon}\boldsymbol{\zeta}\times\boldsymbol{H} + \frac{2\mu'\varepsilon}{\varepsilon + m}(\boldsymbol{v}\cdot\boldsymbol{H})\boldsymbol{v}\times\boldsymbol{\zeta} + \frac{2\mu m + 2\mu'\varepsilon}{\varepsilon + m}\boldsymbol{\zeta}\times(\boldsymbol{E}\times\boldsymbol{v}), \tag{16}$$

where $\boldsymbol{\zeta}$ is the spin vector of the particle in its instantaneous rest frame K, and where $\mu' = \mu + e/2m$ is the anomalous part of the magnetic moment μ of the particle.

If we take into account eqs. (9) and after a straightforward calculation, the eq. (16) can be written in the form of the following equations:

$$\frac{\mathrm{d}\zeta_x}{\mathrm{d}\tau} = \varrho(\nu)\frac{\mathrm{d}\nu}{\mathrm{d}\tau}\zeta_y, \tag{17}$$

$$\frac{\mathrm{d}\zeta_y}{\mathrm{d}\tau} = -\varrho(\nu)\frac{\mathrm{d}\nu}{\mathrm{d}\tau}\zeta_x, \tag{18}$$

$$\frac{\mathrm{d}\zeta_z}{\mathrm{d}\tau} = 0, \tag{19}$$

where

$$\varrho(\nu) = \left(1 + \frac{\nu^2}{4}\right)^{-1} - 2\frac{m}{e}\mu'.$$

If we define now the complex quantity

$$\chi = \zeta_x + i\zeta_y \qquad (i = \sqrt{-1}),$$

eqs. (17) and (18) can be written in the form

$$\frac{\mathrm{d}\chi}{\mathrm{d}\tau} = -i\varrho(\nu)\frac{\mathrm{d}\nu}{\mathrm{d}\tau}\chi.$$

(6) V. B. Berestetskii, E. M. Lifshitz and L. P. Pitaevskii: *Relativistic Quantum Theory*, Part I, Sect. **41**, eq. (41.9) (Oxford, 1971).

The solution of the above equation is obviously

$$\chi(\tau) = (\zeta_{0_x} + i\zeta_{0_y}) \exp[-i\psi'(\tau)], \tag{20}$$

where

$$\psi'(\tau) = 2 \operatorname{arctg} \frac{\nu}{2} - 2 \frac{m}{e} \mu' \nu, \tag{21}$$

and where $\zeta_{0_x}, \zeta_{0_y}, \zeta_{0_z}$ were the components of the vector $\boldsymbol{\zeta}(\tau)$ when the potentials were put equal to zero, that is before the field was switched on.

Obviously eq. (19) gives

$$\zeta_z = \zeta_{0_z}, \tag{22}$$

while eq. (20) yields

$$\begin{cases} \zeta_x(\tau) = \operatorname{Re} \chi(\tau) = \zeta_{0_x} \cos\psi'(\tau) + \zeta_{0_y} \sin\psi'(\tau), \\ \zeta_y(\tau) = \operatorname{Im} \chi(\tau) = -\zeta_{0_x} \sin\psi'(\tau) + \zeta_{0_y} \cos\psi'(\tau). \end{cases} \tag{23}$$

The relations (22) and (23) show clearly that the particle spin rotates in the plane (x, y), around the vector $\boldsymbol{H}(\tau)$ with angle $-\psi'(\tau)$ given by (21).

Let us now neglect the anomalous part μ' of the magnetic moment μ of the particle, *i.e.* let us now consider a charged particle (say an electron) with a gyromagnetic ratio $g = 2$ (as for a Dirac particle) $(\mu = -ge/4m)$.

In this case, from (21) we have

$$\cos\psi'(\tau) = \frac{1 - \nu^2/4}{1 + \nu^2/4} \quad \text{and} \quad \sin\psi'(\tau) = \frac{\nu}{1 + \nu^2/4}.$$

The motion of spin of a classical Dirac particle in the field of a plane wave is then given by

$$\boldsymbol{\zeta}(\tau) = \mathscr{R}(-\nu)\boldsymbol{\zeta}_0 = \mathscr{R}^{-1}(\nu)\boldsymbol{\zeta}_0,$$

or

$$\boldsymbol{\zeta}_0 = \mathscr{R}(\nu)\boldsymbol{\zeta}(\tau),$$

where $\mathscr{R}(\nu)$ is just the operator given by (14).

One can therefore conclude that *the passage from the frame where an electron was at rest before the field of a plane wave was switched on, to the frame where the electron and its spin are at rest at each moment of time, is made by an operator of Lorentz transformation which has the remarkable property of being a gauge transformation.*

4. – Conclusion.

We have therefore shown that for the problem of a charged particle classically interacting with the field of a plane wave, it is possible to derive an operator which cannot be physically understood without introducing the notion of electron spin. The operator in question $\mathscr{R}(\nu)$ represents exactly the motion of the spin of a particle the magnetic moment of which is precisely that of a Dirac particle. As was shown, another value of the magnetic moment would provide another operator than the one which was previously derived.

This result seems to indicate that there is a link between relativistic invariance, gauge invariance and electron spin on the classical level.

However, we must not be too optimistic: the results presented in this paper do not imply that the spin «must exist». The Lorentz force law is *a priori* valid for any charged particle (neglecting radiation reaction) and can therefore be used in the problem of a charged pion which is classically interacting with the field of a plane wave. But since this particle has no spin, a physical interpretation of the operator $\mathscr{R}(\nu)$ does not exist in the case of the pion.

For such a particle, from an aesthetic point of view we can only come to an irritating conclusion.

● RIASSUNTO (*)

Senza alcuna ipotesi oltre la teoria dell'elettrodinamica classica applicata al problema di una particella carica interagente con il campo di un'onda elettromagnetica piana, si deduce un operatore che non ha altro significato che il moto dello spin di una particella di Dirac, dato dalla soluzione dell'equazione di Bargmann-Michel-Telegdi. Si dimostra che il moto di un elettrone ed il moto del suo spin in un'onda piana sono dati da un operatore del tipo di Lorentz che ha la lodevole proprietà di essere equivalente a trasformazioni di gauge.

(*) *Traduzione a cura della Redazione.*

Существет ли связь между калибровочной инвариантностью, релятивистской инвариантностью и спином электрона?

Резюме (*). — Применяя теорию классической электродинамики к проблеме заряженной частицы, взаимодействующей с полем плоской электромагнитной волны, выводится оператор, описывающий движение спина частицы Дирака, которое задается решением уравнения Баргмана-Мишеля-Телегди. Показывается, что движение электрона и движение его спина в плоской волне определяется оператором лорентцевского типа, который обладает интересным свойством: этот оператор эквивалентен калибровочному преобразованию.

(*) *Переведено редакцией.*

Reprinted from AMERICAN JOURNAL OF PHYSICS, Vol. 38, No. 2, 218–222, February 1970
Printed in U. S. A.

Special Relativity and Diagonal Transformations

LEONARD PARKER AND GLENN M. SCHMIEG
Department of Physics, University of Wisconsin–Milwaukee, Milwaukee, Wisconsin 53201
(Received 13 June 1969; revision received 7 September 1969)

We discuss the form of the special Lorentz transformation, and the corresponding transformation of the electromagnetic field, in which the transformation matrix is diagonal. We derive the diagonal form of the special Lorentz transformation directly, in a simple way, and show that it is sometimes more convenient to apply than the algebraically equivalent conventional form of the transformation. The convenience is especially evident in deriving the linear Doppler effect, and the relativistic addition of more than two parallel velocities. By writing Maxwell's equations in terms of linear combinations of coordinates which have simple transformation properties, we arrive at the transformation equations of the Maxwell fields in a diagonal form, as well as at the plane wave solutions, in a natural manner. The derivations and applications described above should be of use in a course on relativity because of their simplicity and directness.

INTRODUCTION

We wish to present here an alternate derivation of the special Lorentz transformation, and several of its consequences. These derivations are appropriate for use in a course on special relativity because of their simplicity and directness. The authors have used these techniques in the classroom, and it is hoped that they will prove useful to others.

The derivation of the special Lorentz transformation given here has, to the best of our knowledge, not appeared before. It arrives directly at what we shall call the diagonal form of the special Lorentz transformation, that is, the form in which the transformation matrix is diagonal. That form of the Lorentz transformation has appeared before in the literature, and its usefulness has been noted.[1] However, explicit applications seem to be lacking in the elementary literature.[2] Therefore, we have presented a number of simple applications in which the use of the diagonal form of the special Lorentz transformation is most direct. These include the linear Doppler effect, aberration, and the relativistic addition of any number of parallel velocities. We also apply the diagonal form of the transformation to Maxwell's equations in free space, in order to arrive at an analogous form of the transformation of the electromagnetic field. The diagonal form of the Lorentz transformation is useful in applications involving light propagation and Maxwell's equations because the combinations of x and t appearing in the transformation vanish on the light cone. Such linear combinations of x and t are known as null coordinates.[3]

I. THE LORENTZ TRANSFORMATION

The special Lorentz transformation is the transformation connecting the space–time coordinates of two inertial frames S and S' in standard configuration. The velocity of the origin of S' with respect to S is v, and it is directed along the positive x axis. The axes of S' are parallel to those of S, and the spatial origins of the two systems coincide when the clocks there read zero in each system. By assuming the invariance of the velocity of light, the linearity of the transformation (or homogeneity), and isotropy, many authors arrive at the following equations[4]:

$$y'=y, \qquad z'=z, \tag{1}$$

$$(x-ct)(x+ct)=(x'-ct')(x'+ct'). \tag{2}$$

It is in the further derivation of the equations connecting x and t with x' and t' that we make a simplification. From Eq. (2) it follows that a space–time event with $x=ct$ in S must have $x'=\pm ct'$ in S'. The minus sign can be eliminated by imagining a light pulse originating from the common space–time origin and traveling along the positive x axis. The x' as well as the x coordinate of the pulse must be positive because the origin of S' moves at a velocity v which is less than c with respect to S. Thus, if the coordinates of a space–time event satisfy in S the equation $x-ct=0$, then they satisfy in S' the equation $x'-ct'=0$. Similarly, if $x+ct=0$, then $x'+ct'=0$. It now follows from the presumed linearity of the transformation, and Eq. (2), that

$$x-ct=A(v)(x'-ct')$$
$$x+ct=A^{-1}(v)(x'+ct'), \tag{3}$$

Reprinted from *Am. J. Phys.* **38**, 218 (1970).

where $A(v)$ is independent of the coordinates, but can depend on the relative velocity v.

The quantity $A(v)$ is determined by dividing the first of Eqs. (3) by the second, and noting that the origin of S' has the coordinates $(x=vt, t)$ in S and $(x'=0, t')$ in S'. Thus,

$$(vt-ct)/(vt+ct)=A^2(v)(-ct'/ct'),$$

whence

$$A(v)=(c-v)^{1/2}/(c+v)^{1/2}. \tag{4}$$

The positive square root is chosen because if an event on the x axis satisfies the condition $x>ct$, then also $x'>ct$, since a light signal originating at the common space–time origin could not have reached the event in question relative to either frame.[5]

We have thus arrived at the special Lorentz transformation in the form

$$x-ct=A(v)(x'-ct')$$
$$x+ct=A^{-1}(v)(x'+ct')$$
$$y=y', \qquad z=z'. \tag{5}$$

We call this the diagonal form of the special Lorentz transformation because the matrix transforming the variables $\xi=x+ct$, $\eta=x-ct$, y, z into $\xi'=x'+ct'$, $\eta'=x'-ct'$, y', z' is a diagonal matrix. The conventional form of the special Lorentz transformation is readily obtained from Eq. (5) using the identities

$$A+A^{-1}=2(1-v^2/c^2)^{-1/2}$$
$$A-A^{-1}=-2(1-v^2/c^2)^{-1/2}v/c. \tag{6}$$

We now turn to some applications in which the direct use of Eqs. (5) is very convenient.

II. ELEMENTARY APPLICATIONS

A. Linear Doppler Effect

Consider a plane light wave traveling in the positive x direction, represented by

$$A_0 \sin 2\pi\nu(t-x/c).$$

For such a wave the phase is invariant[6]:

$$\nu(t-x/c)=\nu'(t'-x'/c).$$

Comparison with Eqs. (5) immediately yields the linear Doppler effect

$$\nu'=A(v)\,\nu. \tag{7}$$

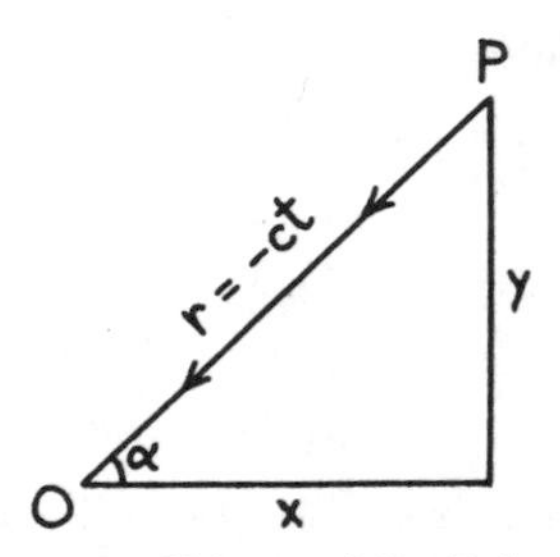

Fig. 1. Light ray relative to S.

B. Aberration

Consider a light ray in the x–y plane, as shown in Fig. 1. Suppose that a pulse on the light ray passes through the coincident origins of S and S', represented by O in the figure, at $t=t'=0$. At an earlier time $t=-r/c$, that pulse was at point P with coordinates x and y in S, as shown in Fig. 1. Using the trigonometric identity

$$\tan(\alpha/2)=\csc\alpha-\cot\alpha,$$

we obtain the equation

$$\tan(\alpha/2)=-(x+ct)/y.$$

Similarly,

$$\tan(\alpha'/2)=(x'+ct')/y',$$

where α' is the angle that the light ray makes with the x' axis in S', and x', t' are the coordinates in S' of the pulse which has coordinates x, t in S. Application of Eqs. (5) to the last two Eqs. yields the aberration formula[7]:

$$\tan(\alpha'/2)=A(v)\tan(\alpha/2). \tag{8}$$

C. Relativistic Addition of Several Parallel Velocities

Consider three inertial frames S, S', and S'' in standard configuration. Let v_1 be the velocity of S' relative to S, and v_2 be the velocity of S'' relative to S'. Let $u=dx/dt$ be the velocity of a particle moving in the x direction relative to S, and $u''=dx''/dt''$ be the corresponding velocity relative to S''. The problem is to relate u'' to u.

The coordinates of the particle relative to S

and S' satisfy

$$x+ct=A^{-1}(v_1)(x'+ct'),$$

$$x-ct=A(v_1)(x'-ct'),$$

while relative to S' and S'' they satisfy

$$x'+ct'=A^{-1}(v_2)(x''+ct''),$$

$$x'-ct'=A(v_2)(x''-ct'').$$

Eliminating $x'\pm ct'$ yields the relation between the coordinates of S and S'',[8]

$$x+ct=A^{-1}(v_1)A^{-1}(v_2)(x''+ct''),$$

$$x-ct=A(v_1)A(v_2)(x''-ct''). \tag{9}$$

Forming the differentials of Eqs. (9) and dividing yields at once the relation

$$A(u)=A(v_1)A(v_2)A(u''). \tag{10}$$

This is a form of the relativistic velocity addition formula.[9] The generalization when any number of intermediate inertial frames are involved is obvious. From Eq. (10) it is easy to see that when $u''=c$, then also $u=c$, and when $u''<c$, then also $u<c$. That is not so easy to deduce, even when only two inertial frames are involved, if the more conventional form of the velocity addition formula is used.[10]

III. MAXWELL'S EQUATIONS

The simple transformation properties of the variables

$$\xi\equiv x+ct,$$

$$\eta\equiv x-ct \tag{11}$$

under special Lorentz transformation make them very useful in dealing with Maxwell's equations. Using the variables ξ and η, we obtain the transformation equations of the Maxwell fields under special Lorentz transformation directly in a form analogous to Eq. (5). The derivation is certainly no more difficult, and probably is somewhat simpler, than the corresponding derivations which also do not use tensors, but use the variables x and t.[11] We also show that the most general plane wave solution of Maxwell's equations in free space follows immediately from the first-order Maxwell equations written in terms of ξ and η, without introduction of the second-order wave equation.

A. Transformation Properties of the Fields

Consider the homogeneous Maxwell equations in vacuum[12]:

$$\nabla\times\mathbf{E}+c^{-1}(\partial\mathbf{B}/\partial t)=0 \tag{12a}$$

$$\nabla\cdot\mathbf{B}=0. \tag{12b}$$

Derivatives in x and t can be replaced by derivatives in ξ and η by using the identities

$$\partial/\partial x=(\partial/\partial\xi)+(\partial/\partial\eta),$$

$$c^{-1}(\partial/\partial t)=(\partial/\partial\xi)-(\partial/\partial\eta).$$

This procedure is advantageous because of the simple transformation properties

$$\partial/\partial\xi=A(\partial/\partial\xi'),$$

$$\partial/\partial\eta=A^{-1}(\partial/\partial\eta'), \tag{13}$$

which follow from Eqs. (5).

The y and z components of Eq. (12a) then have the explicit forms

$$\frac{\partial}{\partial z}E_x-\frac{\partial}{\partial\xi}(E_z-B_y)-\frac{\partial}{\partial\eta}(E_z+B_y)=0$$

$$-\frac{\partial}{\partial y}E_x+\frac{\partial}{\partial\xi}(E_y+B_z)+\frac{\partial}{\partial\eta}(E_y-B_z)=0. \tag{14}$$

By subtracting Eq. (12b) from the x component of Eq. (12a), one obtains

$$\frac{\partial}{\partial y}(E_z-B_y)-\frac{\partial}{\partial z}(E_y+B_z)-2\frac{\partial}{\partial\eta}B_x=0. \tag{15}$$

Substituting Eqs. (13) and $\partial/\partial y=\partial/\partial y'$, $\partial/\partial z=\partial/\partial z'$ into Eq. (14), we get the equations

$$\frac{\partial}{\partial z'}E_x-\frac{\partial}{\partial\xi'}A(E_z-B_y)-\frac{\partial}{\partial\eta'}A^{-1}(E_z+B_y)=0$$

$$-\frac{\partial}{\partial y'}E_x+\frac{\partial}{\partial\xi'}A(E_y+B_z)+\frac{\partial}{\partial\eta'}A^{-1}(E_y-B_z)=0. \tag{16}$$

Clearly, the requirement of form invariance, i.e., that Eqs. (16) have the same form as Eqs. (14) in the primed system, is satisfied if

$$E_x'=E_x \tag{17a}$$

$$E_z'-B_y'=A(E_z-B_y) \tag{17b}$$

$$E_z'+B_y'=A^{-1}(E_z+B_y) \tag{17c}$$

$$E_y'+B_z'=A(E_y+B_z) \tag{17d}$$

$$E_y'-B_z'=A^{-1}(E_y-B_z). \tag{17e}$$

One then readily finds, with the aid of Eqs. (17b) and (17d), that Eq. (15) is form invariant if

$$B_x' = B_x. \tag{17f}$$

Equations (17a)–(17f) are analogous to Eqs. (5) in that they only involve multiplication by the factor A.[13]

The above considerations arrive at Eqs. (17) in a natural and simple manner, and are therefore suitable for use in an introductory course. In fact, analogous derivations of the more conventional form of the transformation equations are given in the texts cited in Ref. 11. However, it should be noted that the considerations given show only that form-invariance is satisfied if the correct transformation equations of the Maxwell fields hold. Those considerations alone are not sufficient to arrive uniquely at the transformation equations, since six field quantities are involved, while less than six equations are used. Furthermore, since the equations used were all homogeneous, the transformation equations arrived at could obviously all be multiplied by a common factor dependent on v. In the paper cited in Ref. 11, Einstein avoids the above difficulties by considering six of the Maxwell equations in the absence of charges, thereby deriving uniquely the transformation of the fields, to within a common factor $\psi(v)$. He then shows by simple considerations that $\psi(v) = 1$.

We can overcome the uniqueness difficulty in our derivation by noting that Eqs. (12a) and (12b) become the remaining Maxwell equations in the absence of charges, if one makes the substitutions $\mathbf{E} \rightarrow \mathbf{B}$ and $\mathbf{B} \rightarrow -\mathbf{E}$. That substitution in Eqs. (14) and (15) then gives three new equations, which are form-invariant if Eqs. (17) with $\mathbf{E} \rightarrow \mathbf{B}$ and $\mathbf{B} \rightarrow -\mathbf{E}$ hold. However, the full set of Eqs. (17) are simply interchanged among themselves by that substitution. Hence, if Eqs. (17) hold, then the six independent equations given by Eqs. (14), (15), and the equations obtained from them under the substitution $\mathbf{E} \rightarrow \mathbf{B}$ and $\mathbf{B} \rightarrow -\mathbf{E}$ are form-invariant. Consequently, form invariance of Maxwell's equations in the absence of charges uniquely determines Eqs. (17) to within a common factor $\psi(v)$. We refer the reader to Einstein's paper cited in Ref. 11, for the proof that $\psi(v) = 1$.

One can easily show that Eqs. (17) are equivalent to the more conventional form of the transformation equations by using Eqs. (6). However, as with Eqs. (5), the transformation Eqs. (17) are often useful in their original form. For example, multiplication of Eq. (17b) by Eq. (17c), and Eq. (17d) by Eq. (17e), followed by addition yields the invariance of $\mathbf{E}^2 - \mathbf{B}^2$. Similarly, multiplication of Eq. (17c) by Eq. (17d), and Eq. (17b) by Eq. (17e) yields, upon subtraction, the invariance of $\mathbf{E} \cdot \mathbf{B}$. The invariance of these quantities can immediately be generalized to arbitrary Lorentz transformations because the quantities are clearly invariant under three-dimensional rotations.

B. Plane Wave Solutions

It is worth noting how simply one can extract the general plane wave solution from Maxwell's equations written in terms of ξ and η. We seek solutions of Maxwell's equations which are independent of y and z. It is then evident in the usual way that the x component of Eq. (12a) together with Eq. (12b) implies that B_x is a constant, which we set equal to zero. Similarly one finds that E_x is a constant, which we put equal to zero. It is easy to show that the remaining four components of Maxwell's equations are equivalent to

$$(\partial/\partial\xi)(E_z - B_y) = 0 \qquad (\partial/\partial\eta)(E_z + B_y) = 0$$
$$(\partial/\partial\xi)(E_y + B_z) = 0 \qquad (\partial/\partial\eta)(E_y - B_z) = 0. \tag{18}$$

The solutions, to within additive constants, which we put equal to zero, are

$$E_z - B_y = 2f_1(\eta) \qquad E_z + B_y = 2g_1(\xi)$$
$$E_y + B_z = 2f_2(\eta) \qquad E_y - B_z = 2g_2(\xi). \tag{19}$$

Hence

$$E_y = f_2(\eta) + g_2(\xi) \qquad E_z = f_1(\eta) + g_1(\xi)$$
$$B_y = -f_1(\eta) + g_1(\xi) \qquad B_z = f_2(\eta) - g_2(\xi). \tag{20}$$

Equations (20) involve four arbitrary differentiable functions, and represent the most general plane wave solution of Maxwell's equations (the x axis has been chosen to be along the direction of the wave motion, without significant loss of generality). The solution clearly consists of a superposition of a wave traveling at velocity c in the positive x direction (the η dependence), with a wave traveling at velocity c in the negative x direction (the ξ dependence). By comparing Eqs. (19) with Eqs. (17) and Eqs. (5), it is evident that the plane wave solutions in one inertial frame transform into similar plane wave solutions under special Lorentz transformations.

IV. CONCLUSIONS

We feel that the diagonal form of the special Lorentz transformation, and the corresponding transformation of the electromagnetic field, can be advantageously employed when covering certain material in a relativity course. A derivation which arrives directly at Eqs. (5) is evidently somewhat simpler than the more conventional derivations of the Lorentz transformation. Equations (5) can be applied directly to the Doppler effect, aberration, and the addition of any number of parallel velocities.[14]

By expressing the vacuum form of Maxwell's equations in terms of the variables $x+ct$ and $x-ct$ we arrived at the transformation Eqs. (17). Equations (17) are clearly analogous to Eqs. (5) in that they involve only multiplication by factors of A. We also pointed out that the plane wave solutions follow naturally from the above form of Maxwell's equations.

In group theoretic language, the variables $x-ct$, $x+ct$, y, and z form the basis of the diagonal representation of the special Lorentz transformation, which is given by Eqs. (5). Similarly, the linear combinations of the fields appearing in Eqs. (17) form the components of an antisymmetric tensor which transforms like the direct product of two basis vectors of the diagonal representation of the special Lorentz transformation. To go more deeply into such matters would take us beyond the scope of this article.

[1] W. Rindler, *Special Relativity* (Wiley–Interscience, Inc., New York, 1966), p. 22.

[2] However, H. Bondi's k-calculus is related to the diagonal form of the Lorentz transformation because of its connection with the linear Doppler effect [see the derivation of Eq. (7)]. Bondi emphasizes the reflection of radar signals in his applications, whereas our viewpoint is generally quite different. Also, our considerations with regard to Maxwell's equations are unrelated to the k-calculus. For the k-calculus, see H. Bondi, *Relativity and Common Sense* (Doubleday & Co., Inc., Garden City, N. Y., 1964).

[3] The use of null coordinates in treating radiation is well known. See, for example, R. Penrose in *Relativity, Groups and Topology*, C. and B. DeWitt, Eds. (Gordon and Breach Science Publ., Inc., New York, 1964), p. 565.

[4] See, for example Ref. 1, or A. P. French, *Special Relativity* (W. W. Norton & Company, Inc., New York, 1968); R. Resnick, *Introduction to Special Relativity* (John Wiley & Sons, Inc., New York, 1968); W. G. V. Rosser, *An Introduction to the Theory of Relativity* (Butterworths Scientific Publications Ltd., London, 1964), as well as many others.

[5] Or because $A(v)$ must clearly reduce to unity as v approaches zero.

[6] Invariance of the phase follows from a consideration of the counting of wave crests. See for example, W. G. V. Rosser, Ref. 4, p. 154, or C. Møller, *Theory of Relativity* (Oxford University Press, London, 1952), p. 7.

[7] The derivation of Eq. (8) was suggested by a problem in Ref. 1, p. 54.

[8] The fact that Eqs. (9) have the form of a special Lorentz transformation illustrates the group property of the special Lorentz transformations.

[9] The general form of Eq. (10) has been given in a group theoretic context by P. Malvaux, Compt. Rend. **235,** 1009 (1952).

[10] D. Bohm, *Special Relativity* (W. A. Benjamin, Inc., New York, 1965), p. 68; N. D. Mermin, *Space and Time in Special Relativity* (McGraw-Hill Book Co., New York, 1968), p. 132.

[11] A. Einstein, Ann. Physik **17,** 891 (1905); translation in *Principle of Relativity* (Dover Publications, Inc., New York, 1923), p. 37; R. Resnick, Ref. 4, pp. 178–181; W. G. V. Rosser, *Introductory Relativity* (Plenum Press, Inc., New York, 1967), pp. 224–227.

[12] We use Heaviside–Lorentz units. To change the formulas in this paper into rationalized MKS units, simply replace **B** by c**B**. (That procedure works for the formulas in this paper, but not in general.)

[13] We are not aware of any reference where it is pointed out that the transformation of the electromagnetic field can be written in the form of Eqs. (17).

[14] For a further application of Eqs. (5) to accelerating frames of reference, see R. T. Jones, Amer. J. Phys. **28,** 109 (1960).

Reprinted from AMERICAN JOURNAL OF PHYSICS, Vol. 38, No. 11, 1298–1302, November 1970
Printed in U. S. A.

A Useful Form of the Minkowski Diagram

LEONARD PARKER AND GLENN M. SCHMIEG
Department of Physics, University of Wisconsin, Milwaukee, Wisconsin 53201
(Received 2 April 1970; revision received 22 May 1970)

We give a diagrammatic representation of the diagonal form of the special Lorentz transformation. The null coordinates $x \pm ct$ are plotted along a single set of orthogonal axes. Special Lorentz transformations are then represented only by a change of scale along those orthogonal axes. This diagram, which we call a null coordinate diagram, and the Minkowski diagram are closely connected. To demonstrate the use of the null coordinate diagram, we apply it to the linear Doppler effect, time dilation, and Lorentz contraction.

In a previous paper,[1] we considered a form of the special Lorentz transformation, the so-called diagonal form, which is especially convenient in certain applications. Our methods were mainly algebraic in nature. In this paper, we will show that the diagonal form of the Lorentz transformation can be conveniently represented on a diagram, and we will apply this geometrical representation to several elementary problems. This representation is equivalent to the Minkowski diagram. However, our version of the diagram has the favorable feature that only one set of orthogonal axes are used to represent any number of frames related by special Lorentz transformation. A special Lorentz transformation is represented simply by a change of scale of the orthogonal axes. Because our version of the Minkowski diagram uses null coordinates $x \pm ct$, we will call it a null coordinate diagram to distinguish it from the customary form of Minkowski diagram.[2]

To demonstrate its use, we apply the null coordinate diagram to the linear Doppler shift, the time dilation, and the length contraction. We feel that the null coordinate diagram can serve as a complement to the other diagrammatic representations of the special Lorentz transformation,[3] each representation being most convenient in particular applications.

I. THE NULL COORDINATE DIAGRAM

A. Explanation of the Diagram

Let S' and S be two inertial frames in standard configuration, with S' moving at velocity v in the positive x direction relative to S. The special Lorentz transformation relating the coordinates of S and S' can be written in the so-called diagonal form[1]:

$$\xi' = A(v)\xi,$$
$$\eta' = A^{-1}(v)\eta,$$
$$y' = y, \qquad z' = z, \tag{1}$$

where

$$\xi = x + ct, \qquad \xi' = x' + ct',$$
$$\eta = x - ct, \qquad \eta' = x' - ct', \tag{2}$$

and

$$A(v) = (c-v)^{1/2}/(c+v)^{1/2}. \tag{3}$$

Reprinted from *Am. J. Phys.* **38**, 1298 (1970).

The coordinates ξ and η are known as null coordinates.

In a null coordinate diagram, we draw the ξ and η axes as shown in Fig. 1. By convention, the unprimed coordinates ξ and η are always plotted with equal scales along the two axes. The coordinates ξ and η of a space–time point are obtained by orthogonal projection onto the axes. The primed coordinates ξ' and η' are plotted along the same axes, respectively, as ξ and η. However, in accordance with the transformation given in Eq. (1), one unit of ξ' is equal to $A^{-1}(v)$ units of ξ, and one unit of η' is equal to $A(v)$ units of η. For positive v, since $A(v) < 1$, it follows that the units of ξ' will be dilated with respect to those of ξ, and the units of η' will be contracted with respect to those of η. The ξ' and η' coordinates of a space–time point are also obtained by orthogonal projection. A given event corresponds to a single point on the diagram, regardless of which coordinate system we refer to.

In practice, as will become evident in the applications, it is sufficient to know that the primed and unprimed coordinates are related by Eq. (1), so that it is not actually necessary to plot the various scales along the axes. If more than two inertial frames in standard configuration are involved, the additional null coordinates are also plotted along the single set of orthogonal axes, with scales which depend on the relation of the additional null coordinates to ξ and η. Note that only a single orthogonal projection is necessary to read the null coordinates of a given event in any Lorentz frame.

B. Equivalence with the Minkowski Diagram

To illustrate that the null coordinate diagram is equivalent to the Minkowski diagram, we plot

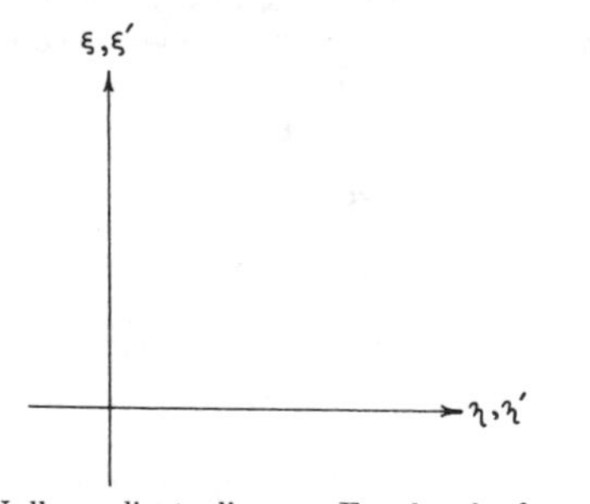

FIG. 1. Null coordinate diagram. Equal scales for ξ and η. Unequal scales for ξ' and η'.

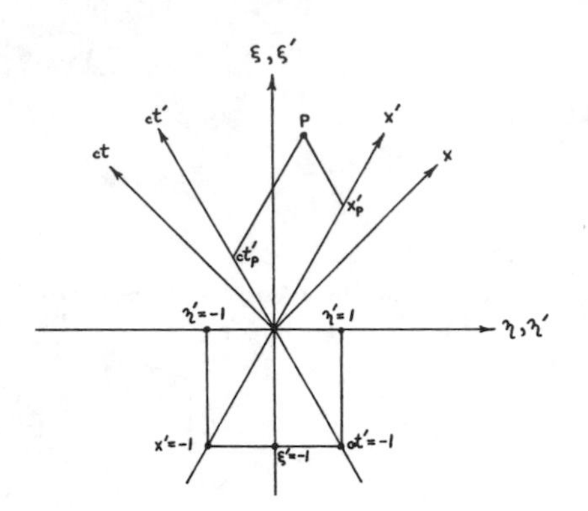

FIG. 2. Equivalence of null coordinate and Minkowski diagrams. Relation of coordinate scales is shown, as well as the x' and ct' coordinates of a point P.

the various position and time axes on a null coordinate diagram and determine how the position and time coordinates of a space–time point can be obtained using those axes rather than the null coordinate axes. The x axis is the locus of points with $t=0$ or $\xi=\eta$. Similarly, the x' axis is the line $\xi'=\eta'$. The ct axis is the locus of points with $x=0$ or $\xi=-\eta$. Similarly, the ct' axis is the locus of points with $\xi'=-\eta'$. The direction of the axes can be determined by noting that as x or ct increase, ξ also increases. The various position and time axes are plotted in Fig. 2 for the case when $v>0$. From Eq. (2), it follows that the lines of constant x' are parallel to the ct' axis, and the lines of constant ct' are parallel to the x' axis. Hence, to find the x' and ct' coordinates of a point, such as the point P shown in Fig. 2, one projects lines parallel to the ct' and x' axes, respectively. Thus, our diagram is equivalent to a Minkowski diagram rotated through 45°.[4] The null coordinate diagram also provides a convenient method of determining the scales on the x' and ct' axes as shown in Fig. 2. The examples given below will demonstrate the use of null coordinates diagrams and will show that they are very convenient in some cases.

II. APPLICATIONS

A. Linear Doppler Effect

Suppose that a source at the origin of the inertial frame S emits pulses periodically in the $+x$ direction with a proper period τ. Figure 3 shows the

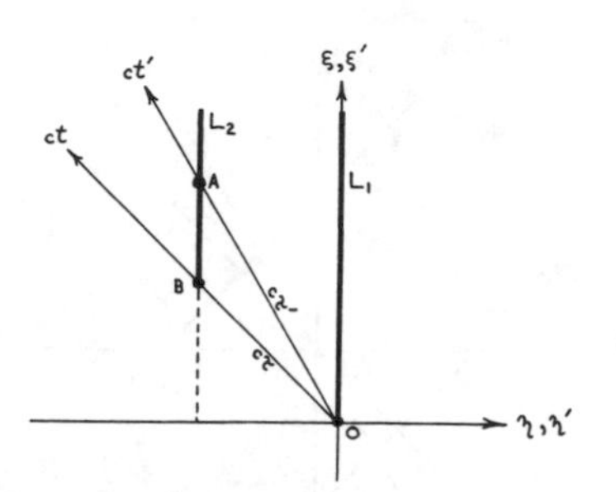

FIG. 3. Doppler effect. The first pulse L_1 is emitted (and received) at $t=t'=0$.

world lines L_1 and L_2 of two such pulses,[5] as well as the time axes of S and S', which are the world lines, respectively, of the source at the origin of S and an observer at the origin of S'. Because $x'=0$ for both the events A and O, it follows that[6]

$$\Delta\eta_{AO}'=c\tau'. \tag{4}$$

Similarly, because $x=0$ for both the events B and O, we have

$$\Delta\eta_{BO}=c\tau. \tag{5}$$

By projection onto the η, η' axis, we see that

$$\Delta\eta_{AO}=\Delta\eta_{BO}. \tag{6}$$

Thus, making use of Eqs. (1) and (6), we have

$$\Delta\eta_{AO}'=A^{-1}(v)\,\Delta\eta_{AO}=A^{-1}(v)\,\Delta\eta_{BO}. \tag{7}$$

This gives the linear Doppler effect:

$$\tau'=A^{-1}(v)\tau$$

or

$$\nu'=A(v)\nu, \tag{8}$$

where ν and ν' are the frequencies relative to S and S', respectively.

B. Time Dilation

To illustrate further the usefulness of the null coordinate diagram, consider a clock[7] consisting of two mirrors at the ends of a rod, with a light pulse bouncing back and forth between the mirrors. The period of the clock is the round trip time of the light pulse. Figure 4 shows the world line of the light pulse in one such round trip. The rod is at rest in the inertial system S. The light starts from the mirror at space–time point A, is reflected from the other mirror at space–time point B, and returns to the original mirror at space–time point C.

For any light pulse moving in the $+x$ direction we have $\Delta\xi=2c\Delta t=2\Delta x$, while for any light pulse moving in the $-x$ direction we have $\Delta\eta=2c\Delta t=2\Delta x$, as is easily obtained from Eq. (2). The analogous equations hold in the primed coordinates. Thus, the distances and times of travel of light pulses are proportional to the lengths of their world lines, as measured in the appropriate null coordinate units.

Therefore, we can see directly from Fig. 4 that

$$\Delta t_{AB}'=A(v)\,\Delta t_{AB} \tag{9}$$

and

$$\Delta t_{BC}'=A^{-1}(v)\,\Delta t_{BC}. \tag{10}$$

The factors of A and A^{-1} appear because of the different scales for the primed and unprimed null coordinates, in accordance with Eq. (1). In the proper system of the rod, the time intervals Δt_{AB} and Δt_{BC} are clearly equal, and will be denoted by Δt. The period τ' of the clock with respect to S' is therefore

$$\tau'=\Delta t_{AB}'+\Delta t_{BC}'=\tfrac{1}{2}[A(v)+A^{-1}(v)]\tau, \tag{11}$$

where $\tau=2\Delta t$ is the proper period of the clock. Using Eq. (3), the expression for τ' becomes

$$\tau'=[1-(v^2/c^2)]^{-1/2}\tau, \tag{12}$$

which is the familiar form of the time dilation equation. The ease with which Eqs. (9) and (10) were obtained from Fig. 4 illustrates the utility of null coordinate diagrams in problems involving light pulses.

C. Length Contraction

Suppose that a rod of proper length l is at rest along the x axis of S, with its center at $x=0$. If a light flash is emitted at the center of the rod, it

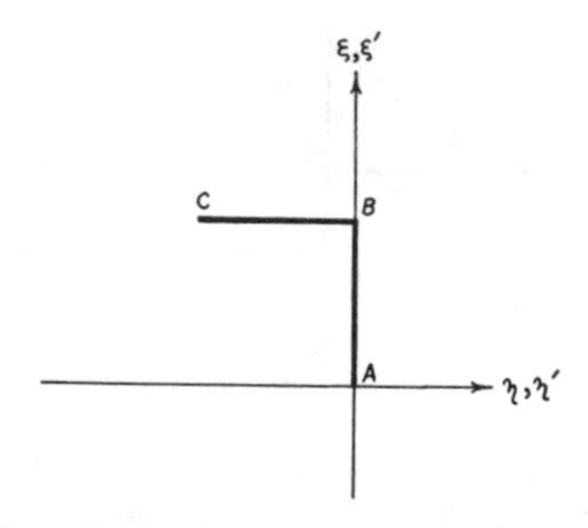

FIG. 4. Round trip of light pulse bouncing between two mirrors.

will reach the ends of the rod simultaneously relative to S, and the pulses in the $+x$ and $-x$ directions will each travel a distance $l/2$. The world lines L_1 and L_2 of the ends of the rod and the world lines of the light pulses (emitted at $t=0$) are shown in Fig. 5.

Let a different light flash be emitted from another source on the rod, such that the flash will reach the ends of the rod simultaneously relative to S'. Then, relative to S', the initial position of the source must be halfway between the final positions of the ends of the rod at the time when the flash reaches them.[8] Therefore, relative to S', the wave front of the flash travels to reach the ends of the rod a distance $l'/2$, where l' is the length of the rod. The world lines AD and DE of the flash which reaches the ends of the rod simultaneously with respect to S' are also shown in Fig. 5.[9]

We have

$$\Delta\xi_{AB}=\Delta\eta_{BC}=l \tag{13}$$

and

$$\Delta\xi_{AD}'=\Delta\eta_{DE}'=l'. \tag{14}$$

Since the world line L_1 makes a 45° angle with the $-\eta$ axis, it follows that

$$\Delta\xi_{CF}=\Delta\eta_{EF}$$

or

$$\Delta\xi_{AD}-\Delta\xi_{AB}=\Delta\eta_{BC}-\Delta\eta_{DE}. \tag{15}$$

But

$$\Delta\xi_{AD}=A^{-1}(v)\,\Delta\xi_{AD}' \tag{16}$$

and

$$\Delta\eta_{DE}=A(v)\,\Delta\eta_{DE}'. \tag{17}$$

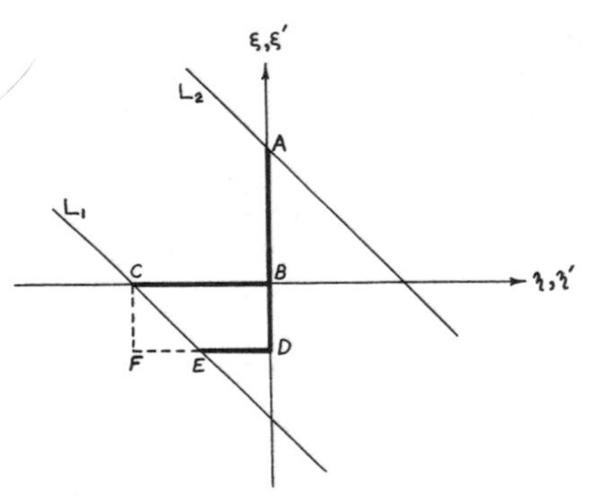

FIG. 5. World lines L_1 and L_2 of ends of rod. World lines AB and BC of flash emitted from center of rod. World lines AD and DE of flash emitted from position such that events A and E are simultaneous in S'.

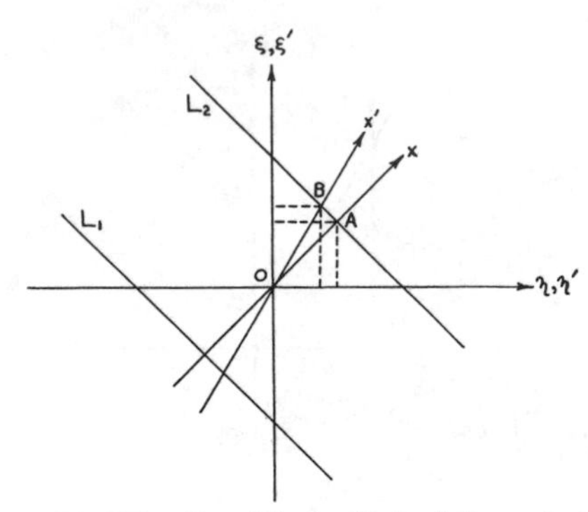

FIG. 6. World lines L_1 and L_2 of ends of rod, the x axis of S, and the x' axis of S'.

Substituting (16) and (17) into (15), and using (13) and (14), we obtain

$$A^{-1}(v)l'-l=l-A(v)l'$$

or

$$l=\tfrac{1}{2}[A(v)+A^{-1}(v)]l'. \tag{18}$$

Hence,

$$l'=[1-(v^2/c^2)]^{1/2}l. \tag{19}$$

The length contraction can also be obtained directly without the use of light rays. Figure C shows the world lines of the same rod at rest with respect to S with its center at $x=0$. As in the previous derivation, because the world line L_1 makes a 45° angle with the $-\eta$ direction, it follows that the perimeters of the two dotted figures with vertices at A and B, respectively, are equal. Thus,

$$\Delta\xi_{OA}+\Delta\eta_{OA}=\Delta\xi_{OB}+\Delta\eta_{OB}$$

or

$$\Delta\xi_{OA}+\Delta\eta_{OA}=A^{-1}(v)\,\Delta\xi_{OB}'+A(v)\,\Delta\eta_{OB}'. \tag{20}$$

From Eq. (2), we have

$$\Delta\xi_{OA}+\Delta\eta_{OA}=l$$

and

$$\Delta\xi_{OB}'+\Delta\eta_{OB}'=l'.$$

But since A lies on the x axis and B lies on the x' axis, we must have

$$\Delta\xi_{OA}=\Delta\eta_{OA}=l/2 \tag{21}$$

and

$$\Delta\xi_{OB}'=\Delta\eta_{OB}'=l'/2. \tag{22}$$

Then Eq. (20) yields the length contraction

$$l=\tfrac{1}{2}[A^{-1}(v)+A(v)]l'.$$

The time dilation can be derived analogously.

The above examples should be sufficient to familiarize the reader with the use of the null coordinate form of the Minkowski diagram.

[1] L. Parker and G. M. Schmieg, Amer. J. Phys. **38**, 218 (1970).

[2] Although diagrams involving null coordinates are not new, we have not previously seen null coordinate diagrams used to represent special Lorentz transformations, nor have we seen them applied to elementary problems in special relativity.

[3] The Minkowski, Brehme, and Loedel diagrams are described in A. Shadowitz, *Special Relativity* (Saunders, Philadelphia, Pa., 1968).

[4] The question of whether to draw the ξ and η axes vertically and horizontally, or rotated through a 45° angle, is purely a matter of taste.

[5] The world lines of light pulses are always parallel to the ξ or η axis.

[6] We use the convention that $\Delta\eta_{PQ}$ denotes the positive change in η between events P and Q. Similarly, in our notation Δ followed by any variable always denotes the positive increment in that variable. An alternate procedure would be to use coordinates $ct \pm x$ in Eq. (2). If this were done, many examples could be easily discussed in the first quadrant of the ξ, η plane.

[7] Often called a "Feynman clock."

[8] Since the rod is moving relative to S', the source at the time of emission must be closer to the leading end of the rod.

[9] For convenience, in Fig. 5 both flashes are drawn so that they reach the end of the rod at $x = l/2$ simultaneously.

PHYSICAL REVIEW A VOLUME 33, NUMBER 6 JUNE 1986

SU(2) and SU(1,1) interferometers

Bernard Yurke, Samuel L. McCall, and John R. Klauder
AT&T Bell Laboratories, Murray Hill, New Jersey 07974
(Received 30 October 1985)

A Lie-group-theoretical approach to the analysis of interferometers is presented. Conventional interferometers such as the Mach-Zehnder and Fabry-Perot can be characterized by SU(2). We introduce a class of interferometers characterized by SU(1,1). These interferometers employ active elements such as four-wave mixers or degenerate-parametric amplifiers in their construction. Both the SU(2) and SU(1,1) interferometers can in principle achieve a phase sensitivity $\Delta\phi$ approaching $1/N$, where N is the total number of quanta entering the interferometer, provided that the light entering the input ports is prepared in a suitable quantum state. SU(1,1) interferometers can achieve this sensitivity with fewer optical elements.

I. INTRODUCTION

In a conventional interferometer such as the Mach-Zehnder[1] depicted in Fig. 1 light is fed into one of the input ports. The light beam is split into two beams which propagate along different paths and suffer a phase shift relative to each other of ϕ. The light beams are combined and interfere with each other at a second beam splitter. The relative phase shift ϕ can be determined by measuring the position of the interference fringes in the output beams. Such an interferometer can achieve a phase sensitivity

$$\Delta\phi=\frac{1}{\sqrt{N}}\ , \tag{1.1}$$

where N is the total number of photons that have passed through the interferometer during the measurement time. Caves[2] has pointed out that by feeding suitably constructed squeezed states into both input ports of the interferometer the phase sensitivity can approach

$$\Delta\phi=\frac{1}{N}\ . \tag{1.2}$$

Bondurant and Shapiro[3,4] and Ni[5] have also investigated the use of squeezed states in increasing interferometer sensitivity.

The interferometers considered by Caves[2] and Bondurant and Shapiro[3,4] were primarily passive lossless devices with two input ports and two output ports. We will show that the group SU(2) naturally characterizes such interferometers and present group-theoretical arguments indicating the ultimate sensitivity that can be achieved by such devices. We will then introduce a class of active lossless interferometers characterized by the group SU(1,1). In these devices the interference arises not from recombining light beams via beam splitters, but from the phase-sensitive response of active elements such as degenerate-parametric amplifiers and four-wave mixers. In contrast to SU(2) interferometers, SU(1,1) interferometers can achieve a phase sensitivity of $1/N$ with only vacuum fluctuations entering the input ports and coherent light pumping the active devices. SU(1,1) interferometers can achieve a phase sensitivity of $1/N$ with fewer optical elements than the SU(2) interferometers and hence present a more practical way of doing sensitive interferometry, once sufficiently low-noise parametric amplifiers or four-wave mixers become available.

II. SU(2) CHARACTERIZATION OF PASSIVE LOSSLESS DEVICES WITH TWO INPUT AND TWO OUTPUT PORTS

In this section the connection between a linear lossless passive device having two input ports and two output ports and the group SU(2) is presented. Since SU(2) is equivalent to the rotation group in three dimensions, this will allow one to visualize the operations of beam splitters and phase shifters as rotations in 3-space. This insight will be exploited in the next section to discuss the performance of the Mach-Zehnder interferometer.

Let a_1 and a_2 denote the annihilation operators for two light beams which may be, for example, the two light beams entering a beam splitter or the two light beams leaving a beam splitter. These operators and their Hermitian conjugates satisfy the boson commutation relations:

$$\begin{aligned}&[a_i,a_j]=[a_i^\dagger,a_j^\dagger]=0\ ,\\&[a_i,a_j^\dagger]=\delta_{ij}\ ,\end{aligned} \tag{2.1}$$

where i and j take on the values 1 and 2. One can introduce the Hermitian operators

$$\begin{aligned}J_x&=\tfrac{1}{2}(a_1^\dagger a_2+a_2^\dagger a_1)\ ,\\J_y&=-\frac{i}{2}(a_1^\dagger a_2-a_2^\dagger a_1)\ ,\\J_z&=\tfrac{1}{2}(a_1^\dagger a_1-a_2^\dagger a_2)\ ,\end{aligned} \tag{2.2}$$

and

$$N=a_1^\dagger a_1+a_2^\dagger a_2\ . \tag{2.3}$$

The operators (2.2) satisfy the commutation relations for

Reprinted from *Phys. Rev. A* 33, 4033 (1986).

the Lie algebra of SU(2):

$$\begin{aligned}[J_x, J_y] &= iJ_z ,\\ [J_y, J_z] &= iJ_x ,\\ [J_z, J_x] &= iJ_y .\end{aligned} \tag{2.4}$$

The Casimir invariant for this group, using (2.2) and (2.3), can be put into the form

$$J^2 = \frac{N}{2}\left[\frac{N}{2}+1\right] , \tag{2.5}$$

in fact, N itself commutes with all the operators of (2.2).

Why one should want to characterize a lossless passive device with two input ports and two output ports with the operators (2.2) and (2.3) will now be explained. Let $a_{1\,\mathrm{in}}$ and $a_{2\,\mathrm{in}}$ denote the annihilation operators for the light entering the two input ports and similarly let $a_{1\,\mathrm{out}}$ and $a_{2\,\mathrm{out}}$ denote the annihilation operators for two light beams leaving the two output ports. The scattering matrix for the device will have the form

$$\begin{bmatrix} a_{1\,\mathrm{out}} \\ a_{2\,\mathrm{out}} \end{bmatrix} = \begin{bmatrix} U_{11} & U_{12} \\ U_{21} & U_{22} \end{bmatrix} \begin{bmatrix} a_{1\,\mathrm{in}} \\ a_{2\,\mathrm{in}} \end{bmatrix} . \tag{2.6}$$

Since the creation and annihilation operators for the two input beams and the two output beams must satisfy (2.1) the matrix

$$U = \begin{bmatrix} U_{11} & U_{12} \\ U_{21} & U_{22} \end{bmatrix} \tag{2.7}$$

must be unitary. Such a transformation will in general transform J_x, J_y, and J_z among themselves.

How $\mathbf{J}=(J_x, J_y, J_z)$ transforms under U will now be determined for some common optical elements.

Consider a beam splitter with the scattering matrix

$$U = \begin{bmatrix} \cos\frac{\alpha}{2} & -i\sin\frac{\alpha}{2} \\ -i\sin\frac{\alpha}{2} & \cos\frac{\alpha}{2} \end{bmatrix} . \tag{2.8}$$

This transformation will transform $\mathbf{J}$ according to

$$\begin{bmatrix} J_x \\ J_y \\ J_z \end{bmatrix}_{\mathrm{out}} = \begin{bmatrix} 1 & 0 & 0 \\ 0 & \cos\alpha & -\sin\alpha \\ 0 & \sin\alpha & \cos\alpha \end{bmatrix} \begin{bmatrix} J_x \\ J_y \\ J_z \end{bmatrix}_{\mathrm{in}} . \tag{2.9}$$

That is, the abstract angular momentum vectors are rotated about the x axis by an angle α. This transformation can be expressed in the form

$$\begin{bmatrix} J_x \\ J_y \\ J_z \end{bmatrix}_{\mathrm{out}} = e^{i\alpha J_x} \begin{bmatrix} J_x \\ J_y \\ J_z \end{bmatrix} e^{-i\alpha J_x} , \tag{2.10}$$

where the angular momentum operators on the right-hand side are evaluated for the input beams $a_{1\,\mathrm{in}}$ and $a_{2\,\mathrm{in}}$. The equivalence of (2.9) and (2.10) can be checked using the operator identity

$$e^{\xi A} B e^{-\xi A} = B + \xi[A,B] + \frac{\xi^2}{2!}[A,[A,B]] + \cdots . \tag{2.11}$$

One can alternatively work in a Schrödinger picture where the operators J_x, J_y, and J_z remain unchanged but the state vector, after interacting with the beam splitter, becomes

$$|\,\mathrm{out}\rangle = e^{-i\alpha J_x}\,|\,\mathrm{in}\rangle , \tag{2.12}$$

where $|\,\mathrm{in}\rangle$ is the state vector for the light before it has interacted with the beam splitter. Throughout this paper we will hop back and forth between the Heisenberg picture where $\mathbf{J}$ is rotated while the state vector remains fixed and the Schrödinger picture where $\mathbf{J}$ remains fixed depending on which picture is most convenient for the discussion at hand.

Another realizable scattering matrix for a beam splitter is

$$U = \begin{bmatrix} \cos\frac{\beta}{2} & -\sin\frac{\beta}{2} \\ \sin\frac{\beta}{2} & \cos\frac{\beta}{2} \end{bmatrix} . \tag{2.13}$$

At radio frequencies devices with the scattering matrix Eq. (2.8) and Eq. (2.13) would be distinguished, respectively, as 90° and 180° couplers. The scattering matrix Eq. (2.13) transforms $\mathbf{J}$ according to

$$\begin{bmatrix} J_x \\ J_y \\ J_z \end{bmatrix}_{\mathrm{out}} = \begin{bmatrix} \cos\beta & 0 & \sin\beta \\ 0 & 1 & 0 \\ -\sin\beta & 0 & \cos\beta \end{bmatrix} \begin{bmatrix} J_x \\ J_y \\ J_z \end{bmatrix}_{\mathrm{in}} . \tag{2.14}$$

This transformation represents a rotation of $\mathbf{J}$ about the y axis by an angle β. This transformation can be written as

$$\begin{bmatrix} J_x \\ J_y \\ J_z \end{bmatrix}_{\mathrm{out}} = e^{i\beta J_y} \begin{bmatrix} J_x \\ J_y \\ J_z \end{bmatrix} e^{-i\beta J_y} . \tag{2.15}$$

Hence in the Schrödinger picture where $\mathbf{J}$ remains fixed the state vector for the light after interacting with the beam splitter is

$$|\,\mathrm{out}\rangle = e^{-i\beta J_y}\,|\,\mathrm{in}\rangle . \tag{2.16}$$

How $\mathbf{J}$ transforms under a phase shift or change in optical path length is now determined. Let light beams 1 and 2 incur a phase shift γ_1 and γ_2, respectively. The unitary matrix associated with this process is

$$U = \begin{bmatrix} e^{i\gamma_1} & 0 \\ 0 & e^{i\gamma_2} \end{bmatrix} . \tag{2.17}$$

Under this transformation $\mathbf{J}$ transforms as

$$\begin{bmatrix} J_x \\ J_y \\ J_z \end{bmatrix}_{\mathrm{out}} = \begin{bmatrix} \cos(\gamma_2-\gamma_1) & -\sin(\gamma_2-\gamma_1) & 0 \\ \sin(\gamma_2-\gamma_1) & \cos(\gamma_2-\gamma_1) & 0 \\ 0 & 0 & 1 \end{bmatrix} \begin{bmatrix} J_x \\ J_y \\ J_z \end{bmatrix} . \tag{2.18}$$

This represents a rotation about the z axis by the angle

$\gamma_2-\gamma_1$ corresponding to the relative phase shift between the two light beams. This transformation can be expressed as

$$\begin{bmatrix} J_x \\ J_y \\ J_z \end{bmatrix}_{\text{out}} = e^{i(\gamma_2-\gamma_1)J_z} \begin{bmatrix} J_x \\ J_y \\ J_z \end{bmatrix} e^{-i(\gamma_2-\gamma_1)J_z} . \quad (2.19)$$

Hence in the Schrödinger picture this represents a transformation of the incoming state vector $|\text{in}\rangle$ according to

$$|\text{out}\rangle = e^{-i(\gamma_2-\gamma_1)J_z}|\text{in}\rangle . \quad (2.20)$$

It is worth noting that under the full transformation Eq. (2.17) the incoming state transforms as $|\text{out}\rangle = e^{i(\gamma_1+\gamma_2)N/2}e^{-i(\gamma_2-\gamma_1)J_z}|\text{in}\rangle$ but since N commutes with $\mathbf{J}$ the operator $e^{i(\gamma_1+\gamma_2)N/2}$ gives rise to phase factors which do not contribute to the expectation values or moments of number-conserving operators such as $\mathbf{J}$ and N. In fact, it is the insensitivity of photodetectors (photon counters) to the extra phase $e^{i(\gamma_1+\gamma_2)N/2}$ that allows one to fully characterize an interferometer by the SU(2) transformations described above. It has now been shown that the transformations the beam splitters and phase shifters perform on the two incoming light beams can be visualized as rotations of the vector $\mathbf{J}$. Further, since the operator ($a_1^\dagger a_1$ or $a_2^\dagger a_2$) characterizing the number of photons counted by a photodetector placed in one of the light beams can be expressed in terms of the operator N and J_z, interferometry can be visualized as the process of measuring rotations of $\mathbf{J}$. The operators giving rise to the mode transformations of Eqs. (2.12), (2.16), and (2.20) have also been recently discussed by Schumaker in Ref. 6 where they are referred to as two-mode mixing operators.

III. THE MACH-ZEHNDER INTERFEROMETER

The formalism of the last section will now be applied to the Mach-Zehnder[7] interferometer. This interferometer is

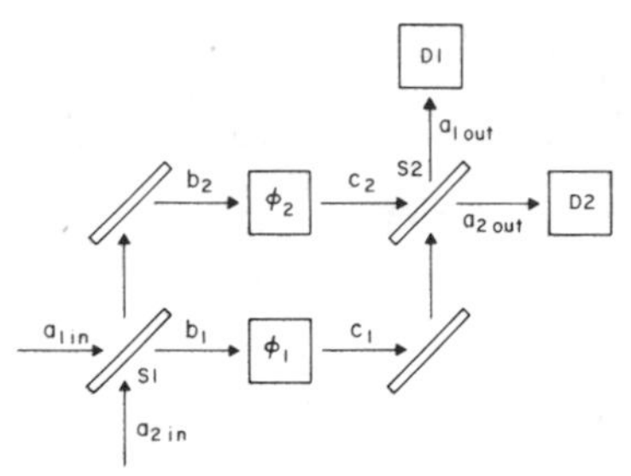

FIG. 1. A Mach-Zehnder interferometer. Light entering one of the two input ports $a_{1\text{in}}$ or $a_{2\text{in}}$ is split into two beams by beam splitter S1. The two light beams b_1 and b_2 accumulate a phase shift ϕ_1 and ϕ_2, respectively, before entering beam splitter S2. The photons leaving the interferometer are counted by detectors D1 and D2.

depicted in Fig. 1. It consists of two 50-50 beam splitters S1 and S2. The relative phase shift $\phi=\phi_2-\phi_1$ is measured by observing the interference fringes in the light leaving S2. Here, as depicted in Fig. 1, the case will be considered where the photodetector is placed in each of the two output beams $a_{1\text{out}}$ and $a_{2\text{out}}$. By counting the number of photoelectrons generated by each detector, D_1 and D_2, separately, one measures the operators $N_1=a_{1\text{out}}^\dagger a_{1\text{out}}$ and $N_2=a_{2\text{out}}^\dagger a_{2\text{out}}$. From Eqs. (2.2) and (2.3) one sees that this is equivalent to measuring both N_{out} and $J_{z\,\text{out}}$.

A geometrical picture of the operation of the interferometer will now be developed. For definiteness the beam splitters S1 and S2 will be chosen to have scattering matrices of the form (2.8). For a 50-50 beam splitter α must take on the value $\pi/2$ or $-\pi/2$. For the beam splitter S1 we take $\alpha=+\pi/2$, for the beam splitter S2 we take $\alpha=-\pi/2$. Let $|\text{in}\rangle$ denote the state vector for the light in the two light beams entering the interferometer. From Eq. (2.12) the state $|\psi\rangle$ of the light

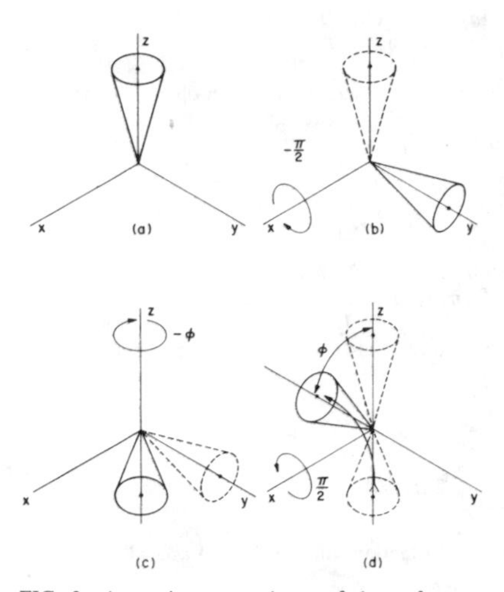

FIG. 2. A rotation-group picture of the performance of a Mach-Zehnder interferometer. When light enters only one input port of the interferometer the input state has the form $|j,m\rangle=|j,j\rangle$ in a fictitious (J_x,J_y,J_z) space and can be represented by a cone centered along the z axis with height j (a). The first beam splitter performs a $-\pi/2$ rotation about the x axis. The cone now lies along the y axis (b). The phase shifts accumulated by the two light beams in the interferometer correspond to a rotation $-\phi$ about the z axis (c). The second beam splitter performs a $\pi/2$ rotation about the x axis (d). Since J_z is proportional to the difference in the number of photons counted by the two photodetectors in the interferometer output beam the interferometer can resolve states whose overall rotation is sufficiently far from the z axis so that on average the J_z measured will differ from j by one. In order for this to be the case the cone must be rotated by approximately the width of its base which is $\sqrt{j}$. Hence the minimum detectable ϕ is of order $1/\sqrt{j}$.

upon leaving the S1 is

$$|\psi\rangle = e^{-i(\pi/2)J_x}|\text{in}\rangle \tag{3.1}$$

which amounts to a rotation of the state vector about the x axis by an amount $-\pi/2$. This is depicted in Figs. 2(a) and 2(b) where for definiteness $|\text{in}\rangle$ was chosen to be the state $|j,m=j\rangle$, that is, **J** lies on the circle surrounding the base of the cone in Fig. 2(a). With a $-\pi/2$ rotation about the x axis this cone now lies along the y axis.

Upon reaching the input ports of S2 one light beam c_1 has undergone a phase shift of ϕ_1 while the other c_2 has undergone a phase shift ϕ_2. Thus, from Eq. (2.19), upon arriving at S2 the light is in the state $|\psi'\rangle$:

$$\begin{aligned}|\psi'\rangle &= e^{-i(\phi_2-\phi_1)J_z}|\psi\rangle \\ &= e^{-i(\phi_2-\phi_1)J_z}e^{-i(\pi/2)J_x}|\text{in}\rangle .\end{aligned} \tag{3.2}$$

Hence, as depicted in Fig. 2(c), the phase shift rotates the state vector about the z axis by an amount $-\phi=-(\phi_2-\phi_1)$.

The second beam splitter rotates the state vector $|\psi'\rangle$ about the x axis by an amount $\pi/2$. The state vector $|\text{out}\rangle$ for the light leaving the interferometer is thus

$$\begin{aligned}|\text{out}\rangle &= e^{i(\pi/2)J_x}|\psi'\rangle \\ &= e^{i(\pi/2)J_x}e^{-i(\phi_2-\phi_1)J_z}e^{-i(\pi/2)J_x}|\text{in}\rangle .\end{aligned} \tag{3.3}$$

As shown in Fig. 2(d), the net result of this sequence of rotations is a rotation of the initial state vector about the y axis by an amount ϕ.

As pointed out earlier, by placing photodetectors in both of the output beams one can measure both N (the total number of photons passing through the interferometer) and J_z (the difference in the number of photons arriving at each detector divided by 2). Because N commutes with **J** it by itself gives one no useful information about ϕ. It does, however, give one useful information about $|\text{in}\rangle$, in particular the total number of photoelectrons n counted after the light has passed through the interferometer tells one that $|\text{in}\rangle$ was in an eigenstate of N:

$$N|\text{in}\rangle = n|\text{in}\rangle . \tag{3.4}$$

From this, using (2.5) one concludes

$$j(j+1) = \frac{n}{2}\left[\frac{n}{2}+1\right] \tag{3.5}$$

or

$$j=\frac{n}{2} , \tag{3.6}$$

that is, $|\text{in}\rangle$ was an eigenstate of J^2.

The other variable measured, J_z, allows one to infer what the value of ϕ was. In particular,

$$\langle\text{out}|J_z|\text{out}\rangle = \langle\text{in}|e^{i(\pi/2)J_x}e^{i\phi J_z}e^{-i(\pi/2)J_x}J_z e^{i(\pi/2)J_x}e^{-i\phi J_z}e^{-i(\pi/2)J_x}|\text{in}\rangle . \tag{3.7}$$

One can show

$$e^{i(\pi/2)J_x}e^{i\phi J_z}e^{-i(\pi/2)J_x}J_z e^{i(\pi/2)J_x}e^{-i\phi J_z}e^{-i(\pi/2)J_x} = -(\sin\phi)J_x+(\cos\phi)J_z . \tag{3.8}$$

Hence

$$\begin{aligned}\langle J_z\rangle &= \langle\text{out}|J_z|\text{out}\rangle \\ &= -\sin\phi\langle\text{in}|J_x|\text{in}\rangle+\cos\phi\langle\text{in}|J_z|\text{in}\rangle\end{aligned} \tag{3.9}$$

and

$$\begin{aligned}\langle J_z^2\rangle &= \langle\text{out}|J_z^2|\text{out}\rangle \\ &= \sin^2\phi\langle\text{in}|J_x^2|\text{in}\rangle \\ &\quad -\sin\phi\cos\phi\langle\text{in}|J_xJ_z+J_zJ_x|\text{in}\rangle \\ &\quad +\cos^2\phi\langle\text{in}|J_z^2|\text{in}\rangle .\end{aligned} \tag{3.10}$$

To proceed further one needs additional information on $|\text{in}\rangle$. Let us suppose the interferometer is operated in the usual manner where light enters the interferometer only along one of the input beam paths, say a_1. Then from the total number of photons n counted by D1 and D2 one knows that there were n photons in the incoming light beam. Hence $|\text{in}\rangle$ is an eigenstate of J_z:

$$J_z|\text{in}\rangle = \frac{n}{2}|\text{in}\rangle . \tag{3.11}$$

From (3.6) and (3.11) one concludes that the incoming light beam was in an eigenstate of $|j,m\rangle$

$$|\text{in}\rangle = |j=n/2,m=n/2\rangle . \tag{3.12}$$

Hence the incoming light is in the eigenstate that was depicted in Fig. 2.

Intuitively the smallest ϕ that can be measured is one where the cones of Fig. 2(d) do not appreciably overlap. The distance from the apex of one of the cones to a point on the circle of the cone's base is the square root of the eigenvalue of J^2 or $\sqrt{j(j+1)}$. The distance from the apex of one of the cones to the center of its base is the eigenvalue of J_z or j. Hence the radius of the base of one of the cones is $[j(j+1)-j^2]^{1/2}=\sqrt{j}$. The minimum detectable ϕ is thus of order $\phi_{\min}\simeq j^{-1/2}$, and since from Eq. (3.12) $j=n/2$,

$$\phi_{\min}\simeq n^{-1/2} . \tag{3.13}$$

Hence the sensitivity of an interferometer operated in the mode where light enters only one of the two input ports has a sensitivity that goes as the square root of the number of photons passing through the interferometer.

Equation (3.13) is now made more rigorous by a direct calculation from Eq. (3.9) and Eq. (3.10). For the state (3.12) the mean value of J_z is

$$\bar{J}_z = \frac{n}{2}\cos\phi . \tag{3.14}$$

The mean-square fluctuation $(\Delta J_z)^2$ about this value is

$$(\Delta J_z)^2 \equiv \overline{J_z^2} - \bar{J}_z^2 = \frac{n}{4}\sin^2\phi \ . \tag{3.15}$$

The mean-square noise in ϕ is thus

$$(\Delta\phi)^2 = \frac{(\Delta J_z)^2}{\left|\frac{\partial \bar{J}_z}{\partial \phi}\right|^2} = \frac{1}{n} \ . \tag{3.16}$$

Hence the rms fluctuation of ϕ due to photon noise goes as $n^{-1/2}$,

$$\Delta\phi = n^{-1/2} \tag{3.17}$$

in agreement with the intuitive argument based on Fig. 2. We will refer to (3.17) as the "standard noise limit" for an interferometer.

Note that no assumption was made about the quantum statistics of the source of light entering the interferometer. The total number of photons n entering the interferometer completely characterizes the ultimate sensitivity that can be achieved with an interferometer in which light is fed into only one input port. If instead of using photodetectors in both output ports and measuring N and J_z one chooses to use only one photodetector or to measure only J_z, then one is throwing away information. In this case knowledge about the photon statistics of the source becomes important. For this situation the performance of the interferometer will generally degrade although for some particular values of $\phi_1 - \phi_2$ the $n^{1/2}$ phase sensitivity can still be achieved.

As will be pointed out in Sec. V, the sensitivity of an interferometer can be dramatically improved if photons are allowed to enter both input ports provided the photons are prepared in the right quantum state.

IV. THE FABRY-PEROT INTERFEROMETER

In the last section a geometrical picture of the operation of the Mach-Zehnder interferometer was presented in terms of rotations of the operators (J_x, J_y, J_z) defined by Eq. (2.2). Photodetectors placed in the output beam of the interferometer measure the operator J_z. A relative phase shift between two optical beams produces a rotation of $\mathbf{J}$ about the z axis, see Eq. (2.19). A measurement of J_z, however, is only sensitive to rotations in a plane containing the z axis. The function of the two 50-50 beam splitters is thus to convert a rotation about the z axis into a rotation in a plane containing the z axis. For the particular set of beam splitters chosen in the last section this corresponds to a net rotation in the x-z plane as depicted in Fig. 2.

The Fabry-Perot interferometer,[7] by employing semitransparent mirrors, also converts a rotation about the x axis into a rotation lying in a plane containing the z axis. One can also take advantage of the multiple passes of the light between the two mirrors to enhance the sensitivity of the interferometer, but at the expense of the interferometer's bandwidth. Here expressions are obtained for the phase sensitivity of a Fabry-Perot.

A Fabry-Perot interferometer is depicted in Fig. 3(a). It consists of semitransparent mirrors M1 and M2. This interferometer measures the phase shift ϕ suffered by light as it propagates from one mirror to the other. This device has two input ports $a_{1\,\mathrm{in}}$ and $a_{2\,\mathrm{in}}$, and two output ports $a_{1\,\mathrm{out}}$ and $a_{2\,\mathrm{out}}$. Although $a_{1\,\mathrm{in}}$ and $a_{2\,\mathrm{out}}$ and $a_{2\,\mathrm{in}}$ and $a_{1\,\mathrm{out}}$ are collinear they can be separated with optical circulators as shown in Fig. 3(b). In this manner one can place photodetectors in both beams $a_{1\,\mathrm{out}}$ and $a_{2\,\mathrm{out}}$ without obstructing the light injected into $a_{1\,\mathrm{in}}$ or $a_{2\,\mathrm{in}}$. Hence one is allowed to measure N and J_z for the two output beams.

An analysis of the Fabry-Perot interferometer is now carried out. The mirrors M1 and M2 will be taken to have scattering matrices of the form Eq. (2.8). In particular for the mirror M1 we take

$$\begin{aligned} b_1 &= \cos(\tfrac{1}{2}\beta)a_{1\,\mathrm{in}} + i\sin(\tfrac{1}{2}\beta)b_2 \ , \\ a_{2\,\mathrm{out}} &= +i\sin(\tfrac{1}{2}\beta)a_{1\,\mathrm{in}} + \cos(\tfrac{1}{2}\beta)b_2 \ , \end{aligned} \tag{4.1}$$

and for the mirror M2

$$\begin{aligned} a_{1\,\mathrm{out}} &= \cos(\tfrac{1}{2}\beta)c_1 - i\sin(\tfrac{1}{2}\beta)a_{2\,\mathrm{in}} \ , \\ c_2 &= -i\sin(\tfrac{1}{2}\beta)c_1 + \cos(\tfrac{1}{2}\beta)a_{2\,\mathrm{in}} \ . \end{aligned} \tag{4.2}$$

In writing (4.1) and (4.2) it has been assumed that both mirrors have the same transmission coefficient

$$T = \cos^2(\tfrac{1}{2}\beta) \ . \tag{4.3}$$

The phase shift θ sustained by the light as it propagates between the two mirrors is given by

$$\begin{aligned} c_1 &= e^{i\theta}b_1 \ , \\ b_2 &= e^{i\theta}c_2 \ . \end{aligned} \tag{4.4}$$

Equations (4.1)–(4.4) can be solved to obtain $a_{1\,\mathrm{out}}, a_{2\,\mathrm{out}}$ in terms of $a_{1\,\mathrm{in}}$ and $a_{2\,\mathrm{in}}$. One finds

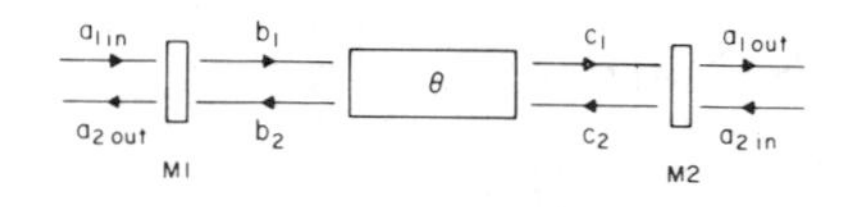

(a)

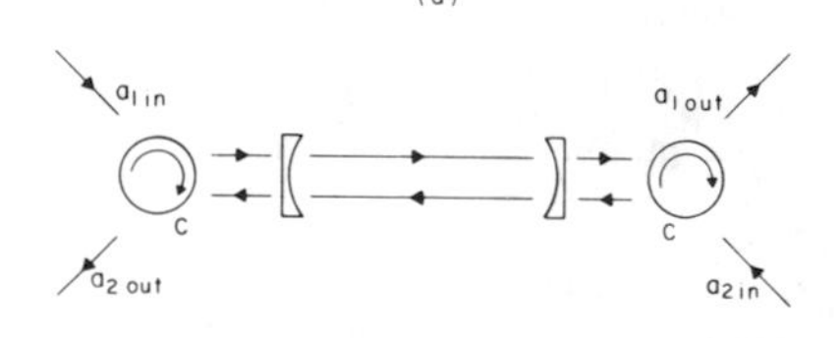

(b)

FIG. 3. A Fabry-Perot interferometer. The device (a) consists of two semitransparent mirrors M1 and M2. The light propagating between M1 and M2 suffers a one-way phase shift θ. In (b) circulators C have been placed behind the mirrors to physically separate the incoming and outgoing light beams. It is evident that the Fabry-Perot interferometer has two input ports and two output ports.

$$\begin{pmatrix} a_{1\,\text{out}} \\ a_{2\,\text{out}} \end{pmatrix} = \begin{pmatrix} \mu & -\nu \\ \nu & \mu \end{pmatrix} \begin{pmatrix} a_{1\,\text{in}} \\ a_{2\,\text{in}} \end{pmatrix}, \tag{4.5}$$

where

$$\mu = \frac{\cos^2(\frac{1}{2}\beta)e^{i\theta}}{1-\sin^2(\frac{1}{2}\beta)e^{2i\theta}}, \quad \nu = -\frac{i\sin(\frac{1}{2}\beta)(e^{2i\theta}-1)}{1-\sin^2(\frac{1}{2}\beta)e^{2i\theta}}. \tag{4.6}$$

The scattering matrix of Eq. (4.5) is unitary. Using Eq. (2.2) one can determine how **J** transforms under this unitary transformation. One finds

$$\begin{pmatrix} J_x \\ J_y \\ J_x \end{pmatrix}_{\text{out}} = \begin{pmatrix} \cos\phi & 0 & \sin\phi \\ 0 & 1 & 0 \\ -\sin\phi & 0 & \cos\phi \end{pmatrix} \begin{pmatrix} J_x \\ J_y \\ J_z \end{pmatrix}_{\text{in}}, \tag{4.7}$$

where

$$\cos\phi = |\mu|^2 - |\nu|^2, \quad \sin\phi = \mu^*\nu + \nu^*\mu. \tag{4.8}$$

So the Fabry-Perot interferometer, for the mirrors chosen, performs a rotation of **J** about the y axis. Hence, following the same line of reasoning as in the last section, if light enters the Fabry-Perot in only one input port the ultimate phase sensitivity $\Delta\phi$ is given by

$$\Delta\phi = n^{-1/2}. \tag{4.9}$$

In order to determine what this implies for the ultimate phase sensitivity $\Delta\theta$ one needs to evaluate $|d\phi/d\theta|$. With Eq. (4.6), Eq. (4.8) becomes

$$\cos\phi = \frac{\cos^4(\frac{1}{2}\beta) - 4\sin^2(\frac{1}{2}\beta)\sin^2\theta}{\cos^4(\frac{1}{2}\beta) + 4\sin^2(\frac{1}{2}\beta)\sin^2\theta}, \tag{4.10}$$

$$\sin\phi = +\frac{4\cos^2(\frac{1}{2}\beta)\sin(\frac{1}{2}\beta)\sin\theta}{\cos^4(\frac{1}{2}\beta) + 4\sin^2(\frac{1}{2}\beta)\sin^2\theta}, \tag{4.11}$$

and thus ϕ is given by

$$\phi = \arctan\left[\frac{4\cos^2(\frac{1}{2}\beta)\sin(\frac{1}{2}\beta)\sin\theta}{\cos^4(\frac{1}{2}\beta) - 4\sin^2(\frac{1}{2}\beta)\sin^2\theta}\right]. \tag{4.12}$$

Differentiating this equation with respect to θ one obtains

$$\frac{d\phi}{d\theta} = \frac{4\cos^2(\frac{1}{2}\beta)\sin(\frac{1}{2}\beta)\cos\theta}{\cos^4(\frac{1}{2}\beta) + 4\sin^2(\frac{1}{2}\beta)\sin^2\theta}. \tag{4.13}$$

The Fabry-Perot is most sensitive for those angles θ for which $|d\phi/d\theta|$ is maximized. From Eq. (4.13) one sees that the sensitivity is greatest when $|\cos\theta| = 1$ and $\sin\phi = 0$. So

$$\left|\frac{d\phi}{d\theta}\right|_{\max} = \frac{4|\sin(\frac{1}{2}\beta)|}{\cos^2(\frac{1}{2}\beta)}, \tag{4.14}$$

or in terms of the transmission coefficient T for the mirrors

$$\left|\frac{d\phi}{d\theta}\right|_{\max} = \frac{4(1-T)^{1/2}}{T}. \tag{4.15}$$

Hence the smallest rms fluctuations in $\Delta\theta$ achieved by a Fabry-Perot is

$$\Delta\theta_{\min} = \frac{\Delta\phi}{\left|\frac{d\phi}{d\theta}\right|_{\max}} = \frac{T\Delta\phi}{4(1-T)^{1/2}}. \tag{4.16}$$

For mirrors with a small transmission coefficient T, and using (4.9)

$$\Delta\theta_{\min} \simeq \frac{T}{4n^{1/2}}. \tag{4.17}$$

Hence, as with the Mach-Zehnder, the sensivity of the interferometer scales as $n^{-1/2}$ where n is the total number of photons entering the interferometer. As with the derivation of (3.17), Eq. (4.17) is based on the assumption that light enters only one port of the interferometer.

In the next section it is shown that the sensitivity of an interferometer can be greatly enhanced if light, prepared in a suitable quantum state, is allowed to enter both ports of the interferometer. Although the arguments will be applied to the Mach-Zehnder, with the tools developed in this section, they can be applied to the Fabry-Perot interferometer as well.

V. SURPASSING THE STANDARD NOISE LIMIT

In the last two sections it was shown that an interferometer can be regarded as a device which performs rotations on the operators (J_x, J_y, J_z) defined by Eq. (2.2). Photodetectors placed in the output beam of the interferometer measure the operator J_z. Hence the overall rotation must lie in a plane containing the z axis. For the choice of beam splitters used in the Mach-Zehnder of Sec. III and the mirrors used in the Fabry-Perot of Sec. IV this rotation was in the x-z plane. [See Eqs. (3.8) and (4.7)].

For the Mach-Zehnder the sequence of rotations performed is depicted in Fig. 2. A cone was used to represent a J_z eigenstate. Based on how such an object transforms under the rotations performed by the interferometer a minimum detectable phase shift of order $n^{-1/2}$ was derived. As will be shown, a J_z eigenstate is not the optimum eigenstate for interferometry. In particular, by forming a linear superposition of J_z eigenstates near $m=0$, one might imagine constructing a squashed cone or "fan-shaped" state lying in the x-y plane as depicted in Fig. 4(a). Such a state constructed from a superposition of J_z eigenstates near $m=1$ would have an extent along the z axis of order unity. Figure 4 indicates how such a geometrical object would transform under the rotations performed by the interferometer. Since the extent of the state along the z axis is ~ 1 and the distance from the origin to the edge of the cone is $\sim j$, Fig. 4(d) would indicate that the minimal detectable ϕ is $\phi_{\min} \sim 1/j$. Or, from Eq. (3.6),

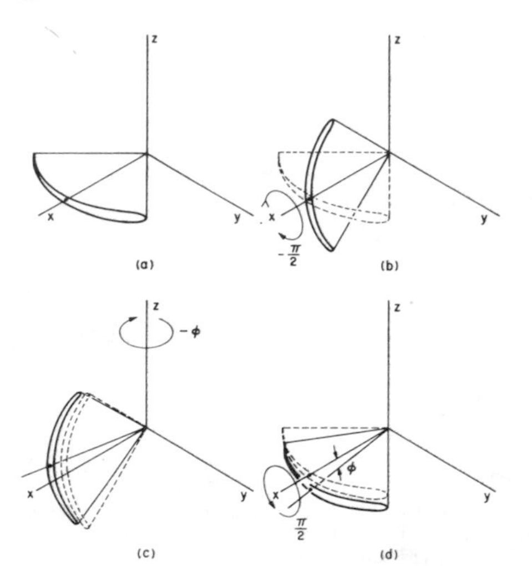

FIG. 4. The performance of a Mach-Zehnder interferometer in which an input state, of length j when depicted in the (J_x, J_y, J_z) space, is a flattened cone whose width along the z axis is of order unity. The sequence of rotations performed by the interferometer is the same as that of Fig. 2. In contrast to the state depicted in Fig. 2, an overall rotation $\phi \sim 1/j$ can be resolved with the state depicted here.

$$\phi_{\min} \sim 1/n \ . \tag{5.1}$$

Hence, by choosing the appropriate incoming state $|\,\text{in}\rangle$, an interferometer's sensitivity can be greatly improved over the $n^{-1/2}$ sensitivity of Eq. (3.17) or Eq. (4.9).

The above discussion is now made rigorous by explicitly exhibiting a state with the properties described above. Consider the state

$$|\,\text{in}\rangle = \frac{1}{\sqrt{2}}\,|\,j,0\rangle + \frac{1}{\sqrt{2}}\,|\,j,1\rangle \ . \tag{5.2}$$

From this equation one can immediately show

$$\langle \text{in}\,|\,J_z^k\,|\,\text{in}\rangle = \tfrac{1}{2} \ . \tag{5.3}$$

Hence this state lies close to the x-y plane and has a mean-square height of order unity,

$$(\Delta J_z)^2 = \tfrac{1}{4} \ . \tag{5.4}$$

It is also straightforward to show that

$$\begin{aligned} \langle \text{in}\,|\,J_x \text{in}\rangle &= \tfrac{1}{2}[j(j+1)]^{1/2} \ , \\ \langle \text{in}\,|\,J_y\,|\,\text{in}\rangle &= 0 \ , \end{aligned} \tag{5.5}$$

and

$$\begin{aligned} \langle \text{in}\,|\,J_x^2\,|\,\text{in}\rangle &= \tfrac{1}{2}[j(j+1) - \tfrac{1}{2}] \ , \\ \langle \text{in}\,|\,J_y^2\,|\,\text{in}\rangle &= \tfrac{1}{2}[j(j+1) - \tfrac{1}{2}] \ . \end{aligned} \tag{5.6}$$

So the mean-square uncertainties in J_x and J_y are

$$\begin{aligned} (\Delta J_x)^2 &= \tfrac{1}{4}[j(j+1) - 1] \ , \\ (\Delta J_y)^2 &= \tfrac{1}{2}[j(j+1) - \tfrac{1}{2}] \ . \end{aligned} \tag{5.7}$$

Equations (5.5) and (5.7) indicate that the state is oriented along the x axis and is very broad along the x and y axes. This state could thus be represented by a geometrical object similar to that depicted in Fig. 4(a). One can also show that

$$\langle \text{in}\,|\,J_x J_z\,|\,\text{in}\rangle = \tfrac{1}{4}[j(j+1)]^{1/2} \tag{5.8}$$

and consequently

$$\langle \text{in}\,|\,J_x J_z + J_z J_x\,|\,\text{in}\rangle = \tfrac{1}{2}[j(j+1)]^{1/2} \ . \tag{5.9}$$

The rms fluctuations in ϕ for this state will now be determined. Substituting Eqs. (5.3) and (5.5) into Eq. (3.9), $\bar{J}_z$, half the mean differenced photocurrent, is given by

$$\bar{J}_z = -\tfrac{1}{2}[j(j+1)]^{1/2}\sin\phi + \tfrac{1}{2}\cos\phi \ . \tag{5.10}$$

Substituting Eqs. (5.6) and (5.9) into Eq. (3.10) one has

$$\overline{J_z^2} = \tfrac{1}{2}[j(j+1) - \tfrac{1}{2}]\sin^2\phi + \tfrac{1}{2}\cos^2\phi \ . \tag{5.11}$$

The mean-square fluctuation in J_z is then

$$(\Delta J_z)^2 = \tfrac{1}{4}[j(j+1) - 1]\sin^2\phi + \tfrac{1}{4}\cos^2\phi \ . \tag{5.12}$$

The mean-square fluctuation in ϕ is given by

$$(\Delta\phi)^2 = \frac{(\Delta J_z)^2}{\left[\dfrac{d\bar{J}_z}{d\phi}\right]^2} \tag{5.13}$$

or

$$(\Delta\phi)^2 = \frac{[j(j+1) - 1]\sin^2\phi + \cos^2\phi}{\{[j(j+1)]^{1/2}\cos\phi + \sin\phi\}^2} \ . \tag{5.14}$$

This quantity has its minimum value when $\sin\phi = 0$, then

$$(\Delta\phi)^2_{\min} = \frac{1}{j(j+1)} \tag{5.15}$$

or in terms of the number of photons passing through the interferometer, since $j = n/2$,

$$(\Delta\phi)^2_{\min} = \frac{4}{n(n+2)} \ . \tag{5.16}$$

Hence when the state Eq. (5.2) is fed into the input ports of an interferometer a minimum rms fluctuation $\Delta\phi_{\min}$ in the phase of order n^{-1} can be achieved:

$$\Delta\phi_{\min} \simeq \frac{2}{n} \ . \tag{5.17}$$

This maximum sensitivity is however achieved only at particular values of ϕ satisfying $\sin\phi = 0$. For other values of ϕ the sensitivity of the interferometer is degraded. Since $\phi = \phi_1 - \phi_2$, ϕ_1 may be tracked as a function of time with the precision Eq. (5.17) by controlling ϕ_2 with a feedback loop which maintains $\phi_1 - \phi_2$ at zero. The error signal for this loop is the differenced photodetector current $2J_z$. The use of feedback loops with be further discussed in Sec. VII.

A state $|\text{in}\rangle$ which allows an interferometer to achieve phase uncertainty of order n^{-1} has now been presented. How one prepares light in such a state, or a state similar to it, is the topic of the next section. Here we simply point out some properties of the state $|\text{in}\rangle$ of Eq. (5.2). It is a superposition of the states $|j,0\rangle$ and $|j,1\rangle$. For the state $|j,0\rangle$, N has the eigenvalue $n=2j$ and J_z has the eigenvalue $m=0$. Equations (2.2) and (2.3) allow one to recognize this state as one in which exactly j photons enter each of the two input ports of the interferometer. For the state $|j,1\rangle$, N has the eigenvalue $n=2j$ and J_z has the eigenvalue $m=1$. This state can be recognized as one in which exactly $j+1$ photons enter the input port $a_{1\text{in}}$ while exactly $j-1$ photons enter the input port $a_{2\text{in}}$.

VI. THE TWO-MODE FOUR-WAVE MIXER

In the last section it was shown that the sensitivity of an interferometer could be greatly improved provided one could prepare the light delivered to the input ports of the interferometer in a state which consists of a superposition of two states, one in which exactly j photons enter each of the two input ports in the interferometer and a state in which $j+1$ photons enter one port while $j-1$ photons enter the other port. In this section it is shown that states similar to this can be generated with two-mode four-wave mixers. For the analysis of such a device it will be convenient to introduce a set of operators whose commutation relations are those for the generators of the group SU(1,1).

In particular we introduce the Hermitian operators

$$\begin{aligned} K_x &= \tfrac{1}{2}(a_1^\dagger a_2^\dagger + a_1 a_2)\,, \\ K_y &= -\frac{i}{2}(a_1^\dagger a_2^\dagger - a_1 a_2)\,, \\ K_z &= \tfrac{1}{2}(a_1^\dagger a_1 + a_2 a_2^\dagger)\,. \end{aligned} \tag{6.1}$$

The commutation relations for these operators,

$$\begin{aligned} [K_x,K_y] &= -iK_z\,, \\ [K_y,K_z] &= iK_x\,, \\ [K_z,K_x] &= iK_y\,, \end{aligned} \tag{6.2}$$

can be recognized as those belonging to the group[6,8] SU(1,1). It is also useful to introduce the raising and lowering operators

$$\begin{aligned} K_+ &= K_x + iK_y = a_1^\dagger a_2^\dagger\,, \\ K_- &= K_x - iK_y = a_1 a_2 \end{aligned} \tag{6.3}$$

which satisfy the commutation relations[9]

$$\begin{aligned} [K_-,K_+] &= 2K_z\,, \\ [K_z,K_\pm] &= \pm K_\pm\,. \end{aligned} \tag{6.4}$$

The Casimir invariant K^2 is

$$K^2 = K_z^2 - K_x^2 - K_y^2 \tag{6.5}$$

which upon the substitution of Eq. (6.1) becomes

$$K^2 = J_z(J_z+1)\,, \tag{6.6}$$

where J_z is given in Eq. (2.2). In fact, the operator J_z commutes with all the K_i.

There has been a considerable amount of theoretical work, beginning with Yuen and Shapiro,[10] on four-wave mixers as possible sources of squeezed states. The reader is directed to Reid and Walls[11] and references therein for work that has been done on four-wave mixers. For the purposes of this paper, a four-wave mixer will be regarded as a device with two input ports $a_{1\text{in}},a_{2\text{in}}$ and two output ports $a_{1\text{out}},a_{2\text{out}}$ which performs the mode transformation of the form[12,13]

$$\begin{bmatrix} a_{1\text{out}} \\ a_{2\text{out}}^\dagger \end{bmatrix} = \begin{bmatrix} S_{11} & S_{12} \\ S_{21} & S_{22} \end{bmatrix} \begin{bmatrix} a_{1\text{in}} \\ a_{2\text{in}}^\dagger \end{bmatrix}. \tag{6.7}$$

Both backward degenerate four-wave mixing in which two counter propagating pump beams pass through the nonlinear medium, and forward four-wave mixing, in which the pump beam propagates in only one direction through the nonlinear medium, perform mode transformations[14] of the form (6.7). Since the incoming and outgoing creation and annihilation operators must satisfy (2.1), the following restrictions are placed on the S_{ij}:

$$\begin{aligned} &|S_{11}|^2 - |S_{12}|^2 = 1\,, \\ &|S_{22}|^2 - |S_{21}|^2 = 1\,, \\ &S_{11}S_{21}^* = S_{12}S_{22}^*\,. \end{aligned} \tag{6.8}$$

From these relationships one can show

$$\begin{aligned} &|S_{11}|^2 = |S_{22}|^2\,, \\ &|S_{12}|^2 = |S_{21}|^2\,. \end{aligned} \tag{6.9}$$

The phases of the S_{ij} are controlled by the pump phase.

How the operators (6.1) transform under the scattering matrix (6.7)

$$S = \begin{bmatrix} S_{11} & S_{12} \\ S_{21} & S_{22} \end{bmatrix} \tag{6.10}$$

will now be determined for some particular examples. A possible realization of S is

$$S = \begin{bmatrix} \cosh(\tfrac{1}{2}\beta) & e^{-i\delta}\sinh(\tfrac{1}{2}\beta) \\ e^{i\delta}\sinh(\tfrac{1}{2}\beta) & \cosh(\tfrac{1}{2}\beta) \end{bmatrix}, \tag{6.11}$$

where δ is controlled by the phase of the pump light relative to some master clock and β is related to the reflectivity R of the four-wave mixer (when it is used as a phase-conjugating mirror) via $\sinh^2(\tfrac{1}{2}\beta)=R$.

When the pump phase is set such that $\delta=\pi/2$ Eq. (6.11) becomes

$$S = \begin{bmatrix} \cosh(\tfrac{1}{2}\beta) & -i\sinh(\tfrac{1}{2}\beta) \\ i\sinh(\tfrac{1}{2}\beta) & \cosh(\tfrac{1}{2}\beta) \end{bmatrix}. \tag{6.12}$$

Under this transformation, the vector $\mathbf{K}=(K_x,K_y,K_z)$ transforms as

$$\begin{bmatrix} K_x \\ K_y \\ K_z \end{bmatrix}_{\text{out}} = \begin{bmatrix} 1 & 0 & 0 \\ 0 & \cosh\beta & \sinh\beta \\ 0 & \sinh\beta & \cosh\beta \end{bmatrix} \begin{bmatrix} K_x \\ K_y \\ K_z \end{bmatrix}_{\text{in}} \tag{6.13}$$

which represents a Lorentz boost along the y axis, where z transforms as time. This transformation can be expressed in the form

$$\begin{bmatrix} K_x \\ K_y \\ K_z \end{bmatrix}_{\text{out}} = e^{i\beta K_x} \begin{bmatrix} K_x \\ K_y \\ K_z \end{bmatrix} e^{-i\beta K_x} . \tag{6.14}$$

Equivalently, in the Schrödinger picture where $\mathbf{K}$ remains fixed the state vector transforms as

$$|\,\text{out}\,\rangle = e^{i\beta K_x}\,|\,\text{in}\,\rangle . \tag{6.15}$$

When the pump phase is set at $\delta=0$, the scattering matrix (6.11) becomes

$$S = \begin{bmatrix} \cosh(\frac{1}{2}\beta) & \sinh(\frac{1}{2}\beta) \\ \sinh(\frac{1}{2}\beta) & \cosh(\frac{1}{2}\beta) \end{bmatrix} . \tag{6.16}$$

Under this transformation $\mathbf{K}$ transforms as

$$\begin{bmatrix} K_x \\ K_y \\ K_z \end{bmatrix}_{\text{out}} = \begin{bmatrix} \cosh\beta & 0 & \sinh\beta \\ 0 & 1 & 0 \\ \sinh\beta & 0 & \cosh\beta \end{bmatrix} \begin{bmatrix} K_x \\ K_y \\ K_z \end{bmatrix}_{\text{in}} . \tag{6.17}$$

This transformation has the form of a Lorentz boost along the x axis and can be expressed in the form

$$\begin{bmatrix} K_x \\ K_y \\ K_z \end{bmatrix}_{\text{out}} = e^{-i\beta K_y} \begin{bmatrix} K_x \\ K_y \\ K_z \end{bmatrix} e^{i\beta K_y} . \tag{6.18}$$

In the Schrödinger picture the state vector is transformed as

$$|\,\text{out}\,\rangle = e^{i\beta K_y}\,|\,\text{in}\,\rangle . \tag{6.19}$$

The operators performing the transformations of Eqs. (6.15) and (6.19) are two-mode squeeze operators.[6,12,13]

At this point it will be useful to determine how $\mathbf{K}$ transforms when the two input light beams sustain phase shifts. Letting $a_{1\text{in}}$ undergo a phase shift of ϕ_1 and $a_{2\text{in}}$ undergo a phase shift of ϕ_2, then

$$S = \begin{bmatrix} e^{i\phi_1} & 0 \\ 0 & e^{-i\phi_2} \end{bmatrix} . \tag{6.20}$$

Under this transformation, $\mathbf{K}$ transforms as

$$\begin{bmatrix} K_x \\ K_y \\ K_z \end{bmatrix}_{\text{in}} = \begin{bmatrix} \cos(\phi_1+\phi_2) & \sin(\phi_1+\phi_2) & 0 \\ -\sin(\phi_1+\phi_2) & \cos(\phi_1+\phi_2) & 0 \\ 0 & 0 & 1 \end{bmatrix} \begin{bmatrix} K_x \\ K_y \\ K_z \end{bmatrix}_{\text{out}} \tag{6.21}$$

which can be recognized as a rotation about the z axis by an angle $\phi = -(\phi_1+\phi_2)$. This transformation may be expressed as

$$\begin{bmatrix} K_x \\ K_y \\ K_z \end{bmatrix}_{\text{out}} = e^{-i(\phi_1+\phi_2)K_z} \begin{bmatrix} K_x \\ K_y \\ K_z \end{bmatrix} e^{i(\phi_1+\phi_2)K_z} . \tag{6.22}$$

In the Schrödinger picture the state vector is transformed as

$$|\,\text{out}\,\rangle = e^{i(\phi_1+\phi_2)K_z}\,|\,\text{in}\,\rangle . \tag{6.23}$$

The transformation (6.11) can be factorized into the form

$$S(\delta)S(\beta)S(-\delta) = \begin{bmatrix} \cosh(\frac{1}{2}\beta) & e^{-i\delta}\sinh(\frac{1}{2}\beta) \\ e^{i\delta}\sinh(\frac{1}{2}\beta) & \cosh(\frac{1}{2}\beta) \end{bmatrix} , \tag{6.24}$$

where

$$S(\delta) = \begin{bmatrix} e^{-i\delta} & 0 \\ 0 & 1 \end{bmatrix} , \tag{6.25}$$

$$S(\beta) = \begin{bmatrix} \cosh(\frac{1}{2}\beta) & \sinh(\frac{1}{2}\beta) \\ \sinh(\frac{1}{2}\beta) & \cosh(\frac{1}{2}\beta) \end{bmatrix} , \tag{6.26}$$

$$S(-\delta) = \begin{bmatrix} e^{i\delta} & 0 \\ 0 & 1 \end{bmatrix} . \tag{6.27}$$

From (6.20) the transformation $S(-\delta)$ can be recognized as a rotation about the z axis by an angle $-\delta$. $S(\beta)$ represents a Lorentz boost along the x axis, and $S(\delta)$ represents a rotation about the z axis by the angle δ. The product of transformations Eq. (6.24) thus represent a Lorentz transformation along a direction making an angle δ with respect to the x axis. Hence, in the Schrödinger picture, after the incoming light $|\,\text{in}\,\rangle$ has passed through a four-wave mixer, it will be in the state

$$|\,\text{out}\,\rangle = e^{-i\delta K_z} e^{i\beta K_y} e^{i\delta K_z}\,|\,\text{in}\,\rangle . \tag{6.28}$$

It has now been demonstrated that a four-wave mixer performs Lorentz transformations on the vector $\mathbf{K}$, the direction of the Lorentz boost being determined by the pump phase which is at the experimenter's control. Since J_z commutes with $\mathbf{K}$, it remains unchanged under the transformations performed by the four-wave mixer. From Eq. (2.2) one sees that this invariant is equal to half the difference in the number of photons entering the input port of the four-wave mixer. This invariant has been noted by Graham[15] and Reid and Walls.[16]

Let us now consider the case when no light enters the input ports of the four-wave mixer. The state delivered to the output is then given by Eq. (6.28) where $|\,\text{in}\,\rangle$ is the vacuum state $|\,0\,\rangle$.

The probability amplitude that n_1 photons will appear in the output beam $a_{1\text{out}}$ and n_2 photons in the beam $a_{2\text{out}}$ is

$$\langle\, n_1,n_2\,|\,\text{out}\,\rangle = \langle\, n_1,n_2\,|\,e^{-i\delta K_z} e^{i\beta K_y} e^{i\delta K_z}\,|\,0\,\rangle , \tag{6.29}$$

where the state $|\,n_1,n_2\,\rangle$ is

$$|n_1,n_2\rangle=\frac{(a_{1\,\text{out}}^\dagger)^{n_1}(a_{2\,\text{out}}^\dagger)^{n_2}}{\sqrt{n_1!n_2!}}|0\rangle\ . \tag{6.30}$$

From (6.1) one sees that K_z can be put in the form

$$K_z=\tfrac{1}{2}(N_1+N_2+1)\ , \tag{6.31}$$

where $N_1=a_{1\,\text{out}}^\dagger a_{1\,\text{out}}$ and $N_2=a_{2\,\text{out}}^\dagger a_{2\,\text{out}}$ are the number operators for output beams 1 and 2, respectively. With this equation it is readily apparent that

$$e^{i\delta K_z}|0\rangle=e^{i\delta/2}|0\rangle \tag{6.32}$$

and

$$e^{i\delta K_z}|n_1,n_2\rangle=e^{i(\delta/2)(n_1+n_2+1)}\ . \tag{6.33}$$

So Eq. (6.29) simplifies to

$$\langle n_1,n_2|\text{out}\rangle=e^{-i(\delta/2)(n_1+n_2)}\langle n_1,n_2|e^{i\beta K_y}|0\rangle\ . \tag{6.34}$$

In order to simplify things further we make use of the identity[9]

$$\exp(\tau K_+-\tau^*K_-)=\exp\left[\left[\frac{\tau}{|\tau|}\tanh|\tau|\right]K_+\right]\times\exp[-2(\ln\cosh|\tau|K_z]\times\exp\left[-\left[\frac{\tau^*}{|\tau|}\tanh|\tau|\right]K_-\right]\ . \tag{6.35}$$

Hence, noting (6.1) and (6.3), $e^{i\beta K_y}$ can be put into the form

$$e^{i\beta K_y}=\exp[i\tanh(\tfrac{1}{2}\beta)K_+]\exp[-2\ln\cosh(\tfrac{1}{2}\beta)K_z]\times\exp[i\tanh(\tfrac{1}{2}\beta)K_-]\ . \tag{6.36}$$

From (6.3) $K_-=a_1a_2$, hence

$$\exp[i\tanh(\tfrac{1}{2}\beta)K_-]|0\rangle=|0\rangle\ . \tag{6.37}$$

Making use of Eq. (6.31) one has

$$\exp[-2\ln\cosh(\tfrac{1}{2}\beta)K_z]|0\rangle=\text{sech}(\tfrac{1}{2}\beta)|0\rangle\ . \tag{6.38}$$

Finally, using (6.3)

$$\exp[i\tanh(\tfrac{1}{2}\beta)K_+]|0\rangle=\sum_{n=0}^{\infty}[i\tanh(\tfrac{1}{2}\beta)]^n|n,n\rangle\ , \tag{6.39}$$

where $|n,n\rangle$ is defined by (6.30).

Collecting the results, Eq. (6.34), (6.37), (6.38), and (6.39), one has

$$\langle n_1,n_2|\text{out}\rangle=\delta_{n_1,n_2}e^{-i(\delta/2)(n_1+n_2)}\times\text{sech}(\tfrac{1}{2}\beta)[i\tanh(\tfrac{1}{2}\beta)]^{n_1}\ . \tag{6.40}$$

The probability $P(n_1,n_2)$ that the beam $a_{1\,\text{out}}$ will contain n_1 photons and the beam $a_{2\,\text{out}}$ n_2 photons is thus

$$P(n_1,n_2)=\delta_{n_1,n_2}\text{sech}^2(\tfrac{1}{2}\beta)[\tanh^2(\tfrac{1}{2}\beta)]^{n_1}\ . \tag{6.41}$$

From this equation one sees that $P(n_1,n_2)$ is zero if $n_1\neq n_2$. For the vacuum state one has

$$J_z|0\rangle=0\ . \tag{6.42}$$

Since J_z is an invariant for the four-wave mixing process, when there are n_1 photons in beam 1 there must be n_1 photons in the second beam as well, that is, the photons are emitted in correlated pairs. These photons are in fact more highly correlated than allowed classically.[15,16]

From (6.41) the mean $\langle n\rangle$ and the mean-square $\langle n^2\rangle$ number of photons emitted by the four-wave mixer can be computed:

$$\langle n\rangle=\sum_{n_1n_2}^{\infty}(n_1+n_2)P(n_1,n_2)=2\sinh^2(\tfrac{1}{2}\beta)\ , \tag{6.43}$$

$$\langle n^2\rangle=\sum_{n_1n_2}^{\infty}(n_1+n_2)^2P(n_1,n_2)=4[\sinh^2(\tfrac{1}{2}\beta)\cosh^2(\tfrac{1}{2}\beta)+\sinh^4(\tfrac{1}{2}\beta)]\ . \tag{6.44}$$

The mean-square fluctuation in n is rather large:

$$(\Delta n)^2=4\sinh^2(\tfrac{1}{2}\beta)\cosh^2(\tfrac{1}{2}\beta) \tag{6.45}$$

so

$$\frac{(\Delta n)^2}{\langle n\rangle^2}=\coth^2(\tfrac{1}{2}\beta)\geq 1\ . \tag{6.46}$$

Hence the light emitted is super-Poissonian but approaches Poissonian as β becomes large.

We are now in a position to argue that a four-wave mixer can generate states similar to the one described in the last section, Eq. (5.2). The case when no photons enter the input port of the four-wave mixer has already been discussed. The eigenvalue m of J_z is 0. Hence if a total of n photons were measured coming out of the four-wave mixer one can infer that the light leaving the four-wave mixer was in the state $|j,0\rangle$ where, from Eq. (2.5), $j=n/2$.

If instead the state

$$|2,0\rangle=\frac{a_{1\,\text{in}}^{\dagger 2}}{\sqrt{2}}|0\rangle \tag{6.47}$$

is fed into the input ports, that is, if two photons are forced to enter the input port $a_{1\,\text{in}}$ of the four-wave mixer, then the eigenvalue m for J_z is 1. If one thus measures a total of n photons leaving the four-wave mixer one can infer that the light leaving the four-wave mixer is in the state $|j,1\rangle$ where again $j=n/2$.

If now the light entering the input port is a superposition of a vacuum state and the $|2,0\rangle$ state of Eq. (6.47), say,

$$|\text{in}\rangle=\frac{1}{\sqrt{2}}|0\rangle+\frac{1}{\sqrt{2}}|2,0\rangle\ , \tag{6.48}$$

then upon measuring n photons leaving the four-wave mixer all one can infer is that the state leaving the four-mixer is in a superposition of the states $|j,0\rangle$ and $|j,1\rangle$. Hence a four-wave mixer can generate states of the form (5.2) provided the state (6.48) is fed into its input.

From a practical point of view the state Eq. (6.48) may be hard to generate. It would be more practical to attenuate laser light until on average there is one photon per unit coherence time propagating along the beam. The input state generated by feeding this coherent state into the input port $a_{1\,\mathrm{in}}$ of the four-wave mixer will have a strong overlap with states $|n_1,n_2\rangle$ only when n_1 is small and n_2 is 0. Consequently, $\langle J_z\rangle$ and ΔJ_z for the light fed into the four-wave mixer will still be of order unity. Such light when passed through the four-wave mixer should still produce "fan-shaped" states that will allow an interferometer to reach a phase sensitivity $\Delta\phi$ of order $1/n$. Interferometry, using the light coming from a four-wave mixer fed with coherent states, will be discussed in detail in the next section.

VII. ACHIEVING A PHASE SENSITIVITY OF $1/N$

The device to be considered in this section is depicted in Fig. 5. It consists of a Mach-Zehnder interferometer whose input ports are fed by the output beams b_1 and b_2 of a four-wave mixer. The four-wave mixing medium is pumped with a laser. Part of the laser light is split off of the main beam phase shifted, attenuated and then fed into the input port a_1 of the four-wave mixer. The other input port a_2 is terminated with a cold blackbody absorber so that no light enters the four-wave mixer from this port. The pump light's phase is controlled with the phase shifter δ.

Letting a_1 and a_2 denote the creation operators for the light beams fed into the input ports of the four-wave mixer, the state vector for this light $|\alpha\rangle$ is defined by

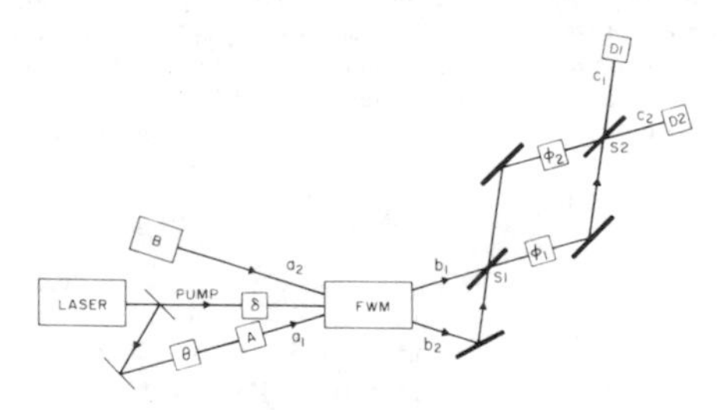

FIG. 5. A method by which the state depicted in Fig. 4 can be generated and fed into an interferometer. The state is generated via a degenerate four-wave mixer (FWM) pumped via a laser. A small fraction of the pump light is split off of the pump beam, phase shifted by θ, attenuated by A and then fed into one of the FWM inputs, a_1. The input port a_2 is terminated with a cold blackbody absorber B. The two output ports b_1 and b_2 of the four-wave mixer are fed into the input ports of the Mach-Zehnder interferometer. δ is a phase shifter for the pump light before it enters FWM1.

$$a_1|\alpha\rangle=\alpha|\alpha\rangle \tag{7.1}$$

and

$$a_2|\alpha\rangle=0\;, \tag{7.2}$$

that is, $|\alpha\rangle$ is a coherent state for a_1 and a vacuum state for a_2.

Since J_z is an invariant under the transformation (6.7) performed by the four-wave mixer, its expectation values can be computed at the input port,

$$J_z=\tfrac{1}{2}(a_1^\dagger a_1-a_2^\dagger a_2)\;. \tag{7.3}$$

One can readily show

$$\langle J_z\rangle=\tfrac{1}{2}|\alpha|^2 \tag{7.4}$$

and

$$(\Delta J_z)^2=\tfrac{1}{4}|\alpha|^2\;. \tag{7.5}$$

For $|\alpha|^2$ of order unity such a state will lie near the x-y plane and have a spread along the z axis of order unity.

In order to compute the expectation values of J_x and J_y at the output of the four-wave mixer it is necessary to express J_x and J_y in terms of the operators a_1 and a_2. Again we choose the scattering matrix for the four-wave mixer to be given by Eq. (6.11),

$$\begin{bmatrix} b_1 \\ b_2^\dagger \end{bmatrix}=\begin{bmatrix} \cos(\tfrac{1}{2}\beta) & e^{-i\delta}\sinh(\tfrac{1}{2}\beta) \\ e^{i\delta}\sinh(\tfrac{1}{2}\beta) & \cosh(\tfrac{1}{2}\beta) \end{bmatrix}\begin{bmatrix} a_1 \\ a_2^\dagger \end{bmatrix}\;. \tag{7.6}$$

Using this transformation, the output J_x and J_y expressed in terms of the input creation and annihilation operators are

$$\begin{aligned} J_x=&\tfrac{1}{4}\cos\delta\sinh\beta(a_1^\dagger a_1^\dagger+a_1a_1+a_2^\dagger a_2^\dagger+a_2a_2) \\ &-\frac{i}{4}\sin\delta\sinh\beta(a_1^\dagger a_1^\dagger-a_1a_1+a_2^\dagger a_2^\dagger-a_2a_2) \\ &+\tfrac{1}{2}\cosh\beta(a_1^\dagger a_2+a_2^\dagger a_1)\;, \end{aligned} \tag{7.7}$$

$$\begin{aligned} J_y=&-\frac{i}{4}\cos\delta\sinh\beta(a_1^\dagger a_1^\dagger-a_1a_1-a_2^\dagger a_2^\dagger+a_2a_2) \\ &-\tfrac{1}{4}\sin\delta\sinh\beta(a_1^\dagger a_1^\dagger+a_1a_1-a_2^\dagger a_2^\dagger-a_2a_2) \\ &-\frac{i}{2}\cosh\beta(a_1^\dagger a_2-a_2^\dagger a_1)\;. \end{aligned} \tag{7.8}$$

The expectation values of J_x and J_y for the state $|\alpha\rangle$ can now be readily evaluated. Writing

$$\alpha=|\alpha|e^{-i\theta} \tag{7.9}$$

one has

$$\begin{aligned} \langle\alpha|J_x|\alpha\rangle&=\tfrac{1}{2}|\alpha|^2\sinh\beta\cos(2\theta-\delta)\;, \\ \langle\alpha|J_y|\alpha\rangle&=\tfrac{1}{2}|\alpha|^2\sinh\beta\sin(2\theta-\delta)\;. \end{aligned} \tag{7.10}$$

The mean-square fluctuation in J_x and J_y is independent of ϕ and δ:

$$\begin{aligned} (\Delta J_x)^2&=(\Delta J_y)^2 \\ &=\frac{|\alpha|^2}{2}(\sinh^2\beta+\tfrac{1}{2})+\tfrac{1}{8}\sinh^2\beta\;. \end{aligned} \tag{7.11}$$

One also has

$$\langle \alpha | J_x J_z + J_z J_x | \alpha \rangle = \frac{|\alpha|^2}{2}(|\alpha|^2+1)\sinh\beta\cos(2\theta-\delta) . \quad (7.12)$$

Since from Eq. (3.8) $J_{z\,\text{out}}$, measured at the output of the interferometer, is

$$J_{z\,\text{out}} = -(\sin\phi)J_x + (\cos\phi)J_z , \quad (7.13)$$

one can now evaluate

$$\langle J_{z\,\text{out}} \rangle = -\tfrac{1}{2}|\alpha|^2 \sin\phi \sinh\beta\cos(2\theta-\delta) + \tfrac{1}{2}|\alpha|^2\cos\phi \quad (7.14)$$

and

$$(\Delta J_{z\,\text{out}})^2 = \frac{|\alpha|^2}{2}\sin^2\phi\sinh^2\beta - \frac{|\alpha|^2}{2}\sin\phi\cos\phi\sinh\beta\cos(2\theta-\delta) + \frac{1}{8}\sin^2\phi\sinh^2\beta + \frac{|\alpha|^2}{4} . \quad (7.15)$$

The mean-square phase uncertainty $\Delta\phi$ can be evaluated via

$$(\Delta\phi)^2 = \frac{(\Delta J_{z\,\text{out}})^2}{\left|\dfrac{\partial\langle J_{z\,\text{out}}\rangle}{\partial\phi}\right|^2} . \quad (7.16)$$

This quantity is minimized with respect to the pump phase δ when $2\theta-\delta=0$, that is, $\cos(2\phi-\delta)=1$. This quantity is also near its minimum value when $\phi=0$. When ϕ is set to zero Eq. (7.16) reduces to

$$(\Delta\phi)^2 = \frac{1}{|\alpha|^2\sinh^2\beta} . \quad (7.17)$$

The parameter β will now be expressed in terms of the mean number of photons $\langle N \rangle$ leaving the four-wave mixer and $|\alpha|$. This will allow us to optimize $\Delta\phi$ holding $\langle N \rangle$ fixed. From Eq. (6.31) the number operator N can be expressed in terms of K_z,

$$N = 2K_z - 1 . \quad (7.18)$$

Upon leaving the interferometer the state $|\alpha\rangle$ has been transformed, according to Eq. (6.28), into the state $|\psi\rangle$,

$$|\psi\rangle = e^{-i\delta K_z} e^{i\beta K_y} e^{i\delta K_z} |\alpha\rangle . \quad (7.19)$$

From Eq. (6.18) and Eq. (6.22) one has

$$e^{-i\delta K_z} e^{-i\beta K_y} e^{i\delta K_z} K_z e^{-i\delta K_z} e^{i\beta K_y} e^{i\delta K_z} = (\cos\phi)(\sinh\beta)K_x + (\sin\phi)(\sinh\beta)K_y + (\cosh\beta)K_z . \quad (7.20)$$

The mean number of photons leaving the four-wave mixer can then easily be shown to be

$$\overline{N} = (|\alpha|^2+1)\cosh\beta - 1 . \quad (7.21)$$

This equation is easily solved for $\sinh^2\beta$. Equation (7.17) then becomes

$$(\Delta\phi)^2 = \frac{(|\alpha|^2+1)^2}{|\alpha|^2[(\overline{N}+1)^2-(|\alpha|^2-1)^2]} . \quad (7.22)$$

This expression can be optimized for $|\alpha|$ holding N fixed. One finds that for large N $(\Delta\phi)^2$ is smallest when $|\alpha|^2$ is close to 1, hence

$$\Delta\phi \simeq \frac{2}{\overline{N}} . \quad (7.23)$$

This equation implies that the interferometer of Fig. 5 can achieve a phase sensitivity approaching $1/N$, and that photons are most economically used by the interferometer when the coherent state $|\alpha\rangle$ fed into the input port a_1 has its intensity reduced to $|\alpha|^2 \simeq 1$, that is, on average only one photon per unit coherence time of the four-wave mixer enters the input port a_1.

The sensitivity $\Delta\phi$ of Eq. (7.23) with the particular numerical coefficient 2 cannot be achieved in practice. The reason for this is now indicated. Equation (7.23) holds only for $\phi \lesssim 1/\overline{N}$. Hence a practical interferometer employed to measure ϕ_1 must incorporate a feedback loop which adjusts ϕ_2 to follow ϕ_1 such that $\phi=\phi_2-\phi_1=0$. However, for angles ϕ outside the narrow range $|\phi| \lesssim 1/\overline{N}$ the uncertainty in ϕ, defined by Eq. (7.16), becomes

$$\Delta\phi = \left[\frac{4|\alpha|^2+1}{2|\alpha|^4}\right]^{1/2} \tan\phi \quad (7.24)$$

which for small ϕ and $|\alpha|^2=1$ becomes

$$\Delta\phi \simeq (\tfrac{5}{2})^{1/2}\phi . \quad (7.25)$$

That is, the uncertainty in ϕ is greater than ϕ itself.

A feedback loop presented with a measurement of ϕ whose uncertainty is greater than ϕ will generally not be able to adjust ϕ_2 properly to drive ϕ to zero. The uncertainty $\Delta\phi$ can be decreased by increasing $|\alpha|$ or by averaging several[4] successive measurements of ϕ. In the next section the problem of locking an interferometer to $\phi=0$ will be discussed in more detail and the maximum uncertainty in $\Delta\phi$ that a feedback loop can tolerate will be determined.

VIII. TRACKING THE PHASE

In the last section an interferometer capable of achieving a phase sensitivity $\Delta\phi$ approaching $1/N$, where N is the number of photons passing through the interferometer per unit measurement time, was discussed. This sensitivity is only achieved, however, for a small range of angles within $1/N$ of $\phi=0$. The interferometer can be made to follow the phase ϕ_1 with a sensitivity approaching $1/N$ provided a feedback loop is employed to adjust a controllable phase shifter ϕ_2 such that $\phi=\phi_2-\phi_1$ is maintained at zero. The operation of the interferometer of the last section with a feedback loop will now be discussed in more detail.

The parameters θ, δ, $|\alpha|$, and β are under the control of the experimenter. It will be assumed that $2\theta-\delta=0$ and that $|\alpha|$ and β are known. Then the quantum statistics of the light entering the interferometer is well characterized and in particular one knows the numbers $\langle J_z \rangle$ and $\langle J_x \rangle$, which according to Eq. (7.4) and (7.10) are

$$\langle J_z \rangle = \tfrac{1}{2} |\alpha|^2 , \quad \langle J_x \rangle = \tfrac{1}{2} |\alpha|^2 \sinh\beta . \tag{8.1}$$

The differenced photocurrent is measured at the output of the interferometer, that is, the photodetectors measure $2J_z$. A sequence of measurements will generate a string of numbers, each of which is an eigenvalue of $2J_z$. One is free to process these numbers and in particular one can subtract $\langle J_z \rangle$ from them and divide them by $-2\langle J_x \rangle$. Then the sequence of numbers $\{d_1, d_2, \ldots\}$ are eigenstates of the operator D

$$D = \frac{(\sin\phi) J_x}{\langle J_x \rangle} - \frac{(\cos\phi) J_z - \langle J_z \rangle}{\langle J_x \rangle} . \tag{8.2}$$

For simplicity it will be assumed that ϕ is small so that the approximations $\sin\phi \simeq \phi$ and $\cos\phi \simeq 1$ can be made. Then one can write

$$D = \phi A - B , \tag{8.3}$$

where

$$A = \frac{J_x}{\langle J_x \rangle} , \quad B = \frac{J_z - \langle J_z \rangle}{\langle J_z \rangle} . \tag{8.4}$$

Since

$$\langle A \rangle = 1 , \quad \langle B \rangle = 0 \tag{8.5}$$

it is immediately evident that

$$\langle D \rangle = \phi , \tag{8.6}$$

that is, the sequence of numbers $\{d_1, d_2, \ldots\}$ are estimates of ϕ.

The phase shifter ϕ_2 of Fig. 5 will be taken to be controllable. A feedback algorithm that will track ϕ_1 maintaining $\phi = \phi_2 - \phi_1$ at zero will now be described. Let $\phi_2(i)$ be the setting of ϕ_2 during the ith measurement. The measurement provides the estimate of $\phi(i) = \phi_2(i) - \phi_1(i)$, d_i, which is an eigenvalue of

$$D_i = \phi(i) A_i - B_i . \tag{8.7}$$

The feedback loop then adjusts ϕ_2 to the new setting

$$\phi_2(i+1) = \phi_2(i) - \lambda d_i ,$$

or in operator form

$$\phi_2(i+1) = \phi_2(i) - \lambda A_i [\phi_2(i) - \phi_1(i)] + \lambda B_i , \tag{8.8}$$

where λ is a feedback parameter.

It is now assumed that the successive measurements are performed on a time scale equal to the characteristic coherence time of the four-wave mixer so that the ith operators A_i and B_i are independent of the j operators A_j and B_j. Then Eq. (8.8) can readily be iteratively substituted into itself. For convenience $\phi_1(i)$ will be held fixed to $\phi_1(0)$. It will be determined how rapidly ϕ_2 approached $\phi_1(0)$ given the feedback algorithm (8.8). Equation (8.8) iteratively substituted into itself yields

$$\phi(n) = \sum_{k=0}^{n-1} (1 - \lambda A_k) \phi(0) + \lambda \sum_{k=0}^{n-1} B_K \prod_{m=k+1}^{n-1} (1 - \lambda A_m) , \tag{8.9}$$

where the product is defined in the usual way, except that

$$\prod_{m=n}^{n-1} F(m) \equiv 1 . \tag{8.10}$$

The mean value $\langle \phi(n) \rangle$ is, using Eq. (8.5),

$$\langle \phi(n) \rangle = (1-\lambda)^n \langle \phi(0) \rangle . \tag{8.11}$$

It is apparent that the mean value of $\phi(n)$ will converge to zero only if $|1-\lambda| < 1$. Hence the feedback parameter is restricted to the range

$$0 < \lambda < 2 . \tag{8.12}$$

The mean-square value of $\phi(n)$ can be obtained by squaring (8.9) and then taking the expectation value. One obtains

$$\begin{aligned} \langle [\phi(n)]^2 \rangle = {} & \langle (1-\lambda A)^2 \rangle^n [\phi(0)]^2 \\ & + \lambda^2 \langle B^2 \rangle \sum_{k=0}^{n-1} \langle (1-\lambda A)^2 \rangle^k \\ & - \lambda^2 \langle AB + BA \rangle \phi(0) \\ & \times \sum_{k=0}^{n-1} \langle (1-\lambda A)^2 \rangle^{n-1-k} \langle 1-\lambda A \rangle^k . \end{aligned} \tag{8.13}$$

The sums can be evaluated to yield

$$\begin{aligned} \langle [\phi(n)]^2 \rangle = {} & \langle (1-\lambda A)^2 \rangle^n [\phi(0)]^2 \\ & + \lambda^2 \langle B^2 \rangle \frac{1 - \langle (1-\lambda A^2) \rangle^n}{1 - \langle (1-\lambda A)^2 \rangle} \\ & - \lambda^2 \langle AB + BA \rangle \frac{\langle (1-\lambda A)^2 \rangle^n - \langle 1-\lambda A \rangle^n}{\langle (1-\lambda A)^2 \rangle - \langle 1-\lambda A \rangle} . \end{aligned} \tag{8.14}$$

The expectation values $\langle (1-\lambda A)^2 \rangle$ and $\langle 1-\lambda A \rangle$ can be written, keeping in mind Eq. (8.5), as

$$\langle 1-\lambda A \rangle = 1-\lambda , \quad \langle (1-\lambda A)^2 \rangle = (1-\lambda)^2 + \lambda^2 (\Delta A)^2 . \tag{8.15}$$

Substituting these expressions into (8.14) one finally has

$$\begin{aligned} \langle [\phi(n)]^2 \rangle & \\ = {} & [(1-\lambda)^2 + \lambda^2 (\Delta A)^2]^n [\phi(0)]^2 \\ & + \lambda^2 \langle B^2 \rangle \frac{1 - [(1-\lambda)^2 + \lambda^2 (\Delta A)^2]^n}{1 - [(1-\lambda)^2 + \lambda^2 (\Delta A)^2]} \\ & - \lambda^2 \langle AB + BA \rangle \frac{[(1-\lambda)^2 + \lambda^2 (\Delta A)^2]^n - (1-\lambda)^n}{[(1-\lambda)^2 + \lambda^2 (\Delta A)^2] - (1-\lambda)} . \end{aligned} \tag{8.16}$$

From this equation one sees that in order for $\langle[\phi(n)]^2\rangle$ to converge one must, in addition to (8.12), have $[(1-\lambda)^2+\lambda^2(\Delta A)^2]^n<1$. This expression yields the restriction

$$\lambda\langle A^2\rangle<2\,. \tag{8.17}$$

As a particular example, consider the case when $\lambda=1$, then

$$\langle[\phi(n)]^2\rangle=(\Delta A)^{2n}[\phi(0)]^2+\langle B^2\rangle\frac{1+(\Delta A)^{2n}}{1-(\Delta A)^2}-\langle AB+BA\rangle(\Delta A)^{2n-1}\,. \tag{8.18}$$

The mean-square value of ϕ converges to

$$\langle\phi^2\rangle_{\min}=\frac{\langle B^2\rangle}{1-(\Delta A)^2}\,. \tag{8.19}$$

As ϕ is driven to zero, the characteristic number of measurements $\bar{n}$ that must be made to reduce ϕ^2 to $1/e$ of its original value is

$$\bar{n}=-\frac{1}{\ln(\Delta A)^2}\,. \tag{8.20}$$

We now substitute the results of Sec. VII into these expressions. One has

$$(\Delta A)^2=\frac{2|\alpha|^2(\sinh^2\beta+\frac{1}{2})+\frac{1}{2}\sinh^2\beta}{|\alpha|^4\sinh^2\beta}\,, \tag{8.21}$$

$$\langle B^2\rangle=\frac{1}{|\alpha|^2\sinh^2\beta}\,, \tag{8.22}$$

and

$$\langle AB+BA\rangle=\frac{2}{|\alpha|^2\sinh\beta}\,. \tag{8.23}$$

In the large-β limit Eq. (8.21) reduces to

$$(\Delta A)^2=\frac{2|\alpha|^2+\frac{1}{2}}{|\alpha|^4}\,. \tag{8.24}$$

In order that the sensitivity $(\Delta\phi)^2$ not be degraded too much from its minimum value $(\Delta\phi)^2=\langle B^2\rangle$ [see Eq. (8.19)], let us choose $(\Delta A)^2=\frac{1}{2}$. Then the characteristic number of measurements necessary to reduce ϕ^2 to $1/e$ of its original value is $\bar{n}=1.44$. From (8.24) one sees that $|\alpha|^2$ has the value $|\alpha|^2=2+\sqrt{5}$. Using (7.21), Eq. (8.19) becomes, for large $\bar{N}$,

$$\langle\phi^2\rangle_{\min}\simeq\frac{2(|\alpha|^2+1)^2}{|\alpha|^2\bar{N}^2}\simeq\frac{12.9}{\bar{N}^2}\,. \tag{8.25}$$

Suppose ϕ_1 is stationary so that ϕ_2 has settled down and ϕ fluctuates with the mean-square value of (8.25), i.e.,

$$\Delta\phi\simeq\frac{3.6}{\bar{N}}\,. \tag{8.26}$$

Then if a small disturbance should come along to displace ϕ_1 by an amount $\Delta\phi$ from its quiescent value it will take approximately 1.44 measurements in order for ϕ_2 to adjust itself to the new ϕ_1. Hence, on the average, the total number of photons $\bar{N}_T$ used to detect this displacement is of the order $\bar{N}_T=1.44\bar{N}$ and $\Delta\phi$ in terms of the total number of photons used is

$$\Delta\phi\simeq\frac{5}{\bar{N}_T}\,. \tag{8.27}$$

Hence, by increasing the number of photons fed into the four-wave mixer from 1 to $|\alpha|^2\simeq2+\sqrt{5}\simeq4.24$, the interferometer can be operated stably in a feedback mode and ϕ_2 can detect changes in ϕ_1 as small as that given by (8.27).

Consider now the case where, instead of choosing $|\alpha|^2$ large enough so that the feedback loop would be stable with the feedback parameter λ set to unity, one chose $|\alpha|^2=1$ as was done in Sec. VII in order to optimize the sensitivity (7.22) with $\bar{N}$ fixed. In this case

$$(\Delta A)^2=\frac{5}{2}\,,\quad \langle B^2\rangle=\frac{1}{\sinh^2\beta}\,,\quad \langle AB+BA\rangle=\frac{2}{\sinh\beta}\,. \tag{8.28}$$

Since $(\Delta A)^2>1$ it is apparent from (8.18) that the feedback loop cannot be operated stably with the feedback parameter set to unity. In fact, from Eq. (8.17) it follows that λ must be less than $\frac{4}{7}$ if the feedback loop is to be operated stably. In order to make the e-folding time for ϕ^2 as short as possible we choose the value of λ which minimizes

$$[(1-\lambda)^2+\lambda^2(\Delta A)^2]$$

of Eq. (8.16), that is,

$$\lambda=\frac{1}{1+(\Delta A)^2}=\frac{2}{7}\,. \tag{8.29}$$

If $\phi_1(0)$ is held fixed $\langle\phi^2\rangle$ settles to a steady-state value

$$\langle\phi^2\rangle_{\min}=\frac{\lambda^2\langle B^2\rangle}{1-[(1-\lambda)^2+\lambda^2(\Delta A)^2]}\,. \tag{8.30}$$

The characteristic number of measurements which must be made to reduce ϕ^2 to $1/e$ of its initial value is

$$\bar{n}=-\frac{1}{\ln[(1-\lambda)^2+\lambda^2(\Delta A)^2]}\,. \tag{8.31}$$

For $\lambda=\frac{2}{7}$ and $(\Delta A)^2=\frac{5}{2}$ one has, upon using Eq. (7.21),

$$\langle\phi^2\rangle_{\min}=\frac{1.14}{\bar{N}^2} \tag{8.32}$$

and

$$\bar{n}=3.0\,. \tag{8.33}$$

Again consider the case where ϕ_1 has remained constant for a long time so that the rms fluctuations in ϕ have settled down to the value determined by (8.32), $\Delta\phi\simeq1.07/\bar{N}$. Suppose now that ϕ_1 is displaced instantaneously to a new value a distance $\Delta\phi$ from its old value. It takes characteristically three measurements for ϕ_2 to adjust itself to the new ϕ_1. Hence, on the average, the total number of photons N_T used to detect the change in ϕ_1 is $N_T=3\overline{N}$. The sensitivity of the interferometer, operated with $|\alpha|^2=1$ and $\lambda=\frac{2}{7}$, is thus expressed in terms of the total number of photons needed to observe the change as

$$\Delta\phi=\frac{3.2}{N_T}\ , \tag{8.34}$$

a number that is somewhat better than Eq. (8.27).

In this section it has been shown that by using suitable feedback loops the interferometer of Sec. VII can track changes in ϕ_1 in a stable manner and can achieve a phase sensitivity of order $1/N$. Hence the two problems encountered in Sec. VII, namely the fact that the interferometer achieves its optimum sensitivity only for a small range of phases, $\phi \lesssim 1/N$, and that the fluctuations in $J_{z\,\text{out}}$, the interferometer's output, are greater than $\langle J_{z\,\text{out}}\rangle$ for $|\alpha|^2$ set at its optimum value, can be overcome be operating the interferometer with $|\alpha|^2$ slightly degraded or by choosing the response of the feedback loop to be such that it averages enough successive measurements of ϕ that a useful error signal can be generated.

In the literature[2-4] a number of schemes for achieving interferometer sensitivities of $1/N$ have been described. All of these schemes employ standard interferometers into which light from degenerate-parametric amplifiers or four-wave mixers is injected. In the next section we will describe a novel set of interferometers which dispense with beam splitters and use the SU(1,1) boosts to convert phase shifts into light amplitude changes rather than the SU(2) rotations employed by a conventional interferometer.

IX. AN SU(1,1) MACH-ZEHNDER INTERFEROMETER

In Sec. III it was shown how the operation of a Mach-Zehnder interferometer could be viewed in terms of rotations of the vector **J** under the rotation group SU(2).

In this picture relative phase shifts between two light beams correspond to rotations about the z axis while photodetectors are sensitive to rotations in a plane containing the z axis. The function of the beam splitters was to convert a rotation about the z axis into one perpendicular to the z axis.

In this section an interferometer whose operation can be viewed in terms of transformations of the vector **K** Eq. (6.1), under the Lorentz group SU(1,1) is considered. From Eq (6.21) one sees that the common mode phase shift of two light beams corresponds to a rotation of **K** about the z axis. But from Eq. (6.1) one sees that photodetectors placed in the two light beams will be sensitive only to transformations perpendicular to the z axis.

Again, a device is required which will convert rotations about the z axis into transformations perpendicular to this axis. The four-wave mixers described in Sec. VI can carry out such transformations. These transformations consist of Lorentz boosts.

As a specific example, consider the device of Fig. 6. The phase shifter ψ in the pump beam is adjusted such that four-wave mixer FWM2 performs the inverse of the transformation performed by four-wave mixer FWM1. In

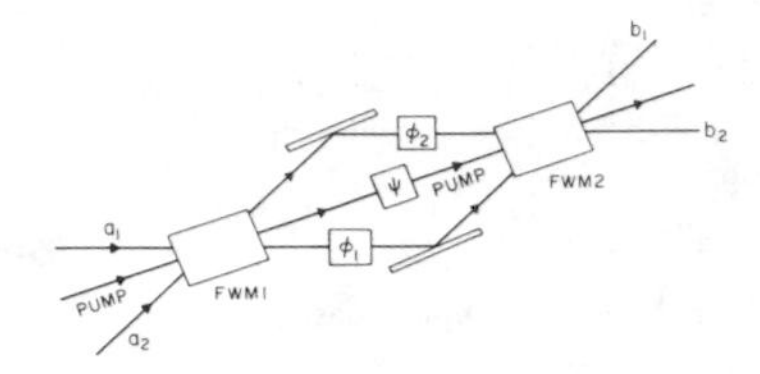

FIG. 6. An SU(1,1) interferometer. The beam splitters of a conventional interferometer have been replaced by the four-wave mixers FWM1 and FWM2. The light pumping FWM2 is phase shifted from the light pumping FWM1 by the angle ψ.

particular let FWM1 have the scattering matrix

$$S(-\beta)=\begin{bmatrix}\cosh(\tfrac{1}{2}\beta) & +i\sinh(\tfrac{1}{2}\beta)\\ -i\sinh(\tfrac{1}{2}\beta) & \cosh(\tfrac{1}{2}\beta)\end{bmatrix}. \tag{9.1}$$

As can be seen from Eq. (6.13), **K** transforms as a Lorentz boost $L(-\beta,y)$ along the $-y$ axis under this scattering matrix:

$$L(-\beta,y)=\begin{bmatrix}1 & 0 & 0\\ 0 & \cosh\beta & -\sinh\beta\\ 0 & -\sinh\beta & \cosh\beta\end{bmatrix}. \tag{9.2}$$

The scattering matrix for FWM2 is

$$S(\beta)=\begin{bmatrix}\cosh(\tfrac{1}{2}\beta) & -i\sinh(\tfrac{1}{2}\beta)\\ i\sinh(\tfrac{1}{2}\beta) & \cosh(\tfrac{1}{2}\beta)\end{bmatrix}. \tag{9.3}$$

This scattering matrix transforms **K** as a Lorentz boost $L(\beta,y)$ along the $+y$ axis. The transformation performed by the phase shifters ϕ_1 and ϕ_2 is, from Eq. (6.20),

$$S(\phi)=\begin{bmatrix}e^{i\phi_1} & 0\\ 0 & e^{-i\phi_2}\end{bmatrix}. \tag{9.4}$$

Under this scattering matrix **K** transforms as a rotation $R(\phi,z)$ about the z axis by an angle $\phi=-(\phi_1+\phi_2)$,

$$R(\phi,z)=\begin{bmatrix}\cos\phi & -\sin\phi & 0\\ \sin\phi & \cos\phi & 0\\ 0 & 0 & 1\end{bmatrix}. \tag{9.5}$$

The overall scattering matrix for the device of Fig. 6 is

$$S=S(\beta)S(\phi)S(-\beta)\ , \tag{9.6}$$

and the overall transformation performed on **K** is

$$\mathbf{K}_{\text{out}}=L(\beta,y)R(\phi,z)L(-\beta,y)\mathbf{K}_{\text{in}}\ . \tag{9.7}$$

It will be useful to reexpress this transformation as follows:

$$L(\beta,y)R(\phi,z)L(-\beta,y)=R(\theta,z)L(\gamma x)R(\theta,z)\ , \tag{9.8}$$

where $L(\gamma,x)$ denotes a Lorentz boost along the x axis,

$$L(\gamma,x)=\begin{bmatrix}\cosh\gamma & 0 & \sinh\gamma\\ 0 & 1 & 0\\ \sinh\gamma & 0 & \cosh\gamma\end{bmatrix} . \quad (9.9)$$

Equation (9.8) holds when θ and γ are chosen such that

$$\sin\theta=\frac{(1-\cos\phi)\cosh\beta}{[\sin^2\phi+(1-\cos\phi)^2\cosh^2\beta]^{1/2}} , \quad (9.10)$$

$$\cos\theta=\frac{\sin\phi}{[\sin^2\phi+(1-\cos\phi)^2\cosh^2\beta]^{1/2}} , \quad (9.11)$$

$$\cosh\gamma=(1-\cos\phi)\cosh^2\beta+\cos\phi , \quad (9.12)$$

$$\sinh\gamma=\sinh\beta[\sin^2\phi+(1-\cos\phi)^2\cosh^2\beta]^{1/2} . \quad (9.13)$$

Hence the transformation performed by the device of Fig. 6 on **K** can be regarded as a rotation θ about the z axis, followed by a Lorentz boost along the x axis, followed by a second rotation θ about the z axis.

Let us now consider the operation of this device when no light enters the input ports, that is when $|\text{in}\rangle$ is the vacuum state $|0\rangle$. The vacuum state is both an eigenstate of K_z and J_z:

$$K_z|0\rangle=\tfrac{1}{2}|0\rangle , \quad (9.14)$$

$$J_z|0\rangle=0 . \quad (9.15)$$

Consequently, from Eq. (6.6) the invariant K^2 is zero, that is, we can think of **K** for the vacuum state as lying on the light cone. In the spirit of Fig. 2, the vacuum state is depicted in Fig. 7(a) as a cone whose base intersects the z axis at $\frac{1}{2}$. The Lorentz boost $L(-\beta,y)$ is equivalent, in the Schrödinger picture, to a boost of the state vector in the opposite direction. The Lorentz boost performed by the first four-wave mixer is depicted in Fig. 7(b). The mean value of K_z in terms of the mean number of photons $\langle N\rangle$ emitted by the four-wave mixer is, from Eq. (6.31),

$$\langle K_z\rangle=\tfrac{1}{2}(\langle N\rangle+1) .$$

The phase shifts ϕ_1 and ϕ_2 encountered by the two light beams leaving the four-wave mixer then rotate the state vector about the z axis by an angle $-\phi=\phi_1+\phi_2$. This is depicted in Fig. 7(c). A second Lorentz boost with the same rapidity, but in the opposite direction, is then performed. If $\phi=0$ the final state will be a vacuum state and no photons will be detected by the photodetectors in the output beams. If ϕ is nonzero the state of the light delivered to the photodetectors will be a Lorentz-boosted vacuum, the rapidity parameter being determined by Eq. (9.12) or (9.13).

In Fig. 7(b) the projected ellipse lying in the x-y plane has a width of $\frac{1}{2}$ and the distance from the origin to its center is $\langle K_z\rangle=\frac{1}{2}(\langle N\rangle+1)$. Hence Fig. 7(c) suggests that the minimum detectable phase $\phi_{\min}$ is of the order

$$\phi_{\min}=\frac{1}{\langle N\rangle+1} , \quad (9.16)$$

that is, this detector can achieve a phase sensitivity approaching $1/N$.

That this is the case will now be demonstrated with an

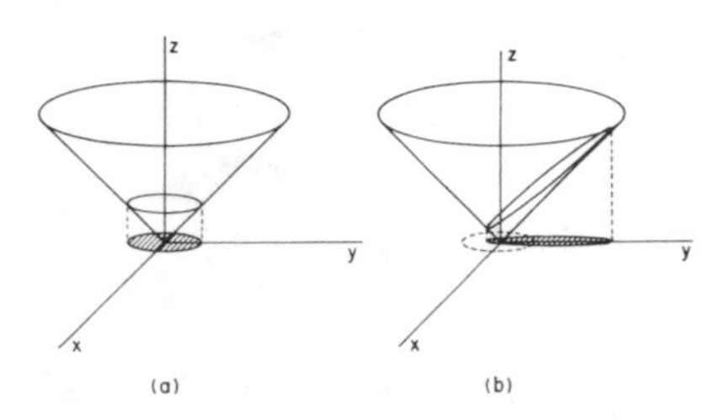

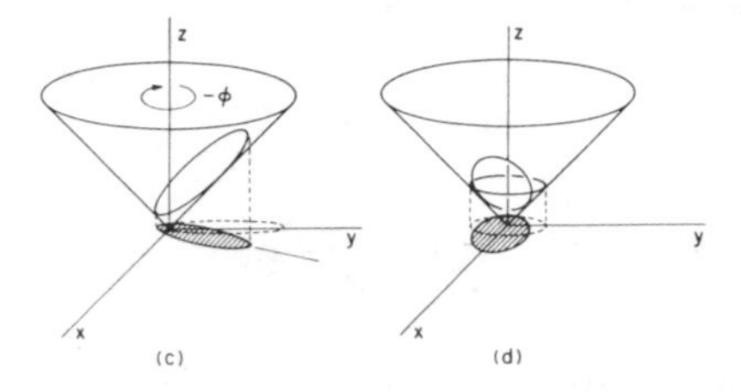

FIG. 7. A geometrical view of the performance of an SU(1,1) interferometer. (a) The input state consisting of the vacuum state is depicted in the (K_x,K_y,K_z) space. where K_x and K_y are regarded as space coordinates and K_z as a time coordinate, as a circle on the light cone. (b) The first four-wave mixer performs a Lorentz boost along the positive y axis. (c) The phase shifts accumulated by the light beams propagating in the interferometer result in a rotation in the xy plane. (d) The second four-wave mixer performs a Lorentz transformation along the negative y axis. The total number of photons leaving the interferometer is a linear function of K_z.

explicit calculation. From Eq. (9.6) and Eqs. (6.15) and (6.23) the incoming state vector $|\text{in}\rangle$ is transformed as

$$|\text{out}\rangle=e^{-i\beta K_x}e^{-i\phi K_z}e^{i\beta K_x}|\text{in}\rangle , \quad (9.17)$$

but from (9.8) this is equivalent to the transformation

$$|\text{out}\rangle=e^{i\theta K_z}e^{i\gamma K_y}e^{-i\theta K_z}|\text{in}\rangle . \quad (9.18)$$

The operator N_d for the total number of photons detected by the photodetectors placed in the output beam is from Eq. (6.31)

$$N_d=2K_z-1 . \quad (9.19)$$

Hence in order to evaluate $\langle N_d\rangle$ and ΔN_d one needs to evaluate $\langle\text{out}|K_z|\text{out}\rangle$ and $\langle\text{out}|K_z^2|\text{out}\rangle$. From Eq. (9.18) one has

$$\langle\text{out}|K_z|\text{out}\rangle=\langle\text{in}|e^{i\theta K_z}e^{-i\gamma K_y}K_z e^{i\gamma K_y}e^{-i\theta K_z}|\text{in}\rangle . \quad (9.20)$$

Since $|\text{in}\rangle$ is the vacuum state, one has

$$e^{-i\theta K_z}|\text{in}\rangle=e^{-i\theta/2}|\text{in}\rangle . \quad (9.21)$$

Equation (9.20) thus simplifies to

$$\langle \text{out} \mid K_z \mid \text{out} \rangle = \langle 0 \mid e^{-\gamma K_y} K_z e^{i\gamma K_y} \mid 0 \rangle . \tag{9.22}$$

From Eqs. (6.18) and (6.17),

$$e^{-i\gamma K_y} K_z e^{i\gamma K_y} = (\sinh\gamma) K_x + (\cosh\gamma) K_z , \tag{9.23}$$

but

$$\langle 0 \mid K_z \mid 0 \rangle = 0 \tag{9.24}$$

so

$$\langle \text{out} \mid K_z \mid \text{out} \rangle = \tfrac{1}{2}\cosh\beta . \tag{9.25}$$

From Eq. (9.19), the mean number of photons $\langle N_d \rangle$ detected by the photodetectors is

$$\langle N_d \rangle = \cosh\gamma - 1 . \tag{9.26}$$

In a similar manner one can show

$$(\Delta N_d)^2 = \sinh^2\gamma . \tag{9.27}$$

The dependence of γ on ϕ and β is given by Eqs. (9.12) and (9.13). Hence Eqs. (9.26) and (9.27) can be rewritten as

$$\begin{aligned} \langle N_d \rangle &= (1-\cos\phi)\sinh^2\beta , \\ (\Delta N_d)^2 &= [\sin^2\phi + (1-\cos\phi)^2\cosh^2\beta]\sinh^2\beta . \end{aligned} \tag{9.28}$$

The mean-square fluctuation in ϕ due to the photon statistics is thus

$$\begin{aligned} (\Delta\phi)^2 &= \frac{(\Delta N_d)^2}{\left[\dfrac{\partial \langle N_d \rangle}{\partial \phi} \right]^2} \\ &= \frac{\sin^2\phi + (1-\cos\phi)^2\cosh^2\beta}{\sin^2\phi \sinh^2\beta} . \end{aligned} \tag{9.29}$$

The quantity $(\Delta\phi)^2$ is minimized when $\phi = 0$, then

$$(\Delta\phi)^2_{\min} = \frac{1}{\sinh^2\beta} . \tag{9.30}$$

Expressed in terms of the mean number of photons $\langle N \rangle$ emitted by the first four-wave mixer, $\sin^2\beta = \langle N \rangle(\langle N \rangle - 2)$, so

$$(\Delta\phi)^2_{\min} = \frac{1}{\langle N \rangle(\langle N \rangle - 2)} . \tag{9.31}$$

Hence it has now been shown that the SU(1,1) interferometer depicted in Fig. 6 can achieve a phase sensitivity approaching $1/N$. This sensitivity is achieved when no light is fed into the input ports. A comparison with Fig. 5 shows that SU(1,1) interferometers achieving a sensitivity of $1/N$ require fewer optical elements than an SU(2) interferometer achieving the same sensitivity. Further, at $\phi = 0$ no light is delivered to the photodetectors, that is, the pairs of pump photons converted into pairs of four-wave-mixer output photons are absorbed by the second four-wave mixer and converted back into pump photons. Hence an SU(1,1) interferometer can be very economical with photons. It will absorb pump power only when ϕ is nonzero. It is also worth pointing out that the beams 1 and 2 need not be at the same frequency, as long as they are placed symmetrically about the pump frequency, ω_0, that is, with beam 1 at the frequency $\omega_0 + \Delta\omega$ and beam 2 at the frequency $\omega_0 - \Delta\omega$; the scattering matrix[6,12,13] for the four-wave mixer will still have the form (6.11).

By using techniques similar to those used in deriving Eq. (6.41) one can show that the probability $P(N)$ of detecting a total of N photons leaving the output ports of the interferometer is

$$P(N) = \begin{cases} 0 & \text{if } N \text{ is odd} \\ \dfrac{2}{(1-\cos\phi)\sinh^2\beta + 2} \left[\dfrac{(1-\cos\phi)\sinh^2\beta}{(1-\cos\phi)\sinh^2\beta + 2} \right]^{N/2} & \text{if } N \text{ is even} . \end{cases} \tag{9.32}$$

Let N_i denote the number of photons counted during the ith measurement in a sequence of measurements. One is free to take the square root of each of these numbers. Hence it is meaningful to talk about the average and rms value of $\sqrt{N}$. The motivation for investigating the statistics of $\sqrt{N}$ stems from the fact that $\langle N \rangle$ is an even function of ϕ. Now

$$\langle \sqrt{N} \rangle = \sum_{N \text{ even}} \sqrt{N} P(N) \tag{9.33}$$

can be put into the form

$$\begin{aligned} \langle \sqrt{N} \rangle = &\frac{2^{3/2}}{(1-\cos\phi)\sinh^2\beta + 2} \\ &\times \sum_{k=0}^{\infty} \sqrt{k} \left[\frac{(1-\cos\phi)\sinh^2\beta}{(1-\cos\phi)\sinh^2\beta + 2} \right]^k . \end{aligned} \tag{9.34}$$

The sum has the form

$$\sum_{k=0}^{\infty} \sqrt{k} \left[\frac{1}{a} \right]^k$$

which can be approximated by the integral

$$\int_0^\infty x^{1/2} \left[\frac{1}{a} \right]^x dx = \frac{\sqrt{\pi}}{2(\ln a)^{3/2}} . \tag{9.35}$$

Hence

$$\begin{aligned} \langle \sqrt{N} \rangle \simeq &\frac{\sqrt{2\pi}}{(1-\cos\phi)\sinh^2\beta + 2} \\ &\times \left[\ln \left[\frac{(1-\cos\phi)\sinh^2\beta + 2}{(1-\cos\phi)\sinh^2\beta} \right] \right]^{-3/2} . \end{aligned} \tag{9.36}$$

Since this approximation holds reasonably well for

$$(1-\cos\beta)\sinh^2\beta \gtrsim 4 \tag{9.37}$$

the logarithm can be approximated via $\ln(1+x)\simeq x$ and one has

$$\langle \sqrt{N}\,\rangle \simeq \frac{\sqrt{\pi}}{2}(1-\cos\phi)^{1/2}\sinh\beta\ . \tag{9.38}$$

Approximating $(1-\cos\phi)$ as $\phi^2/2$ one finally has

$$\langle \sqrt{N}\,\rangle = \frac{1}{2}\left[\frac{\pi}{2}\right]^{1/2} |\,\phi\,|\sinh\beta\ . \tag{9.39}$$

Hence the norm of the phase $\tilde{\phi}$ inferred from a measurement of N is

$$|\,\tilde{\phi}\,| = \frac{2\left[\dfrac{2}{\pi}\right]^{1/2}\sqrt{N}}{\sinh\beta}\ . \tag{9.40}$$

From (9.39) one has

$$\langle\,|\,\tilde{\phi}\,|\,\rangle = |\,\phi\,|\ , \tag{9.41}$$

that is, the mean value of $|\,\tilde{\phi}\,|$ inferred from a measurement of N is equal to the norm of the actual phase setting ϕ of the interferometer. Now

$$\langle\,|\,\tilde{\phi}\,|^2\rangle = \frac{8}{\pi}(1-\cos\phi) \simeq \frac{4}{\pi}\phi^2\ , \tag{9.42}$$

so

$$(\Delta\,|\,\tilde{\phi}\,|^2) = \frac{4-\pi}{\pi}\phi^2 \simeq 0.273\phi^2\ . \tag{9.43}$$

Hence for $N \gtrsim 4$ it has been shown that the uncertainty in the inferred norm of the phase $\tilde{\phi}$ is to a good approximation

$$\Delta\,|\,\tilde{\phi}\,| = 0.522\,|\,\tilde{\phi}\,|\ . \tag{9.44}$$

It is also instructive to ask what the probability $P(\,|\,\tilde{\phi}\,|: (1-\alpha)\,|\,\phi\,| < |\,\tilde{\phi}\,| < (1+\alpha)\,|\,\phi\,|\,)$ is that a measured $|\,\tilde{\phi}\,|$ will lie in the range

$$(1-\alpha)\,|\,\phi\,| < |\,\tilde{\phi}\,| < (1+\alpha)\,|\,\phi\,|\ .$$

From (9.40) this is equivalent to determining the probability $P(N\colon\ N_1 < N < N_2)$ that N lies in the range $N_1 < N < N_2$ where

$$N_1 = \frac{\pi}{8}(1-\alpha)^2\phi^2\sinh^2\beta\ ,$$
$$N_2 = \frac{\pi}{8}(1+\alpha)^2\phi^2\sinh^2\beta\ . \tag{9.45}$$

One can show rigorously

$$P(N\colon N_1 < N < N_2) = \left[\left[\frac{(1-\cos\phi)\sinh^2\beta}{(1-\cos\phi)\sinh^2\beta+2}\right]^{N_1/2} - \left[\frac{(1-\cos\phi)\sinh^2\beta}{(1-\cos\phi)\sinh^2\beta+2}\right]^{N_2/2}\right]\ . \tag{9.46}$$

So from Eq. (9.45)

$$P(\,|\,\tilde{\phi}\,|: (1-\alpha)\,|\,\phi\,| < |\,\tilde{\phi}\,| < (1+\alpha)\,|\,\phi\,|\,) = \left[\left[\frac{(1-\cos\phi)\sinh^2\beta}{(1-\cos\phi)\sinh^2\beta+2}\right]^{(\pi/16)(1-\alpha)^2\phi^2\sinh^2\beta} - \left[\frac{(1-\cos\phi)\sinh^2\beta}{(1-\cos\phi)\sinh^2\beta+2}\right]^{(\pi/16)(1+\alpha)^2\phi^2\sinh^2\beta}\right]\ . \tag{9.47}$$

Approximating $1-\cos\phi$ by $\phi^2/2$, this expression can be put into the form

$$P(\,|\,\tilde{\phi}\,|: (1-\alpha)\,|\,\phi\,| < |\,\tilde{\phi}\,| < (1+\alpha)\,|\,\phi\,|\,) = \left[\left[\frac{x}{x+4}\right]^{(\pi/16)(1-\alpha)^2x} - \left[\frac{x}{x+4}\right]^{(\pi/16)(1+\alpha)^2x}\right]\ , \tag{9.48}$$

where $x = \phi^2\sinh^2\beta$.

Now

$$\lim_{x\to\infty}\left[\frac{x}{x+4}\right]^x = e^{-4}\ . \tag{9.49}$$

This limiting value is not a bad approximation for $[x/(x+4)]^x$ even for x as low as 10, the level at which on average five photons are counted in the interferometer output beams. Hence

$$P(\,|\,\tilde{\phi}\,|: (1-\alpha)\,|\,\phi\,| < |\,\tilde{\phi}\,| < (1+\alpha)\,|\,\phi\,|\,) \simeq e^{-(\pi/2)(1-\alpha)^2} - e^{-(\pi/2)(1+\alpha)^2}\ . \tag{9.50}$$

As an example, let $\alpha = \frac{1}{2}$, then

$$P(\,|\,\tilde{\phi}\,|:\ |\,\phi\,|/2 < |\,\tilde{\phi}\,| < 3\,|\,\phi\,|/2) \simeq e^{-\pi/8} - e^{-9\pi/8} = 0.646\ . \tag{9.51}$$

Hence from a single measurement of $|\,\tilde{\phi}\,|$ one has 65% confidence that $|\,\phi\,|/2 < |\,\tilde{\phi}\,| < 3\,|\,\phi\,|/2$.

The interferometer described here suffers from drawbacks similar to those of the SU(2) interferometer of Sec. VII. Maximum sensitivity occurs at $\phi=0$ and the sensitivity rapidly degrades as ϕ is adjusted away from zero. It was shown in Sec. VIII that such drawbacks can be overcome with feedback. However, implementing a feedback algorithm for the SU(1,1) interferometer described here is complicated by the fact that $\langle N\rangle$ is an even function of ϕ

and hence the sign of the error signal cannot be determined from the number of photons counted by the photodetector during a single measurement.

The sign of the error signal can be generated by changing (dithering) ϕ_2 between successive measurements and constructing the derivative signal $(N_{i+1}-N_i)/\Delta\phi_2$.

Alternatively, one could implement the feedback algorithm which will now be described. Make repeated measurements of $|\tilde{\phi}|$ until $|\phi|$ is determined to some predetermined precision: $\Delta|\phi|=\alpha|\phi|$ where α is a constant. Then move ϕ_2 according to

$$\phi_2(\text{new})=\phi_2(\text{old})+|\tilde{\phi}| . \tag{9.52}$$

Make repeated measurements of $|\tilde{\phi}|$ at this new setting so that a new $|\phi|$ can be inferred with the precision $\Delta|\phi|=\alpha|\phi|$. If $|\phi|$ inferred for the new setting of ϕ_2 is less than $|\phi|$ inferred for the old setting one assumes that one has moved in the right direction. If, on the other hand, the inferred value of $|\phi|$ for the new setting of ϕ_2 is greater than the inferred $|\phi|$ for the old setting one assumes that one has moved ϕ_2 in the wrong direction and ϕ_2 is then readjusted so that

$$\phi_2(\text{new})=\phi_2(\text{old})-|\tilde{\phi}| . \tag{9.53}$$

The process is then repeated.

If $|\phi|$ is determined to sufficient precision this algorithm will move one closer to $\phi=0$ most of the time. On the occasions when this algorithm moves one in the wrong direction it generally does not move ϕ_2 very far in the wrong direction and the lost ground is regained during the next few iterations of the feedback procedure. Further, since a single measurement already determines $|\phi|$ with a precision $\Delta\phi\approx 0.5|\phi|$ at a 65% confidence level, one does not have to make very many repeated measurements of $|\phi|$ in order for the feedback algorithm to work.

X. SINGLE-MODE SU(1,1) INTERFEROMETERS

In this section interferometers based on devices having the scattering matrix

$$a_{\text{out}}=\cosh(\tfrac{1}{2}\beta)a_{\text{in}}+e^{-i\delta}\sinh(\tfrac{1}{2}\beta)a_{\text{in}}^{\dagger} \tag{10.1}$$

will be described. Such a single-mode device can be regarded as a limiting case of the four-wave mixers of Sec. VI in which the two input and the two output beams are made collinear and are sufficiently close in frequency that they cannot be resolved during the coherence time of the device. Both four-wave mixers and parametric amplifiers configured properly[17,18] are capable of performing the mode transformation Eq. (10.1). Connected with Eq. (10.1) it is convenient to introduce the operators

$$\begin{aligned} L_x&=\tfrac{1}{4}(a^{\dagger}a^{\dagger}+aa) , \\ L_y&=-\frac{i}{4}(a^{\dagger}a^{\dagger}-aa) , \\ L_z&=\tfrac{1}{4}(a^{\dagger}a+aa^{\dagger}) . \end{aligned} \tag{10.2}$$

These operators behave as generators[6,9] of the group SU(1,1) satisfying commutation relations identical with Eq. (6.2):

$$\begin{aligned} [L_x,L_y]&=-iL_z , \\ [L_yL_z]&=iL_x , \\ [L_z,L_x]&=iL_y . \end{aligned} \tag{10.3}$$

Again, it is useful to introduce the raising and lowering operators

$$\begin{aligned} L_+&=L_x+iL_y=\tfrac{1}{2}a^{\dagger}a^{\dagger} , \\ L_-&=L_x-iL_y=\tfrac{1}{2}aa \end{aligned} \tag{10.4}$$

which satisfy the commutation relations

$$\begin{aligned} [L_-,L_+]&=2L_z , \\ [L_z,L_\pm]&=\pm L_\pm . \end{aligned} \tag{10.5}$$

The Casimir invariant

$$L^2=L_z^2-L_x^2-L_y^2 , \tag{10.6}$$

when expressed in terms of the operators a and $a^{\dagger}$, reduces to the number

$$L^2=-\tfrac{3}{16} . \tag{10.7}$$

It is useful to determine how $\mathbf{L}=(L_x,L_y,L_z)$ transforms under specific cases of Eq. (10.1). Under the mode transformation

$$a_{\text{out}}=\cosh(\tfrac{1}{2}\beta)a_{\text{in}}+\sinh(\tfrac{1}{2}\beta)a_{\text{in}}^{\dagger} , \tag{10.8}$$

$\mathbf{L}$ transforms as a boost along the x axis:

$$\begin{pmatrix} L_x \\ L_y \\ L_z \end{pmatrix}_{\text{out}} = \begin{pmatrix} \cosh\beta & 0 & \sinh\beta \\ 0 & 1 & 0 \\ \sinh\beta & 0 & \cosh\beta \end{pmatrix} \begin{pmatrix} L_x \\ L_y \\ L_z \end{pmatrix}_{\text{in}} . \tag{10.9}$$

Under the mode transformation

$$a_{\text{out}}=\cosh(\tfrac{1}{2}\beta)a_{\text{in}}-i\sinh(\tfrac{1}{2}\beta)a_{\text{in}}^{\dagger} \tag{10.10}$$

$\mathbf{L}$ transforms as a boost along the y axis:

$$\begin{pmatrix} L_x \\ L_y \\ L_z \end{pmatrix}_{\text{out}} = \begin{pmatrix} 1 & 0 & 0 \\ 0 & \cosh\beta & \sinh\beta \\ 0 & \sinh\beta & \cosh\beta \end{pmatrix} \begin{pmatrix} L_x \\ L_y \\ L_z \end{pmatrix} . \tag{10.11}$$

A phase shift

$$a_{\text{out}}=e^{-i\phi}a_{\text{in}} \tag{10.12}$$

transforms $\mathbf{K}$ as a rotation about the z axis by the amount 2ϕ:

$$\begin{pmatrix} L_x \\ L_y \\ L_z \end{pmatrix}_{\text{out}} = \begin{pmatrix} \cos2\phi & -\sin2\phi & 0 \\ \sin2\phi & \cos2\phi & 0 \\ 0 & 0 & 1 \end{pmatrix} \begin{pmatrix} L_y \\ L_y \\ L_z \end{pmatrix}_{\text{in}} . \tag{10.13}$$

It can be shown that $\mathbf{K}$ transforms under Eq. (10.1) as a rotation of $\mathbf{K}$ about the z axis by the angle $-\delta/2$ followed by a boost β along the x axis followed by a rotation about the z axis by the angle $\delta/2$. Equivalently, in the Schrödinger picture the state vector transforms according to

$$|\,\text{out}\rangle = e^{-i(\delta/2)L_x} e^{i\beta L_x} e^{i(\delta/2)L_z} |\,\text{in}\rangle\,, \tag{10.14}$$

where $e^{i\beta L_x}$ is a single-mode squeeze operator[19] and $e^{i(\delta/2)L_z}$ has been called a single-mode rotation operator.[6,12,13]

More generally one could consider a device which transforms a state vector according to

$$|\,\text{out}\rangle = e^{i\phi L_z} e^{i\beta L_x} e^{i\theta L_z} |\,\text{in}\rangle\,. \tag{10.15}$$

The probability distribution for the number n of photons in the output beam will now be determined for the case when the input consists of vacuum fluctuations. A more general case, when the input consists of coherent states, has been treated by Yuen.[20] A photodetector in the output beam measures $N = a^\dagger a$, which can, from Eq. (10.2), be written in the form

$$N = 2L_z - \tfrac{1}{2}\,. \tag{10.16}$$

The amplitude that n photons will be counted in the output beam is $\langle n\,|\,\text{out}\rangle$, hence the probability $P(n)$ that n photons will be counted is

$$P(n) = |\langle n\,|\,\text{out}\rangle|^2\,. \tag{10.17}$$

Now for an n-photon state $|\,n\,\rangle$ one has

$$L_z\,|\,n\,\rangle = \left[\frac{n}{2} + \frac{1}{4}\right] |\,n\,\rangle\,. \tag{10.18}$$

Hence one has

$$e^{i\theta L_z}\,|\,0\rangle = e^{i\theta/4}\,|\,0\rangle \tag{10.19}$$

and

$$e^{-i\phi L_z}\,|\,n\,\rangle = e^{-i\phi(n/2+1/4)}\,|\,n\,\rangle\,. \tag{10.20}$$

The probability distribution (10.17) thus reduces to

$$P(n) = |\langle n\,|\,e^{i\beta L_x}\,|\,0\rangle|^2\,. \tag{10.21}$$

Hence $P(n)$ is independent of the phase angles ϕ and θ, i.e., $P(n)$ depends only on the magnitude of the boost. Again, using (6.35) one has

$$e^{i\beta L_x} = \exp[i\tanh(\tfrac{1}{2}\beta)L_+]\exp[-2\ln\cosh(\tfrac{1}{2}\beta)L_z] \times \exp[i\tanh(\tfrac{1}{2}\beta)L_-]\,. \tag{10.22}$$

Hence by using the techniques used to arrive at Eq. (6.41) one can show

$$P(n) = \begin{cases} 0, & n \text{ odd} \\ \dfrac{1}{2^n}\begin{bmatrix} n \\ \dfrac{n}{2}\end{bmatrix} \dfrac{\tanh^2(\frac{1}{2}\beta)}{\cosh(\frac{1}{2}\beta)}, & n \text{ even} \end{cases} \tag{10.23}$$

where

$$\begin{bmatrix} n \\ m \end{bmatrix} = \frac{n!}{(n-m)!m!}\,. \tag{10.24}$$

It is useful to compare this probability distribution with distribution $P_T(n)$ for the probability that a total of n photons will be counted leaving the two-port four-mixer of Sec. VI. From Eq. (6.41)

$$P_T(n) = \begin{cases} 0, & n \text{ odd} \\ \operatorname{sech}^2(\frac{1}{2}\beta)[\tanh^2(\frac{1}{2}\beta)]^2, & n \text{ even}\,. \end{cases} \tag{10.25}$$

With the identity

$$\sum_{k=0}^{n} \begin{bmatrix} 2k \\ k \end{bmatrix} \begin{bmatrix} 2(n-k) \\ n-k \end{bmatrix} = 2^{2n} \tag{10.26}$$

one can show

$$P_T(n) = \sum_{\substack{n_1,n_2 \\ n_1+n_2=n}} P(n_1)P(n_2)\,. \tag{10.27}$$

This equation implies that the statistics of the total number of photons leaving the two-mode four-wave mixer is the same as the statistics of the total number of photons coming from two independent single-mode devices. This observation will allow a simplification of the discussion of feedback loops for the interferometer discussed in this section, since the results of Sec. IX can be made to apply by pairwise averaging successive measurements made with a single-mode device.

The interferometer to be considered in this section is depicted in Fig. 8 where for the sake of definiteness, degenerate-parametric amplifiers DPA1 and DPA2 are used to perform the Lorentz boost. DPA1 will be taken to perform the boost $e^{i\beta L_x}$ on the incoming state vector. The phase shifter is taken to perform a phase shift $e^{-i\phi L_z}$ on the light beam. The last degenerate-parametric amplifier is taken to perform the boost

$$e^{-i\delta L_z} e^{-i\beta L_x} e^{i\delta L_z}\,,$$

where δ is proportional to the phase of the pump light entering DPA2. Letting $|\,\text{in}\rangle$ denote the state vector for the incoming light, the state vector $|\,\text{out}\rangle$ for light leaving the interferometer is

$$|\,\text{out}\rangle = e^{-i\delta L_z} e^{-i\beta L_x} e^{i\delta L_z} e^{-i\phi L_z} e^{i\beta L_x}\,|\,\text{in}\rangle\,. \tag{10.28}$$

The behavior of this device when $|\,\text{in}\rangle$ is the vacuum state will now be considered. Figures analogous to Fig. 7 can be drawn to illustrate the behavior of the interferometer. However, in this case the Casimir invariant Eq. (10.6) has the numerical value $-\frac{3}{16}$. Hence **L** lies on a spacelike hyperboloid instead of the light cone of Fig. 7. The vacuum state is an eigenstate of L_z

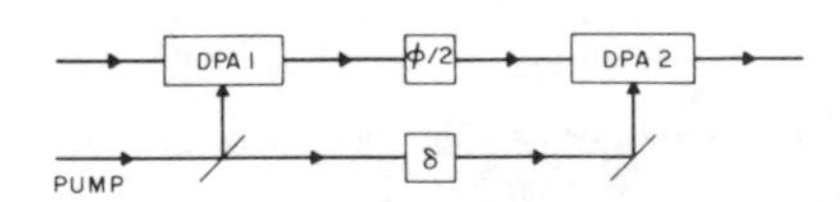

FIG. 8. A single-mode SU(1,1) interferometer. The device employs two degenerate-parametric amplifiers DPA1 and DPA2. The output of the device is sensitive to the difference between the phases ϕ and δ accumulated by the signal and pump beam, respectively.

$$L_z|0\rangle=\tfrac{1}{4}|0\rangle \tag{10.29}$$

and hence could be represented as a circle drawn around the hyperboloid at a height along the z axis of $\frac{1}{4}$.

We now determine the mean and variance in the number of photons counted at the output of the interferometer. Since the number operator N for the total number of photons counted at the output of the detector is linear in L_z, Eq. (10.16), one would like to determine $\langle \text{out}|L_z^k|\text{out}\rangle$. One can readily show from Eq. (10.28) that

$$\langle \text{out}|L_z^k|\text{out}\rangle=\langle \text{in}|e^{-i\beta L_x}e^{i(\phi-\delta)L_z}e^{i\beta L_x}L_z^k e^{-i\beta L_x}e^{-i(\phi-\delta)L_z}e^{i\beta L_x}|\text{in}\rangle . \tag{10.30}$$

Using the techniques of Sec. IX this can be further reduced to

$$\langle \text{out}|L_z^k|\text{out}\rangle=\langle \text{in}|e^{i\theta L_z}e^{-i\gamma L_y}e^{i\theta L_z}L_z^k e^{-i\theta L_z}e^{i\gamma L_y}e^{-i\theta L_z}|\text{in}\rangle=\langle \text{in}|e^{-i\gamma L_y}L_z^k e^{i\gamma L_y}|\text{in}\rangle , \tag{10.31}$$

where in analogy with Eqs. (9.12) and (9.13),

$$\cosh\gamma=[1-\cos(\phi-\delta)]\cosh^2\beta+\cos(\phi-\delta) , \tag{10.32}$$

$$\sinh\gamma=\sinh\beta\{\sin^2(\phi-\delta)+[1-\cos(\phi-\delta)]^2\cosh^2\beta\}^{1/2} .$$

It is straightforward then to show that

$$\langle N\rangle\equiv\langle \text{out}|N|\text{out}\rangle=\tfrac{1}{4}\cosh\gamma , \tag{10.33}$$

and

$$\langle N^2\rangle\equiv\langle \text{out}|N^2|\text{out}\rangle =\tfrac{1}{2}\sinh^2\gamma+\tfrac{1}{4}(\cosh\gamma-1)^2 \tag{10.34}$$

or, using Eq. (10.32)

$$\langle N\rangle=\tfrac{1}{2}[1-\cos(\phi-\delta)](\cosh^2\beta-1) , \tag{10.35}$$

$$(\Delta N)^2=\tfrac{1}{2}\sinh^2\beta\{\sin^2(\phi-\delta)+[1-\cos(\phi-\delta)]^2\cosh^2\beta\} .$$

The mean-square fluctuation in the readings for ϕ, given by

$$(\Delta\phi)^2=\frac{\langle\Delta N\rangle^2}{\left|\dfrac{\partial\langle N\rangle}{\partial\phi}\right|^2} , \tag{10.36}$$

is readily evaluated and has a minimum given by

$$(\Delta\phi_{\min})^2=\frac{1}{2\sinh^2\beta} . \tag{10.37}$$

The mean number of photons $\langle N_I\rangle$ in the light beam passing through the phase shifter ϕ is

$$\langle N_I\rangle=\langle 0|e^{-i\beta L_x}Ne^{i\beta L_x}|0\rangle =\tfrac{1}{2}(\cosh\beta-1) . \tag{10.38}$$

Solving this equation for $\sinh^2\beta$ one finally has

$$(\Delta\phi_{\min})^2=\frac{1}{8\langle N_I\rangle(\langle N_I\rangle-1)} . \tag{10.39}$$

Hence it has been shown that the device of Fig. 8 can indeed achieve a phase sensitivity approaching $1/n$. This minimum sensitivity is achieved when $\phi-\delta=0$. Hence by implementing a feedback loop one can track ϕ maintaining $\phi-\delta=0$ by using the error-correcting signal to adjust the phase shifter δ in the pump beam delivered to DPA2. As was mentioned earlier the statistics of the total number of photons counted in two successive measurements of ϕ are the same as for the total number of photons leaving the interferometer of Sec. IX. Hence the feedback algorithms discussed in Sec. IX will also work for the single-mode device discussed here.

XI. CONCLUSION

A geometric or Lie-group-theoretical approach to the analysis of interferometers was presented. Such an approach facilitates identifying the input states which optimize the interferometer's sensitivity. It was shown that ordinary interferometers are characterized by the group SU(2) which is equivalent to the group of rotations in three dimensions. With suitable input states such an interferometer can achieve a phase sensitivity $\Delta\phi$ approaching $1/N$ where N is the total number of photons passing through the phase-shifting element ϕ of the interferometer. Although this sensitivity can only be achieved for ϕ within $1/N$ of $\phi=0$, it was shown that by employing a feedback loop the interferometer could track phase as a function of time with a precision of $1/N$.

A class of interferometers in which four-wave mixers serve as active analogs of beam splitters was also presented. Such interferometers are characterized by the group SU(1,1) and have the virtue of being able to achieve a phase sensitivity approaching $1/N$ with only vacuum fluctuations entering the input port. SU(1,1) interferometers can consequently achieve the $1/N$ phase sensitivity much more readily and in fact have a simpler construction than SU(2) interferometers. The output of a four-wave mixer depends on the relative phase between the pump and the incoming signal. It is this phase sensitivity which the SU(1,1) interferometers employ. In fact, it was shown that degenerate-parametric amplifiers which are also phase-sensitive devices can be used to construct SU(1,1) interferometers.

ACKNOWLEDGMENT

We would like to thank R. E. Slusher for stimulating discussions on the work presented here.

[1]M. Born and E. Wolf, *Principles of Optics* (Pergamon, Oxford, 1975).
[2]C. M. Caves, Phys. Rev. D **23**, 1693 (1981).
[3]R. S. Bondurant and J. H. Shapiro, Phys. Rev. D **30**, 2548 (1984).
[4]R. S. Bondurant, Ph.D. thesis, Massachusetts Institute of Technology, 1983.
[5]Wei-Tou Ni (unpublished).
[6]B. L. Schumaker, Phys. Rep. (to be published).
[7]W. H. Steel, *Interferometry,* 2nd ed. (Cambridge University, Cambridge, 1983).
[8]K. Wódkiewicz and J. H. Eberly, J. Opt. Soc. Am. B **2**, 458 (1985).
[9]D. R. Truax, Phys. Rev. D **31**, 1988 (1985).
[10]H. P. Yuen and J. H. Shapiro, Opt. Lett. **4**, 334 (1979).
[11]M. D. Reid and D. F. Walls, Phys. Rev. A **31**, 1622 (1985).
[12]C. M. Caves and B. L. Schumaker, Phys. Rev. A **31**, 3068 (1985).
[13]B. L. Schumaker and C. M. Caves, Phys. Rev. A **31**, 3093 (1985).
[14]P. Kumar and J. H. Shapiro, Phys. Rev. A **30**, 1568 (1984).
[15]R. Graham, Phys. Rev. Lett. **52**, 117 (1984).
[16]M. D. Reid and D. F. Walls, Phys. Rev. Lett. **53**, 955 (1984).
[17]B. Yurke, Phys. Rev. A **32**, 300 (1985).
[18]B. Yurke, Phys. Rev. A **29**, 408 (1984).
[19]J. N. Hollenhorst, Phys. Rev. D **19**, 1669 (1979).
[20]H. P. Yuen, Phys. Rev. A **13**, 2226 (1976).

Thomas precession, Wigner rotations and gauge transformations

D Han†, Y S Kim‡ and D Son§

† National Aeronautics and Space Administration, Goddard Space Flight Center (Code 636), Greenbelt, MD 20771, USA

‡ Department of Physics and Astronomy, University of Maryland, College Park, MD 20742, USA

§ Department of Physics, Kyungpook National University, Taegu 635, South Korea

Received 10 March 1987, in final form 1 July 1987

Abstract. The exact Lorentz kinematics of the Thomas precession is discussed in terms of Wigner's O(3)-like little group which describes rotations in the Lorentz frame in which the particle is at rest. A Lorentz-covariant form for the Thomas factor is derived. It is shown that this factor is a Lorentz-boosted rotation matrix, which becomes a gauge transformation in the infinite-momentum or zero-mass limit.

1. Introduction

To most physicists, the Thomas precession is known as an isolated event of the $\frac{1}{2}$ factor in the spin-orbit coupling in the hydrogen atom [1-6]. The purpose of this paper is to point out that the Thomas rotation plays a very important role in studying the internal spacetime symmetries of massive and massless particles. Einstein's $E = mc^2$, which means $E = (p^2 + m^2)^{1/2}$, unifies the momentum-energy relations for massive and massless particles, as is illustrated in table 1. We are interested in the question of whether the internal spacetime symmetries can be unified in a similar manner.

The Thomas precession is caued by the extra rotation the particle in a circular orbit feels *in its own rest frame* [1, 3, 5, 7]. In this paper, we point out first that Wigner's little group [8] is the natural language for the Thomas effect and perform the exact calculation of the precession angle. We shall then examine its implications.

Table 1. Significance of Wigner's little groups. Einstein's formula, commonly known as $E = mc^2$, unifies the energy-momentum relation for massive and massless particles. Likewise, Wigner's little groups unify the internal spacetime symmetries of massive and massless particles. The Thomas rotation occupies an important place in this figure as an element of the O(3)-like little group for a massive particle.

	Massive, slow	Between	Massless, fast
Energy, momentum	$E = p^2/2m$	Einstein's $E = (m^2 + p^2)^{1/2}$	$E = p$
Spin, gauge, helicity	S_3, S_1, S_2	Wigner's little group	S_3, Gauge transformations

0264-9381/87/061777+07$02.50

Reprinted from *Class. Quantum Grav.* **4**, 1777 (1987).

The little group is the maximal subgroup of the Lorentz group which leaves the 4-momentum of a given particle invariant [8]. The little groups for massive and massless particles are locally isomorphic to O(3) and E(2) respectively. The O(3)-like little group for a massive particle becomes the three-dimensional rotation group in the Lorentz frame in which the particle is at rest, which is therefore the Thomas frame.

In the case of massless particles, the rotational degree of freedom of E(2) is associated with the helicity [8, 9] and the translational degrees of freedom correspond to the gauge degrees of freedom [10–12]. It has been established that the O(3)-like little group becomes the E(2)-like little group in the limit of infinite momentum and/or vanishing mass [13–16]. This means that both the O(3)- and E(2)-like little groups are two different manifestations of a single little group, just as $E = P^2/2m$ and $E = cP$ are two different limits of $E = [(cP)^2 + (mc^2)^2]^{1/2}$. We shall show in this paper that the Thomas effect stands between these two limiting cases.

In § 2, the Wigner rotation is constructed for the kinematics of the Thomas precession. In § 3, the Thomas effect is shown to be an element of the O(3)-like little group. In § 4, it is shown that the Thomas factor becomes a gauge transformation in the large-momentum or zero-mass limit.

2. Wigner rotations

Let us consider a system of three Lorentz boosts, as is described in figure 1. We start with a massive particle at rest whose 4-momentum is

$$P_a = (0, 0, m) \tag{1}$$

in the 4-vector convention: $X^\mu = (x, z, t)$, where we omit the y component which is not affected by the transformations discussed in this paper. B_1 boosts the above 4-momentum along the z axis with the boost parameter η:

$$P_b = B_1 P_a = m(0, \sinh\eta, \cosh\eta). \tag{2}$$

The 3×3 matrix for B_1 is

$$B_1 = \begin{pmatrix} 1 & 0 & 0 \\ 0 & \cosh\eta & \sinh\eta \\ 0 & \sinh\eta & \cosh\eta \end{pmatrix}. \tag{3}$$

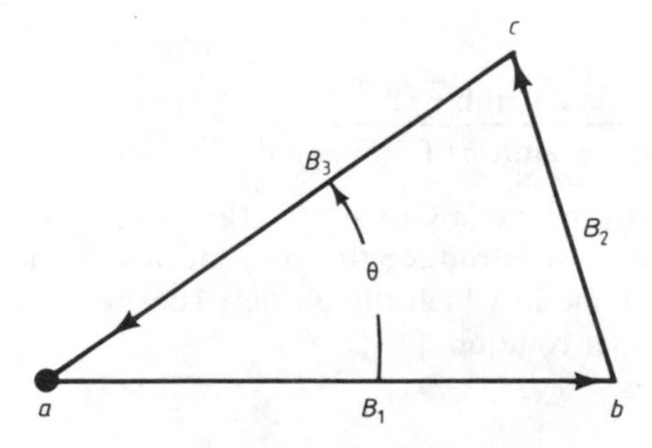

Figure 1. Closed Lorentz boosts. Initially, a massive particle is at rest with 4-momentum P_a. The first boost B_1 brings P_a to P_b. The second boost B_2 transforms P_b to P_c. The third boost B_3 brings P_c back to P_a. The net effect is a rotation around the axis perpendicular to the plane containing these three transformations. We may assume for convenience that P_b is along the z axis, and P_c in the zx plane. The rotation is then made around the y axis.

The second boost B_2 transforms P_b into P_c whose momentum has the same magnitude as that of P_b but makes an angle θ with the direction of P_b:

$$P_c = B_2 P_b = m((\sinh \eta) \sin \theta, (\sinh \eta) \cos \theta, \cosh \eta). \tag{4}$$

From the triangular geometry of figure 1, the direction of the boost for B_2 becomes $(\theta + \pi)/2$ and its boost parameter is

$$\lambda = 2 \tanh^{-1}([\sin(\theta/2)] \tanh \eta). \tag{5}$$

The transformation matrix is

$$B_2 = \begin{pmatrix} 1 + (\sin \theta/2)(\cosh \lambda - 1) & \frac{1}{2} \sin \theta (1 - \cosh \lambda) & (\cos \theta/2) \sinh \lambda \\ \frac{1}{2} \sin \theta (1 - \cosh \lambda) & 1 + (\sin \theta/2)^2 (\cosh \lambda - 1) & -(\sin \theta/2) \sinh \lambda \\ (\cos \theta/2) \sinh \lambda & -(\sin \theta/2) \sinh \lambda & \cosh \lambda \end{pmatrix}. \tag{6}$$

The momentum of P_c is in the zx plane. We can also obtain P_c by rotating P_b around the y axis by θ as is illustrated in figure 1. The third boost B_3 brings P_c back to the state of zero momentum:

$$P_a = B_3 P_c = (0, 0, m). \tag{7}$$

B_3 takes the form:

$$B_3 = \begin{pmatrix} 1 + (\sin \theta)^2 (\cosh \eta - 1) & \frac{1}{2} \sin 2\theta (\cosh \eta - 1) & -(\sin \theta) \sinh \eta \\ \frac{1}{2} \sin 2\theta (\cosh \eta - 1) & 1 + (\cos \theta)^2 (\cosh \eta - 1) & -(\cos \theta) \sinh \eta \\ -(\sin \theta) \sinh \eta & -(\cos \theta) \sinh \eta & \cosh \eta \end{pmatrix}. \tag{8}$$

This means that the above three successive boosts will leave the 4-momentum $P_a = (0, 0, m)$ invariant. Since all the boosts are made in the zx plane, the net effect will be a rotation around the y axis, which does not change P_a:

$$B_3 B_2 B_1 = W \qquad \text{or} \qquad B_3 B_2 B_1 W^{-1} = I \tag{9}$$

where I is the identity matrix. W is a one-parameter matrix representing a rotation around the y axis in the Lorentz frame in which the particle is at rest and its form is

$$W = \begin{pmatrix} \cos \alpha & \sin \alpha & 0 \\ -\sin \alpha & \cos \alpha & 0 \\ 0 & 0 & 1 \end{pmatrix} \tag{10}$$

with

$$\alpha = 2 \sin^{-1} \left(\frac{(\sin \theta)(\sinh \eta/2)^2}{[(\cosh \eta)^2 - (\sinh \eta)^2 (\sin \theta/2)^2]^{1/2}} \right). \tag{11}$$

W is definitely a rotation matrix of Wigner's O(3)-like little group which leaves P_a invariant. Since Wigner was the first to introduce the concept of this little group in terms of rotations in the Lorentz frame in which the particle is at rest [8], it is quite appropriate to call $W(\alpha)$ the 'Wigner rotation' [17].

3. Thomas effect

The Thomas effect is caused by two successive boosts with the same boost parameter in different directions [1–7]. The transformation needed for this case is $(B_3)^{-1}(B_1)^{-1}$

which brings P_b to P_a and then transforms P_a into P_c. However, this is not B_2, but requires an additional matrix T:

$$(B_3)^{-1}(B_1)^{-1} = B_2 T \tag{12}$$

where T is the Thomas factor. According to (9)

$$(B_3)^{-1}(B_1)^{-1} = B_2 B_1 W^{-1}(B_1)^{-1}. \tag{13}$$

If we compare (13) and (14):

$$T = B_1 W^{-1}(B_1)^{-1}. \tag{14}$$

While W is the rotation matrix whose form is given in (10), T is not a rotation matrix but is a Lorentz-boosted rotation matrix. Its form is

$$T = \begin{pmatrix} \cos\alpha & -(\sin\alpha)\cosh\eta & (\sin\alpha)\sinh\eta \\ (\sin\alpha)\cosh\eta & (\cos\alpha)\cosh^2\eta - \sinh^2\eta & (\cosh\eta)[\cosh\eta - (\sinh\eta)\cos\alpha] \\ (\sin\alpha)\sinh\eta & (\sinh\eta)(\cosh\eta)(\cos\alpha - 1) & \cosh^2\eta - (\cos\alpha)\sinh^2\eta \end{pmatrix}. \tag{15}$$

Furthermore, T can be written as

$$T = (B_2)^{-1}(B_3)^{-1}(B_1)^{-1}. \tag{16}$$

This matrix leaves the 4-momentum P_b invariant and is therefore an element of the O(3)-like little group for P_b.

The right-hand side of (12) is different from the decomposition given in the existing literature [2–6, 18–20]. $(B_3)^{-1}(B_1)^{-1}$ can also be written as a boost preceded by a rotation:

$$(B_3)^{-1}(B_1)^{-1} = B'R(-\alpha'). \tag{17}$$

As is illustrated in figure 2, R represents a rotation in the Lorentz frame in which the particle is not at rest and the boost B' is quite different from B_2. The rotation angle for R is

$$\alpha' = 2\tan^{-1}\left(\frac{(\sin\theta)[\sinh(\eta/2)]^2}{[\cosh(\eta/2)]^2 - [\sinh(\eta/2)]^2\cos\theta}\right). \tag{18}$$

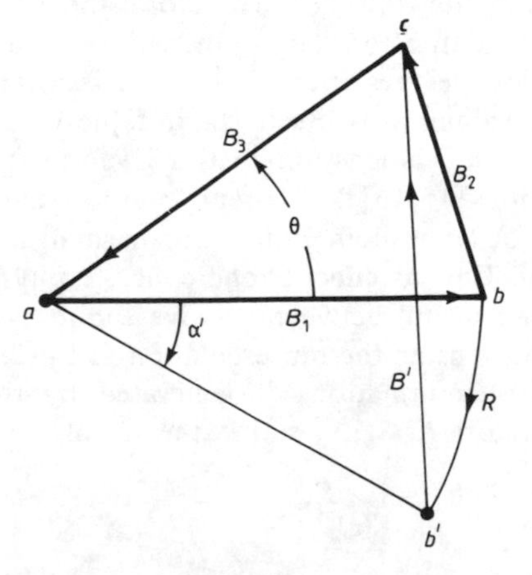

Figure 2. Lorentz boosts which are not closed. Two successive boosts $(B_3)^{-1}(B_1)^{-1}$ result in boost B' preceded by rotation R performed in the Lorentz frame where the particle is *not* at rest. It is quite clear that B' is not B_2.

This angle has been reported in the literature [18–20] but is conceptually quite different from α of equation (11). Figures 1 and 2 will illustrate this difference. α and α' are numerically different.

The Thomas precession of a charged particle in an electromagnetic field and its connection with the Wigner rotation has been thoroughly discussed in a recent review article by Chakrabarti [21]. Our main interest in this paper is the Thomas effect in the large-momentum or zero-mass limit.

4. Large-momentum or zero-mass limit

Let us go back to the T matrix of (15). In the low-energy (small-η) limit, T becomes a rotation matrix and the rotation angle α becomes equal to α' in the same limit. The above discussion therefore does not alter the existing treatment of the Thomas precession in the small-η limit. If η is not small, (15) has a new meaning. In the limit of large η, T takes the form:

$$T = \begin{pmatrix} 1 & 0 & -u & u \\ 0 & 1 & 0 & 0 \\ u & 0 & 1-u^2/2 & u^2/2 \\ u & 0 & -u^2/2 & 1+u^2/2 \end{pmatrix} \tag{19}$$

with

$$\alpha = 2u\, \mathrm{e}^{-\eta}.$$

For a finite value of u, α vanishes as η becomes very large. We have restored the y coordinate in order to compare this form with those given in the literature [8, 10, 11]. The T matrix is now applicable to the 4-vector (x, y, z, t).

The above expression for the T matrix represents an element of the E(2)-like little group which leaves invariant the 4-momentum of a massless particle moving in the z direction [8]. Furthermore, when applied to the 4-potential of a plane electromagnetic wave, it performs a gauge transformation [10–12]. Therefore, the Lorentz-boosted rotation given in (19) becomes a gauge transformation. Indeed, the Thomas rotation stands between slow particles with spin degrees of freedom and massless particles with the helicity and gauge degrees of freedom, as is illustrated in table 1.

Using techniques different from the Thomas kinematics given in this paper, we have reported in our earlier publications [15, 16] that Lorentz-boosted rotations become gauge transformations in the limit of large momentum and/or small mass. What is new in the present paper is that the Thomas effect is one concrete physical example of the Lorentz-boosted rotation which stands between massive and massless particles.

Let us illustrate what we did above using the representation of SL(2, c) for spin-$\frac{1}{2}$ particles. The group of Lorentz transformations is generated by three rotation generators S_i and three boost generators K_i. They satisfy the Lie algebra:

$$\begin{aligned} &[S_i, S_j] = \mathrm{i}\varepsilon_{ijk} S_k \qquad [S_i, K_j] = \mathrm{i}\varepsilon_{ijk} K_k \\ &[K_i, K_j] = -\mathrm{i}\varepsilon_{ijk} S_k. \end{aligned} \tag{20}$$

Since these commutation relations remain invariant when K_i are replaced by $-K_i$, we have to consider both signs of the boost generators.

For spin-$\frac{1}{2}$ particles, the generators of rotations take the form

$$S_i = \tfrac{1}{2}\sigma_i \tag{21}$$

and the rotation matrix W of (10) is translated into

$$W = \begin{pmatrix} \cos\alpha/2 & -\sin\alpha/2 \\ \sin\alpha/2 & \cos\alpha/2 \end{pmatrix}. \tag{22}$$

Since there are two different sets of boost generators:

$$K_i^{(+)} = \tfrac{1}{2}\mathrm{i}\sigma_i \qquad K_i^{(-)} = -\tfrac{1}{2}\mathrm{i}\sigma_i \tag{23}$$

there are two different boost matrices for B_1:

$$B_1^{(\pm)} = \begin{pmatrix} \exp(\pm\eta/2) & 0 \\ 1 & \exp(\mp\eta/2) \end{pmatrix}. \tag{24}$$

Consequently, there are two different forms of T:

$$T^{(+)} = B_1^{(\pm)} W^{-1} (B_1^{(\pm)})^{-1}. \tag{25}$$

In the limit of large η

$$T^{(+)} = \begin{pmatrix} 1 & -u \\ 0 & 1 \end{pmatrix} \qquad T^{(-)} = \begin{pmatrix} 1 & 0 \\ u & 1 \end{pmatrix}. \tag{26}$$

This reflects the fact that, in the SL(2, c) regime, there are two different representations of the E(2)-like little group for a given direction [22]. The 4×4 gauge transformation matrix of (19) can be constructed from the direct product of the above 2×2 matrices [16, 22].

Acknowledgment

We would like to thank Professor Eugene P Wigner for very illuminating discussions on his little groups and table 1.

References

[1] Thomas L H 1927 *Phil. Mag.* **3** 1
[2] Moller C 1952 *The Theory of Relativity* (Oxford: Oxford University Press)
[3] Corben H C and Stehle P 1960 *Classical Mechanics* (New York: Wiley) 2nd edn
[4] Taylor E F and Wheeler J A 1966 *Spacetime Physics* (San Francisco: Freeman)
[5] Jackson J D 1975 *Classical Electrodynamics* (New York: Wiley) 2nd edn
[6] Goldstein H 1980 *Classical Mechanics* (Reading, MA: Addison-Wesley) 2nd edn
[7] Hestenes D 1974 *J. Math. Phys.* **15** 1768
[8] Wigner E P 1939 *Ann. Math.* **40** 149
[9] Wigner E P 1957 *Rev. Mod. Phys.* **29** 255
[10] Weinberg S 1964 *Phys. Rev.* **135** B1049
[11] Kupersztych J 1976 *Nuovo Cimento* B **31** 1
[12] Han D and Kim Y S 1981 *Am. J. Phys.* **49** 348
[13] Inonu E and Wigner E P 1953 *Proc. Natl. Acad. Sci. USA* **39** 510
[14] Talman J D 1968 *Special Functions, A Group Theoretical Approach Based on Lectures by E P Wigner* (New York: Benjamin)
[15] Han D, Kim Y S and Son D 1983 *Phys. Lett.* **131B** 327

[16] Han D, Kim Y S and Son D 1986 *J. Math. Phys.* **27** 2228
[17] Chakrabarti A 1964 *J. Math. Phys.* **5** 1747
[18] Hestenes D 1966 *Space-time Algebra* (New York: Gordon and Breach)
[19] Ben-Menahim A 1985 *Am. J. Phys.* **53** 62
[20] Salingaros N 1986 *J. Math. Phys.* **27** 157
[21] Chakrabarti A 1987 *Preprint* Centre de Physique Theorique de l'Ecole Polytechnique, Palaiseau, A.767.0187
[22] Kim Y S and Noz M E 1986 *Theory and Applications of the Poincaré Group* (Dordrecht: Reidel)

PHYSICAL REVIEW A VOLUME 37, NUMBER 3 FEBRUARY 1, 1988

Linear canonical transformations of coherent and squeezed states in the Wigner phase space

D. Han
National Aeronautics and Space Administration, Goddard Space Flight Center (Code 636), Greenbelt, Maryland 20771

Y. S. Kim
Department of Physics and Astronomy, University of Maryland, College Park, Maryland 20742

Marilyn E. Noz
Department of Radiology, New York University, New York, New York 10016
(Received 8 September 1987)

It is shown that classical linear canonical transformations are possible in the Wigner phase space. Coherent and squeezed states are shown to be linear canonical transforms of the ground-state harmonic oscillator. It is therefore possible to evaluate the Wigner functions for coherent and squeezed states from that for the harmonic oscillator. Since the group of linear canonical transformations has a subgroup whose algebraic property is the same as that of the (2+1)-dimensional Lorentz group, it may be possible to test certain properties of the Lorentz group using optical devices. A possible experiment to measure the Wigner rotation angle is discussed.

I. INTRODUCTION

Coherent and squeezed states now form the basic language for quantum optics.[1-3] They preserve the minimum-uncertainty product in the phase space consisting of phase and intensity. The Wigner phase space, which was initially formulated in 1932,[3,4] is also becoming the standard scientific language in many branches of physics, including quantum optics.[5,6] It is therefore of interest to formulate the coherent and squeezed states within the framework of the Wigner phase-space representation.

The Wigner distribution function for the coherent states has been discussed in the literature.[7] The Wigner function for the squeezed states has also been studied recently by Schleich and Wheeler for the deformation along the "x" or "p" axis caused by real or purely imaginary parameters.[5] However, the deformation in phase space of squeezed states with complex parameters has not been systematically studied.

In this paper we shall study the squeezed states with complex parameters. It will be shown that for a complex value of the squeeze parameter, the deformation is along the direction of the phase angle of the squeeze parameter. We shall achieve this purpose not by performing a direct calculation but by studying transformation properties in phase space.

Classical mechanics can be effectively formulated in terms of the Poisson brackets and canonical transformations.[8] Although the Poisson brackets become Heisenberg's uncertainty relations in quantum mechanics, it is cumbersome to use canonical transformations in quantum mechanics because the translation operators in phase space, which are x and p, do not commute with each other.[9,10]

The basic advantage of the Wigner function is that these operators commute with each other in phase space. In this paper we study coherent and squeezed states in the Wigner phase space. We shall show that these states are canonically transformed states of the ground-state harmonic oscillator. A subset of these transformations form a group whose algebraic properties are identical to that of the (2 + 1)-dimensional Lorentz group. It may therefore be possible to design an optical experiment to test the properties of the Lorentz group.

In Sec. II we briefly review the linear canonical transformations in classical mechanics. In Sec. III we discuss the canonical transformations of the Wigner distribution function in phase space. The canonical transformation of the Wigner function is much simpler than the conventional Weyl transformation applicable to the Schrödinger picture. In Sec. IV the Wigner phase-space formalism is discussed in detail for the harmonic oscillators.

In Sec. V we discuss coherent and squeezed states in terms of canonical transformations in phase space. It is possible from this formalism to determine the Wigner function for the squeezed state with a complex parameter. It is noted in Sec. VI that the algebra of squeezed and coherent states is the same as that for the (2 + 1)-dimensional Lorentz group. This enables us to discuss a possible experiment to measure the Wigner rotation angle using optical devices.

II. LINEAR CANONICAL TRANSFORMATIONS IN CLASSICAL MECHANICS

The group of linear canonical transformations consists of translations, rotations, and squeezes in phase space.[8,10,11] These operations preserve the area element

Reprinted from *Phys. Rev. A* **37**, 807 (1988).

in phase space. We present in this section a short formalism which will be useful for studying coherent and squeezed states in quantum optics.

In order to define the word "squeeze" in phase space, let us consider a circle around the origin in the coordinate system of x and p. If we elongate the x axis by multiplying it by a real number greater than 1 and contract the p axis by dividing it by the same real number, the circle becomes an ellipse. The area of the ellipse remains the same as that of the circle. This is precisely an act of squeeze. If we combine this operation with rotation around the origin, the squeezing can be done in every possible direction in phase space.

The coordinate transformation representing translations,

$$x'=x+u, \quad p'=p+v , \tag{1}$$

can be written as

$$\begin{bmatrix} x' \\ p' \\ 1 \end{bmatrix} = \begin{bmatrix} 1 & 0 & u \\ 0 & 1 & v \\ 0 & 0 & 1 \end{bmatrix} \begin{bmatrix} x \\ p \\ 1 \end{bmatrix} . \tag{2}$$

The matrix performing the rotation around the origin by $\theta/2$ takes the form

$$R(\theta) = \begin{bmatrix} \cos\frac{\theta}{2} & -\sin\frac{\theta}{2} & 0 \\ \sin\frac{\theta}{2} & \cos\frac{\theta}{2} & 0 \\ 0 & 0 & 1 \end{bmatrix} . \tag{3}$$

The matrix which squeezes along the x axis is

$$S_x(\eta) = \begin{bmatrix} e^{\eta/2} & 0 & 0 \\ 0 & e^{-\eta/2} & 0 \\ 0 & 0 & 1 \end{bmatrix} . \tag{4}$$

The elongation along the x axis is necessarily the contraction along the p axis.

Since a canonical transformation followed by another one is a canonical transformation, the most general form of the transformation matrix is a product of the above three forms of matrices. We can simplify these mathematics by using the generators of the transformation matrices. If we use $T(u,v)$ for the translation matrix given in Eq. (2), it can be written as

$$T(u,v) = e^{-i(uN_1+vN_2)} , \tag{5}$$

where

$$N_1 = \begin{bmatrix} 0 & 0 & i \\ 0 & 0 & 0 \\ 0 & 0 & 0 \end{bmatrix}, \quad N_2 = \begin{bmatrix} 0 & 0 & 0 \\ 0 & 0 & i \\ 0 & 0 & 0 \end{bmatrix} . \tag{6}$$

The rotation matrix is generated by

$$L = \begin{bmatrix} 0 & -i/2 & 0 \\ i/2 & 0 & 0 \\ 0 & 0 & 0 \end{bmatrix} , \tag{7}$$

and

$$R(\theta) = e^{-i\theta L} . \tag{8}$$

The squeeze matrix can be written as

$$S_x = e^{-i\eta B_1} , \tag{9}$$

where

$$B_1 = \begin{bmatrix} i/2 & 0 & 0 \\ 0 & -i/2 & 0 \\ 0 & 0 & 0 \end{bmatrix} . \tag{10}$$

In addition, if we introduce the matrix B_2 defined as

$$B_2 = \begin{bmatrix} 0 & i/2 & 0 \\ i/2 & 0 & 0 \\ 0 & 0 & 1 \end{bmatrix} , \tag{11}$$

which generates the squeeze along the direction which makes 45° with the x axis, then the matrices L, B_1, and B_2 satisfy the following commutation relations:

$$[B_1,B_2] = -iL, \quad [B_1,L] = -iB_2, \quad [B_2,L] = iB_1 . \tag{12}$$

This set of commutation relations is identical to that for the generators of the (2+1)-dimensional Lorentz group.[12] The group generated by the above three operators is known also as the symplectic group Sp(2),[13,14] and its connection with the Lorentz group has been extensively discussed in the literature.[15]

If we take into account the translation operators, the commutation relations become

$$\begin{aligned} &[B_1,N_1] = (i/2)N_1, \quad [B_1,N_2] = (-i/2)N_2 , \\ &[B_2,N_1] = (i/2)N_2, \quad [B_2,N_2] = (i/2)N_1 , \\ &[N_1,L] = (i/2)N_2, \quad [N_1,L] = (-i/2)N_1 , \\ &[N_1,N_2] = 0 . \end{aligned} \tag{13}$$

These commutators, together with those of Eq. (12), form the set of closed commutation relations (or Lie algebra) of the group of canonical transformations. This group is the inhomogeneous symplectic group in the two-dimensional space or ISp(2).[11]

The translations form an Abelian subgroup generated by N_1 and N_2. Since their commutation relations with all the generators result in N_1, N_2, or 0, the translation subgroup is an invariant subgroup. The translations and the rotation form the two-dimensional Euclidean group generated by N_1,N_2, and L, which have closed commutation relations. This group also has been extensively discussed recently in connection with the internal space-time symmetries of massless particles.[16,17]

Indeed, it is of interest to see how the representations of the Lorentz group can be useful in optical sciences. It is also of interest to see how the experimental resources in optical science can be helpful in understanding some of the "abstract" mathematical identities in group theory.

III. LINEAR CANONICAL TRANSFORMATIONS IN THE WIGNER PHASE SPACE

If $\psi(x)$ is a solution of the Schrödinger equation, the Wigner distribution function in phase space is defined as

$$W(x,p)=(1/\pi)\int \psi^*(x+y)\psi(x-y)e^{2ipy}dy \ . \quad (14)$$

This is a function of x and p which are c numbers. This function is real but is not necessarily positive everywhere in phase space. The properties of this function have been extensively discussed in the literature.[4-7]

When we make linear canonical transformations of this function in phase space, the infinitesimal generators are

$$N_1=-i\frac{\partial}{\partial x}, \quad N_2=-i\frac{\partial}{\partial p} ,$$
$$L=-\frac{i}{2}\left[x\frac{\partial}{\partial p}-p\frac{\partial}{\partial x}\right] , \quad (15)$$
$$B_1=\frac{i}{2}\left[x\frac{\partial}{\partial x}-p\frac{\partial}{\partial p}\right], \quad B_2=\frac{i}{2}\left[x\frac{\partial}{\partial p}+p\frac{\partial}{\partial x}\right] .$$

These operators satisfy the commutation relations given in Eqs. (12) and (13). We can therefore derive the algebraic relations involving the above differential forms using the matrix representation discussed in Sec. II.

The rotation of the translation operators takes the form

$$R(\theta)N_1R(-\theta)=\left[\cos\frac{\theta}{2}\right]N_1-\left[\sin\frac{\theta}{2}\right]N_2 ,$$
$$R(\theta)N_2R(-\theta)=\left[\sin\frac{\theta}{2}\right]N_1+\left[\cos\frac{\theta}{2}\right]N_2 . \quad (16)$$

Under the same rotation, the squeeze generators become

$$R(\theta)B_1R(-\theta)=(\cos\theta)B_1+(\sin\theta)B_2 ,$$
$$R(\theta)B_2R(-\theta)=-(\sin\theta)B_1+(\cos\theta)B_2 . \quad (17)$$

Likewise, we can derive all the algebraic relations using matrix algebra. The important point is that the group of canonical transformations in the Wigner phase space is identical to that for classical mechanics.

Next, let us consider the above transformations in terms of operators applicable to the Schrödinger wave function. From the expression of Eq. (14) it is quite clear that the operation e^{-ivx} on the wave function leads to a translation along the p axis by v. The operation of $\exp[-u(\partial/\partial x)]$ on the wave function leads to a translation of the above distribution function along the x axis by u.

Likewise, the operation in the Wigner phase space of $ix(\partial/\partial p)$ and $ip(\partial/\partial x)$ become $x^2/2$ and $\frac{1}{2}(\partial/\partial x)^2$, respectively. Thus, the transformations in phase space can be generated from the operators applicable to the wave function. The generators applicable to the wave function are

$$\tilde{N}_1=-i\frac{\partial}{\partial x}, \quad \tilde{N}_2=x ,$$
$$\tilde{L}=\frac{1}{4}\left[\left[\frac{\partial}{\partial x}\right]^2-x^2\right] , \quad (18)$$
$$\tilde{B}_1=-i\left[\frac{x}{2}\right]\frac{\partial}{\partial x}, \quad \tilde{B}_2=\frac{1}{4}\left[x^2+\left[\frac{\partial}{\partial x}\right]^2\right] .$$

These operators satisfy the commutation relations given in Eqs. (12) and (13), except the last one. The operators $\tilde{N}_1$ and $\tilde{N}_2$ do not commute with each other, and

$$[\tilde{N}_1,\tilde{N}_2]=-i \ . \quad (19)$$

Therefore, it appears that the operators applicable to the Schrödinger wave function do not satisfy the same set of commutation relations as that for classical phase space.[9,10]

Let us consider the translation along the x axis followed by the translation along the p axis, and the operation in the opposite order. From the Baker-Campbell-Hausdorff formula for two operators,[9,10]

$$(e^{-iuN_1})(e^{-ivN_2})=(e^{iuv})(e^{-ivN_2})(e^{-iuN_1}) . \quad (20)$$

The interchange of the above two translations results in a multiplication of the wave function by a constant factor of unit modulus.

However, this factor disappears when the Wigner function W is constructed according to the definition of Eq. (14). Therefore, the translation along the x direction and the translation along the p direction commute with each other in the Wigner phase space. This means that the commutation relation $[N_1,N_2]=0$ in the Wigner phase space and the Heisenberg relation $[N_1,N_2]=-i$ are perfectly consistent with each other. The basic advantage of the Wigner phase-space representation is that its canonical transformation property is the same as that of classical mechanics.

We now have three sets of operators. The first set consists of the three-by-three matrices in Eqs. (6), (7), (10), and (11), and this set is for classical mechanics. The differential operators in two-dimensional phase space form the second set, and they are for the Wigner function. The third set consists of the differential operators of Eq. (18) applicable to the Schrödinger wave function. The first and second sets are the same. While both the second set of double-variable operators and the third set of single-variable operators are extensively used in the literature,[11,14,18] it is interesting to see that the connection between these two sets can be established through the Wigner function.

The transformations discussed in this section constitute the basic language for coherent and squeezed states in quantum optics. The relevance of the translation in phase space to coherent states has been noted before.[1] The word squeeze comes from quantum optics. It has been also noted that its mathematics is like that of (2 + 1)-dimensional Lorentz transformations. As was emphasized in the literature,[12,17] combining translations with Lorentz transformations is not a trivial problem. We shall discuss the problem in Secs. V and VI.

IV. HARMONIC OSCILLATORS

The one-dimensional harmonic oscillator occupies a unique place in the physics of phase space. For the Hamiltonian of the form

$$H=\tfrac{1}{2}(p^2+x^2)\ , \tag{21}$$

the Wigner function is a function only of (p^2+x^2), and is thus invariant under rotations around the origin in phase space. The Wigner function for the ground-state harmonic oscillator is[7,19]

$$W(x,p)=\frac{1}{\pi}\exp[-(x^2+p^2)]\ . \tag{22}$$

This function is localized within the circular region whose boundary is defined by the equation

$$x^2+p^2=1\ . \tag{23}$$

Therefore, the study of the Wigner function for the harmonic oscillator is the same as the study of a circle on the two-dimensional plane. The canonical transformation consists of rotations, translations, and area-preserving elliptic deformations of this circle. These transformations are straightforward.

Under the translation by r along the x axis, the above circle becomes

$$(x-r)^2+p^2=1\ . \tag{24}$$

This circle is centered around the point $(r,0)$. We can rotate the above circle around the origin. Then the resulting Wigner function is

$$R(\theta)T(r,0)W(x,p)=\frac{1}{\pi}\exp\left\{\left[\left[x-r\cos\frac{\theta}{2}\right]^2+\left[p-r\sin\frac{\theta}{2}\right]^2\right]\right\}, \tag{25}$$

where $T(r,0)$ and $R(\theta)$ are the translation and rotation operators. Because the circle of Eq. (23) is invariant under rotations around the point where $x=r$ and $p=0$, the above Wigner function is the same as the translated Wigner function,

$$T\left[r\cos\frac{\theta}{2},r\sin\frac{\theta}{2}\right]W(x,p)=R(\theta)T(r,0)W(x,p)\ . \tag{26}$$

Let us next elongate the translated circle of Eq. (24) along the x direction. The circle will be deformed into

$$e^{-\eta}(x-r')^2+e^{\eta}p^2=1\ , \tag{27}$$

where

$$r'=re^{\eta/2}\ .$$

If we rotate this ellipse, the resulting Wigner function will be

$$R(\theta)S_1(\eta)T(r,0)W(x,p)$$
$$=\frac{1}{\pi}\exp\left\{-\left[e^{-\eta}\left[x\cos\frac{\theta}{2}+p\sin\frac{\theta}{2}-r'\cos\frac{\theta}{2}\right]^2+e^{\eta}\left[x\sin\frac{\theta}{2}-p\cos\frac{\theta}{2}-r'\sin\frac{\theta}{2}\right]^2\right]\right\} \tag{28}$$

This transformation is illustrated in Fig. 1. As we shall see in Sec. V, the translated and deformed Wigner functions will be useful for studying coherent and squeezed states, respectively.

In the meantime, let us observe other useful properties of the harmonic oscillator. We noted above that, in order to study the harmonic oscillator, we can start with a circle in phase space. How does this rotational invariance manifest itself in the Schrödinger picture? The generator of rotations is

$$L=\frac{1}{4}\left[\left[\frac{\partial}{\partial x}\right]^2-x^2\right]=\tfrac{1}{2}(-H)\ . \tag{29}$$

If the wave function is a solution of the time-independent Schrödinger equation with the above Hamiltonian, the application of the rotation operator $\exp(-i\theta L)$ will only generate a constant factor of unit modulus. This is the reason why the Wigner function for the above Hamiltonian system is invariant under rotations in phase space.

In order to study rotations more carefully in the Schrödinger picture, let us use a and $a^\dagger$, defined in this case as

$$a=(1/\sqrt{2})\left[x+\frac{\partial}{\partial x}\right],\qquad a^\dagger=(1/\sqrt{2})\left[x-\frac{\partial}{\partial x}\right]. \tag{30}$$

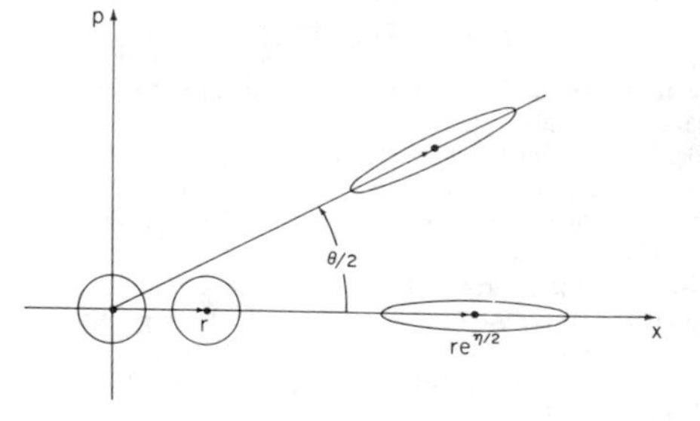

FIG. 1. Coherent and squeezed states in the Wigner phase space. The circle centered around the origin describes the ground-state harmonic oscillator. The circle around $(r,0)$ is for the coherent state. This coherent state can be squeezed to ellipse along the x axis, with a real value of the squeeze parameter. When the squeeze parameter becomes complex then the ellipse is rotated around the origin in the Wigner phase space.

These operators serve two distinct purposes in physics. They are step-up and step-down operators for the one-dimensional harmonic oscillator in nonrelativistic quantum mechanics.

On the other hand, in quantum-field theory, they serve as the annihilation and creation operators. We are here interested in the creation and annihilation of photons. Then, what is the physics of the phase space spanned by x and p variables? Indeed, the concept of creation and annihilation comes from the commutation relation

$$[a,a^\dagger]=1 . \tag{31}$$

This form of uncertainty relation states also that the area element in phase space cannot be smaller than Planck's constant. The area element in the Cartesian coordinate system is $(\Delta x)(\Delta p)$. It is also possible to write the area element in the polar coordinate system. If this area is described in the polar-coordinate system, the uncertainty relation is the relation between phase and intensity.[20] This is the uncertainty relation we are discussing in this paper. We are particularly interested in the minimum-uncertainty states.

In both Eq. (25) and Eq. (28) the rotation plays the essential role. Let us see how the operators a and $a^\dagger$ can be rotated. For two operators A and B, we note the relation[21]

$$e^{A}Be^{-A}=B+[A,B]+\tfrac{1}{2}[A,[A,B]] + \tfrac{1}{6}[A,[A,[A,B]]]+\cdots , \tag{32}$$

and

$$[L,a]=-\tfrac{1}{2}a, \quad [L,a^\dagger]=\tfrac{1}{2}a^\dagger . \tag{33}$$

Since $R(\theta)=e^{-i\theta L}$,

$$\begin{aligned} R(\theta)aR(-\theta)&=(e^{-i\theta/2})a , \\ R(\theta)a^\dagger R(-\theta)&=(e^{i\theta/2})a^\dagger . \end{aligned} \tag{34}$$

In terms of the a and $a^\dagger$ operators, the generators of canonical transformations take the form

$$\begin{aligned} &\tilde{N}_1=(-i/\sqrt{2})(a-a^\dagger), \quad \tilde{N}_2=(1/\sqrt{2})(a+a^\dagger) , \\ &\tilde{L}=\tfrac{1}{4}(aa^\dagger+a^\dagger a) , \\ &\tilde{B}_1=\tfrac{1}{4}(aa-a^\dagger a^\dagger), \quad \tilde{B}_2=\tfrac{1}{4}(aa+a^\dagger a^\dagger) . \end{aligned} \tag{35}$$

We can rotate these operators using Eq. (34). In particular, the rotations given in Eq. (17) can now be written as

$$\begin{aligned} R(\theta)aaR(-\theta)&=e^{-i\theta}aa , \\ R(\theta)a^\dagger a^\dagger R(-\theta)&=e^{i\theta}a^\dagger a^\dagger . \end{aligned} \tag{36}$$

These relations will be useful in evaluating the Wigner function for the squeezed state.

V. COHERENT STATES AND SQUEEZED STATES

In terms of the a and $a^\dagger$ operators, the coherent state is defined as

$$|\alpha\rangle=[\exp(-|\alpha|^2/2)]\sum_{n=0}^{\infty}(\alpha^n/n!)(a^\dagger)^n|0\rangle . \tag{37}$$

We can obtain this state by applying the translation operator to the ground state,

$$|\alpha\rangle=T(\alpha)|0\rangle , \tag{38}$$

where

$$T(\alpha)=\exp(\alpha a^\dagger-\alpha^* a) .$$

The translation operator in the phase space depends on two real parameters. In the above case, the parameter α is a complex number containing two real parameters.

It is possible to evaluate the Wigner function from the above expression to obtain the form given in Eq. (25),[7,9] with

$$r\cos\frac{\theta}{2}=\sqrt{2}[\mathrm{Re}(\alpha)], \quad r\sin\frac{\theta}{2}=\sqrt{2}[\mathrm{Im}(\alpha)] . \tag{39}$$

It is also possible to obtain the Wigner function starting from a real value of α by rotation. From the rotation properties of the a and $a^\dagger$ operators given in Sec. IV, the rotation of this operator becomes

$$R(\theta)T(r)R(-\theta)=T(\alpha) , \tag{40}$$

with

$$\alpha=(e^{-i\theta/2})r .$$

This means that we can make α complex starting from a real number r by rotation.

The squeezed state $|\xi,\alpha\rangle$ is defined to be[2,3,5,18,22]

$$|\xi,\alpha\rangle=S(\xi)|\alpha\rangle=S(\xi)T(\alpha)|0\rangle , \tag{41}$$

where

$$S(\xi)=\exp\left[\frac{\xi}{2}a^\dagger a^\dagger-\frac{\xi^*}{2}aa\right] . \tag{42}$$

Here again the parameter ξ is complex and contains two real numbers for specifying the direction and the strength of the squeeze.

If ξ is real, it is possible to evaluate the Wigner function by direct evaluation of the integral. If, on the other hand, ξ is complex, the present authors were not able to manage the calculation. We can, however, overcome this difficulty by using the method of canonical transformation developed in this paper. We can make ξ complex starting from a real value of η by rotating the above squeeze operator using the rotation properties of the a and $a^\dagger$ operators.

Let us start from a real value of ξ for which the evaluation is possible.[5] For the real value η, the squeeze operator becomes

$$S(\eta)=\exp\left[-\frac{\eta}{2}\left[x\frac{\partial}{\partial x}\right]\right] . \tag{43}$$

This operator makes the scale change of x to $(e^{-\eta/2})x$. It is therefore possible to visualize the deformation of the circle into an ellipse in the phase space. Let us next rotate this ellipse. From Eqs. (36) and (42),

$$R(\theta)S(\eta)R(-\theta)=S(\xi) , \tag{44}$$

where

$$\xi = (e^{-i\theta})\eta \ .$$

The operator $S(\xi)$, when applied to the wave function, leads to the Wigner function which is elongated along the $\theta/2$ direction in the phase space. It is indeed possible to evaluate the Wigner function for the squeezed state with a complex value of ξ simply by rotating the ellipse elongated along the x direction.

Table I describes how we can determine the Wigner functions for coherent and squeezed states. Figure 1 illustrates how the above calculation can be carried out. The translated circle in phase space describes the coherent state. This circle can be elongated along the x direction. The resulting ellipse is for the squeezed state with a real parameter. This ellipse can be rotated. This rotated ellipse corresponds to the squeezed state with a complex parameter.

VI. POSSIBLE MEASUREMENT OF THE WIGNER ROTATION

We have noted in Sec. II that the transformation group contains the subgroup Sp(2) which is locally isomorphic to the (2 + 1)-dimensional Lorentz group. It may therefore be possible to design experiments in optics to test the mathematical identities in the Lorentz group. The Wigner rotation is a case in point. Two successive applications of Lorentz boosts in different directions is not a Lorentz boost, but is a boost preceded by a rotation which is commonly called the Wigner rotation.[23–26]

TABLE I. How to evaluate the Wigner function for coherent and squeezed states.

	Coherent states	Squeezed states
Direct computation	Possible	Not known
Canonical transformation	Possible	Possible

This effect exhibits itself in the Thomas effect in atomic physics.[25]

Since the mathematics of squeeze is the same as that of Lorentz boost, we can discuss the possibility of measuring the effect of the Wigner rotation in optical experiments. In order to illustrate how the Wigner rotation comes into this subject, let us start with a circle of unit radius centered around the origin in the Cartesian-coordinate system with the coordinate variables x and p, whose equation is given in Eq. (23). If we squeeze this circle by elongating along the x axis, the squeeze matrix applicable to the vector (x,p) is

$$S(0,\lambda) = \begin{pmatrix} e^{\eta/2} & 0 \\ 0 & e^{-\eta/2} \end{pmatrix} . \tag{45}$$

This will deform the circle into the ellipse

$$(e^{-\eta})x^2 + (e^{\eta})p^2 = 1 \ . \tag{46}$$

If we squeeze the circle centered around the origin along the $\theta/2$ direction with the deformation parameter η, the squeeze matrix is

$$S(\theta,\lambda) = \begin{pmatrix} \cosh\frac{\lambda}{2} + \left(\sinh\frac{\lambda}{2}\right)\cos\theta & \left(\sinh\frac{\lambda}{2}\right)\sin\theta \\ \left(\sinh\frac{\lambda}{2}\right)\sin\theta & \cosh\frac{\lambda}{2} - \left(\sinh\frac{\lambda}{2}\right)\cos\theta \end{pmatrix} , \tag{47}$$

and the circle is deformed into the ellipse

$$e^{-\lambda}\left[x\cos\frac{\theta}{2} + p\sin\frac{\theta}{2}\right]^2 + e^{\lambda}\left[x\sin\frac{\theta}{2} - p\cos\frac{\theta}{2}\right]^2 = 1 \ . \tag{48}$$

In order to understand the squeeze mechanism thoroughly, we should know how to squeeze an ellipse. We can achieve this goal by studying two successive squeezing properties. Let us therefore consider the squeeze $S(\theta,\lambda)$ of the circle centered around the origin preceded by $S(0,\eta)$. This will result in another ellipse,

$$e^{-\xi}\left[x\cos\frac{\alpha}{2} + p\sin\frac{\alpha}{2}\right]^2 + e^{\xi}\left[x\sin\frac{\alpha}{2} - p\cos\frac{\alpha}{2}\right]^2 = 1 \ . \tag{49}$$

where

$$\cosh\xi = (\cosh\eta)\cosh\lambda + (\sinh\eta)(\sinh\lambda)\cos\theta \ ,$$

and

$$\tan\alpha = \frac{(\sin\theta)[\sinh\lambda + (\tanh\eta)(\cosh\lambda - 1)\cos\theta]}{(\sinh\lambda)\cos\theta + (\tanh\eta)[1 + (\cosh\lambda - 1)(\cos\theta)^2]} \ .$$

This is an ellipse elongated along the $\alpha/2$ direction with the parameter ξ.

The above calculation gives an indication that two successive squeezes become one squeeze. This is not true. The product of the matrices $S(\theta,\lambda)S(0,\eta)$ does not result in $S(\alpha,\xi)$. Instead, it becomes[16,17,23–25]

$$S(\theta,\lambda)S(0,\eta) = S(\alpha,\xi)R(\phi) \ , \tag{50}$$

where

$$\tan\left(\frac{\phi}{2}\right) = \frac{(\sin\theta)[\tanh(\lambda/2)][\tanh(\eta/2)]}{1 + [\tanh(\lambda/2)][\tanh(\eta/2)](\cos\theta)} \ .$$

The right-hand side of the above equation is a squeeze preceded by a rotation, which may be called the Wigner rotation.[23–26] Although Eq. (49) does not show the effect of this rotation which leaves the initial circle centered around the origin invariant, we need the derivation of Eq. (49) in order to determine α, ξ, and eventually ϕ.

The study of coherent states representations requires transformations of a circle not centered around the origin. If we squeeze this circle by applying $S(0,\eta)$, the circle is transformed into the ellipse given in Eq. (27). If we squeeze this ellipse by applying $S(\theta,\lambda)$, the net effect is the squeeze $S(\alpha,\xi)$ preceded by the Wigner rotation $R(\phi)$. If we apply this rotation to the circle of Eq. (24),

$$\left[x - r\cos\frac{\phi}{2}\right]^2 + \left[p - r\sin\frac{\phi}{2}\right]^2 = 1 . \quad (51)$$

The effect of this rotation is illustrated in Fig. 2.

Next, if we apply the squeeze $S(\alpha,\xi)$ to the above circle, the resulting ellipse is

$$e^{-\xi}\left[(x-a)\cos\frac{\alpha}{2} + (y-b)\sin\frac{\alpha}{2}\right]^2 + e^{\xi}\left[(x-a)\sin\frac{\alpha}{2} - (y-b)\cos\frac{\alpha}{2}\right]^2 = 1 , \quad (52)$$

where

$$a = r\left\{\left[\cosh\frac{\xi}{2} + \left[\sinh\frac{\xi}{2}\right]\cos\alpha\right]\cos\frac{\phi}{2} + \left[\sinh\frac{\xi}{2}\right](\sin\alpha)\sin\frac{\phi}{2}\right\} ,$$

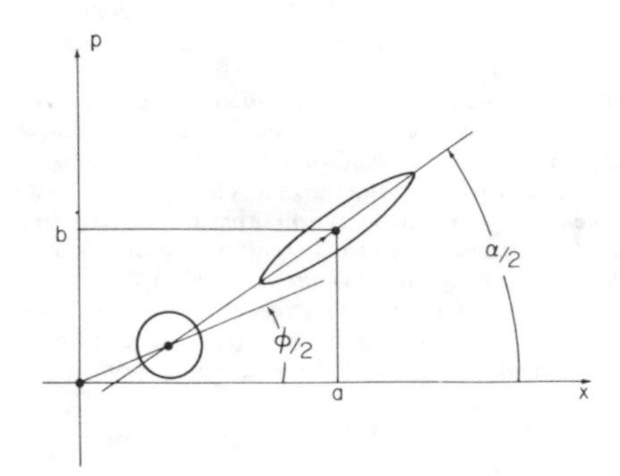

FIG. 2. Two repeated squeezes resulting in one squeeze preceded by one rotation. The circle around $(r,0)$ in Fig. 1 is rotated around the origin by $\phi/2$ and is then elongated along the $\alpha/2$ direction.

$$b = r\left\{\left[\sinh\frac{\xi}{2}\right](\sin\alpha)\cos\frac{\phi}{2} + \left[\cosh\frac{\xi}{2} - \left[\sinh\frac{\xi}{2}\right]\cos\alpha\right]\sin\frac{\phi}{2}\right\} .$$

The effect of this squeeze is also illustrated in Fig. 2.

The Wigner rotation angle ϕ can now be determined from a,b, which can be measured. In terms of these parameters,

$$\tan\left[\frac{\phi}{2}\right] = \frac{b\left[\cosh\frac{\xi}{2} + \left[\sinh\frac{\xi}{2}\right]\cos\alpha\right] - a\left[\sinh\frac{\xi}{2}\right]\sin\alpha}{a\left[\cosh\frac{\xi}{2} - \left[\sinh\frac{\xi}{2}\right]\cos\alpha\right] - b\left[\sinh\frac{\xi}{2}\right]\sin\alpha} \quad (53)$$

The parameters ξ and α can be measured or determined from Eq. (49). The angle ϕ determined from the above expression can be compared with the angle calculated from η,λ, and α according to the expression given in Eq. (50).

Indeed, if the parameters of the coherent and squeezed states can be determined experimentally, the Wigner rotation can be measured in optical laboratories. The question is then whether this experiment can be carried out with the techniques available at the present time. While the analysis presented in this section is based on single-mode squeezed states, the squeezed states that have been generated to date are two-mode states.[3,18] Hence, in order to be directly applicable to experiment, the present work has to be extended to the two-mode case, unless the single-mode squeezed state can be generated in the near future. In the meantime, the present work indicates that some of optical experiments may serve as analog computers for the (2 + 1)-dimensional Lorentz group.

VII. CONCLUDING REMARKS

It is quite clear from this paper that the coherent and squeezed states can be described by circles and ellipses in the Wigner phase space. One circle or ellipse can be transformed into another by area-preserving transformations. The group governing these transformations is the inhomogeneous symplectic group ISp(2).

We studied the generators of these transformations both for phase space and for the Schrödinger representation. It has been shown that the connection between these two sets of operators can be established through the Wigner function.

We also studied in detail rotations in the Wigner phase space and their counterparts in the Schrödinger representation. It is now possible to evaluate the Wigner function for a squeezed state with a complex parameter.

The correspondence (local isomorphism) between Sp(2) and the (2 + 1)-dimensional Lorentz group allows us to study quantum optics using the established language of the Lorentz group. At the same time it allows us to look into possible experiments in optical science to study some of mathematical formulas in group theory. It may be possible to measure the Wigner rotation angle in optical laboratories.

ACKNOWLEDGMENTS

We are grateful to Professor Eugene P. Wigner for very helpful discussions on the subject of canonical transformations in quantum mechanics and for maintaining his interest in the present work. We would like to thank Mr. Seng-Tiong Ho and Dr. Yanhua Shi for explaining to us the experimental techniques available at the present time.

---

[1]J. R. Klauder, Ann. Phys. (N.Y.) **11**, 123 (1960); R. J. Glauber, Phys. Rev. Lett. **10**, 84 (1963); F. T. Arechi, *ibid.* **15**, 912 (1965); E. Goldin, *Waves and Photons* (Wiley, New York, 1982); J. R. Klauder and B. S. Skagerstam, *Coherent States* (World Scientific, Singapore, 1985).

[2]D. Stoler, Phys. Rev. D **1**, 3217 (1970); H. P. Yuen, Phys. Rev. A **13**, 2226 (1976).

[3]For some of the recent papers on the squeezed state, see C. M. Caves, Phys. Rev. D **23**, 1693 (1981); D. F. Walls, Nature (London) **306**, 141 (1983); R. S. Bondourant and J. H. Shapiro, Phys. Rev. D **30**, 2584 (1984); M. D. Reid and D. F. Walls, Phys. Rev. A **31**, 1622 (1985); B. Yurke, *ibid.* **32**, 300 (1985); **32**, 311 (1985); R. E. Slusher, L. W. Hollberg, B. Yurke, J. C. Mertz, and J. F. Valley, Phys. Rev. Lett. **55**, 2409 (1985); C. M. Caves and B. L. Schumaker, Phys. Rev. A **31**, 3068 (1985); B. L. Schumaker and C. M. Caves, *ibid.* **31**, 3093 (1985); M. J. Collet and D. F. Walls, *ibid.* **32**, 2887 (1985); J. R. Klauder, S. L. McCall, and B. Yurke, *ibid.* **33**, 3204 (1986); S. T. Ho, P. Kumar, J. H. Shapiro, *ibid.* **34**, 293 (1986); **36**, 3982 (1987); Z. Y. Ou, C. K. Hong, and L. Mandel, *ibid.* **36**, 192 (1987).

[4]E. P. Wigner, Phys. Rev. **40**, 749 (1932). For review articles on this subject, see E. P. Wigner, in *Perspective in Quantum Theory,* edited by W. Yourgrau and A. van der Merwe (MIT Press, Cambridge, MA, 1971); M. Hillery, R. F. O'Connell, M. O. Scully, and E. P. Wigner, Phys. Rep. **106**, 121 (1984); N. L. Balazs and B. K. Jennings, *ibid.* **104C**, 347 (1984).

[5]W. Schleich and J. A. Wheeler, in *Proceedings of the First International Conference on the Physics of Phase Space, College Park, Maryland 1986,* edited by Y. S. Kim and W. W. Zachary (Springer-Verlag, Heidelberg, 1987); Nature (London) **326**, 574 (1987).

[6]W. H. Louisell, *Quantum Statistical Properties of Radiation* (Wiley, New York, 1973); B. V. K. Vijaya Kumar and C. W. Carroll, Opt. Eng. **23**, 732 (1984); R. L. Easton, A. J. Ticknor, and H. H. Barrett, *ibid.* **23**, 738 (1984); R. Procida and H. W. Lee, Opt. Commun. **49**, 201 (1984); N. S. Subotic and B. E. A. Saleh, *ibid.* **52**, 259 (1984); A. Conner and Y. Li, Appl. Opt. **24**, 3825 (1985); S. W. McDonald and A. N. Kaufman, Phys. Rev. A **32**, 1708 (1985); O. T. Serima, J. Javanainen, and S. Varro, *ibid.* **33**, 2913 (1986); K. J. Kim, Nucl. Instrum. A **246**, 71 (1986); H. Szu, in *Proceedings of the First International Conference on the Physics of Phase Space, College Park, Maryland, 1986,* edited by Y. S. Kim and W. W. Zachary (Springer-Verlag, Heidelberg, 1987).

[7]P. Carruthers and F. Zachariasen, Rev. Mod. Phys. **55**, 245 (1983).

[8]H. Goldstein, *Classical Mechanics,* 2nd ed. (Addison-Wesley, Reading, MA, 1980).

[9]H. Weyl, *The Theory of Groups and Quantum Mechanics,* 2nd ed. (Dover, New York, 1950).

[10]K. B. Wolf, Kinam **6**, 141 (1986).

[11]A. Perelomov, *Generalized Coherent States* (Springer-Verlag, Heidelberg, 1986).

[12]E. P. Wigner, Ann. Math. **40**, 149 (1939); V. Bargmann, *ibid.* **48**, 568 (1947); V. Bargmann and E. P. Wigner, Proc. Nat. Acad. Sci. U.S.A. **34**, 211 (1948).

[13]H. Weyl, *Classical Groups,* 2nd ed. (Princeton University, Princeton, N.J., 1946).

[14]V. Guillemin and S. Sternberg, *Symplectic Techniques in Physics* (Cambridge University Press, Cambridge, England, 1984).

[15]Y. S. Kim and M. E. Noz, Am. J. Phys. **51**, 368 (1983).

[16]D. Han, D. Son, and Y. S. Kim, Phys. Rev. D **26**, 3717 (1982).

[17]Y. S. Kim and M. E. Noz, *Theory and Applications of the Poincaré Group* (Reidel, Dordrecht, 1986).

[18]B. Yurke, S. McCall, and J. R. Klauder, Phys. Rev. A **33**, 4033 (1986).

[19]S. Shlomo, J. Phys. A **16**, 3463 (1983).

[20]W. Heitler, *Quantum Theory of Radiation,* 3rd ed. (Oxford University Press, London, 1954).

[21]W. Miller, *Symmetry Groups and Their Applications* (Academic, New York, 1972); D. R. Truax, Phys. Rev. D **31**, 1988 (1985).

[22]This definition is equivalent to $|\xi,\alpha\rangle = T(\alpha')S(\xi)|0\rangle$, with a suitably adjusted value of α, since the translation subgroup is an invariant subgroup which enables us to write and $ST=STS^{-1}S$ with $ST^{-1}=T'$, which is another element in the translation subgroup.

[23]V. I. Ritus, Zh. Eksp. Teor. Fiz. **40**, 352 (1961) [Sov. Phys.—JETP **13**, 240 (1961)].

[24]A. Chakrabarti, J. Math. Phys. **5**, 1747 (1964).

[25]D. Han, Y. S. Kim, and D. Son, Class. Quantum Grav. **4**, 1777 (1987).

[26]A. Ben-Menahim, Am. J. Phys. **53**, 62 (1985).

Of Related Interest

Theory and Applications of the Poincaré Group

by

Y. S. Kim

Department of Physics and Astronomy, University of Maryland, U.S.A.

and

Marilyn E. Noz

Department of Radiology, New York University, U.S.A.

Special relativity and quantum mechanics are likely to remain the two most important languages in physics for many years to come. The underlying language for both disciplines is group theory. Eugene P. Wigner's 1939 paper on the Poincaré group laid the foundation for unifying the concepts and algorithms of quantum mechanics and special relativity. This book systematically presents physical examples which can best be explained in terms of Wigner's representation theory. The examples include the relativistic quark model, hadronic mass spectra, the Lorentz-Dirac deformation of hadrons, the form factors of nucleons, Feynman's parton picture and the proton structure function, the kinematical origin of the gauge degrees of freedom for massless particles, the polarization of neutrinos as a consequence of the gauge invariance, and massless particles as the (small-mass/large-momentum) limits of massive particles.

This book is intended mainly as a teaching tool directed toward those who desire a deeper understanding of group theory in terms of examples applicable to the physical world and/or of the physical world in terms of the symmetry properties which can best be formulated in terms of group theory. Each chapter contains problems and solutions, and this makes it potentially useful as a textbook. Graduate students and researchers interested in space-time symmetries of relativistic particles will find the book of interest.

ISBN 90–277–2141–6 FTP 17

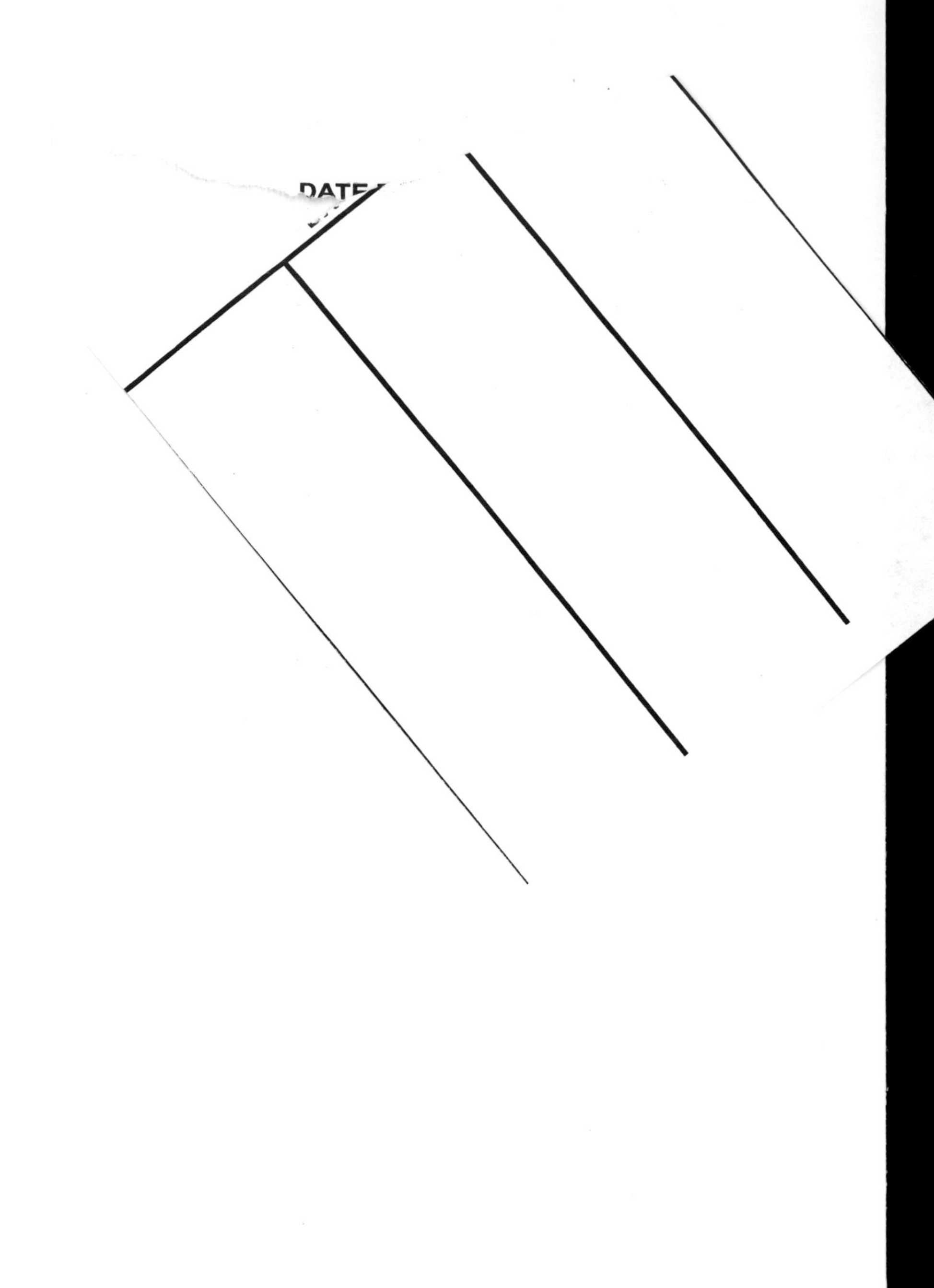